ESSENTIALS OF FIRE FIGHTING

THIRD EDITION

Validated by
The International Fire Service Training Association

Published by
Fire Protection Publications
Oklahoma State University

COVER PHOTOS COURTESY OF:
BOB ESPOSITO — PENNSBURG, PA
TOM McCARTHY — CHICAGO, IL
MICHAEL A. WIEDER — STILLWATER, OK

ISBN 0-87939-101-4
Library of Congress 92-73546

Third Edition
Eighth Printing

Printed in the United States of America

Oklahoma State University in compliance with Title VI of the Civil Rights Act of 1964 and Title IX of the Educational Amendments of 1972 (Higher Education Act) does not discriminate on the basis of race, color, national origin or sex in any of its policies, practices or procedures. This provision includes but is not limited to admissions, employment, financial aid and educational services.

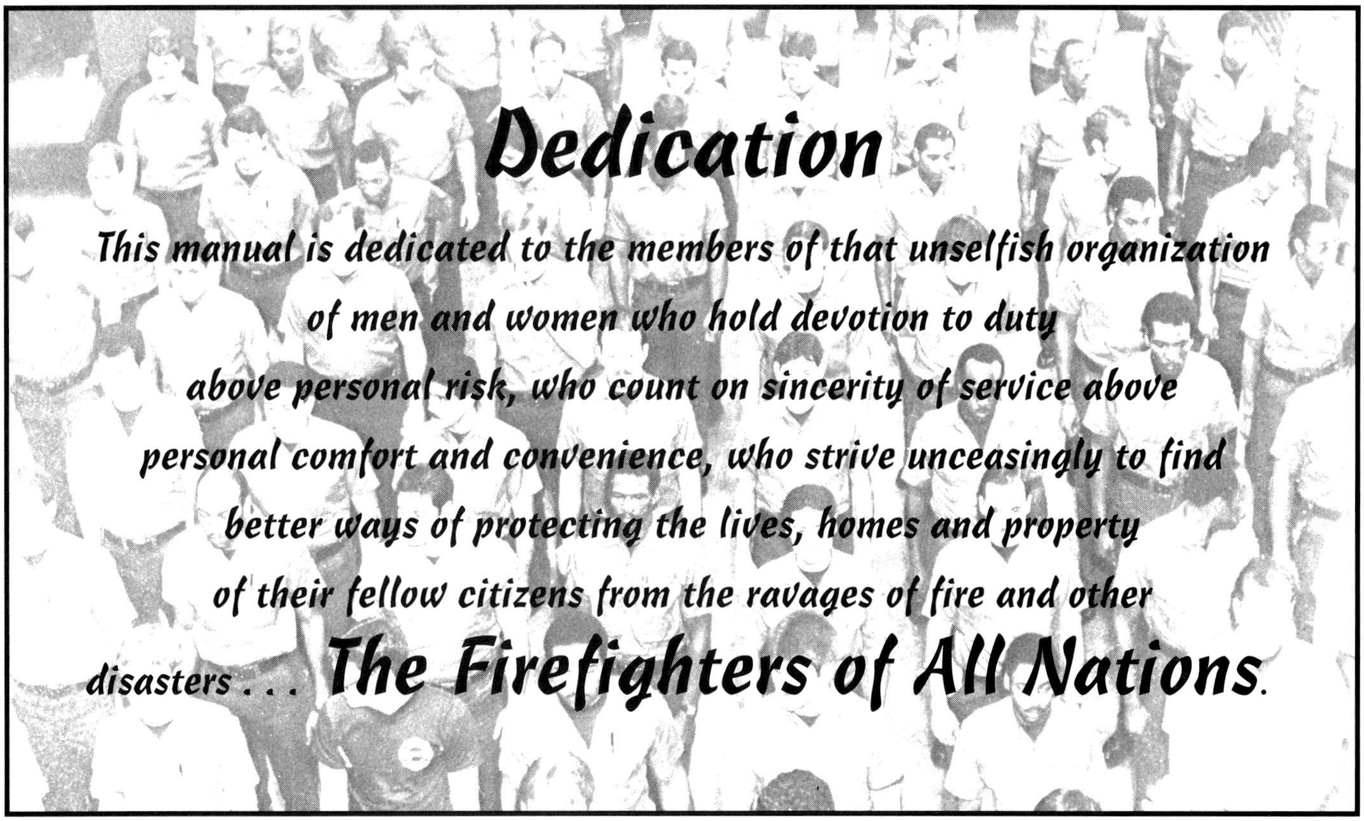

Dedication

This manual is dedicated to the members of that unselfish organization of men and women who hold devotion to duty above personal risk, who count on sincerity of service above personal comfort and convenience, who strive unceasingly to find better ways of protecting the lives, homes and property of their fellow citizens from the ravages of fire and other disasters . . . ***The Firefighters of All Nations.***

Dear Firefighter:

The International Fire Service Training Association (IFSTA) is an organization that exists for the purpose of serving firefighters' training needs.Fire Protection Publications is the publisher of IFSTA materials. Fire Protection Publications staff members participate in the National Fire Protection Association and the International Association of Fire Chiefs.

If you need additional information concerning our organization or assistance with manual orders, contact:

Customer Services
Fire Protection Publications
Oklahoma State University
Stillwater, OK 74078-0118
1 (800) 654-4055

For assistance with training materials, recommended material for inclusion in a manual, or questions on manual content, contact:

Technical Services
Fire Protection Publications
Oklahoma State University
Stillwater, OK 74078-0118
(405) 744-5723

THE INTERNATIONAL FIRE SERVICE TRAINING ASSOCIATION

The International Fire Service Training Association (IFSTA) was established as a "nonprofit educational association of fire fighting personnel who are dedicated to upgrading fire fighting techniques and safety through training." This training association was formed in November 1934, when the Western Actuarial Bureau sponsored a conference in Kansas City, Missouri. The meeting was held to determine how all the agencies interested in publishing fire service training material could coordinate their efforts. Four states were represented at this initial conference. Because the representatives from Oklahoma had done some pioneering in fire training manual development, it was decided that other interested states should join forces with them. This merger made it possible to develop training materials broader in scope than those published by individual agencies. This merger further made possible a reduction in publication costs, because it enabled each state or agency to benefit from the economy of relatively large printing orders. These savings would not be possible if each individual state or department developed and published its own training material.

To carry out the mission of IFSTA, Fire Protection Publications was established as an entity of Oklahoma State University. Fire Protection Publications' primary function is to publish and disseminate training texts as proposed and validated by IFSTA. As a secondary function, Fire Protection Publications researches, acquires, produces, and markets high-quality learning and teaching aids as consistent with IFSTA's mission. The IFSTA Executive Director is officed at Fire Protection Publications.

IFSTA's purpose is to validate training materials for publication, develop training materials for publication, check proposed rough drafts for errors, add new techniques and developments, and delete obsolete and outmoded methods. This work is carried out at the annual Validation Conference.

The IFSTA Validation Conference is held the second full week in July at Oklahoma State University or in the vicinity. Fire Protection Publications, the IFSTA publisher, establishes the revision schedule for manuals and introduces new manuscripts. Manual committee members are selected for technical input by Fire Protection Publications and the IFSTA Executive Secretary. Committees meet and work at the conference addressing the current standards of the National Fire Protection Association and other standard-making groups as applicable.

Most of the committee members are affiliated with other international fire protection organizations. The Validation Conference brings together individuals from several related and allied fields, such as:

- Key fire department executives and training officers
- Educators from colleges and universities
- Representatives from governmental agencies
- Delegates of firefighter associations and industrial organizations
- Engineers from the fire insurance industry

Committee members are not paid nor are they reimbursed for their expenses by IFSTA or Fire Protection Publications. They come because of commitment to the fire service and its future through training. Being on a committee is prestigious in the fire service community, and committee members are acknowledged leaders in their fields. This unique feature provides a close relationship between the International Fire Service Training Association and other fire protection agencies, which helps to correlate the efforts of all concerned.

IFSTA manuals are now the official teaching texts of most of the states and provinces of North America. Additionally, numerous U.S. and Canadian government agencies as well as other English-speaking countries have officially accepted the IFSTA manuals.

Table of Contents

Tables

Preface

This third edition of **Essentials of Fire Fighting** is intended to serve as a primary text for the firefighter candidate or as a reference text for fire fighting personnel who are already on the job. This book addresses most of the fire fighting objectives found in NFPA 1001, *Standard on Fire Fighter Professional Qualifications*. This updated version of **Essentials** contains new and revised information, two new chapters, and completely new photographs and artwork.

Acknowledgement and special thanks are extended to the members of the validating committee who contributed their time, wisdom, and knowledge to this manual.

Chairman
Jesse V. Jackson
Maryland Fire & Rescue Institute
College Park, MD

Secretary
Russell J. Strickland
Maryland Fire & Rescue Institute
College Park, MD

James P. Bugler
Nassau County Fire Service Academy
Old Bethpage, NY

Michael Callan
Connecticut State Fire School
Meriden, CT

David F. Clark
Illinois Fire Service Institute
Champaign, IL

Ronald C. Cody
Redondo Beach Fire Department
Redondo Beach, CA

Jeffrey B. Coffman
Fairfax County Fire & Rescue Department
Fairfax, VA

Daniel Contini
Dan's Fire House
Palmer, AK

Eric Haussermann
Fire Service Training
Stillwater, OK

George Lyon
Arlington County Fire Department
Arlington, VA

George Malik
Robert J. Quinn Fire Academy
Chicago, IL

Robert H. Noll
Yukon Fire Department
Yukon, OK

Adam Piskura
Minnesota Technical Institute System
St. Paul, MN

Frederick S. Richards
Office of Fire Prevention & Control
Albany, NY

Frederick C. Stark
Fire Standards & Training
Ocala, FL

Gerald L. Straus
Kitchener Fire Department
Kitchener, Ont., Canada

Paul C. Smith
Fire Academy
Vancouver, B.C., Canada

Special recognition is given to the Plano (TX) Fire Department whose help and cooperation was essential in obtaining a large number of the photographs needed to illustrate this manual. Helping with the project were:

Fire Chief William Peterson
Operations Chief Kirk Owen
Communications Supervisor William T. Jones, Jr.

Also instrumental in helping us obtain photos was the Tulsa (OK) Fire Department. Chief Tom Baker and Firefighter/Communications Specialist Frank Mason provided the assistance.

The Stillwater (OK) Fire Department, under the direction of Chief Jim Smith, also provided assistance. In particular, the following crew members at Station 3(B) assisted with numerous pictures:

Captain Steve Burdick
Firefighter Mike Eytcheson
Firefighter Tom Low

Oklahoma State Fire Service Training provided their facilities and personnel for a number of the photographs needed to finish the project. The following personnel assisted:

Assistant Director Nancy J. Trench
Kevin Smith
Thom Jones
Craig Woods

Special thanks to Henry Morse of Lakeland, Florida; Jon Jones of the NFPA; and Professor Emeritus O. C. Dermer of the OSU Chemistry Department who provided technical assistance in the revision of the Fire Behavior chapter of this manual.

The following individuals and organizations contributed information, photographs, and other help that was instrumental in the completion of this manual:

Harvey Eisner, Tenafly, NJ
Ron Jeffers, New Jersey Metro Fire Photographers Assoc.
Joseph Marino, New Britain, CT
Joel Woods, Maryland Fire & Rescue Institute
Bob Norman, Singerly (MD) Volunteer Fire Department
Fire Marshal Donald Tully, Buena Park, CA
Dennis Cranor, Kansas City (KS) Fire Department
Oklahoma City (OK) Fire Department:
    Assistant Chief Gary Marrs
    Station 14 Personnel
Phoenix (AZ) Fire Department:
    Kevin Roche
    Elaine Hutchings
Yukon (OK) Fire Department
Pennsburg (PA) Fire Company No. 1
Norristown (PA) Fire Department
East Greenville (PA) Fire Company No. 1
Rocky Hill (CT) Fire Department
New Haven (CT) Fire Department
Gainesville (FL) Fire-Rescue:
    George Braun, Public Information Officer
Claremore (OK) Fire Department
Edmond (OK) Fire Department
Jim Cox, Ansul Fire Protection
Oklahoma State University Safety Department
OSU Campus Fire Station Volunteer Firefighters
Fire and Rescue Training Institute at the University of Missouri - Columbia
Hale Fire Pump Co. — Hurst Rescue Tool Division
Lukas of America, Inc.
ALACO Ladder Company, Chino, CA:
    Kurt Kenworth, President
Montgomery County (PA) Fire Academy
Saulsbury Fire Apparatus, Tully, NY
Superior Flame Fighter Corp.

Bob Esposito, Pennsburg, PA
Tom McCarthy, Chicago (IL) Fire Department
Emergency One Fire Apparatus:
Jim Weigle
Shawn and Ryan Schoonover
Fred Myers
Ed Prendergast, Chicago (IL) Fire Department
Reliable Fire Equipment
Oberlin World Colorpress, Stillwater, OK
Southwest Loss Control, Inc.
Jim Hanson, Stillwater, OK
Oklahoma State University School of Fire Protection and Safety Engineering Technology
Electronic Data Systems, Plano, TX:
Tom Bennett, Security Support Services
Santa Rosa (CA) Fire Department
Ted and Janet Maker
Robert Fleischner

Last, but not least, gratitude is also extended to the following members of the Fire Protection Publications staff whose contributions made the final publication of this manual possible:

Michael A. Wieder, Senior Publications Editor
Carol M. Smith, Associate Editor
Cynthia S. Brakhage, Publications Specialist
Carl Goodson, Senior Publications Editor
Barbara Adams, Publications Specialist
Marsha Sneed, Publications Specialist
Ann Moffat, Senior Graphic Designer
Lori Schoonover, Graphic Designer
Desa Porter, Graphic Designer
Karen Murphy, Phototypesetting Technician
Don Davis, Publications Production Coordinator
Andy Maxey, Photographer
Kimberly Edwards, Photographic Technician
Joel A. Naspinski, Research Technician
Richard Windham, Research Technician
Kevin Kolb, Research Technician
Michael Bryant, Research Technician
Pat Agostinelli, Research Technician
Jason Goetz, Research Technician

Lynne C. Murnane
Managing Editor

Introduction

In the early 1970s, the leaders of the various major fire service organizations saw the need for a standard of professional competence for firefighters. A committee was formed to develop this standard; the results of which became known to the fire service as the National Fire Protection Association (NFPA) Standard 1001, *Standard on Fire Fighter Professional Qualifications.* The purpose of this standard is to specify, in terms of performance-based objectives, the minimum levels of competence required of a person who wishes to serve as either a paid or volunteer firefighter. NFPA 1001 has become widely accepted as the standard of measurement for all firefighters in North America and beyond.

The original edition of NFPA 1001 contained three levels of competence. These levels were Fire Fighter I, II, and III. These levels were maintained through the first several revisions on the standard. (In keeping with NFPA principles, all standards are revised every three to five years.) However, the latest edition of NFPA 1001, adopted in May of 1992, includes a major change being made. After reviewing the existing edition of the standard and realigning information as required, the committee determined that the information left in the Fire Fighter III section was repetitive of many of the objectives in the lower levels of NFPA 1021, *Standard on Fire Officer Professional Qualifications.* Thus, it was decided that NFPA 1001 would be limited to two levels, Fire Fighter I and II. It is also important to note that the new standard addresses several major topics that were not covered in the past. These include personal protective equipment, building construction, and firefighter safety. This manual addresses these issues with new chapters not contained in the previous editions.

It is also important to understand the committee's philosophy on the delineation between Fire Fighter I and Fire Fighter II. A Fire Fighter I is a person who is minimally trained to function safely and effectively as a member of a fire fighting team under direct supervision. A person meeting the requirements of Level I is by no means considered a "complete" firefighter. This is not accomplished until the objectives of both Levels I and II have been satisfied. The Fire Fighter II may operate under general supervision and may be expected to lead a group of equally or lesser trained personnel through the performance of a specified task.

It must be pointed out that NFPA 1001 is a **minimum** standard. Local jurisdictions may desire to exceed the minimum requirements in any or all areas of the standard. This is acceptable. It is not acceptable for local jurisdictions to weaken any portion of the standard to suit their needs.

The acceptance and recognition of a national standard provides a baseline for professionalism in the international fire service. It is the intent of Fire Protection Publications and the International Fire Service Training Association to promote this professionalism by providing excellence in the materials used to prepare fire personnel to meet the objectives of these standards.

PURPOSE AND SCOPE

The **Essentials of Fire Fighting** manual is designed to provide the firefighter candidate with the information needed to meet the fire-related performance objectives in NFPA 1001, Levels I and II. This manual also serves as an excellent reference for seasoned fire service personnel. The methods shown throughout this text have been approved by the International Fire Service Training Association as accepted methods for accomplishing each task. However, they are *not* to be interpreted as the only methods to accomplish a given task. Other specific methods for achieving any performance task may be specified by a local authority having jurisdiction. For guidance in seeking additional
ods for performing a given task, the student
may consult any of the IFSTA expand
(such as **Hose Practices** or **Fire S**
Ladder Practices) for more in-depth

In order to meet all of the objectives of NFPA 1001, the student will require several other IFSTA texts. These texts include **Fire Service First Responder**, **Fire Service Orientation and Indoctrination**, and **Hazardous Materials for First Responders**. The scope of the objectives under these particular topics made it impossible for us to cover them in one manual.

For ease of organization and presentation, the **Essentials** manual has been divided into chapters that correspond with the chapters in NFPA 1001. Because of the interrelatedness of the information, it was not possible to delineate between Level I and Level II information within each chapter. When possible, the chapters follow the order of the objectives of the standard; however, this is not always possible. It should be noted that the standard itself does not require the objectives to be mastered in the order in which they appear. Local jurisdictions may decide the order in which the material is to be presented.

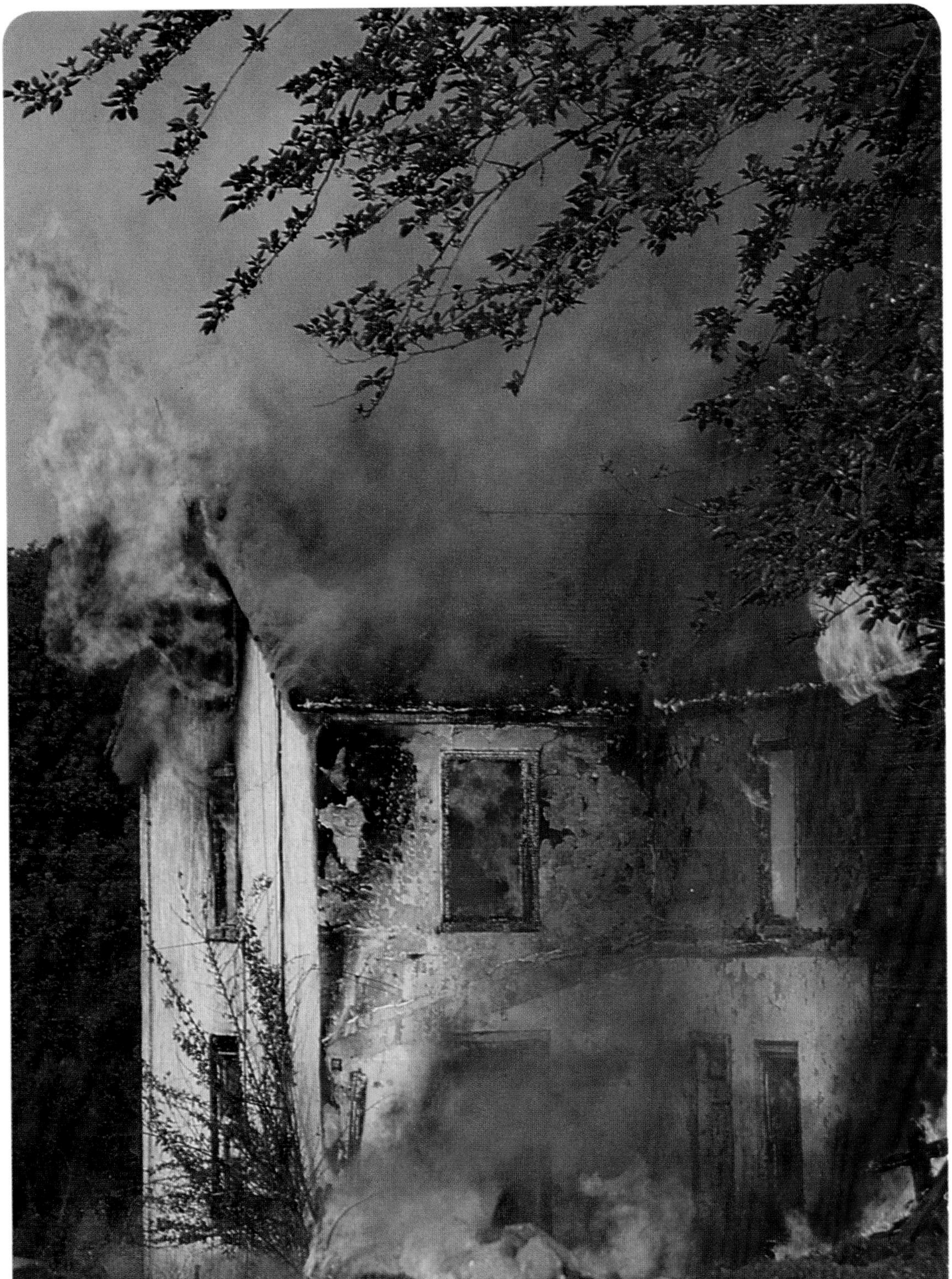

1

Fire Behavior

This chapter provides information that addresses performance objectives described in NFPA 1001, *Standard for Fire Fighter Professional Qualifications* (1992), particularly those referenced in the following sections:

3-5 Fire Behavior

3-5.1

3-5.2

3-5.3

3-5.4

3-5.5

3-5.6

3-5.7

3-5.8

4-5 Fire Behavior

4-5.1

4-5.2

4-5.3

4-5.4

Chapter 1
Fire Behavior

People first learn about fire as children. They know that fire consumes fuel, needs air, and gives off heat and light. Normally, that degree of understanding is all that one needs. Firefighters, however, have to take their understanding of this process a step or two further. In particular, they have to know more about the chemical process that goes on, the methods of heat transfer a fire can use, the makeup and nature of the fuels, and the environment the fire needs. It is this knowledge that arms the firefighter to fight fire and win.

Fire is actually a by-product of a larger process called combustion. Fire and combustion are two words used interchangeably by most people; however, firefighters should understand the difference. *Combustion* is the self-sustaining process of rapid oxidation of a fuel, which produces heat and light. *Fire* is the result of a rapid combustion reaction (Figure 1.1).

Fuels are most commonly oxidized by the oxygen in air; although, in some cases, other oxidizers, such as those inherent in a specific material, are used. The normal oxygen content in air is 21 percent; nitrogen makes up 78 percent of air; and the remaining 1 percent is made up of other gases such as water vapor, neon, and carbon dioxide.

Figure 1.1 Fire can be a rapid process that quickly destroys an entire structure. *Courtesy of Harvey Eisner.*

When oxygen is not oxidizing a fire, chlorine or chemicals with chlorine are oxidizing the fire. Some fuels do not need any oxidizer at all, because they have their own oxidizer tied up in their chemical formula. (This phenomenon can be compared to a firefighter with his or her own air supply functioning in an atmosphere that could not support breathing.) These self-oxidizing fuels can normally be recognized by the word oxide or oxalate in their name such as organic peroxide.

Combustion is defined as the process of rapid oxidation (resulting in fire). But oxidation is not always rapid. It may be very slow, or it may be instantaneous. Neither of these extremes produce fire (combustion) as we know it, but they are common occurrences in themselves. Very slow oxidation is more commonly known as rusting. (A light film of oil placed on metal prevents rusting because it keeps air and its oxygen off the metal so that it cannot react and oxidize the metal.) Instantaneous oxidation is an explosion, such as what occurs inside the casing of a bullet cartridge when the primer is ignited.

Rapid oxidation (combustion) can occur in two forms: smoldering fires and steady-state fires. Steady-state fires are sometimes also called free-burning fires.

IMPORTANT TERMS

In order to understand the material in this chapter, it is important that the reader know some basic terms that pertain to fire behavior. Knowledge of terms denoted by an asterisk [*] is required by NFPA 1001, *Standard for Fire Fighter Professional Qualifications.*

Boiling point — The temperature of a substance when the vapor pressure exceeds atmospheric pressure. At this temperature, the rate of evaporation exceeds the rate of condensation. At this point, more liquid is turning into gas than gas is turning back into a liquid.

**British thermal unit (Btu)* — The amount of heat needed to raise the temperature of one pound of water one degree Fahrenheit.

**Calorie* — The amount of heat needed to raise the temperature of one gram of water one degree Celsius.

**Celsius (Centigrade)* — Metric unit of temperature measurement. On the Celsius scale, 0 is the melting point of ice; 100 degrees is the boiling point of water.

**Fahrenheit* — Unit of temperature measurement primarily used in the United States. On the Fahrenheit scale, 32 degrees is the melting point of ice; 212 degrees is the boiling point of water.

**Fire point* — The temperature at which a liquid fuel will produce vapors sufficient to support continuous combustion once ignited. The fire point is usually a few degrees above the flash point.

Flame spread — The movement of flame away from the ignition source.

Flammable or explosive limits — The percentage of a substance (vapor) in air that will burn once it is ignited. Most substances have an upper (too rich) and a lower (too lean) flammable limit.

**Flash point* — The minimum temperature at which a liquid fuel gives off sufficient vapors to form an ignitable mixture with the air near the surface. At this temperature, the ignited vapors will flash but will not continue to burn.

Heat — The form of energy that raises temperature. Heat can be measured by the amount of work it does; for example, the amount of heat needed to make a column of mercury expand inside a glass thermometer.

**Ignition temperature* — The minimum temperature to which a fuel in air must be heated to start self-sustained combustion without a separate ignition source.

Oxidation — The complex chemical reaction of organic materials with oxygen or other oxidizing agents resulting in the formation of more stable compounds. Organic materials are substances containing carbon such as all plant and animal material and hydrocarbon fuels. More stable compounds are simply those with less bound-up chemical energy. They become more stable by releasing some of their energy as heat and light during combustion.

Vapor pressure — A measure of the tendency of a substance to evaporate.

SOURCES OF HEAT ENERGY

Heat is a form of energy that may be described as a condition of "matter in motion" caused by the movement of molecules. All matter contains some heat, regardless of how low the temperature, because molecules are constantly moving. When a body of matter is heated, the speed of the molecules increases, and thus the temperature also increases. Anything that sets the molecules of a material in faster motion produces heat in that material. The five general categories of heat energy are as follows:

- Chemical
- Electrical
- Mechanical
- Nuclear
- Solar

Chemical Heat Energy

Chemical heat energy is generated as the result of some type of chemical reaction. The four types of chemical reactions that result in heat production are heat of combustion, spontaneous heating, heat of decomposition, and heat of solution.

HEAT OF COMBUSTION

Heat of combustion is the amount of heat generated by the combustion (oxidation) reaction. The amount of heat generated by burning materials will vary depending on the material. This phenomenon is why some materials are said to burn "hotter" than others (Figure 1.2).

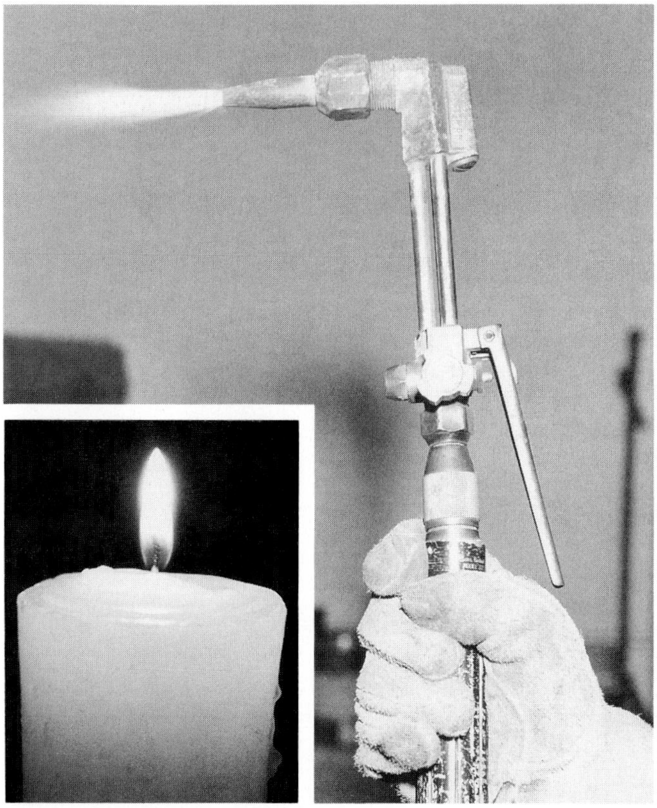

Figure 1.2 The heat of combustion varies depending on the fuel and other conditions. The flame on a candle will not burn as "hot" as the flame on a cutting torch.

SPONTANEOUS HEATING

Spontaneous heating is the heating of an organic substance without the addition of external heat. Spontaneous heating occurs most frequently where sufficient air is not present and insulation prevents dissipation of heat — heat that is produced by a low-grade chemical breakdown process. An example would be oil-soaked rags that are rolled into a ball and thrown into a corner. If there is not enough ventilation to let the heat drift off, eventually the heat will become sufficient to cause ignition of the rags (Figure 1.3). The speed of a heating reaction doubles with each 18°F (10°C) temperature increase.

HEAT OF DECOMPOSITION

Heat of decomposition is the release of heat from decomposing compounds, usually due to bacterial action. In some cases, these compounds may be unstable and release their heat very quickly; they may even detonate. In other cases, the reaction and resulting release of heat is much slower. This reaction can be commonly seen when viewing a compost pile (Figure 1.4). The decomposition of organic materials creates heat that can be seen on cold days when holes are poked into the pile. The heated vapors and steam can be seen rising from the holes in the pile.

Figure 1.3 Spontaneous heating of oily rags is prevented when they are stored in an oxygen-limiting container.

Figure 1.4 Heat of decomposition occurs in composted organic materials.

HEAT OF SOLUTION

Heat of solution is the heat released by the solution of matter in a liquid. Some acids, when dissolved in water, can produce violent reactions, spewing hot water and acid with explosive force.

Electrical Heat Energy

It is not uncommon to see "electrical" listed as the cause for a fire that damaged a building or automobile. Electricity has the ability to generate high temperatures that are capable of igniting any combustible materials near the heated area. Electrical heating can occur in a variety of ways. This section highlights some of the more common ways.

RESISTANCE HEATING

Resistance heating refers to the heat generated by passing an electrical current through a conductor such as a wire or an appliance. Resistance heating is increased if the wire is not large enough in diameter for the amount of current. Fires are caused when a simple extension cord is overloaded with too many appliances plugged into it. Resistance heating can also be increased if the conductor is tightly wound. Fires are often started by cheap electrical extension cords that are wound around the leg of a table or chair to take up the slack (Figure 1.5).

DIELECTRIC HEATING

Dielectric heating, used in microwave ovens, occurs as a result of the action of pulsating either direct current (DC) or alternating current (AC) at high frequency on a nonconductive material (Figure 1.6). The nonconductive material is not heated by the dielectric heating; however, it is heated by being in constant contact with electricity. This is somewhat similar to bombarding an object with many little lightning bolts.

Figure 1.5 Resistance heating may occur when an electrical cord is wound tightly around an object.

Figure 1.6 Dielectric heating is mainly used in microwave ovens.

LEAKAGE CURRENT HEATING

Leakage current heating occurs when a wire is not insulated well enough to contain all the current. Some current leaks out into the surrounding material such as inside the wall of a structure. This current causes heat and can cause a fire.

HEAT FROM ARCING

Heat from arcing is a type of electrical heating that occurs when the current flow is interrupted. Interruption may be from an open switch or a loose connection.

Arc temperatures are extremely high and may even melt the conductor. A common arc used in industrial applications is the arc welder (Figure 1.7). Here the welding rod (conductor) melts away as metals are joined together.

Figure 1.7 A welding machine uses heat created by arcing.

Figure 1.8 Flammable liquid containers should be bonded together before transferring liquid between the two.

STATIC ELECTRICITY

Static electricity is the buildup of positive charge on one surface and negative charge on another surface. The charges are naturally attracted to each other and seek to become evenly charged again. This condition is shown when the two surfaces come close to each other, as in the case of a person's finger and a metal doorknob, and an arc occurs producing the familiar spark and small shock.

Static electricity is often to blame when a fire occurs as flammable liquids are transferred between containers that are not properly electrically bonded together (Figure 1.8). This is why airplanes and fuel tankers have to be grounded and bonded by a wire during fuel transfer operations. Electric bonding is not necessary when filling the fuel tank of an automobile, because gasoline has special additives that make the fuel act as a ground and the metal nozzle is in contact with the filling pipe.

Heat generated by lightning is static electricity on a very large scale. The heat generated by the discharge of billions of volts from either earth to cloud, cloud to cloud, or cloud to ground can be in excess of 60,000°F (33 300°C).

Mechanical Heat Energy

Mechanical heat is generated two ways — by friction and compression. Heat of friction is created by the movement of two surfaces against each other. This movement results in heat and/or sparks being generated. Heat of compression is generated when a gas is compressed. Diesel engines ignite fuel vapor without a spark plug by the use of this principle. It is also the reason that SCBA bottles feel warm to the touch after they have been filled.

Nuclear Heat Energy

Nuclear heat energy is generated when atoms are either split apart (fission) or combined (fusion). In a controlled setting, fission is used to heat water to drive steam turbines and produce electricity. Currently, fusion cannot be controlled and has no commercial use.

Solar Heat Energy

The energy transmitted from the sun in the form of electromagnetic radiation is called solar heat energy.

Typically, solar energy is distributed fairly evenly over the face of the earth and in itself is not really capable of starting a fire. However, when solar energy is concentrated on a particular point, as through the use of a lens, it may ignite combustible materials.

HEAT TRANSFER

A number of the natural laws of physics are involved in the transmission of heat. One is called the Law of Heat Flow; it specifies that heat tends to flow from a hot substance to a cold substance. The colder of two bodies in contact will absorb heat until both objects are at the same temperature. Heat can travel throughout a burning building by one or more of three methods: conduction, convection, and radiation. The following sections describe how this transfer takes place.

Conduction

Heat may be conducted from one body to another by direct contact of the two bodies or by an intervening heat-conducting medium (Figure 1.9). An example of this type of heat transfer is a basement fire that heats pipes enough to ignite the wood inside walls several rooms away. The amount of heat that will be transferred and its rate of travel depend upon the conductivity of the material through which the heat is passing. Not all materials have the same heat conductivity. Aluminum, copper, and iron are good conductors; however, fibrous materials, such as felt, cloth, and paper, are poor conductors.

Liquids and gases are poor conductors of heat because of the movement of their molecules, and air is a relatively poor conductor. This factor is why double building walls and storm windows that contain an air space provide additional insulation from outside air temperatures. Certain solid materials, such as fiberglass, shredded into fibers and packed into batts make good insulation because the material itself is a poor conductor and there are air pockets within the batting.

Convection

Convection is the transfer of heat by the movement of air or liquid (Figure 1.10). When water is heated in a glass container, the movement within the vessel can be observed through the glass. If sawdust is added to the water, the movement is more apparent. As the water is heated, it expands and grows lighter, hence, the upward movement. In the same manner, as air near a steam

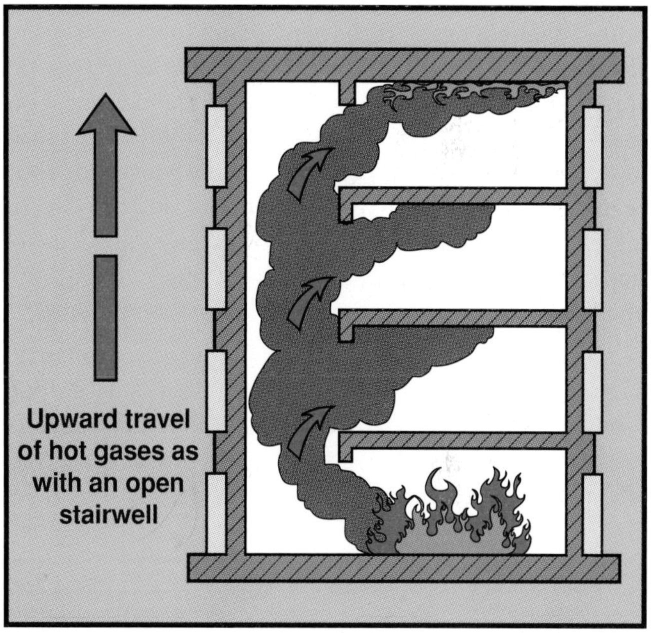
Upward travel of hot gases as with an open stairwell

Figure 1.10 An example of convection.

Figure 1.9 An example of conduction is a basement fire that heats pipes enough to ignite the wood inside walls several rooms away.

radiator becomes heated by conduction, it expands, becomes lighter, and moves upward. As the heated air moves upward, cooler air takes its place at the lower levels. When liquids and gases are heated, they begin to move within themselves. This movement is different from the molecular motion discussed in conduction of heat and is responsible for heat transfer by convection.

Heated air in a building will expand and rise. For this reason, fire spread by convection is mostly in an upward direction; however, air currents can carry heat in any direction. Convection currents are generally the cause of heat movement from floor to floor, from room to room, and from area to area. The spread of fire through corridors, up stairwells and elevator shafts, between walls, and through attics is caused mostly by the convection of heat currents. If the convecting heat encounters a ceiling or other barrier that keeps it from rising, it will spread out laterally (sideways) along the ceiling. If it runs out of ceiling space, it will travel down the wall toward the floor, being pushed by more heated air that is rising behind it. Convected heat encountering a ceiling is commonly referred to as mushrooming. Convection has more influence upon the positions for fire attack and ventilation than either radiation or conduction.

Although often mistakenly thought to be a separate form of heat transfer, direct flame contact is actually a form of convective heat transfer. When a substance is heated to the point where flammable vapors are given off, these vapors may be ignited, creating a flame. As other flammable materials come in contact with the burning vapors, or flame, they may be heated to a temperature where they, too, will ignite and burn.

Radiation

Although air is a poor conductor, it is obvious that heat can travel where matter does not exist. The warmth of the sun reaches us even though it is not in direct contact with us (conduction), nor is it heating up gases that travel to us (convection). This method of heat transmission is known as radiation of heat waves (Figure 1.11). Heat and light waves are similar in nature, but they differ in length per cycle. Heat waves are longer than light waves, and they are sometimes called infrared rays. Radiated heat will travel through space until it reaches an opaque object. As the object is exposed to heat radiation, it will in return radiate heat from its surface. Radiated heat is one of the major sources of fire spread to exposures, and its importance as a source of fire spread demands immediate attention at locations where radiation exposure is severe.

PRINCIPLES OF FIRE BEHAVIOR

Fuel may be found in any of three states of matter: solid, liquid, or gas (Figure 1.12). Only gases burn. The initiation of combustion of a liquid or solid fuel requires their conversion into a gaseous state by heating. Fuel gases are evolved from solid fuels by pyrolysis. Pyrolysis is the chemical decomposition of a substance through the action of heat.

Figure 1.11 Radiated heat is one of the major sources of fire spread to exposures.

Figure 1.12 Fuel comes in solid, liquid, and gaseous states.

Fuel gases are evolved from liquids by vaporization. This process is the same for water evaporating by boiling or water in a container evaporating in sunlight. In both cases, heat causes the liquid to vaporize. Generally, the vaporization process of liquid fuels requires less heat input than does the pyrolysis process for solid fuels. This places considerable restraints on the control and extinguishment of gas fuel fires because their reignition is much more likely.

Gaseous fuels can be the most dangerous, because they are already in the natural state required for ignition. No pyrolysis or vaporization will be needed to ready the fuel. These fuels are also the most difficult to contain.

Fuel Characteristics

Solid fuels have a definite shape and size that significantly affect the ignitability of the fuel. Of primary consideration is the surface-to-mass ratio, which is the ratio of the surface area of the fuel to the mass of the fuel. As this ratio increases, the fuel particles become smaller and more finely divided (i.e., sawdust as opposed to logs), and the ignitability increases tremendously (Figure 1.13). As the surface area increases, heat transfer is easier and the material heats more rapidly, thus speeding pyrolysis.

Figure 1.13 Sawdust will ignite much quicker than will a log, even though they are composed of the same material.

The physical position of a solid fuel is also of great concern to fire fighting personnel. If the solid fuel is in a vertical position, fire spread will be more rapid than if it is in a horizontal position. The rapidity of the fire spread is due to increased heat transfer through convection as well as conduction and radiation.

Liquid fuels have physical properties that increase the difficulty of extinguishment and the hazard to personnel. A liquid will assume the shape of its container. When a spill occurs, the liquid will assume the shape of the ground (flat), and it will flow and accumulate in low areas.

Density is a measure of how tightly the molecules of a substance are packed together. Dense materials are heavy. Imagine two boats of equal size, one with two people in it and the other with twenty people (Figure 1.14 on next page). The twenty people would have to be packed together tightly. Although the two boats are the same size, the boat with the twenty people would be heavier. That boat would be more dense.

Figure 1.14 The boat containing twenty people is more dense than the boat containing two people.

The density of liquids in relation to water is known as specific gravity. Water is given a value of one. Liquids with a specific gravity less than one are lighter than water, while those with a specific gravity greater than one are heavier than water. If the other liquid also has a density of one, it will mix evenly with water. It is interesting to note that most flammable liquids have a specific gravity of less than one. This means that if a firefighter is confronted with a flammable liquid fire and flows water on it improperly, the whole fire can just float away on the water and ignite everything in its path.

The solubility of a liquid fuel in water is also an important factor. Alcohols and other polar solvents dissolve in water. If large volumes of water are used, alcohols and other polar solvents may be diluted to the point where they will not burn. As a rule, hydrocarbon liquids (nonpolar solvents) will not dissolve in water and will float on top of water. This is why water alone cannot wash oil off our hands; the oil does not dissolve in the water. In addition to the water, soap must be used to dissolve the oil.

Consideration must be given to which extinguishing agents are effective on hydrocarbons (insoluble) and which affect polar solvents and alcohols (soluble). Today, multipurpose foams are available that will work on both types of liquid fuels (Figure 1.15).

The volatility or ease with which the liquid gives off vapor influences fire control. All liquids give off vapors to a greater or lesser degree in the form of simple evaporation. Liquids that give off large quantities of flammable or combustible vapors can be dangerous because they may be easily ignited.

Vapor density is the density of gas or vapor in relation to air. Vapor density is of concern with volatile liquids and with gaseous fuels. Gases tend to assume the shape of their container but have no specific volume. If a vapor is less dense than air (air is given a value of one), it will rise and tend to dissipate. If a gas or vapor is heavier than air, which is more common than the previous case, it will tend to hug the ground and travel as directed by terrain and wind.

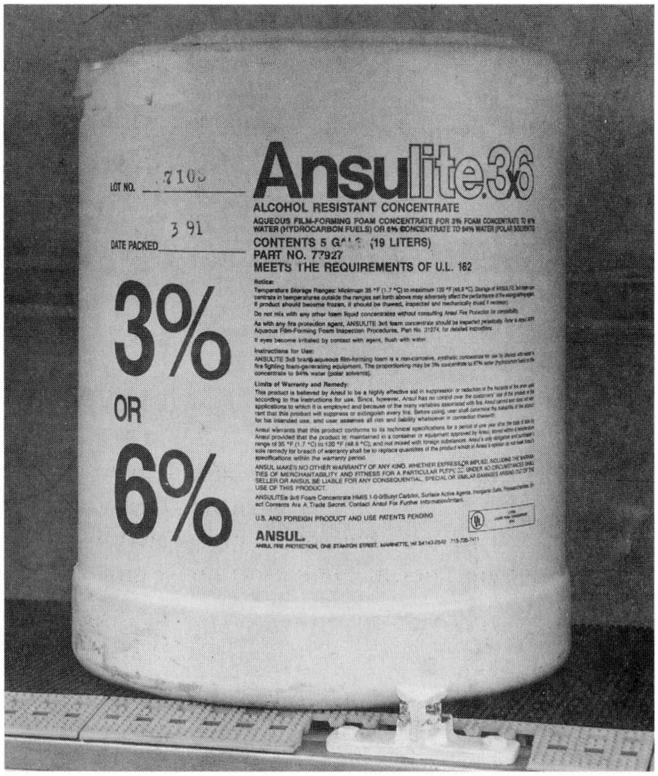

Figure 1.15 Multipurpose foams are used on both polar solvent and hydrocarbon fires.

It is important for all firefighters to know that every hydrocarbon except the lightest one, methane, has a vapor density greater than one, will sink and hug the ground, will flow into low lying areas, and poses a real hazard for that reason. Common gases such as ethane, propane, and butane are examples of hydrocarbon gases that are heavier than air.

Fuel-To-Air Mixture

Once a fuel has been converted into a gaseous state, it must mix with an oxidizer to burn. The usual oxidizer is the oxygen in air. The mixture of the fuel vapor and the oxidizer must be within the flammable limits for that fuel (Table 1.1). That is, there must be enough but not too

TABLE 1.1
EXAMPLES OF FLAMMABLE RANGES

Fuel	*Lower Limit (%)*	*Upper Limit (%)*
Gasoline Vapor	1.4	7.6
Methane (Natural Gas)	5.0	17
Propane	2.2	9.5
Hydrogen	4.0	75
Acetylene	2.5	100

much fuel vapor for the amount of oxidizer. If there is too much fuel vapor, the mixture is too rich to burn. If there is not enough, it is too lean to burn. The flammable limits of how rich or how lean a fuel vapor mixture can be and still burn are recorded in handbooks and are usually reported for temperatures of 70°F (21°C). These limits change slightly with temperature.

The Burning Process

When the proper fuel vapor/air mixture has been achieved, it must then be raised to its ignition temperature or the point at which self-sustained combustion will continue. Fire burns in two basic modes: flaming or surface combustion. The flaming mode of combustion, such as the burning of logs in a fireplace, is represented by the fire tetrahedron (fuel, temperature, oxygen, and the uninhibited chemical chain reaction) (Figure 1.16). The surface, or smoldering mode of combustion, is represented by the fire triangle (fuel, temperature, and oxygen) (Figure 1.17).

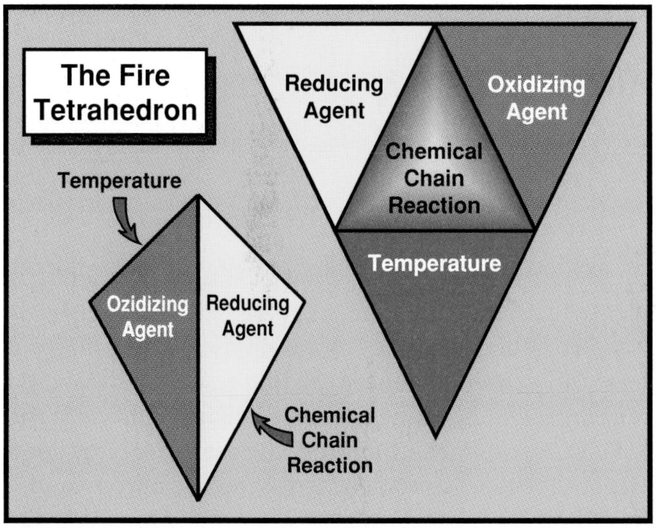

Figure 1.16 The fire tetrahedron represents the flaming mode of combustion.

The fuel segment of both diagrams is any solid, liquid, or gas that can combine with oxygen in the chemical reaction known as oxidation. A fuel with a sufficiently high temperature will ignite if an oxidizing agent is being liberated. Combustion will continue as long as enough energy or heat is present. Under most conditions, the oxidizing agent will be the oxygen in air. However, some materials, such as sodium nitrate and potassium chlorate, release their own oxygen during combustion and can cause fuels to burn in an oxygen-free atmosphere. While scientists only partially understand what happens in the combustion chemical chain reaction, they do know that heating a fuel can produce vapors that contain substances that will combine with oxygen and burn. These substances must continue to be produced for combustion to continue.

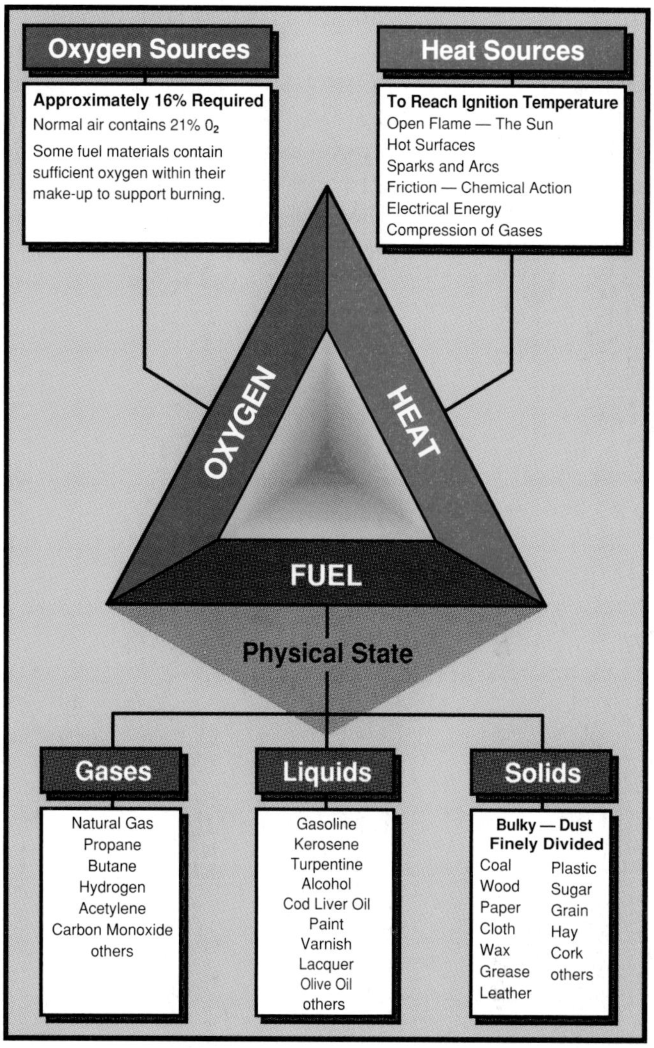

Figure 1.17 The fire triangle represents the smoldering mode of combustion.

A self-sustaining combustion reaction of solids and liquids depends on radiative feedback. Radiative feedback is radiant heat providing energy for continued vaporization. When sufficient heat is present to maintain or increase this feedback, the fire will remain constant or will grow, depending on the heat produced. When heat is fed back to the fuel, this is known as a positive heat balance. If heat is dissipated faster than it is generated, a negative heat balance is created. A positive heat balance is required to maintain combustion.

The amount of oxygen available to support combustion is important. As stated previously, air contains about 21 percent oxygen under normal circumstances. However, firefighters may frequently encounter situations where less than 21 percent oxygen is present. The three most common situations where this would be found are in empty storage tanks, oxygen-limiting silos, and structural fires (Figure 1.18). Often, storage tanks are being purged or are filled with residual gases or vapors that result in an oxygen-deficient atmosphere. If a worker or rescuer enters the tank without breathing equipment, the person will soon become unconscious and may die.

Figure 1.18 A firefighter entering an oxygen-limiting silo must always wear personal protective breathing apparatus.

The oxygen-deficient atmosphere often found in structural fires is a result of the fire's need for oxygen to continue the burning process. If the structure is tightly closed, the fire will use much of the available oxygen. As we will see later in this chapter, this can create even more serious problems in addition to the inability to support life.

In either case, oxygen concentrations below the normal 21 percent adversely affect both fire production and life safety. Table 1.2 lists the symptoms associated with operating in an oxygen-deficient atmosphere from a life safety standpoint. From a fire standpoint, the intensity of the fire will begin to decrease below 18 percent oxygen concentration. Oxygen concentrations below 15 percent will not support combustion.

TABLE 1.2
PHYSIOLOGICAL EFFECTS OF REDUCED OXYGEN (HYPOXIA)

Oxygen in Air (Percent)	*Symptoms*
21	None — normal conditions
17	Some impairment of muscular coordination; increase in respiratory rate to compensate for lower oxygen content
12	Dizziness, headache, rapid fatigue
9	Unconsciousness
6	Death within a few minutes from respiratory failure and concurrent heart failure

NOTE: These data cannot be considered absolute because they do not account for difference in breathing rate or length of time exposed.

These symptoms occur only from reduced oxygen. If the atmosphere is contaminated with toxic gases, other symptoms may develop.

Phases Of Fire

Fires may start at any time of the day or night if a hazard exists. If the fire happens when the area is occupied and/or protected by automatic suppression and detection systems, chances are that it will be discovered and controlled in the beginning (incipient) phase. If the fire occurs when the building is closed, deserted, and without fixed protection systems, the fire may go undetected until it has gained major headway. The phase of a fire in a closed building is of chief importance when determining ventilation requirements.

Fire in a confined room or building has two particularly important characteristics. The first characteristic is that there is a limited amount of oxygen. This differs from an outdoor fire, where the oxygen supply is unlimited. The second characteristic is that the fire gases that are given off are trapped inside the structure and build up, unlike outdoors where they can dissipate. When fire is confined in a building or room, the situation requires carefully thought-out and executed ventilation procedures if further damage is to be prevented and danger reduced. Fire confined to a building or room can be best understood by an investigation of its three main progressive phases: incipient, steady-state burning, and hot smoldering. A firefighter may be confronted by one or all of the phases of fire at any time; therefore, a working knowledge of these phases is important for understanding ventilation procedures. Firefighters must also be aware of the variety of potentially hazardous conditions that may be intertwined within the three main phases. These hazards include rollover, flashover, and backdraft.

INCIPIENT PHASE

The incipient phase is the earliest phase of a fire beginning with the actual ignition. The fire is limited to the original materials of ignition. In the incipient phase, the oxygen content in the air has not been significantly reduced, and the fire is producing water vapor (H_2O), carbon dioxide (CO_2), perhaps a small quantity of sulfur dioxide (SO_2), carbon monoxide (CO), and other gases. Some heat is being generated, and the amount will increase as the fire progresses. The fire may be producing a flame temperature well above 1,000°F (537°C), yet the temperature in the room at this stage may be only slightly increased (Figure 1.19).

Rollover

Rollover, sometimes referred to as flameover, takes place when unburned combustible gases released during the incipient or early steady-state phase accumulate at the ceiling level (Figure 1.20). These superheated gases are pushed, under pressure, away from the fire area and into uninvolved areas where they mix with oxygen. When their flammable range is reached, they ignite and a fire front develops, expanding very rapidly and rolling over the ceiling (Figure 1.21 on the next page). This is one of the reasons firefighters must stay low when advancing hoselines. Rollover differs from flashover in that only the gases are burning and not the contents of the room. The rollover will continue until its fuel is eliminated. This is done by extinguishing the main body of fire. The rollover will cease when the fire itself stops producing the flammable gases that are feeding the rollover.

STEADY-STATE BURNING PHASE

For purposes of simplicity, the steady-state burning phase (sometimes referred to as the free-burning phase)

Figure 1.19 The incipient phase is the earliest phase of a fire.

Figure 1.20 Prerollover occurs when unburned combustible gases accumulate at the ceiling level.

Figure 1.21 An example of rollover.

can generally be considered the phase of the fire where sufficient oxygen and fuel are available for fire growth and open burning to a point where total involvement is possible. During the early portions of this phase, oxygen-rich air is drawn into the flame, as convection (the rise of heated gases) carries the heat to the uppermost regions of the confined area. The heated gases spread out laterally from the top downward, forcing the cooler air to seek lower levels, and eventually igniting all the combustible material in the upper levels of the room (Figure 1.22). This early portion of the steady-state burning phase is often called the flame-spread phase. The presence of this heated air is one of the reasons firefighters are taught to keep low and use protective breathing equipment. One breath of this superheated air can sear the lungs. At this point, the temperature in the upper regions can exceed 1,300°F (700°C).

If conditions are perfect, and they rarely are, the fire may achieve what is commonly referred to as "clear burning." Clear burning is accompanied by high temperatures and complete combustion. Little or no smoke is given off. This fire is usually seen only when very clean fuels, such as methanol-based race car fuels, burn. Thermal columns will normally occur with rapid air movements upward from the base of the fire. As the fire progresses (in a confined space) through the latter portions of the steady-state burning phase, the fire continues to consume the free oxygen until it reaches the point where there is insufficient oxygen to react with the fuel. The fire is then reduced to the smoldering phase, but this fire needs only a fresh supply of oxygen to burn rapidly.

Figure 1.22 The steady-state burning phase.

Flashover

Flashover occurs when flames flash over the entire surface of a room or area (Figure 1.23). The actual cause of flashover is attributed to the buildup of heat from the fire itself. As the fire continues to burn, all the contents of the fire area are gradually heated to their ignition temperatures. When they reach their ignition point, simultaneous ignition occurs, and the area becomes fully involved in fire. This actual ignition is almost instantaneous and can be quite dramatic. A flashover can usually be avoided by directing water toward the ceiling level and the room contents to cool materials below their ignition temperatures.

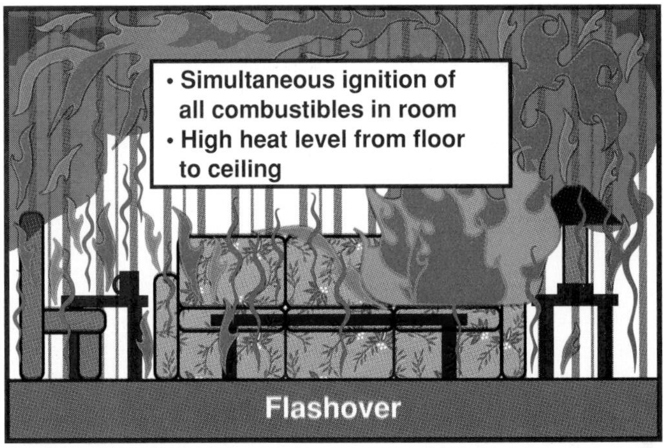

Figure 1.23 A flashover can usually be avoided by directing water toward the ceiling and the room contents to cool materials below their ignition temperatures.

HOT-SMOLDERING PHASE

After the steady-state burning phase, flames may cease to exist if the area of confinement is sufficiently airtight. In this instance, burning is reduced to glowing embers. As the flames die down, the room becomes completely filled with dense smoke and gases. Air pressure from gases being given off may build to the extent that smoke and gases are forced through small cracks (Figure 1.24). Room temperatures in excess of 1,000°F (537°C) are possible. The intense heat will have liberated the lighter fuel fractions, such as methane, from the combustible material in the room. These fuel gases will be added to those produced by the fire and will further increase the hazard to the firefighter and create the possibility of a backdraft if air is improperly introduced into the room. If air is not introduced into the room, the fire will eventually burn out, leaving totally incinerated contents.

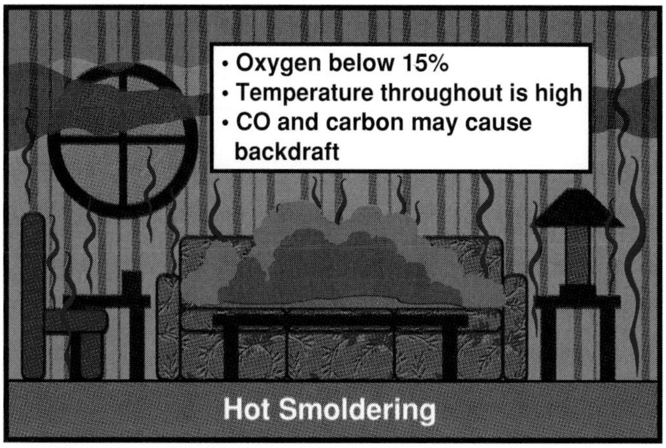

Figure 1.24 The hot-smoldering phase.

Backdraft

Firefighters responding to a confined fire that is late in the steady-state burning phase or in the hot-smoldering phase risk causing a backdraft (also known as a smoke explosion) if the science of fire is not considered in opening the structure.

In the hot-smoldering phase of a fire, burning is incomplete because of insufficient oxygen to sustain the fire (Figure 1.25). However, the heat from the steady-state burning phase remains, and the unburned carbon particles and other flammable products of combustion are available for instantaneous combustion when more oxygen is supplied. Improper ventilation, such as opening a door or breaking a window, supplies the dangerous missing link — oxygen. As soon as the needed oxygen rushes in, the stalled combustion resumes; it can be devastating in its speed, truly qualifying as an explosion (Figure 1.26 on the next page). Backdraft can be the most hazardous condition a firefighter will ever face.

Combustion is oxidation, and oxidation is a chemical reaction in which oxygen combines with other elements. Carbon is a naturally abundant element present in wood and most plastics, among other things. When wood burns, carbon combines with oxygen to form carbon dioxide (CO_2) or carbon monoxide (CO), depending on the availability of oxygen. When enough oxygen is no longer available, large quantities of free carbon are released in the smoke. Thus, a warning sign of possible backdraft is dense, black (carbon-filled) smoke.

The following characteristics may indicate the potential for a backdraft to occur:

- Pressurized smoke exiting small openings
- Black smoke becoming dense gray yellow
- Confinement and excessive heat
- Little or no visible flame
- Smoke leaving the building in puffs or at intervals
- Smoke-stained windows

Figure 1.25 There is insufficient oxygen to sustain the fire in the prebackdraft stage.

- Muffled sounds
- Sudden rapid movement of air inward when opening is made

This situation can be made less dangerous by proper ventilation. If the room or building is opened at the highest point involved, the heated gases and smoke will be released, reducing the possibility of an explosion.

THERMAL LAYERING OF GASES

The thermal layering of gases is the tendency of gases to form into layers, according to temperature. Other terms sometimes used to describe this layering of gases by heat are heat stratification and thermal balance. The hottest gases tend to be in the top layer, while the cooler ones form the bottom layer. Smoke is a heated mixture of air, gases, and particles, and it rises. If a hole is made in the roof, the smoke will rise from the building or room to the outside.

Thermal layering is critical to fire fighting activities. As long as the hottest air and gases are allowed to rise, the lower levels will be safer for firefighters (Figure 1.27). This normal layering of the hottest gases to the top and out the ventilation opening can be disrupted if water is improperly applied.

If water is improperly applied to the fire area and the area is not ventilated, the water will cool and condense the steam generated by the initial fire attack. This reaction causes the smoke and steam to circulate within all levels of the fire area. This swirling of smoke and steam is the result of disrupted normal thermal layering (Figure 1.28). This process is sometimes referred to as disrupting the thermal balance or creating a thermal imbalance. Many firefighters have been needlessly burned when thermal layering was disrupted. Once the normal layering is disrupted, forced ventilation procedures must be used to clear the area.

Figure 1.26 Backdraft can be the most hazardous condition a firefighter will ever face.

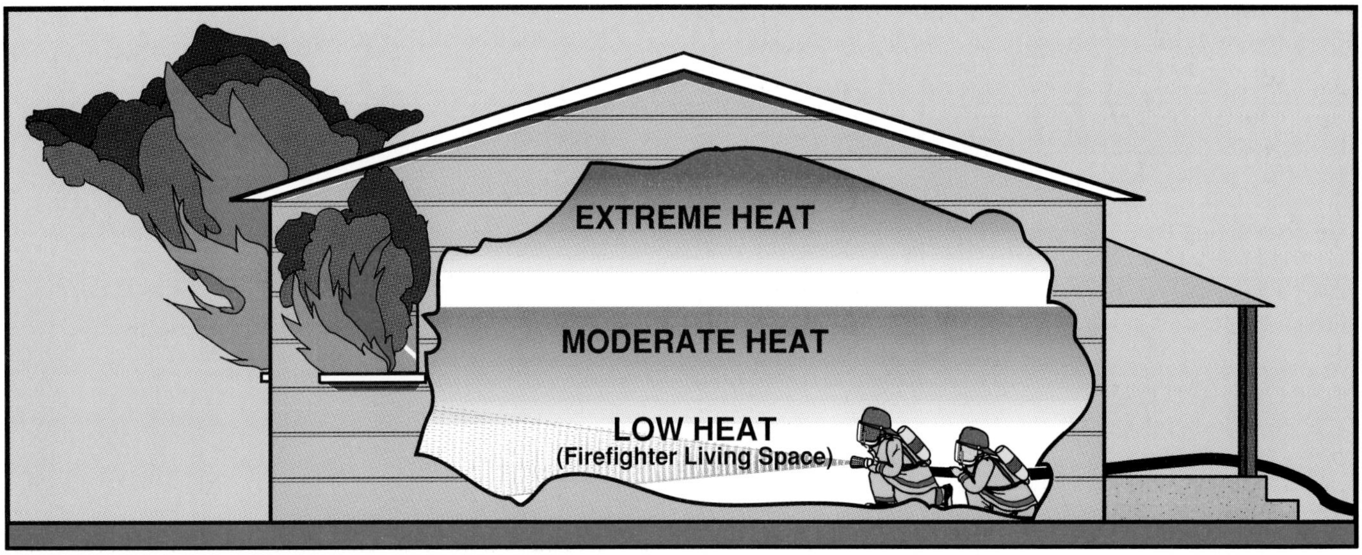

Figure 1.27 Under normal fire conditions in a closed structure, the highest levels of heat will be found at ceiling level, and the lowest level of heat will be at floor level.

Figure 1.28 Improper application of water creates a thermal imbalance.

PRODUCTS OF COMBUSTION

Incomplete combustion, of course, also leaves behind some unburned or charred fuel. When a material (fuel) burns, it undergoes a chemical change. None of the elements making up the material are destroyed in the process, but all of the material is transformed into another form or state. For example, when a piece of paper burns, the gases and moisture contained within the paper are liberated. The remaining solids take on the appearance of carbonized, charred flakes. Although it was once thought that the weight of the various by-products was the same as the original weight of the fuel, it is now known that a tiny amount of the fuel is indeed converted into energy, so the by-products weigh slightly less than the fuel did.

When a fuel burns, there are four products of combustion: heat, light, smoke, and fire gases (Figure 1.29). Heat is a form of energy that is measured in degrees of temperature to signify its intensity. Heat is the product of combustion that is responsible for the spread of fire. It is also the direct cause of burns, dehydration, heat exhaustion, and injury to the respiratory tract.

Flame is the visible, luminous body of a burning gas. When a burning gas is mixed with the proper amounts of oxygen, the flame becomes hotter and less luminous. The loss of luminosity is caused by a more complete combustion of the carbon. For these reasons, flame is considered to be a product of combustion. Of course, it is not present in those types of combustion, such as smoldering fires, that do not produce a flame.

The smoke encountered at most fires consists of a mixture of oxygen, nitrogen, carbon dioxide, carbon monoxide, finely divided carbon particles (soot), and a miscellaneous assortment of products that have been released from the material involved. The contents of the smoke will vary depending on the exact material that is burning. Some materials give off more smoke than others. Liquid fuels generally give off dense, black smoke. Oil, tar, paint, varnish, rubber, sulfur, and many plastics also give off dense smoke.

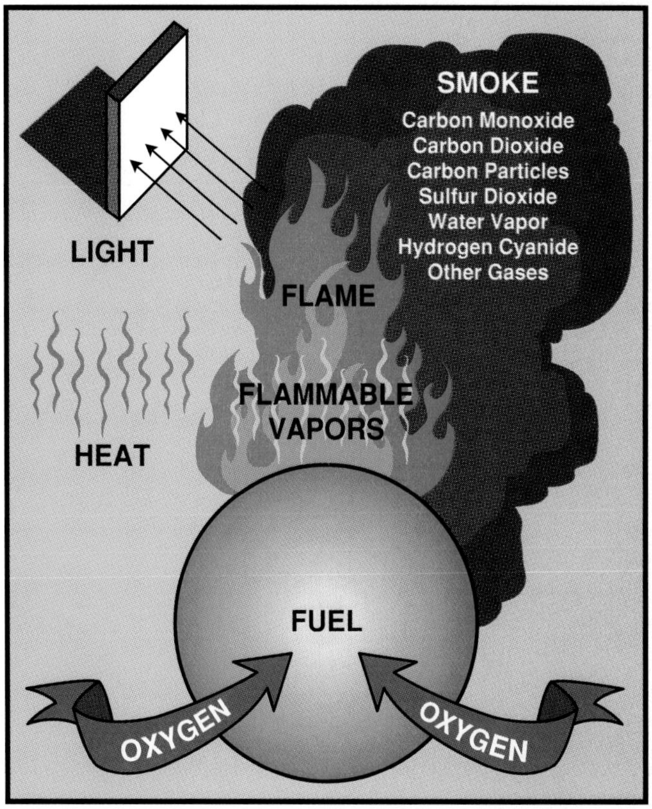

Figure 1.29 The four products of combustion are heat, light, smoke, and fire gases.

FIRE EXTINGUISHMENT THEORY

The extinguishment of fire is carried out by limiting or interrupting one or more of the essential elements in the combustion process. With flaming combustion, the fire may be extinguished by reducing temperature, eliminating fuel or oxygen, or by stopping the uninhibited chemical chain reaction (Figure 1.30). If a fire is in the smoldering mode of combustion, only three extinguishment options exist: reduction of temperature, elimination of fuel, or elimination of oxygen.

Extinguishment By Temperature Reduction

One of the most common methods of extinguishment is by cooling with water. This process of extinguishment is dependent on reducing the temperature of the fuel to a point where it does not produce sufficient vapor to burn. Solid fuels and liquid fuels with high flash points can be extinguished by cooling. Fires involving low flash point liquids and flammable gases cannot be extinguished by cooling with water, because vapor production cannot be sufficiently reduced. Reduction of temperature is dependent on the application of an adequate flow in proper form to establish a negative heat balance.

Extinguishment By Fuel Removal

In some cases, a fire is effectively extinguished by removing the fuel source. Removal of the fuel sources may be accomplished by stopping the flow of liquid or

Figure 1.30 With flaming combustion, there are four methods of extinguishment.

gaseous fuel or by removing solid fuel in the path of the fire (Figure 1.31). Another method of fuel removal is to allow the fire to burn until all fuel is consumed.

Figure 1.31 Fuel removal is the preferred method for extinguishing fires involving pressurized flammable and combustible liquids and gases.

Extinguishment By Oxygen Dilution

Reducing the oxygen content in an area also puts out the fire. Reduction of the oxygen content can be done by flooding an area with an inert gas, such as carbon dioxide, which displaces the oxygen; or the oxygen can be reduced by separating the fuel from the air such as by blanketing it with foam. Of course, neither of these methods work on those rare fuels that are self-oxidizing.

Extinguishment By Chemical Flame Inhibition

Some extinguishing agents, such as dry chemicals and halogenated hydrocarbons (Halons), interrupt the flame-producing chemical reaction and stop flaming. This method of extinguishment is effective on gas and liquid fuels, because they must flame to burn. Smoldering fires are not easily extinguished by this method, because the moment the Halon is shut off, air once again has access to the smoldering fuel and it continues to burn. Cooling is the only practical way to extinguish a smoldering fire.

CLASSIFICATION OF FIRES AND EXTINGUISHMENT METHODS

Class A Fires

Class A fires are fires involving ordinary combustible materials such as wood, cloth, paper, rubber, and many plastics (Figure 1.32). Water is used in a cooling or quenching effect to reduce the temperature of the burning material below its ignition temperature. The addition of Class A foams (sometimes referred to as wet water) may enhance water's ability to extinguish Class A fires, particularly those that are deep seated in bulk materials. This is because the Class A foam agent reduces the water's surface tension, allowing it to penetrate more easily into piles of the material.

Figure 1.32 Class A fires involve ordinary combustibles.

Class B Fires

Class B fires involve flammable and combustible liquids and gases such as gasoline, oil, lacquers, paints, mineral spirits, and alcohols (Figure 1.33). The smother-

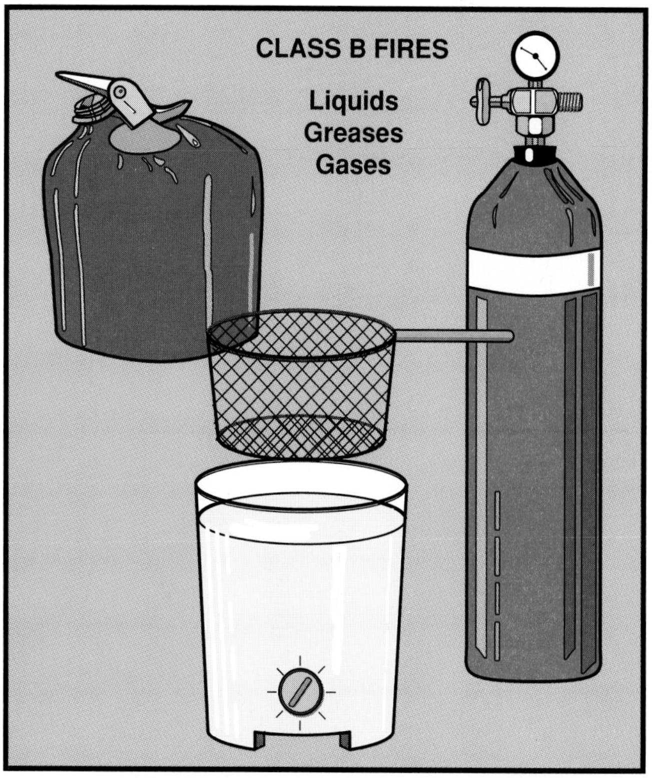

Figure 1.33 The smothering or blanketing effect of oxygen exclusion is most effective for extinguishment of Class B fires.

ing or blanketing effect of oxygen exclusion is most effective for extinguishment. Other extinguishing methods include removal of fuel and temperature reduction when possible.

Class C Fires

Fires involving energized electrical equipment are Class C fires (Figure 1.34). Household appliances, computers, transformers, and overhead transmission lines are examples of these. These fires can sometimes be controlled by a nonconducting extinguishing agent such as Halon, dry chemical, or carbon dioxide. The safest extinguishment procedure is to first deenergize high voltage circuits and then to treat the fire as a Class A or Class B fire depending upon the fuel involved.

Class D Fires

Class D fires involve combustible metals such as aluminum, magnesium, titanium, zirconium, sodium, and potassium (Figure 1.35). These materials are particularly hazardous in their powdered form. Proper airborne concentrations of metal dusts can cause powerful explosions given a suitable ignition source. The extremely high temperature of some burning metals makes water and other common extinguishing agents ineffective. There is no single agent available that will effectively control fires in all combustible metals. Special extinguishing agents are available for control of fire in each of the metals and are marked specifically for that metal. These agents are used to cover up the burning material and smother the fire.

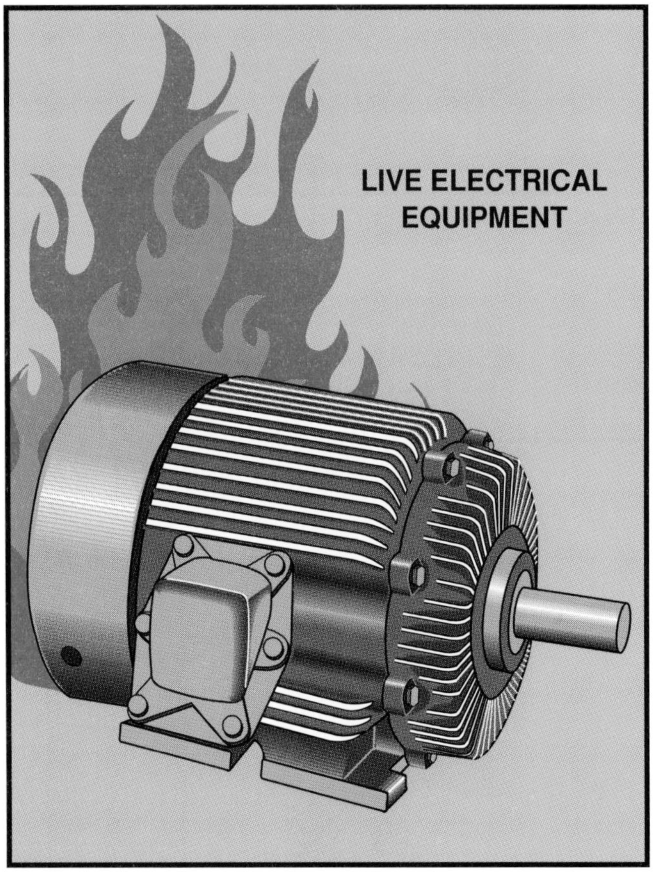

Figure 1.34 Fires involving energized electrical equipment are Class C fires.

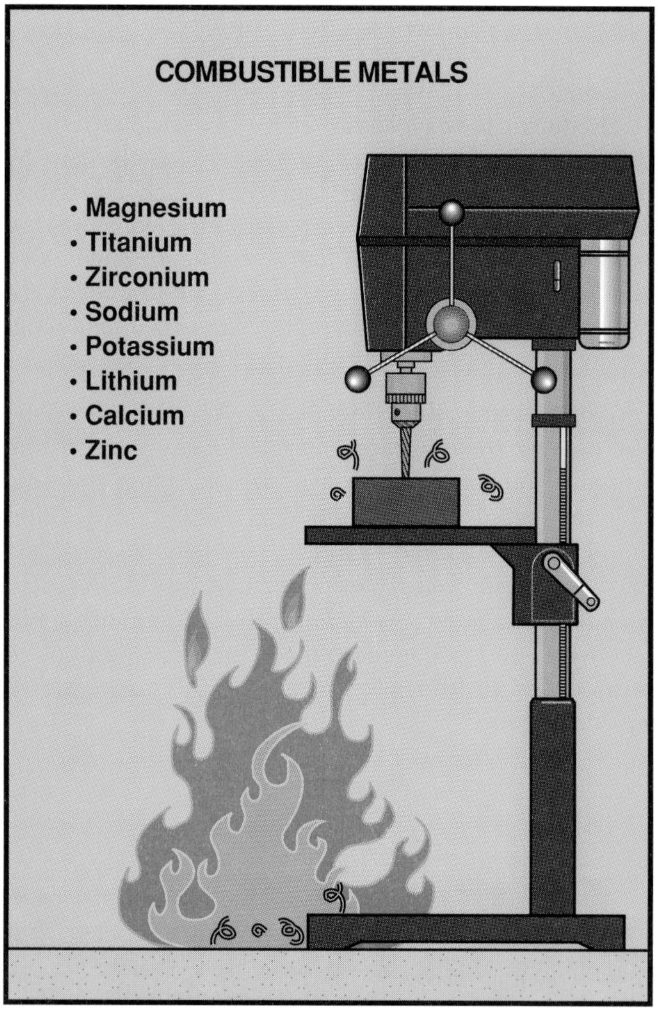

Figure 1.35 Class D fires involve combustible metals.

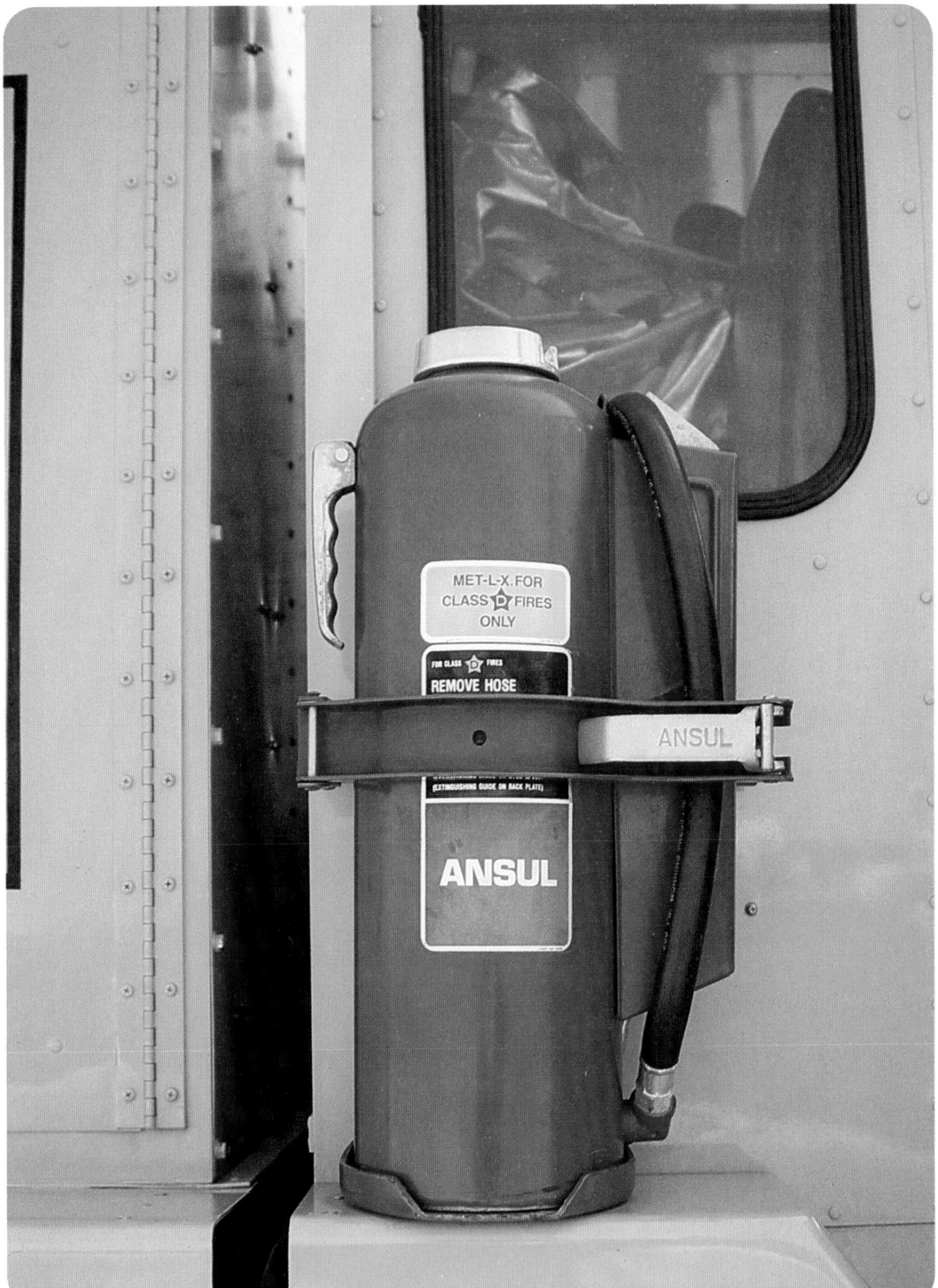

2

Portable Extinguishers

This chapter provides information that addresses performance objectives described in NFPA 1001, *Standard for Fire Fighter Professional Qualifications* (1992), particularly those referenced in the following sections:

3-6 Portable Extinguishers

3-6.1

3-6.2

3-6.3

3-6.4

Chapter 2

Portable Extinguishers

The portable fire extinguisher, one of the most common fire protection appliances in use today, is found in fixed facilities and on fire apparatus (Figure 2.1). A portable fire extinguisher is excellent to use on incipient fires. In many cases, a portable extinguisher can extinguish a fire in much less time than it would take to deploy a hoseline.

It is important that firefighters be knowledgeable in the different types of portable fire extinguishers and their correct usage. This chapter covers the various types of portable fire extinguishers that firefighters are likely to encounter. Also covered is information on the rating, selection, and inspection of portable fire extinguishers. All portable fire extinguishers should meet the requirements set forth in NFPA 10, *Standard for Portable Fire Extinguishers.*

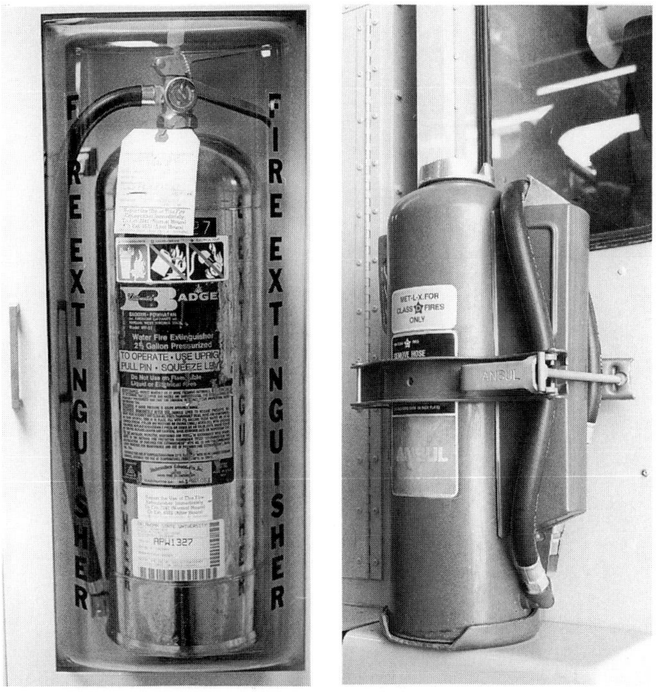

Figure 2.1 Fire extinguishers may be located at fixed positions in an occupancy or mounted on the apparatus.

EXTINGUISHER RATING SYSTEM

Portable fire extinguishers are classified according to their intended use on the four classes of fire (A, B, C, and D) (Figure 2.2 on next page). In addition to the letter classification, Class A and Class B extinguishers receive a numerical rating. The numerical rating system is based on tests conducted by Underwriter's Laboratories, Inc. (UL) and Underwriter's Laboratories of Canada (ULC). These tests are designed to determine the extinguishing potential for each size and type of extinguisher. Extinguishers for use on Class C fires receive only the letter rating because Class C fires are essentially Class A or Class B fires involving energized electrical equipment. The Class C designation just confirms that the extinguishing agent is nonconductive. Class D extinguishers, likewise, do not contain a numerical rating. The effectiveness of the extinguisher on Class D metals is detailed on the faceplate (Figure 2.3 on next page). Multiple letters or numeral-letter ratings are used on extinguishers that are effective on more than one class of fire (Figure 2.4 on next page).

Class A Ratings

Class A portable fire extinguishers are rated from 1-A through 40-A, depending on their size. For a 1-A rating, 1¼ gallons (5 L) of water are required. A 2-A rating requires 2½ gallons (10 L) or twice the 1-A capacity (Figure 2.5 on next page). Therefore, a dry chemical extinguisher rated 10-A is equivalent to five 2½ gallon (10 L) water extinguishers.

To receive a 1-A through 6-A rating, the extinguisher must be capable of extinguishing three types of Class A fires:

- Wood crib
- Wood panel
- Excelsior

Although all three are Class A combustibles, each presents a substantially different type of burning. Por-

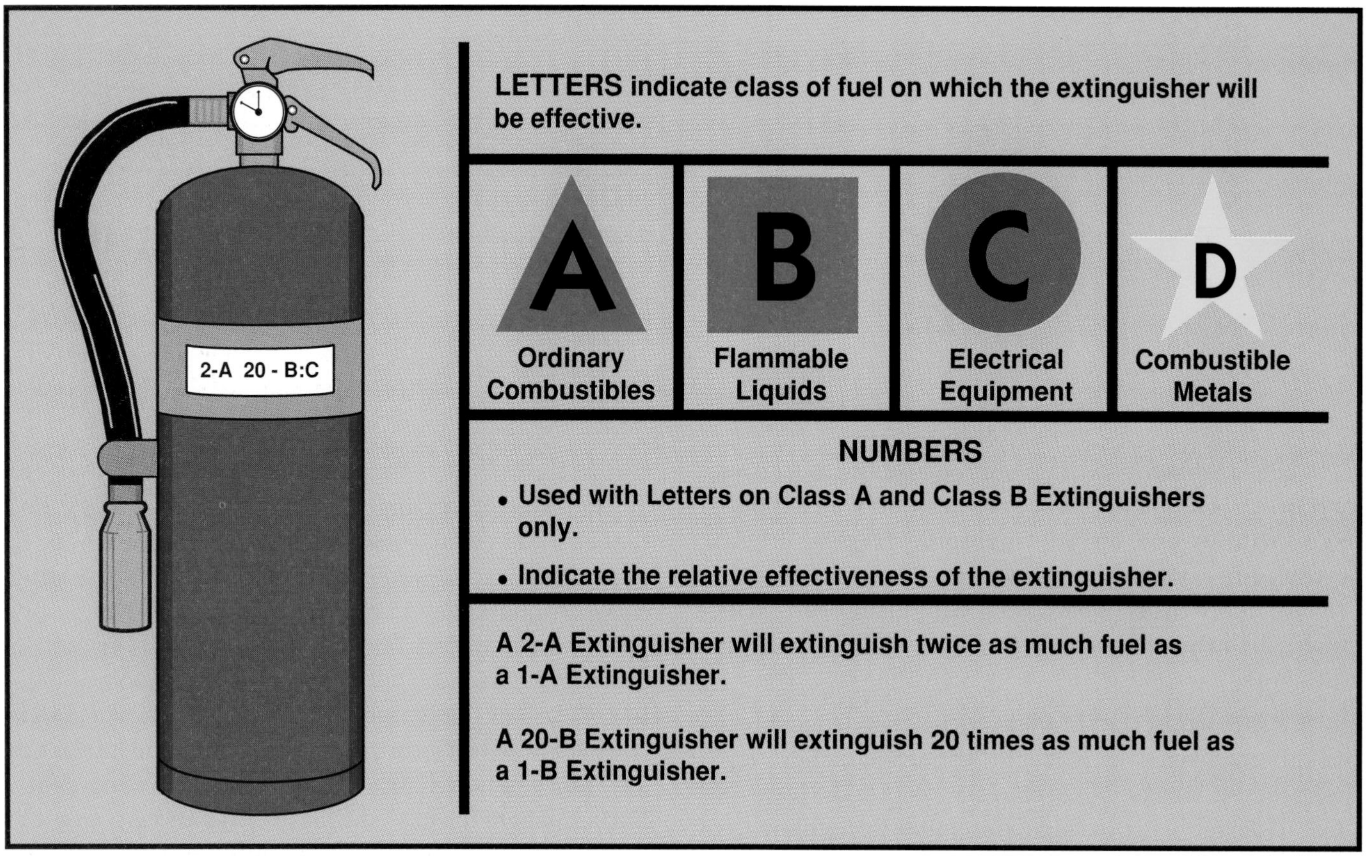
2-A 20 - B:C
LETTERS indicate class of fuel on which the extinguisher will be effective.
A
Ordinary Combustibles
B
Flammable Liquids
C
Electrical Equipment
D
Combustible Metals
NUMBERS
• Used with Letters on Class A and Class B Extinguishers only.
• Indicate the relative effectiveness of the extinguisher.
A 2-A Extinguisher will extinguish twice as much fuel as a 1-A Extinguisher.
A 20-B Extinguisher will extinguish 20 times as much fuel as a 1-B Extinguisher.

Figure 2.2 Portable fire extinguishers are classified according to their intended use on the four classes of fire.

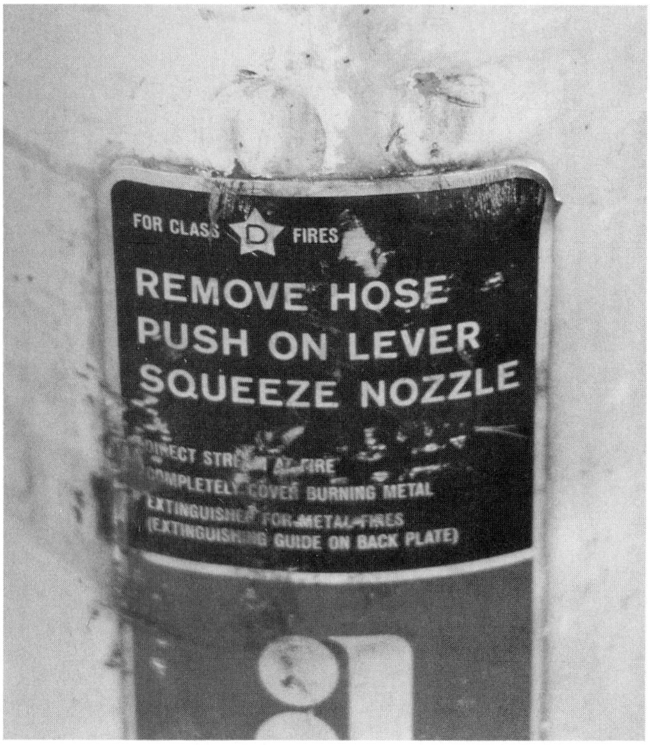

Figure 2.3 The Class D faceplate does not contain a numerical rating.

Figure 2.4 The numeral-letter rating on this faceplate signifies that this extinguisher is effective on more than one class of fire.

Figure 2.5 The typical air-pressurized water fire extinguisher holds 2½ gallons (10 L) of water and has a 2-A rating.

table fire extinguishers that are tested for larger than a 6-A rating are only subjected to the wood crib test.

When tests are conducted on Class A fires, there is little difference in the size of fire extinguished by expert operators and nonexpert operators (Figure 2.6). According to the rating board, technique and application are not as important on a Class A fire as they are on a Class B fire.

Figure 2.6 There is no differentiation between expert and nonexpert operators when rating Class A extinguishers.

Class B Ratings

Extinguishers suitable for use on Class B fires are classified with numerical ratings ranging from 1-B through 640-B (Figure 2.7). The test used by UL to determine the rating of Class B extinguishers consists of burning the flammable liquid n-Heptane in square steel pans. The rating is based on the approximate square foot (square meter) area of a flammable liquid fire that a nonexpert operator can extinguish.

Figure 2.7 A wheeled Class B extinguisher, as they are commonly found on an airport tarmac.

Underwriters' Laboratories always use an expert operator in conducting the tests. However, the numerical rating of the extinguisher is applied under the assumption that the extinguisher will be used by a nonexpert or untrained operator. To determine the amount of fire a nonexpert can extinguish, a working rating is determined to be 40 percent of the fire area an expert operator can consistently extinguish in the tests. For example, an expert operator, using a unit rated 60-B, can extinguish a flammable liquid fire in a 150-square-foot (14 square meter) pan. The nonexpert operator can be expected to extinguish 40 percent of the 150 square feet (14 square meters) or 60 square feet (5.6 square meters) of fire (Figure 2.8). Consequently, the rating of 60-B is applied to the unit.

A common-sized extinguisher, such as the multipurpose extinguisher rated 4-A 20-B:C, can be reviewed to better understand the rating system (Figure 2.9). This extinguisher should extinguish a Class A fire that is 4 times larger than a 1-A fire, extinguish approximately 20 times as much Class B fire as a 1-B extinguisher, and extinguish a deep-layer flammable liquid fire of a 20-square-foot (2 square meter) area. It is also safe to use on fires involving energized electrical equipment.

Figure 2.9 A typical multipurpose extinguisher.

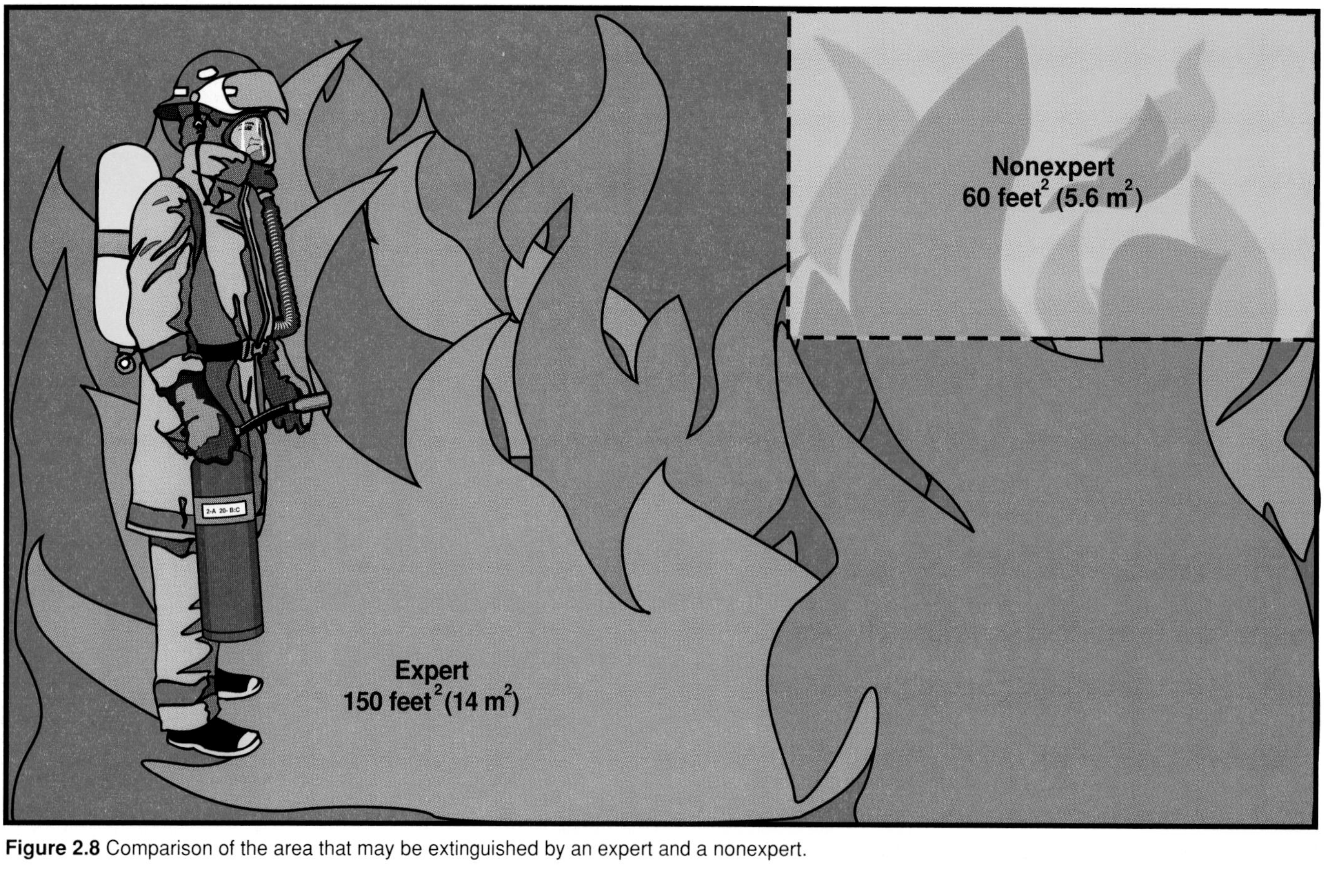

Figure 2.8 Comparison of the area that may be extinguished by an expert and a nonexpert.

Class C Ratings

There are no fire tests specifically conducted for Class C ratings. In assigning a Class C designation, the extinguishing agent is only tested for electrical nonconductivity. If the agent meets the test requirements, the Class C rating is provided in conjunction with a rating previously established for Class A and/or Class B fires.

Class D Ratings

Test fires for establishing Class D ratings vary with the type of combustible metal being tested. The following are several factors considered during each test.

- Reactions between the metal and the agent
- Toxicity of the agent
- Toxicity of the fumes produced and the products of combustion
- The possible burnout of the metal instead of extinguishment

When an extinguishing agent is determined to be safe and effective for use on a metal, the details of instruction are included on the facepiece of the extinguisher, although no numerical rating is applied. Class D agents cannot be given a multipurpose rating to be used on other classes of fire.

Multiple Markings

Extinguishers suitable for more than one class of fire should be identified by multiples of the symbols A, B, and/or C. The three most common combinations are Class A-B-C, Class A-B, and Class B-C (Figure 2.10). There is no extinguisher with a Class A-C rating. A new portable fire extinguisher must be labeled with its appropriate marking. If a new extinguisher is not properly marked, the seller should be requested to supply the proper decals.

There are two methods of labeling portable fire extinguishers. The first system uses specifically colored geometric shapes with the class letter shown within the shape (Figure 2.11). The second system is a *picture-symbol* labeling system that is designed to make the

Figure 2.11 Two systems are used to label portable fire extinguishers. One system uses specifically colored geometric shapes with the class letter shown within the shape.

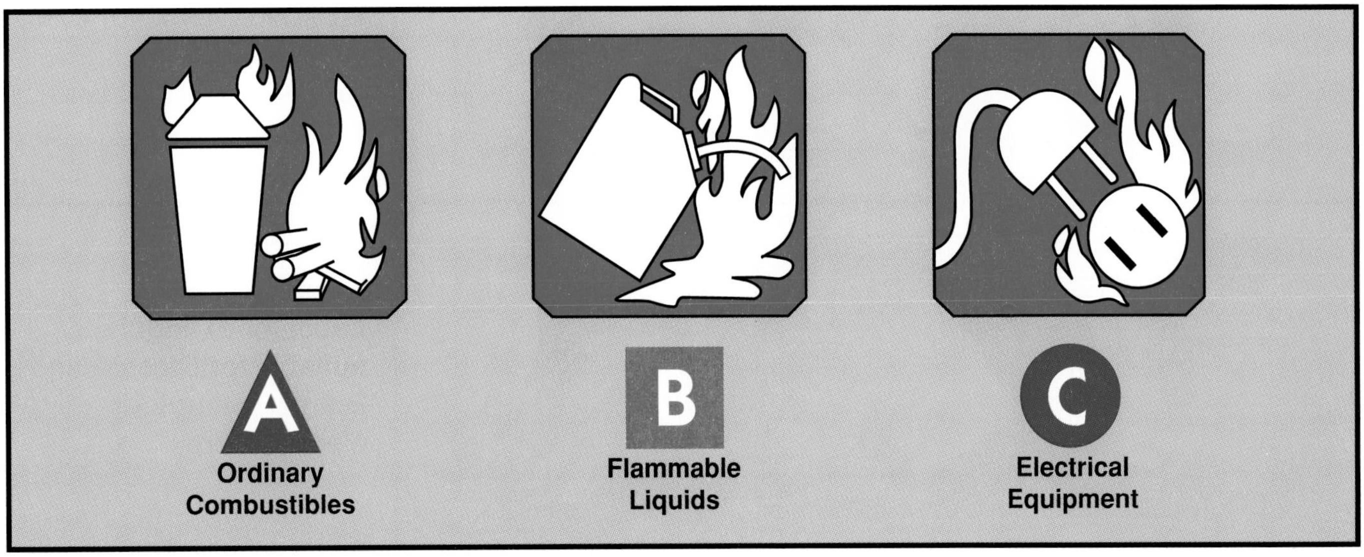

Figure 2.10 This extinguisher is effective on Class A, B, and C fires.

selection of fire extinguishers more effective and safe to use (Figure 2.12). The picture-symbol system also emphasizes when not to use an extinguisher on certain types of fires. Regardless of which system is used, it is important that the marking be clearly visible.

SELECTING AND USING PORTABLE FIRE EXTINGUISHERS

Selection of the proper portable fire extinguisher depends on numerous factors:

- Hazards to be protected
- Severity of the fire
- Atmospheric conditions
- Personnel available
- Ease of handling extinguisher
- Any life hazard or operational concerns

Extinguishers should be chosen that minimize the risk to life and property, yet that are effective in extinguishing the fire. Hopefully, these considerations will be addressed when choosing a fire extinguisher to be mounted in a particular area (Figure 2.13). For example, it would be unwise to place dry chemical extinguishers with a corrosive agent in areas where highly sensitive computer equipment is located. In these particular areas, Halon or carbon dioxide extinguishers would be a better choice (Figure 2.14).

Portable extinguishers come in many shapes, sizes, and types. While the operating procedures of each type of extinguisher are similar, operators should become familiar with the detailed instructions found on the label of the extinguisher. In an emergency, every second is of great importance; therefore, everyone should be acquainted with the following general instructions applicable to most portable fire extinguishers. The general operating instructions follow the letters P-A-S-S.

P Pull the pin at the top of the extinguisher that keeps the handle from being pressed (Figure

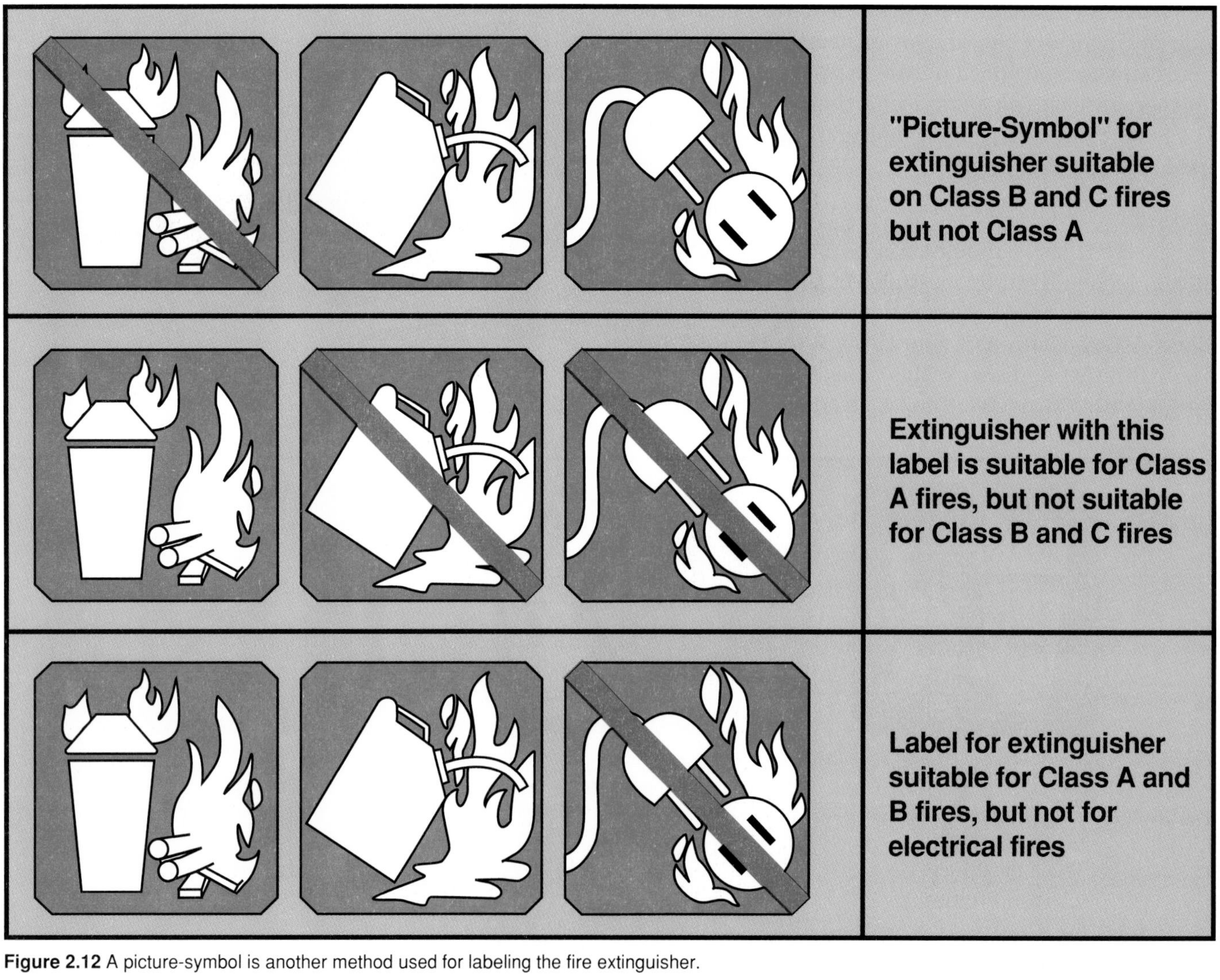

"Picture-Symbol" for extinguisher suitable on Class B and C fires but not Class A
Extinguisher with this label is suitable for Class A fires, but not suitable for Class B and C fires
Label for extinguisher suitable for Class A and B fires, but not for electrical fires

Figure 2.12 A picture-symbol is another method used for labeling the fire extinguisher.

Figure 2.14 Halon extinguishers are used in computer areas.

Figure 2.13 Careful consideration of surroundings should be given to the type of extinguisher used for permanent placement.

Figure 2.16 Aim at the base of the fire.

2.15). Break the plastic or thin wire inspection band as the pin is pulled.

A Aim the nozzle or outlet toward the fire (Figure 2.16). Some hose assemblies are clipped to the extinguisher body. Release the hose and point.

S Squeeze the handle above the carrying handle to discharge the agent (Figure 2.17). The handle can be released to stop the discharge at any time. Before approaching the fire, try a very short test burst to ensure proper operation.

S Sweep the nozzle back and forth at the base of the flames to disperse the extinguishing agent (Figure 2.18). After the fire is out, watch for remaining smoldering hot spots or possible reflash of flammable liquids. *Make sure that the fire is out.*

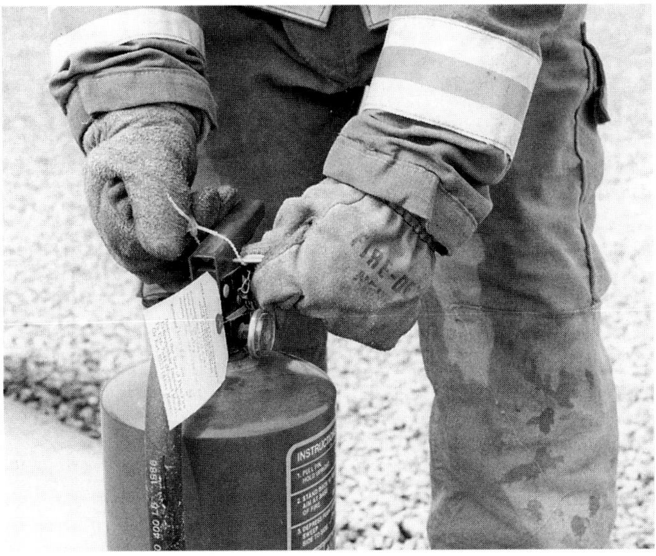

Figure 2.15 Pull the pin from the handle.

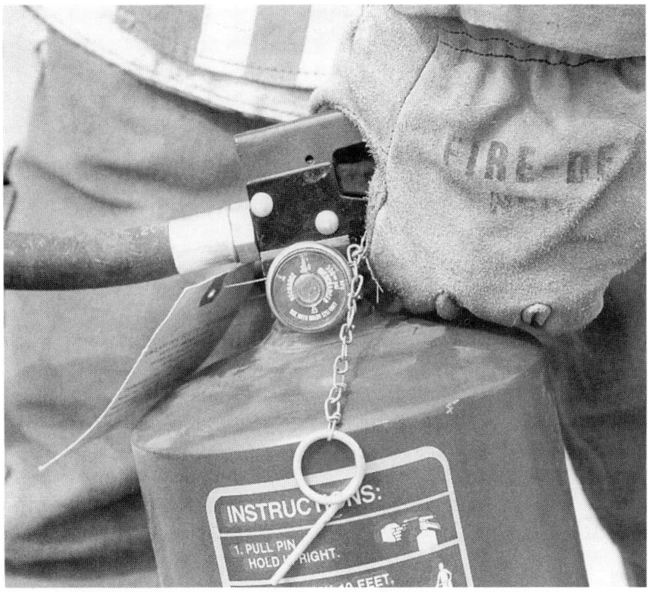

Figure 2.17 Squeeze the handle.

Figure 2.18 Sweep back and forth.

Modern extinguishers are designed to be carried to the fire in an upright position. When instructing the general public in the use of extinguishers, emphasize that extinguishers are operated in an upright position (Figure 2.19). (**NOTE:** Only obsolete soda-acid extinguishers, foam extinguishers, and cartridge-operated water extinguishers are designed to be turned upside down; however, these extinguishers are no longer acceptable for use.) Make sure that the fire is within reach before discharging the extinguishing agent (Figure 2.20). Otherwise, the agent will be wasted. Smaller extinguishers require closer approach to the fire because they have less of a stream reach than do larger extinguishers.

Figure 2.19 Firefighters will frequently be asked to demonstrate extinguishing techniques to the public.

Figure 2.20 Make sure that the agent is hitting the fire.

TYPES OF PORTABLE FIRE EXTINGUISHERS

There are many different types of portable fire extinguishers. This section highlights some of the common extinguishers encountered by fire service personnel.

Pump Tank Water Extinguishers

Pump tank water extinguishers are applicable to Class A fires only. They are available in sizes of 1½ to 5 gallons (6 L to 20 L) (Figure 2.21). Under normal conditions, their stream reach is 30 to 40 feet (9 m to 12 m). The discharge time is 45 seconds to 3 minutes, depending on the size of the unit. NFPA 10 requires that these extinguishers be protected against freezing if they are going to be exposed to temperatures less than 40°F (4°C). Freeze protection is accomplished by the addition of an antifreeze solution to the water.

There are several kinds of pump tank extinguishers. Some are designed to be hand carried and set down at the point of operation, and others are designed to be used while worn on the operator's back. Both may be equipped with a double-action pump that delivers a continuous stream of water. However, some backpack tanks are equipped with a combination piston pump/nozzle that provides bursts of water as the nozzle is pumped (Figure 2.22).

Figure 2.21 A typical pump tank water extinguisher.

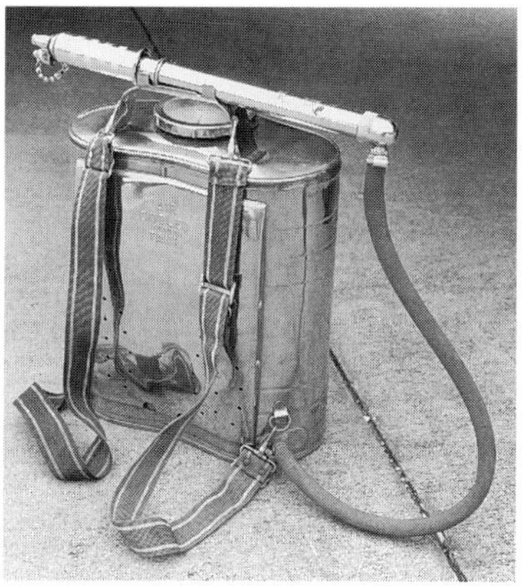

Figure 2.22 This backpack tank is equipped with a combination piston pump/nozzle that provides bursts of water as the nozzle is pumped.

HAND-CARRIED PUMP TANK WATER EXTINGUISHERS

To operate the hand-carried pump tank water extinguisher, carry the extinguisher by the handle to the fire (Figure 2.23). Operate the pump with an up-and-down (or side-to-side) stroke (Figure 2.24). Short, continuous strokes of the pump handle may provide a better stream than long strokes. Some nozzles are provided with two tips, one for straight stream and the other for a broken spray stream (Figure 2.25). The filler cap is provided with a tiny vent that must be kept clear so that air may replace the water as it is discharged from the tank. In some cases, it may be desirable to place one finger into the discharge stream at the nozzle to break the solid stream into a broken stream (Figure 2.26).

Figure 2.23 The pump tank extinguisher is carried to the location where it is needed.

Figure 2.24 Pump the extinguisher with one hand and direct the stream with the other.

Figure 2.25 Some pump tank extinguisher nozzles will have both straight stream and spray tips.

Figure 2.26 The stream may be broken by placing the thumb or another finger partially over the discharge orifice.

BACKPACK PUMP TANK WATER EXTINGUISHERS

A backpack fire extinguisher consists of either a metal or rubber tank, a hose, and a combination nozzle/pump device that is worn on the firefighter's back (Figure 2.27). This type of extinguisher usually has a capacity of 5 gallons (20 L) of water. When a firefighter is wearing a backpack extinguisher, both hands are free. This enables firefighters to use their hands for balance. When full of water, these extinguishers are quite heavy and require physical stamina if they are to be worn for long periods.

Backpack extinguishers have several useful purposes. Because they are so portable and carry a fair amount of water, they are very useful for small grass and brush fires that fire apparatus are unable to access. Backpack extinguishers are also useful at structure fires. They make excellent tools for overhauling smoldering debris such as insulation, furniture, and wallboard.

Figure 2.27 Backpack tanks are most commonly used to fight wildland fires.

Stored-Pressure Water Extinguishers

Stored-pressure water extinguishers, also known as air-pressurized water extinguishers (APWs), are only used on Class A fires (Figure 2.28). These extinguishers are useful for all types of small Class A fires and are often used for hitting confined hot spots during overhaul operations, as well as for extinguishing chimney flue fires. They are available in sizes of 1¼ to 2½ gallons (5 L to 10 L). Under normal conditions, their stream reach is 30 to 40 feet (9 m to 12 m). The discharge time is 30 to

60 seconds, depending on the size of the unit. NFPA 10 requires that these extinguishers be protected against freezing if they are going to be exposed to temperatures less than 40°F (4°C). Freeze protection is accomplished by the addition of an antifreeze solution to the water.

The water is discharged by compressed air that is stored in the tank with the water. A gauge located on the top of the tank shows when the extinguisher is properly pressurized (Figure 2.29). With this type of extinguisher, the pressure is ready to release the extinguishing agent at any time. When the shutoff device is released, a stream of water is expelled through the hose. Stored-pressure water extinguishers are designed to be carried to the fire in an upright position. The hose is held in one hand and the shutoff device in the other hand. When water is needed, the shutoff handle is squeezed and the water directed at the target (Figure 2.30).

Figure 2.29 The pressure gauge clearly shows the operable range.

Figure 2.28 A typical stored-pressure water extinguisher.

Figure 2.30 Squeeze the handle to direct water at the target.

Aqueous Film Forming Foam (AFFF) Extinguishers

Aqueous film forming foam extinguishers are suitable for use on Class A and Class B fires (Figure 2.31). They are particularly useful in combating fires or suppressing vapors on small liquid fuel spills. The most common AFFF extinguishers are 2½ gallons (10 L) in size. Their stream reach is 20 to 25 feet (6 m to 7.5 m). The discharge time is about 50 seconds. NFPA 10 requires that these extinguishers be protected against freezing if they are going to be exposed to temperatures less than 40°F (4°C). Freeze protection is accomplished by the addition of an antifreeze solution to the water. This antifreeze must be compatible with the AFFF concentrate.

AFFF extinguishers are very similar to the stored-pressure water extinguishers. The main difference is that in addition to water, the tank contains a proportionate amount of AFFF concentrate. The other primary difference is that an AFFF extinguisher has a special aspirating nozzle that provides a better quality of foam than a regular nozzle provides (Figure 2.32).

The water/AFFF solution is expelled by compressed air stored in the tank with the solution. When the shutoff device is released, the solution is expelled through the hose and aspirating nozzle where air mixes with the solution and forms the finished foam. The foam is applied with a side-to-side sweeping motion across the

Figure 2.31 A typical AFFF portable extinguisher.

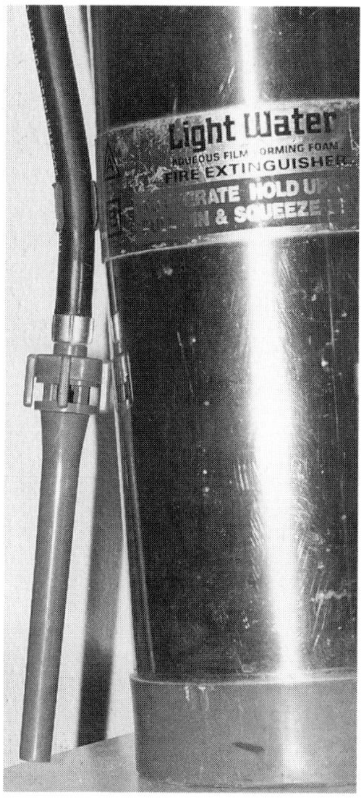

Figure 2.32 AFFF hand extinguishers have special aerating nozzles.

entire width of the fire (Figure 2.33). (**NOTE:** The operator must be careful to avoid splashing liquid fuels. This special foam has the ability to make water float on fuels that are lighter than water.) The vapor seal that is created extinguishes the flame and prevents reignition. The foam also has good wetting and penetrating properties on Class A fires.

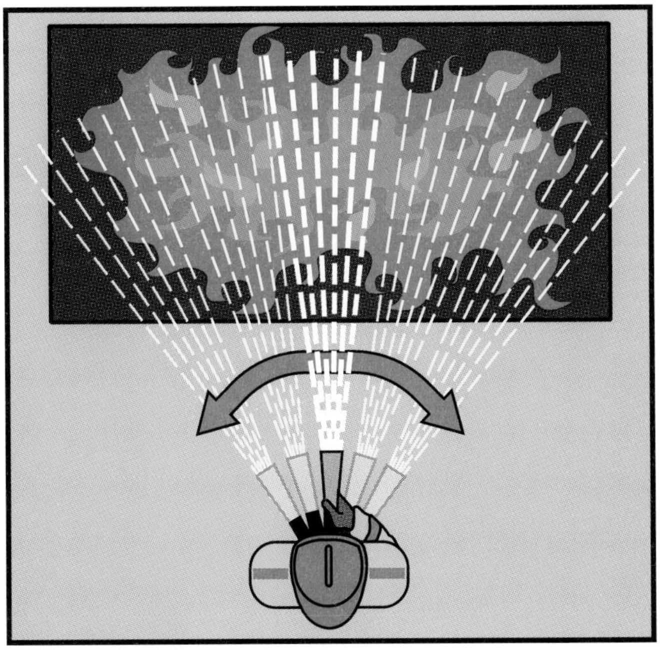

Figure 2.33 Sweep the foam across the entire surface area of the fuel/fire.

Figure 2.34 A hand-carried Halon extinguisher.

Figure 2.35 A wheeled Halon extinguisher. *Courtesy of Ansul Fire Protection.*

Figure 2.36 The Halon agent is discharged in a solid stream.

Halon 1211 Extinguishers

Halon 1211 (bromochlorodifluoromethane) extinguishers are primarily designed for Class B and Class C fires. However, Halon 1211 extinguishers greater than 9 pounds (4 kg) in capacity will also have a small Class A rating (1-A to 4-A, depending on the size). Hand-carried extinguishers are available in sizes from 2½ to 22 pounds (1 kg to 10 kg) (Figure 2.34). Larger Halon 1211 extinguishers are available on wheeled units up to 150 pounds (68 kg) (Figure 2.35). These extinguishers have a stream reach of about 8 to 18 feet (2.5 m to 5.5 m). The total discharge time is 8 to 18 seconds. The inherent features of Halon 1211 extinguishers do not require special freeze protection.

Halon 1211 is stored in the extinguisher as a liquefied compressed gas. Nitrogen is added to give the Halon 1211 added pressure when discharged. When the shutoff handle is squeezed, the Halon 1211 is released in a clear liquid stream, giving it greater reach ability than a gaseous agent (Figure 2.36). How Halon 1211 or any of the other Halon agents extinguish fires is not clearly known. Research indicates that they interrupt the chain reaction of the combustion process.

Halon 1211 extinguishers are designed to be carried by the top handle. They have a limited range, and the discharge of the extinguishing agent may be affected by draft and wind. The initial application should be made close to the fire, and the discharge should be directed at the base of the flames. The discharge should be applied to the burned surface even after the flames are extinguished. Best results will be obtained on flammable liquid fires if the discharge is directed to sweep the flame from the burning surface (Figure 2.37 on next page). Apply the discharge first at the near edge of the fire and gradually progress forward while moving from side to side.

Halon 1301 Extinguishers

Halon 1301 (bromotrifluoromethane) is also used in small portable fire extinguishers. These extinguishers are comparable to the Halon 1211 extinguishers with a few exceptions. They are limited to 2½ pounds (1 kg) and their stream is discharged in a nearly invisible gaseous form (Figure 2.38). Because a gas stream has less reach than a liquid stream, the effective reach of a Halon 1301 extinguisher is only about 4 to 6 feet (1.3 m to 2 m).

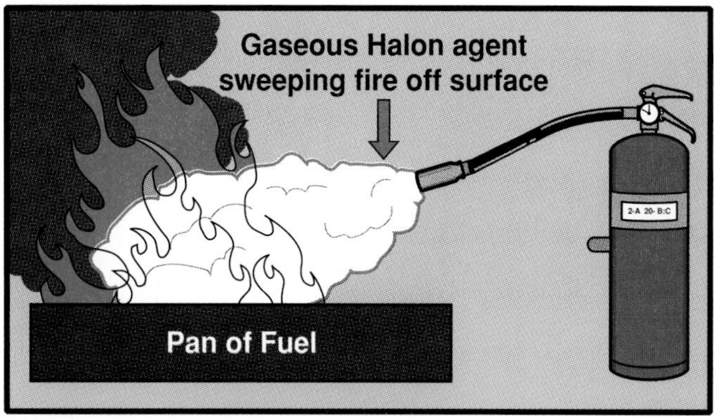

Figure 2.37 Sweep the flames off the surface of the fuel with the Halon stream.

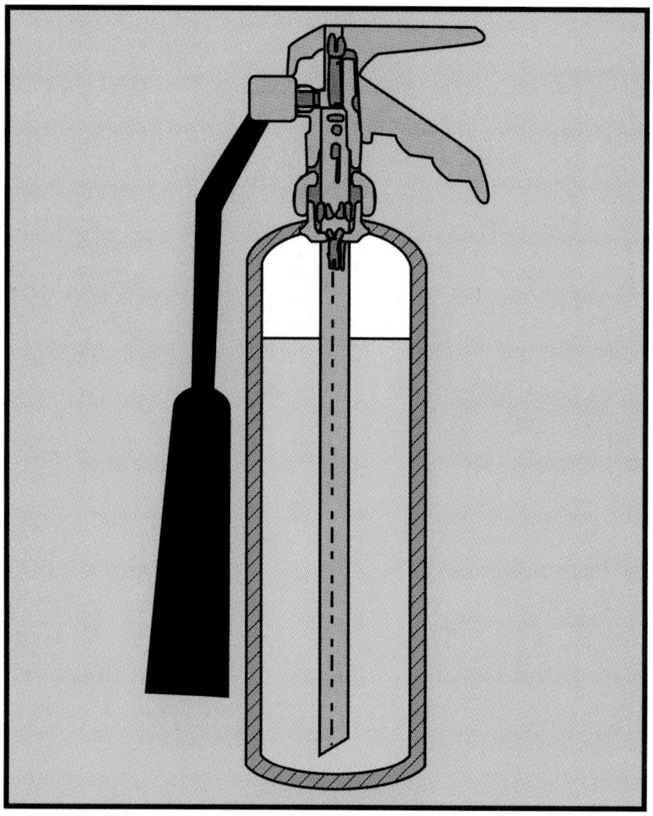

Figure 2.38 The parts of a Halon 1301 portable extinguisher.

Carbon Dioxide Extinguishers (Hand Carried)

Carbon dioxide (CO_2) extinguishers are effective in extinguishing Class B and Class C fires. Hand-carried units are available in sizes from 2 to 20 pounds (1 kg to 9 kg) (Figure 2.39). Because their discharge is in the form of a gas, they have a limited reach of only 3 to 8 feet (1 m to 2.5 m). Their total discharge time ranges from 8 to 30 seconds, depending on the size of the extinguisher. They do not require any special freeze protection.

Carbon dioxide is stored as a liquefied compressed gas. When the shutoff handle is squeezed, the carbon dioxide is discharged in a gaseous form to smother the fire. The carbon dioxide is stored under its own pressure and ready for release at any time. The gaseous discharge is usually accompanied by little dry ice crystals or "snowflakes" (Figure 2.40). These flakes sublime into a gaseous form shortly after discharge.

Carbon dioxide extinguishers are designed to be carried by the top handle. The discharge expels through a horn on the end of a hose or short metal fitting (Figure 2.41). (**CAUTION:** Often, a "frost" residue will form on the nozzle horn. Contact with the skin could result in frostbite.) The discharge horn should be pointed at the

Figure 2.39 A portable CO_2 extinguisher.

Figure 2.40 CO_2 is discharged as a gas, but moisture may appear in the form of snowflakes.

base of the fire (Figure 2.42). Application should continue even after the flames are extinguished to prevent a possible reflash of the fire. On flammable liquid fires, best results are obtained when the discharge from the extinguisher is employed to sweep the flame off the burning surface. Apply the discharge first at the near edge of the fire and gradually progress forward, moving the discharge cone very slowly from side to side (Figure 2.43).

Carbon Dioxide Wheeled Units

Carbon dioxide wheeled units are similar to the hand-carried units, except they are considerably larger (Figure 2.44). These units are to be used only on Class B and Class C fires. Wheeled units range in size from 50 to 100 pounds (23 kg to 45 kg). Because of their size, they have a longer stream reach. The stream reach under normal conditions is 8 to 20 feet (2 m to 6 m). The discharge time is usually no more than 30 seconds.

Wheeled units are most commonly found in industrial facilities (Figure 2.45). The principle of operation is

Figure 2.41 The agent is expelled through a conical horn.

Figure 2.42 Aim the horn at the base of the fire. **NOTE:** Do not touch the horn with bare hands.

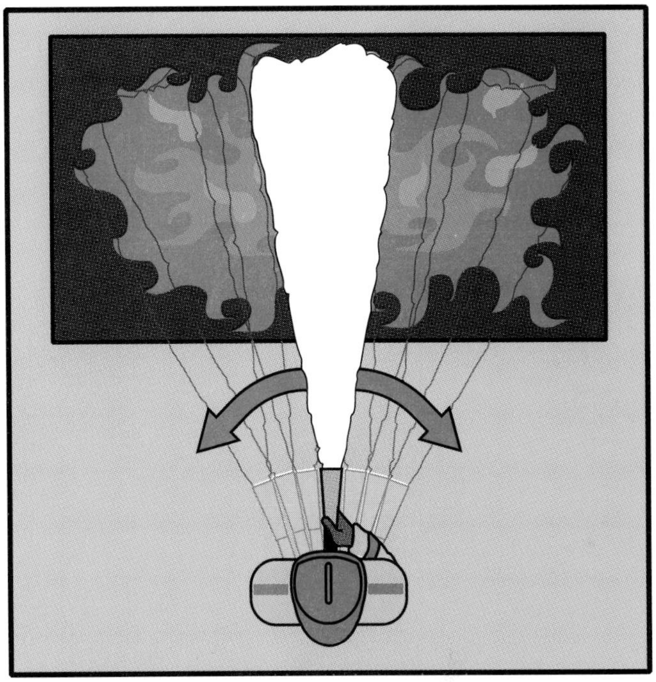

Figure 2.43 Sweep the CO_2 stream across the fuel's surface.

Figure 2.44 A typical wheeled CO_2 unit.

Figure 2.45 Wheeled CO_2 units are often found at airports.

the same for these larger units as for the smaller CO_2 extinguishers. They are designed to be wheeled to the fire and operated according to the instructions on the extinguisher. They have a short hose, typically less than 15 feet (4.5 m), that must be unraveled before use (Figure 2.46).

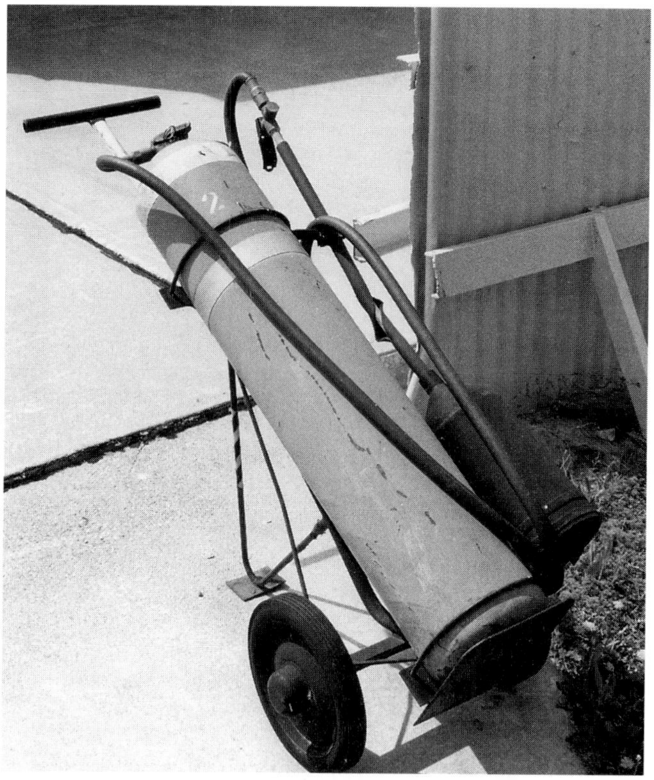

Figure 2.46 Make sure that the hose is in good condition.

Dry Chemical Extinguishers (Hand Carried)

Dry chemical extinguishers are among the most common portable fire extinguishers in use today. There are two basic types of dry chemical extinguishers: ordinary and multipurpose (Figure 2.47). Ordinary dry chemical extinguishers are rated for Class B and Class C fires, while multipurpose dry chemical extinguishers are rated for Class A, Class B, and Class C fires. Unless specifically noted in this section, the characteristics and operation of both types are the same.

Hand-carried dry chemical fire extinguishers are available in sizes from 2½ to 30 pounds (1.5 kg to 11 kg). The stream reach under normal conditions is 5 to 20 feet (2 m to 6 m), although it is easily adversely affected by wind. The total discharge time is 10 to 25 seconds. Because they are dry chemicals, no freeze protection is required.

The following dry chemicals are commonly used in ordinary base and multipurpose agent fire extinguishers:

Ordinary Base
- Sodium bicarbonate
- Potassium bicarbonate
- Potassium chloride

2/94

Multipurpose Agent
- Monoammonium phosphate
- Barium sulfate
- Ammonium phosphate

4-93

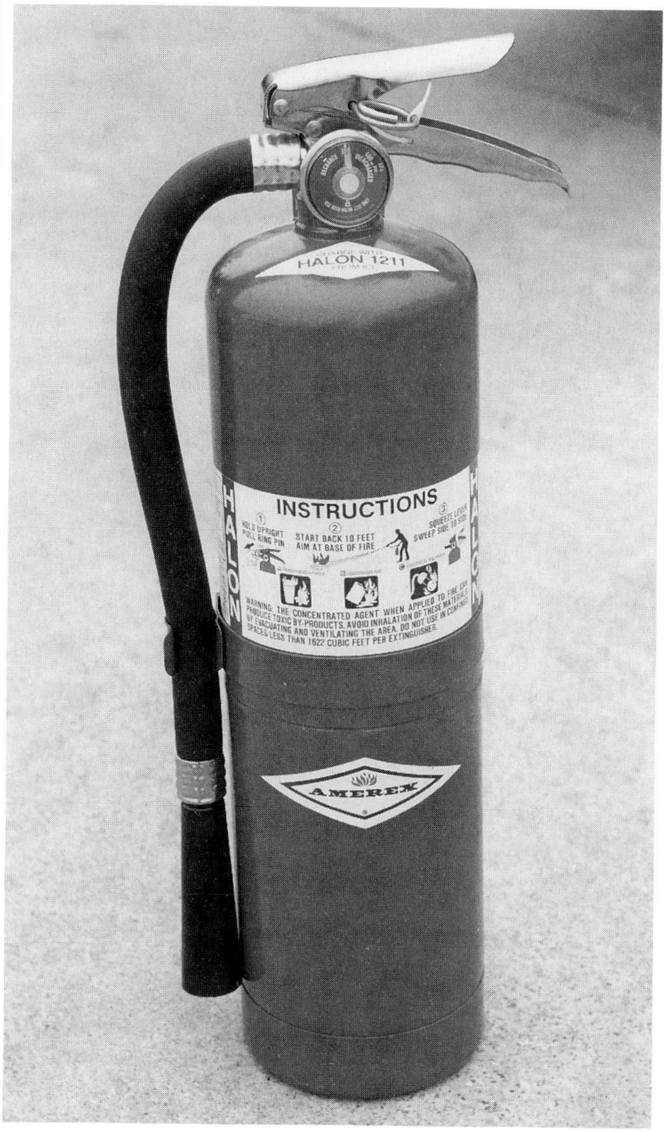

Figure 2.47 A hand-carried dry chemical fire extinguisher rated for Class A, B, and C fires.

During manufacture, these chemicals are mixed with small amounts of additives that prevent the agent from caking, and this allows the agent to be discharged easily. Care should be taken to avoid mixing or contaminating ordinary agents with multipurpose agents and vice versa.

NOTE: The terms *dry chemical* and *dry powder* are often incorrectly interchanged. Dry chemical agents are for use on Class A-B-C fires and/or B-C fires. Dry powder agents are for Class D fires only.

There are two basic designs for hand-carried dry chemical extinguishers: stored pressure and cartridge operated. The stored-pressure type is similar in design to the pressurized-water extinguisher (Figure 2.48). The agent storage tank is maintained under a constant pressure. This pressure is commonly in the area of 200 psi (1 400 kPa). Cartridge-operated extinguishers maintain a separate agent tank and pressure cylinder (Figure 2.49). The agent tank is not pressurized until a plunger is pushed to release the gas from the cartridge (Figure 2.50). Both types of extinguishers primarily use nitrogen as the pressurizing gas.

Figure 2.48 A stored-pressure dry chemical extinguisher.

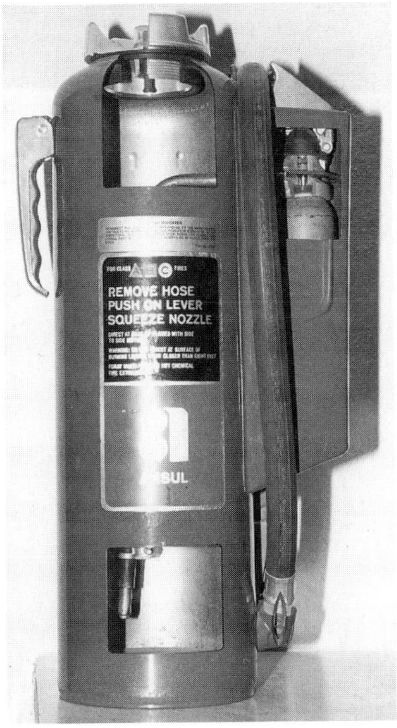

Figure 2.49 A cartridge-operated dry chemical extinguisher. This cutaway shows the position of the nitrogen cylinder in relation to the larger dry chemical tank.

Figure 2.50 The cartridge plunger breaks the seal on the nitrogen cylinder and pressurizes the dry chemical tank.

The method of operation will be slightly different, depending on whether a stored-pressure or cartridge-operated extinguisher is used. To operate the stored-pressure type, simply pull the pin and squeeze the shutoff handle to discharge the agent (Figure 2.51). To operate the cartridge type, first remove the hose from its storage position. In the stored position, the hose prevents the activation plunger from being accidentally pushed. Once the hose is removed, point the top of the extinguisher away from people and depress the activation plunger (Figure 2.52). Within a second or two, the agent tank should be fully charged, and the agent can then be discharged by squeezing the control handle on the nozzle.

Figure 2.51 Stored-pressure dry chemical extinguishers operate similar to stored-pressure water extinguishers.

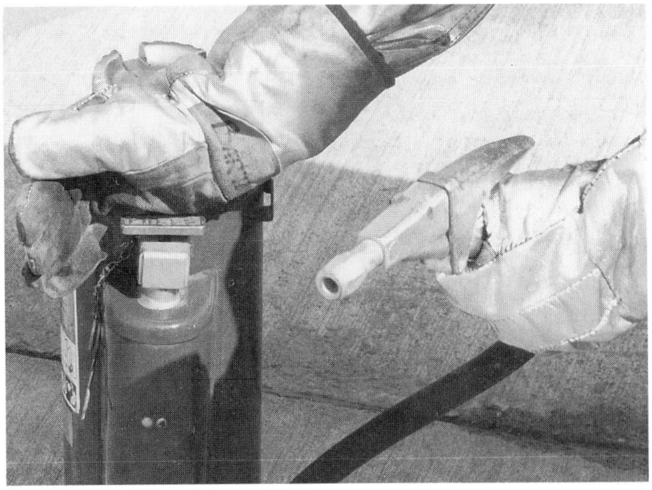

Figure 2.52 To charge the extinguisher, push the cartridge plunger.

Regardless of which type extinguisher is used, operation expels a cloud of dry chemical, and the discharge is controlled by a shutoff valve. Best results are obtained by attacking the near edge of the fire and progressing forward while moving the nozzle in a side-to-side sweeping motion (Figure 2.53). The discharge should be applied to the burning surface even after the flames are extinguished. This process will prevent possible reflash by coating the hot surfaces and any glowing materials present. When used on electrical fires, care should be taken around open electrical contacts to prevent costly cleaning. It may be better to use other agents, such as Halon or carbon dioxide, in such cases.

Figure 2.53 Aim the dry chemical stream at the base of the fire.

Dry Chemical Wheeled Units

Dry chemical wheeled units are similar to the hand-carried units, but on a larger scale. They are rated for Class A, Class B, and Class C fires. Wheeled units range in size from 75 to 350 pounds (34 kg to 160 kg) (Figure 2.54). They are capable of shooting a stream up to 45 feet (14 m). The total discharge time ranges from 20 seconds to 120 seconds. They are usually equipped with hoses that are 50 to 100 feet (15 m to 30 m) in length.

Operating the wheeled dry chemical extinguisher is similar to operating the hand-carried, cartridge-type, dry chemical extinguisher. The extinguishing agent is kept in one tank and the pressurizing gas is stored in a separate cylinder. When the extinguisher is needed, the first thing that should be done is to completely remove the hose from the storage position, because removal could become difficult after the unit is charged (Figure 2.55). The gas is released to enter the agent tank by turning a hand wheel on the top of the gas tank or by activating a quick-pressurization device (Figure 2.56). The agent tank should be totally pressurized within seconds. The agent is applied as described for the hand-carried extinguishers.

On Class A fires, the discharge should be directed at the burning surfaces to cover them with chemical. When the flames have been extinguished, the chemical discharge should be intermittently directed on any glowing areas (Figure 2.57). A careful watch should be maintained for hot spots that may develop. Additional agent is applied to those surfaces as required to adequately coat them with the extinguishing agent.

Figure 2.54 A wheeled dry chemical extinguisher. *Courtesy of Ansul Fire Protection.*

Figure 2.55 Remove the hose from the bracket.

Figure 2.56 A quick actuating device.

Figure 2.57 Wheeled multipurpose dry chemical extinguishers may be used to attack Class A fires.

Extinguishers And Powder Extinguishing Agents For Metal Fires

Normal extinguishing agents generally should not be used on metal (Class D) fires. Specialized techniques and extinguishing agents have been developed to control and extinguish metal fires (Figure 2.58). A given agent does not, however, necessarily control or extinguish all metal fires. Some agents are valuable in working with several metals; others are useful for combating only one type of metal fire. Some of the agents are intended to be applied by means of a hand shovel or scoop, others by means of portable fire extinguishers designed for use with dry powders.

Figure 2.58 Cover the fire with the dry powder. *Courtesy of Tulsa (OK) Fire Department.*

Portable fire extinguishers for Class D fires come in both hand-carried and wheeled models. Hand-carried units have a capacity of 30 pounds (14 kg) (Figure 2.59). Wheeled units are found in 150- and 350-pound (70 kg and 158 kg) sizes (Figure 2.60). The agent used in most of these extinguishers is sodium chloride. Added to the sodium chloride are flow enhancers and a thermoplastic material that enhances crusting after the material is discharged onto the fire.

The application of the agents should be of sufficient depth to adequately cover the fire area and provide a smothering blanket. The agent should be applied gently on metal fires to avoid breaking any crust that may have formed over the burning metal (Figure 2.61). If the crust is broken, the fire may flare up and expose more raw material to combustion. Additional applications may be necessary to cover any hot spots that may develop. The material should be left undisturbed, and disposal should not be attempted until the mass has cooled. Care should be taken to avoid scattering the burning metal. If the burning metal is on a combustible surface, the fire should first be covered with powder, then a 1- or 2-inch (25 mm to 50 mm) layer of powder spread out nearby, and the burning metal shoveled onto this layer with more powder added as needed (Figure 2.62). Reference

Figure 2.59 A Class D extinguisher.

Figure 2.60 A wheeled Class D extinguisher looks much like a wheeled Class B extinguisher.

Figure 2.61 Dry powder from a Class D extinguisher should be used to cover the burning metal.

Figure 2.62 Class D agent may be shoveled onto the fire.

should be made to the manufacturer's recommendations for use and special techniques for extinguishing fires in various combustible metals.

DAMAGED EXTINGUISHERS

Leaking, corroded, or otherwise damaged extinguisher shells or cylinders should be discarded or returned to the manufacturer for repair (Figure 2.63). Damaged extinguishers can fail when operation is attempted. This can result in serious injury to the user or people standing nearby. (**CAUTION:** Never try to repair the shell or cylinder of a fire extinguisher. Contact the manufacturer for instructions on where to have it repaired or replaced.)

If an extinguisher shows only slight damage or corrosion and it is questionable whether it is safe to use, it should be given a hydrostatic test by the manufacturer or a qualified testing agency. Leaking hoses, gaskets, nozzles, and inner chambers can be replaced by firefighters.

Figure 2.63 Never rely on a damaged fire extinguisher.

OBSOLETE EXTINGUISHERS

Occasionally, firefighters encounter older fire extinguishers with which they are not familiar. These are often found in older buildings, particularly factories and educational facilities. In most cases, these extinguishers will be types that are no longer recommended for use; therefore, it is important that firefighters recognize obsolete extinguishers (Figure 2.64). Knowing that they are obsolete allows the firefighters to notify the occupants and fire department superiors that a change must be made. All obsolete extinguishers should be removed from service and replaced with new extinguishers that meet NFPA 10.

One of the most common obsolete extinguishers still found in service is the type designed to operate after being inverted. The inverting types include the following:

- The soda-acid extinguisher is the most common inverting-type extinguisher (Figure 2.65). When this extinguisher is inverted, acid from a bottle within the cylinder mixes with a soda-and-water solution and produces a gas that expels the liquid. The pressure on an acid-corroded shell has exploded many soda-acid extinguishers.

Figure 2.64 Obsolete extinguishers should be removed from service.

Figure 2.65 Soda-acid extinguishers are no longer acceptable for use.

- Inverting foam extinguishers look like soda-acid extinguishers. Inverting mixes solutions from two chambers forming a foam that expands at a 1:89 ratio and a gas that expels the foam to fight Class A and Class B fires. These units have been replaced by stored-pressure AFFF units described earlier in this chapter.
- Cartridge-operated water extinguishers are used by inverting and bumping the units to puncture a

CO_2 cylinder. The pressure of the gas released from the cartridge expels the water.

- Liquid carbon tetrachloride extinguishers have been obsolete since the 1960s. These extinguishers resemble pump-type insecticide sprayers (Figure 2.66). When carbon tetrachloride comes in contact with heat, it releases a highly toxic phosgene gas.

Many of these obsolete models were made of copper or brass joined by soft solder or rivets. It is estimated that there are still several million in use, despite the fact that NFPA standards ruled them out in 1978. The disadvantages of these types of extinguishers include the following:

- They cannot be turned off once activated.
- The agent is corrosive.
- They are potentially dangerous. If the discharge hose is blocked, these extinguishers can build up pressures in excess of 300 psi (2 100 kPa) and explode causing serious injury or death. They can also explode if the cylinder is damaged or corroded.

INSPECTION OF FIRE EXTINGUISHERS

Fire extinguishers must be inspected regularly to ensure that they are accessible and operable. This is done by verifying that the extinguisher is in its designated location, that it has not been actuated or tampered with, and there is no obvious physical damage or condition present that will prevent its operation (Figure 2.67). Servicing of portable fire extinguishers (or any other privately owned fire suppression or detection equipment) is the responsibility of the property owner or building occupant.

Although it is usually performed by the building owner or the owner's designate, fire inspectors should include extinguisher inspections in their building inspection and pre-incident planning programs (Figure 2.68). During the inspection, the inspector should remember that there are three important factors that determine the value of a fire extinguisher: its serviceability, its accessibility, and the user's ability to operate it.

The following procedures should be part of every fire extinguisher inspection:

- Check to ensure that the extinguisher is in a proper location and that it is accessible (Figure 2.69).
- Inspect the discharge nozzle or horn for obstructions. Check for cracks and dirt or grease accumulations.
- Check to see if the operating instructions on the extinguisher nameplate are legible (Figure 2.70).

Figure 2.68 Pre-incident planning should include the inspection of fire extinguishers.

Figure 2.66 An old carbon tetrachloride extinguisher.

Figure 2.67 Note extra extinguishers that might be kept on the premises.

Figure 2.69 Make sure that the extinguisher is not blocked.

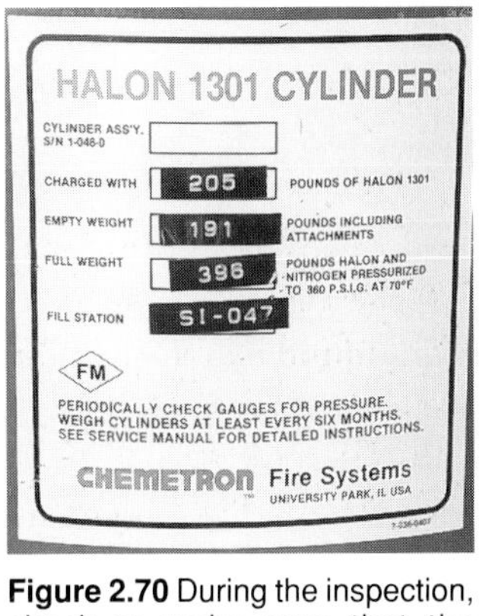

Figure 2.70 During the inspection, check to make sure that the instructions are legible.

- Check the lock pins and tamper seals to ensure that the extinguisher has not been tampered with (Figure 2.71).
- Determine if the extinguisher is full of agent and/or fully pressurized by checking the pressure gauge, weighing the extinguisher, or inspecting the agent level. If an extinguisher is found to be deficient in weight by 10 percent, it should be removed from service and replaced.
- Check the inspection tag for the date of the previous inspection, maintenance, or recharging (Figure 2.72).
- Examine the condition of the hose and its associated fittings.

If any of the items listed are deficient, the extinguisher should be removed from service and repaired as required. The extinguisher should be replaced with an extinguisher that has an equal or greater rating.

EXTINGUISHER MAINTENANCE REQUIREMENTS AND PROCEDURES

All maintenance procedures should include a thorough examination of the three basic parts of an extinguisher: mechanical parts, extinguishing agents, and expelling means. Building owners should keep accurate and complete records of all maintenance and inspections that include the month, year, type of maintenance, and date of the last recharge. The inspector is responsible for reviewing these records.

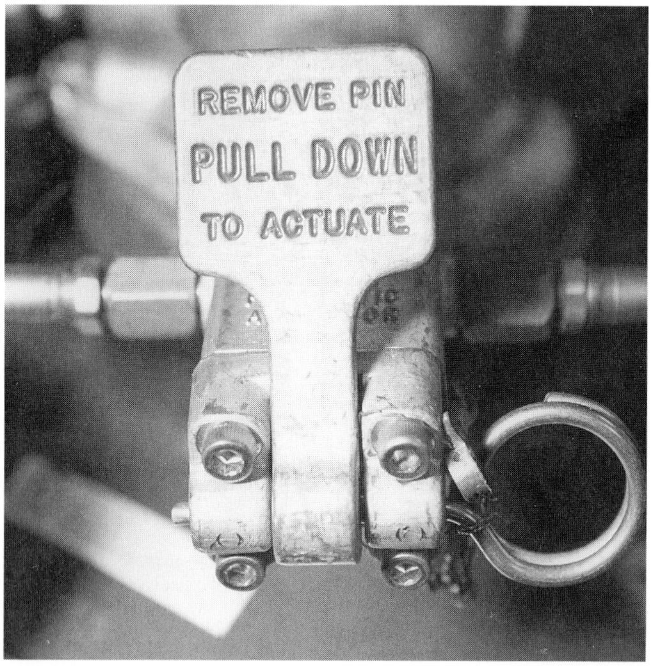

Figure 2.71 Check the lock pins and tamper seals to ensure that the extinguisher has not been tampered with.

Figure 2.72 Make sure that the inspection tag is up to date.

Fire extinguishers should be thoroughly inspected at least once a year. Such an inspection is designed to provide maximum assurance that the extinguisher will operate effectively and safely. A thorough examination of the extinguisher determines if any repairs are necessary or if the extinguisher should be replaced.

Stored-pressure extinguishers containing a loaded stream agent should be disassembled for complete maintenance. Before disassembly, the extinguisher should be discharged to check the operation of the discharge valve and pressure gauge.

Stored-pressure extinguishers that require a twelve-year hydrostatic test must be emptied every six years for complete maintenance. Extinguishers having nonrefillable disposable containers are exempt.

All carbon dioxide hose assemblies should have a conductivity test. Hoses found to be nonconductive must be replaced. Hoses must be conductive because they act as bonding devices to prevent the generation of static electricity.

TABLE 2.1
CHARACTERISTICS OF EXTINGUISHERS

Extinguishing Agent	*Method of Operation*	*Capacity* U.S.	Metric	*UL or ULC Classification*
Water	Stored Pressure	2½ gal	9.5 L	2-A
Water	Pump Tank	1½ gal	5.7 L	1-A
	Pump Tank	2½ gal	9.5 L	2-A
	Pump Tank	4 gal	15.1 L	3-A
	Pump Tank	5 gal	18.9 L	4-A
Water (Antifreeze Calcium Chloride)	Cartridge or Stored Pressure	1¼, 1½ gal	4.73, 5.7 L	1-A
	Cartridge or Stored Pressure	2½ gal	9.5 L	2-A
	Cylinder	33 gal	125 L	20-A
Water (Wetting Agent)	Stored Pressure	1½ gal	5.7 L	2-A
	Carbon Dioxide Cylinder	25 gal (wheeled)	94.6 L	10-A
	Carbon Dioxide Cylinder	45 gal (wheeled)	170.3 L	30-A
	Carbon Dioxide Cylinder	60 gal (wheeled)	227 L	40-A
Water (Soda Acid)	Chemically Generated Expellant	1¼, 1½ gal	4.73, 5.7 L	1-A
	Chemically Generated Expellant	2½ gal	9.5 L	2-A
	Chemically Generated Expellant	17 gal (wheeled)	64.3 L	10-A
	Chemically Generated Expellant	33 gal (wheeled)	125 L	20-A
Water (Loaded Stream)	Stored Pressure	2½ gal	9.5 L	2 to 3-A:1-B
	Cartridge or Stored Pressure	33 gal (wheeled)	125 L	20-A
AFFF	Stored Pressure	2½ gal	9.5 L	3-A:20-B
	Nitrogen Cylinder	33 gal (wheeled)	125 L	20-A:160-B
Carbon Dioxide	Self-Expellant	2 to 5 lb	.9 to 2.26 kg	1 to 5-B:C
	Self-Expellant	10 to 15 lb	4.5 to 6.8 kg	2 to 10-B:C
	Self-Expellant	20 lb	9 kg	10-B:C
	Self-Expellant	50 to 100 lb (wheeled)	22.6 to 45.3 kg	10 to 20-B:C
Dry Chemical (Sodium Bicarbonate)	Stored Pressure	1 lb	.45 kg	1 to 2-B:C
	Stored Pressure	1½ to 2½ lb	.68 to 1.13 kg	2 to 10-B:C
	Cartridge or Stored Pressure	2¾ to 5 lb	1.24 to 2.26 kg	5 to 20-B:C
	Cartridge or Stored Pressure	6 to 30 lb	2.7 to 13.6 kg	10 to 160-B:C
	Nitrogen Cylinder or Stored Pressure	75 to 350 lb (wheeled)	34 to 158.7 kg	40 to 320-B:C
Dry Chemical (Potassium Bicarbonate)	Stored Pressure	1 to 2 lb	.45 to .9 kg	1 to 5-B:C
	Cartridge or Stored Pressure	2¼ to 5 lb	1.02 to 2.26 kg	5 to 20-B:C
	Cartridge or Stored Pressure	5½ to 10 lb	2.4 to 4.5 kg	10 to 80-B:C

TABLE 2.1 (Continued)
CHARACTERISTICS OF EXTINGUISHERS

Extinguishing Agent	*Method of Operation*	*Capacity* *U.S.*	*Metric*	*UL or ULC Classification*
	Cartridge or Stored Pressure	16 to 30 lb	7.2 to 13.6 kg	40 to 120-B:C
	Cartridge	48 lb	21.7 kg	120-B:C
	Nitrogen Cylinder or Stored Pressure	125 to 315 lb (wheeled)	56.7 to 142.8 kg	80 to 640-B:C
Dry Chemical (Potassium Chloride)	Stored Pressure	2 to 2• lb	.9 to 1.1 kg	5 to 10-B:C
	Stored Pressure	5 to 9 lb	2.26 to 4 kg	20 to 40-B:C
	Stored Pressure	10 to 20 lb	4.5 to 9 kg	40 to 60-B:C
	Stored Pressure	135 lb	61.2 kg	160-B:C
Dry Chemical (Ammonium Phosphate)	Stored Pressure	1 to 5 lb	.45 to 2.26 kg	1 to 2-A and 2 to 10-B:C
	Stored Pressure or Cartridge	2•to 8• lb	1.13 to 3.8 kg	1 to 4-A and 10 to 40-B:C
	Stored Pressure or Cartridge	9 to 17 lb	4 to 7.7 kg	2 to 20-A and 10 to 80-B:C
	Stored Pressure or Cartridge	17 to 30 lb	7.7 to 13.6 kg	3 to 20-A and 30 to 120-B:C
	Cartridge	45 lb	20.4 kg	20-A and 80-B:C
	Nitrogen Cylinder or Stored Pressure	110 to 315 lb (wheeled)	49.9 to 142.8 kg	20 to 40-A and 60 to 320-B:C
Dry Chemical (Foam Compatible)	Cartridge or Stored Pressure	4≠ to 9 lb	2.1 to 4 kg	10 to 20-B:C
	Cartridge or Stored Pressure	9 to 27 lb	4 to 12.2 kg	20 to 30-B:C
	Cartridge or Stored Pressure	18 to 30 lb	8.2 to 13.6 kg	40 to 60-B:C
	Nitrogen Cylinder or Stored Pressure	150 to 350 lb	68 to 158.7 kg	80 to 240-B:C
Dry Chemical (Potassium Chloride)	Cartridge or Stored Pressure	(wheeled) 2• to 5 lb	1.13 to 2.26 kg	10 to 20-B:C
	Cartridge or Stored Pressure	9• to 20 lb	4.3 to 9 kg	40 to 60-B:C
	Cartridge or Stored Pressure	19• to 30 lb	8.8 to 13.6 kg	60 to 80-B:C
	Stored Pressure	125 to 200 lb (wheeled)	56.7 to 90.7 kg	160-B:C
Dry Chemical	Stored Pressure	5 to 11 lb	2.26 to 4.98 kg	40 to 80-B:C
(Potassium Bicarbonate Urea Base)	Stored Pressure	9 to 23 lb	4 to 10.4 kg	60 to 160-B:C
	Stored Pressure	175 lb	79.3 kg	480-B:C
Bromotrifluoromethane	Stored Pressure	2• lb	1.3 kg	2-B:C
Bromochlorodifluro-methane	Stored Pressure	2 to 4 lb	.9 to 1.8 kg	2 to 5-B:C
	Stored Pressure	5• to 9 lb	2.49 to 4 kg	1-A and 10-B:C
	Stored Pressure	13 to 22 lb	5.90 to 9.97 kg	1 to 4-A and 20 to 80-B:C

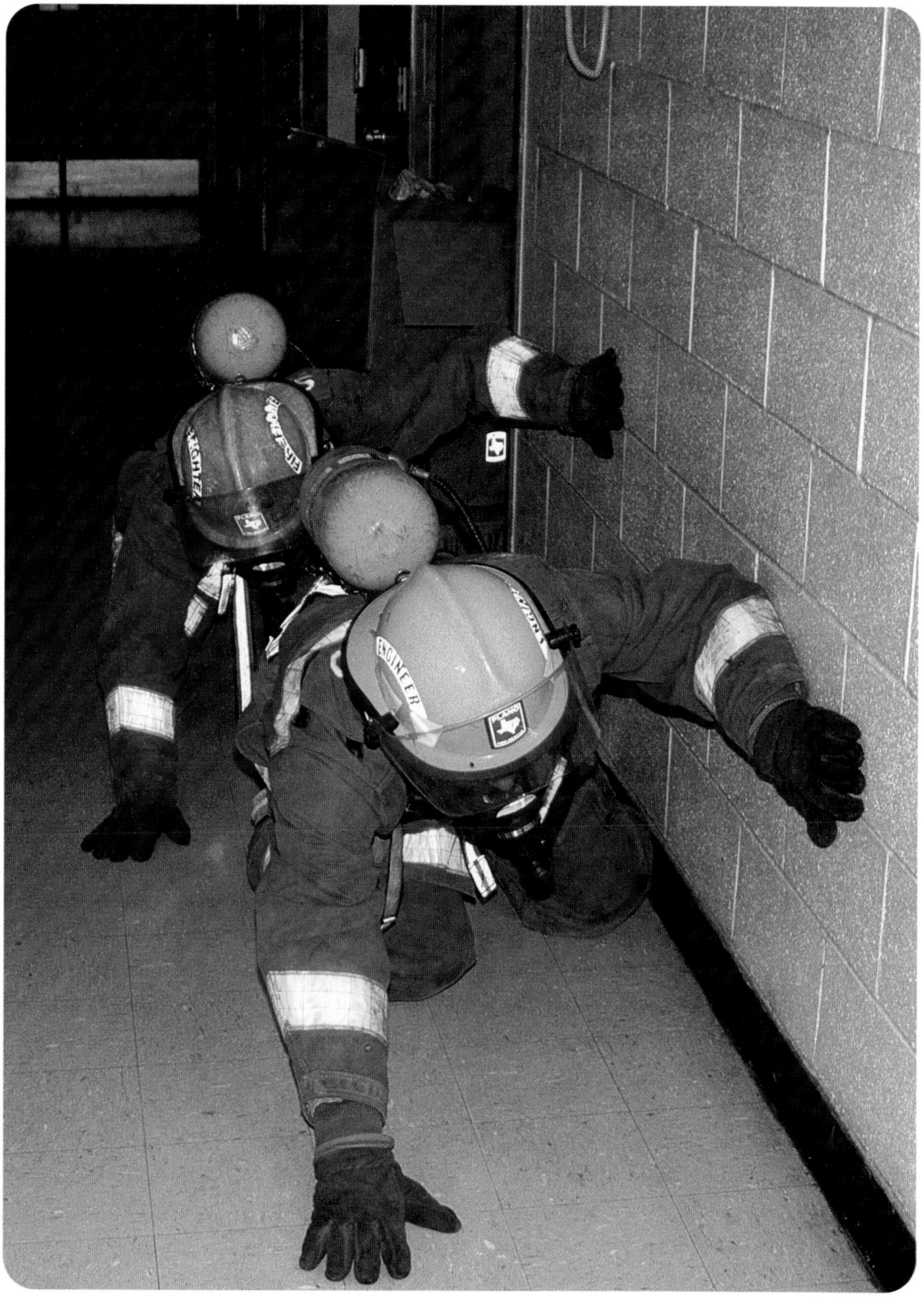

3

Firefighter Personal Protective Equipment

This chapter provides information that addresses performance objectives described in NFPA 1001, *Standard for Fire Fighter Professional Qualifications* (1992), particularly those referenced in the following sections:

3-3 Safety

3-3.12

3-7 Personal Protective Equipment

3-7.1

3-7.2

3-7.3

3-7.4

3-7.5

3-7.6

3-7.7

3-7.8

3-7.9

3-7.10

3-7.11

3-7.12

3-7.13

3-7.14

3-7.15

3-7.16

3-7.17

Chapter 3

Firefighter Personal Protective Equipment

Firefighters require the best personal protective equipment available because of the environment in which they perform their duties (Figure 3.1). Injuries can be reduced and prevented if firefighters have and use good protective clothing and breathing apparatus. However, providing and using quality protective equipment will not necessarily guarantee firefighter safety. All protective equipment has inherent limitations that must be recognized so that firefighters will not overextend the item's range of protection. Extensive training in the use and maintenance of equipment is required to ensure that the equipment will provide optimum protection.

The first part of this chapter discusses overall protective clothing. Eye protection, hearing protection, and work uniforms are discussed, in addition to standard turnout gear. The second part of this chapter gives an extensive overview of protective breathing equipment. Included is information on the different types of protective breathing equipment, reasons why protective breathing equipment must be worn, and the procedures for donning, doffing, inspecting, and refilling. Personal alert safety systems (PASS) are also discussed.

Figure 3.1 Fire fighting exposes personnel to a hostile environment. *Courtesy of Ron Jeffers.*

PERSONAL PROTECTIVE CLOTHING

Personal protective clothing refers to the garments firefighters need to wear while performing on the job. When possible, all firefighters operating at an emergency scene should wear full protective clothing suitable to that incident (Figure 3.2). Full protective clothing for structural fire fighting consists of the following:

- Helmet — protects the head from impact and puncture injuries as well as from scalding water.

Figure 3.2 A firefighter working at a structural fire should always wear full personal protective equipment, which includes an SCBA and a PASS device.

- Protective hood — protects portions of the firefighter's face, ears, and neck not covered by the helmet or coat.
- Protective coat and trousers — protect trunk and limbs against cuts, abrasions, and burn injuries (resulting from radiant heat), and provide limited protection from corrosive liquids.
- Gloves — protect the hands from cuts, wounds, and burn injuries.
- Safety shoes or boots — protect the feet from burn injuries and puncture wounds.
- Eye protection (goggles or faceshields) — protects the wearer's eyes from flying solid particles or liquids. May be a part of or separate from the helmet.
- Hearing protection — limits noise-induced damage to the firefighter's ears when loud noise situations cannot be avoided.
- SCBA — protects the face and lungs from toxic smoke and products of combustion.
- Personal Alert Safety System (PASS) — provides life-safety protection by emitting a loud shriek if the firefighter should collapse or remain motionless for approximately 30 seconds.

The quality of protective clothing must be thoroughly researched before selection and purchase. All equipment purchased should meet current applicable standards. There is a variety of protective clothing available that has been tested to determine different characteristics. Proximity, entry, and approach suits are also available for special situations (Figure 3.3).

Figure 3.3 The outer shell of this proximity suit is made of aluminized materials.

Special chemical-resistant suits must be considered for handling chemical emergencies. For more information on these special types of protective clothing, see IFSTA's **Firefighter Occupational Safety** and IFSTA's **Hazardous Materials for First Responders** manuals and the following National Fire Protection Association Standards:

NFPA 1991, *Standard On Vapor-Protective Suits for Hazardous Chemical Emergencies*

NFPA 1992, *Standard On Liquid Splash-Protective Suits for Hazardous Chemical Emergencies*

NFPA 1993, *Standard On Support Function Garments for Hazardous Chemical Operations*

The firefighter must understand the design and purpose of the various types of protective clothing and be especially aware of each garment's inherent limitations. The following sections highlight some of the important features of specific types of firefighter personal protective clothing.

Helmets

Head protection was one of the first concerns for firefighters. The traditional function of the helmet was to shed water, not to protect from heat, cold, or impact. The wide brim, particularly where it extends over the back of the neck, was designed to prevent hot water and embers from reaching the ears and neck. Newer helmet designs perform this function as well as provide the following additional benefits:

- Protect the head from impact
- Provide protection from heat and cold
- Provide faceshields to protect the face and eyes when SCBA is not required (Figure 3.4)

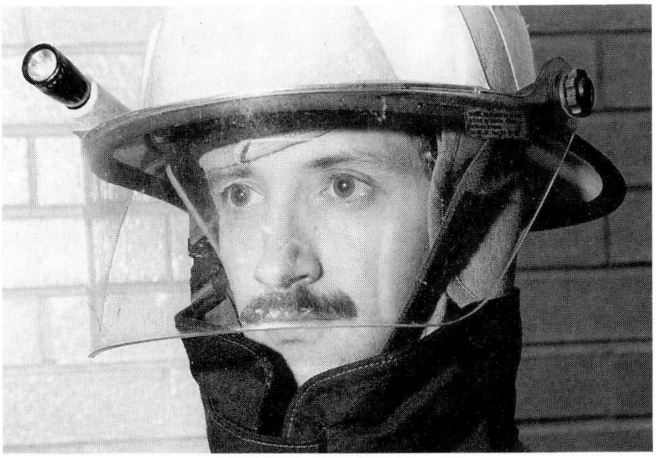

Figure 3.4 Flip-up faceshields attached to the helmet provide face and eye protection when SCBA is not required.

Helmet design will probably remain a subject of debate for years to come. Every type of helmet provides some protection, but whether a helmet is accepted by firefighters and officers depends on comfort and tradition (Figure 3.5). All helmets in service today should meet the specifications set forth in NFPA 1972, *Standard on Helmets for Structural Fire Fighting*. A fire department's failure to heed such a consensus standard may cause greater jeopardy in the event of a lawsuit/liability claim for negligence.

Figure 3.5 Firefighters may have several different types of helmets from which to choose. This photo shows both fire fighting and rescue models.

Figure 3.6 An information sticker will be placed on the inside of helmets that meet the specifications in the standard.

Helmets that meet the specifications in the standard will be marked with the following (Figure 3.6):

- Name of the manufacturer
- Model number of design
- Date of manufacture
- Lot number
- Weight
- Recommended cleaning procedure
- Helmet size or range

Each helmet will be durably marked with the following warning: **"This Helmet Must Be Properly Adjusted And Secured To The Head With All Components In Place To Provide Designed Protection."**

Helmets must have earflaps, which should always be used during fire fighting (Figure 3.7). Chin straps ensure that helmets stay in place upon impact (Figure 3.8). Some helmets are equipped with flashlights that free the hands as well as provide good visibility (Figure 3.9).

Eye Protection

Perhaps one of the most common injuries on the fireground is injury to the eyes. Eye injuries are not

Figure 3.7 All helmets must have earflaps to protect the firefighter's ears during fire fighting activities.

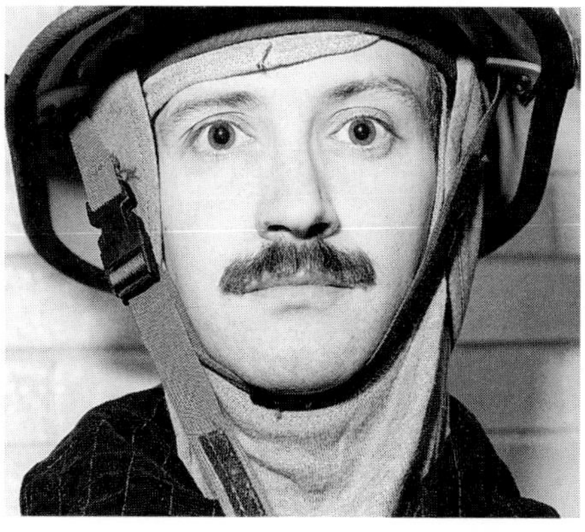

Figure 3.8 Chin straps ensure that the helmet will remain in place.

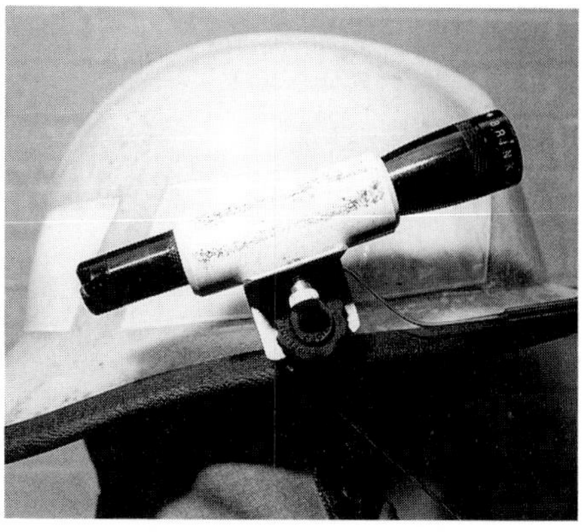

Figure 3.9 Helmet-mounted flashlights free the firefighter's hands and provide visibility.

always reported because they are not always debilitating. However, eye injuries can be serious, and they are fairly easy to prevent. It is important to protect the eyes on the fireground and while performing duties around the station. Eye protection for the firefighter comes in many forms such as safety glasses, safety goggles, helmet faceshields, and SCBA masks (Figure 3.10).

Consider the durability of eye protection. Select materials that will absorb impact without shattering, are not subject to easy scratching, and will not melt at high temperatures. Some of the best materials available for eye protection are made of polycarbonates. Fire departments should prepare for frequent replacements as a basic cost of operations and should maintain an ample supply of faceshields and facepiece replacement parts. For more information on the requirements for eye protection, reference ANSI Standard Z87.1.

For face and eye protection, many fire departments depend on faceshields that attach to the helmet (Figures 3.11 a and b). Most of these flip up and out of the field of vision fairly easily and are generally acceptable to firefighters. Most assemblies do not interfere with protective breathing equipment.

Many of the current faceshields are made of polycarbonate plastics. Many of these shields start to lose structural stability at temperatures around 250°F to 300°F (121°C to 149°C) — melting faceshields should not necessarily be considered a drawback. A melting faceshield can indicate to firefighters that they are in extremely hot and dangerous environments.

WARNING:

When the faceshield begins to distort from heat, the firefighter is in danger and should leave the area immediately.

When selecting faceshields, give consideration to the height of the shield and to the degree of side protection it offers. The most common sizes for firefighter helmet faceshields are 4 and 6 inches (100 mm and 150 mm). Faceshields attached to a helmet must be approved by the manufacturer of the helmet. Only manufacturer-approved mounting hardware and techniques should be used to attach the shield to the helmet.

Firefighters may encounter a variety of situations where eye protection, other than that afforded by a helmet faceshield or SCBA mask, is required. These situations include station operations such as welding, grinding, or cutting (Figure 3.12). Other situations that require eye protection include operating at vehicle extrications or brush fires and conducting inspections in industrial occupancies (Figure 3.13). NFPA 1500, *Standard on Fire Department Occupational Safety and Health Program*, requires eye protection when hazards to the eyes, such as flying debris or chemical splashing, are present.

Safety glasses and goggles will protect against approximately 85 percent of all eye hazards. Several styles are available, including some that fit over prescription

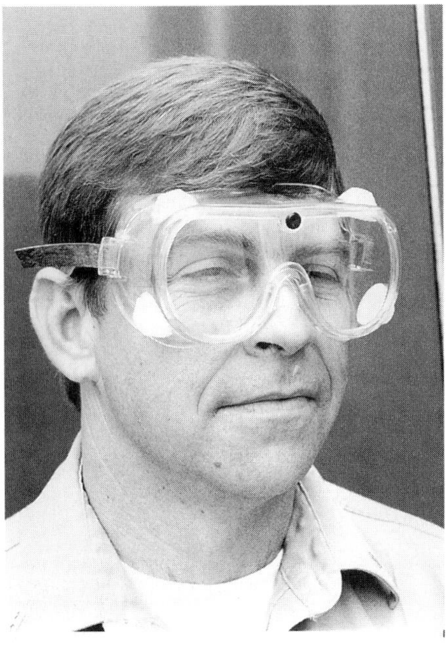

Figure 3.10 Any time that there is a danger to the eyes, safety glasses or goggles must be worn.

Figure 3.11a A 6-inch (150 mm) faceshield.

Figure 3.11b A 4-inch (100 mm) faceshield.

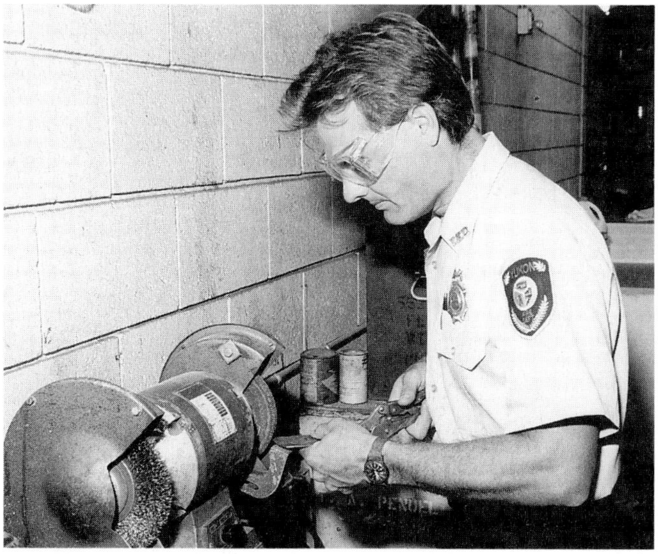

Figure 3.12 Many station operations, such as grinding, cutting, or welding, require eye protection.

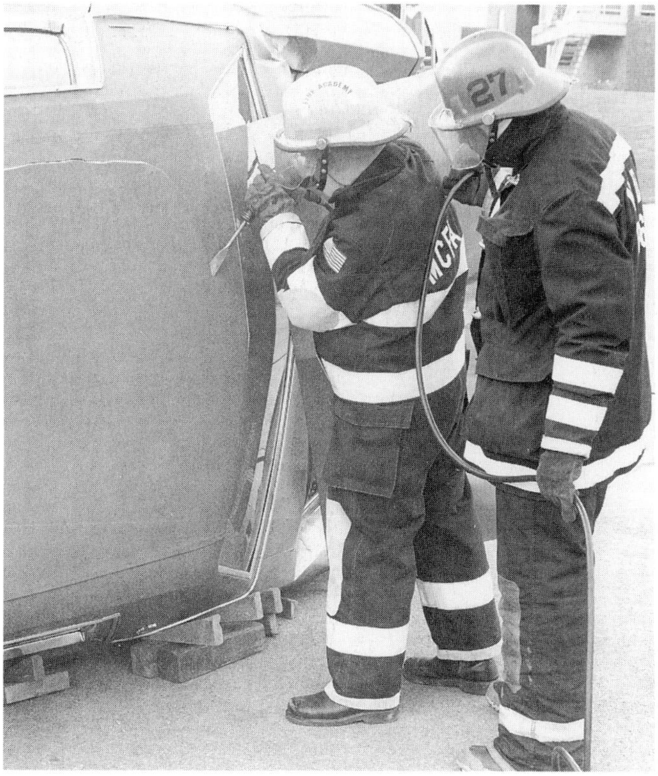

Figure 3.13 Firefighters are wearing eye protection at a vehicle extrication.

glasses (Figure 3.14). Firefighters who wear prescription safety eyeglasses should select frames and lenses that meet ANSI Standard Z87.1 for severe exposure to impact and heat.

Warning signs should be posted near operations requiring eye protection (Figure 3.15). Use of eye protection must be required through departmental standard operating procedures (SOPs) and enforced by supervisory personnel.

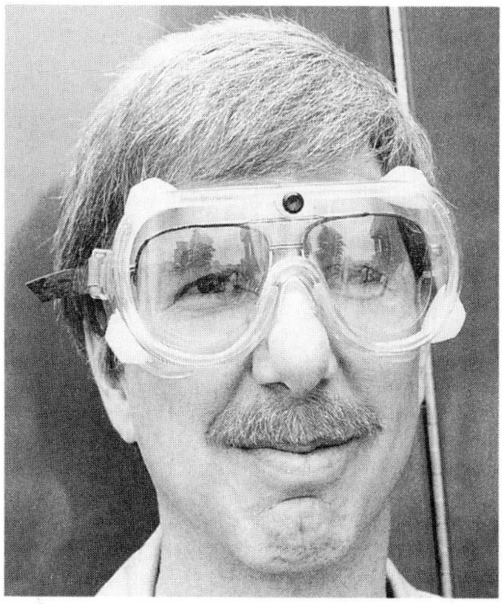

Figure 3.14 Goggles that fit over prescription eyeglasses are available.

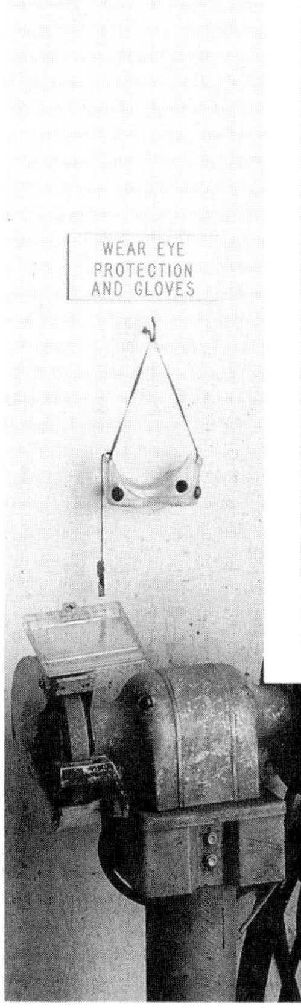

Figure 3.15 Warning signs should be posted in areas where goggles are required.

Hearing Protection

Firefighters are exposed to a variety of sounds in the station, en route to the scene, and on the fireground. Often, exposure to these sounds or a combination of sounds can produce permanent hearing loss. To prevent exposure to unacceptably high levels of noise, it is necessary for the department to initiate a hearing protection program to identify, control, and reduce the potentially harmful noise and/or provide protection from it (Figure 3.16).

Figure 3.16 Earmuffs provide hearing protection.

Eliminating or reducing the noise level is the best solution. However, this is often not possible. Therefore, acceptable hearing protection should be provided to firefighters, and they should be required to use it. Earplugs or earmuffs should be used for hearing protection (Figure 3.17). If earplugs are to be used, each firefighter should carry a set as part of the standard turnout gear. There are potential hazards associated with earplugs and earmuffs. For example, in a structural fire fighting situation, earmuffs can compromise protection of the face by making it awkward to use SCBA and hoods. Earplugs may melt when exposed to intense heat. For these reasons, wearing hearing protection during structural fire fighting is impractical. The most common use for hearing protection will be for firefighters who ride apparatus that exceed maximum noise exposure levels (Figure 3.18).

Intercom/ear protection systems provide a dual benefit because of their ability to reduce the amount of noise the ear is exposed to and at the same time allow the crew to communicate or monitor the radio at a normal volume (Figure 3.19). These systems can be worn during response by each member riding the apparatus or by the officer in charge and then handed over to the apparatus operator when the unit is on the scene (Figure 3.20).

Protective Hoods

Protective hoods are an important part of the total protective clothing system. Protective hoods are designed to protect the firefighter's ears, neck, and face from exposure to extreme heat. These hoods will also cover areas not otherwise protected by the SCBA facepiece, earflap, or coat collar. Hoods are typically made of Kevlar® aramid fibers, Nomex® fire resistant material, or PBI® polybenzimidazole and are available in long or short styles (Figures 3.21 a and b). Protective hoods used in conjunction with the SCBA facepiece provide effective protection. However, care must be taken to ensure that the hood does not interfere with the facepiece-to-face seal (Figure 3.22).

Firefighters who have grown accustomed to interior structural fire fighting without the use of a protective hood will be required to perform live-fire training with the hood in place. This training is necessary for the firefighters to develop a new sense of the level of heat in a given area. This awareness will prevent them from overextending themselves in an extremely hot atmosphere.

Firefighter Protective Coats (Turnout Coats)

Turnout coats are used for protection in structural fire fighting and other fire department activities. A

Figure 3.17 Earplugs are inexpensive and easy to transport.

Figure 3.18 Firefighters riding in the jump seat area on the apparatus should be required to wear hearing protection.

Figure 3.19 Intercom/ear protection allows the crew to communicate and at the same time reduces the amount of noise to which they are exposed.

Figure 3.20 The apparatus operator should wear ear protection when operating at the pump panel.

Figure 3.21a Longer protective hoods extend over the shoulders and chest.

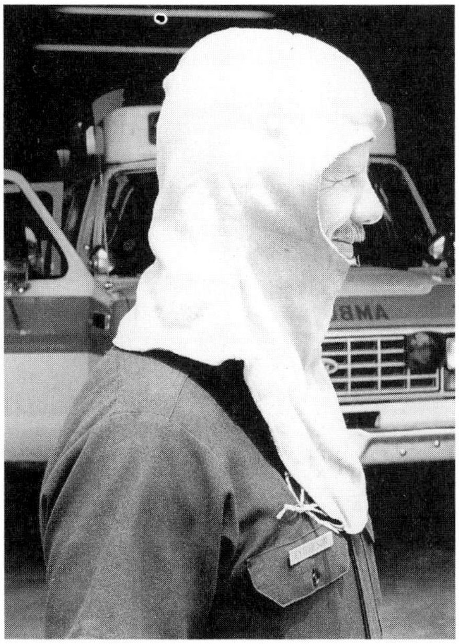

Figure 3.21b The short protective hood covers the ears, neck, and face from exposure to extreme heat.

Figure 3.22 Proper placement of the protective hood will not interfere with the facepiece-to-face seal.

variety of coats are available from numerous manufacturers (Figures 3.23 a and b). All turnout coats and trousers should meet the standards set forth in NFPA 1971, *Standard on Protective Clothing for Structural Fire Fighting*.

Coats that meet the NFPA 1971 standard have labels attached to each separable layer. The labels state that the garment meets the requirements of the standard, explains the limitations of the garment, and gives basic recommendations for proper care of the garment (Figure 3.24).

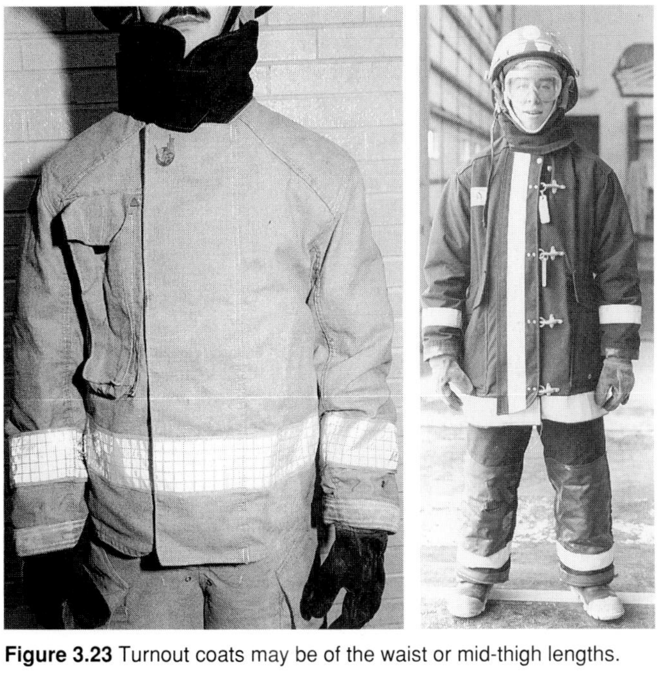

Figure 3.23 Turnout coats may be of the waist or mid-thigh lengths.

Figure 3.24 Coats that meet NFPA 1971 will have labels attached to each layer.

NFPA 1971 requires that all turnout coats be made of three layers: outer shell, moisture barrier, and thermal barrier (Figure 3.25 on the next page). These barriers serve to trap insulating air that inhibits the transfer of heat from the outside to the firefighter's body. They also protect the firefighter from direct flame contact, cold temperatures, hot water and vapors, and any number of environmental hazards. Clearly, the construction and function of each layer are of great importance to the firefighter's safety.

The type of barrier material has a great effect upon the amount of protection given. Some fabric combinations provide better protection against heat, while other combinations provide better resistance against cuts and abrasions. Some materials are heavier than others. All these factors must be considered when selecting coats.

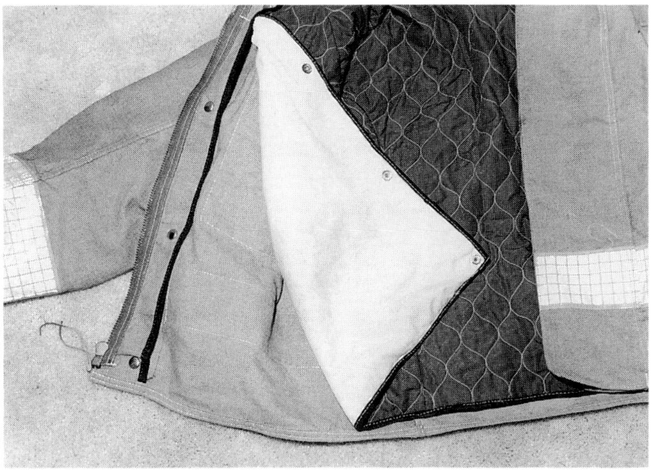

Figure 3.25 The moisture barrier protects the firefighters from water, steam, hot vapors, or corrosive liquids.

WARNING:

All inner liners of the protective coat must be in place and used during any fire fighting operation. Failure to wear the entire coat and liner system during fire conditions may expose the firefighter to severe heat that could result in serious injury or death.

Turnout coats have many features that provide additional protection and convenience to the wearer. Collars can be turned up to protect the neck and throat of the wearer (Figure 3.26). Wristlets prevent water, embers, and other foreign debris from rolling down inside the sleeves (Figure 3.27). The double-layered storm flap closure system on the front of every coat prevents water or fire products from entering through gaps between the snaps or clips (Figure 3.28). A variety of pocket styles are available, and they can be specially designed to carry tools, SCBA facepieces, flashlights, portable radios, and other important equipment (Figure 3.29).

Turnout coats are available in many colors including yellow, black, white, red, olive drab, orange, khaki or natural, silver, and brown. The most practical color has been the subject of much research. Dark coats exposed to the sun absorb more heat than lighter colors; therefore, they may be hotter to wear. However, when dark coats and coats of lighter colors of the same material are exposed to the radiant heat found in ordinary fires, there is little measurable difference during extended exposure.

Dark colors are harder to see, especially at night. Lighter colors, especially white and yellow, are easier to see both at night and during the day. A complaint about lighter colors is that they are more difficult to keep clean than the darker colors. Actually, dark coats get just as dirty, the dirt is just disguised. A dirty coat will absorb more heat than a clean coat. Thus, the ability to more readily see dirt on the lighter colored coat is a safety advantage in that it requires the firefighter to more thoroughly clean the coat. Turnout coats that meet NFPA standards are designed to be laundered.

Reflective trim should be sewn onto the coat before purchase. To meet NFPA standards, each coat must have at least 325 square inches (209 700 mm^2) of reflective tape. All trim should be at least 2 inches (50 mm) wide and should encircle both sleeves and the hem of the

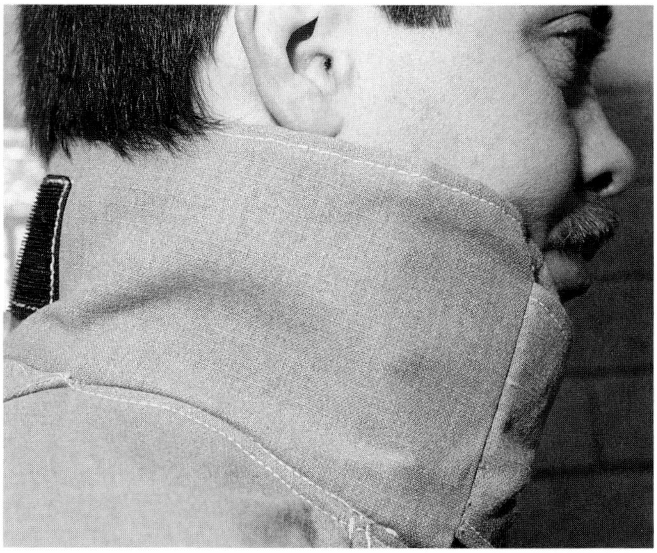

Figure 3.26 Some coat collars are made of the same fabric as the outer shell.

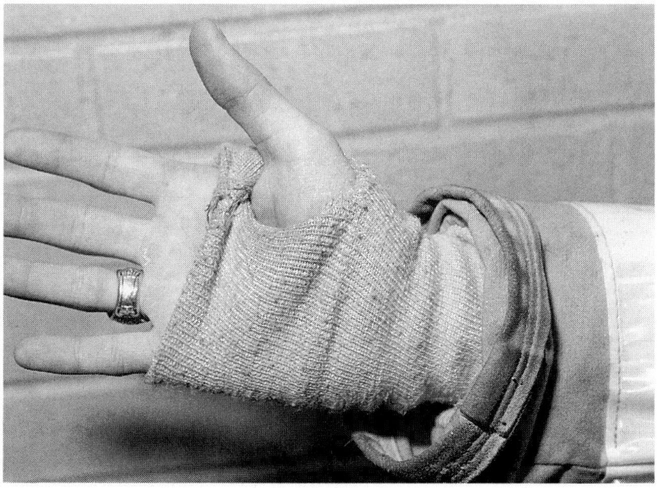

Figure 3.27 The wristlet that extends over the hand should have a thumbhole to prevent it from sliding up the wrist.

coat. It is preferable that the trim be of the two-tone type (Figure 3.30). Trim should not be obscured by pockets, patches, or storm flaps.

There are numerous other protective coat construction features that must be considered when specifying new coats. Pocket corners and sleeve bottoms can be reinforced with leather or additional layers of outer shell fabric for longer wear (Figure 3.31). A hanger loop may be provided inside the neck of the coat. Shoulder padding and leather wear protectors for shoulders are also beneficial.

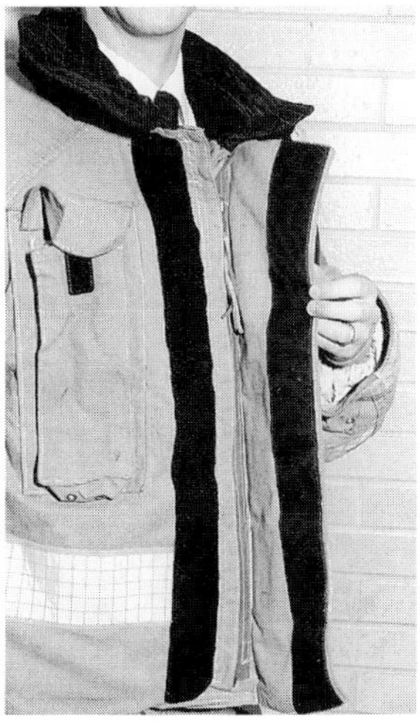

Figure 3.28 The storm flap covers the closure area and prevents steam, water, and fire products from entering the gaps between the closures.

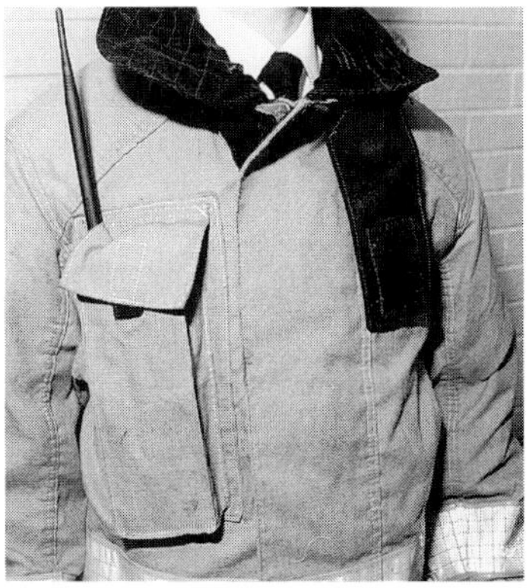

Figure 3.29 Some coats have special pockets designed to carry items such as radios.

Figure 3.30 Two-tone reflective trim.

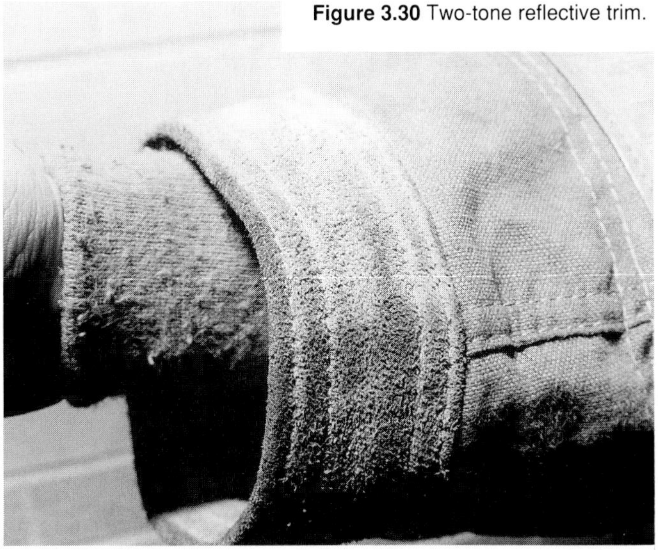

Figure 3.31 Leather sewn to the outer cuff helps protect the cuff from damage.

Firefighter Protective Trousers

Protective trousers, often called bunker pants or night hitches, are required by NFPA 1500 (Figure 3.32). Three-quarter boots and long coats alone do not provide adequate protection for the lower torso or extremities and are no longer permissible according to NFPA 1500. When selecting trousers, consider the same concepts of fabric selection, moisture barriers, and the other considerations used to select protective coats. The layering principles that apply to coats also apply to trousers. Protective trousers should also meet the requirements specified in NFPA 1971, *Standard on Protective Clothing for Structural Fire Fighting*. Options, such as reinforced knees and leather cuffs, may increase the durability of protective trousers (Figure 3.33). Suspenders should be the heavy-duty type so that pants will not sag when they become wet (Figure 3.34). Protective trousers must have 80 square inches (51 600 mm^2) of trim. Each leg should have a band of trim encircling it between the hem and the knee (Figure 3.35).

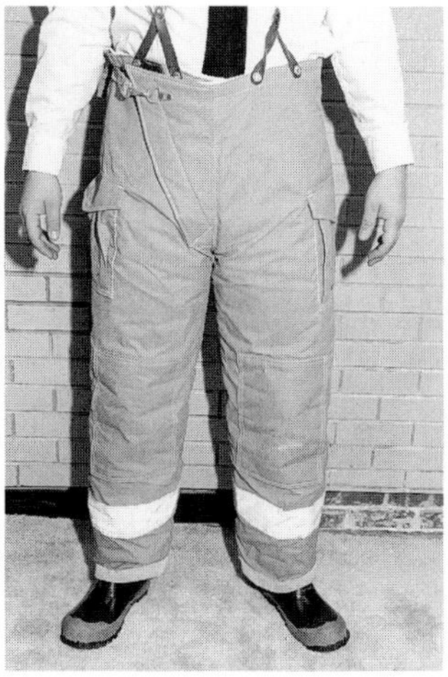

Figure 3.32 Protective trousers are required by NFPA 1500.

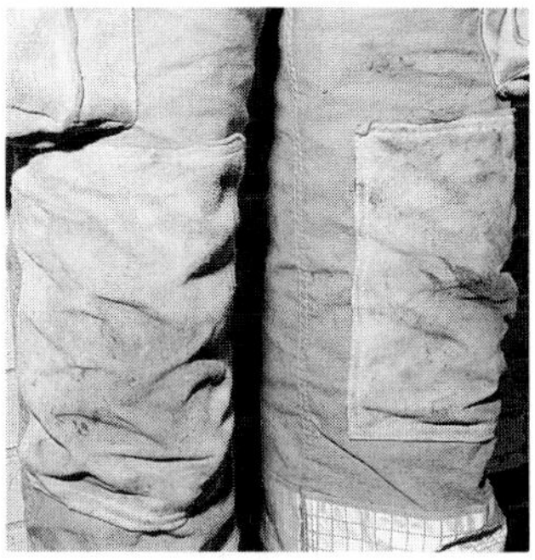

Figure 3.33 Reinforced knees prolong the life of protective trousers.

Figure 3.34 Heavy-duty suspenders keep trousers from sagging when they become wet.

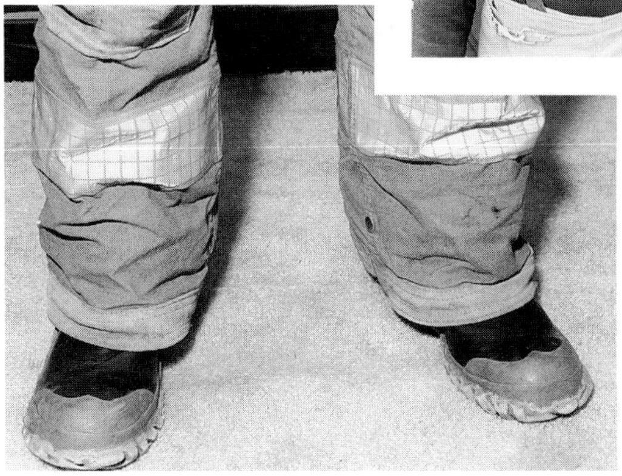

Figure 3.35 According to NFPA 1971, each leg of the protective trouser should have a band of reflective trim between the hem and the knee.

Hand Protection

The most important characteristics of gloves are the protection they provide against heat or cold penetration and their resistance to cuts, punctures, and liquid absorption (Figure 3.36). Gloves must allow enough dexterity and tactile feel for the firefighter to perform the job effectively. If the gloves are too awkward and bulky, the firefighter may be unable to do fine manipulative work (Figure 3.37). Gloves must fit properly and be designed to provide protection as well as to allow dexterity. Unfortunately, in order to provide protection, this dexterity is often reduced. In winter or extremely cold areas, fire fighting mittens may be worn (Figure 3.38). However, these have even less dexterity than gloves.

All firefighters who engage in or are exposed to structural fire fighting should be provided with gloves that meet the requirements of NFPA 1973, *Standard on Gloves for Structural Fire Fighting*. The standard specifies that the gloves contain a label stating that they meet the requirements of the standard (Figure 3.39).

Gloves coated with a rubberlike plastic are not recommended for fire fighting. These gloves are usually orange or red and are commonly referred to as "redball" gloves (Figure 3.40). The main concern about plastic-coated gloves is their reaction to radiant heat. Exposed surfaces of the glove will sometimes begin to soften; and once the heat migrates to the skin, the gloves become uncomfortable. If the heat is sufficient, hand perspiration can turn to steam and cause burns. Plastic-coated gloves may be useful during overhaul and hose pickup operations, especially in wet and freezing conditions. Gloves may be stored on apparatus for these purposes.

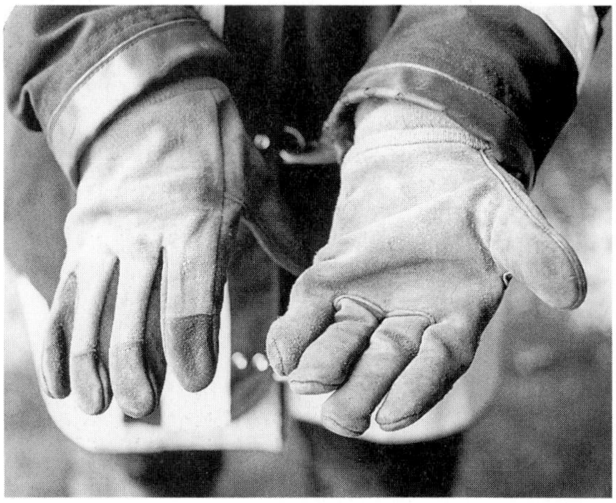

Figure 3.36 Newer styles of fire fighting gloves incorporate aramid fibers with leather palms and fingers to provide a lighter, more flexible glove.

Figure 3.37 Extrication situations often require the firefighter to wear a glove that is less bulky than a normal fire fighting glove. *Courtesy of Bob Norman, Elkton (MD) V.F.D.*

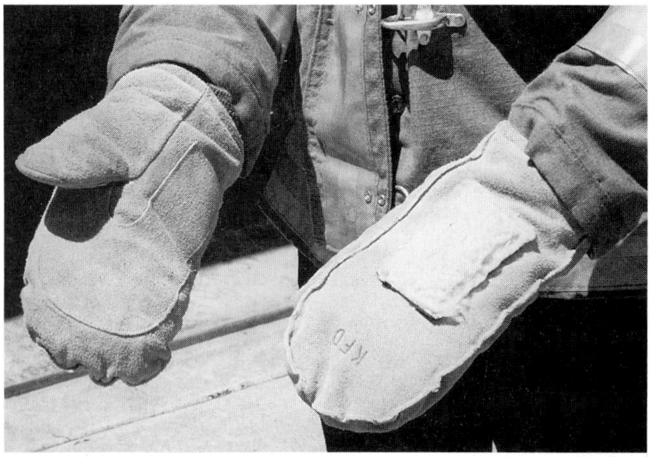

Figure 3.38 Mittens keep the fingers and hands warmer in low temperatures, but offer less dexterity than gloves.

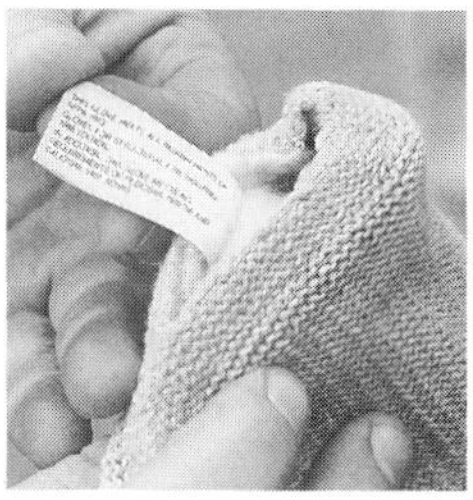

Figure 3.39 Fire fighting gloves should have an NFPA 1973 label inside them.

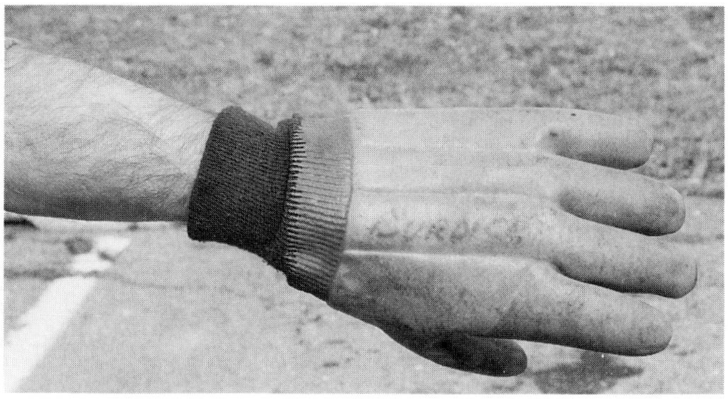

Figure 3.40 "Redball" gloves are not suitable for fire fighting operations.

Foot Protection

There are numerous hazards to the feet on the fire scene. Firebrands, falling objects, and nails are examples of commonly encountered hazards (Figure 3.41). Appropriate foot protection should be selected to ensure that the risk of injury from these hazards is minimized. Because of the nature of their work, firefighters will need to have two different kinds of foot protection:

- Turnout boots for fire fighting and emergency activities (Figure 3.42)
- Safety shoes for station wear and other fire department activities that include inspections, emergency medical responses, and similar activities (Figure 3.43)

Turnout boots must meet the requirements set forth in NFPA 1974, *Standard for Protective Footwear for Structural Fire Fighting*. Puncture resistance should be provided by a stainless steel midsole plate about 0.018 inch (0.46 mm) thick. If there is doubt about midsole protection, X-ray the boot. Some fire departments require insulation laminated into the rubber (Figure 3.44). The only disadvantage to this requirement is that the added weight tends to increase firefighter fatigue. Select a boot lining that will not break up and cause blisters and discomfort. There are also turnout boots with shin pads to reduce the strain from leg locks and crawling (Figure 3.45). Boots should have well-secured pull loops (Figure 3.46 on the next page).

Boots that meet the requirements of the standard will have labels stating: **"Meets All Requirements of NFPA 1974 Protective Footwear for Structural Fire Fighting (1987 Edition)."** Some leather boots will meet the requirements of this standard, as will some boots constructed of fire-resistant materials (Figure 3.47 on the next page). Each firefighter should be fitted as accurately as possible. Half sizes are available in both men's and women's boots. Firefighters should not share turnout boots because this practice is unsanitary. When boots are reissued, sanitize them with procedures recommended by an industrial hygienist.

Safety shoes or boots should be worn while conducting inspections or while doing work around the station. Some departments require safety shoes or boots as a part of their daily uniform. Safety shoes usually have safety toes, puncture-resistant soles, or special inserts (Figure 3.48 on the next page). These shoes provide good support for climbing, give increased physical agility, and are generally less fatiguing than turnout boots. The leather fire fighting boot can be used for work at the station, inspections, and fire fighting operations.

Figure 3.41 Fire scenes have numerous foot and tripping hazards. *Courtesy of Bob Esposito.*

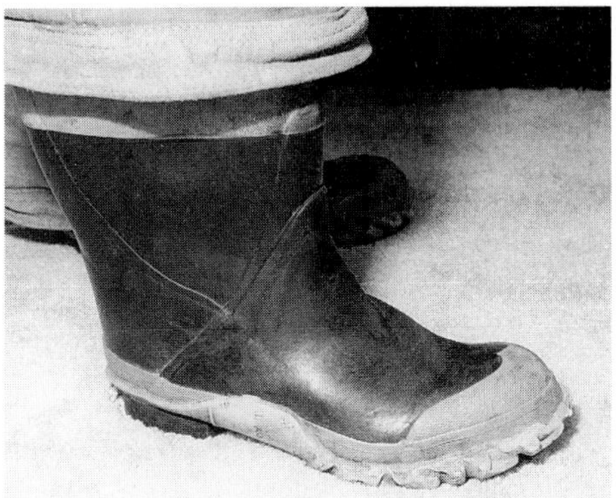

Figure 3.42 The rubber boot is one style of turnout boot. All turnout boots must meet the requirements set forth in NFPA 1974.

Figure 3.43 Newer style leather turnout boots can also serve as station safety shoes. *Courtesy of the Warrington Group, LTD.*

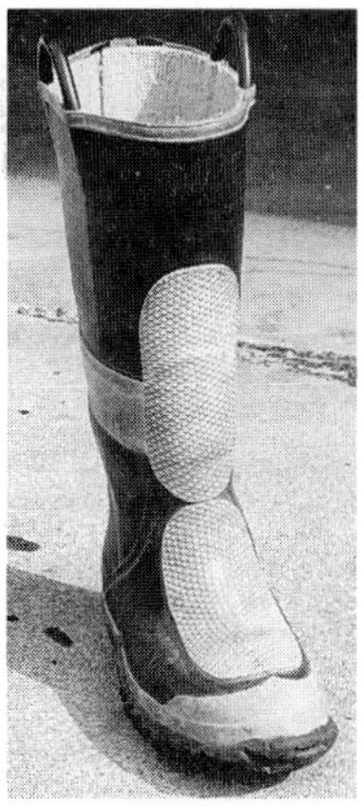

Figure 3.45 Boots with shin pads will help reduce the strain caused by leg locks and crawling.

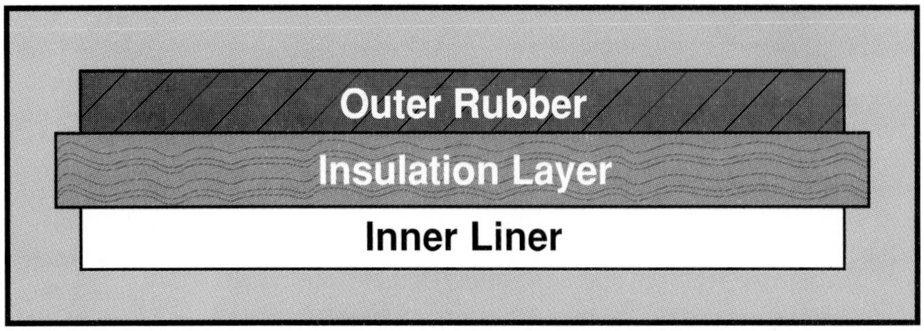

Figure 3.44 Some fire departments require insulation laminated into the rubber of the turnout boot.

Figure 3.46 Well-secured pull loops on fire fighting boots facilitate donning.

Figure 3.47 A new style boot uses both aramid fibers and leather. *Courtesy of the Warrington Group, LTD.*

Figure 3.48 Safety shoes should be worn when working at the station. They provide good support for climbing and are usually less fatiguing than fire fighting boots.

Station/Work Uniforms

Firefighter accident statistics show that certain types of clothing can contribute to on-the-job injuries. Certain synthetic fabrics, such as polyester, can be especially hazardous because they can melt during exposure to high temperatures. Some of the materials that have high temperature resistance are as follows:

- Organic fibers such as wool, cotton, and rayon
- Synthetic fibers such as Kevlar® aramid fibers, Nomex® fire resistant material, PBI® polybenzimidazole, Kynol® phenolic resins, Gore-Tex® water repellant fabric, Orlon® acrylic fiber, neoprene, Teflon® fluorocarbon resins (nonstick coatings), silicone, and panotex

All firefighter station and work uniforms should meet the requirements set forth in NFPA 1975, *Standard on Station / Work Uniforms for Fire Fighters*. The purpose of the standard is to provide work wear that will not contribute to firefighter injury and negate the effects of the outer protective clothing. Garments falling under this standard include trousers, shirts, jackets, and coveralls, but not underwear (Figure 3.49). Underwear made of 100 percent cotton is recommended.

Figure 3.49 Station uniforms should provide additional protection to the firefighter.

The main part of the standard requires that no components of garments burn or melt when exposed to heat at 482°F (250°C) for five minutes. A garment meeting all parts of the standard will have a notice to that effect permanently attached. It is important to note that while this clothing is designed to be fire resistant, it is not designed to be worn for fire fighting operations. Standard structural fire fighting clothing must always be worn over these garments when a firefighter is engaged in fire fighting activities, with the exception of wildland fire fighting (Figure 3.50). Special clothing designed for wildland fire fighting is more appropriate in these situations.

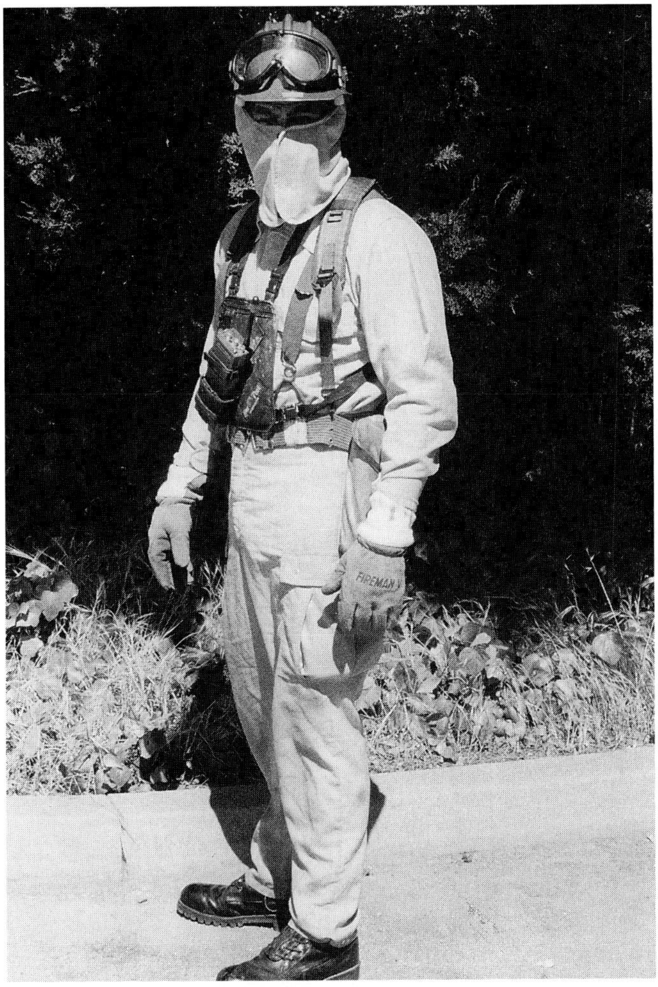

Figure 3.50 Wildland firefighters should be equipped with personal protective equipment designed especially for wildland fire conditions. *Courtesy of Monterey (CA) Fire Department.*

Care Of Personal Protective Clothing

In order for personal protective clothing to perform properly, it must be maintained within the manufacturer's specifications. Each piece of protective clothing has a particular maintenance procedure that should be followed to ensure it is ready for service.

Helmets should be properly cleaned and maintained to ensure their durability and maximum life expectancy. The following are guidelines for their proper care and maintenance.

- Remove dirt from the shell (Figure 3.51). Dirt will absorb heat faster than the shell itself, thus exposing the wearer to more severe conditions.
- Remove chemicals, oils, and petroleum products from the shell as soon as possible (Figure 3.52). These agents may soften the shell material and reduce its impact and dielectric protection. See the manufacturer's instructions for suggested cleansers.
- Repair or replace helmets that do not fit properly (Figure 3.53). A poor fit will reduce the helmet's ability to resist the transmission of force.
- Repair or replace helmets that are damaged (Figure 3.54 on the next page). This includes leather helmets that have become cracked or brittle with age.

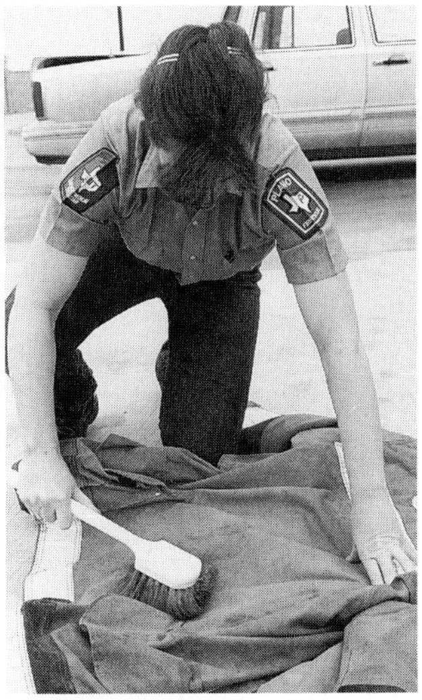

Figure 3.51 Clean dirt from the coat shell.

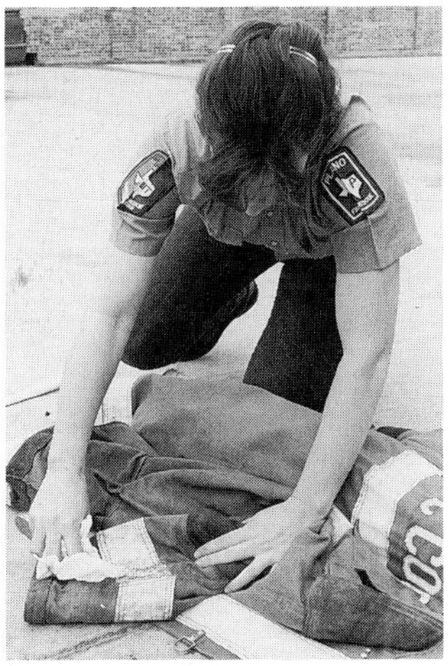

Figure 3.52 Remove any type of substance, such as chemicals, oils, and petroleum products, from the shell as soon as possible.

Figure 3.53 The firefighter should try on the helmet to ensure a proper fit. Some may be adjusted to correct a loose or tight fit.

Figure 3.54 Damaged helmets should be removed from service.

- Check for adequate separation between the suspension web and the outer shell. The proper distance is at least 1¼ inches (32 mm) between the top of the firefighter's head and the crown of the shell (Figure 3.55). Tie any suspension lace with a square knot.
- Inspect suspension systems frequently to detect deterioration. Replace if necessary.
- Consult the helmet manufacturer if a helmet needs repainting. Manufacturers can inform the departments about the choice of paints available for a particular shell material.
- Polycarbonate helmets that have come into contact with hydraulic oil from a rescue tool should be removed from service and checked. Some oils attack the polycarbonate material and will weaken the helmet.

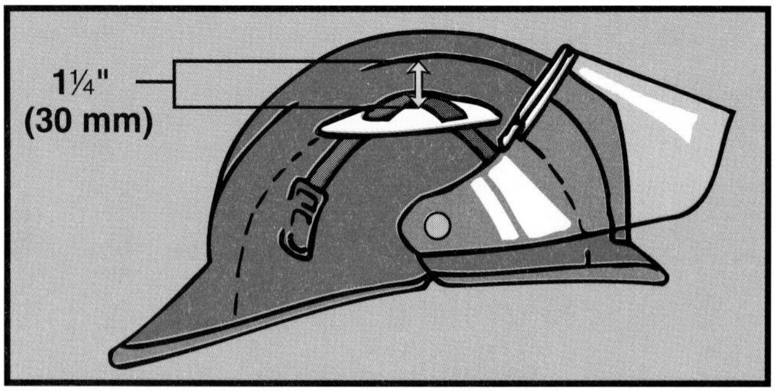

Figure 3.55 Check for adequate separation between the suspension web and the outer shell.

Cleanliness also affects the performance of protective coats, trousers, and hoods. The outer shell should be cleaned regularly. Clean outer shells have better fire resistance; dirty turnouts absorb more heat. Follow the manufacturer's directions for cleaning. The directions are usually contained on a tag sewn to the coat. However, most coats and trousers can be cleaned by laying them out on concrete, blacktop, or on a large table and hand-scrubbing them with a mild soap and water (Figure 3.56). The coat should be rinsed and then hung to dry.

Gloves should also be cleaned according to the manufacturer's instructions. They can usually be treated in the same way as protective coats and trousers. Some may be machine washed if the manufacturer approves.

The following guidelines are suggested for care and maintenance of fire fighting boots.

- Wash oil, grease, chemicals, and debris from the boot because they deteriorate rubber (Figure 3.57).
- Store rubber boots in a cool, dark place because ozone will deteriorate rubber and cause the boots to lose their protective quality (Figure 3.58).
- Replace worn, cut, or punctured boots that cannot be repaired.
- Apply the recommended dressing to leather boots at the required intervals.

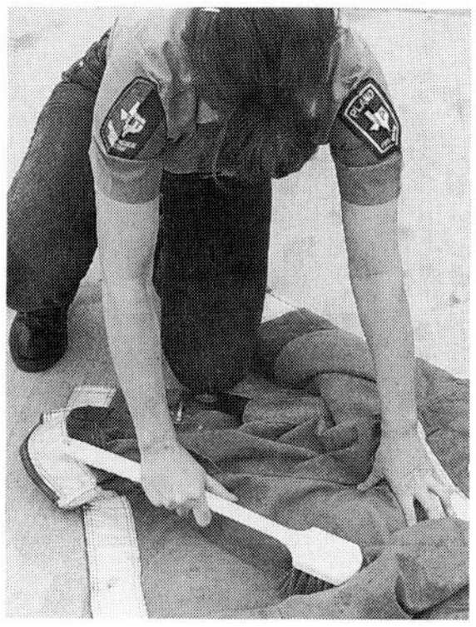

Figure 3.56 The turnout coat can be cleaned by laying it on concrete and hand-scrubbing with soap and water.

Figure 3.57 Cleaning helps prevent the boot from deteriorating.

Figure 3.58 Store boots in a cool, dark place.

PROTECTIVE BREATHING APPARATUS

Protective breathing apparatus is extremely crucial to the well-being of the firefighter. The lungs and respiratory tract are more vulnerable to injury than any other body area, and the gases encountered in association with fires are, for the most part, dangerous in one way or another. It should be a fundamental rule in fire fighting that no one be permitted to enter any potentially toxic atmosphere, such as an interior or exterior fire attack, below-grade rescue, or hazardous materials emergency, unless equipped with protective breathing apparatus (Figure 3.59). Failure to use this equipment could lead to failed rescue attempts, firefighter injuries, or fatalities. The well-trained firefighter should be knowledgeable of the requirements for wearing protective breathing apparatus, the procedures for donning and doffing the apparatus, and the proper care and maintenance of the equipment.

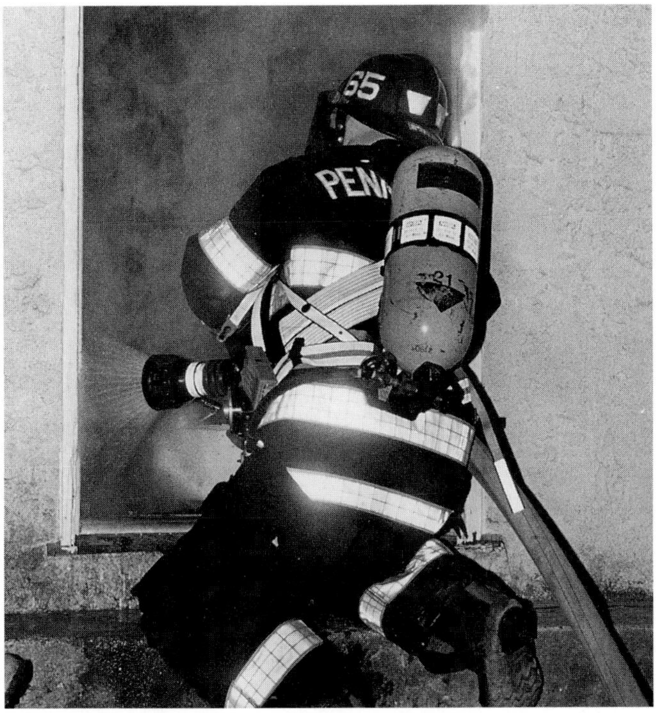

Figure 3.59 Firefighters should always wear their SCBA while performing a fire attack. *Courtesy of Bob Esposito.*

Respiratory Hazards

There are four common hazardous atmospheres associated with fires or other emergencies. These atmospheres include the following:

- Oxygen deficiency
- Elevated temperatures
- Smoke
- Toxic atmospheres (with or without fire)

OXYGEN DEFICIENCY

The combustion process consumes oxygen while producing toxic gases that either physically displace oxygen or dilute its concentration (Figure 3.60). When oxygen concentrations are below 18 percent, the human body responds by increasing the respiratory rate. Symptoms of oxygen deficiency by percentage of available oxygen are shown in Table 3.1. Remember, oxygen deficiency can also occur in below-grade locations, chemi-

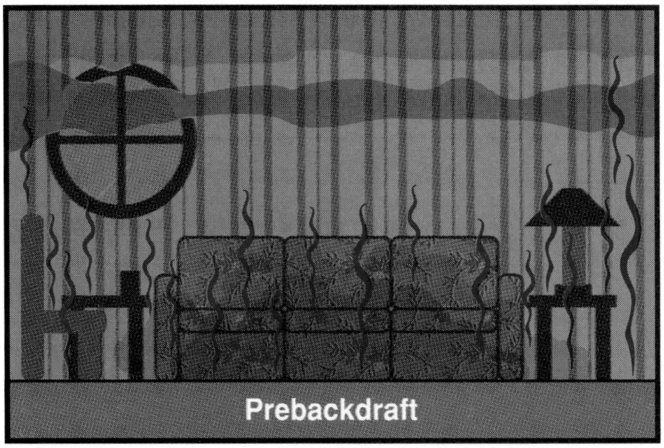

Figure 3.60 The combustion process consumes oxygen while producing toxic gases that either physically displace oxygen or dilute its concentration.

TABLE 3.1
PHYSIOLOGICAL EFFECTS OF REDUCED OXYGEN (HYPOXIA)

Oxygen in Air (Percent)	*Symptoms*
21	None — normal conditions
17	Some impairment of muscular coordination; increase in respiratory rate to compensate for lower oxygen content
12	Dizziness, headache, rapid fatigue
9	Unconsciousness
6	Death within a few minutes from respiratory failure and concurrent heart failure

NOTE: These data cannot be considered absolute because they do not account for difference in breathing rate or length of time exposed.

These symptoms occur only from reduced oxygen. If the atmosphere is contaminated with toxic gases, other symptoms may develop.

cal storage tanks, and grain bins and silos. Another area of potential hazard would be a room protected by a total-flooding carbon dioxide extinguishing system after discharge. A person who has lost consciousness in a confined space should be a warning that the atmosphere may not have enough oxygen to support life. OSHA requires workers to wear protective breathing equipment whenever entering an atmosphere that has less than a 19½ percent oxygen concentration (Figure 3.61).

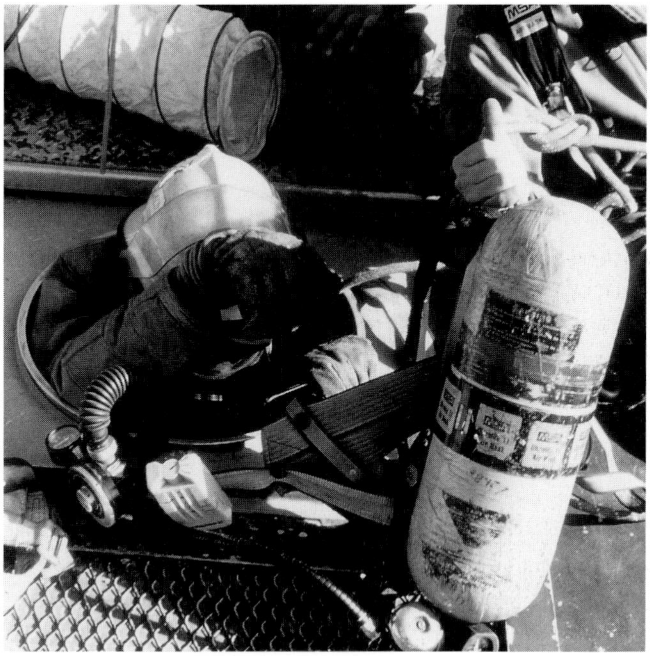

Figure 3.61 SCBA must be worn whenever entering a potentially oxygen-deficient confined space. *Courtesy of Joel Woods, University of Maryland Fire & Rescue Institute.*

ELEVATED TEMPERATURES

Exposure to heated air can damage the respiratory tract, and if the air is moist, the damage can be much worse. Excessive heat (temperatures exceeding 120°F to 130°F [49°C to 54°C]) taken quickly enough into the lungs can cause a serious decrease in blood pressure and failure of the circulatory system. Inhaling heated gases can cause pulmonary edema (accumulation of fluids in the lungs and associated swelling), which can cause death from asphyxiation. The tissue damage from inhaling hot air is not immediately reversible by introducing fresh, cool air.

SMOKE

The smoke at a fire is a suspension of small particles of carbon, tar, and dust floating in a combination of heated gases (Figure 3.62). The particles provide a means for the condensation of some of the gaseous products of combustion, especially aldehydes and organic acids formed from carbon. Some of the suspended particles in smoke are merely irritating, but others may be lethal. The size of the particle will determine how deeply into the unprotected lungs it will be inhaled.

Figure 3.62 Some fires generate large quantities of smoke. *Courtesy of Ron Jeffers.*

TOXIC ATMOSPHERES ASSOCIATED WITH FIRE

The firefighter should remember that a fire means exposure to combinations of the irritants and toxicants that cannot be predicted accurately. In fact, the combination can have a synergistic effect in which the combined effect of two or more substances is more toxic or more irritating than the total effect would be if each were inhaled separately.

Inhaled toxic gases may have several harmful effects on the human body (Figure 3.63). Some of the gases

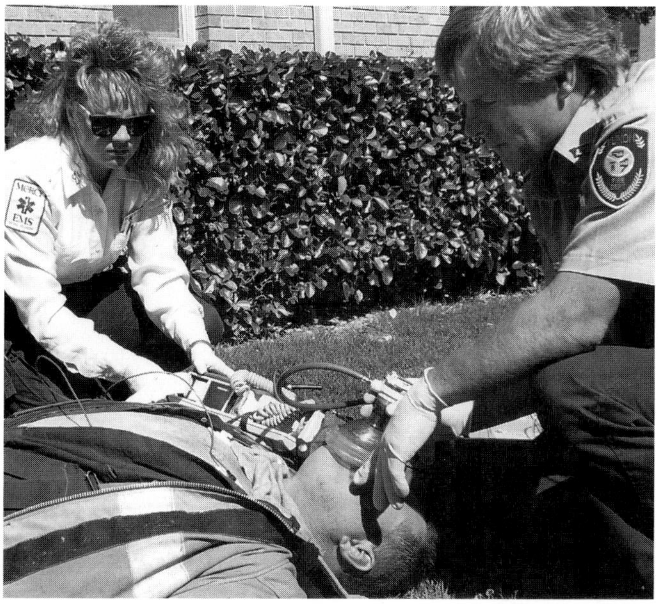

Figure 3.63 Oxygen therapy being given to a downed firefighter who was exposed to a toxic atmosphere.

directly cause disease of the lung tissue and impair its function. Other gases have no directly harmful effect on the lungs but pass into the bloodstream and to other parts of the body and impair the oxygen-carrying capacity of the red blood cells.

The particular toxic gases given off at a fire vary according to four factors:

- Nature of the combustible
- Rate of heating
- Temperature of the evolved gases
- Oxygen concentration

The following sections address some of the most commonly found gases on the fire scene.

Carbon Monoxide

More fire deaths occur from carbon monoxide (CO) than from any other toxic product of combustion. This colorless, odorless gas is present with every fire. The poorer the ventilation and the more inefficient the burning, the greater the quantity of carbon monoxide formed. A rule of thumb, although subject to much variation, is that darker smoke means higher carbon monoxide levels (Figure 3.64). Black smoke is high in particulate carbon and carbon monoxide because of incomplete combustion.

Figure 3.64 Some materials, such as rubber and plastic, give off a particularly black smoke. *Courtesy of Harvey Eisner.*

The blood's hemoglobin combines with and carries oxygen in a loose chemical combination called *oxyhemoglobin*. The most significant characteristic of carbon monoxide is that it combines with the blood's hemoglobin so readily that the available oxygen is excluded. The loose combination of oxyhemoglobin becomes a stronger combination called *carboxyhemoglobin* (COHb). In fact, carbon monoxide combines with hemoglobin about 200 times more readily than does oxygen. The carbon monoxide does not act on the body, but crowds oxygen from the blood and leads to eventual hypoxia of the brain and tissues, followed by death, if the process is not reversed.

Concentrations of carbon monoxide in air above five hundredths of one percent (0.05 percent) (500 ppm) can be dangerous. When the level is more than 1 percent, unconsciousness and death can occur without physiological signs. Even at low levels, the firefighter should not use signs and symptoms for safety factors. Headaches, dizziness, nausea, vomiting, and cherry-red skin can occur at many concentrations, based on an individual's dose and exposure (Figure 3.65). Therefore, these signs and symptoms are not good indicators of safety. Table 3.2 on the next page shows the toxic effects of different levels of carbon monoxide in air. These effects are not absolute because they do not take into account variations in breathing rate or length of exposure. Such factors could cause toxic effects to occur more quickly.

Measurements of carbon monoxide concentrations in air are not the best way to predict rapid physiological effects, because the actual reaction is from the concentration of carboxyhemoglobin in the blood, causing oxygen starvation. High oxygen user organs, such as the heart and brain, are damaged early. The combination of carbon monoxide with the blood will be greater when the concentration in air is greater. An individual's general physical condition, age, degree of physical activity, and length of exposure all affect the actual carboxyhemoglobin level in the blood. Studies have shown that it takes years for carboxyhemoglobin to dissipate from the bloodstream. People frequently exposed to carbon monoxide develop a tolerance to it, and they can function asymptomatically (without symptoms) with residual levels of serum carboxyhemoglobin that would produce significant discomfort in the average adult. The bottom line is that firefighters may be suffering the effects of CO exposure even though they are asymptomatic.

Experiments have provided some comparisons relating air and blood concentrations to carbon monoxide.

Figure 3.65 Carbon monoxide exposure at many concentrations can cause headaches and red skin.

TABLE 3.2
TOXIC EFFECTS OF CARBON MONOXIDE

Carbon Monoxide (CO) (ppm)	*Carbon Monoxide (CO) in Air (Percent)*	*Symptoms*
100	0.01	No symptoms — no damage
200	0.02	Mild headache; few other symptoms
400	0.04	Headache after 1 to 2 hours
800	0.08	Headache after 45 minutes; nausea, collapse, and unconsciousness after 2 hours
1,000	0.10	Dangerous — unconsciousness after 1 hour
1,600	0.16	Headache, dizziness, nausea after 20 minutes
3,200	0.32	Headache, dizziness, nausea after 5 to 10 minutes; unconsciousness after 30 minutes
6,400	0.64	Headache, dizziness, nausea after 1 to 2 minutes; unconsciousness after 10 to 15 minutes
12,800	1.26	Immediate unconsciousness, danger of death in 1 to 3 minutes

ppm - parts per million

A 1 percent concentration of carbon monoxide in a room will cause a 50 percent level of carboxyhemoglobin in the bloodstream in 2½ to 7 minutes. A 5 percent concentration can elevate the carboxyhemoglobin level to 50 percent in only 30 to 90 seconds. Because the newly formed carboxyhemoglobin may be traveling through the body, a person previously exposed to a high level of carbon monoxide may react later in a safer atmosphere. A person so exposed should not be allowed to use breathing apparatus or resume fire control activities until the danger of toxic reaction has passed. Even with protection, a toxic condition could be endangering consciousness. A hardworking firefighter may be incapacitated by a 1 percent concentration of carbon monoxide. The stable combination of carbon monoxide with the blood is only slowly eliminated by normal breathing. Administering pure oxygen is the most important element in first aid care. After an uneventful convalescence from a severe exposure, signs of nerve or brain injury may appear any time within three weeks. This is why an overcome firefighter who quickly revives should not be allowed to reenter a smoky atmosphere.

Hydrogen Chloride

Hydrogen chloride (HCl) is colorless but is easily detected by its pungent odor and intense irritation of the eyes and respiratory tract. Hydrogen chloride causes swelling and obstruction of the upper respiratory tract. Breathing is labored and suffocation can result. This gas is more commonly present in fires because of the increased use of plastics, such as polyvinyl chloride (PVC), containing chlorine (Figure 3.66).

In addition to the usual presence of plastics in homes, firefighters can expect to encounter plastics containing chlorine in drug, toy, and general merchandise stores (Figure 3.67). The overhaul stage is especially dangerous because breathing apparatus is often removed, and toxic fumes linger in a room (Figure 3.68). Heated concrete can remain hot enough to decompose the plastics in telephone or electrical cables and release more hydrogen chloride.

Figure 3.66 PVC piping.

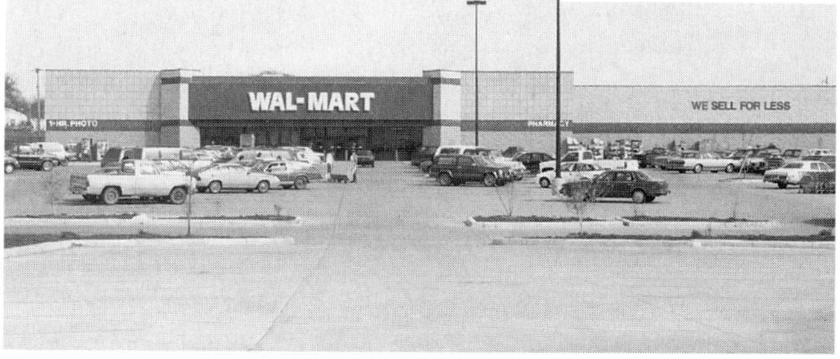

Figure 3.67 Firefighters can expect to encounter plastics containing chlorine in a general merchandise store such as Wal-Mart.

Figure 3.68 SCBA must be worn during overhaul, unless the area has been monitored and proven to have acceptable carbon monoxide levels. *Courtesy of Ron Jeffers.*

Hydrogen Cyanide

Hydrogen cyanide (HCN) interferes with respiration at the cellular and tissue level. The proper exchange of oxygen and carbon dioxide is hampered, so hydrogen cyanide is classified as a chemical asphyxiant. The gas inhibits the enzymes by which the tissues take up and use oxygen. Hydrogen cyanide also may be absorbed through the skin.

Materials that give off hydrogen cyanide include wool, nylon, polyurethane foam, rubber, and paper. Unusually high concentrations of hydrogen cyanide might be found at fires in clothing stores, rug shops, or fires involving aircraft cabins (Figure 3.69). Exposure to this colorless gas, which has a noticeable almond odor, might cause gasping respirations, muscle spasms, and increased heart rate, possibly up to 100 beats per minute. Collapse is often sudden. An atmosphere containing 135 parts per million (ppm) (0.0135 percent) is fatal within 30 minutes; a concentration of 270 ppm is fatal almost immediately.

Figure 3.69 Materials used in aircraft interiors can give off hydrogen cyanide when they burn. *Courtesy of Parker Drilling, Inc., Tulsa, Oklahoma.*

Carbon Dioxide

Although carbon dioxide (CO_2) is generally thought of as an extinguishing agent, it is also a common gas given off by the fire itself. It is an end product of the complete combustion of carboniferous materials. Carbon dioxide is nonflammable, colorless, and odorless. Free-burning fires form more carbon dioxide than smoldering fires. Normally, the presence of carbon dioxide in the air and the exchange of carbon dioxide from the bloodstream into the lungs stimulates the respiratory center of the brain. Air normally contains about 0.03 percent carbon dioxide. At a 5 percent concentration of carbon dioxide in air there is a marked increase in respiration, along with headache, dizziness, sweating, and mental excitement. Concentrations of 10 to 12 percent cause death within a few minutes from paralysis of the brain's respiratory center. Unfortunately, increased breathing increases the inhalation of other toxic gases. As the gas increases, the initially stimulated breathing rate becomes depressed before total paralysis takes place.

Firefighters should anticipate high carbon dioxide levels when a carbon dioxide total-flooding system has been activated (Figure 3.70). These systems are de-

Figure 3.70 Carbon dioxide total flooding systems can leave an area oxygen deficient. *Courtesy of Reliable Fire Equipment Co.*

signed to extinguish a fire by excluding the oxygen, and they will have the same effect on a firefighter. According to the American Conference of Government Industrial Hygienists, exposure for even a short time to carbon dioxide concentrations greater than 15,000 ppm should be avoided. Because CO_2 is colorless and odorless, meters should be used in any suspected atmosphere (Figure 3.71).

Figure 3.71 This photo dramatically illustrates the need to have carbon dioxide monitoring devices in place where total flooding systems are in use. It is possible that even the area above the visible cloud could be hazardous. *Courtesy of Reliable Fire Equipment Co.*

Nitrogen Oxides

There are two dangerous oxides of nitrogen: nitrogen dioxide (NO_2) and nitric oxide (NO). Nitrogen dioxide is the most significant because nitric oxide readily converts to nitrogen dioxide in the presence of oxygen and moisture. Nitrogen dioxide is a pulmonary irritant that has a reddish-brown color. Nitrogen dioxide is commonly called silo gas and is frequently the culprit when farmers are overcome in or around silos and grain bins (Figure 3.72). Nitrogen dioxide is also liberated when pyroxylin plastics decompose. This is common in office supply store fires because many drafting tools (triangles, T-squares, etc.) are made of pyroxylin plastics (Figure 3.73). When inhaled in sufficient concentrations, it causes pulmonary edema that blocks the body's natural respiration processes and leads to death by suffocation.

Additionally, all oxides of nitrogen are soluble in water and react in the presence of oxygen to form nitric and nitrous acids. These acids are neutralized by the alkalies in the body tissues and form nitrites and nitrates. These substances chemically attach to the blood and can lead to collapse and coma. Nitrates and nitrites can also cause arterial dilation, variation in blood pressure, headaches, and dizziness. The effects of nitrites and nitrates are secondary to the irritant effects of nitrogen dioxide but can become important under certain circumstances and cause delayed physical reactions.

Figure 3.72 Farmers working around grain bins are often overcome by nitrogen dioxide.

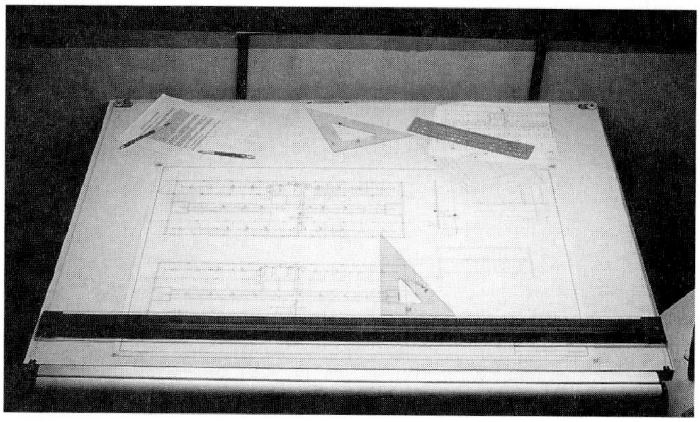

Figure 3.73 Many drafting tools are made of pyroxylin plastics.

Nitrogen dioxide is an insidious gas because its irritating effects in the nose and throat can be tolerated even though a lethal dose is being inhaled. Therefore, the hazardous effects from its pulmonary irritation action or chemical reaction may not become apparent for several hours after exposure.

Phosgene

Phosgene ($COCl_2$) is a colorless, tasteless gas with a disagreeable odor. It may be produced when refrigerants, such as freon, contact flame (Figure 3.74). It is a strong lung irritant, and the full poisonous effect is not evident for several hours after exposure. The musty-hay odor of phosgene is perceptible at 6 ppm, although lesser amounts cause coughing and eye irritation. Twenty-five ppm is deadly. When phosgene contacts water, it decomposes into hydrochloric acid. Because the lungs and bronchial tubes are always moist, phosgene forms hydrochloric acid in the lungs when inhaled.

Figure 3.74 Phosgene gas may be produced when refrigeration systems burn.

TOXIC ATMOSPHERES NOT ASSOCIATED WITH FIRE

Hazardous atmospheres can be found in numerous situations in which fire is not involved. Many industrial processes use extremely dangerous chemicals to make ordinary items (Figure 3.75). For example, quantities of carbon dioxide would be stored at a facility where wood alcohol, ethylene, dry ice, or carbonated soft drinks are manufactured. Any other specific chemical could be traced to numerous, common products.

Figure 3.75 Hazardous atmospheres can be found in numerous situations such as this large factory.

Many refrigerants are toxic and may be accidentally released, causing a rescue situation to which firefighters may respond. Ammonia and sulfur dioxide are two dangerous refrigerants that irritate the respiratory tract and eyes. Sulfur dioxide reacts with moisture in the lungs to form sulfuric acid. Other gases also form strong acids or alkalies on the delicate surfaces of the respiratory system.

An obvious location where a chlorine gas leak may be encountered is at a manufacturing plant; a not so obvious location is at a swimming pool (Figure 3.76). Incapacitating concentrations can be found at either location. Chlorine is also used in manufacturing plastics, foam, rubber, and synthetic textiles and is commonly found at water and sewage treatment plants.

Figure 3.76 Swimming pools may have large quantities of chlorine on the property. These chlorine tanks are installed near an oudoor pool. Firefighters must be prepared to handle emergencies involving this and other chemicals.

Sometimes the leak is not at the manufacturing plant but occurs during transportation of the chemical. Train derailments have resulted in container failures, exposing the public to toxic chemicals and gases (Figure 3.77). The large quantities involved can travel long distances.

Because of the likelihood of the presence of toxic gas, rescues in sewers, caves, trenches, storage tanks, tank cars, bins, silos, manholes, pits, and other confined places require the use of self-contained breathing appa-

Figure 3.77 Train derailments can quickly elevate into major hazardous materials incidents.

ratus (Figure 3.78). Workers have been overcome by harmful gases in large tanks during cleaning or repairs; unprotected personnel have also been overcome while attempting a rescue. In addition, the atmosphere in many of these areas is oxygen deficient and will not support life even though there may be no toxic gas.

Figure 3.78 SCBA must be worn when performing a rescue at a confined space such as a storage tank.

The manufacture and transport of hazardous materials has made virtually every area a potential site for a hazardous materials incident. Hazardous materials are routinely transported by truck, rail, water, air, and car. A firefighter needs to be able to recognize when a chemical spill or incident is hazardous and to wear protective breathing apparatus (Figure 3.79). The United States Department of Transportation (DOT) defines a hazardous material as "any substance which may pose an unreasonable risk to health and safety of operating or emergency personnel, the public, and/or the environment if it is not properly controlled during handling, storage, manufacture, processing, packaging, use, disposal, or transportation." Hazardous materials can range from chemicals in liquid or gas form to radioactive materials to etiologic (disease-causing) agents. Fire may complicate the hazards and pose an even greater danger. Many times a response to an industrial site may deal with hazardous materials. Positive-pressure self-contained breathing apparatus should be a mandatory piece of protective equipment when dealing with hazardous materials situations.

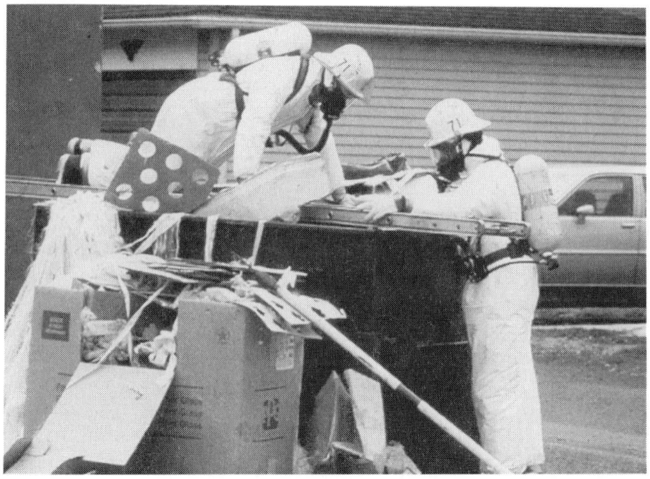

Figure 3.79 Appropriate personal protective equipment must be worn at hazardous materials incidents. *Courtesy of Bob Esposito.*

When responding to a vehicle accident involving a truck, the placard on the truck should serve as a warning that the atmosphere may be toxic and that self-contained breathing apparatus should be worn. In industrial facilities, placards may also be placed on containers warning of the dangerous materials inside. It is safer to attempt to view these placards through binoculars from a distance before moving in close to them (Figure 3.80).

Do not limit self-contained breathing apparatus to transportation hazardous materials incidents. Common calls, such as natural gas leaks or carbon monoxide (CO) poisonings, may also require the use of self-contained breathing apparatus. *When in doubt, wear self-contained breathing apparatus!*

For more information on hazardous materials, see IFSTA's **Hazardous Materials for First Responders** manual or Fire Protection Publications' **Hazardous Materials: Managing the Incident**.

Figure 3.80 Firefighter viewing placard before moving close to the accident.

Protective Breathing Apparatus Limitations

To operate effectively, the firefighter must be aware of the limitations of protective breathing apparatus. These include limitations of the wearer, equipment, and air supply.

One issue of frequent debate is the use of contact lenses while wearing a protective breathing apparatus facepiece. Presently, OSHA and ANSI standards prohibit firefighters from wearing contact lenses while using a respirator. However, this regulation has been repeatedly challenged by users. In 1985, OSHA reevaluated its position after funding a study conducted by Robert A. da Roza and Catherine Weaver of the Lawrence Livermore National Laboratory, Livermore, California. Weaver and da Roza surveyed 9,100 firefighters in the United States and Canada regarding their use of contact lenses with SCBA. Of the 403 respondents who wore contacts in spite of the regulation, 6 reported that contact-lens-related problems had caused them to remove the facepiece in an environment in which the facepiece would normally have been worn. However, 30 respondents reported safety-related problems regarding mask-mounted eyeglasses (which are legal) in SCBA. The researchers concluded that the prohibition against wearing contacts while using SCBA should be discontinued.

Based on the results of the Livermore study and the review of other reports and studies, OSHA intends to modify the regulation (1910.134) to permit firefighters to use soft, extended-wear contact lenses with SCBA. The issue with nongas-permeable, hard contact lenses will also be resolved in the revision effort, which is presently underway. Until the revised regulation is printed, OSHA will follow an interim enforcement policy that states:

"Violations of the respirator standard involving the use of gas-permeable and soft contact lenses shall continue to be documented in the case file and recorded as *de minimis* [the smallest infraction]; citations shall not be issued."

Limitations Of Wearer

Several factors affect the firefighter's ability to use SCBA effectively. These factors include physical, medical, and mental limitations.

Physical

- Physical condition — The wearer must be in sound physical condition in order to maximize the work that can be performed and to stretch the air supply as far as possible.
- Agility — Wearing a protective breathing apparatus restricts the wearer's movements and affects his or her balance. Good agility will overcome these obstacles.
- Facial features — The shape and contour of the face will affect the wearer's ability to get a good facepiece-to-face seal (Figure 3.81).

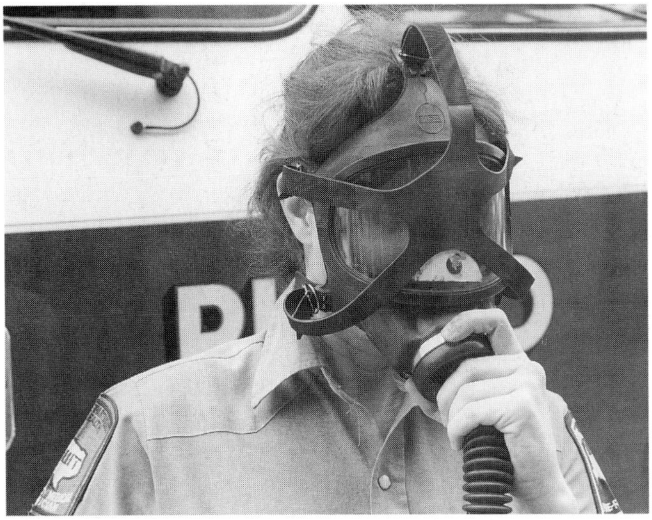

Figure 3.81 Facial features can affect the fit of an SCBA mask.

Medical

- Neurological functioning — Good motor coordination is necessary for operating in protective breathing equipment. The firefighter must be of sound mind to handle emergency situations that may arise.
- Muscular/skeletal condition — The firefighter must have the physical strength and size required to wear the protective equipment and to perform necessary tasks.
- Cardiovascular conditioning — Poor cardiovascular conditioning can result in heart attacks, strokes, or other related problems during strenuous activity.
- Respiratory functioning — Proper respiratory functioning will maximize the wearer's operation time in a self-contained breathing apparatus.

Mental

- Adequate training in equipment use — The firefighter must be knowledgeable in every aspect of protective breathing apparatus use.
- Self-confidence — The firefighter's belief in his or her ability will have an extremely positive overall effect on the actions that are performed.

- Emotional stability — The ability to maintain control in an excited or high stress environment will reduce the chances of a serious mistake being made.

In addition to being concerned about the limitations of the wearer, firefighters must also be cognizant of the limitations of the equipment and the air supply. These limitations are highlighted below:

Limitations Of Equipment

- Limited visibility — The facepiece reduces peripheral vision, and facepiece fogging can reduce overall vision.
- Decreased ability to communicate — The facepiece hinders voice communication.
- Increased weight — Depending on the model, the protective breathing equipment adds 25 to 35 pounds (11 kg to 16 kg) of weight to the firefighter.
- Decreased mobility — The increase in weight and the splinting effect of the harness straps reduce the firefighter's mobility (Figure 3.82).

Figure 3.82 SCBA decreases the firefighter's mobility; for example, working above one's head can become difficult. *Courtesy of Bob Esposito.*

Limitations Of Air Supply

- Physical condition of user — If the wearer is in poor physical condition, the air supply will be expended faster.
- Degree of physical exertion — The harder the firefighter exerts himself or herself, the faster the air supply is expended (Figure 3.83).
- Emotional stability — A person who becomes excited will increase his or her respirations and use air faster.
- Condition of apparatus — Minor leaks and poor adjustment of regulators result in excess air loss.
- Cylinder pressure before use — If the cylinder is not filled to capacity, the amount of working time will be reduced proportionately.
- Training and experience — Properly trained and highly experienced personnel will be able to draw the maximum air supply from a cylinder.

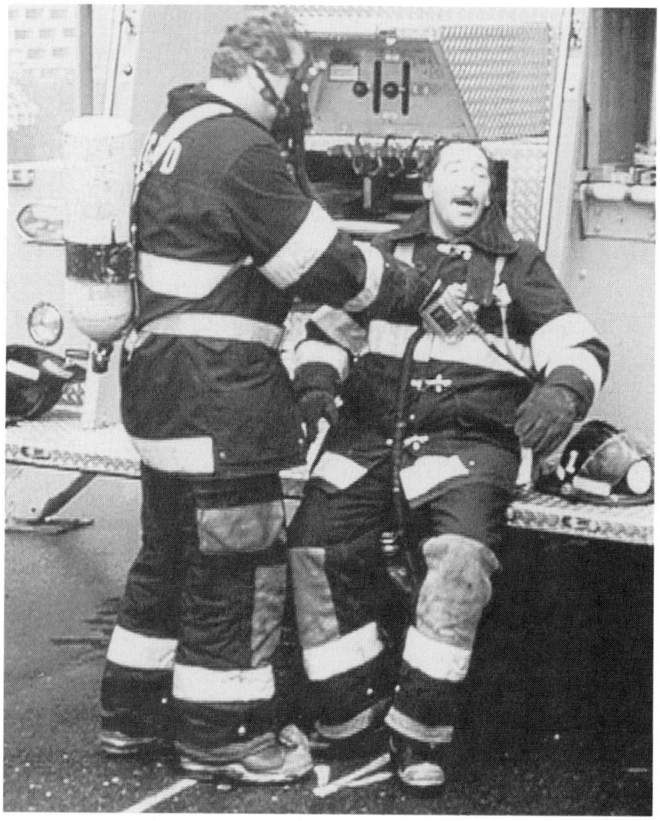

Figure 3.83 Firefighters are subject to extreme exertion. *Courtesy of Ron Jeffers.*

Types Of Breathing Apparatus

There are two types of self-contained breathing apparatus used in the fire service: open-circuit and closed-circuit. Open-circuit SCBA is used much more frequently than closed-circuit SCBA. Open-circuit SCBA uses compressed air; closed-circuit uses compressed or liquid oxygen. Closed-circuit SCBA is also known as "rebreather" apparatus because the user's exhaled air stays within the system for reuse. Closed-circuit breathing apparatus is rarely used in today's fire service. It is only used in some extended hazardous materials operations. Regardless of the type of SCBA used, training in its use is essential.

OPEN-CIRCUIT SELF-CONTAINED BREATHING APPARATUS

Open-circuit SCBA are the most commonly used protective breathing apparatus in the fire service (Figure 3.84). The air supply in open-circuit systems is compressed air. Exhaled air is vented to the outside atmosphere.

Until July 1, 1983, two types of open-circuit SCBA were used by the fire service: demand and positive pressure. Since that date, demand units are no longer in compliance with OSHA requirements and should not be used (Figure 3.85). NFPA and ANSI standards also require that only positive-pressure breathing apparatus be used in the fire service. (**NOTE**: Any departments still using demand apparatus should be aware that these units can and should be converted to positive-pressure units as rapidly as possible.)

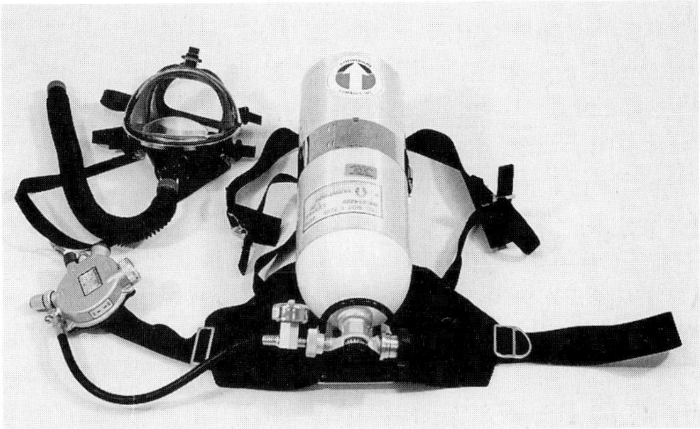

Figure 3.84 Most commonly used SCBA in the fire service are the open-circuit type.

Figure 3.85 Older, nonpositive-pressure SCBA must be removed from service.

The main reason for the change to positive pressure is the greater protection factor afforded by the positive-pressure units. Positive-pressure SCBA maintains a slightly increased pressure (above atmospheric) in the user's facepiece. This positive pressure helps prevent contaminants from entering the facepiece if a leak develops (Figure 3.86).

Figure 3.86 A positive-pressure facepiece seal provides protection from contaminants entering the facepiece if a leak develops.

Several companies manufacture open-circuit SCBA, each with different design features or mechanical construction. Certain parts, such as cylinders and backpacks, are interchangeable; however, such substitution voids National Institute for Occupational Safety and Health (NIOSH) and Mine Safety and Health Administration (MSHA) certification and is not a recommended practice. Substituting different parts may also void warranties and leave the department or firefighter liable for any injuries incurred.

There are four basic SCBA component assemblies:

- Backpack and harness assembly
- Air cylinder assembly — includes cylinder, valve, and pressure gauge
- Regulator assembly — includes high-pressure hose and low-pressure alarm
- Facepiece assembly — includes low-pressure hose (breathing tube), exhalation valve (for SCBA with harness-mounted regulator), and head harness

Backpack And Harness Assembly

The backpack assembly is designed to hold the air cylinder on the firefighter's back as comfortably and securely as possible. Adjustable harness straps provide a secure fit for whatever size the individual requires. The waist straps are designed to help properly distribute the weight of the cylinder or pack (Figure 3.87). One problem is that waist straps are often not used or are removed. Remember that NIOSH and MSHA certify the entire SCBA unit, and removal of waist straps could void warranties.

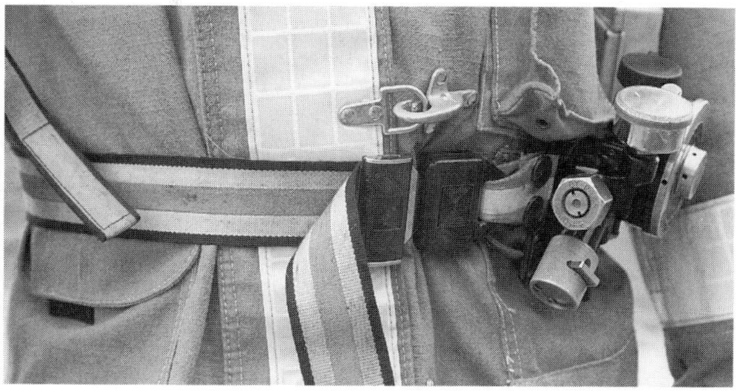

Figure 3.87 The waist strap helps to distribute the weight of the cylinder.

Air Cylinder Assembly

Air cylinders come in different sizes and with a variety of high-pressure hose connections. Because the cylinder must be strong enough to safely contain the high pressure of the compressed air, it constitutes the main weight of the breathing apparatus (Figure 3.88).

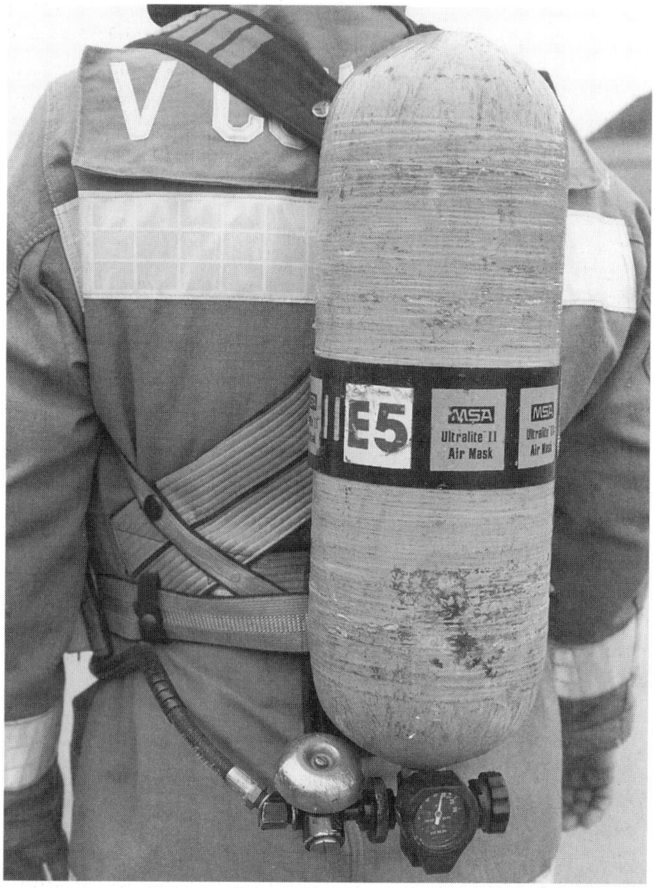

Figure 3.88 On an average, a composite cylinder weighs about 16 pounds (7.2 kg).

When full, a 2,216 psi (15 290 kPa), 30-minute cylinder ranges from 9.6 pounds (4.3 kg) for a composite cylinder to 23.8 pounds (10.7 kg) for a steel cylinder. On the average, composite cylinders weigh about 16 pounds (7.2 kg), while steel and aluminum cylinders weigh about 20 pounds (9 kg). Lightweight aluminum and combination fiberglass-wrapped aluminum cylinders are the most common cylinder types and are 10 percent stronger than steel.

Fully charged, the typical 30-minute-rated cylinder contains 45 cubic feet (1 275 L) of breathing air at 2,216 psi (15 290 kPa). Cylinders rated for 60 minutes contain 88 cubic feet (2 490 L) of breathing air at 4,500 psi (31 000 kPa). In most cases, particularly during periods of heavy exertion, the actual time of use will be less than the time the cylinder is rated for.

Regulator Assembly

Air from the cylinder travels through the high-pressure hose to the regulator. The regulator reduces the pressure of the cylinder air to slightly above atmospheric pressure and controls the flow of air to meet the respiratory requirements of the wearer (Figure 3.89). When the firefighter inhales, a partial vacuum is created in the regulator. The apparatus diaphragm moves inward, tilting the admission valve so that low-pressure air can flow into the facepiece. The diaphragm is then held open, which creates the positive pressure. Exhalation moves the diaphragm back to the "closed" position. Some SCBA units have regulators that fit into the facepiece (Figure 3.90). On other units, the regulator is on the firefighter's chest or waist strap (Figure 3.91).

On many models, two external knobs, differing in color, shape, and location, control operating and emergency functions. These are the mainline valve and the

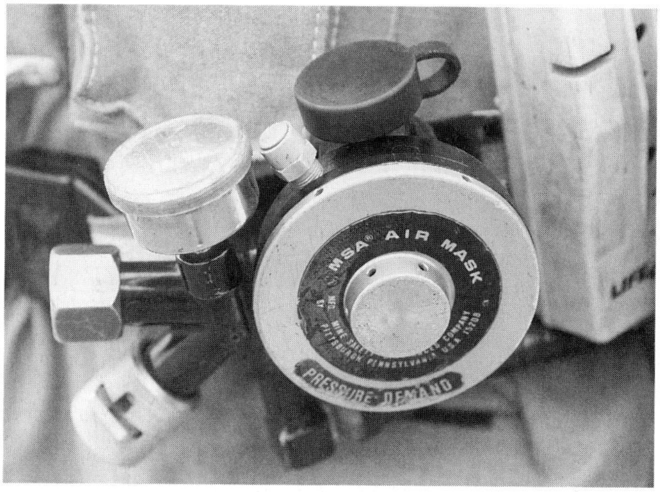

Figure 3.89 The regulator controls the flow of air to meet the respiratory requirements of the user.

Figure 3.90 This regulator connects directly to the facepiece.

Figure 3.91 The regulator of this unit is on the firefighter's waist strap.

bypass valve (Figure 3.92). During normal operation, the mainline valve is fully open and locked, if there is a lock. The bypass valve is closed. On some SCBA, the bypass valve controls a direct airline from the cylinder in the event that the regulator fails. Once the valves are set in their normal operating position, they should not be changed unless the emergency bypass is needed (Figure 3.93).

Figure 3.92 The mainline valve (bottom) and the bypass valve (top).

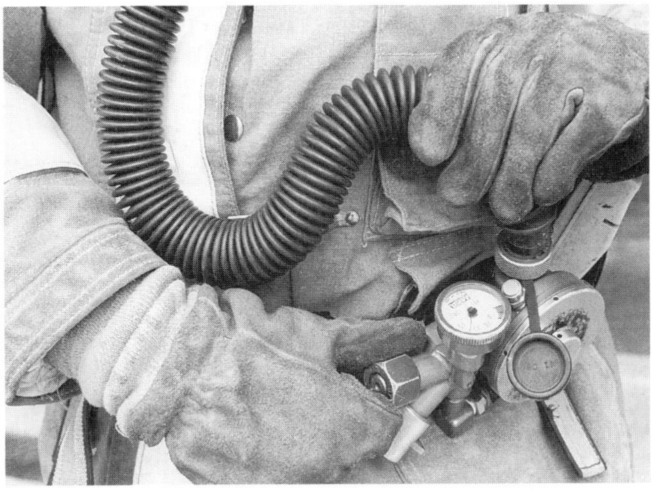

Figure 3.93 Firefighter operating the bypass valve.

A pressure gauge that shows the air pressure remaining in the cylinder is usually mounted on or near the regulator or on the high-pressure hose (Figure 3.94). The regulator pressure gauge should read within 100 psi (700 kPa) of the cylinder gauge if increments are in psi (kPa). If increments are shown in other measurements, such as percents or fractions, both measurements should be the same (Figure 3.95 on the next page). These pressure readings are most accurate at or near the upper range of the gauge's rated working pressures. Low pressures in the cylinder may cause inconsistent readings between the cylinder and regulator gauges. If they are not consistent, rely on the lower reading and check the equipment for any needed repair before using it

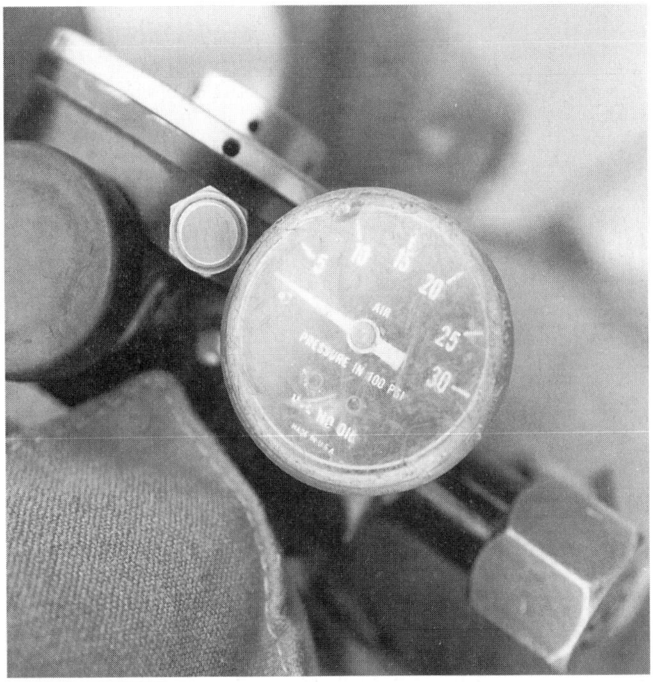

Figure 3.94 The pressure gauge is mounted close to the regulator.

again. All modern units have an audible alarm that sounds when the cylinder pressure decreases to a preset level: 450 to 550 psi (3 100 kPa to 3 795 kPa), depending on the manufacturer. SCBA teams should leave the fire area *immediately* after the first firefighter's alarm sounds (Figure 3.96).

Figure 3.95 A closeup of the pressure gauge on a cylinder.

Figure 3.96 This alarm sounds when the firefighter needs to leave the area because of a low air supply.

Figure 3.97 This low-pressure hose carries the air from the regulator to the facepiece.

Figure 3.98 A head harness with adjustable straps.

Facepiece Assembly

The facepiece assembly consists of the facepiece lens, an exhalation valve, and if the regulator is separate, a low-pressure hose to carry the air from the regulator to the facepiece (Figure 3.97). The facepiece lens is made of clear safety plastic and is connected to a flexible rubber mask. The facepiece is held snugly against the face by a head harness with adjustable straps, net, or some other arrangement (Figure 3.98). The lens should be protected from scratches during use and storage. Some facepieces have a speech diaphragm to make communication easier (Figure 3.99).

The low-pressure hose brings air from the regulator into the facepiece; therefore, it must be kept free of kinks and away from contact with abrasive surfaces. The hose is usually corrugated to prevent collapse when a person is working in close quarters, breathing deeply, or leaning against a hard surface (Figure 3.100). Like the facepiece, this hose is oil-resistant rubber, neoprene, silicone, or plastic resin, all of which are referred to generically as elastomer.

The exhalation valve at the chin of the facepiece is a simple, one-way valve that releases an exhaled breath without admitting any of the contaminated outside atmosphere (Figure 3.101). Dirt or foreign materials can cause the valve to become partially opened, which may permit the contaminated outside atmosphere to enter the facepiece. Therefore, it is important that the valve be kept clean and free of foreign material. It is also important that the exhalation valve be tested by the firefighter during facepiece-fit tests and before entering a hazardous atmosphere (Figure 3.102).

A facepiece provides some protection from facial burns and holds in the cool breathing air. The protective

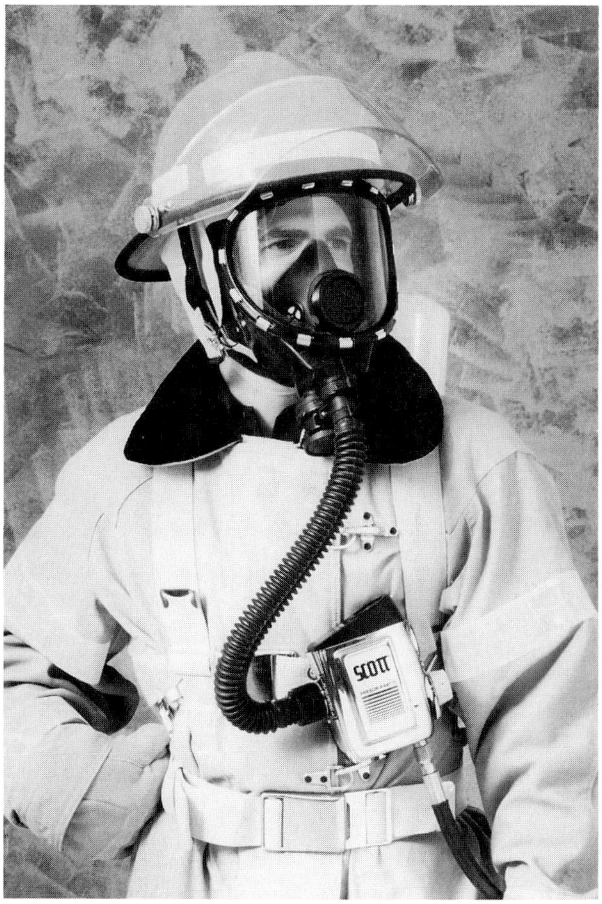

Figure 3.99 A speaking diaphragm mounted in the facepiece allows easier communication among firefighters. *Courtesy of Scott Aviation.*

Figure 3.100 The low-pressure hose must be kept free of kinks.

hood completes the protective envelope for the head and neck area, along with the helmet and earflaps (Figure 3.103).

An improperly sealed facepiece or a fogged lens can cause problems for the wearer. The different temperatures inside and outside the facepiece, where the exhaled air or outside air is moist, can cause the facepiece lens to fog, which hampers vision (Figure 3.104). Internal fogging occurs when the lens is cool, causing the

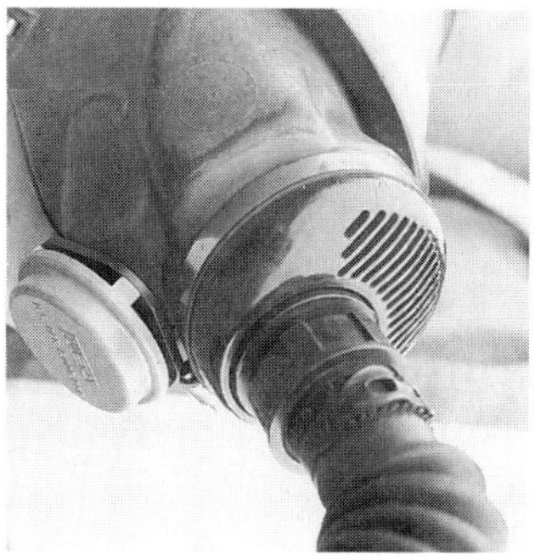

Figure 3.101 The exhalation valve is a simple, one-way valve that releases an exhaled breath without admitting any of the contaminated outside atmosphere.

Figure 3.102 Test the exhalation valve during fit testing.

Figure 3.103 The protective hood protects the firefighter's head and neck area.

Figure 3.104 A fogged facepiece can hamper the firefighter's vision.

highly humid exhaled breath to condense. As the cooler dry air from the cylinder passes over the facepiece lens, it often removes the condensation. External fogging occurs when condensation collects on the relatively cool lens during interior fire fighting operations. External fogging can be removed by wiping the lens.

One of the following methods can be used to prevent or control internal fogging of a lens:

- Releasing cylinder air — Quickly opening and closing the bypass valve to release a brief flow of cylinder air removes internal fogging; however, *this technique uses valuable air and its use should be restricted.*
- Using a nosecup — Facepieces can be equipped with a nosecup that deflects exhalations away from the lens (Figure 3.105). However, if the nosecup does not fit well, it will permit exhaled air to leak into the facepiece and condense on the lens.
- Applying an antifogging chemical — Special antifogging chemicals recommended by the manufacturer can be applied to the lens of the facepiece. Some SCBA facepieces are permanently impregnated with an antifogging chemical.

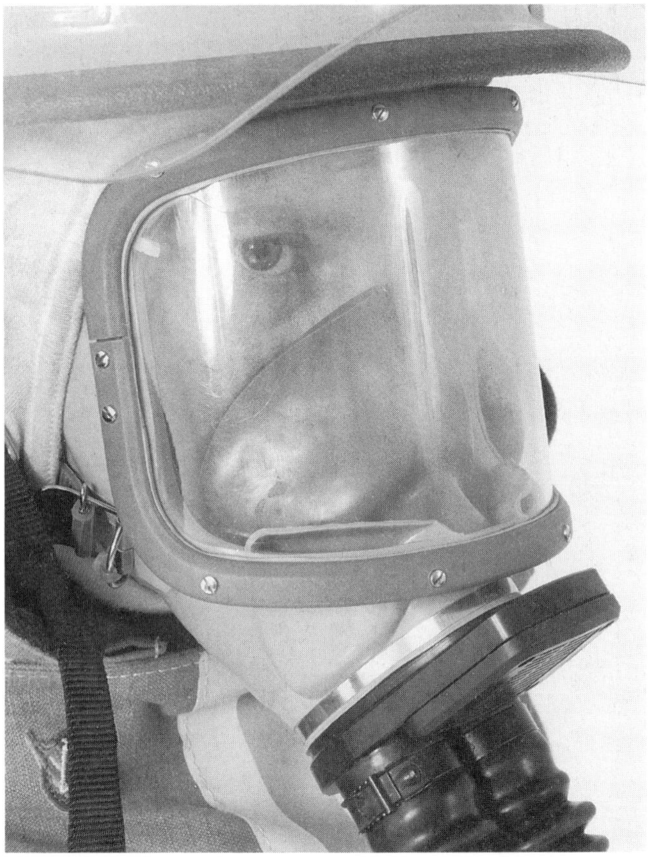

Figure 3.105 Facepiece nosecups help reduce mask fogging.

OPEN-CIRCUIT AIRLINE EQUIPMENT

Incidents involving hazardous materials or rescues often require a longer air supply than can be obtained from standard open-circuit SCBA. In these situations, an airline attached to one or several large air cylinders can be connected to an open-circuit facepiece, regulator, and egress cylinder (Figure 3.106). Airline equipment enables the firefighter to travel up to 300 feet (90 m) from the *regulated* air supply source (Figure 3.107). This type of respiratory protection enables the firefighter to work for several hours without the encumbrance of a backpack. If greater mobility is needed, the firefighter can also wear a standard SCBA with an airline option. The firefighter can then temporarily disconnect from the airline supply, using the SCBA to provide breathing air, and perform necessary tasks beyond the range of the airline equipment.

The number and sizes of air cylinders used with airline equipment vary from several small bottles to large, cascade-type cylinders (Figure 3.108). Air supply is available as long as there are spare full cylinders. All airline units should be capable of using more than one cylinder in order to provide a continuous air source to the wearer. A small 45-cubic-foot (1 275 L) cylinder at 2,216 psi (15 290 kPa) is rated for 30 minutes. A large 240-cubic-foot (6 792 L) cylinder at 2,400 psi (16 560 kPa) is rated for approximately 2½ hours. The smallest cylinders are rated for 5 minutes and are to be used only for escape purposes.

Figure 3.106 Airline breathing equipment is useful for extended duration operations. *Courtesy of MSA.*

Small SCBA cylinders are easily and readily transported, but have a limited capacity and require frequent changing (Figure 3.109). (**NOTE:** A small cylinder here refers to a 45-cubic-foot (1 275 L) cylinder, not the 5-minute escape cylinder of some airline units that is worn on the hip.) The larger cylinders are difficult to maneuver in small spaces and may not be readily available, but they permit extended operations without cylinder changes. Any airline respirator that is used in a hazardous atmosphere must provide enough breathing air for the wearer to escape the atmosphere in the event the airline is severed. This requirement is usually accomplished by attaching the very small breathing cylinders, rated for 5 minutes, to the airline unit (Figure 3.110). Almost all airline units used in rescue situations will require the 5-minute escape cylinder. The 5-minute escape cylinders must not be disconnected from the air supply line for untethered work—they are for escape only! To perform untethered work, a 30- or 60-minute SCBA that can be augmented by an airline should be used.

Figure 3.107 A typical airline assembly. *Courtesy of ISI.*

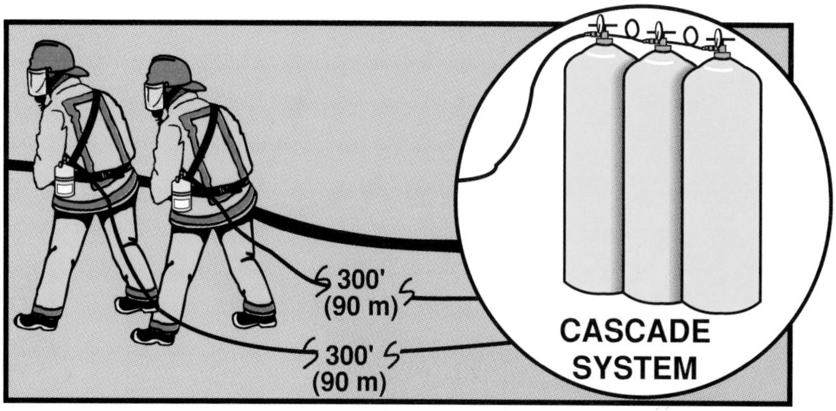

Figure 3.108 Cascade cylinders supplying air to airline units.

Figure 3.109 Air cylinders are available in various sizes, volumes, and compositions.

Figure 3.110 Note the emergency escape cylinder on the airline assembly. *Courtesy of MSA.*

CLOSED-CIRCUIT BREATHING APPARATUS

Closed-circuit breathing apparatus are not used in the fire service as commonly as open-circuit breathing apparatus. However, they are sometimes used for hazardous materials incidents because of their longer air supply duration (Figure 3.111). Closed-circuit SCBA are available with durations of 30 minutes to 4 hours and usually weigh less than open-circuit units of similarly rated service time. They weigh less because a smaller cylinder containing pure oxygen is used. For more information on closed-circuit breathing apparatus, see IFSTA's **Self-Contained Breathing Apparatus** manual.

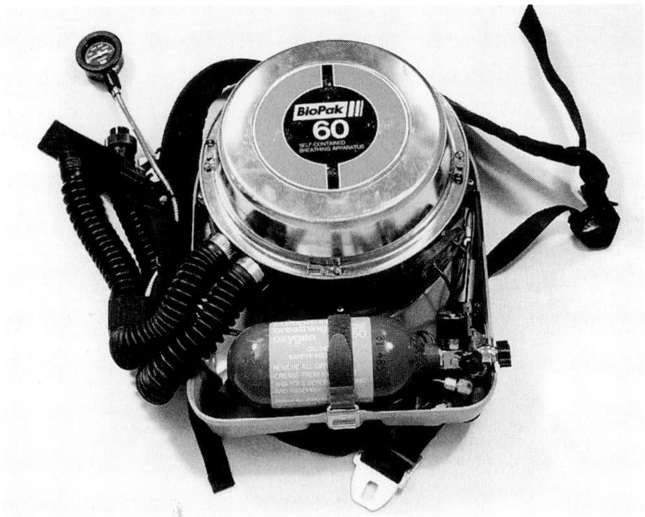

Figure 3.111 A typical closed-circuit SCBA.

Mounting Protective Breathing Apparatus

Methods of storing self-contained breathing apparatus vary from department to department. Each department should use the most appropriate method to facilitate quick and easy donning (Figures 3.112 a through c). SCBA can be placed on the apparatus in seat mounts, side mounts, and compartment mounts, and stored in cases. If placed in seat mounts, the SCBA should be arranged so that it may be donned without the firefighter having to remove his or her seat belt.

Figure 3.112a Seat-mounted SCBA are shown in the fire fighting vehicle.

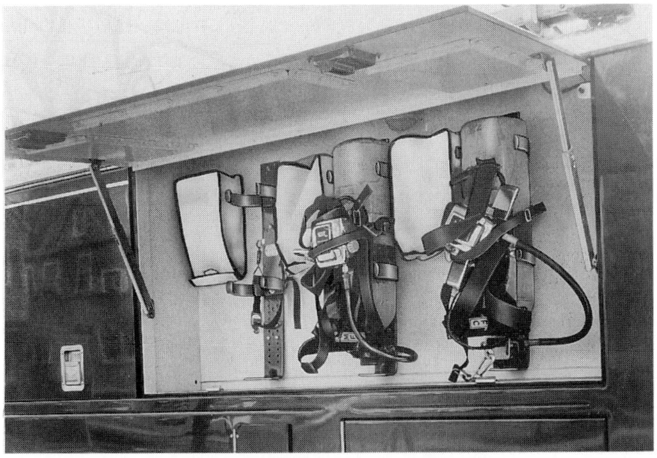

Figure 3.112b These SCBA are stored in a closed compartment. While providing good protection, this compartment may not be the correct height for proper donning. *Courtesy of Ziamatic Corporation.*

Figure 3.112c Some SCBA may be stored in special cases carried on the apparatus.

Donning And Doffing Protective Breathing Apparatus

Several methods can be used to don self-contained breathing apparatus, depending on how the apparatus is stored. The methods used in the fire service include the over-the-head method, the coat method, and donning from a seat, rear, or compartment mount. The steps needed to get the apparatus onto the body differ with each method, but once the apparatus is on the body, the method of securing the unit will be the same for any one model (there are different steps for securing different makes and models). Regardless of the method used, once the breathing apparatus has been donned, the wearer should follow manufacturer's instructions during its operation.

Regardless of the SCBA model or method of donning, several precautionary safety checks should be made when preparing to don the SCBA:

- Check the air cylinder gauge to ensure that the cylinder is full.
- Check the regulator gauge and cylinder gauge to ensure that they read within 100 psi of the same pressure. Gauges not marked in increments of 100 psi should read relatively close to each other.
- Check the harness assembly to ensure that it is fully extended.
- Check the regulator valves to ensure that they are in the proper setting.

Once these checks are complete, the protective breathing apparatus may be donned using the most appropriate of the following methods.

OVER-THE-HEAD-METHOD

SCBA must be stored ready to don. The backpack straps should be arranged so that they do not interfere with grasping the cylinder or backplate. The firefighter should put the protective hood on, pull it back, button the turnout coat, and turn the collar up so that the shoulder straps do not hold the collar down. The procedures for donning a backpack using the over-the-head method are as follows:

Step 1: Check the unit.

a. Crouch or kneel at the end opposite the cylinder valve, regardless of whether the unit is in its case or on the ground (Figure 3.113).

b. Check the cylinder gauge to make sure that the air cylinder is full (Figure 3.114). Open the cylinder valve slowly and listen for the audible alarm as the system pressurizes. Then, open the cylinder valve fully. If the audible alarm does not sound, or if it sounds but does not stop, place the unit out of service by tagging it and notifying an officer; use another unit.

c. Check the regulator gauge; both the cylinder and regulator gauges should register within 100 psi of each other — if increments are in psi — when the cylinder is pressurized to its rated capacity (Figure 3.115). If increments are in other measurements, such as fractions or minutes, they should correspond. If the unit has a donning switch (for units with facepiece-mounted regulators), leave the cylinder valve open and the unit in the donning mode. If the unit is positive pressure only, refer to the manufacturer's instructions concerning the cylinder valve.

NOTE: Positive-pressure SCBA that can supply air in a demand-type mode do not meet the requirements of NFPA 1981, *Standard on Open-Circuit Self-Contained Breathing Apparatus for Fire Fighters.*

Figure 3.113 Crouch or kneel at the end opposite the cylinder valve, regardless of whether the unit is in its case or on the ground.

Figure 3.114 Check the cylinder gauge to make sure that the air cylinder is full.

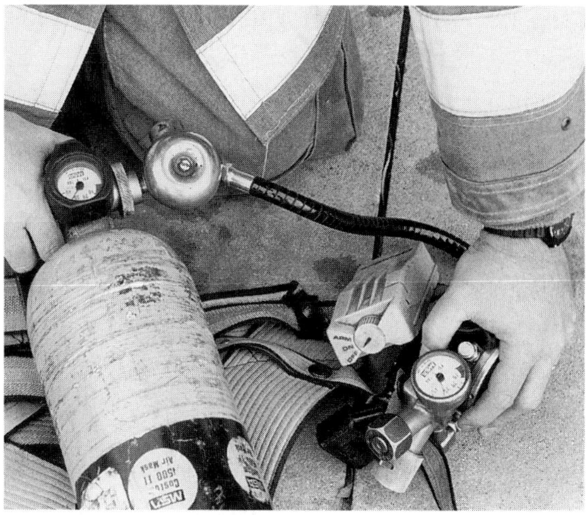

Figure 3.115 Check the regulator gauge. It should be relatively the same as the cylinder gauge.

Step 2: Spread the harness straps out to their respective sides.

Step 3: Grasp the backplate or cylinder with both hands, one at each side (Figure 3.116). Make sure that the cylinder valve is pointed away from you. There should be no straps between your hands.

Step 4: Lift the cylinder, and let the regulator and harness hang freely (Figure 3.117).

Step 5: Raise the cylinder overhead, and let your elbows find their respective loosened harness shoulder strap loops (Figure 3.118). Keeping your elbows close to your body, tuck your chin and grasp the shoulder straps as the SCBA begins to slide down your back. Let the straps slide through your hands as the backpack lowers into place.

Step 6: Lean forward to balance the cylinder on your back and partially tighten the shoulder straps by pulling them outward and downward (Figure 3.119). (**NOTE:** It is sometimes necessary to lean forward with a quick jumping motion to properly position the SCBA on the back while tightening the straps.)

Step 7: Continue leaning forward and then fasten the chest buckle if the unit has a chest strap (Figure 3.120). (**NOTE:** Depending upon the firefighter's physique, it may be more comfortable to fasten the chest buckle before tightening the shoulder straps.)

Step 8: Fasten and adjust the waist strap until the unit fits snugly (Figure 3.121).

Step 9: Don the facepiece. (**NOTE:** This procedure is covered later in this chapter.)

Some departments have removed waist straps from SCBA. Without a waist strap fastened, the SCBA wearer suffers undue stress from side-to-side shifting of the unit and from improper weight distribution of the unit. Even more important, removing waist straps permit the SCBA to be used in a nonapproved manner, which may violate NIOSH certification of the equipment and may void the manufacturer's warranty.

CROSSED-ARMS COAT METHOD

Self-contained breathing apparatus can be donned like a coat. The equipment should be arranged so that both shoulder straps can be grasped for lifting.

Step 1: Check the unit.

a. Crouch or kneel at the cylinder valve end of the unit, regardless of whether the unit is in its case or on the ground.

b. Check the cylinder gauge to make sure that the air cylinder is full (Figure 3.122). Open the cylinder valve slowly and listen for the audible alarm as the system pressurizes. Then, open the cylinder valve fully. If the audible alarm does not sound, or if it sounds but does not stop, place the unit out of service by tagging it and notifying an officer; use another unit.

Figure 3.116 With straps spread, grasp the cylinder through the shoulder straps and lift.

Figure 3.117 Lift the cylinder, and let the straps and regulator hang freely.

Figure 3.118 Raise the cylinder overhead, and allow your elbows to slip through straps.

Figure 3.119 While leaning forward, tighten straps by pulling them outward and downward.

Figure 3.120 Fasten the chest buckle.

Figure 3.121 Fasten and adjust the waist strap.

c. Check the regulator gauge; both the cylinder and regulator gauges should register within 100 psi of each other — if increments are in psi — when the cylinder is pressurized to its rated capacity. If increments are in other measurements, such as fractions or minutes, they should correspond. If the unit has a donning switch, leave the cylinder valve open and the unit in the donning mode. If the unit is positive pressure only, refer to the manufacturer's instructions concerning the cylinder valve.

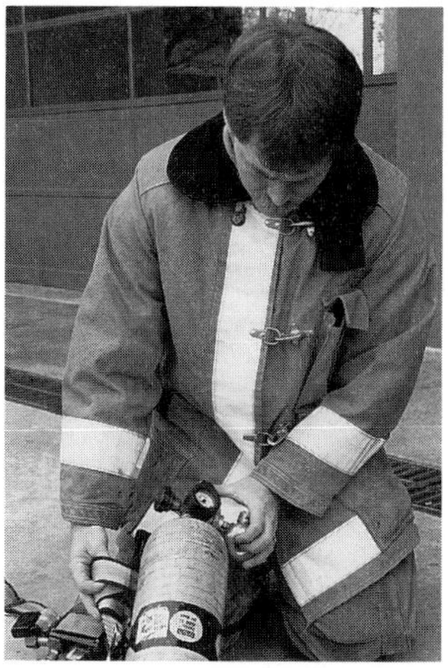

Figure 3.122 Always check the cylinder gauge before donning the SCBA. Open the valve fully before donning the unit.

Figure 3.123 Kneel and grasp the shoulder straps with your arms crossed, left over right.

NOTE: Positive-pressure SCBA that can supply air in the demand-type mode do not meet the requirements of NFPA 1981, 1987 edition.

Step 2: Spread the harness straps out to their respective sides. Cross your arms, left over right. Grasp the shoulder straps at the top of the harness, left hand holding the left strap and right hand holding the right strap (Figure 3.123).

Step 3: Lift the SCBA. Using both arms, swing the unit around your right shoulder, and raising your left arm, continue bringing the unit behind your head and onto your back. Both hands should still be grasping the shoulder straps high on the harness (Figures 3.124 a and b on the next page).

Step 4: Maintaining a firm grip on the straps, slide your hands down the straps to the shoulder strap buckles. Your elbows should be between the straps and the backpack (Figure 3.125 on the next page).

Step 5: Lean slightly forward to balance the cylinder on your back; tighten the shoulder straps by pulling them outward and downward (Figure 3.126 on the next page). (**NOTE:** It is sometimes necessary to lean forward with a quick jumping motion to properly position the SCBA on the back while tightening the straps.)

Figure 3.124a While rising to a standing position, keep your wrists together.

Figure 3.124b Swing the unit around your right shoulder, allowing your wrists to separate.

Figure 3.125 Let the harness straps slide through your hands until the SCBA is positioned on your back.

Figure 3.126 Tighten the shoulder straps by pulling them outward and downward.

Step 6: Continue leaning forward, and fasten the chest buckle if the unit has a chest strap. (NOTE: It may be necessary to fasten the chest strap before completely tightening the shoulder straps.)

Step 7: Fasten and adjust the waist strap until the unit fits snugly.

Step 8: Recheck all straps to see that they are correctly adjusted.

Step 9: Don the facepiece. (NOTE: This procedure is covered later in this chapter.)

REGULAR COAT METHOD

Self-contained breathing apparatus can be donned like a coat, putting one arm at a time through the shoulder strap loops. The unit should be arranged so that *either* shoulder strap can be grasped for lifting.

Step 1: Check the unit.

a. Crouch or kneel at the cylinder valve end of the unit, regardless of whether the unit is in its case or on the ground.

b. Check the cylinder gauge to make sure that the air cylinder is full. Open the cylinder valve slowly and listen for the audible alarm as the system pressurizes. Then, open the cylinder valve fully. If the audible alarm does not sound, or if it sounds but does not stop, place the unit out of service by tagging and notifying an officer; use another unit.

c. Check the regulator gauge; both the cylinder and regulator gauges should register within 100 psi of each other — if increments are in psi — when the cylinder is pressurized to its rated capacity. If increments are in other measurements, such as fractions or minutes, they should correspond. If the unit has a donning switch, leave the cylinder valve open and the unit in the donning mode. If the unit is positive pressure only, refer to the manufacturer's instructions concerning the cylinder valve.

NOTE: Positive-pressure SCBA that can supply air in a demand-type mode do not meet the requirements of NFPA 1981, 1987 edition.

Step 2: Spread the straps out to their respective sides, and position the upper portion of the straps over the top of the backplate (Figure 3.127). (NOTE: By doing this, the straps are less likely to fall, and your arms can go through the straps with less difficulty.)

NOTE: This procedure is written for those harnesses having the regulator attached to the left side of the harness. There are some SCBA that have the regulator mounted on the right. For these types, the right strap should be grasped with your right hand, and the backpack should be donned following the instructions in the next steps, but using directions opposite those indicated.

Step 3: At the top of the harness, grasp the left strap with your left hand; grasp the lower portion of the same strap with your right hand (Figure 3.128). (NOTE: When kneeling at the cylinder valve end, the left harness strap will be to your right.)

Step 4: Lift the unit; swing it around your left shoulder and onto your back. Both hands should still be grasping the shoulder strap (Figure 3.129).

Step 5: Continue to hold the strap with your left hand, release your right hand, and insert your right arm between the right shoulder strap and the backpack frame (Figure 3.130).

NOTE: An alternate way of applying the regular coat method as described above is the following:

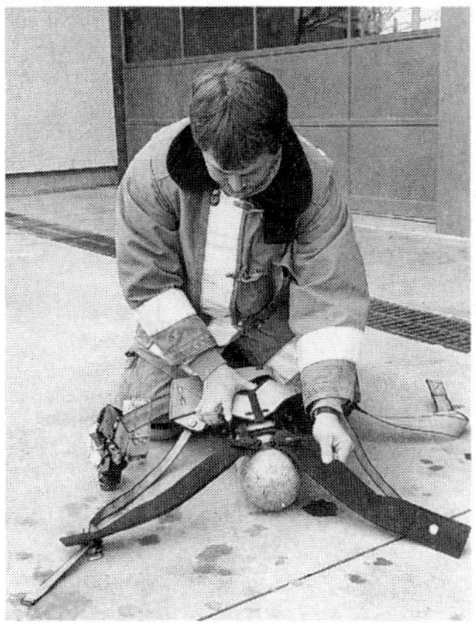

Figure 3.127 Spread the straps out and position the upper part of the straps over the top of the backplate.

Figure 3.128 Grasp the top of the left shoulder strap with your left hand. Grasp the lower part of the same strap with your right hand.

Figure 3.129 Lifting the SCBA by the straps, swing the unit around your left shoulder and onto your back while maintaining a firm grasp of the straps.

Figure 3.130 Holding the strap with your left hand, insert your right arm between the right shoulder strap.

Step 3a: Grasp the top of the left shoulder strap with your left hand; grasp the *regulator* with your right hand.

Step 4a: Lift the unit; swing it around your left shoulder and onto your back, maintaining control of the regulator with your right hand.

Step 5a: Transfer the regulator to your left hand; insert your right arm through the right shoulder strap; grasp the end of the waist strap with your right hand and *loosely* connect the waist strap.

Step 6: Lean slightly forward to balance the cylinder on your back (Figure 3.131); tighten the shoulder straps by pulling them outward and downward (Figure 3.132). (**NOTE:** It is sometimes necessary to lean forward with a quick jumping motion to properly position the SCBA on the back while tightening the straps.)

Step 7: Continue leaning forward, and fasten the chest buckle if the unit has a chest strap (Figure 3.133). Tighten the shoulder straps further if necessary.

Step 8: Fasten and adjust the waist strap until the unit fits snugly (Figure 3.134).

Figure 3.131 While leaning forward, tighten the shoulder straps.

Figure 3.132 Lean forward to balance the cylinder on your back; tighten the shoulder straps by pulling them outward and downward.

Figure 3.133 Fasten the chest strap.

Figure 3.134 Fasten and adjust the waist strap until the unit fits snugly.

Step 9: Recheck all straps to see that they are correctly adjusted.

Step 10: Don the facepiece. (**NOTE:** This procedure is covered later in this chapter.)

DONNING FROM A SEAT MOUNT

Valuable time can be saved if the SCBA is mounted on the back of the firefighter's seat in the vehicle (Figure 3.135). By having a seat mount, firefighters can don SCBA while en route to an incident. If the SCBA is not needed upon arrival, it can be removed quickly and remain mounted in its support. Donning from a seat mount should only be done, however, if it can be safely accomplished without the firefighter having to remove his or her safety belt.

Seat-mounting hardware comes in three main types: lever clamp, spring clamp, or flat hook. A drawstring or other quick-opening bag should enclose the facepiece to keep it clean and to protect it from dust and scratches. (**NOTE:** Do not keep the facepiece connected to the regulator during storage. These parts must be separate to check for proper facepiece seal.)

Donning en route is accomplished by inserting the arms through the straps while sitting with the seat belt on, then adjusting the straps for a snug fit (Figures 3.136 and 3.137).

The cylinder's position should match the proper wearing position for the firefighter. The visible seat-mounted SCBA reminds and even encourages personnel to check the equipment more frequently. Because it is exposed, checks can be made more conveniently. When exiting the fire apparatus, be sure to adjust the straps for a snug and comfortable fit.

WARNING:

Never stand up to don SCBA while the vehicle is moving. Standing places both you and other firefighters in danger of serious injury in the event of a fall. NFPA 1500 requires firefighters to remain seated and belted at all times while the emergency vehicle is in motion.

SIDE OR REAR MOUNT

Although it does not permit donning en route, the side- or rear-mounted SCBA may be desirable (Figure 3.138). This type of mount saves time because the following steps are eliminated: removing the equipment case from the fire apparatus, placing it on the ground, opening the case, and picking up the unit. However, because the unit is exposed to weather and physical damage, a canvas cover is desirable (Figure 3.139).

If the mounting height is right, firefighters can don SCBA with little effort. Having the mount near the running boards or near the tailboard allows the firefighter to don the equipment while sitting. The donning steps are essentially the same as those for seat-mounted SCBA.

COMPARTMENT OR BACKUP MOUNT

SCBA stored in a closed compartment can be ready for rapid donning by using any number of mounts (Figure 3.140). A mount on the inside of a compartment presents the same advantages as does side-mounted

Figure 3.135 Seat-mounted SCBA are shown in the fire fighting vehicle.

Figure 3.136 When donning en route, the first step is to insert the arms through the straps.

Figure 3.137 The second step is to adjust the straps for a snug fit.

Figure 3.138 Some apparatus have SCBA mounted on the sides for easy donning. *Courtesy of Joel Woods, University of Maryland Fire & Rescue Institute.*

Figure 3.139 Canvas covers help protect the SCBA from excess dirt and moisture.

equipment. Some compartment doors, however, may not allow a firefighter to stand fully while donning SCBA. Other compartments may be too high for the firefighter to don the SCBA properly.

Other compartment mounts feature a telescoping frame that holds the equipment out of the way inside the compartment when it is not needed (Figure 3.141). One type of compartment mount telescopes outward, then upward or downward to proper height for quick donning.

The backup mount provides quick access to SCBA (some high-mounted SCBA must be removed from the vehicle and donned using the over-the-head or coat method). The procedure for donning SCBA using the backup method, with slight variation for mounts from which the SCBA can be donned while seated, is as follows:

Step 1: Uncover the SCBA. Remove the facepiece and place it nearby.

Step 2: Check the unit.

a. Open the cylinder valve slowly and listen for the audible alarm as the system pressurizes. Open the cylinder valve fully. If the audible alarm does not sound, or if it sounds but does not stop, place the unit out of service by tagging it and notifying an officer; use another unit.

b. Check the regulator gauge; both the cylinder and regulator gauges should register within 100 psi of each other — if increments are in psi — when the cylinder is pressurized to its rated capacity. If increments are in other measurements, such as fractions or minutes, they should correspond. If the unit has a donning switch, leave the cylinder valve open and the unit in the donning mode. If the unit is positive pressure only, refer to the manufacturer's instructions concerning the cylinder valve.

NOTE: Positive-pressure SCBA that supply air in the demand-type mode do not meet the requirements of NFPA 1981, 1987 edition.

Step 3: Back up against the cylinder backplate, and place your arms through the harness straps (Figure 3.142). As you lean slightly forward to

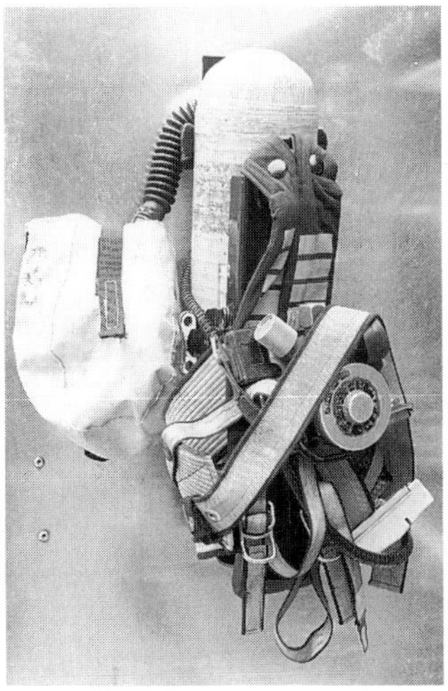

Figure 3.140 Compartment mount installation.

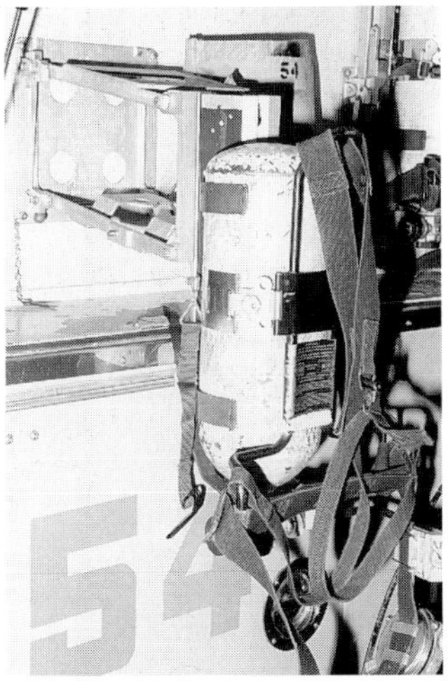

Figure 3.141 A compartment mount featuring a telescoping frame to hold the equipment inside the compartment provides the proper height for donning.

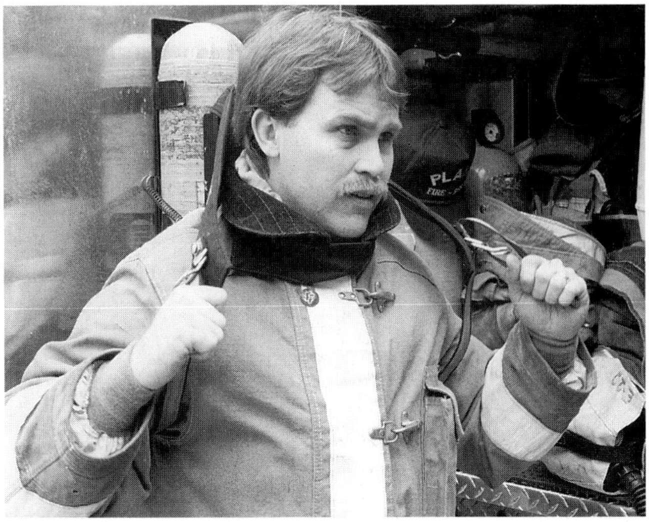

Figure 3.142 Back up against the cylinder backplate, and place your arms through the harness straps.

will temporarily seal any leak and will give a false sense of a proper seal). If there is evidence of leaking, adjust or redon the facepiece.

Step 7: Check the exhalation valve. Inhale, seal the end of the low-pressure hose with the palm of your hand, and exhale. If the exhalation escapes at the edges of the facepiece and does not go through the exhalation valve, keep the low-pressure hose sealed, press the facepiece against your face, and exhale to free the valve (Figure 3.152). Use caution when exhaling against a sealed facepiece in order to prevent discomfort and possible damage to the inner ear from exhaling forcefully. If you cannot get the exhalation valve free, remove the facepiece from service and have it checked.

Step 8: Put your helmet on, first inserting the low-pressure hose through the helmet's chin strap. The helmet should rest on your shoulder until the SCBA is completely donned (Figure 3.153). (**NOTE:** Helmets with straps that completely disconnect may be donned as a last step.)

Step 9: Connect the low-pressure hose to the regulator. If the unit has a donning switch, turn it to the PRESSURE, USE, or ON position. If the unit does not have a donning switch, open the mainline valve.

NOTE: Positive-pressure SCBA that can supply air in a demand-type mode do not meet the requirements of NFPA 1981, 1987 edition.

Step 10: Check for positive pressure. Gently break the facepiece seal by inserting two fingers under the edge of the facepiece (Figure 3.154). You should be able to feel air moving past your fingers. If you cannot feel air movement, remove the unit and have it checked.

Step 11: Pull the protective hood into place, making sure that all exposed skin is covered and that vision is unobscured (Figure 3.155). Check to see that no portion of the hood is located between the facepiece and your face.

Step 12: Place the helmet on your head and tighten the chin strap (Figure 3.156).

An alternative method is to wear the helmet while donning the SCBA. After donning the backpack, loosen the chin strap, allow the helmet to rest on the air cylinder or on your shoulder, and then don the facepiece (Figure 3.157). When the facepiece straps have been tightened and the hood is on, lift the helmet back onto your head and tighten the chin strap (Figure 3.158).

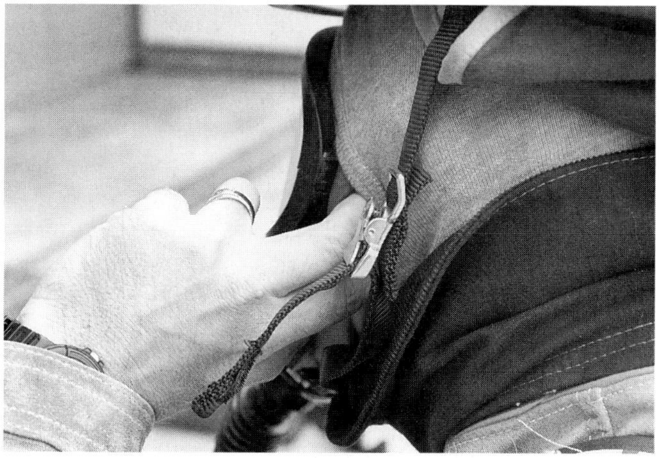

Figure 3.154 Check for positive pressure by inserting two fingers under the edge of the facepiece.

Figure 3.152 To check the exhalation valve, inhale, seal the end of the low-pressure hose with the palm of your hand, and then exhale.

Figure 3.153 Insert the low-pressure hose through the helmet's chin strap.

Figure 3.155 Pull the protective hood into place.

Figure 3.156 Once the helmet is in place, tighten the chin strap.

Figure 3.157 The helmet should rest on the air cylinder until the SCBA is completely donned.

Figure 3.158 Place the helmet on your head, and assure the chin strap is under your chin and not tangled with the facepiece or hood.

Donning The Facepiece With Facepiece-Mounted Regulator

Step 1: Pull the protective hood back and down so that the face opening is around your neck. Turn the collar of your turnout coat up (Figure 3.159). (**NOTE:** Depending upon the style of helmet used, it may be necessary to don the helmet now and allow it to rest on the shoulder.)

Figure 3.159 Pull the protective hood back and down so that the face opening is around your neck. Turn up the collar of the turnout coat.

Step 2: With your thumbs inserted through the straps, grasp the head harness and spread the webbing (Figure 3.160).

Step 3: Stabilize the facepiece with one hand, and use the other hand to remove hair that may be present between your forehead and the sealing surface of the facepiece (Figure 3.161).

Step 4: Center your chin in the chin cup and position the harness so that it is centered at the rear of your head (Figures 3.162 a and b).

Step 5: Tighten the harness straps by pulling them backward (not outward) evenly and simultaneously. Tighten the lower straps first, then the temple straps, and finally the top strap if there is one (Figures 3.163 a through c on the next page). (**NOTE:** For two-strap harnesses, tighten the neck straps, then stroke the harness firmly down the back of your head (Figure 3.164 on the next page). Retighten the straps as necessary.)

Step 6: If the SCBA is so equipped, check the regulator to ensure that the gasket is in place around the regulator outlet port.

Step 7: If the regulator is separated from the facepiece, attach it to the facepiece by positioning it firmly into the facepiece fitting. Lock it into place (Figure 3.165 on the next page). (**NOTE:** This procedure will vary, depending upon the make of SCBA. Always follow the manufacturer's instructions.)

Figure 3.160 With your thumbs inserted through the straps, grasp the head harness and spread the webbing.

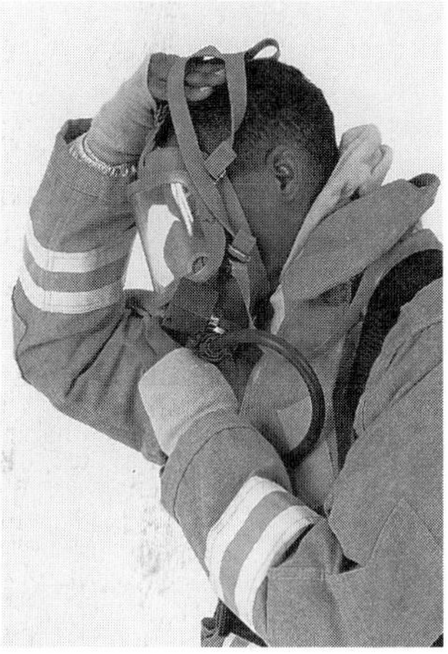

Figure 3.161 Stabilize the facepiece with one hand. Push the web strap back over your head, making sure that hair does not interfere with facepiece seal.

Figure 3.162a Center your chin in the chin cup and position the harness so that it is centered at the rear of your head.

Figure 3.162b Pull the neck strap over your head and onto your neck.

Figure 3.163a Tighten the straps starting with the lower straps, pulling backward not outward.

Figure 3.163b Pull the temple straps to the rear.

Figure 3.163c Tighten the top strap.

Figure 3.164 For a two-strap harness, tighten the neck straps.

Figure 3.165 If separated from the facepiece, attach the regulator to the facepiece by positioning it firmly into the facepiece fitting and locking it in place.

Step 8: Check the facepiece seal. Make sure that the donning switch is in the DON position (positive pressure off) (Figure 3.166). Inhale slowly (not deeply), and hold your breath for 10 seconds. The mask should draw up to your face. Listen for the sound of airflow. There should be no sound and no inward leakage through the exhalation valve or around the facepiece.

Another method for checking facepiece seal is to close the cylinder valve. Continue to breathe slowly until the mask collapses against your face, and hold your breath for 10 seconds. If the mask draws up to your face and no leaks are detected, reopen the cylinder valve. Adjust or redon the facepiece if there is evidence of leaking. If leakage persists, determine and correct the cause of the leakage. If unable to eliminate the leakage, obtain another facepiece and repeat the leak-check procedure. Use care with this method because it uses up some air.

NOTE: Positive-pressure SCBA that can supply air in a demand-type mode do not meet the requirements of NFPA 1981, 1987 edition.

Step 9: Check the exhalation valve. As you exhale during Step 8, make sure that the exhalation goes through the exhalation valve and not to the edges of the facepiece. If it does not go through the valve, the valve may be stuck. To free it, press your facepiece against the sides of your face, and exhale to free the valve (Figure 3.167). Use caution when exhaling against a sealed facepiece in order to prevent discomfort and possible damage to the inner ear from exhaling forcefully. If you cannot get the exhalation valve free, remove the facepiece from service and have it checked.

Figure 3.166 Use the negative-pressure test to check the facepiece seal by inhaling slowly for 10 seconds.

Figure 3.167 Check the exhalation valve by pressing the mask against your face and exhaling sharply.

Figure 3.168 Check for positive pressure by gently breaking the facepiece seal by inserting two fingers under the edge of the facepiece. Air should be felt moving past your fingers.

Figure 3.169 Pull the protective hood into place, making sure that all exposed skin is covered and vision is unobscured.

Figure 3.170 Put the helmet back on and tighten the chin strap.

Step 10: Check for positive pressure. Gently break the facepiece seal by inserting two fingers under the edge of the facepiece (Figure 3.168). You should be able to feel air moving past your fingers. If you cannot feel air movement, remove the unit from service and have it checked.

Step 11: Pull the protective hood into place, making sure that all exposed skin is covered and that vision is unobscured (Figure 3.169). Check to see that no portion of the hood is located between the facepiece and your face.

Step 12: Put your helmet back on your head and tighten the chin strap (Figure 3.170). Be sure to get the helmet strap under your chin. (**NOTE:** Helmets with a breakaway strap can be donned at this point.)

An alternative method is to leave the helmet on while donning the backpack, then loosen the chin strap and allow the helmet to rest on the air cylinder while donning the facepiece (Figure 3.171). This method will have to be used if the helmet is not equipped with a chin strap that is detachable.

DOFFING WITH HARNESS-MOUNTED REGULATOR

When you are in a safe atmosphere, take the following steps to remove SCBA having a harness-mounted regulator:

Step 1: Close the mainline valve and disconnect the low-pressure hose from the regulator. (**NOTE:** If the unit has a donning switch, make sure that it is in the donning mode.)

Step 2: Take off your helmet or loosen it and push it and your hood back off your head.

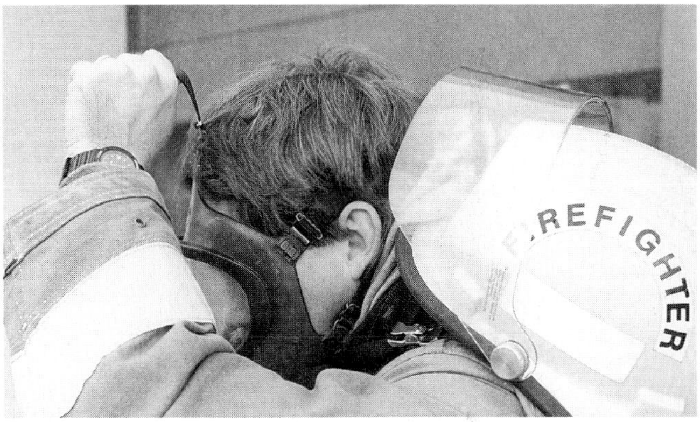

Figure 3.171 Some firefighters prefer to keep their helmet slung back on the SCBA cylinder during donning.

Step 3: Loosen the facepiece harness strap buckles. Either rub them toward your face or lift the buckles slightly to loosen them and to disentangle them from your hair (Figures 3.172 a and b). Take off the facepiece (Figure 3.173 on the next page), extend the harness straps fully,

Figure 3.172a Loosen the facepiece harness buckles by rubbing them toward your face.

Figure 3.172b Lifting the buckles also loosens the facepiece.

and prepare it for inspection, cleaning, sanitizing, and storage (Figure 3.174).

Step 4: Unbuckle the waist belt and fully extend the adjustment (Figure 3.175).

Step 5: Disconnect the chest buckle if the unit has a chest strap.

Step 6: Lean forward; release shoulder strap buckles and hold them open while fully extending the straps (Figure 3.176).

Figure 3.173 Take off the facepiece.

Figure 3.174 Extend the harness straps fully.

Figure 3.175 Unbuckle the waist belt and fully extend the adjustment.

Figure 3.176 Lean forward; release shoulder strap buckles and hold them open while fully extending the straps.

Step 7: Grasp the shoulder straps firmly with the respective hands. Slip off the shoulder strap from your shoulder opposite the regulator, and remove your arm from the shoulder strap. Grasp the regulator with your free hand, allow the other strap to slide off your shoulder, and lower the SCBA to the ground (Figures 3.177 a through c). As you remove the unit, do not drop the regulator or allow it to strike anything.

Step 8: Close the cylinder valve, then relieve the excess pressure from the regulator. If the regulator has been removed from the facepiece, recouple it, hold the facepiece against your face, and breathe until the remaining pressure is depleted. Another method is to open the mainline valve and allow the excess pressure to vent (Figure 3.178). (**NOTE:** Do not use the bypass valve to relieve excess pressure.)

Step 9: Remove the facepiece from the regulator and extend the straps fully. Prepare the facepiece for inspection, cleaning, sanitizing, and storage.

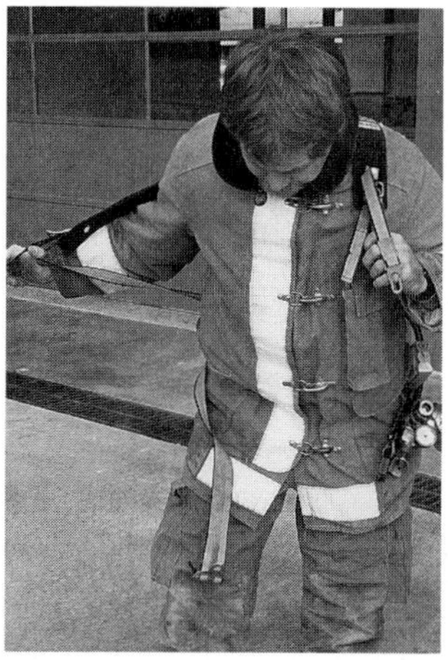

Figure 3.177a Slip off the shoulder strap from your shoulder opposite the regulator, and remove your arm from the shoulder strap.

Figure 3.177b Grasp the regulator with your free hand.

Figure 3.177c With your hand on the regulator, allow the other strap to slide off your shoulder, and lower the SCBA to the ground.

Figure 3.178 Open the mainline valve and allow the excess pressure to vent.

DOFFING WITH FACEPIECE-MOUNTED REGULATOR

When you are in a safe atmosphere, take the following steps to remove SCBA having a facepiece-mounted regulator:

Step 1: Take off your helmet or loosen it and push it and your hood back off your head.

Step 2: If the unit has a donning switch, turn the positive pressure off or place it in donning mode (Figure 3.179).

Step 3: Depending upon the make of SCBA and manufacturer's instructions, disconnect the regulator from the facepiece (Figure 3.180).

Step 4: Loosen the facepiece harness strap buckles. Either rub them toward your face or lift the buckles slightly to loosen them. Take off the facepiece and prepare it for inspection, cleaning, sanitizing, and storage. Extend the harness straps fully (Figures 3.181 a and b).

Step 5: Unbuckle the waist belt and fully extend the adjustment (Figure 3.182).

Step 6: Disconnect the chest buckle if the unit has a chest strap.

Step 7: If the unit is so equipped, attach the regulator to the harness clip; or, control the regulator by holding it while performing the next steps.

Step 8: Lean forward; release shoulder strap buckles and hold them open while fully extending the straps (Figure 3.183).

Step 9: Grasp the shoulder straps firmly with the respective hands. Slip off the shoulder strap from your shoulder opposite the regulator, and remove your arm from the shoulder strap (Figures 3.184 a and b). Grasp the regulator with your free hand, allow the other strap to slide off your shoulder, and lower the SCBA to the ground. As you remove the unit, do not drop the regulator or allow it to strike anything (Figure 3.185 on the next page).

Step 10: Close the cylinder valve and breathe down the pressure from the regulator by holding the facepiece against your face and breathing until the pressure is depleted (Figure 3.186 on the next page). (**NOTE:** Do not bleed off air by operating the bypass valve.)

Figure 3.179 Place the unit in the donning mode (turn positive pressure off).

Figure 3.180 Disconnect the regulator from the facepiece if required.

Figure 3.182 Unbuckle the waist belt and fully extend the adjustment.

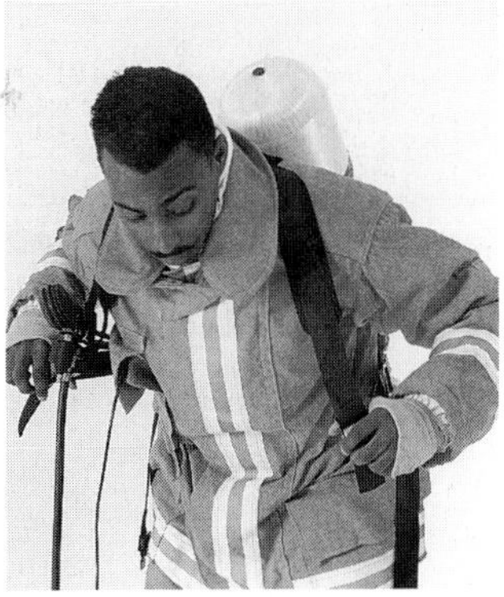

Figure 3.183 Lean forward to loosen shoulder straps.

Figure 3.181a Loosen the straps by briskly rubbing the buckles toward your face.

Figure 3.181b Lift the facepiece from your face, extend the straps, and prepare for cleaning and inspection.

Figure 3.184a Slip the shoulder strap from your shoulder opposite the regulator.

Figure 3.184b Remove your arm from the strap, and with your free hand grasp the regulator.

Figure 3.185 Lower SCBA to the ground while controlling the regulator.

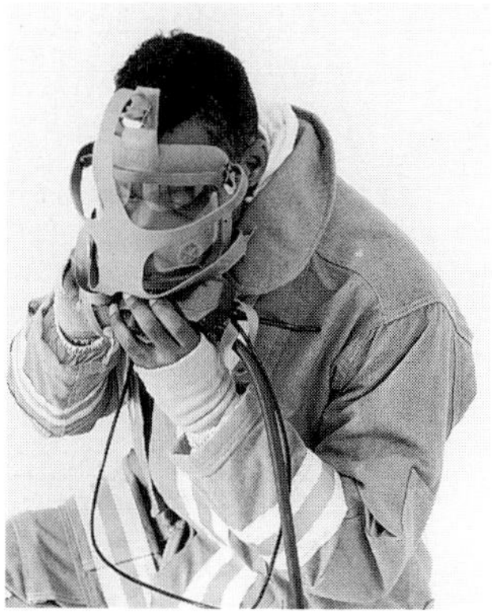

Figure 3.186 Place the facepiece against your face and breathe until the excess pressure is depleted.

Figure 3.187 Disconnect the low-pressure hose from the regulator.

Figure 3.188 Disconnect the high-pressure coupling from the cylinder.

Changing Cylinders

With care and caution, a firefighter can change an air cylinder at the scene of an emergency so that the equipment can be used again as soon as possible. Changing cylinders can be either a one- or two-person job.

The one-person method for changing an air cylinder is described in detail as follows:

Step 1: Doff the unit using the procedures described earlier.

Step 2: Obtain a full air cylinder and have it ready.

Step 3: Disconnect the regulator from the facepiece or disconnect the low-pressure hose from the regulator (Figure 3.187).

Step 4: Close the cylinder valve on the used bottle and release the pressure from the high-pressure hose. On some units, the pressure must be released by breathing down the regulator or opening the mainline valve. Refer to the manufacturer's instructions for the correct method for the particular unit.

NOTE: If the pressure is not released, the high-pressure coupling will be difficult to disconnect.

Step 5: Disconnect the high-pressure coupling from the cylinder (Figure 3.188). (NOTE: If more than hand force is required to disconnect the coupling, repeat Step 4 and then again attempt to disconnect the coupling.) Lay the hose coupling on the ground, directly in line with the cylinder outlet, as a reminder so that the replacement cylinder can be aligned correctly and easily. Be sure that grit or liquids will not enter the end of the unprotected high-pressure hose prior to attaching it to the cylinder outlet valve.

Step 6: Release the cylinder clamp and remove the empty cylinder (Figure 3.189).

Step 7: Place the new cylinder into the backpack, position the cylinder outlet, and lock the cylinder into place (Figure 3.190). (NOTE: For some cylinders, it may be necessary to rotate the cylinder one-eighth turn to the left; this protects the high-pressure hose by lessening the angle of the hose and preventing twisting.)

Step 8: Check the cylinder valve opening and the high-pressure hose fitting for debris and the condition of the O-ring (Figure 3.191). Clear any debris from the cylinder valve opening by quickly opening and closing the cylinder valve or by wiping the debris away. If the O-ring is distorted or damaged, replace it.

Step 9: Connect the high-pressure hose to the cylinder valve opening. (NOTE: Do not overtighten; hand tightening is sufficient.)

Step 10: Open the cylinder valve and check the gauges on the cylinder and the regulator (Figure 3.192). Both gauges should register within 100 psi of

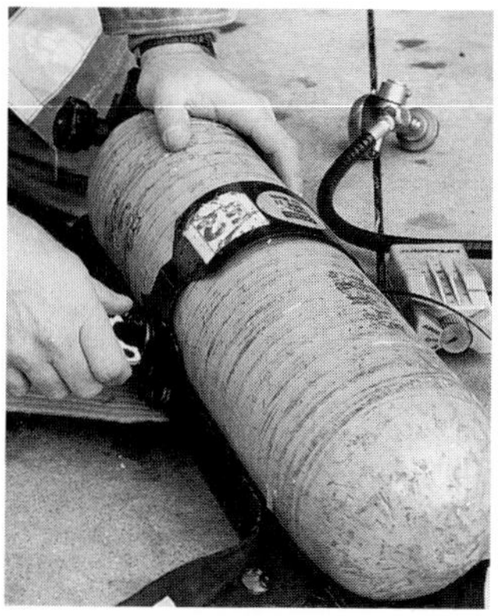

Figure 3.189 Release the cylinder clamp.

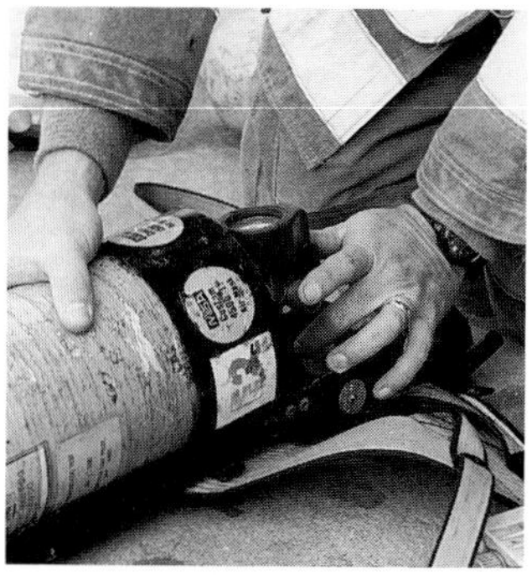

Figure 3.190 Place the new cylinder into the backpack, position the cylinder outlet, and lock the cylinder into place.

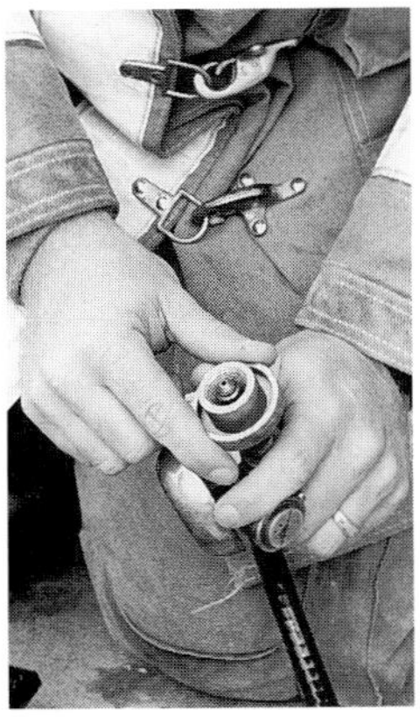

Figure 3.191 Check the O-ring for damage.

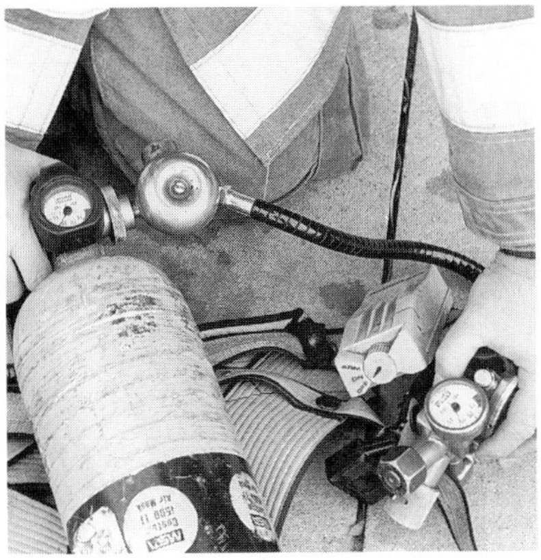

Figure 3.192 Open the cylinder valve and check the gauges on the cylinder and the regulator.

each other — if increments are in psi — when the cylinder is pressurized to its rated capacity. If increments are in other measurements, such as fractions or minutes, they should correspond.

NOTE: Some units require that the mainline valve on the regulator be opened in order to obtain a gauge reading. Seal the regulator outlet port by placing one hand over it. On a positive-pressure regulator, the port must be sealed for an accurate regulator gauge reading.

When there are two people, the firefighter with an empty cylinder simply positions the cylinder so that it can be easily changed by the other firefighter. Two methods for two people are shown in Figures 3.193 and 3.194.

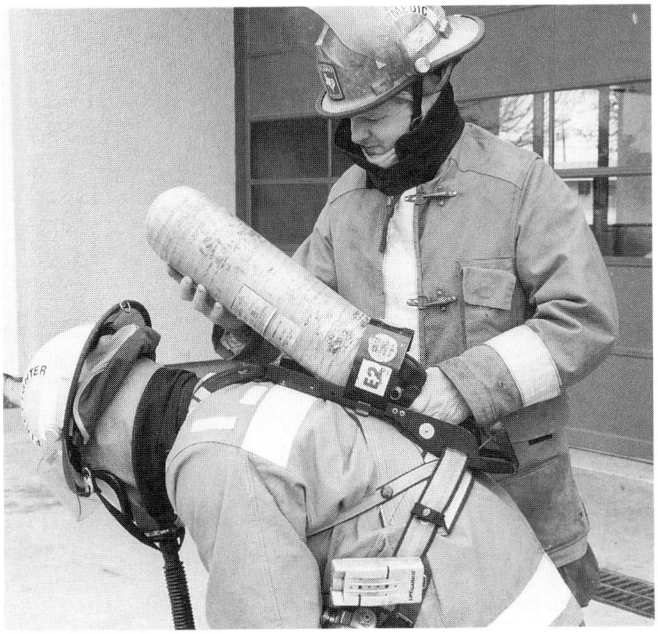

Figure 3.193 One firefighter slides a full cylinder into the backpack assembly while the other firefighter braces to remain steady.

Figure 3.194 The firefighter receiving a full cylinder may choose to kneel while the cylinder is being replaced.

Inspection And Care Of Protective Breathing Apparatus

DAILY INSPECTIONS

Self-contained breathing apparatus requires proper care and inspection before and after each use to provide complete protection. Proper care by paid personnel should include making a daily inspection as soon as possible after reporting for duty (Figure 3.195 on the next page). Volunteer organizations may not be able to check the units every day. In this case, volunteers should check the SCBA after each use and during regular training sessions. The following is a list of things that should be checked:

- Check to ensure that the cylinder is at least 90 percent full.
- Check to ensure that all gauges work. The cylinder gauge and the regulator gauge should read within 100 psi of each other. Gauges not marked in increments of 100 psi should just read relatively close to each other.
- Check to ensure that the low-pressure alarm is in working condition. The alarm should sound briefly when the cylinder valve is turned on and again as the pressure is relieved.

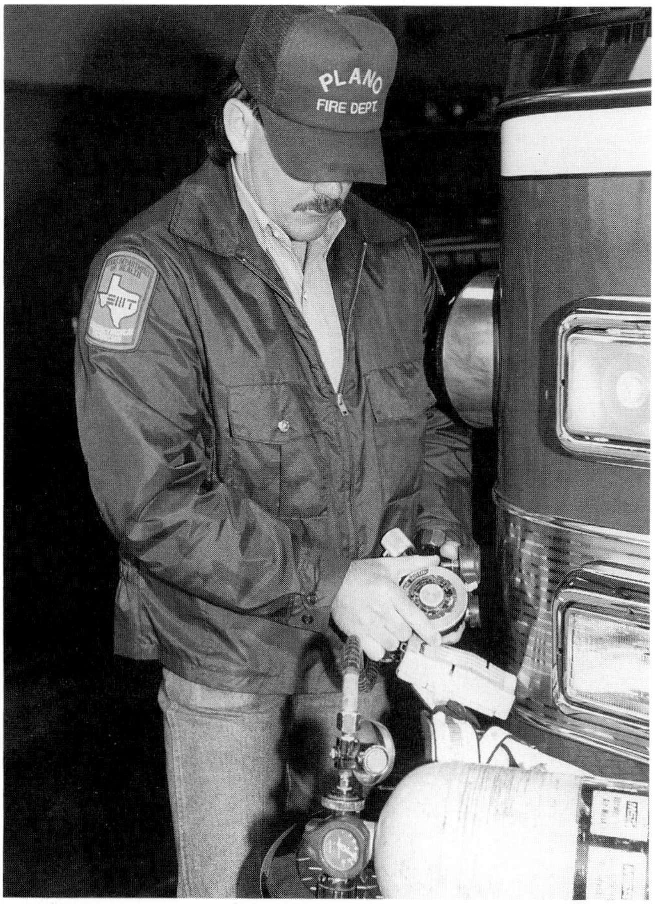

Figure 3.195 Personnel should inspect SCBA on a daily basis.

- Check to ensure that all hose connections are tight and free of leaks.
- Check to ensure that the facepiece is in good condition and clean.
- Check to ensure that the harness system is in good condition and the straps are in the fully extended position.
- Check to ensure that the bypass and mainline valves are operational. After checking the bypass valve, make sure that it is fully closed.

Breathing apparatus should be cleaned and sanitized immediately after each use. Moving parts that are not clean may malfunction. A facepiece that has not been cleaned and sanitized may contain an unpleasant odor and can spread germs to other department members who may wear the mask at a later time. An air cylinder with less air than prescribed by the manufacturer renders the apparatus inefficient, if not useless.

The facepiece should be thoroughly washed with warm water containing any mild commercial disinfectant, and then it should be rinsed with clear, warm water. Special care should be given to the exhalation valve to ensure proper operation. The air hose should be inspected for cracks or tears. The facepiece should be dried with a lint-free cloth or air dried. (**CAUTION:** Do not use paper towels to dry the lens as the paper towel will scratch the plastic lens.)

Many departments now issue personal facepieces to each firefighter. This eliminates the risk of spreading germs from one wearer to the next. Even though the firefighter has his or her own facepiece, it is still important that it be cleaned after each use.

Periodic Inspection And Care

After each three-month period, it is advisable to remove the equipment from service and check the valves, pressure regulator, gauges, harness, and facepiece. The following functional test and inspection should then be made:

- Check the facepiece, hose, and exhalation valve by inhaling slowly with your thumb or palm over the end of the hose connection (Figure 3.196).
- Connect the low-pressure hose to the regulator and check the performance of the regulator. Inhale deeply and quickly. The regulator should supply a full flow to give the user all the air that is needed.

If, on slow inhalation, a "honking" sound is heard in the regulator, it can usually be stopped by inhaling faster. The sound is caused by the bellows vibrating and

Figure 3.196 Check for a good facepiece seal.

in no way affects the performance or safety of the regulator. If the bellows vibrate continuously or excessively, the regulator should be overhauled by competent technicians recommended by the manufacturer. If the demand valve sticks open slightly (this condition may be caused by the diaphragm being cold), the breathing air will continue to flow when the wearer is not inhaling. This condition can usually be corrected by "blowing back" on the regulator. Operate the regulator several minutes to exercise the diaphragm and valves before taking the regulator out of service. With the low-pressure hose out of the regulator connection, close off the cylinder valve (Figure 3.197). With 1,980 psi (13 650 kPa) indicated on the regulator gauge, the regulator and the regulator hose assembly should hold the trapped-in pressure.

Figure 3.197 Close off the cylinder valve.

After every two and one-half years, the regulator with regulator hose should be returned to the factory or to their representative for testing and/or repair. Each cylinder must be stamped or labeled with the date of manufacture and the date of the last hydrostatic test (Figure 3.198). Steel and aluminum cylinders must be tested every five years; composite cylinders every three years. This procedure is necessary to meet the requirements of the United States Department of Transportation. Always empty cylinders before returning them for servicing and testing.

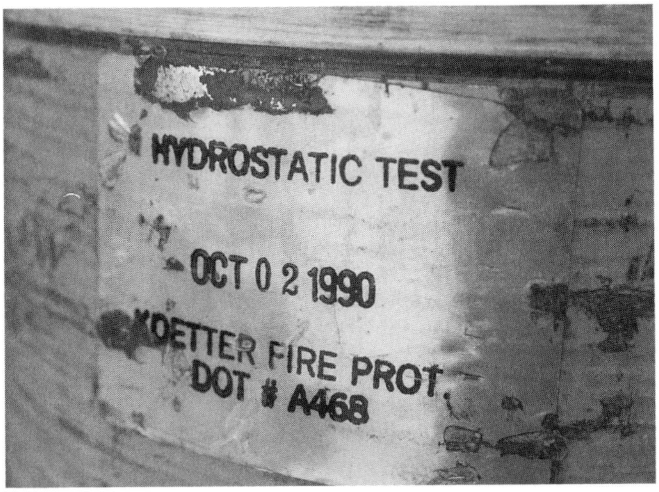

Figure 3.198 Each cylinder must be stamped with the date of the most recent hydrostatic test.

Recharging Self-Contained Breathing Apparatus Cylinders

Air cylinders for self-contained breathing apparatus are filled from either a cascade system (a series of at least three, 300-cubic-foot [8 490 L] cylinders) or directly from a compressor purification system (Figures 3.199 a and b). No matter how the cylinders are filled, the same safety precautions apply: put the cylinders into a shielded

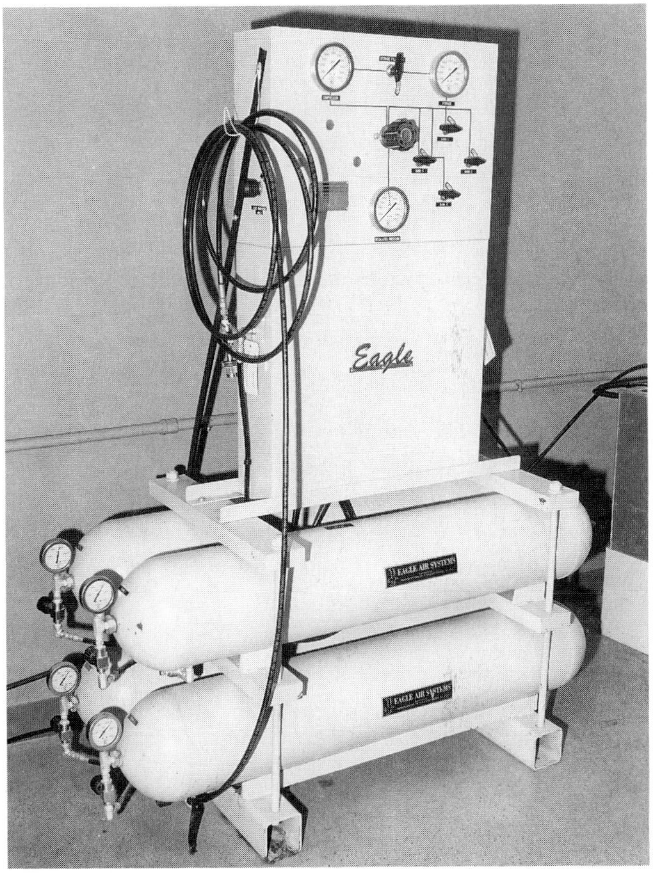

Figure 3.199a Cylinders can be refilled from a cascade system.

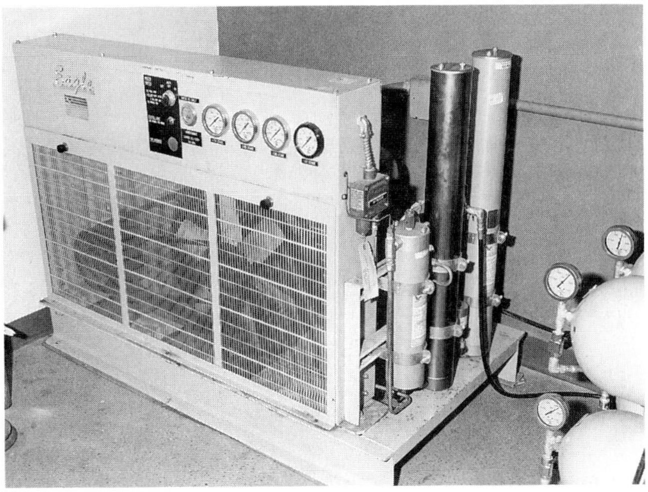

Figure 3.199b Cylinders can also be filled from a compressor purification system.

charging station, prevent cylinder overheating, and be sure that the cylinder is fully charged but not overpressurized.

Although some departments use a water-bath fill station, it is not a recommended practice. **Under no circumstances should a composite bottle be placed in water during a recharging operation.**

FILLING FROM A CASCADE SYSTEM

The following is a recommended procedure for refilling SCBA cylinders from a cascade system:

Step 1: Check the hydrostatic test date, and inspect the SCBA cylinder for damage such as deep nicks, cuts, gouges, or discoloration from heat. If the cylinder is damaged or is out of hydrostatic test date, remove the cylinder from service and tag it for further inspection and hydrostatic testing. (CAUTION: Never attempt to fill a cylinder that is damaged or that is out of hydrostatic test date.)

Step 2: Place the SCBA cylinder in a fragment-proof charging station, and connect the charging hose to the cylinder (Figure 3.200). If the charging hose has a bleed valve, make sure that the bleed valve is closed.

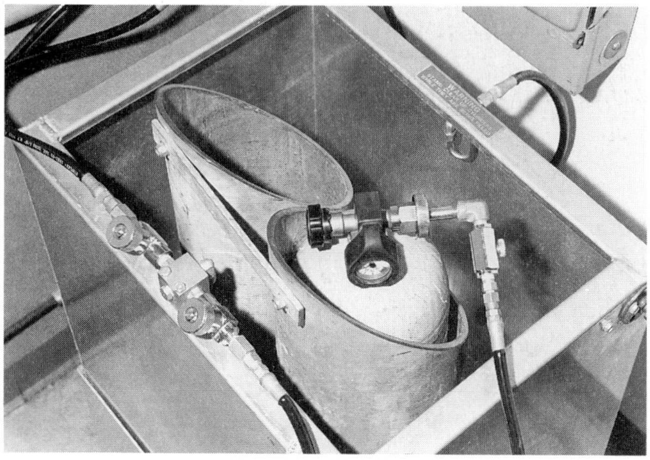

Figure 3.200 Place the SCBA cylinder in a fragment-proof charging station, and connect the charging hose to the cylinder.

Step 3: Open the SCBA cylinder valve.

Step 4: Open the valve at the charging hose, at the cascade system manifold, or the valves at both locations, if the system is so equipped. (NOTE: Some cascade systems may have a valve at the charging hose, at the manifold, or at both places.)

Step 5: Open the valve of the cascade cylinder that has the least pressure but that has more pressure than the SCBA cylinder. The airflow from the cascade cylinder must be slow enough to avoid "chatter" or excessive heating of the cylinder being filled. Watch to see that the cylinder gauge needle rises slowly by about 300 to 600 psi (2 070 kPa to 4 140 kPa) per minute. Your hand should be able to rest on the SCBA cylinder without undue discomfort from the heating of the cylinder.

Step 6: When the pressures of the SCBA and the cascade cylinder equalize, close the cascade cylinder valve. If the SCBA cylinder is not yet fully charged, open the valve on the cascade cylinder with the next highest pressure.

Step 7: Repeat Step 6 until the SCBA cylinder is fully charged.

Step 8: Close the valve or valves at the cascade system manifold and/or charging line, if the system is so equipped.

Step 9: Close the SCBA cylinder valve.

Step 10: Open the hose bleed valve to bleed off excess pressure between the cylinder valve and the valve on the charging hose. (CAUTION: Failure to do so could result in O-ring damage.)

Step 11: Disconnect the charging hose from the SCBA cylinder.

Step 12: Remove the SCBA cylinder from the charging stand, and return the cylinder to proper storage.

FILLING FROM A COMPRESSOR/PURIFIER

NOTE: Always check the compressor/purifier manufacturer's instructions before attempting to fill any cylinders.

Step 1: Check the hydrostatic test date, and inspect the SCBA cylinder for damage such as deep nicks, cuts, gouges, or discoloration from heat. If the cylinder is damaged or out of hydrostatic test date, remove the cylinder from service and tag it for further inspection and hydrostatic testing. (CAUTION: Never attempt to fill a cylinder that is damaged or that is out of hydrostatic test date.)

Step 2: Place the SCBA cylinder in a fragment-proof charging station, and connect the charging hose to the cylinder. Make sure that the hose bleed valve is closed.

Step 3: Open the SCBA cylinder valve.

Step 4: Set the cylinder pressure adjustment on the compressor (if applicable) to the desired full-cylinder pressure (Figure 3.201). (**NOTE:** If there is no cylinder pressure adjustment, you must watch the pressure gauge on the cylinder during charging to determine when it is full.)

Step 5: Open the fill valve on the compressor/purifier unit. Airflow should be slow (300 to 600 psi [2 070 kPa to 4 140 kPa] per minute) to avoid excessive heating of the cylinder.

Step 6: When the cylinder is full, close the fill valve on the compressor/purifier.

Step 7: Close the SCBA cylinder valve.

Step 8: Open the hose bleed valve to bleed off excess pressure between the cylinder valve and valve on the charging hose. (**CAUTION:** Failure to do so could result in O-ring damage.)

Step 9: Disconnect the charging hose from the SCBA cylinder.

Step 10: Remove the SCBA cylinder from the charging stand, and return the cylinder to proper storage.

Figure 3.201 Set the cylinder pressure adjustment on the compressor (if applicable) to the desired full-cylinder pressure.

RAPID REFILLING OF CYLINDER WHILE WEARING SCBA

Sometimes it will be necessary to refill cylinders while they are still being worn. The following section describes how to do this:

Step 1: Remove the protective caps from the refill connection on the SCBA and from the refill hose connected to the cascade system (Figure 3.202).

Step 2: Connect the refill hose to the SCBA refill connection (Figure 3.203).

Step 3: Fill the cylinder, following SCBA manufacturer's instructions, and monitor the refilling pressure until the cylinder is charged.

Step 4: Disconnect the refill hose from the SCBA refill connection.

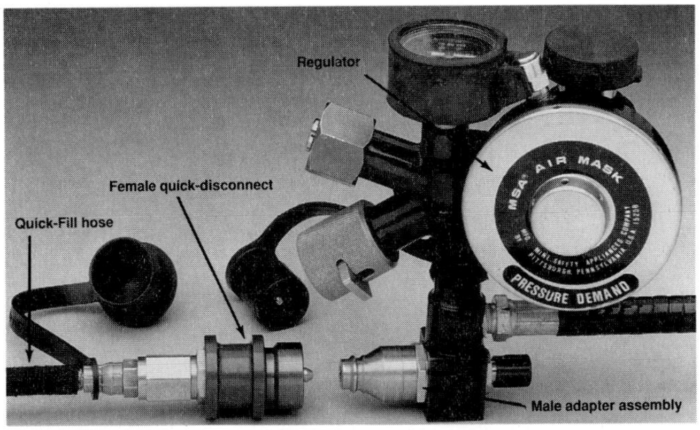

Figure 3.202 This shows the position of the quick-fill connection. *Courtesy of MSA.*

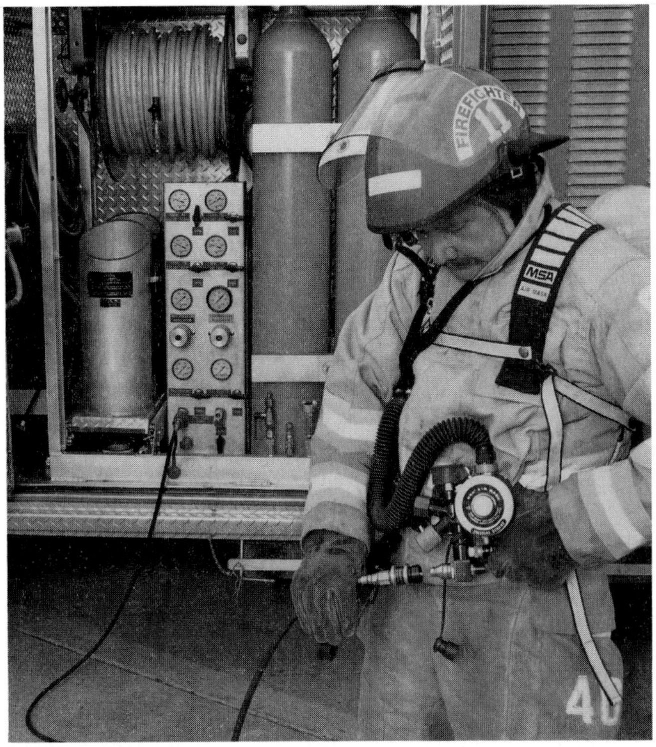

Figure 3.203 Connect the refill hose to the appropriate fill connection. *Courtesy of MSA.*

Step 5: Replace the protective caps on the SCBA refill connection and on the refill hose.

Safety Precautions For SCBA Use

Fire fighting is a strenuous, demanding activity, so firefighters need to be in good physical condition. Their protective gear can work against them while at the same time protecting them. The basic required turnout coat can be a virtual sweat box. It builds up body heat and hinders movement which increases firefighter exhaustion. This condition is intensified when self-contained breathing equipment is used under emergency conditions. The difference between the weight of ordinary street clothes and fire fighting gear plus the mask unit has been measured at an extra 47 pounds (21 kg); the breathing unit alone can weigh from 25 to 35 pounds (11 kg to 16 kg), depending on size and type. All firefighters should be aware of the signs and symptoms of heat-related illnesses (Figure 3.204). For more information on heat-related illnesses, consult IFSTA's **Fire Service First Responder** manual.

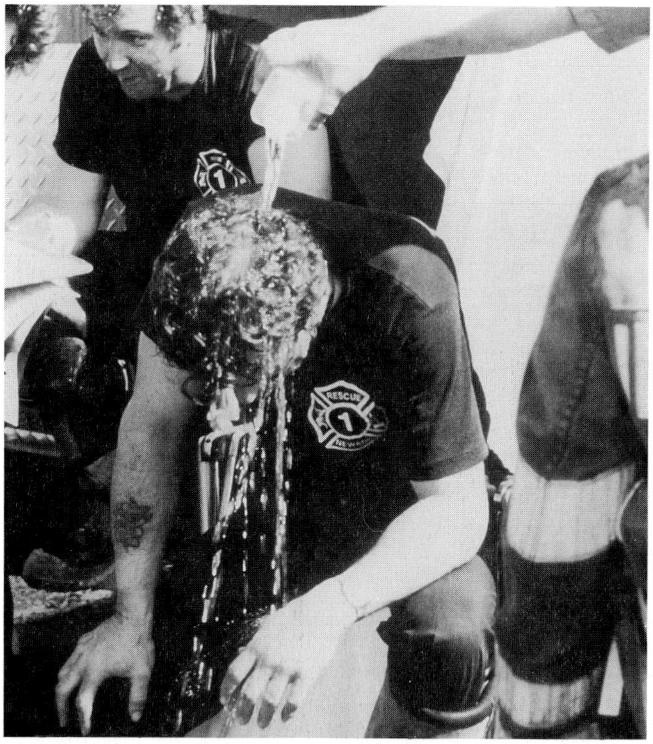

Figure 3.204 This firefighter is experiencing heat stress. *Courtesy of Ron Jeffers.*

When using self-contained breathing apparatus, the following items should be remembered and observed for maximum safety:

- All firefighters who wear SCBA should be certified as physically fit by a physician using criteria established by the fire department.
- Breathing apparatus should not be used immediately after performing strenuous work.
- Air supply duration will vary with:
  - Firefighter conditioning
  - The task performed
  - Level of training
  - Operational environment
  - Degree of excitement
  - Other variables

Know your own limitations and abilities!

- Make sure that the mainline valve is opened and locked (Figure 3.205).
- Use the bypass valve for emergencies only.
- Close the mainline valve when the bypass valve must be used.
- Once entering a contaminated area, do not remove your breathing apparatus until you have left the contaminated area. Improved visibility does not ensure that the area is free from contamination.
- When wearing SCBA, work in groups of two or more.

Figure 3.205 Make sure that the mainline valve is opened and locked.

Emergency Situations

The emergencies created by the malfunction of protective breathing apparatus can be overcome in several ways. In all of these emergencies, the conservation of air and immediate withdrawal from the hazardous atmosphere are of the utmost importance. The following is a list of suggestions that can effectively correct an emergency situation:

- Do not panic! Panicking causes rapid breathing, using more valuable air.
  - Control breathing while crawling.
- Stop and think. How did you get to where you are? Downstairs? Upstairs? Left turns?
- Listen:
  - For noise from other personnel.
  - For hose and equipment operation.
  - For sounds that indicate the location of fire.
- Different methods to use to find a way out:
  - Follow the hoseline out if possible (male coupling is closest to exit, female is closest to the fire) (Figure 3.206).

4-93

— Crawl in a straight line (hands flat on floor, move knee to hand).

— Once in contact with the wall, crawl in one direction (all left-hand turns, all right-hand turns).

— Call for directions, call out or make noise for other firefighters to assist you.

— If possible, break a window or breach a wall to escape.

— Activate your Personal Alert Safety System device.

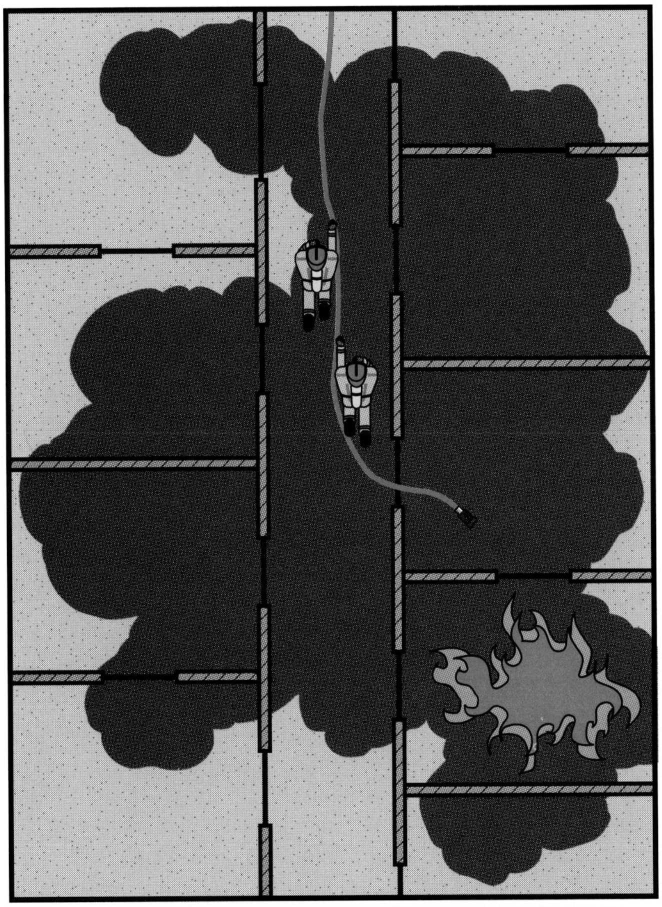

Figure 3.206 In an emergency situation where you need to exit the structure, remember that the male couplings lead away from the fire.

Firefighters should practice controlled breathing when using SCBA. However, when air supply is low, they may practice skip breathing. Skip breathing is an emergency breathing technique used to extend the use of the remaining air supply. To use this technique, the firefighter inhales (as during regular breathing), holds the breath as long as it would take to exhale, and then inhales once again before exhaling. The firefighter should take normal breaths and exhale slowly to keep the carbon dioxide in the lungs in proper balance.

Although a regulator usually works as designed, it can malfunction. One method of using SCBA when the regulator becomes damaged or malfunctions is to open the bypass valve. During normal SCBA operation, the mainline valve is fully open while the bypass valve is fully closed. If needed in an emergency, the firefighter can close the mainline valve and open the bypass valve to provide a flow of air into the facepiece (Figures 3.207). The bypass valve should be closed after the firefighter takes a breath and then opened each time the next breath is needed.

If the facepiece fails, it may be necessary to breathe directly from the low-pressure hose (after disconnecting it from the facepiece) or the regulator (Figure 3.208). In

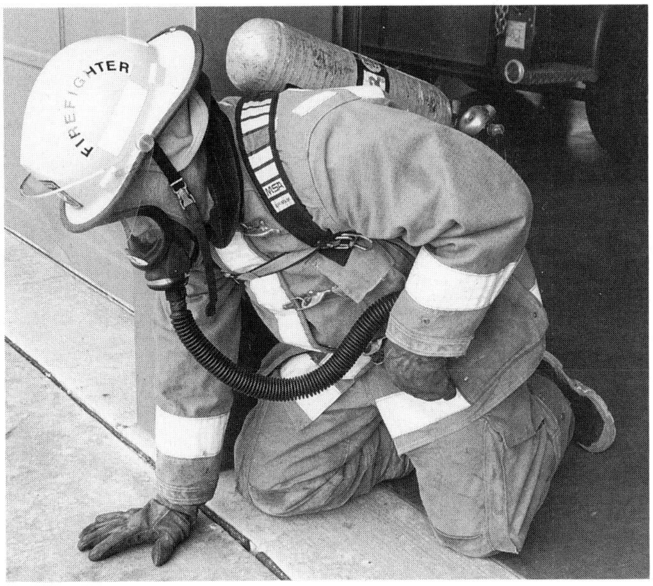

Figure 3.207 This firefighter is using the bypass valve because of regulator malfunction or damage.

Figure 3.208 If the facepiece fails, it may be necessary to breathe directly from the regulator.

either event, the firefighter must make sure that the hose or regulator opening is held close to the mouth to avoid breathing in fire gases when inhaling. Additional information on emergency breathing situations can be found in IFSTA's **Self-Contained Breathing Apparatus** manual.

Using Protective Breathing Apparatus

In addition to being familiar with the donning, operation, and doffing of protective breathing apparatus, the firefighter must also be trained in techniques that accompany the use of the apparatus. In order to operate at maximum efficiency, the firefighter must be able to operate effectively in areas of obscured vision and negotiate tight passages without having to completely shed the breathing apparatus. The following sections address techniques for accomplishing these tasks.

OPERATING IN AREAS OF OBSCURED VISION

In many instances where protective breathing apparatus will be required, firefighters will be operating in an area of obscured visibility. Most interior fire attacks, and many exterior attacks, will present the firefighters with heavy smoke conditions that may reduce visibility to zero. Firefighters must learn techniques for moving about and performing critical tasks when vision is diminished.

Figure 3.209 Firefighter using the "duck" walk.

The primary method of moving about in areas of obscured visibility is by crawling. Crawling is beneficial for several reasons. First, it allows firefighters to remain close to the floor and avoid the higher heat found closer to ceiling level. Second, crawling allows firefighters to feel in front of themselves as they move along. This prevents the firefighters from falling through holes burned in the floor, falling through stair or elevator shafts, or running into objects in front of them. Crawling also allows the firefighters to feel for victims who may be lying on the floor or furniture. If firefighters can see the floor, they may be able to move about using a crouched or "duck" walk (Figure 3.209). This method will be slightly faster than crawling but is more dangerous unless the firefighters can clearly see the floor in front of them.

Figure 3.210 Whenever possible, a guideline should be used to assist rescuers who are making a search.

When entering an area of obscured visibility, firefighters must always operate in teams of at least two, and they should always have some sort of guideline that will lead them back to the point of entrance if necessary. The guideline may be a hoseline, rope, or electrical cord (Figure 3.210). In the event it becomes necessary to evacuate the structure in a hurry, the firefighters can turn around and follow the guideline to safety. If for some reason the team does not have a guideline, or becomes separated from it, they should proceed to a wall and follow it until a door or window is found.

ENTERING AREAS WITH RESTRICTED OPENINGS

Occasionally, firefighters will need to enter an opening that is too small to allow them to pass through in a normal manner while wearing SCBA. These situations include making entry into attics, storage tanks, manholes, silos, and other confined spaces. In these instances, airline breathing equipment will provide the optimum maneuverability (Figure 3.211).

If an airline is not available, it will be necessary to enter the confined space with a standard SCBA cylinder and backpack. The opening of many confined spaces will

Figure 3.211 Manholes are tough to enter while wearing SCBA.

not permit the entry of a rescuer wearing a backpack; therefore, it will be necessary to remove the backpack, enter the confined space, and redon the backpack (Figure 3.212). The following procedures can be used for making a vertical entry with SCBA:

Step 1: Don an approved Class III full-body harness (Figure 3.213).

Step 2: Don the SCBA and test for correct facepiece seal and positive pressure (Figure 3.214).

Step 3: Remove the backpack and attach it to the body harness by one of the D-rings or other hardware, using an approved knot or prepared lanyard. This procedure allows the backpack to be lowered at the same time as the rescuer and helps prevent accidental removal of the facepiece.

Step 4: Attach the rescue rope to the body harness D-ring (Figure 3.215).

Step 5: Enter the confined space, maintaining contact with the SCBA backpack (Figure 3.216).

Step 6: Once inside the confined space, redon the backpack if possible.

Step 7: If necessary, adjust the facepiece taking care not to compromise the seal.

Step 8: At all times, maintain an awareness of the approximate length of time needed to remove a victim from the confined space, as well as the point-of-no-return figure of the rescuer.

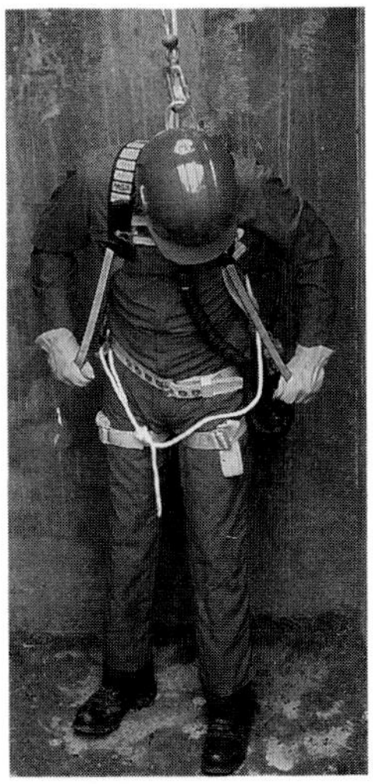

Figure 3.212 Sometimes it is necessary to enter the space without the cylinder on the wearer's back and then redon it once inside.

Figure 3.213 A rescuer dons an approved Class III, full-body harness before entering a confined space.

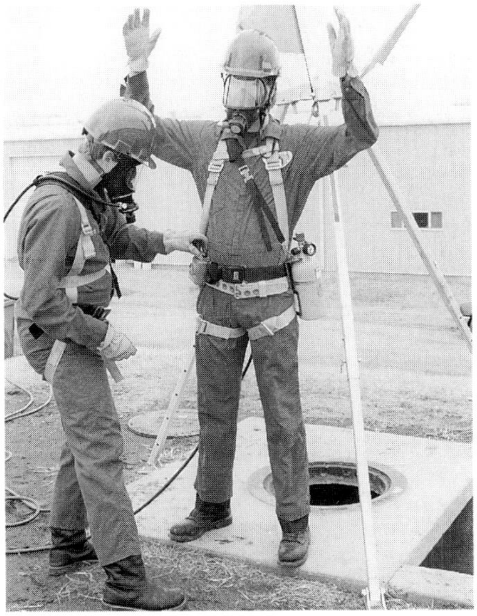

Figure 3.214 The rescuer's partner inspects the rescuer's airline unit to ensure that it has been properly donned.

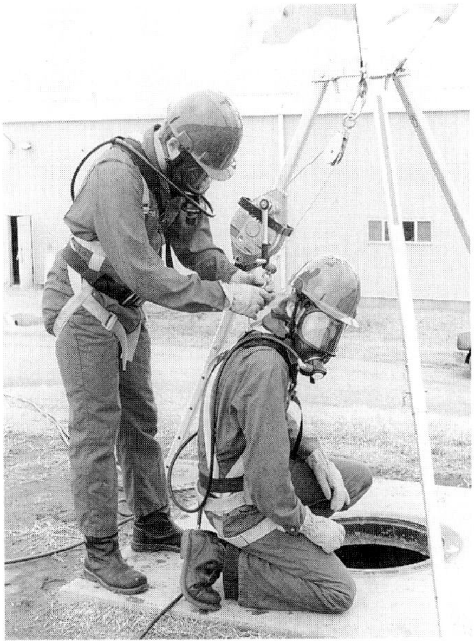

Figure 3.215 The partner attaches a rescue rope to the rescuer's body harness D-ring.

Figure 3.216 A rescuer, properly outfitted, enters the confined space.

For a horizontal entry with SCBA, the following procedures can be used:

Step 1: Don an approved Class III, full-body harness.

Step 2: Don the SCBA and test for correct facepiece seal and positive pressure (Figure 3.217).

Step 3: Remove the backpack and attach a short tag line or lanyard between the backpack and the full-body harness. The tag line should be shorter than the breathing tube or regulator hose. Attaching a tag line prevents the apparatus from falling away and pulling the facepiece from the rescuer's face (Figure 3.218).

Step 4: Attach the rescue line to the body harness D-ring, and enter the space while a team member or attendant maintains control of the backpack.

Step 5: After entering the confined space, have the team member outside pass the backpack assembly through the opening to the rescuer (Figure 3.219.)

Step 6: If applicable, redon the backpack, taking care not to get tangled in the rescue rope.

Step 7: Adjust the facepiece if necessary.

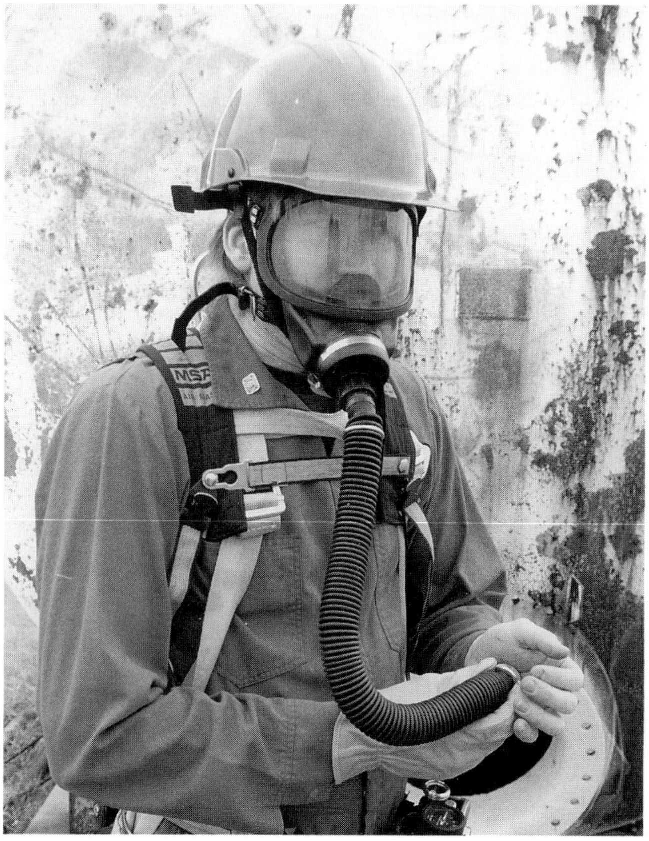

Figure 3.217 The rescuer dons the SCBA and checks for proper facepiece seal and positive pressure.

Step 8: At all times, maintain an awareness of the length of time needed to remove a victim from the confined space, as well as the point-of-no-return figure of the rescuer.

NOTE: In fire fighting situations, the use of a life safety harness and tag line may not be practical. In this case a rope or hoseline may serve in place of the tag line.

Figure 3.218 A rescuer enters the confined space while an attendant maintains control of the SCBA backpack. Note that the rescue rope is attached to the body harness D-ring and the SCBA backpack is attached to a short tag line.

Figure 3.219 After the rescuer enters the confined space, the partner passes the SCBA backpack to the rescuer for redonning.

PERSONAL ALERT SAFETY SYSTEMS

A downed or disoriented firefighter inside a structure poses a severe rescue problem. Personal alert safety system (PASS) devices are designed to assist rescuers attempting to locate the firefighter, even in dense smoke. (The acronym PAD [personal alert device] is also used). The device, about the size of a portable transistor radio, is worn on the firefighter's self-contained breathing apparatus or coat, and a switch is turned on before entering a structure (Figure 3.220). If the firefighter should collapse or remain motionless for approximately 30 seconds, the PASS device will emit a loud, pulsating shriek. It can also be activated manually. Either way, rescuers can follow the sound to locate the lost or downed firefighter. The use of PASS devices by all firefighters and rescuers is mandatory under NFPA 1500.

PASS devices can save lives, but they must be used and maintained properly. The user must remember to turn on and test the device before entering a structure (Figure 3.221). Training classes should be conducted on techniques to be used when attempting rescue of a lost

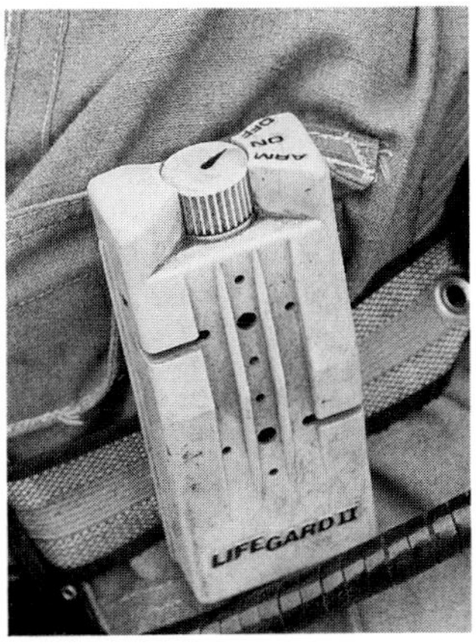

Figure 3.220 A PASS device worn on the firefighter's coat.

Figure 3.221 The PASS device must be turned on before entering a structure.

firefighter. Tests performed by the Mesa (Arizona) Fire Department showed that locating even the loud shriek of a PASS device in poor visibility conditions can be more difficult than expected. One reason was the sound reflecting off walls, ceilings, and floors made it difficult to locate. Noise from SCBA operation and muffled hearing because of protective hoods added to the difficulty. Rescuers had a tendency to sidestep established search procedures when they thought they could tell the location of the alarm sound. The Mesa tests resulted in several recommendations for use of PASS devices:

- Make sure that the system selected meets the requirements of NFPA 1982, *Standard on Personal Alert Safety Systems (PASS) for Fire Fighters*.
- Test the PASS at least weekly and maintain in accordance with manufacturer's instructions.
- Conduct practical training with the PASS under realistic conditions.
- Check the PASS device calibration during training, and return units for recalibration if necessary.
- Retrain every six months with PASS devices.
- Train firefighters to always turn on and test the device before entering a hazardous atmosphere.
- Train rescuers to listen for the distress sound by stopping in unison, controlling breathing, and lifting hood or earflaps away from ears.
- When a downed firefighter is located, turn off the PASS device or communications will be impossible.

4

Ropes and Knots

This chapter provides information that addresses performance objectives described in NFPA 1001, *Standard for Fire Fighter Professional Qualifications* (1992), particularly those referenced in the following sections:

3-10 Ropes

3-10.1

3-10.2

3-10.3

3-10.4

3-10.5

3-10.6

4-10 Ropes

4-10.1

4-10.2

Chapter 4
Ropes And Knots

Rope is one of the oldest tools used by the fire service. Rope is very valuable for applications such as hauling tools, accomplishing rescues from areas of different elevations, stabilizing vehicles, and cordoning off areas. Firefighters must be knowledgeable of the different types of rope so that the correct one will be chosen to do the required job. The ability to tie proper knots is crucial to the safety of rope maneuvers. The knots discussed in this chapter are limited to only those basic knots that NFPA 1001, *Standard for Fire Fighter Professional Qualifications*, requires firefighters to know. Local policies may require firefighters to know additional knots or to use different methods than are shown in this chapter — this knowledge is not discouraged. However, any knots that deviate from the standard should be thoroughly tested under controlled conditions before use in life safety applications.

One section of this chapter deals with the proper care, inspection, and storage of fire service ropes, and the last section deals with hoisting tools and equipment. For expanded information on ropes and knots, consult IFSTA's **Fire Service Rescue Practices** manual.

ROPE MATERIALS

The materials used to construct fire service rope can be divided into two basic categories: natural fibers and synthetic fibers. Each has its own advantages and disadvantages. This section describes the features of the major natural and synthetic fibers used for rope construction.

Natural Fibers

For many years natural fiber rope was the primary type of rope used for rescue. However, after much testing and evaluation, natural fiber rope is no longer accepted for use in life safety applications. It is acceptable to use natural fiber rope for utility purposes; however, it must not be used for specific rescue purposes. The following sections describe the most common fibers used for natural fiber rope.

MANILA

The manila fiber, grown in Manila in the Philippines, is a strong, hard fiber that comes from the leaf stems of the abaca plant's trunk. The oily-waxy feeling of manila rope comes from special preserving and lubricating oils added during manufacturing.

There are several types of manila rope available; however, the best type to purchase is Type #1. Type #1 is made from the innermost part of the plant and is generally identified with a colored string twisted into the fibers (Figure 4.1). Type #1 manila rope is comparable in price to nylon, but its tensile strength is much less. This rope is biodegradable — it is subject to normal rot and decay from various environmental sources.

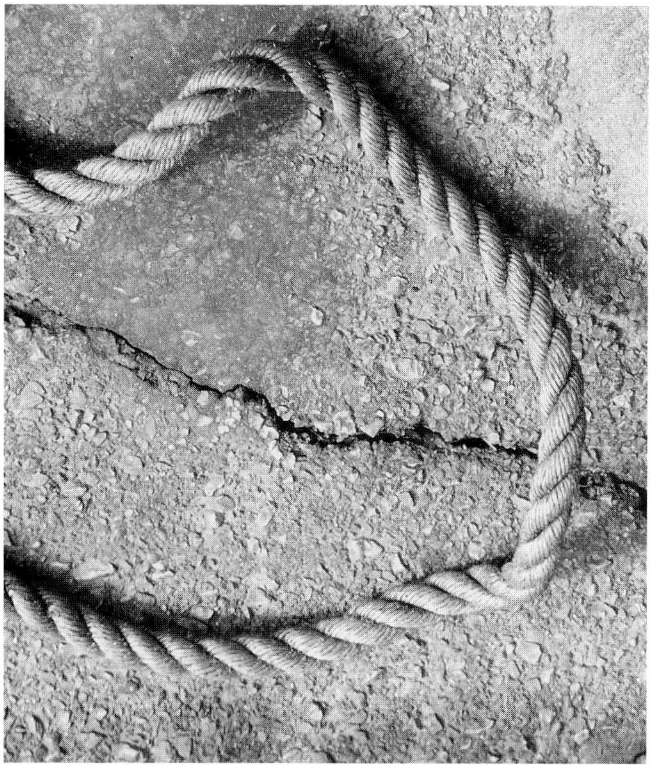

Figure 4.1 Note the lack of a colored strand on this manila rope. This is not Type 1 manila rope.

Manila rope is susceptible to the following types of damage:

- Deterioration — Manila is a natural fiber that deteriorates rapidly over the years. Its tensile strength is reduced proportionately. Manila rope should be considered "used" if it is six months old and has not been employed.
- Water — Once manila rope gets wet, it loses one-half of its tensile strength. Even though it has been a common practice, DO NOT soak manila before use. This practice dramatically decreases the strength of the line.
- Humidity — Rope stored in a humid atmosphere loses one-half its strength in one year.
- Abrasion and chemicals — Manila rope is severely affected by chemicals and abrasion. Once it has been exposed to these hazards, it should be removed from service.
- Charring — Manila rope chars at 380°F (190°C) and loses strength at 180°F (82°C).

SISAL

Sisal fiber is the most common substitute for manila. Sisal is a hard fiber with about three-fourths the tensile strength of manila. Its most common use is in binder's twine, but it is sometimes used in larger ropes.

COTTON

Cotton fiber is used when a soft, pliable rope is needed. Cotton's tensile strength is slightly less than that of sisal and considerably less than that of manila. Cotton rope is the most susceptible to physical abrasion and damage. If still in regular use, it should be given careful examination for problems similar to those listed for manila.

Synthetic Fibers

The use of synthetic fiber rope is common in the fire and rescue service. Advances in synthetic rope construction have made the use of synthetic rope preferable to natural fiber rope, especially in life safety applications (Figure 4.2). Synthetic fiber rope has excellent resistance to mildew and rotting, excellent strength, and easy maintenance. Unlike natural fiber rope, which is made of short overlapping strands of fiber, the synthetic rope may feature continuous fibers running the entire length of the rope.

Figure 4.2 It is important that the proper rope be used for life safety applications. *Courtesy of Ron Jeffers.*

NYLON

Nylon is one of the best materials used for ropes. Nylon has a high resistance to abrasion, high tensile strength, and basic properties resistant to moisture and most chemicals. However, acids and ultraviolet rays will harm nylon after repeated or concentrated exposure.

Nylon is one of the strongest synthetic fiber ropes and has about three to three and one-half times the tensile strength of manila rope. The high tensile strength of nylon permits the use of smaller rope to obtain the equivalent strength of larger rope made from different materials. The advantage of using smaller ropes is that they are easy to handle and require less space for storage. Nylon rope is also lightweight.

Nylon resists wear and abrasion and works comparatively well when wet. When wet, nylon rope maintains approximately 75 to 80 percent of its strength. Nylon ropes have melting points around 400° to 500°F (204°C to 260°C) but will begin losing strength and integrity at around 300°F (149°C). Nylon rope cannot be easily formed into solid knots and hitches, and because it stretches under load, is not suitable for vehicle stabilization or similar applications.

POLYPROPYLENE

Polypropylene rope is one of the most lightweight ropes available. This rope's resistance to water damage and its ability to float make it very popular in rescue incidents around water (Figure 4.3).

Polypropylene rope has excellent resistance to rotting, mildew, and abrasion. It has moderate elastic principles and about 60 percent of the energy absorption capacity of nylon but maintains a relatively low breaking strength.

Polypropylene is quickly affected by heat and should not be exposed to any source of heat. Prolonged exposure to direct sunlight will cause polypropylene to deteriorate. (This is particularly true at high altitudes.) Polypropylene begins to lose strength at around 200°F (93°C) and will begin to melt at around 285° to 300°F (140°C to 149°C). Polypropylene rope is difficult to secure into good knots and hitches.

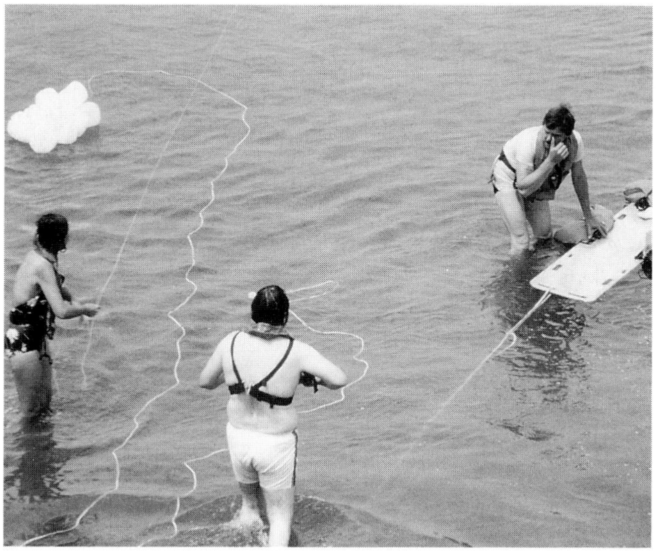

Figure 4.3 Polypropylene is excellent for water incidents because of its ability to float. *Courtesy of Joel Woods, University of Maryland Fire & Rescue Institute.*

POLYETHYLENE

Polyethylene fiber is made from the same synthetic fiber family as polypropylene. Polyethylene is similar to polypropylene in weight, strength, elasticity, and chemical and abrasion resistance. At 230°F (110°C) it shows some degradation from the heat, with its melting point at 285°F (140°C). Polyethylene rope has no moisture absorption and floats indefinitely. The surface of polyethylene rope feels slick and oily and can be manufactured in bright colors for greater visibility (Figure 4.4). Polyethylene is not easily formed into knots and hitches because of its tendency to roll or to give under tension, and it has a relatively low breaking strength. Both polyethylene and polypropylene have relatively high rates of deterioration from sunlight.

Figure 4.4 Polyethylene rope does not form into knots and hitches as easily as some other types of rope.

POLYESTER

Polyester rope is used where a high strength, low-stretch rope is necessary (Figure 4.5). Polyester is NOT subject to damage from water, sunlight, most chemicals, or moderately high temperatures. Polyester can be damaged by 80+ percent acid solutions and 10+ percent basic solutions. Polyester is also fairly resistant to damage caused by flexing or abrasion. Polyester rope begins to lose strength at 300°F (149°C) but does not begin to melt until the temperature reaches 450° to 650°F (232°C to 343°C).

KEVLAR® ARAMID FIBER

Kevlar® has high tensile strength and heat resistance; it will withstand 500°F (260°C) before losing strength. (Kevlar is used in the production of bulletproof vests.) Because Kevlar can be damaged by abrasion, it must be sheathed in another material, such as nylon or polyester, to protect it from abrasion during its use (Figure 4.6). Improvement is needed in Kevlar's shock-absorbing capabilities; therefore, the use of Kevlar rope

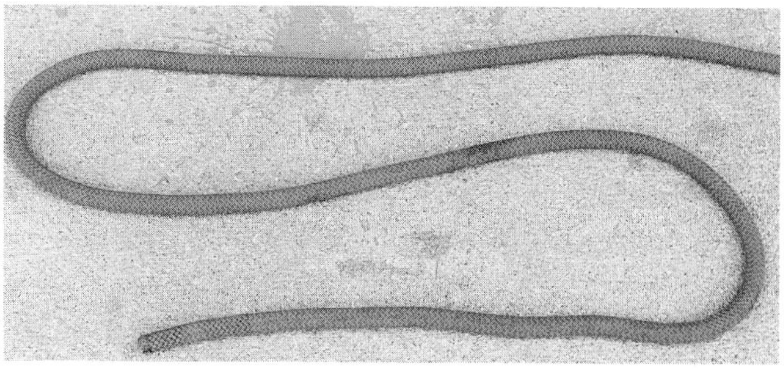

Figure 4.5 Polyester rope.

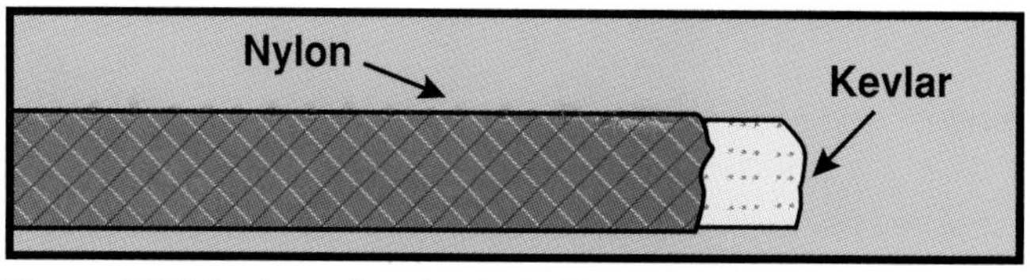

Figure 4.6 Nylon is used to sheath the Kevlar core.

for rescue purposes is limited. This rope is very hard to tie in knots.

ROPE CONSTRUCTION

Ropes fall into one of two categories: static or dynamic. Static rope stretches very little — 1½ to 2 percent — under normal loads. Dynamic lines stretch more than static lines both under weight and shock loads. Because static line has a low stretch factor, it is most often preferred for rescue work. Dynamic rope may be preferred under certain circumstances such as to arrest a fall or absorb the weight of a fall. Both static and dynamic ropes may be constructed in a variety of ways. The most common types of rope construction are laid, braided, braid-on-braid, and kernmantle.

Laid (Twisted) Natural Or Synthetic Rope

All natural fiber ropes and some synthetic fiber ropes are of laid (also called twisted) construction. Hard-laid rope, constructed with only synthetic materials, is twisted tightly to form a rope that is stiff and resists abrasion. (Hard-laid rope was commonly used for mountaineering.) The disadvantage of this type of rope is that it is difficult to form into certain knots and hitches, and it must be tied off with a safety knot. Soft-laid rope is not twisted as tightly and is softer, easily tied, and somewhat stronger.

Laid ropes are constructed by twisting together yarns to form strands (Figure 4.7). Generally, three strands are twisted together to make the final rope. How tightly these ropes are twisted will determine the rope's properties. Twisted rope is susceptible to abrasion and other types of physical damage. Twisting a rope leaves all three load-bearing strands exposed at various points along the rope. Although this exposure allows for easy inspection, it also means that any damage will immediately affect the rope's strength.

Figure 4.7 Laid rope is formed by twisting smaller sections together.

Braided Rope

Although some braided ropes are made from natural fibers, most are of the synthetic variety. Braided rope is constructed by uniformly intertwining strands of rope together (similar to braiding a person's hair) (Figure 4.8). This type of construction is without any type of outer sheath or core. Braided rope reduces or eliminates the twisting common to laid ropes. Because of its construction characteristics, the load-bearing fibers are subject to direct abrasion and damage.

Braid-On-Braid Rope

Because braid-on-braid is a jacketed rope, it is often confused with kernmantle rope. Braid-on-braid rope is just what the name implies: it is constructed with both a braided core and a braided sheath (Figure 4.9). However, this rope does remain a static-type rope. The appearance of the sheath is that of a herringbone pattern.

Braid-on-braid rope is very strong. Half of its strength is in the sheath and the other half of its strength is in the core. A disadvantage of braid-on-braid rope is that it does not resist abrasion as well as the kernmantle rope. It also has a problem with the outer sheath sliding along the inner core.

Kernmantle Rope

Kernmantle, a jacketed rope, is composed of a braided covering or sheath (mantle) over the main load-bearing

Figure 4.8 An example of braided rope.

Figure 4.9 Braid-on-braid rope has a herringbone pattern.

strands (kern) (Figure 4.10). The strands may be twisted or braided together. The rope core runs parallel with the rope covering (Figure 4.11). This increases the rope's stretch resistance and load characteristics. The core (kern) is made of high-strength fibers; these account for about three-fourths of the total strength of the rope with the sheath picking up the balance. With this type of construction, the sheath absorbs most of the abrasion and protects the load-bearing core. Kernmantle rope comes in both dynamic and static types. Dynamic lines stretch more than static lines do, both under weight and shock loads. Dynamic kernmantle is most commonly used as a sport rope for rock or ice climbing. Static kernmantle rope is most commonly used as rescue rope.

Kernmantle has low elastic properties under small loads such as body weight. These elongation figures average somewhere around 1 to 1.5 percent for static kernmantle and 8 to 12 percent for dynamic kernmantle. NFPA 1983, *Standard on Fire Service Life Safety Rope, Harness, and Hardware*, states that the rope breaking elongation should be not less than 15 percent and not more than 55 percent.

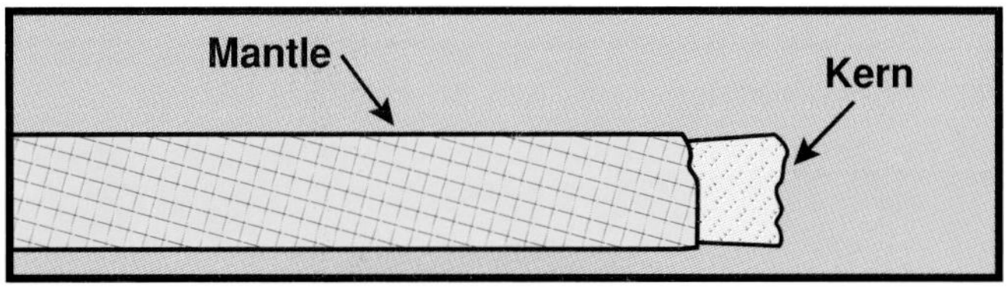

Figure 4.10 The mantle covers the kern.

Figure 4.11 Kernmantle rope is the most common type used for rappelling operations. *Courtesy of Joel Woods, University of Maryland Fire & Rescue Institute.*

LIFE SAFETY AND UTILITY ROPE

Fire service rope falls into two use classifications: life safety rope and utility rope. Life safety rope is used to support rescuers and/or victims (Figure 4.12). Because these situations demand a high degree of safety, the rope used must conform to the standards set forth by NFPA 1983. Life safety rope is defined by the NFPA as "rope dedicated solely for the purpose of constructing lines for supporting people during rescue, fire fighting, or other emergency operations, or during training evolutions." Only rope constructed of continuous filament fiber is suitable for life safety requirements. Rope made of any other materials should not be used in life safety applications. NFPA 1983 requires life safety rope to be used only once and then taken out of service. It may be reused for training purposes only.

Figure 4.12 Rappelling is a life safety application. *Courtesy of Harvey Eisner.*

Utility rope is used in any instance, excluding life safety applications, where the use of a rope is required. Utility rope can be used to hoist equipment, secure unstable objects, or cordon off an area (Figure 4.13). There are no standards set forth for utility rope; however, common sense should prevail in its use. Regularly inspect utility rope to see if it is damaged.

Figure 4.13 Utility rope is often used to stabilize objects.

KNOTS

The ability to tie knots is a vital part of fire and rescue operations. Improperly tied knots can be extremely hazardous to both rescuers and victims. Good knots should be easy to tie, easy to identify, easy to untie, and strong enough to do the required job.

The methods of selecting and tying knots have changed with the advent of synthetic material rope. Knots considered safe and acceptable for many years may no longer be considered safe in some cases. Personnel who were educated in rope work with manila or other natural fiber ropes must reeducate themselves in the proper procedures and knots used with the newer ropes.

The newer synthetic rescue rope has a much smoother, slicker outside surface. The rope is much more likely to slide on itself than natural fibers will. Thus, many older, traditional knots can slip under load. In order to prevent slipping, a single-overhand, double-overhand, half hitch, or double half hitch should be applied to the tail of the working end of the rope.

Elements Of A Knot

Knots weaken a rope because the rope is bent in order to form the knot. The fibers on the outside of the bend are stretched, and the fibers on the inside of the bend are crushed. A knot with sharp bends weakens a rope more than a knot with easy bends. The bends that a rope undergoes in the formation of a knot or hitch are known as the bight, loop, and round turn. Each of these formations is shown in the following figures.

- The bight is formed by simply bending the rope back on itself while keeping the sides parallel (Figure 4.14).
- The loop is made by crossing the side of a bight over the standing part (Figure 4.15).
- The round turn consists of further bending one side of a loop (Figure 4.16).

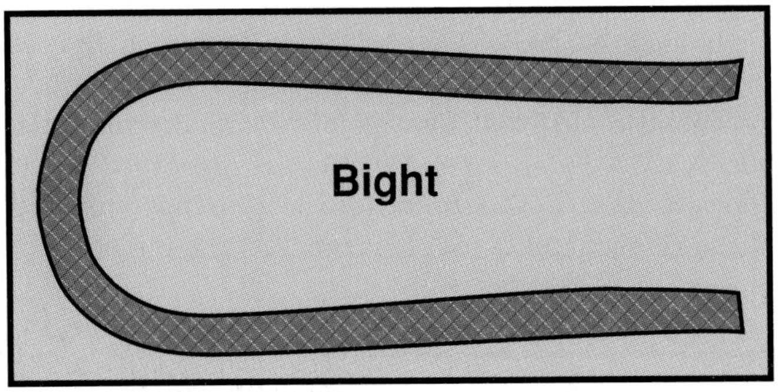

Figure 4.14 A bight.

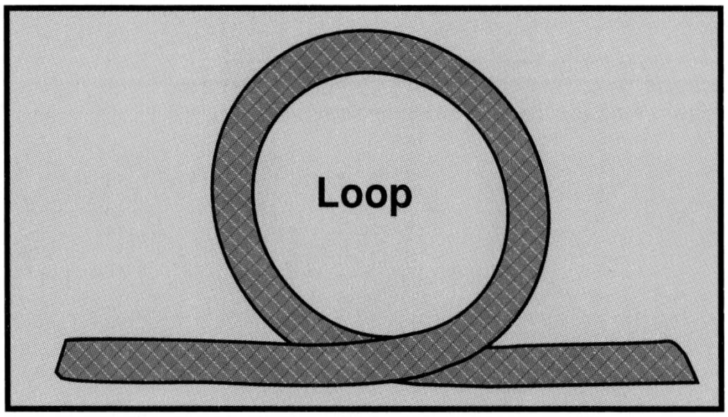

Figure 4.15 A loop.

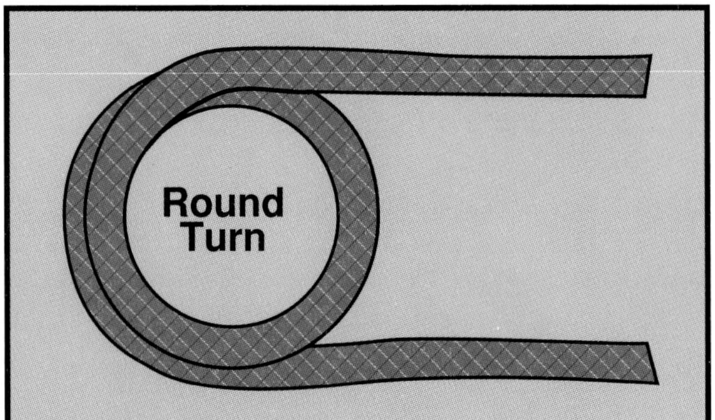

Figure 4.16 A round turn.

Knots and hitches are formed by combining these elements in different ways so that the tight part of the rope bears on the free end to hold it in place. Knots and hitches for fire service use should be those that may be rapidly tied, can be easily untied, are not subject to slippage, and have a minimum of abrupt bends.

Throughout the following descriptions of how to tie knots, the terms *running part*, *working end*, and *standing part* are used (Figure 4.17). The running part is the part of the rope that is to be used for work such as hoisting, pulling, or belaying. The working end is the part of the rope that is to be used in forming the knot (commonly referred to as the "loose end" or "bitter end"). The standing part is that part between the working end and the running part.

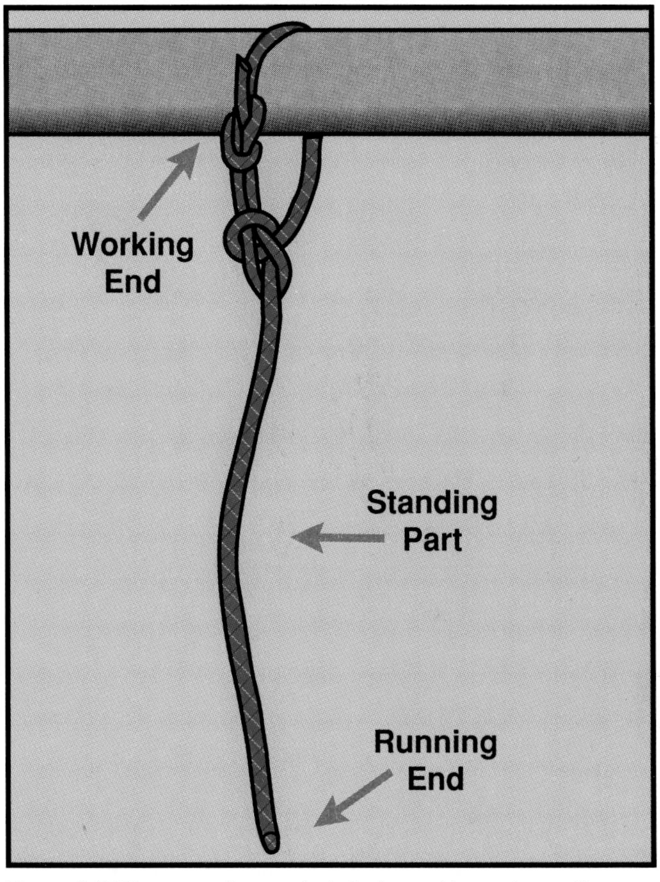

Figure 4.17 The parts of a rope include the working end, standing part, and running part.

The Bowline Knot

The bowline may be easily untied and is a good knot for forming a loop that will not constrict the object it is placed around. Its use in the fire service is extensive, and all firefighters should be able to tie the bowline in the open as well as around an object. (**NOTE:** The bowline should be used only on natural fiber rope.) The bowline is not a secure knot on synthetic fiber rope; therefore, it cannot be used in life safety situations. The following method is one good way of tying the bowline, although other methods may be just as effective.

Step 1: Measure off sufficient rope to form the size of the knot desired, and form an overhand loop in the standing part (Figure 4.18).

Step 2: Pass the working end upward through the loop (Figure 4.19).

Step 3: Pass the working end over the top of the loop under the standing part, and bring the working end completely around the standing part and down through the loop (Figure 4.20).

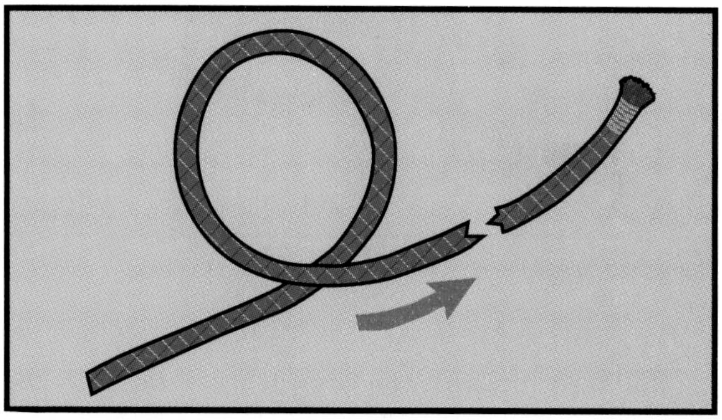

Figure 4.18 First make a loop in the rope.

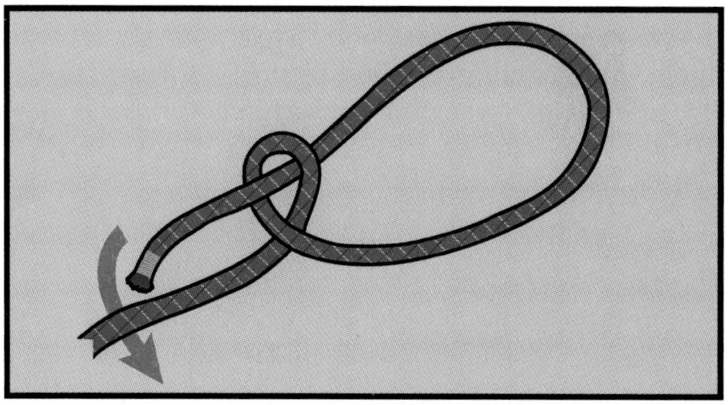

Figure 4.19 Pull the working end through the loop.

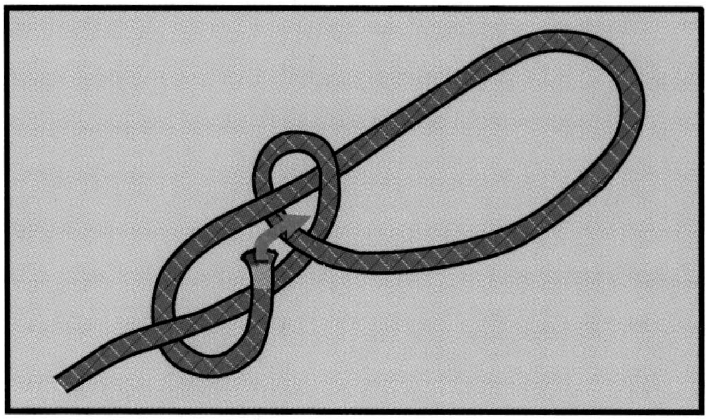

Figure 4.20 Pass the working end around the standing part and pass it back through the loop.

Step 4: Pull the knot snugly into place, forming an *inside* bowline with the working end on the inside of the loop (Figure 4.21).

NOTE: The bowline may be tied with the working end outside the loop. This is known as an outside bowline. The outside bowline is as strong as the inside bowline; however, the preferred method is the inside bowline because of the safety feature of having the working end locked against the object to which it is tied.

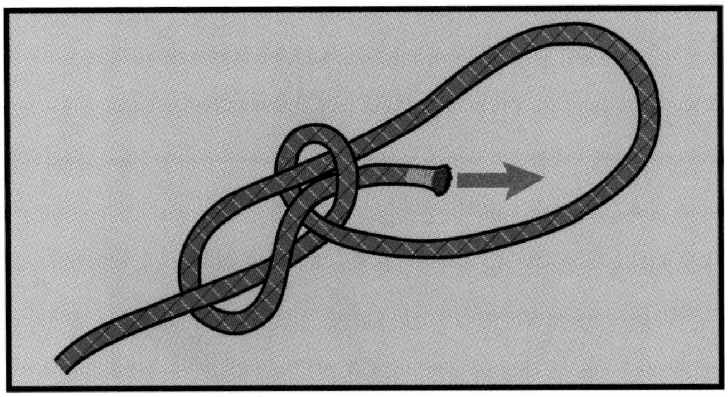

Figure 4.21 Dress the knot when it is complete.

The Clove Hitch

The clove hitch may be formed by several methods. It consists essentially of two half hitches. Its principal use is to attach a rope to an object such as a pole, post, or hose. The clove hitch may be formed anywhere in the rope from one end to the middle. When properly applied, it will stand a pull in either direction without slipping. One method for developing a clove hitch in the open is as follows:

Step 1: Form a loop in your left hand with the working end to the right crossing under the standing part (Figure 4.22).

Step 2: Form another loop in your right hand with the working end crossing under the standing part (Figure 4.23).

Step 3: Slide the right-hand loop on top of the left-hand loop (Figure 4.24). (**NOTE:** This is the important step in forming the clove hitch.)

Step 4: Hold the two loops together at the rope, and thus form the clove hitch (Figure 4.25). Pull the ends in opposite directions to tighten.

The clove hitch, when formed by the method just described, cannot be placed over an object that has no free end such as the center of a hoseline. Therefore, it is necessary to know how to tie the clove hitch around an object. This method is as follows:

Step 1: Make one complete loop around the object, bringing the working end below the standing part (Figure 4.26).

Figure 4.22 Step 1: Form a loop in your left hand with the working end to the right crossing under the standing part.

Figure 4.23 Step 2: Form another loop in your right hand with the working end crossing under the standing part.

Figure 4.25 Step 4: Hold the two loops together at the rope, and thus form the clove hitch. Pull the ends in opposite directions to tighten.

Figure 4.24 Step 3: Slide the right-hand loop on top of the left-hand loop.

Step 2: Cross the working end over the standing part, and complete the *round turn* about the object just above the first loop as shown in (Figure 4.27).

Step 3: Pass the working end under the upper wrap, just above the cross, and properly set the hitch by pulling (Figure 4.28).

To ensure that the clove hitch does not loosen during use, a safety hitch should be applied. To apply a safety hitch, tie a half hitch or overhand knot around the standing part of the rope with the working end (Figure 4.29).

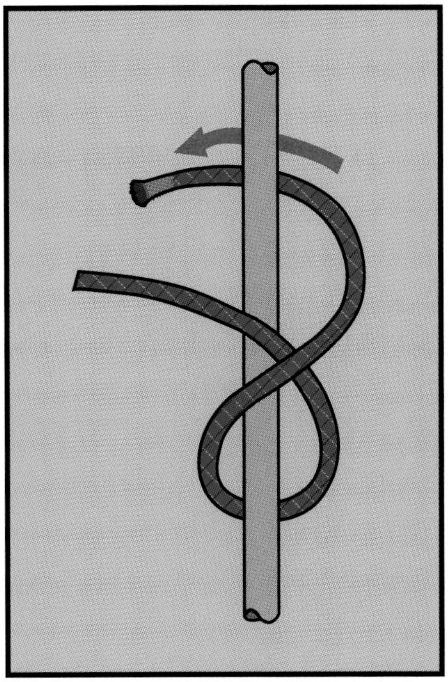

Figure 4.26 Make a loop around the object.

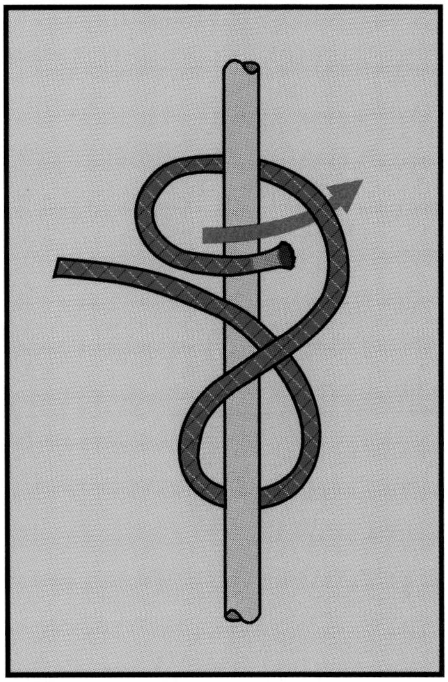

Figure 4.27 Make a second loop above the first one.

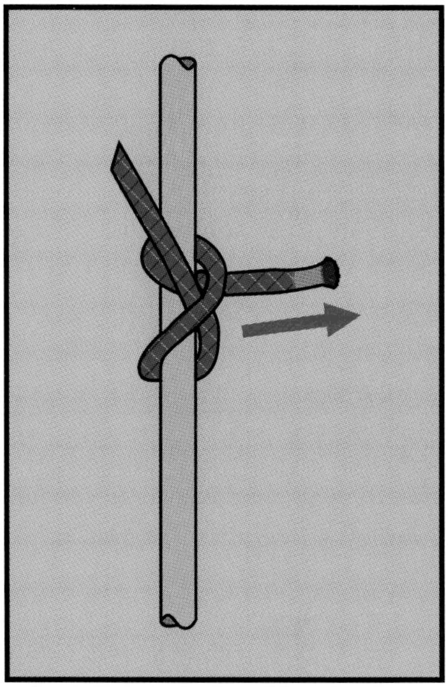

Figure 4.28 The working end and running part should be next to each other.

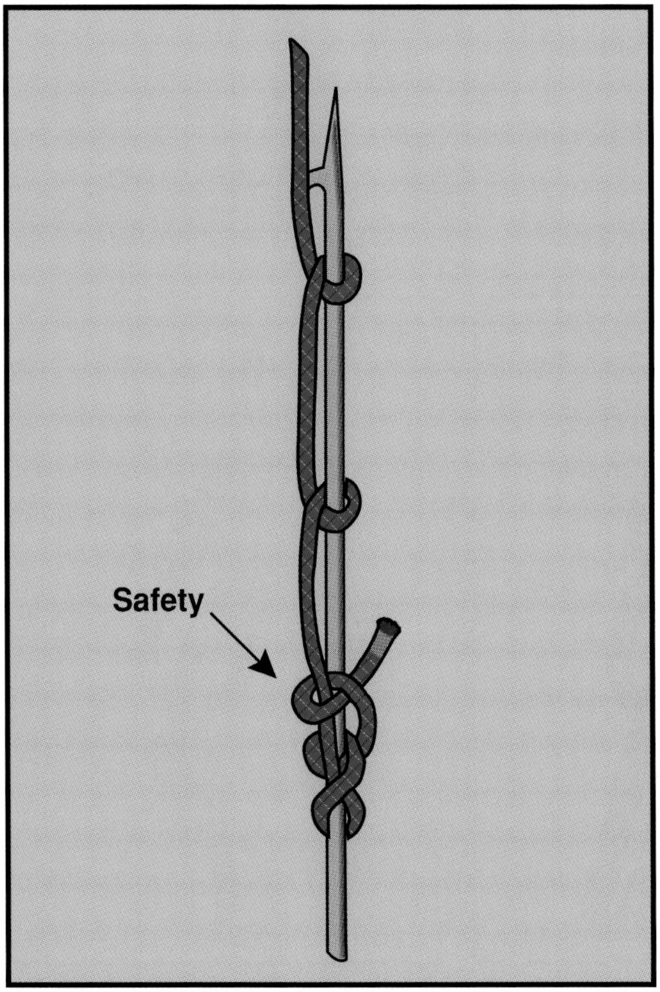

Figure 4.29 Use a safety knot to prevent the main knot from failing.

The Half Hitch

The half hitch is particularly useful in stabilizing tall objects that are being hoisted. The half hitch is always used in conjunction with another knot. For example, when hoisting a pick-head axe, a half hitch is used around the handle; a clove hitch, timber hitch, or girth hitch is tied around the axe head. The half hitch is formed by making a round turn around the object. The standing part of the rope is passed under the round turn on the side opposite the intended direction of pull (Figure 4.30). Several half hitches can be applied in succession if required.

Figure 4.30 A half hitch.

The Figure Of Eight Family Of Knots

Since the introduction of synthetic rope to the fire and rescue service, the figure of eight knot has basically replaced the bowline. Figure of eights are tighter and stronger knots than the bowline. They are also not as apt to damage the synthetic rope as the bowline will. There are several variations of the figure of eight that are commonly used in the rescue service.

DOUBLE FIGURE OF EIGHT KNOT (FIGURE OF EIGHT FOLLOW THROUGH)

This knot is used to tie ropes of equal diameters together. A safety knot, such as the double overhand, should be used in conjunction with this knot. The procedure for tying the double figure of eight knot is as follows:

Step 1: A figure of eight knot is tied on one end of the rope (Figure 4.31).

Step 2: The end of the other rope is fed through the figure of eight knot in reverse. It should follow (hence the name) the exact path of the original knot (Figure 4.32).

Step 3: A safety knot, such as the double overhand, should be used in conjunction with this knot (Figure 4.33).

FIGURE OF EIGHT ON A BIGHT (GUIDE KNOT)

This knot is preferred as the replacement for the bowline when using synthetic rope. It can be tied in the middle of the rope, or if a loop is needed in the end, it can be tied at the end using a bight formed near the end of the rope (Figure 4.34). This variation is used as both an anchoring attachment and as a harness tie-in. The procedure for tying this knot is as follows:

Figure 4.34 A figure of eight on a bight can be tied in the center of the rope.

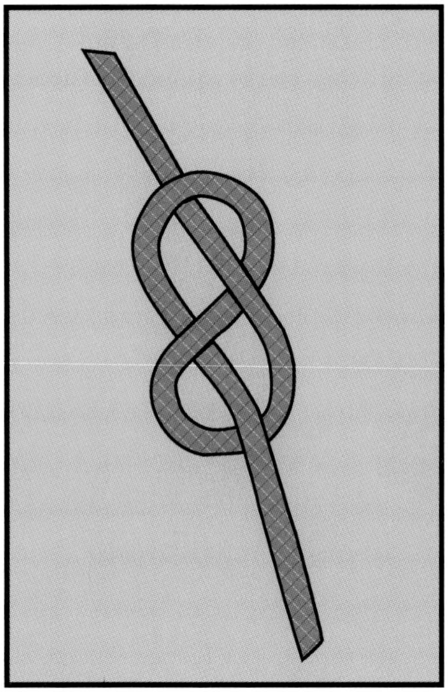

Figure 4.31 Trace the route of the first rope with the second rope.

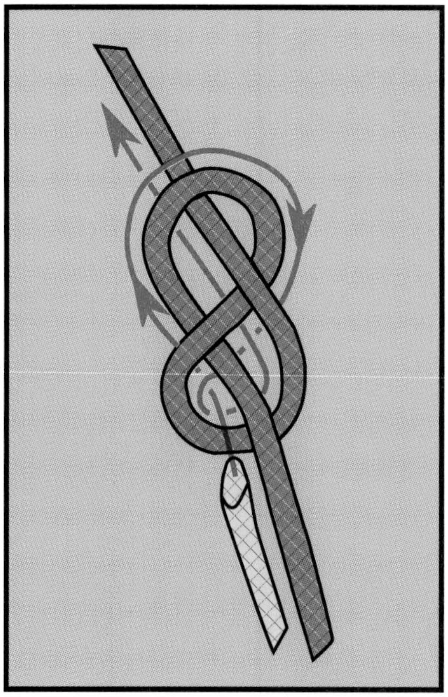

Figure 4.32 Dress the knot when it is complete.

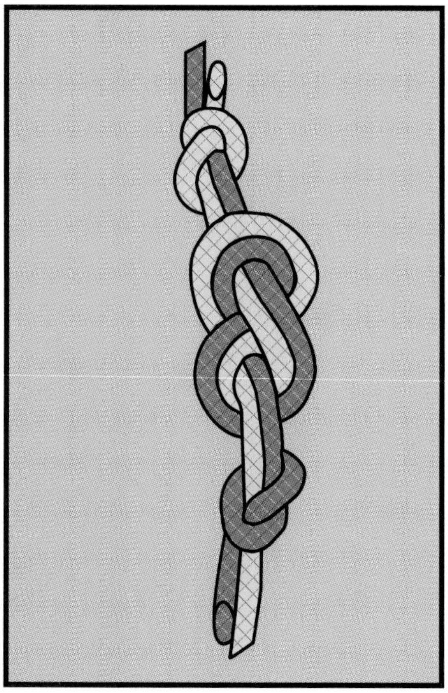

Figure 4.33 Use an overhand safety on both ends to finish the knot.

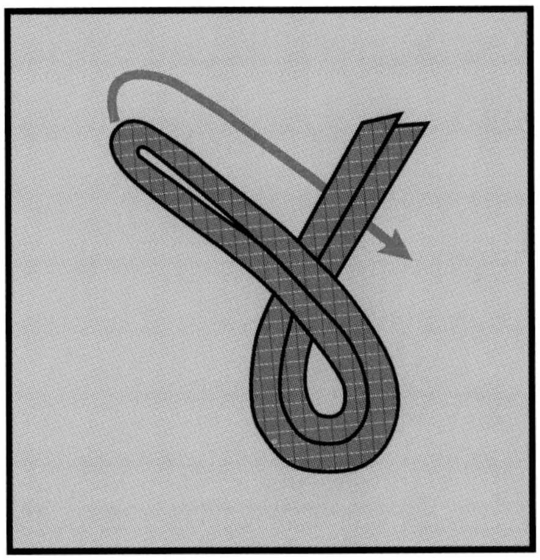

Figure 4.35 Make a loop with the doubled end of the rope.

Figure 4.36 Pass the working end over the loop and through the hole.

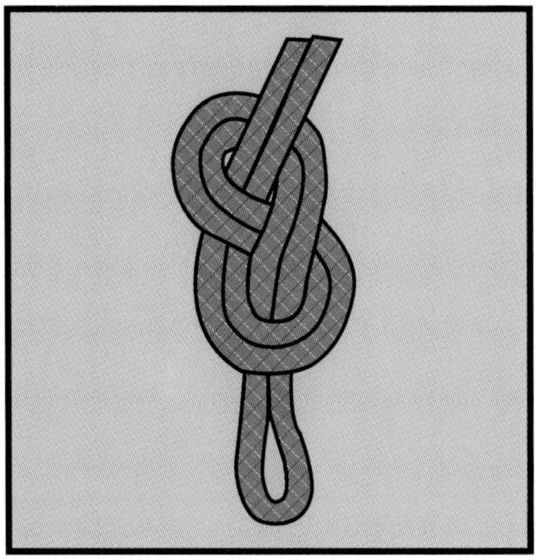

Figure 4.37 The completed knot.

Step 1: Form a bight in the working end and pass it over the standing part to form a loop (Figure 4.35).

Step 2: Pass the bight under the standing part, then over the loop and down through it (Figure 4.36).

Step 3: Having formed the figure of eight, extend the bight through the knot to whatever size working loop is needed and dress the knot (Figure 4.37).

The Becket Or Sheet Bend

The becket or sheet bend is used for joining two ropes and is particularly well suited for joining ropes of unequal diameters or joining a rope and a chain. It is also unlikely to slip when the rope is wet. These advantages make it useful and dependable in fire service rope work. The becket bend is tied as follows:

Step 1: Form a bight in one of the ends to be tied (if two ropes of unequal diameter are being tied, the bight always goes in the larger of the two), and pass the end of the second rope through the bight (Figure 4.38).

Step 2: Bring the loose end around both parts of the bight (Figure 4.39).

Figure 4.38 Pass the smaller rope through the bight in the larger rope.

Step 3: Tuck this end under its own standing part and over the bight (Figure 4.40).

Step 4: Draw the knot down snug (Figure 4.41).

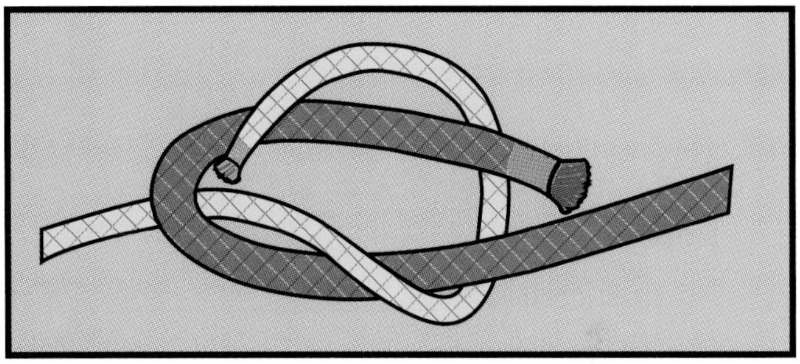

Figure 4.39 The working end of the smaller rope is passed around both sides of the bight.

Figure 4.40 Pass the working end of the smaller rope back under itself.

Figure 4.41 The completed sheet bend.

Overhand Safety Knot

As an added measure of safety, use an overhand safety knot whenever tying any type of knot. Though any properly tied knot should hold, it is always desirable to provide the highest level of safety possible. Use of the overhand safety knot eliminates the danger of the end of
4-93 the rope slipping back through the knot and causing the knot to fail. See Figure 4.42 for examples of correctly tied overhand safety knots.

Figure 4.42 An overhand safety.

Figure 4.43 Inspect the rope carefully after each use.

ROPE CARE AND MAINTENANCE

Inspecting Rope

Inspect all types of rope after each use. Inspect the rope visually and tactilely (by touch) (Figure 4.43). Make periodic inspections, perhaps monthly, during periods of inactivity. In order to better record and evaluate data on these ropes, keep a rope log detailing all information on each rope (Figure 4.44). Numbers for these ropes can be

Oklahoma State University
Fire Service Training
Rope Log

Rope Type: ____________ Rope Size: ____________ Rope #: ____________
Manufacturer: ____________ Model: ____________ Rope Color: ____________
Purchased From: ____________ Date: ____________ Bag Color: ____________

Date:	Sign-Out:	Use:	Possible Damage/Comments:	Sign-In:

Figure 4.44 An example of a rope log.

burned into the fused ends of each rope with a metal stamp. Send any rope that is found to be worn, even in the slightest way, for testing or disposal. If the rope becomes damaged on the scene, label it immediately. Remember that NFPA 1983 specifies that life safety rope only be used once during actual emergencies. It may be reused for training purposes only. When making inspections, use the following methods described for the various ropes and note any observations.

LAID ROPE

Tactilely inspect synthetic laid ropes for soft, crusty, stiff, or brittle spots; areas of excessive stretching; and cuts, nicks, and abrasions. Visually inspect rope for nicks and cuts, dirt, and other obvious flaws. Laid rope should be untwisted and checked internally for the aforementioned flaws (Figure 4.45). The presence of mildew does not necessarily indicate a problem; however, the rope should be cleaned and reinspected. A foul smell might indicate rotting or mildew in manila rope. This rope should also be untwisted and inspected internally.

Figure 4.45 Inspect laid rope by untwisting the sections.

BRAIDED ROPE

Visually inspect braided rope for exterior damage, such as heat sears, caused by friction or fire. Also, visually inspect for cuts and fuzziness. Tactilely inspect braided rope for permanent mushy spots.

BRAID-ON-BRAID ROPE

Inspect braid-on-braid rope for heat sears, nicks, and cuts. Also inspect for the sheath sliding on the core. If sliding is found, cut the end of the rope and pull off the excess skin; then the end should be seared (Figure 4.46). Inspect braid-on-braid rope for lumps which indicate core damage. Carefully examine any type of damage or

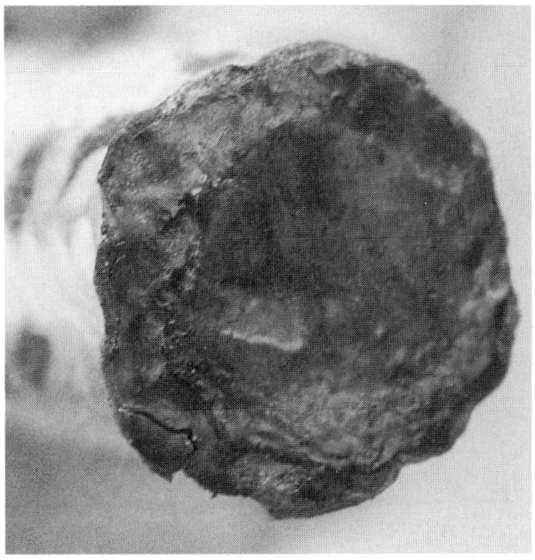

Figure 4.46 The seared end of a braid-on-braid rope.

questionable wear to the sheath; 50 percent of the rope's strength is in the sheath.

KERNMANTLE ROPE

Inspecting kernmantle rope for damage is difficult; however, more than one method can be used. One method used to check kernmantle rope is to carefully inspect the mantle while placing it into its storage bag (Figure 4.47). Another method is to place slight tension on the rope and at the same time feel for any lumps, depressions, or soft mushy spots in the rope (Figure 4.48). Sometimes a temporary soft spot may be felt as a result of hard knots or bends in the rope. In time, if undamaged, this soft spot will return to its original texture. To determine if this is a damaged spot or just normal rope, a careful, slow

Figure 4.47 One method used to inspect kernmantle rope is to check the mantle while placing it in its bag.

Figure 4.48 Another method for inspecting kernmantle rope is to place a slight tension on the rope and feel for any lumps, depressions, or soft mushy spots.

inspection must be conducted on the outer sheath. Any damage to this outer sheath indicates probable damage to the core. Also be aware of irregularities in shape or weave, foul smells, discoloration from chemical action, roughness, abrasion, or fuzziness. (Fuzziness is a natural occurrence and not necessarily a point of concern.) Seventy-five percent of this rope's strength lies in the core. Contrary to some opinions and criticisms, the kernmantle rope's core *can be* damaged without showing evidence of this damage to the outer sheath.

Cleaning Rope

This section gives general guidelines for cleaning rope. Methods for cleaning rope will vary with the manufacturer — contact for specific instructions.

NATURAL FIBERS

Natural fiber rope cannot be cleaned effectively, because water cannot be used in the cleaning process. Water weakens and damages the fiber. Wipe or gently brush the rope to remove as much of the dirt and grit as possible. With the reduction of strength in the rope after one use, serious consideration should be given before placing the section back into service.

SYNTHETIC FIBERS

Contact the manufacturer for specific instructions, because methods of washing and drying synthetic rope will vary. Cool water will have less chance of damaging the rope. Avoid any type of bleach, and make certain that only mild soap solutions are used. At times some rope may appear stiff after washing. This occurs because the finish of the rope is washed out, leaving the rope stiffer than originally made.

Regardless of the water temperature or cleaning agents used, there are three principal ways to clean synthetic fiber rope: hand launder, rope washer, and clothes washer.

- Hand launder — Wipe the rope with a cloth or scrub with a brush and thoroughly rinse (Figure 4.49).
- Special rope washer — Some rope manufacturers market a rope washer that can be connected to a standard faucet. The rope washer provides multidirectional streams of water within the unit. The rope is then fed manually through the washer, thus being cleansed on all sides at the same time (Figure 4.50).
- Regular clothes washing machine — Rope may be washed in any regular washing machine provided the machine does not mechanically damage the rope in the washing process. To protect the rope from damage, either place the rope in a cloth bag before placing in washer or coil the rope in a "bird's nest coil" and place in the washer (Figure 4.51).

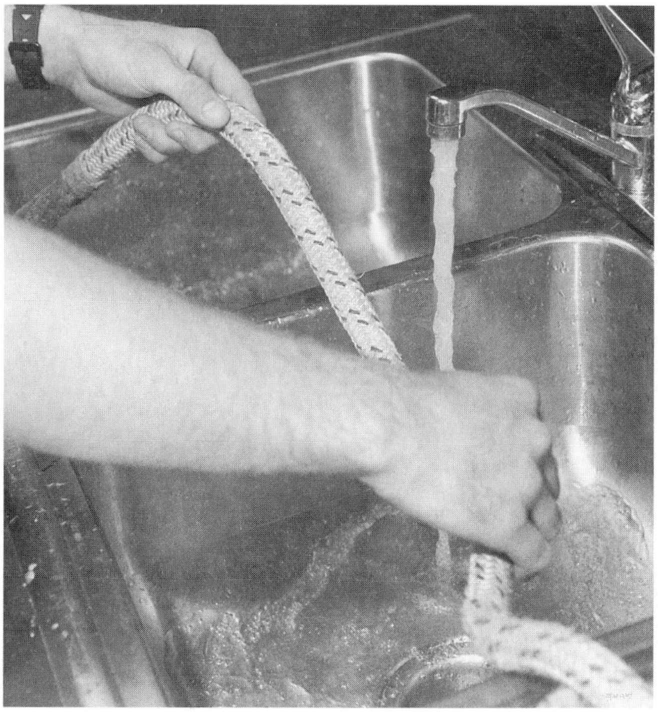

Figure 4.49 One way to clean rope is to hand launder.

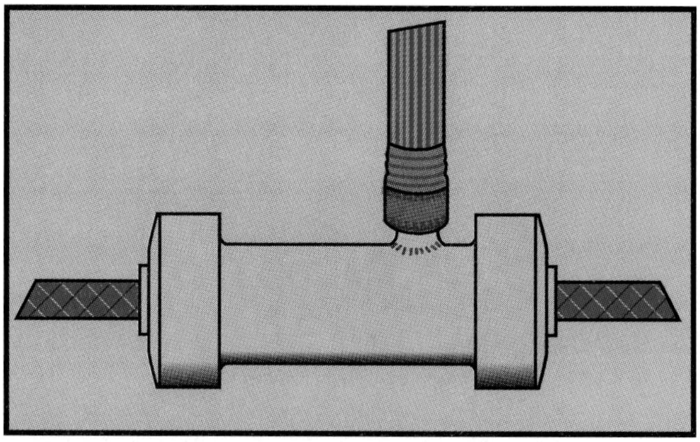

Figure 4.50 The rope is pulled through the washer to clean it.

Figure 4.51 Coil rope in this manner before placing it in a washing machine.

Once the rope has been washed, it may be dried several ways. Rope can be dried in a dryer, provided that the machine has a temperature setting low enough to prevent damage to the rope (Figure 4.52). Another method of drying is to lay the rope on a flat, clean, dry surface out of direct sunlight (Figure 4.53). Turn the rope periodically to thoroughly dry the entire section. Perhaps the best method of drying rope is to handle it as you would hose. Rope can be strung in a hose tower or laid on racks out of direct sunlight for complete drying (Figure 4.54).

Figure 4.54 Another method for drying rope is to place it on racks that are out of direct sunlight.

Figure 4.52 Rope can be dried in a regular dryer that has a low temperature setting.

STORAGE OF ROPES

The proper storage of rope is important for the preservation of the rope and for the fast deployment of the rope when it is needed. The two most common methods of storing rope are by coiling it or by storing it in a canvas bag.

Coiling A Rope For Service

Coiling a rope so that it may be placed into service with a minimum of delay is very essential in the fire service. An improperly coiled handline or lifeline may result in failure of an evolution. A method of coiling a rope is described in the following steps:

Figure 4.53 Rope can also be dried by placing it on a clean, flat surface that is out of the direct sunlight.

Step 1: Select enough rope at the loop end to make a tie around the coil when completed. This amount is usually about three times the distance between standards (Figure 4.55).

Step 2: Wrap the rope around the standards until sufficient width is developed. Coil the rope in two layers to use a sufficient amount of rope. (**NOTE:** This amount can be determined by trial). Avoid making the coils too tight, which would make removal of the finished coil difficult. (Figure 4.56).

Step 3: Coil the remainder of the rope around the loops, and fasten the end securely with a clove hitch (Figure 4.57).

Step 4: Make the tie in the manner illustrated, and the finished coil is secure. (Figure 4.58).

Figure 4.55 Step 1: Select enough rope at the loop end to make a tie around the coil when completed. This amount is usually about three times the distance between standards.

Figure 4.56 Step 2: Wrap the rope around the standards until sufficient width is developed. Coil the rope in two layers to use a sufficient amount of rope.

Figure 4.57 Step 3: Coil the remainder of the rope around the loops and fasten the end securely with a clove hitch.

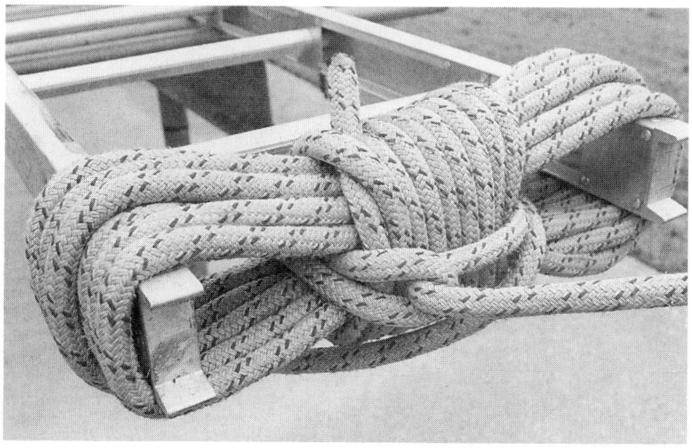

Figure 4.58 Step 4: Make the tie in the manner illustrated, and the finished coil is secure.

A compact, finished coil can be prepared by using the end of a ladder. To uncoil the rope, perform the following: release the tie, grasp the inside, pull out two or three loops to loosen the coil, and drop the coil from the top of a building or from a window (Figure 4.59). If the coil is not carefully prepared, it may not pay out to the ground. Another method of uncoiling the rope is to uncoil it by hand on the roof and then lower it hand over hand (Figure 4.60).

Bagging A Rope For Storage

The best method for storing kernmantle rope and other life safety rope is to place the rope into a storage bag (Figure 4.61). The bag allows easy carrying of the rope and keeps dirt and grime from the rope as well. If the bag is waterproof, a drain hole will need to be added. The bag should be labeled with the type of rope, its length, and its number. The bag should have the same

Figure 4.59 To uncoil the rope, perform the following: release the tie, grasp the inside, and pull out two or three loops to loosen the coil.

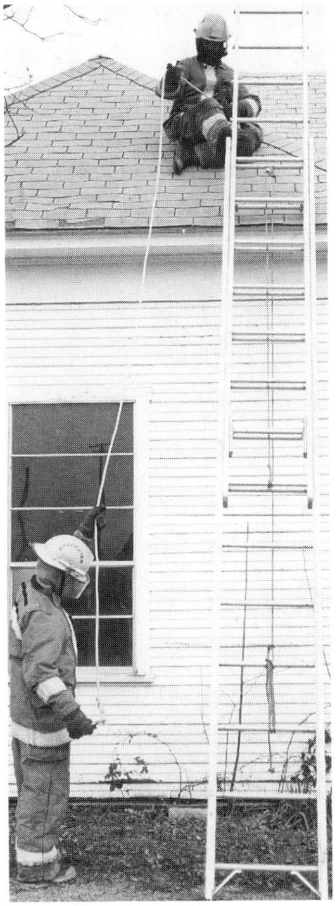

Figure 4.60 Another method of uncoiling the rope is to uncoil it by hand on the roof and then lower it hand over hand.

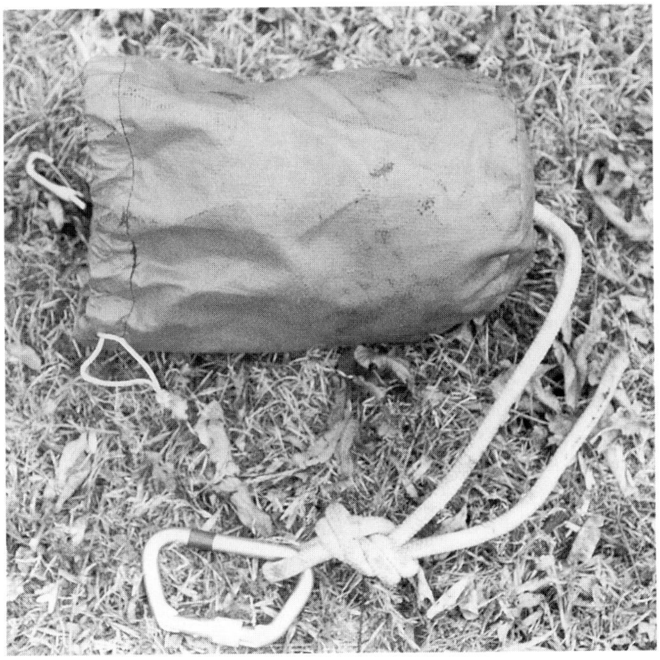

Figure 4.61 A rope storage bag helps to keep the rope clean.

number as the rope. The bag may have a drawstring and shoulder straps for ease in carrying. Burlap-type bags can be used in place of canvas. It is a good idea to have some type of waterproof pouch to hold the rope log. To store rope in a bag, one person should hold the storage bag while the other person stuffs the rope into the bag. Use a figure of eight stopper on both ends of the rope.

HOISTING TOOLS AND EQUIPMENT

A common activity at large-scale fire or rescue incidents is to use ropes to haul various pieces of equipment from one elevation to another (Figure 4.62). Almost any piece of equipment can be hauled by rope. Common sense and knowledge of proper knots will aid in lashing these objects. For example, anything with a closed-type D-ring handle can be raised or lowered using a bowline or figure of eight. However, the hoisting of pressurized cylinders, such as fire extinguishers or SCBA bottles, is not recommended.

Use the proper knots and securing procedures to prevent dropping the equipment. This prevents damage to the equipment and serious injury to anyone standing below. Depending on local policy, a tag line may also be tied to any of these objects (Figure 4.63). The tag line will be guided by firefighters on the ground who will prevent the object from coming in contact with the structure or other objects as it is being raised. The following sections discuss a few of the most commonly hoisted pieces of equipment.

Axe

Use a clove hitch, timber hitch, or a girth hitch (each with a safety) around the head of the axe and at least one half hitch around the handle (Figure 4.64 on next page).

Pike Pole

There are two methods of tying a pike pole to be hoisted. One method is to raise the pike pole, head down,

Figure 4.62 Rope being used to hoist equipment to the roof.

Figure 4.63 A tag line is used to prevent the tool from hitting against the side of the building.

Figure 4.64 The axe is hoisted using a clove hitch with a half hitch on the handle.

by using a clove hitch (with a safety) around the head with at least one, preferably two evenly spaced half hitches around the handle (Figure 4.65). Another method is to raise the pike pole, head up, by placing a clove hitch toward the end of the handle, followed by a half hitch in the middle of the handle and another around the head (Figure 4.66).

Figure 4.65 One method is to raise the pike pole, head down, by using a clove hitch (with a safety around the head) with at least one, preferably two evenly spaced half hitches around the handle.

Figure 4.66 Another method is to raise the pike pole, head up, by placing a clove hitch toward the end of the handle, followed by a half hitch in the middle of the handle and another around the head.

Ladder

Use a bowline or figure of eight knot and slip it first through two of the rungs about one-third of the way down from the top. After pulling that loop through, slip it over the top of the ladder (Figure 4.67).

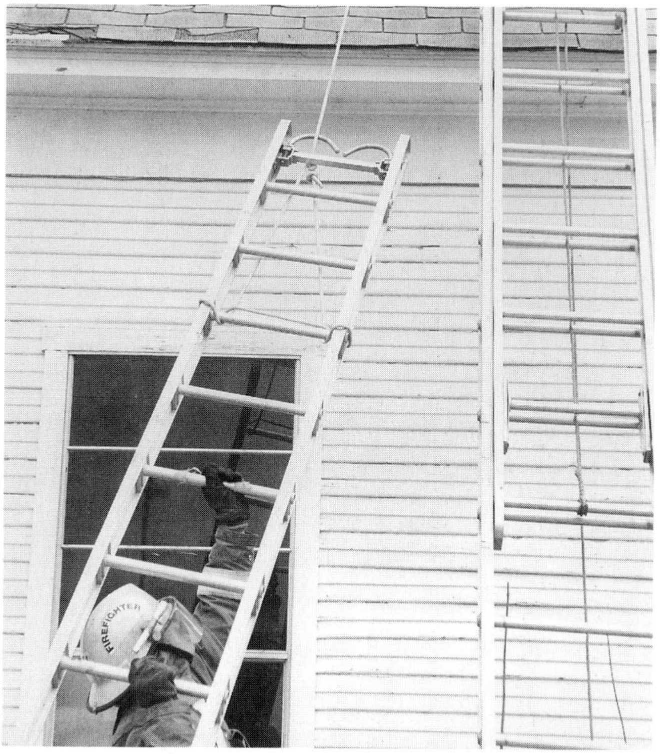

Figure 4.67 To hoist a ladder, use a bowline or figure of eight knot and slip it first through two of the rungs about one-third of the way down from the top. After pulling that loop through, slip it over the top of the ladder.

Hoselines

Both dry and charged hoselines may be hoisted. To hoist a dry hoseline, fold the nozzle and end of the hose back over the rest of the hose so that a bight similar to that described for ropes is formed. This bight should be about 3 to 5 feet (1 m to 1.6 m) long. Tie the nozzle and hose together using a clove hitch with a half hitch safety knot. Then tie at least one half hitch around the bight (Figure 4.68).

To hoist a charged hoseline, tie a clove hitch with a half hitch safety knot around the hose about 1 foot (0.3 m) below the coupling and nozzle. A half hitch should be tied through the nozzle bale so that it will hold the bale in the closed position as the hoseline is hoisted (Figure 4.69).

Smoke Ejector

To securely hoist a smoke ejector, tie a bowline or figure of eight around two of the connecting rods between the front and back plates. This will be the hauling line (Figure 4.70). A second line should be attached to the bottom of the unit (Figure 4.71). This line will be used to keep the ejector from bouncing against the side of the building while it is being raised. It is controlled by personnel on the ground.

Figure 4.68 A dry hoseline being hoisted.

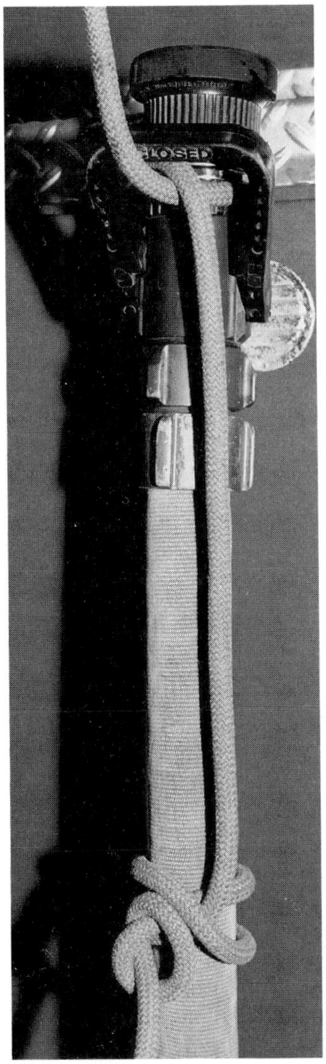

Figure 4.69 A charged hoseline is hoisted with the bale held shut.

Figure 4.70 Use a figure of eight or bowline to hoist a smoke ejector.

Figure 4.71 The tag line prevents the ejector from bouncing against the building.

5

Rescue and Extrication

This chapter provides information that addresses performance objectives described in NFPA 1001, *Standard for Fire Fighter Professional Qualifications* (1992), particularly those referenced in the following sections:

3-18 Rescue

3-18.1

3-18.2

3-18.3

3-18.4

4-18 Rescue

4-18.1

4-18.2

4-18.3

Chapter 5

Rescue And Extrication

While the entire fire service is dedicated to saving lives and property, rescue and extrication deal exclusively with life-threatening situations. Because of such life-threatening situations, firefighters must be thoroughly prepared for any potential rescue and/or extrication situation they encounter. IFSTA makes a definite distinction between rescue and extrication. *Extrication* incidents involve the removal and treatment of victims who are trapped by some type of man-made machinery or equipment. *Rescue* incidents involve the removal and treatment of victims from situations involving natural elements, structural collapse, elevation differences, or any other situation not considered to be an extrication incident.

This chapter covers the basics of rescue equipment and techniques as required by NFPA 1001. For more extensive information on extrication and rescue, see the IFSTA validated manuals **Principles of Extrication** and **Fire Service Rescue Practices.**

RESCUE AND EXTRICATION TOOLS AND EQUIPMENT

The skills and techniques required for rescue and extrication work can be learned only through complete training. Although it is impossible to anticipate every extrication situation, rescue personnel will be best prepared if they are proficient with their equipment. Complete knowledge of tools and equipment allows rescuers to rapidly devise a method for completing almost any rescue or extrication. The following section highlights some of the tools that are more commonly used by firefighters who perform rescue and extrication functions.

Hydraulic Tools

POWERED HYDRAULIC TOOLS

The development of powered hydraulic extrication tools has revolutionized the process of removing victims from various types of entrapments. The wide range of uses, speed, and superior power of these tools has made them the primary tools used in most extrication situations. These tools receive their power from hydraulic fluid pumped through special hoses. Although there are a few pumps that are operated by compressed air, most are powered either by electric motors or by two- or four-cycle gasoline engines. Electric and gas motors may be portable and carried with the tool, or they may be mounted on the vehicle and supply the tool through a hose reel line (Figures 5.1 a through c). Hand-operated pumps are also available in case of a power unit failure (Figure 5.2 on the next page). There are four basic types of powered hydraulic tools used in the rescue service: spreaders, shears, combination spreader/shears, and extension rams.

Figure 5.1a A gas hydraulic rescue tool motor.

Figure 5.1b An electric hydraulic rescue tool motor. *Courtesy of Lukas of America.*

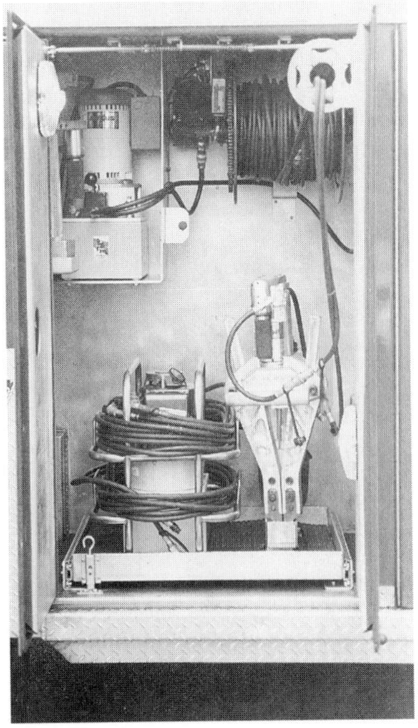

Figure 5.1c Hydraulic rescue tool motors may be mounted on the apparatus. *Courtesy of East Greenville (PA) Fire Co. No. 1.*

Figure 5.2 Hand pumps allow rescuers to operate hydraulic rescue tools without a power unit. *Courtesy of Hale Fire Pump Co., Inc.*

Spreaders

The powered hydraulic spreader was the first powered hydraulic tool to become available to the rescue service (Figure 5.3). It is useful for a variety of different operations involving either pushing or pulling. Depending on the brand, this tool can produce up to 22,000 psi (154 000 kPa) of force at the tips of the tool. The tips of the tool may spread as much as 32 inches (813 mm) apart (Figure 5.4).

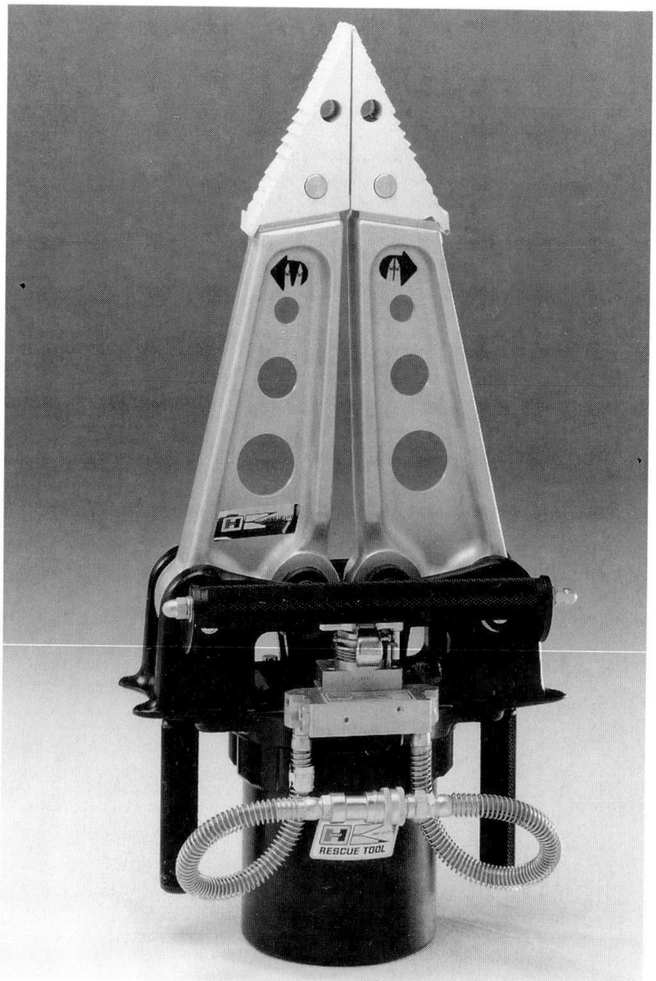

Figure 5.3 A powered hydraulic spreader. *Courtesy of Hale Fire Pump.*

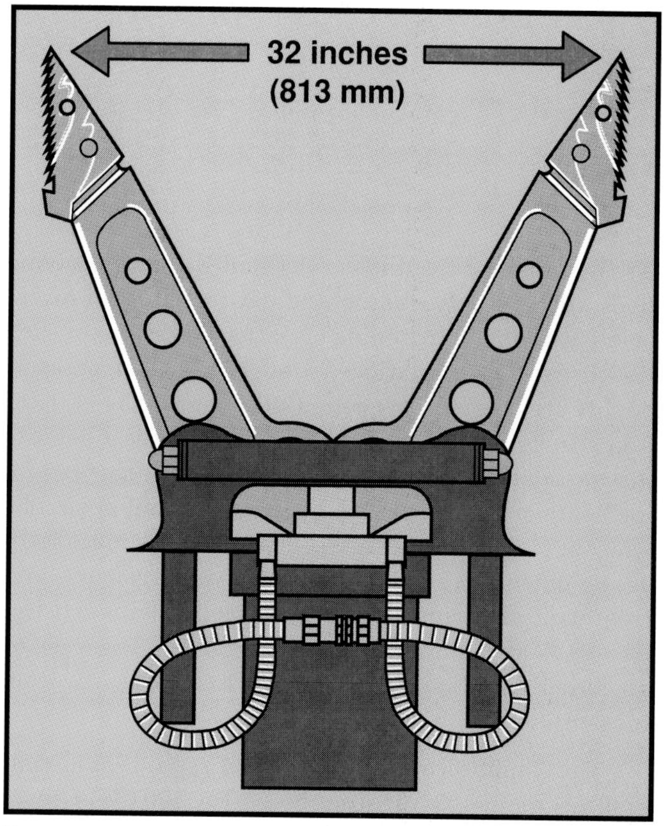

Figure 5.4 Powered hydraulic spreaders may have arms that open as far as 32 inches (813 mm) at the tips.

Shears

In the early days of powered hydraulic extrication tools, cutting was achieved by adding a cutting adapter to the end of a spreader tool. Today, individual hydraulic shears are available to complement the spreader (Figure 5.5). These shears are capable of cutting almost any metal object that can fit between their blades, although some models cannot cut case-hardened steel. The shears may also be used to cut other materials such as plastics or wood. Shears are typically capable of developing about 30,000 psi (206 850 kPa) of cutting force and have an opening spread of up to about 7 inches (175 mm) (Figure 5.6).

Combination Spreader/Shears

Several manufacturers of powered hydraulic extrication equipment now have a combination spreader/ shears tool (Figure 5.7). This tool consists of two arms equipped with spreader tips that can be used for pulling or pushing. The insides of the arms contain cutting

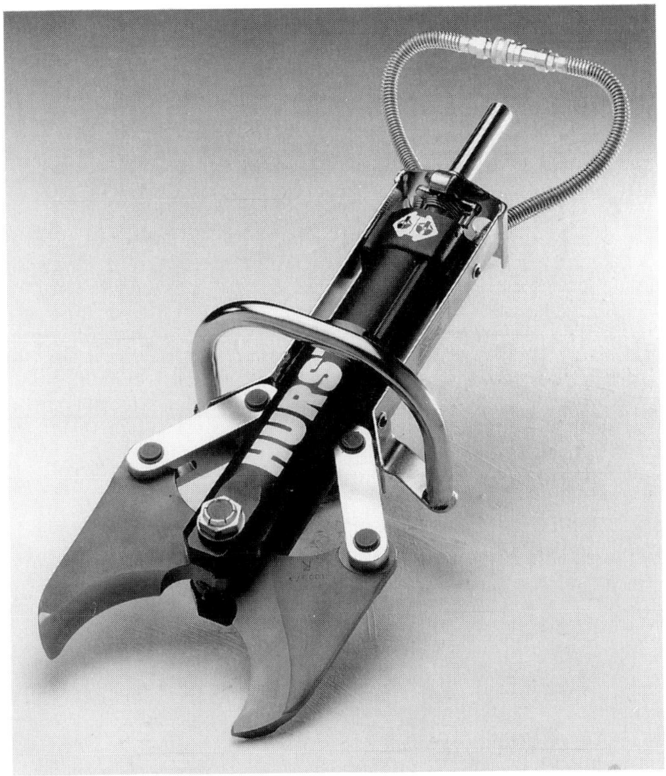

Figure 5.5 Some hydraulic shears have blades that are designed to cut small, high-strength objects. *Courtesy of Hale Fire Pump Co., Inc.*

Figure 5.6 Hydraulic shears in use.

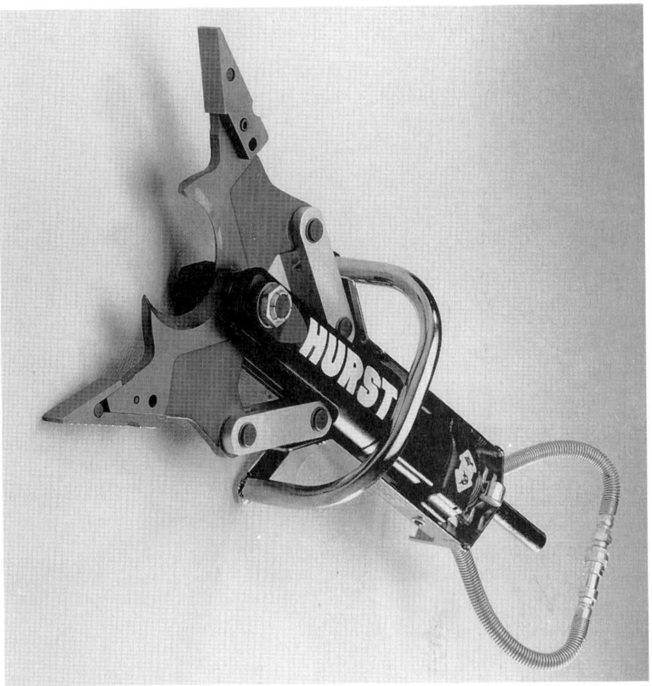

Figure 5.7 This view clearly shows the spreading tips on the outside of the arms and the cutting shears on the inside. *Courtesy of Hale Fire Pump Co., Inc.*

Figure 5.8 A combination spreading/cutting tool in use.

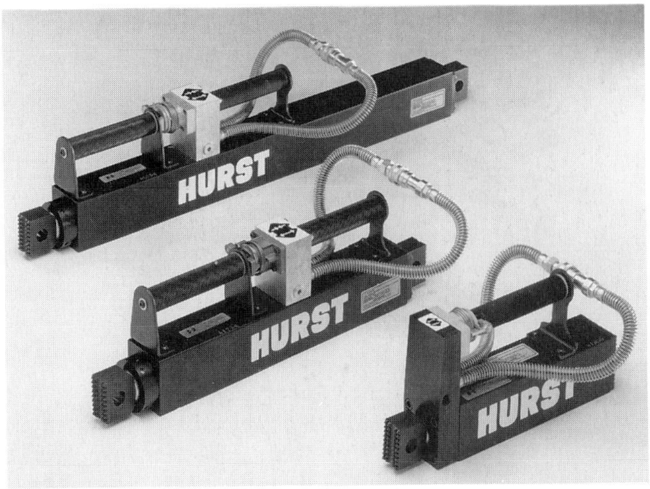

Figure 5.9 Hydraulic rams come in a variety of sizes. *Courtesy of Hale Fire Pump Co., Inc.*

shears similar to those described in the previous paragraph (Figure 5.8). This tool is excellent for use on small initial response vehicles or in areas where limited resources prevent the purchase of larger and more expensive individual spreader and cutting tools. The combination tool's spreading and cutting capabilities are less than those of the individual units.

Extension Rams

Extension rams are the newest addition to powered hydraulic extrication tools (Figure 5.9). These tools are designed primarily for straight pushing operations; however, they are effective at pulling as well. They are especially valuable when it is necessary to push objects farther than the maximum opening distance of the hydraulic spreaders. The largest of these extension rams can extend from a closed length of 36 inches (914 mm) to an extended length of nearly 63 inches (1 600 mm). They open with a pushing force of about 15,000 psi (103 425 kPa). The closing force is about one-half that of the opening force.

MANUAL HYDRAULIC TOOLS

Manual hydraulic tools operate on the same principle as powered hydraulic tools, except that the hydraulic pump is manually powered by a rescuer operating a pump lever. The primary disadvantage of manual hydraulic tools is that they operate more slowly than powered hydraulic tools. The most frequently used manual hydraulic tools are the porta-power system and the hydraulic jack.

Porta-Power

The porta-power tool system is basically an auto body shop tool that has been adapted for use by the rescue service (Figure 5.10). It is operated by transmitting pressure from a hand-pumped compressor through a hydraulic hose to a tool assembly. A number of different tool accessories give the porta-power a variety of applications.

The primary advantage of the porta-power compared to the hydraulic jack is that the porta-power has accessories that allow it to be operated in narrow places in which the jack will not fit or cannot be operated. The primary disadvantage of the porta-power is that assembling complex combinations of accessories and actual operation of the tool are time-consuming.

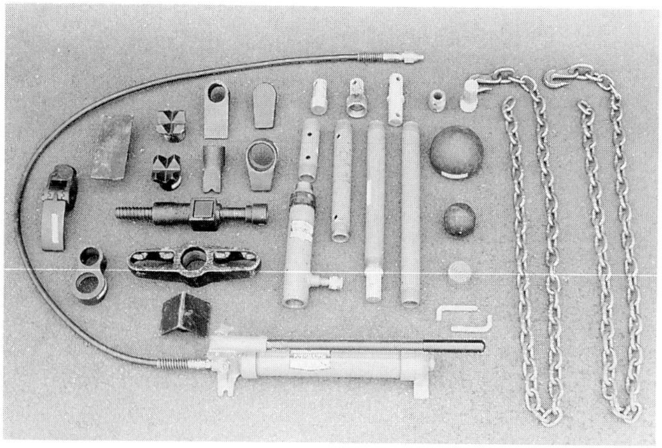

Figure 5.10 Hydraulic porta-power kits have a variety of accessories that enable them to be adapted for many uses.

Hydraulic Jack

The hydraulic jack is good for most heavy lifting situations (Figure 5.11). It is also an excellent compression jack for shoring or stabilizing operations. The hydraulic jack operates on the principle that the pressure of liquids between two chambers of unequal size tends to equalize. A small chamber is used to pump fluid into a larger chamber. The energy applied is multiplied by the surface area differential to do more work in the large chamber. There is a check valve between the chambers that keeps the liquid from flowing back. Hydraulic jacks are commonly available in capacities of up to 20 tons (18 144 kg), although some larger sizes may be found.

Figure 5.11 Hydraulic jacks are manually operated by pumping the lever arm.

When using any type of jack, consider good footing and adequate blocking and cribbing. Never work under a load being supported only by a jack. The load must be blocked or cribbed to make sure that it will not fall if the jack suddenly fails (Figure 5.12). The load will be transmitted to the base of the jack. An expanded surface area must be supplied to distribute the force placed on the jack. The jacking base surface should be solid, flat, and level. If it is not, a flat board or steel plate with wood on top should be put under the jack and shimmed level, making sure there is enough room to place the jack (Figure 5.13).

Figure 5.12 Cribbing should be inserted underneath a vehicle as it is lifted to prevent a total drop of the vehicle if the lifting device fails.

Figure 5.13 Cribbing should be inserted beneath a jack when lifting is to be performed on a potentially unstable surface.

WARNING:

Never work under a load supported only by a jack. If the jack fails, severe injury or death may result. Always make sure that the load is also supported by properly placed cribbing.

Pneumatic (Air-Powered) Tools

Pneumatic tools use the energy created by the release of compressed air for power. This air can be supplied by vehicle-mounted air compressors, apparatus brake system compressors, SCBA bottles, or cascade system cylinders.

WARNING:

Never use compressed *oxygen* supplies to power pneumatic tools. Mixing pure *oxygen* with grease and oils found on the tools will result in fire or violent explosion.

PNEUMATIC CHISELS/HAMMERS

Pneumatic-powered chisels (also called air chisels, pneumatic hammers, or impact hammers) are useful for extrication work. The pneumatic chisel is designed to operate at air pressures between 100 and 150 psi (700 kPa and 1 050 kPa). During periods of normal consumption, it will use about 4 to 5 cubic feet (113 L to 142 L) of air per minute. The pneumatic chisel can be especially effective for auto extrication by cutting through the roof, roof support columns or doorjambs, seat bolts, and door lock assemblies (Figure 5.14). It is good for cutting medium- to heavy-gauge sheet metal and for popping rivets and bolts. Remember, however, that cutting heavier steel or metal requires larger air supplies and higher pressures.

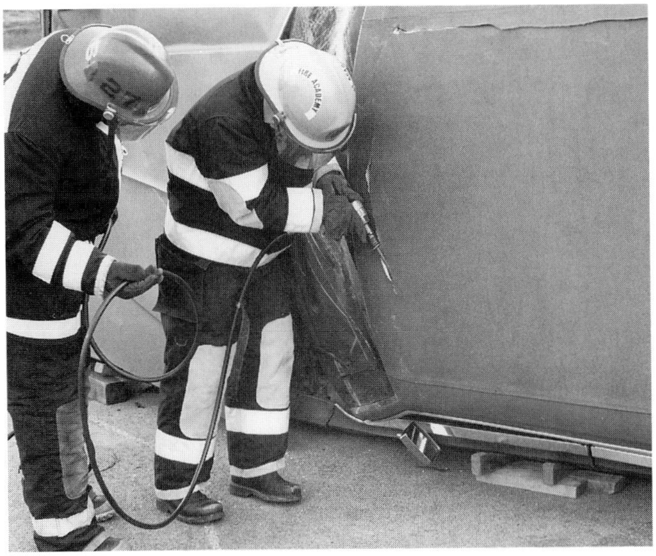

Figure 5.14 Air chisels are well suited for cutting sheet metal.

The pneumatic chisel comes with a variety of tool bits to fit the needs of almost any situation (Figure 5.15). In addition to cutting bits, special bits for such operations as breaking locks or driving in plugs are also available. Exercise caution when using pneumatic chisels in areas containing hazardous atmospheres. The contact between the cutting bit and the metal it is cutting may produce sparks that could provide an ignition source for any flammable vapors that may be present.

CAUTION: Be careful when using pneumatic chisels in areas containing hazardous atmospheres. The sparks produced during cutting may provide an ignition source for flammable vapors.

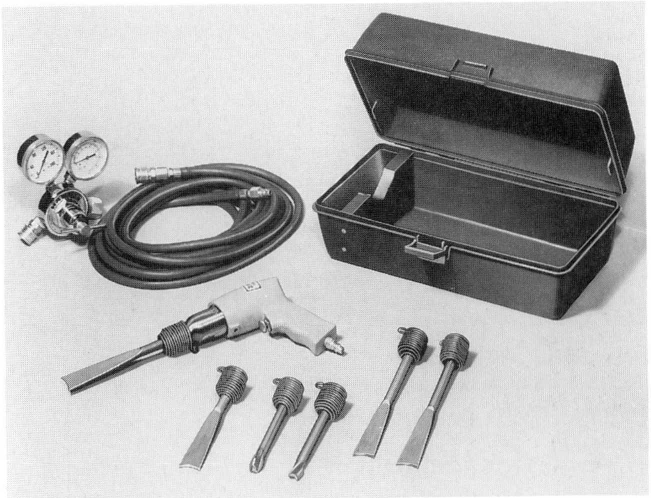

Figure 5.15 Air chisels are equipped with a variety of tool bits for specific applications.

AIR LIFTING BAGS

One of the newest additions to the rescue service has been air lifting bags. These devices give rescuers the ability to lift or displace objects that cannot be lifted with standard extrication equipment. There are three basic types of lifting bags: high pressure, low and medium pressure, and leak-sealing.

The high-pressure bags consist of a rough, neoprene rubber exterior reinforced with either steel wire or Kevlar® aramid fiber (Figure 5.16). Deflated, the bags lie perfectly flat and are about 1 inch (25 mm) thick (Figure 5.17). High-pressure bags come in sizes that range from 6 x 6 inches (150 mm by 150 mm) to 36 x 36 inches (914 mm by 914 mm). Depending on the size of the bags, they may inflate to a height of 20 inches (500 mm). The largest bags can lift around 75 tons (68 040 kg). The inflation pressure of the bags is about 135 psi (931 kPa).

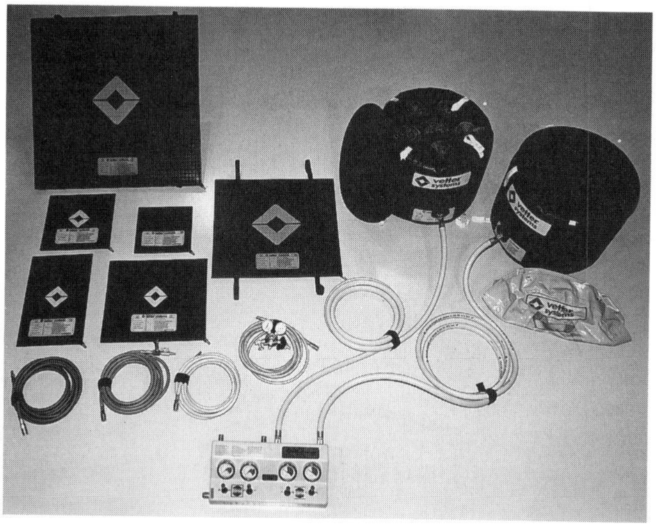

Figure 5.16 Air bags come in a variety of shapes and sizes. *Courtesy of Vetter Systems, Inc.*

Figure 5.17 Air lifting bags can be inserted into spaces as small as 1 inch (25 mm). *Courtesy of Vetter Systems, Inc.*

Low- and medium-pressure bags are considerably larger than high-pressure bags and are most commonly used to lift or stabilize large vehicles or objects (Figure 5.18). Their primary advantage over high-pressure air bags is that they have a much greater lifting distance. Depending on the manufacturer, this distance may be upwards of 6 feet (2 m). Low-pressure bags generally operate on 7 to 10 psi (49 kPa to 70 kPa), while medium-pressure bags use 12 to 15 psi (84 kPa to 105 kPa), again depending on the manufacturer.

Figure 5.18 Low-pressure air lifting bags can lift large loads a greater distance than high-pressure air lifting bags. *Courtesy of Vetter Systems, Inc.*

A third type of air bag is the leak-sealing air bag (Figure 5.19). These bags, along with their accompanying hardware, are designed to be applied to cracks, open ends, or holes in low-pressure liquid storage containers or pipes. These bags are constructed similarly to high-pressure bags but are designed to be inflated at a much lower pressure, usually around 25 psi (175 kPa). Some of these leak-sealing bags may also be used for limited lifting work, but they should not be relied upon for that use.

Figure 5.19 This leak-sealing bag has been inserted into the end of a pipe to stop the flow of liquid. It is not designed for lifting. *Courtesy of Vetter Systems, Inc.*

Air Bag Safety Rules

Operators should use the following safety rules when using air bags:

- Plan the operation before starting the work.
- Be thoroughly familiar with the equipment: its operating principles, methods, and limitations.
- Consult individual operator's manuals and follow the recommendations for the specific system used.
- Keep all components in good operating condition and all safety seals in place.
- Have an adequate air supply and sufficient cribbing before beginning operations.
- Position the bags on or against a solid surface.
- Never inflate the bags against sharp objects.
- Inflate the bags slowly and monitor them continually for any shifting.
- Never work under a load supported only by bags.
- Shore up the load with enough cribbing blocks to more than adequately support the load in case of bag failure.
- Stop the procedure frequently to increase shoring or cribbing.
- Be sure that the top layer is solid when using the box cribbing method; leaving a hole in the center may cause shifting and collapse (Figure 5.20).
- Avoid exposing bags to materials hotter than 220°F (104°C). Insulate the bags with a nonflammable material. A bag should be removed from service if any evidence of heat damage is noticed.
- Never stack more than two bags; center the bags with the smaller bag on top, and inflate the bottom bag first (Figure 5.21).

Figure 5.20 Air lifting bags must be placed on a solid layer of cribbing to ensure stability.

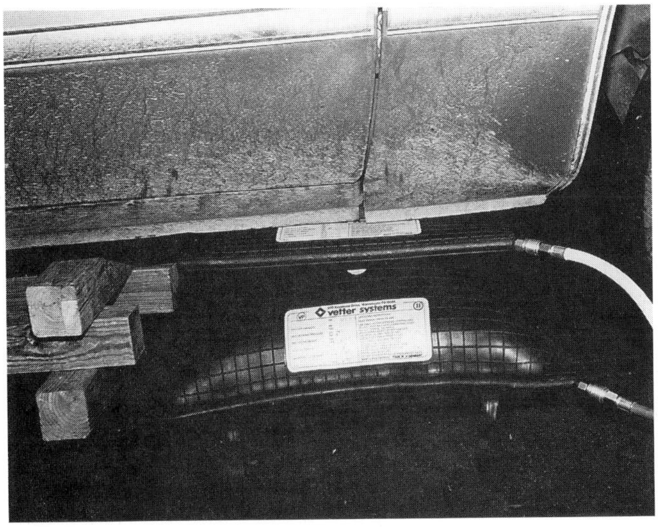

Figure 5.21 When inflating stacked bags, inflate the bottom bag first. *Courtesy of Vetter Systems, Inc.*

Block And Tackle

A block and tackle is useful for lifting or pulling heavy loads because of its mechanical advantage in converting a given amount of pull to a working power greater than the pull. A block is a wooden or metal frame containing one or more pulleys called sheaves (Figure 5.22). The tackle is the assembly of ropes and blocks that the line passes through to multiply the pulling force.

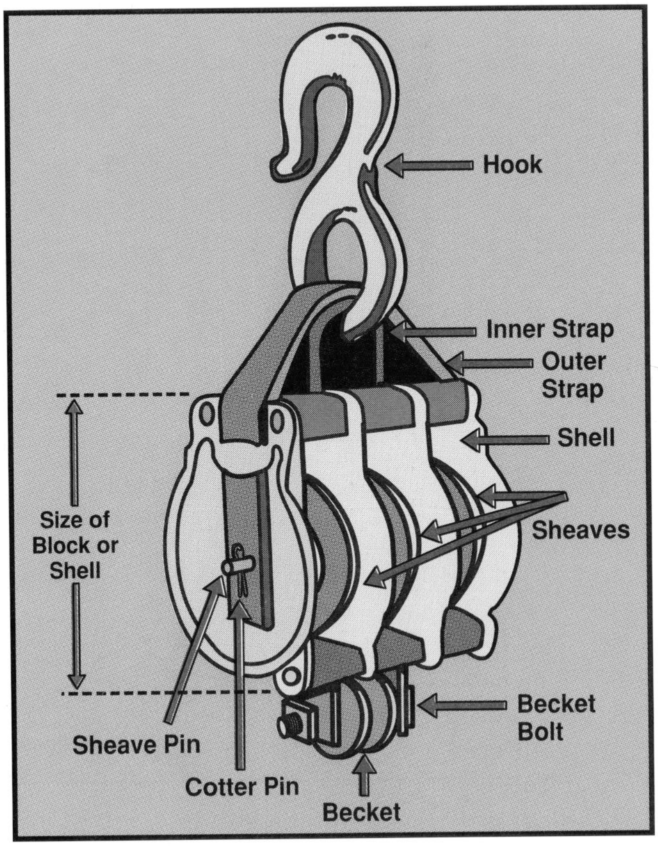

Figure 5.22 The parts of a block and tackle.

Simple tackle is one or more blocks reeved (threaded) with a single rope. Rescuers are most likely to use simple tackle. Compound tackle is two or more blocks reeved with more than one rope. Compound tackle is covered in the IFSTA validated **Fire Service Rescue Practices** manual.

The common terms used in reference to a block and tackle arrangement are:

- Standing block — The block attached to a solid support and from which the fall line leads.
- Running block — The block attached to the load to be moved.
- Leading block — Usually are snatch blocks attached to a solid support that changes the direction of pull without affecting the mechanical advantage. They are used when the fall line leaves the standing block in such a way that applying the motive force is difficult.
- Fall line (Figure 5.23) — The line attached to the power.

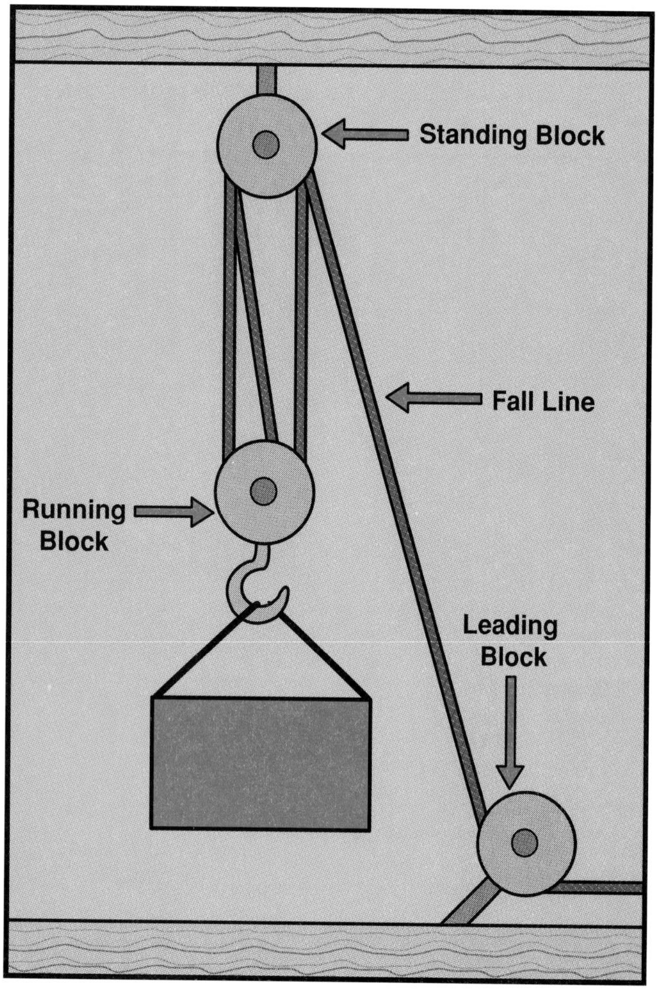

Figure 5.23 The parts of a block and tackle system.

The methods for reeving single, double, and triple blocks are shown in Figures 5.24 through 5.26. After reeving the blocks, the rope should be pulled back and forth through the blocks several times so that it will adjust to the blocks. This prevents the rope from twisting under load. Twisting increases friction, chafes the rope, and may jam the blocks. A method of preventing the running block from twisting is shown in (Figure 5.27).

PRECAUTIONS FOR USING A BLOCK AND TACKLE

Operators should observe the following safety rules when using block and tackle:

- Be sure that the rope is the right size for the weight being lifted and the blocks being used.
- When pulling on a fall line, everyone should exert a steady, simultaneous pull and hold onto the gain.
- Be sure that the supports holding the standing and leading blocks will hold the load and the pull.
- Pull in a direct line with the sheaves, not at an angle to either side.
- Whenever possible, the pull should be downhill.
- Pullers should stand so that they will be out of danger if the tackle or support fails.
- Easing off on a suspended weight should be done gradually without jerking.
- Hooks without safety latches should always be "moused" to prevent slings or ropes from slipping off (Figure 5.28 on the following page). Wrap wire or strong twine eight to ten times around the hook, then take several turns around the mousing, and tie the ends securely.

Cribbing

Rescue vehicles should carry an adequate amount of appropriately sized cribbing. Cribbing is essential in many extrication operations. Its most common use is to stabilize objects; however, it has many other uses. Wood selected for cribbing should be solid, straight, and free of major flaws such as large knots or splits. Various sizes of wood can be used, but the most popular is 4- x 4-inch (100 mm by 100 mm) hardwood timbers. The length of the pieces may vary, but 20 to 22 inches (500 mm to 550 mm) is standard. The ends of the blocks may be painted different colors for easy length identification. Other surfaces of the cribbing should be free of any paint or finish because this can make the wood slippery, especially when it is wet. Individual pieces of wood may have holes cut in the end with a rope loop tied in the holes for easy carrying (Figure 5.29 on the following page).

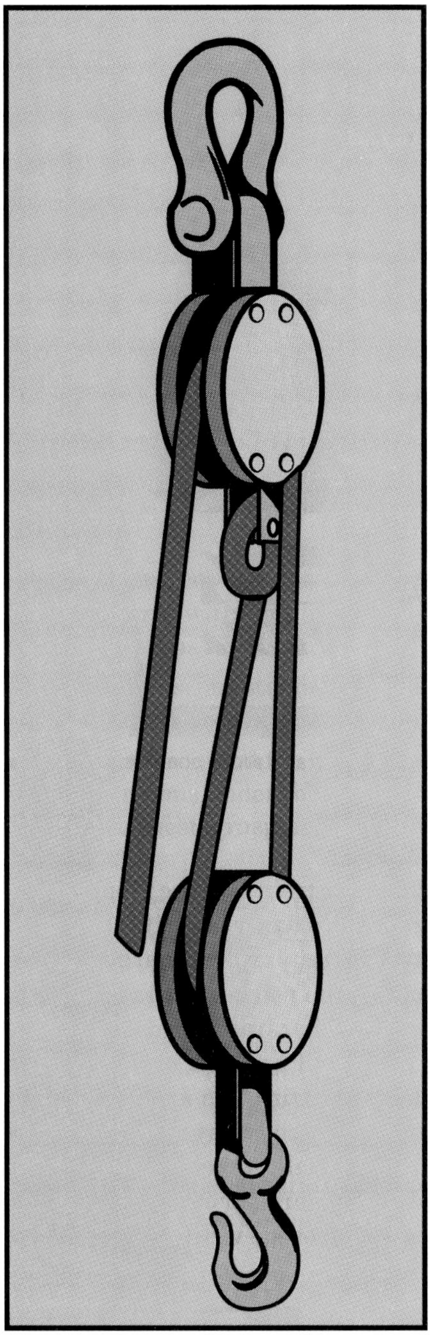
Figure 5.24 Reeving a single block.

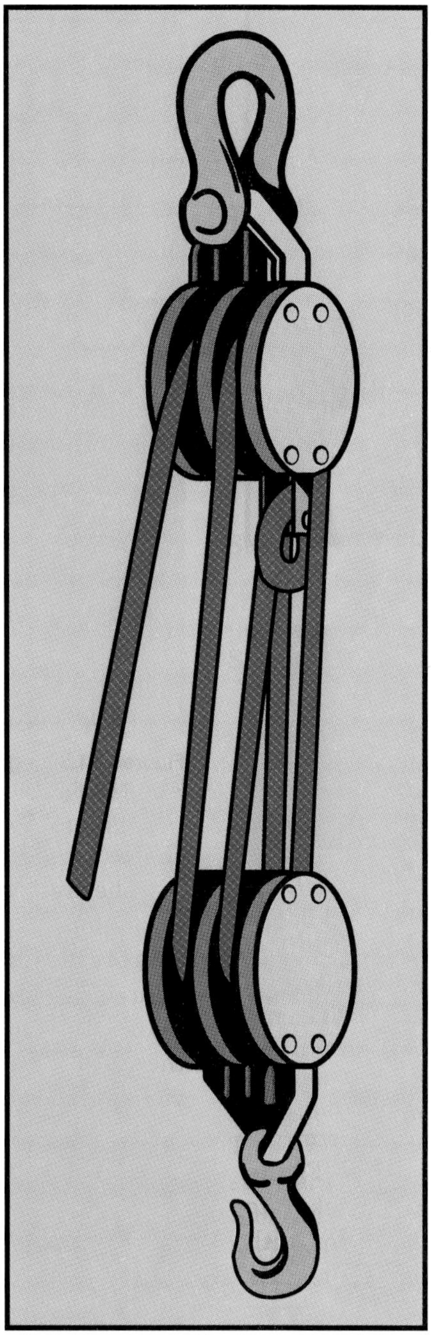
Figure 5.25 Reeving a double block.

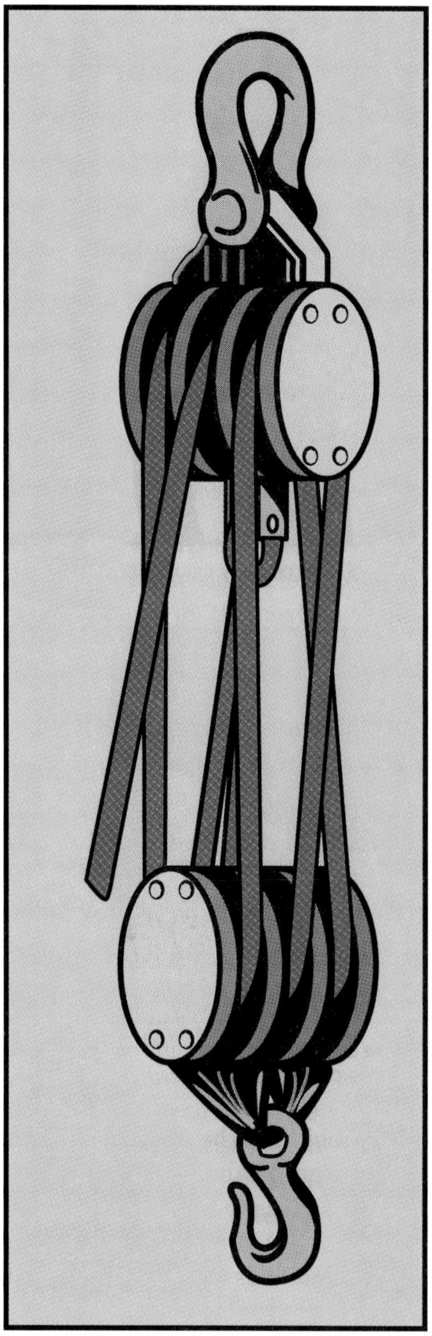
Figure 5.26 Reeving a triple block.

Figure 5.27 A rod may be placed in the system to prevent twisting of the ropes.

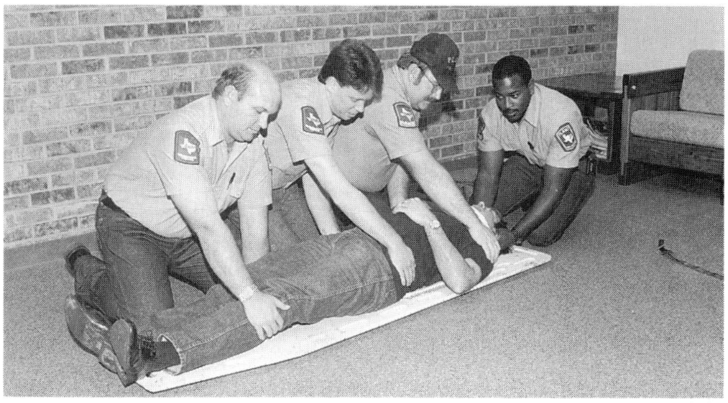

Figure 5.48 Step 8: On the command of the first rescuer, the three rescuers gently push the victim so that the victim is centered on the backboard.

victim's hips, and the third rescuer grasps the victim's legs. On command, all slide the victim and center the victim on the backboard (Figure 5.49).

Step 9: The rescuer who is at the victim's shoulders places rolled towels, blankets, or specially designed immobilization devices on both sides of the victim's head. These items and the victim's head must be secured to the board with a cravat or tape that passes over the forehead. If an immobilizer is used, place the sides in position and secure the chin and forehead straps as shown in (Figure 5.50).

Step 10: Each of the three rescuers fastens the victim to the board with the appropriate strap — one across the chest, one above the hips, and one above the knees (Figure 5.51).

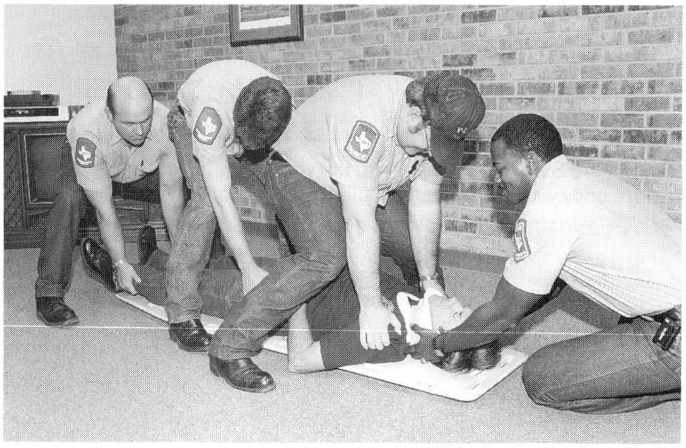

Figure 5.49 While the rescuer at the victim's head maintains axial traction, the other three rescuers stand and straddle the victim. Again, one rescuer grasps the victim's shoulders, one grasps the victim's hips, and the third rescuer grasps the victim's legs. On command, all slide the victim and center the victim on the backboard.

ROPE RESCUE

Every firefighter should be at least minimally familiar with the techniques of using a rope to lower victims from

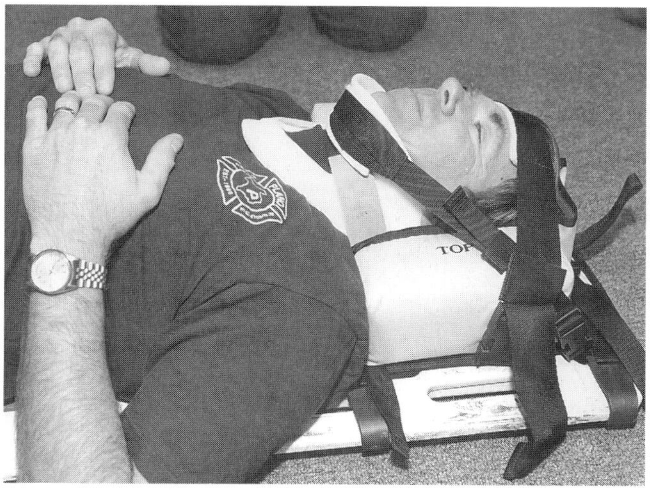

Figure 5.50 When using an immobilizer, place the sides in position and secure the chin and forehead straps.

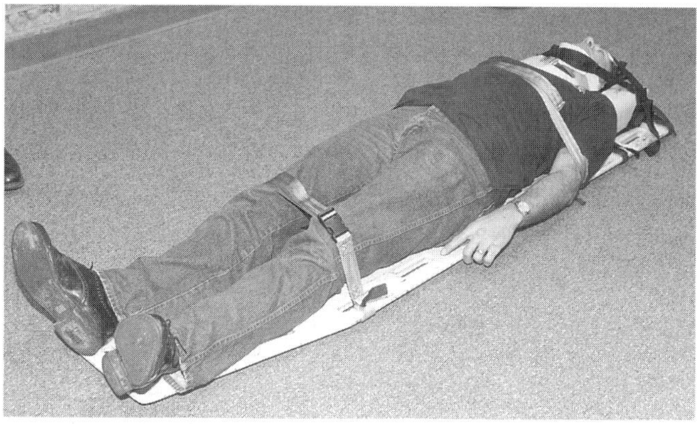

Figure 5.51 Step 10: Each of the three rescuers fastens the victim to the board with the appropriate straps.

an elevated position. This section will present one easy method for accomplishing this task. For more information on rope rescue, see the IFTSA validated **Fire Service Rescue Practices** manual or CMC's *Rope Rescue Manual.*

Life Safety Harnesses

In order to safely accomplish rope rescues, the rescuer must be familiar with the different types of equipment needed. Other than the rope itself, the next most important piece of equipment is the life safety harness that will be attached to the victim. NFPA 1983, *Standard on Fire Service Life Safety Rope, Harness, and Hardware*, lists specific requirements for harnesses that are used in conjunction with life safety ropes. Class I harnesses go around the wearer's waist and are used only to secure firefighters to a ladder or other objects (Figure 5.52). Class I harnesses were traditionally known as the "ladder or pompier belt." However, these are no longer suitable for rope rescue work. Class II harnesses are sit-type harnesses designed to support the weight of two people (victim and rescuer) (Figure 5.53). Class III harnesses

Figure 5.52 Class I harness belt.

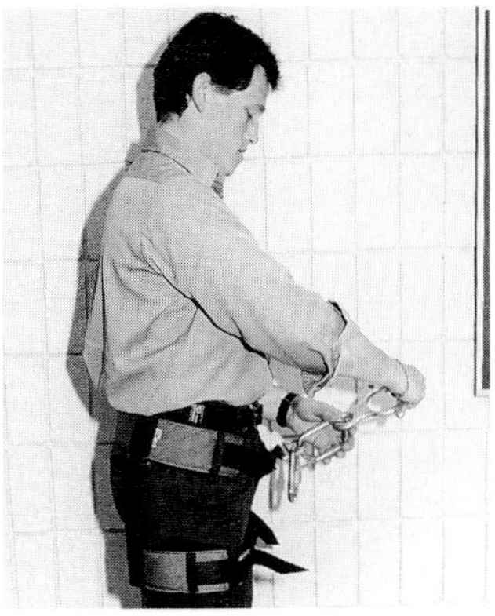

Figure 5.53 Class II harness.

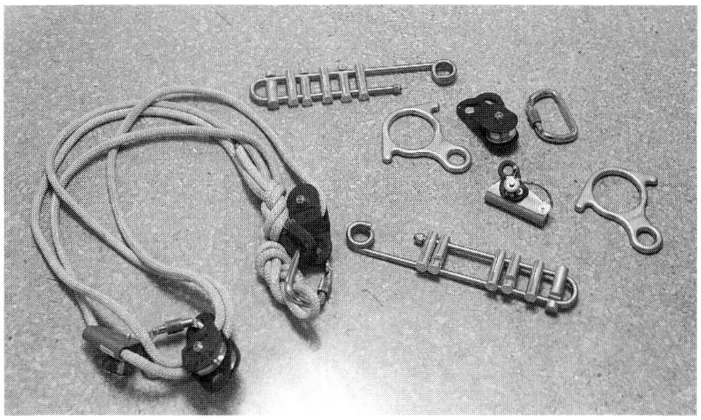

Figure 5.55 Various types of rope hardware.

are also sit-type harnesses designed to support two people (Figure 5.54). However, they have additional support over the shoulders that is designed to prevent the wearer from becoming inverted on the rope. Different types of hardware, such as the figure of eight, carabiner, and ascent and descent devices, are also covered in NFPA 1983 (Figure 5.55). The standard contains design criteria for all these types of hardware and the proper methods of testing each type. For more information on life safety ropes, harnesses, and hardware, consult the standard.

Life safety harnesses should be inspected after every use and periodically between use to make sure that they are in good condition. The harness should be free of cuts, abrasions, and other abnormal wear spots. The presence of fuzz on the webbing is an indication of abnormal wear. Buckles and other metal parts should be checked for cracks, corrosion, or other abnormalities. Rope hardware, such as the carabiner and figure of eight, must also be regularly checked for excessive wear and deformities. These devices should also be subjected to appropriate nondestructive testing on a regular basis to ensure that they are not defective.

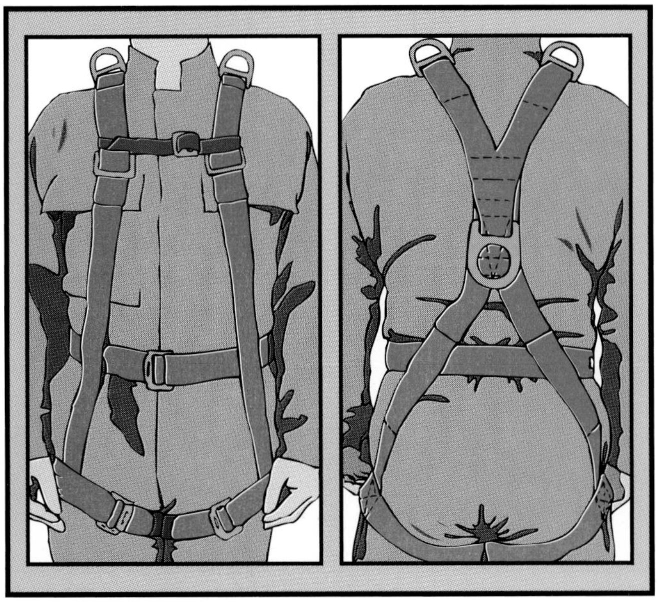

Figure 5.54 A Class III harness.

In the absence of a life safety harness, and under extreme emergency conditions, the victim may be lowered without the use of a life safety harness. In this case, a double loop figure of eight will take the place of a harness. Place a half hitch around the victim's chest with the working line, and secure it in front with an overhand safety to form a loop.

Lowering A Victim From A Window

This evolution may be used when either unconscious or conscious victims have to be lowered from an upper floor. Although it may appear slow compared to carrying the victim down a ladder, this method may be safer under certain conditions. If several people are to be rescued, it may also prove to be more rapid than expected. In performing this evolution, a ladder is raised to a point just above the window where the rescue is to be made. If a double loop figure of eight is going to be used, the knot may be tied on the ground before running the rope up the ladder, or it may be tied in the building after running the rope up the ladder. One firefighter takes the end of the lifeline, passes it underneath the bottom rung to the underside of the ladder, and starts the climb (Figure 5.56 on the next page). The rope is held in either hand so that it will feed up the underside of the ladder. When the windowsill is reached, the rope is threaded back through the rungs toward the firefighter on the ladder, up and over three or four consecutive rungs (Figure 5.57 on the next page). This allows the rope to hang freely, but not too close to the building. Enough rope is then fed through the window to enable the firefighters inside to make the knot or connect to the victim's harness. If a Stokes basket litter is being used, the rope would be attached to the cradle sling on the litter.

Figure 5.56 Run the rope under the bottom rung and up the back side of the ladder.

Once the victim has been secured to the hauling life safety rope, a guideline should be secured to either the harness or the area between the chest hitch and the double loop figure of eight. If a Stokes litter is being used, it may be necessary to use two guidelines, one at each end. A clove hitch is the knot of choice for attaching the guideline(s). The other end of the guideline is lowered out of the window so that a firefighter on the ground can hold the victim away from the building as he or she is being lowered. The victim is lifted out of the window by two or more firefighters inside the structure as the firefighter on the ground, controlling the descent, takes up the slack in the rope (Figure 5.58). This control person at the base of the ladder, with a foot on the bottom rung for support and using a hand-over-hand grasp on the rope, lowers the victim to the ground (Figure 5.59). The firefighter on the guideline should stand at the base of the ladder next to the firefighter doing the lowering to prevent the ladder from being accidentally pulled over, which could occur if the guideline were pulled from a position too far to either side. (**NOTE:** One firefighter at the base of the ladder can easily do the lowering and still support the ladder, but it is a good policy to have the ladder tied in or well anchored for safety reasons.)

Figure 5.57 At windowsill level, thread the rope over the top of the ladder for three to four rungs and then back over the top again.

Figure 5.58 Prepare the victim to be lowered by positioning the victim in the window opening.

Figure 5.59 Gently lower the victim to the ground.

RESCUE FROM BURNING BUILDINGS

Firefighters regularly perform search and rescue operations in burning buildings. These are the most dangerous operations a firefighter will face. Several precautions can be taken to make the job as safe as possible. When firefighters enter a burning building to perform rescue work, they must first consider their own protection. In order to protect the body surface from heat and flame, proper protective clothing should be worn; standard fire fighting clothing is usually adequate. The use of protective breathing equipment is required.

Some type of guideline should be used such as a rope, electric cord, or hoseline; however, the hoseline will drastically slow the rescue operation (Figure 5.60). If it becomes necessary to do the search without a guideline, the rescuers should maintain contact with a wall and try to maintain a sense of where they are in the building (Figure 5.61). They should note the locations of doors and windows in the event that a rapid evacuation from the area is necessary. Firefighters should always stay together in large areas. In smaller areas or rooms, one firefighter may enter a room to conduct the search while the other remains at the doorway. The two firefighters should maintain a running dialogue during the search so that the searcher remains oriented and can thus find the exit quickly if necessary.

Figure 5.60 While performing a rescue, it is helpful for firefighters to use a guideline such as a rope.

Figure 5.61 If a guideline is not available, firefighters should maintain contact with the wall.

WARNING:

Neither interior nor exterior fire attacks (except for wildland fires) should be attempted unless firefighters are wearing full protective clothing and protective breathing apparatus.

Rescuers should carry portable lights, portable radios, and a forcible entry tool to aid entry. A forcible entry tool can be used to allow egress if a means of retreat is cut off or to assist victims trapped in a burning building. In order to reach such victims, fire department ladders may be needed.

Fires occurring in occupied public gathering places may present additional problems. If the normal exits are blocked, occupants must leave through unfamiliar exits. The very fact that fire exists in a building usually unnerves a majority of the people to an extent that panic may result. A public gathering place should be emptied in as orderly a manner as possible. Many modern buildings are constructed with fire-resistant enclosed stairways that are separated from floors by fire-resistant doors. Buildings that are not constructed in this manner should be provided with fire escapes and multiple exits. Institutions, such as hospitals, rest homes, and mental health facilities, present a condition in which the occupants may be unable to walk. Those who perform rescue work must be prepared to move the occupants to safety without further aggravating their condition.

Certain search and rescue procedures must be followed to successfully find and safely remove fire victims. The firefighter conducts a primary search followed by a secondary search. A primary search is the first search of an area in a quick, systematic fashion. Areas that have the highest percentage of chance for finding victims are checked and then the rescuers move on. The secondary search is the slower, more thorough systematic search. The secondary search is usually conducted after the fire has been controlled and the heat and smoke conditions within the building have improved somewhat to permit the more extensive search. The most essential element in rescue work is time. The following guidelines will help ensure a swift and safe rescue operation once the firefighter enters the involved structure (Figure 5.62).

- Wear full protective clothing and protective breathing apparatus when performing search and rescue operations in fire buildings.
- Work in groups of two or more.

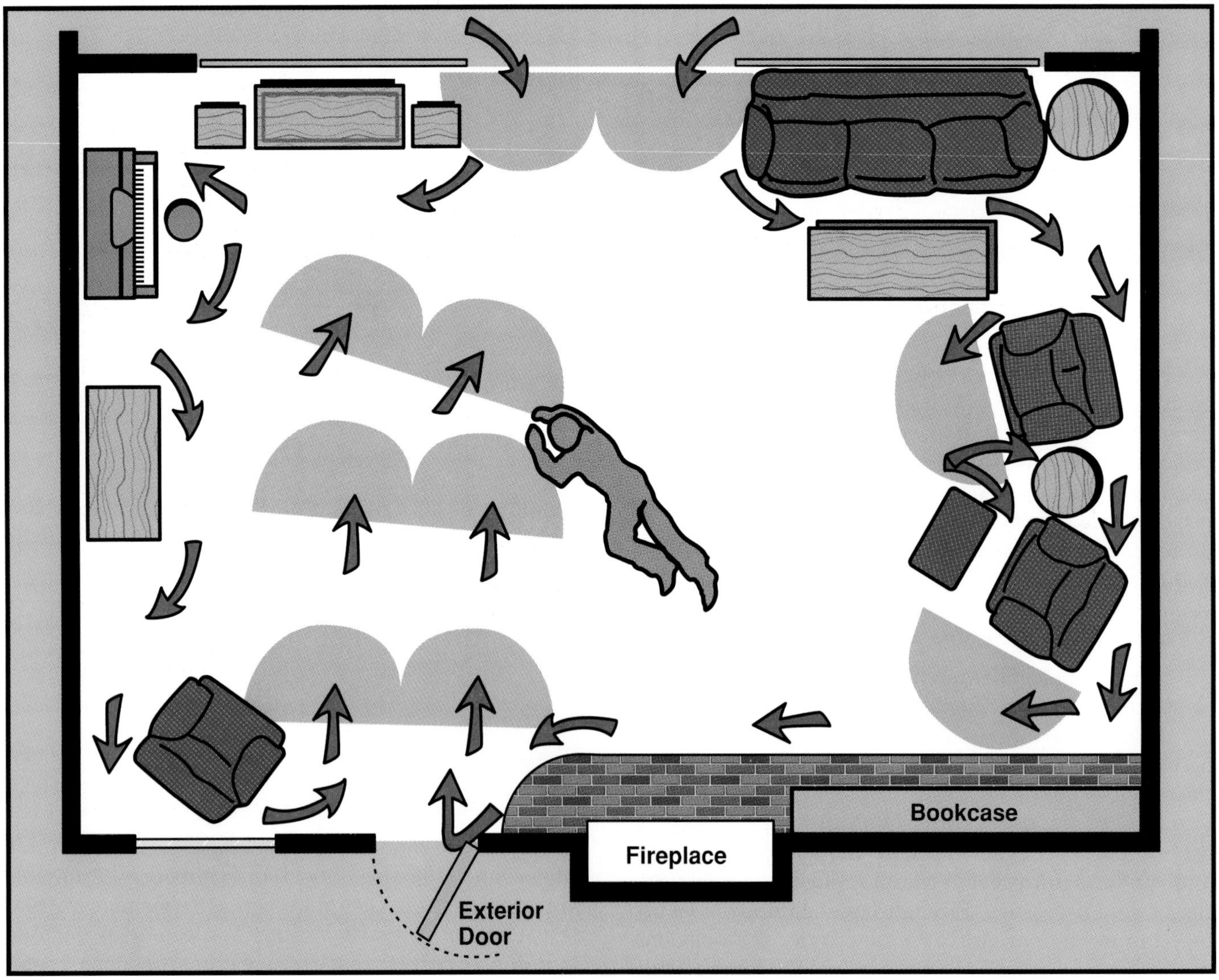

Figure 5.62 Use an organized pattern to search for victims.

- Attempt to locate more than one means of egress before entering the building.
- Search on your hands and knees (Figure 5.63). (Once the building is entered, visibility will be poor at best.) If you cannot see your feet, do not remain standing.

Figure 5.63 Firefighters should stay low while performing a search.

- Search one room completely before moving to the next room.
- Start the search on an outside wall. This will allow you to ventilate by opening windows as soon as possible. Windows should not be opened if positive pressure ventilation is to be used.
- Move all furniture, searching behind and under each piece.
- Search all closets and cupboards including shower stalls.
- Pause occasionally during the search and listen for cries for help or other audible signs or signals.
- Move up and down stairs on your hands and knees; when ascending, proceed head first and when descending, proceed feet first.
- After searching a room, leave a sign or signs indicating that the room has been searched by turning a chair over, folding a mattress sideways on the bed or pulling it off the bed, or opening closet doors. Latch straps or hang tags can be placed over the door knob to leave an indication that the room has been searched. Be sure to close the entry door to the room to prevent the spread of fire.
- Look for extension of fire and report any extension to the commanding officer.
- Reach into the doorway or window with the handle of a tool if rooms or buildings are too hot to enter (Figure 5.64). Frequently, victims will be found just inside the door or windows.
- Once you have successfully removed a conscious victim, place the victim in someone's custody so that he or she will not try to reenter the building for any reason.

Figure 5.64 If a room is too hot to enter, reach through the doorway with the handle of a tool to check for victims.

Any victims that are located should be moved to an area of safety as soon as possible. Firefighters must take into account that the victims do not have the same protective equipment that they have; therefore, it may be necessary to find a safer way to retreat. If the victim is unconscious, the carries and drags discussed earlier in this chapter should be used.

If it becomes necessary to remove a downed firefighter, the rescuers should again use any safe means possible. In most cases, the need to exit the hostile atmosphere will override the need to stabilize injuries before moving the firefighter (Figure 5.65). If the firefighter has a functioning SCBA, carefully move the firefighter so as not to dislodge the mask. If the firefighter does not have a functioning SCBA, either connect the mask to the buddy breathing connection on a rescuer's SCBA or simply quickly remove the victim from the hazardous atmosphere.

WARNING:

At no time should rescuers remove their facepiece or in anyway compromise the proper operation of their SCBA in an attempt to share it with another firefighter or victim.

Figure 5.65 Move the downed firefighter from the building as soon as possible.

A rescuer trapped in a burning building should try to retreat to the ground floor. Conditions are usually worse on upper floors than on lower floors because of heat and smoke buildup. The rescuer who is unable to go down and exit should go to a room that has a window opening to the street. Once in the room, the rescuer should close the door, open a window, straddle the windowsill, and call for help (Figure 5.66). The rescuer should drop available articles, such as a lamp, chair, or pillow, out the window to attract attention. The rescuer must not break his or her protective ensemble of turnout clothes as this is the only protection from the hazards of the fire. If a window cannot be found, the rescuer should consciously stop and relax while considering actions for escape. The rescuer should use shallow breathing to extend the remaining air in the breathing apparatus; this will also calm the rescuer. In rare cases, access to the roof may be available. This can be used if the route to the roof is safe and free of extreme heat.

Figure 5.66 Firefighters in trouble can straddle the windowsill until a ladder is raised to them.

VEHICLE EXTRICATION

The most frequently encountered rescue situations are automobile accidents involving victim entrapment (Figure 5.67). These accidents are the result of collisions with other automobiles, larger motor vehicles, or stationary objects. Because a victim who is trapped may be seriously injured, proper extrication procedures are

Figure 5.67 Automobile extrication incidents are the most common rescue scenario faced by most firefighters. *Courtesy of Bob Norman, Elkton (MD) V.F.D.*

essential to prevent further injury and to speed the victim's removal. It is also critical that firefighters coordinate with emergency medical personnel who are providing first aid to the victim.

In general, there are three methods of gaining access to victims:

- Through a normally operating door
- Through a window
- By compromising the body of the vehicle

The simpler the required operation, the better for all concerned. When complex maneuvers are required to gain access into a vehicle, extrications become long, complicated, and ultimately more dangerous. For example, if a vehicle is not badly damaged, access may be obtained by simply opening an undamaged or operable door. However, when there is severe structural damage, when the car roof is collapsed, or when foreign materials are crushing the passenger compartment, gaining access can be a lengthy and frustrating process.

Assessing The Need For Extrication Activities

Proper accident scene assessment is perhaps the most important step in accomplishing an efficient extrication. Taking several minutes to carefully assess the scene can help avoid confusion, clarify required tasks, prevent further injuries to victims, and prevent injury to personnel.

Assessment should begin as soon as the first emergency vehicle approaches the accident scene. Rescue personnel should be observant as they approach the scene. They should try to determine how many and what type of vehicles are involved, where the vehicles are positioned, whether a fire is involved, and if there are any hazardous materials involved. Personnel should also be checking for any utilities, such as gas or electricity, that may have been damaged and are posing a hazard to the victims and rescue personnel. This and any other pertinent information, including the need for additional resources, should be relayed to the dispatcher as the initial size-up.

At the scene, personnel should make a more thorough assessment of the situation before taking any action. Personnel should assess the immediate area around each vehicle and assess the entire scene in more detail. The rescuer who assesses each vehicle should be concerned with the number of victims in or around the vehicle and the severity of their injuries. The rescuer should also assess the condition of the vehicle, extrication tasks that may be required, and any hazardous conditions that might exist. Ideally, there will be one rescuer to assess each vehicle involved in the incident, but this may not be possible (Figure 5.68). If there is only one rescuer available and more than one vehicle to survey, the rescuer must check each one separately and report the conditions in each vehicle to the incident commander.

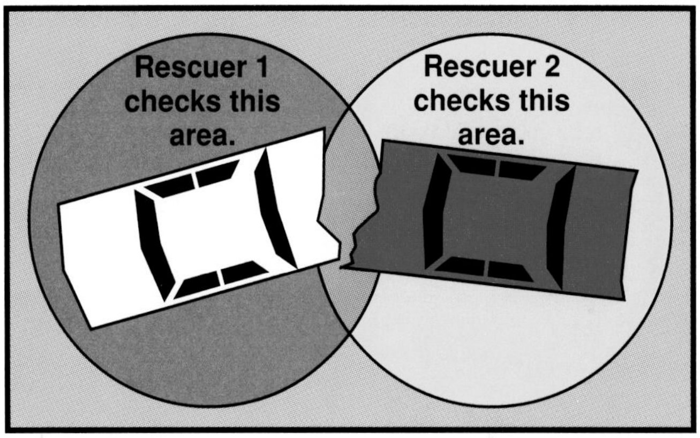

Figure 5.68 Rescuers should search around each of the involved vehicles.

While each vehicle is being checked, another rescuer should be assigned to survey the entire area around the scene (Figure 5.69). This person should check to see if there are any other vehicles involved that may not be readily apparent (over an embankment, for example), any victims who have been thrown clear of the vehicles, any damage to structures or utilities that present a hazard, or any other circumstances that warrant special attention.

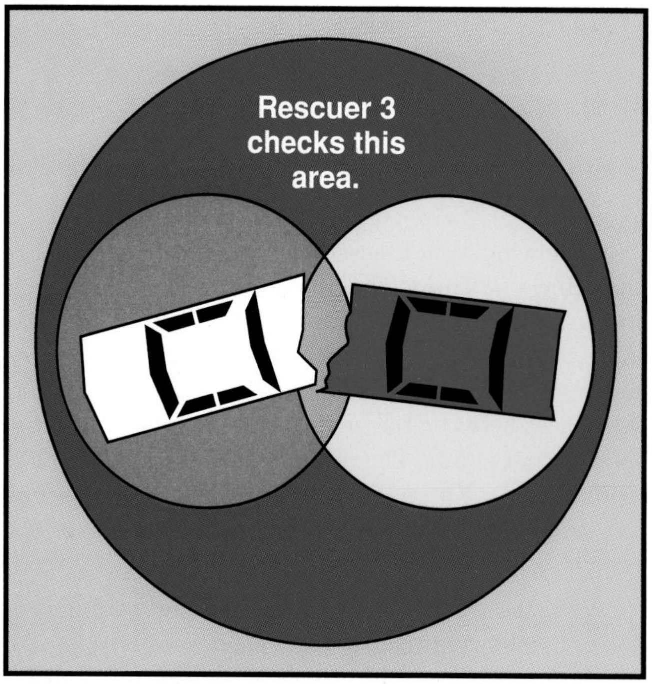

Figure 5.69 A third rescuer should make a general sweep and search of the entire scene. Be on the lookout for victims that may have been thrown or staggered clear of the vehicles.

Rescue personnel who are trained in first aid or more advanced emergency medical techniques should evaluate the condition of the victims (Figure 5.70). They should determine the extent of injury and entrapment. This information will aid the incident commander in determining the order in which victims should be removed. Of course, more seriously injured victims must receive higher priority than those with minor injuries. Victims who are not trapped should be removed first to make more working room for rescuers who are trying to remove those entrapped. As each assessment is completed, the rescuer should report his or her findings to the incident commander.

Figure 5.70 A trained rescuer should assess the victim's condition as soon as possible.

Stabilizing The Vehicle

Following scene assessment, rescuers must stabilize the vehicle(s). This is vital to prevent further damage to the vehicle(s), further injury to victims, or possible injuries to emergency personnel. Proper stabilization refers to the process of providing additional support to key places between the vehicle and the ground or other solid anchor points. The primary goal of stabilization is to maximize the area of contact between the vehicle and the ground to prevent any further movement of the vehicle.

Vehicles can be found in a number of different positions following a collision. Rescuers are often tempted to test the stability of the vehicle in the position in which it is found. Rescuers must be trained to resist this temptation because the slightest push in the wrong place may cause the vehicle to move. This is particularly true of vehicles that are on their sides or resting partially over a cliff or embankment (Figure 5.71).

Figure 5.71 Crib the vehicle on both sides to prevent accidental tipping.

Most vehicles involved in collisions remain upright. Rescuers must realize that even though the vehicle still has all its wheels on the ground, some stabilization is required to ensure maximum stability for extrication operations. The vehicle should be stabilized to prevent vertical and horizontal movement.

Several methods can be used to prevent horizontal motion. The most common method of preventing horizontal movement is to chock the vehicle's wheels. It is most important to chock the wheels on the downhill side of a vehicle that is sitting on a grade (Figure 5.72). If the vehicle is on level ground, chock the wheels in both directions (Figure 5.73). Chocking can be accomplished with standard wheel chocks, pieces of cribbing or other wood, or other appropriately sized objects.

It may also be possible to use one or more of the vehicle's own mechanical systems to assist in stabilization. This will depend on whether or not these systems are still operable. If possible, place automatic transmissions in the park position; place manual transmissions in gear. Set parking or emergency brakes.

CAUTION: Do not rely on mechanical systems, even if they are operable, as the sole source of stabilization. They should be used only with other stabilization procedures.

Figure 5.72 When the vehicle is resting on an incline, chock the wheels in the downhill direction.

Figure 5.73 All tools and cribbing should be in place before actual lifting of the automobile.

There are numerous ways to prevent an automobile from moving vertically. Jacks, air lifting bags, and cribbing are used most frequently for this purpose. Different types of jacks can be used to support the frame of the car. The advantage of jacks is that they can be adjusted to the required height; their disadvantage is that they are time consuming to place. Air lifting bags can also be used for support. To be effective, at least two air lifting bags are needed. They should be positioned either one on each side of the car, or one in the front and one in the rear (Figure 5.74).

Figure 5.74 Air bags can be used to support overturned vehicles; however, they do permit some bouncing movement that a solid box crib would not.

Standard wooden cribbing is also an effective stabilizing tool (Figure 5.75). Cribbing may be built up in a box formation until enough is used to support the car. It may be necessary to use wedges as the top pieces to ensure solid contact between the cribbing and the vehicle (Figure 5.76). Special step blocks can also be used to provide rapid stabilization of the vehicle (Figure 5.77). At least one and preferably two step blocks should be placed on each side of the vehicle.

Figure 5.75 It may be necessary to construct tall box cribs in order to stabilize overturned vehicles.

Figure 5.76 Wedges may be needed to provide maximum contact between the vehicle and the box crib.

Figure 5.77 Insert the step block until it makes solid contact with a portion of the vehicle's undercarriage.

When using any of these methods, the rescuer must take care to avoid placing any part of his or her body under the vehicle while placing the stabilizing device. There is always the possibility that the car may drop unexpectedly, injuring or killing the person beneath it. Handle cribbing on the sides to prevent any crushing hand injuries should a sudden drop occur (Figures 5.78 a and b on the next page).

Figure 5.83 Rescue from a building collapse may require the services of specially trained rescuers who have knowledge of building construction and are proficient in the use of rescue tools. *Courtesy of Harvey Eisner.*

may drop in large sections and form voids. If these sections remain in one piece with support on one side but collapse on the other, they form what is called a *lean-to* collapse (Figure 5.84). The weakening or destruction of bearing walls may cause the floors or the roof to collapse. This collapse may cause the debris to fall as far as the lower floor or basement. A collapse of this nature is referred to as a *pancake* collapse (Figure 5.85). When heavy loads, such as furniture and equipment, are concentrated near the center of a floor, the excess weight may cause the floor to give way. This form of collapse is known as a *V-type* collapse (Figure 5.86). A V-type collapse will result in voids near the walls. Regardless of the type of collapse being dealt with, there are three primary methods of making the situation stable and safe for removing victims. These methods are shoring, cribbing, and tunneling.

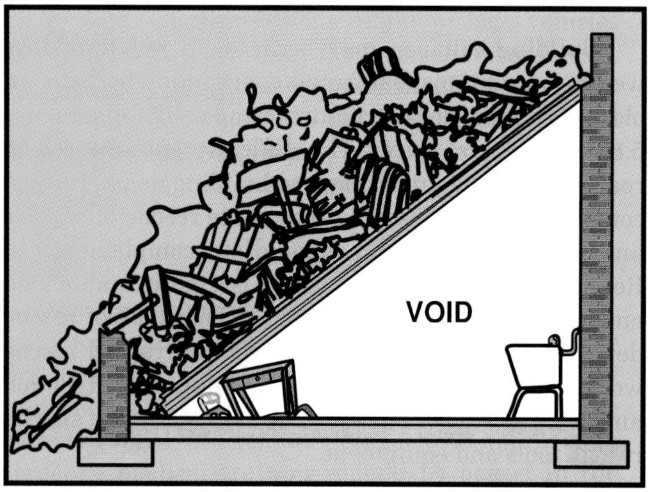

Figure 5.84 Voids may be found near the wall on the high side of a lean-to collapse.

Figure 5.85 Small voids may be found throughout a pancake collapse.

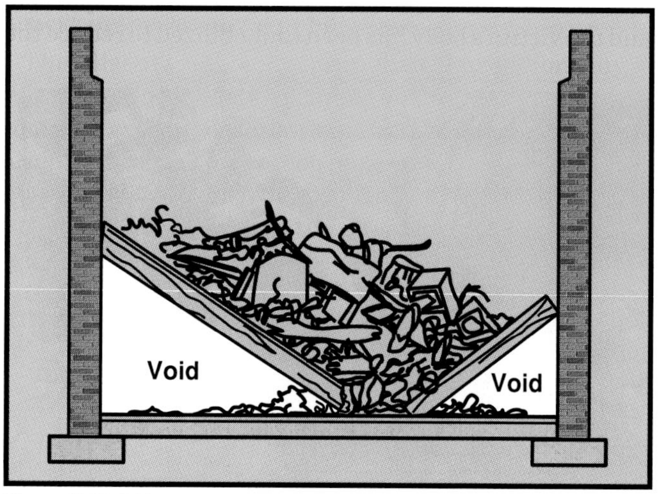

Figure 5.86 A V-type collapse will present voids near the walls.

SHORING

Shoring is a process of erecting a series of timbers or jacks to strengthen a wall or to prevent further collapse of a building or earth opening (Figure 5.87). It is not to be used to restore walls or floors to their original position. Any attempt to force beams, sections of floors, or walls back into place may cause further collapse and damage. Temporary shoring is usually all that is performed by rescue squads. Because bracing, shoring, or supporting walls and floors are basically engineering problems, fire officers should call upon the sources available for assistance when planning and conducting these operations. Shoring is difficult work and requires training and practice to be proficient.

CRIBBING

Cribbing is the process of arranging planks into a cratelike construction; this arrangement usually has separated joints. These planks may sometimes be crossed with the joints staggered. Cribbing, as used in rescue

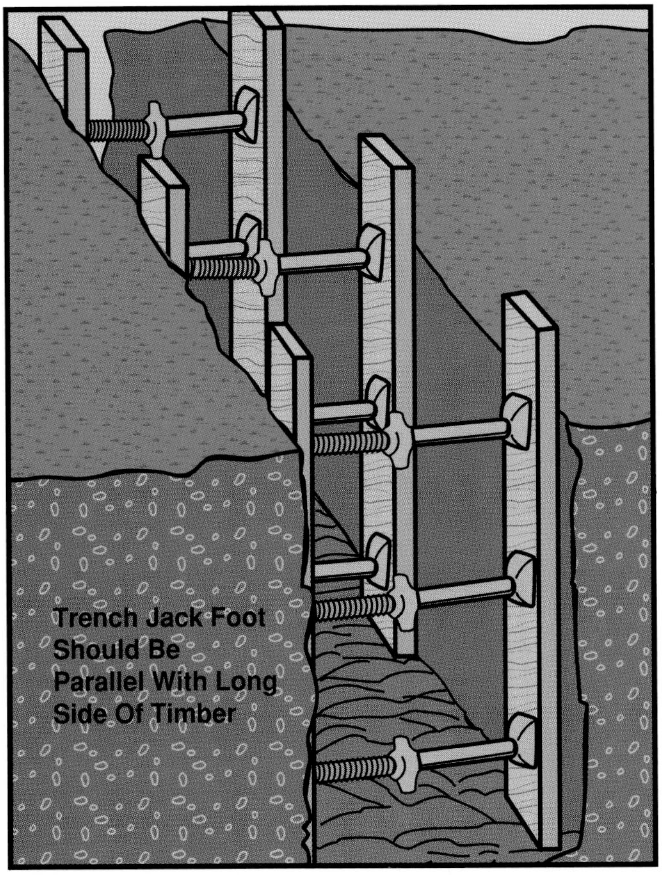

Figure 5.87 Shoring and trench jacks are used to prevent further collapse of trench walls.

work, is usually adapted to roof and ceiling supports, but it can be used on walls when necessary. Tunnel openings and passages are made secure by using the shoring and cribbing processes.

TUNNELING

In order to reach a victim who is buried or trapped beneath debris, it is sometimes necessary to dig a tunnel. Tunneling is a slow, dangerous process and should be undertaken only after all other methods are found impractical. Tunneling should be conducted from the lowest possible level and should not be used for general 3/95 search. It may, however, be used to reach a void under a floor where the search is to be continued. A tunnel must be of sufficient size to permit rescuers to remove victims, and it should not be constructed with abrupt turns (Figure 5.88 on the next page). Tunnels as small as 30 inches (760 mm) wide and 3 feet (1 m) high have proved satisfactory for most rescue work. Tunnels should be driven along a wall whenever possible. This procedure simplifies the framing required to prevent a cave-in.

Rescue From Trench Cave-Ins

A growing number of serious injuries and fatalities are resulting from excavation and trench cave-ins at construction sites. The most common problem that faces trench collapse victims is the reduced ability to breathe. This is a result of being totally covered by dirt and debris or being partially covered but unable to expand the chest cavity to facilitate respirations. The rescuer's first priority must be to uncover enough of the victim's head and chest to allow for sufficient respirations. There are several methods of assisting the victim's breathing. An air hose or a partly opened cylinder can be inserted in a hole dug to the victim's face. In an emergency, air may be directed to a victim through a garden hose. Under most circumstances, regular EMS oxygen equipment can be used to assist the victim.

Although a buried victim must immediately receive air, rescue operations depend on making the site as safe as possible by shoring or cribbing to hold back other weakened earth formations (Figure 5.89 on the following page). Rescuers should not be sent into the trench unless their safety can be reasonably ensured. To ensure rescuer safety, shoring is required to hold unstable trench walls or spoil piles. Meanwhile, rescue apparatus, nonessential personnel, heavy equipment, and spectators should be moved back to avoid causing secondary cave-ins.

While the remaining trench walls are held back, the debris must be removed from around the victim. Heavy equipment should not be used until the exact location and number of victims are known. Likewise, picks should not be used in a search for victims, or they should be used with great care. Remove debris with baskets, buckets, and wheelbarrows to a clear area. Take adequate time for this operation to ensure that the material will not fall back into the cave-in site. Loose debris is unstable.

When removing debris, watch for key timbers or rocks that support other heavy portions of earth or other debris. Improper moving of these key pieces could cause a dangerous collapse or slide. Leaving key pieces in place while debris tunneling is one of the hardest jobs in rescue work and should be undertaken only when other means of gaining access are impractical.

SAFETY PRECAUTIONS

Several safety precautions firefighters and officers must remember when they are involved in cave-ins and excavation rescues are as follows:

- The trench should be entered with proper protective equipment — this equipment includes head, hand, and eye protection. If a toxic or oxygen-deficient atmosphere is suspected, respiratory protection should also be worn.

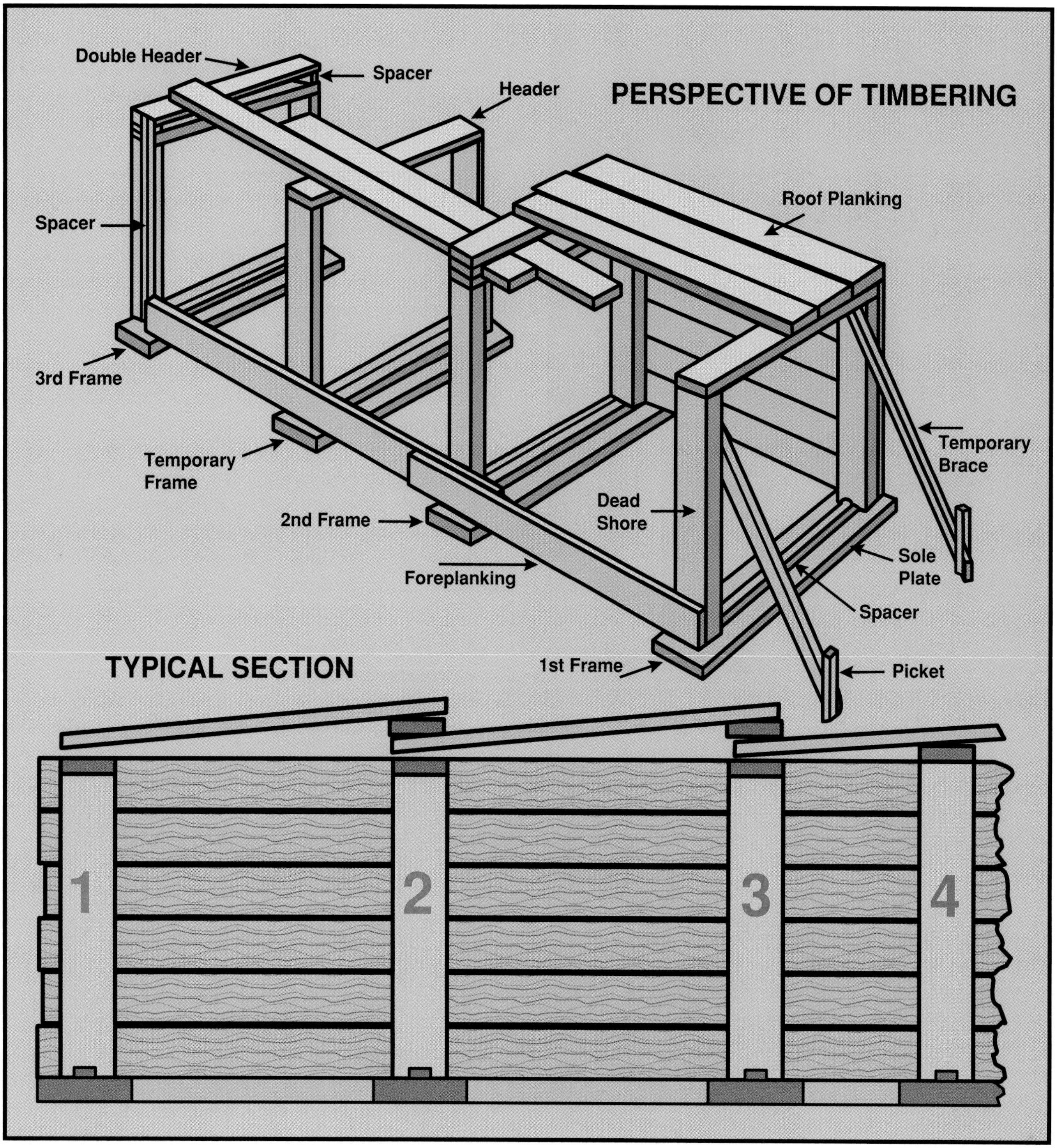

Figure 5.88 It may be necessary to construct tunnels to affect some rescues.

- Exit ladders should be placed in the trench on both sides and ends. Ladders should extend at least 3 feet (1 m) above the top of the trench and be secured in place if possible (Figure 5.90).
- Firefighters should be careful with the tools they use in the trench to avoid injuring each other or the victim.
- Unnecessary fire department personnel and bystanders should be kept out of the trench and away from its edge.
- Rescuers should be aware of any other hazards that might exist at the scene such as underground electrical wiring, water lines, explosives, or toxic or flammable gases.

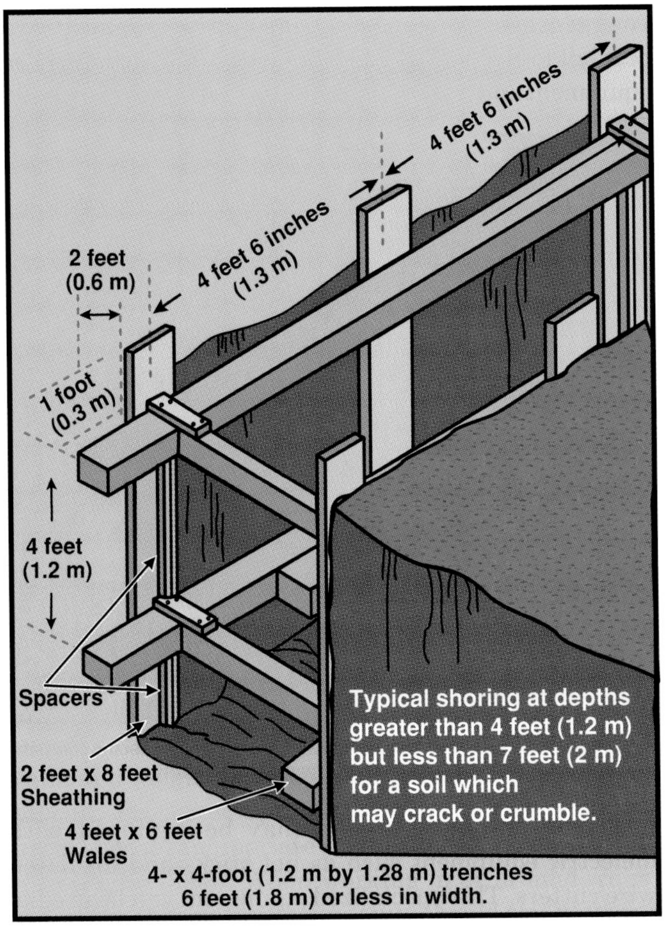

Figure 5.89 Trench shoring must be of appropriate size and quantity.

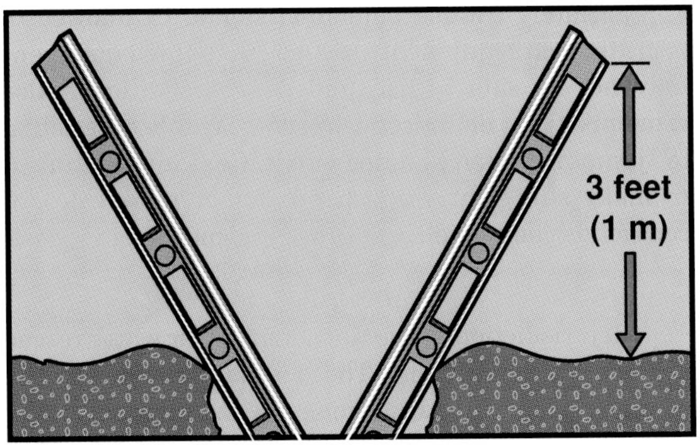

Figure 5.90 Ladders must extend at least three feet beyond the top of the trench.

Rescue From Caves And Tunnels

Fire departments that have caves and tunnels in their jurisdiction must have a plan for locating and removing people who get lost or trapped in them. Operations in caves and tunnels are highly specialized and depending on their capabilities, the fire department may not be the best organization to carry out the operation. Pre-incident planning of these incidents should include the summoning of specially trained cavers, also known as speleologists or "spelunkers," to run or assist the operation. Areas that have numerous caves often have local speleological clubs that can be consulted or relied upon to assist in handling such incidents.

TYPES OF CAVES

The three main types of caves are solution, lava tube, and talus. Solution caves are the most commonly explored. Solution caves have two types of passages: phreatic and vadose. Phreatic passages were formed when the cave was below the water table. These passages are usually large, smooth-walled, level, and dry. Vadose passages are formed by water. They tend to be narrow, irregular, and often contain active streams of water.

Lava tube caves are formed by lava flows that cool on the outside first while the lava inside is still hot and moving, thus leaving a void space. These caves are very unstable, and eventually they will collapse until they are completely filled in.

The talus cave is formed by fallen rock between ridges or at the base of mountains. Although not really caves in the common sense of the word, they do pose the same hazards as other caves. The irregularity of passages and unstable rocks are their greatest hazards.

CAVE RESCUE PROBLEMS AND DANGERS

Most caves have similar characteristics such as darkness, water, passage irregularities and cool temperatures. These characteristics pose rescue problems and dangers to rescuers and victims alike. Darkness will be a given factor in nearly all caves. Rescuers must carry their own light. It is recommended that three sources of light be carried. The primary light should be a cap-mounted, battery-powered lantern similar to those that miners wear. A waterproof flashlight and candle should also be carried. The candle is for use while waiting or when repairing other lights.

Many caves serve as natural conduits for water. Water levels may vary depending on local rainfall, snow melting, and other factors. In most cases, high water must be allowed to subside before it is safe to enter the cave. Be aware that even a small rain shower could cause the cave to fill again. The water temperature in all caves is usually around 55°F (13°C). Water of this temperature poses severe hypothermia potential should the victim or rescuer become immersed. If it becomes necessary to attempt recovery operations in standing water, only experienced cave divers should be allowed in the water.

Passage irregularities can pose many problems. Sharp bends, rises, dropoffs, and restricted openings may all pose challenges to rescuers. In order to traverse

Making personal contact between the rescuer and victim in the water is a last-resort tactic. Panicked victims can injure well-intentioned rescuers and pull them under as well. If it does become necessary to use this method, it is safest to approach and grab the victim from behind.

ICE RESCUE METHODS

Ice rescues require particular caution because of the unstable surface being worked on and the potential for the rescuers becoming quick hypothermia victims. Traversing unstable ice is most safely performed in a flat bottom boat or raft. Rescuers can push themselves across the ice and be protected from the water should the ice fracture further. If a boat is not available, the rescuers should try to distribute their weight by operating from ladders or a sheet of plywood laid on the ice. Wooden ladders are ideal because they will float should the ice fracture. Once the hole is reached, a rope or pole should be extended for the victim to hold on to as he or she is pulled from the water (Figure 5.95).

Figure 5.95 A ladder may be extended to victims for them to grab onto and be pulled to safety.

Industrial Extrication

Industrial extrications are among the most challenging rescue situations that firefighters will ever face. Because there is an endless number of machines that have the potential to entrap victims, it is impossible to list specific techniques for victim removal. When planning for industrial extrications, a process approach must be used. By following a set process, most industrial extrications can be handled in as smooth a manner as possible. The process is as follows:

Step 1: Survey the situation.

Step 2: Seek expert assistance.

Step 3: Neutralize power sources.

Step 4: Stabilize machinery.

Step 5: Survey again.

Step 6: Extricate the victim.

Step 7: Retrieve extrication equipment.

Step 8: Secure the scene.

When surveying the situation, personnel should take into account the following:

- The medical condition and the degree of entrapment of the victim
- The number of rescue personnel required
- The type and amount of extrication equipment needed
- The need for special personnel, equipment, or expert assistance
- The level of fire or hazardous material hazard that is present

This process may pose some very difficult decisions; however, these observations are critical to the rest of the incident. For example, if a victim is seriously entangled and in danger of bleeding to death, amputation by a doctor brought to the scene may be required to save the person's life.

If during the initial survey it becomes obvious that the problem is beyond the expertise of the rescue team, outside expertise is required. In most cases, this will be plant personnel on site who are more familiar with the involved machinery. Plant maintenance personnel are usually good sources of information. In rare cases, it may be necessary to go to off-site sources, such as machinery manufacturers, for help. Ideally, these outside sources will be identified during pre-incident planning.

THE INDUSTRIAL EXTRICATION PROCESS

The first physical step of the extrication process is to neutralize power sources. With almost every machine, the initial action to be taken is to cut the electricity to the machine (Figure 5.96). Once this is done, other sources of energy, such as hydraulic fluids under pressure, compressed air, energy stored in springs, and potential

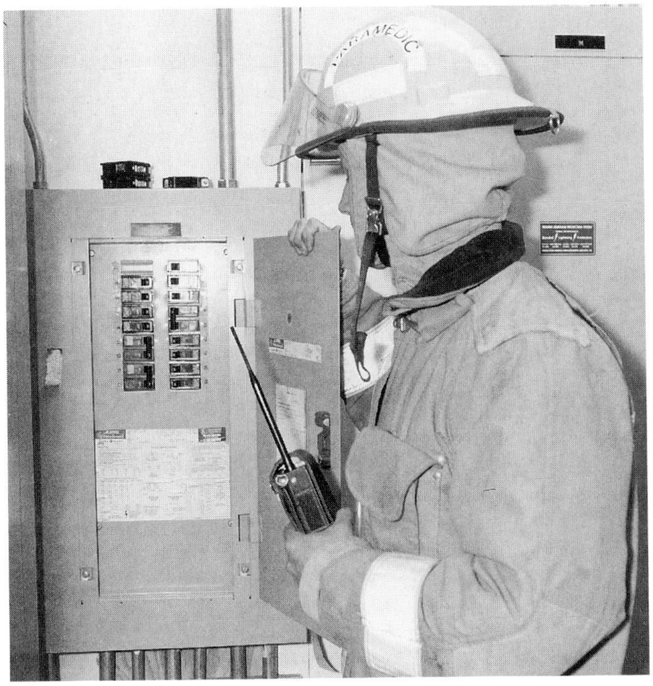

Figure 5.96 Station a firefighter with a portable radio at the electrical panel.

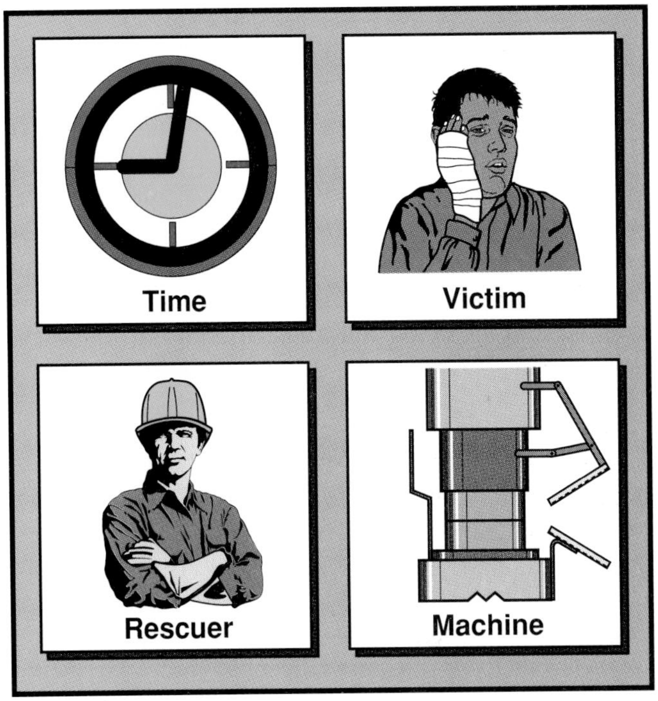

Figure 5.97 There are four critical factors that must be considered when determining a course of action for industrial extrication operations.

energy from suspended parts, should also be neutralized. (**NOTE:** The exception to this rule would be machines that are designed to operate in a full cycle. These machines, such as power printing machines and metal punch presses, may complete the cycle when the power is shut off, inflicting further damage to the victim. The *first* physical extrication step on full-cycle machines is to *stabilize* the machine.)

With the power sources neutralized, the machine can be stabilized to prevent any further unwanted movement. This will generally be accomplished using blocking and cribbing materials that are similar to those used on other extrication incidents. However, it may be necessary to use larger cribbing on industrial machines, because the machines may be capable of cutting cribbing that is normally used.

With all power sources neutralized and the machine stabilized, rescuers should again survey the situation. In some cases, just the process of stabilizing the machine will allow the victim to be removed. If this is not the case, further procedures will be required.

When deciding how to extricate the victim, four critical factors must be addressed (Figure 5.97):

- Time necessary to complete the extrication
- Effects the extrication will have on the victim
- Effects the extrication will have on the rescuers
- Effects (damage) the extrication will have on the equipment involved

There are three basic procedures that can be used to extricate the victim from the machine.

- One method is to manipulate the machine through its normal range of operation to loosen the victim. This approach is generally preferred when the victim is not injured, but merely stuck.
- Another method is to disassemble the machine. (It is very helpful to have plant maintenance personnel who are familiar with the machine and are available to consult.) There are three circumstances when disassembly is the most desirable procedure:
  - — where the victim has minor injuries and quick extrication is not important
  - — where the machinery is highly valuable and possible damage to it is not warranted
  - — where the machine is too strong to be affected by extrication tools
- If the machine cannot be manipulated or disassembled, it will be necessary to force the machine. Care must be used to ensure that parts do not react in a manner that may be injurious to the rescuers or the victim.

Once the extrication is complete, rescue workers can retrieve their tools and place them back in service. The scene should be left in a safe manner that will not promote the possibility of further injury to people who will be in the area.

Elevator Rescue

Firefighters are usually the first people called when an elevator emergency occurs. Most elevator emergencies involve elevators that are stuck between floors due to a mechanical or power failure. Unless there is a medical emergency in the elevator car, the best approach is to reassure the occupants that help is on the way and then wait for the elevator mechanic to arrive and handle the problem. If it does become necessary to take a proactive approach, it is important that firefighters know the basics about elevators and their operation.

ELEVATOR CAR DOORS AND OPENINGS

Elevator doors are powered by an electric motor that rests on the top of the elevator car. The car door does not have locks and can be pushed open by hand; however, electrical interlocks are provided to stop car movement when the doors open. Elevator car doors are designed to open and close automatically when the car receives instructions from the controller to stop at a particular floor landing. The car door, powered by the motor, is also the means employed to open the hoistway door. When the elevator stops at a desired level, the hoistway doors are unlocked by a driving vane attached to the car door (Figure 5.98). As the car door opens, the vane strikes a roller that releases the hoistway door lock, and the car door pushes the hoistway doors completely open. For closing, as a weight forces the hoistway closed, the driving vane moves away from the roller and the hoistway doors are relocked. This action takes place only when the doors are supposed to open and does not happen if the car stops somewhere other than the landing.

Figure 5.98 A common elevator installation.

Access Panels

In addition to the main doors of elevator cars, additional exits are provided for use in emergency situations. These exits are located on the top and at the sides of the elevator car (Figure 5.99).

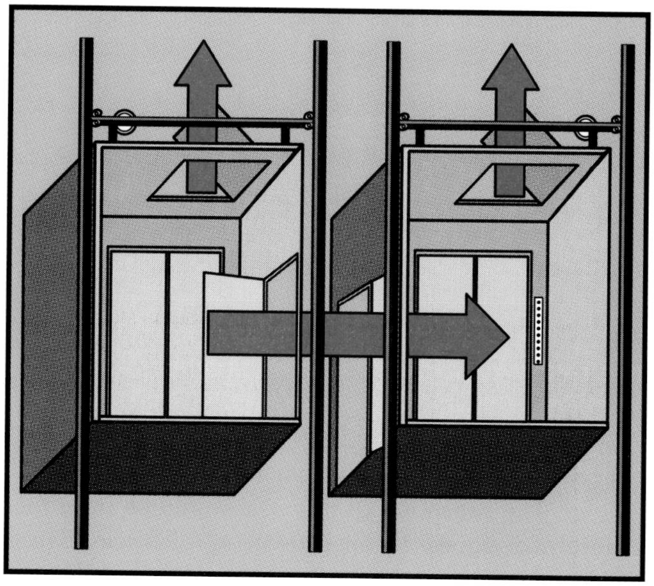

Figure 5.99 Elevator access panels may be found in the sidewall or ceiling of the car.

Top Exit

A top exit is provided on all electric traction elevators. On hydraulic elevators, a top exit may or may not be provided because the system is equipped with a manual lowering valve. Top exits are only required if the elevator does not have this valve, which permits the lowering of the car in the event of trouble. The top exit panel is normally locked from outside the car and is designed to open in that direction. In addition, some elevator cars are provided with electrical interlocks that prevent car movement while the panel is open.

Side Exit

In installations with multiple hoistways, side exits are normally provided so that travel from one car to another is possible for rescue operations. However, it may not be provided on hydraulic installations that have manual lowering valves. Side exits are required to have electrical interlocks to prevent car movement while the panel is open. The panels may be opened from the inside with a special key or from the outside where a permanent handle is provided.

Hoistway Doors

Hoistway doors are equipped with a locking mechanism mounted on one of the panels to prevent the doors from opening if the car is not at the landing. The latching device is shaped on one end so that it hooks around a stop to prevent the doors from opening. In the unlocked position, the shaped end of the latching device is pushed away from the stop and the door can be opened. The latch is attached to a roller or drive block and when an elevator car comes to a landing, a driving vane mounted on the car door is positioned near the roller or block. If the car is not programmed to stop, the vane passes by the roller without contact. When the car does stop, the vane is pushed against the roller, the hoistway latch is released, and the movement of the car door pushes the hoistway door open (Figure 5.100). Hoistway doors are equipped with a weight or spring system that applies a constant force against the door toward the closed position. The force used to open the car door must be sufficient to overcome the force of the hoistway door weights. As the electric motor closes the car doors, the weight or spring on the hoistway door allows it to close simultaneously with the car door. The driving vane then moves away from the roller or block and the latch resets itself, locking the hoistway doors and permitting the car to move on.

Figure 5.100 The hoistway vane is seen here between the doors.

ELEVATOR CONTROL

An elevator car will automatically proceed to the floor from which it receives a call, regardless of what activated the call button. The effects of heat, smoke, and moisture may cause the elevator control mechanism to activate. Because of this activation, it is possible for the elevator car to go to the fire floor, open its doors, and expose the occupants to untenable conditions. However, in emergency situations, it is desirable to have the ability to control the elevator car without interference. The two methods available are known as independent service and emergency service. Both services are controlled by keys that may or may not be readily available; therefore, firefighters should be aware of how to obtain the control keys.

Independent Service

Independent service is a feature available on most elevator installations that allows the car to be controlled manually. The car will respond only to those instructions issued from the operator inside the car, and it will not respond to signals initiated from floor landings. This is an advantage for firefighters because they can move the car at will without having it stop for calls at landings or the fire floor. The service is usually activated by a switch located inside the car in a locked panel and identified by the letters I.S. (Figure 5.101). On some installations, the switch may be located on the control panel in the main lobby. After the car has been switched to independent service, movement and direction of travel are set by pushing a floor selector button. When the floor is reached, the door will open automatically. This control should only be used if it is within the fire department's SOPs to use elevators in fire situations.

Figure 5.101 The fire department can control the elevator by using the appropriate key.

Emergency Service

A more recent innovation is known as the emergency service mode of operation. The operating device is located on the car control panel and is operated by a key. Once on, the elevator will respond only to signals initiated from within the car and disregard those from floor landings. In addition, the doors must be opened and closed by pressing the appropriate button. Both of these operations require the use of keys. Independent service requires a key to unlock the panel that protects the toggle switch. Emergency service requires a key to activate the system.

EVACUATION OF ELEVATORS

Upon arrival at the scene of an elevator emergency, firefighters should have an elevator mechanic dispatched to the scene. The mechanic is trained to make mechanical adjustments to the elevator that may enable passengers to exit from the elevator car in a normal manner. There may be times when a mechanic is unavailable or when the time of arrival will be extremely long. Also, there may be passengers aboard who are in critical medical condition, requiring the fire department to evacuate the car immediately. Elevator rescue requires training in the use of proper rescue techniques, and fire departments should not attempt rescue with untrained personnel. Under no circumstances should firefighters alter the elevator's mechanical system in an attempt to drift the elevator. Adjustments to the mechanical system of the elevator installation should be performed only by the elevator mechanic.

Determine Car Location

After it has been decided that evacuation of the elevator car is necessary, the exact location of the car must be determined. Its position in the hoistway can be located by observing the position indicator panel in the lobby or by opening the hoistway doors at the lobby and looking into the hoistway.

Establish Communication

Communication must be established with the passengers to assure them of their safety and that work is being done to release them. If a telephone or intercom is not available, shouting through the door near the stall location may be sufficient for passing messages back and forth. Communication with the passengers is essential for their morale and mental state and should be established regardless of whether the fire department is going to rescue or wait for the elevator mechanic.

Secure Elevator

Rescue procedures will vary depending upon where the elevator car is located in relation to the nearest landing. Elevators may be stalled near the landing (several inches above or below), below the landing, or above the landing. Each situation requires the use of different techniques for passenger removal.

Firefighters must make certain that the stalled elevator car does not begin to move once rescue operations have begun. The elevator can be secured by opening the main power circuit to the elevator drive motor and applying an electrical lock-out device or removing the fuses. Another method would be to leave a firefighter at the panel. The disconnect switch is in the machine room, and it immobilizes the elevator by removing power and applying the brakes. As a further backup, the car's emergency stop switch should be activated as this also removes power from the drive motor. Passengers can be instructed to activate the switch and when the firefighters enter the car, they should also check to make sure that the switch has been activated.

When working on electrical equipment, specifically elevators, safety lockouts should be used. Approved lockout devices ensure that electrical power will not be inadvertently restored after being shut off. Such lockouts are used in conjunction with padlocks once electricity has been shut off at the control box. Lockouts and/or safety guards should be used during incidents involving elevator rescue, compactor malfunction, industrial process equipment, or other similar situations. Only after it has been verified that ALL operating personnel are in a safe location should the device be removed and the power restored.

Opening Hoistway Doors

The opening of the hoistway doors can easily be accomplished by the use of formed emergency keys, or if necessary, by forcing the doors open. There are three types of keys used for most elevator installations:

- Moon-shaped key — Inserted into the opening a few inches and then pulled downward.
- T-shaped key — Pushed straight into the opening to operate the locking mechanism.
- Drop key — Constructed of two sections hinged together. When the key is inserted into the opening far enough, the front section drops to form a 90-degree angle with the rear section. The key is then rotated to unlock the door.

After the keys have released the locking mechanism, the hoistway doors can be pushed open by hand.

Alternate methods must be used when elevator keys are not available or are unusable. If the elevator is stalled near or within the landing zone, the passengers can be instructed to push open the elevator car door. At this time, if the car is close enough to the landing, the hoistway door will also open. If it is not close enough to the landing, opening the car door will only expose the closed hoistway doors. However, the locking mechanism may also be exposed and the passengers may be able to release it.

Another method, known as poling, can be used with multiple hoistways. This is accomplished by lining up another elevator next to the hoistway door of the stalled elevator. A slender pole or slat can be slipped between the car and wall of the operating elevator and is used to reach the door lock of the stalled elevator.

A less desirable method of opening hoistway doors is to force the door open using forcible entry tools. There is a good possibility of only bending or crimping the door and not opening it. When doors are forced, it should be done at the top, near the lock, and in the direction that the door opens.

Passenger Removal

When an elevator car stalls within the landing zone, the car is close enough to the landing to release the locking mechanism on the hoistway door. Evacuation of the elevator car is simply a matter of pushing the doors open and assisting the passengers from the car.

When the elevator is stalled above or below the landing level, the passengers will be forced to climb up or down to reach the landing. Once the doors are open, a firefighter should enter the car to make sure that the emergency stop switch is activated. A hazard that exists when the car is stalled above the landing is the opening below the car into the hoistway. Precautions should be taken to guard the hole with a ladder to avoid the possibility of someone falling into the hoistway.

Access to the top of a stalled car is accomplished by opening the hoistway door above it. Firefighters lowered by lifelines can then open the top panel, which can only be opened from the outside. Care should be taken so that the elevator ceiling panels do not fall on the passengers. Once the exit panel is removed, a ladder is placed down into the car, and a firefighter should enter the car to check the emergency stop switch. A lifeline should be tied to each passenger as he or she is assisted through the opening to the top of the car. A firefighter located on the landing will then assist the passenger onto the landing and untie the lifeline. If the top of the elevator is not even with the landing, a ladder may be needed.

Side exit removal of passengers is an effective method of rescue if the elevator installation is in a multiple hoistway. To accomplish a rescue, a rescue elevator car (on independent service mode) is moved even with the stalled elevator car. The side exit doors of both cars are opened and rescuers enter the stalled car to check the emergency stop switch. A walking plank is laid between the two cars, and a pike pole is used as a handrail while the passengers are guided into the rescue car. The passengers should be attached to a lifeline rope.

The passengers trapped in the elevator will be very apprehensive. Rescuers must constantly reassure them that they are safe, that there is no need for alarm, and that they will soon be free of the elevator.

Escalator Rescue

Escalators, also called moving stairways, are stairways with electrically powered steps that move continuously in one direction, usually at speeds of either 90 or 120 feet (28 m or 36 m) per minute (Figure 5.102). Each individual step rides a track. The steps are linked together and move around the frame by a step chain. The driving unit is most commonly located under the upper landing and is covered by a landing plate. The handrails also move at the same rate as the stairs.

Many escalators have manual stop switches located on a nearby wall or at a point close to where the handrail

Figure 5.102 Escalators are found in many occupancies.

goes into the newel base (Figure 5.103). Operation of the switch stops the stairs and sets an emergency brake. The stairs should be stopped during rescues or when firefighters are advancing hoselines up or down the moving stairway. Micro switches may be located behind side panels or under the landing plate. These switches allow power to be shut off to the unit.

The design of modern escalators has virtually eliminated the chance of people being caught in them. The most common escalator rescue scenario would be that of a repairperson who gets caught in the linkages during repair operations. In these cases, it is best to get another escalator technician to assist in removing the victim. If the victim is conscious and able to provide information, he or she may be able to direct the firefighters in proper release procedures.

Figure 5.103 Most escalators have emergency stop controls.

6

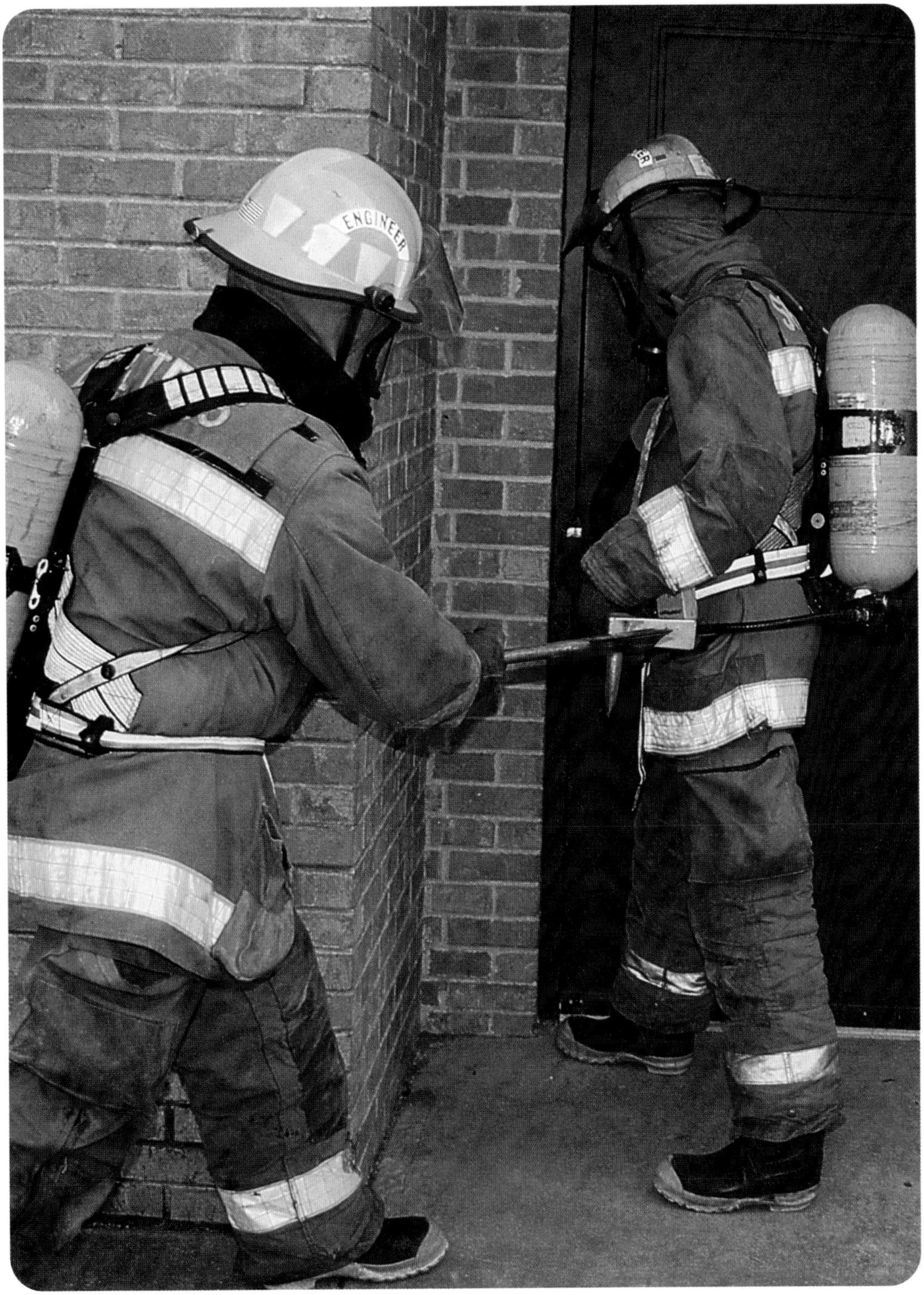

Forcible Entry

This chapter provides information that addresses performance objectives described in NFPA 1001, *Standard for Fire Fighter Professional Qualifications* (1992), particularly those referenced in the following sections:

3-8 Forcible Entry

3-8.1

3-8.2

3-8.3

3-8.4

3-8.5

3-8.6

Chapter 6
Forcible Entry

Firefighters frequently encounter situations where entry to a room or building requires force. In some instances, the firefighter uses physical strength to successfully remove the barriers to entry. However, the job will be made possible or easier by the firefighter's knowledge of the barrier's characteristics, the available tools, and the techniques that should be used. This knowledge allows the firefighter to open the barrier quickly and with less damage than would be caused by using improper tools or techniques. Ultimately, this ability makes the fire fighting job more effective and impresses the general public and building owners.

This chapter begins by highlighting the many tools that can be used for forcible entry operations. Their proper use, care, and maintenance are crucial to the success of the forcible entry operation. Characteristics of the various types of barriers that may have to be forced, such as doors, floors, walls, fences, and windows, are also covered. The opening of roofs is covered in Chapter 7 on ventilation.

FORCIBLE ENTRY TOOLS

The skills and techniques required for forcible entry can be learned only through thorough training. While the basics about common barriers that may require forcible entry can be learned, it is impossible to anticipate every forcible entry situation. Firefighters will be best prepared if they are proficient with their equipment. Complete knowledge of tools and equipment allows firefighters to rapidly devise a method for completing almost any forcible entry operation. Once firefighters develop an appreciation for the tools they work with, they will want to keep the tools properly maintained. Each fire department should set its own rules concerning the care of forcible entry tools.

The safety factor, more than any other consideration, should determine the method by which forcible entry tools are carried. When these tools are carried by hand, precautions should be taken to protect the carrier, other firefighters, and bystanders (Figure 6.1). In atmospheres that could be explosive, extreme caution should be taken in the use of power and hand tools that may cause arcs or sparks. When tools are not in use, they should be kept in their properly designated places on the apparatus (Figure 6.2).

Figure 6.1 Always carry tools in a manner that will avoid harm to yourself and others.

Figure 6.2 When not in use, all tools should be stored in their proper place on the apparatus.

Cutting Tools

Cutting tools are the most diversified of the tool groups. However, most cutting tools are designed to cut only specific types of materials. Misuse occurs when a tool is used to cut material it was not designed to cut. This misuse can destroy the tool and endanger the operator. Cutting tools may be either manual or power tools. The following sections discuss both types.

AXES AND HATCHETS

The fire fighting axe has many uses, but it is designed primarily for cutting. Two of the most popular designs are the flat-head axe and the pick-head axe. The flat-head axe is often used as a striking tool, especially if used in conjunction with another tool during forcible entry (Figure 6.3). The pick-head axe can be used as a puncturing or picking device (Figure 6.4). Both types of axes have tapered blades that make them adaptable for prying. However, handles tend to break when the head is twisted during a prying operation.

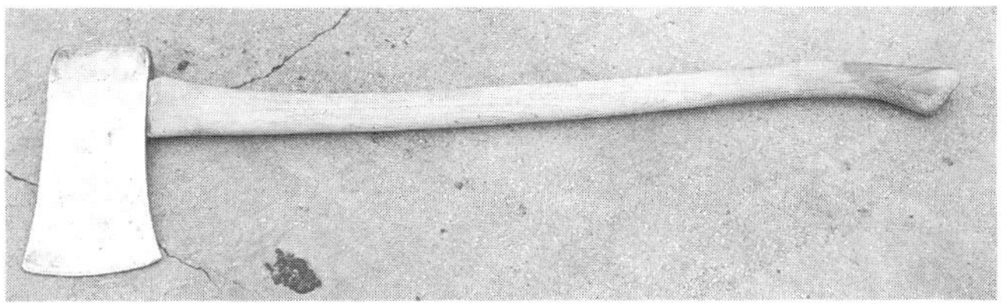

Figure 6.3 A flat-head axe.

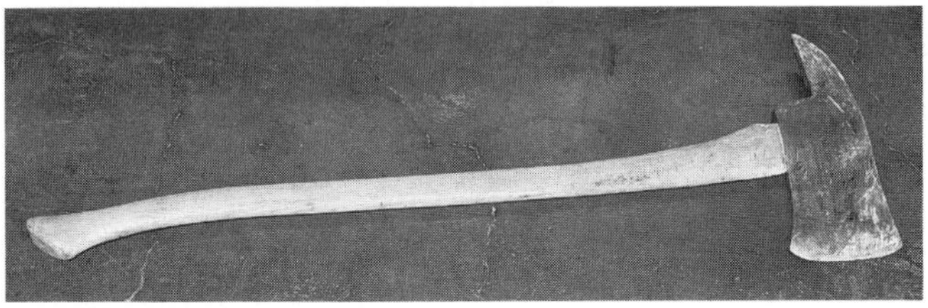

Figure 6.4 A pick-head axe.

Figure 6.5 The firefighter should carry the axe blade away from the body.

Many smaller types of hatchets are also commonly used for forcible entry operations. These include small adzes, pry axes, and standard hatchets. The care and maintenance for these tools is the same as for the full-size axe.

A fire axe should be carried so that there will be little danger of an injury. An axe or any other edged or pointed tool should never be carried on the shoulder. The firefighter should carry the axe blade away from the body. One method of carrying an axe is shown in Figure 6.5. In this carry, the point of the pick is shielded, and the sharp blade protected or pointed away from the body.

HANDSAWS

Generally, power saws are used for forcible entry, and handsaws are used for rescue. However, depending on the situation, firefighters may be required to use handsaws to perform forcible entry.

Handsaws are useful on objects that require a controlled cut but are too big to fit in the jaws of a scissors-type cutter or unsuitable for cutting with a power saw. Handsaws are often more time consuming to use than are power saws or shears. However, handsaws are safer to use when working close to a victim. Handsaws commonly used by firefighters include carpenter's saws, hacksaws, coping saws, and keyhole saws (Figure 6.6). The efficiency of hacksaws can be increased by using two blades, installed in opposite directions, to allow cutting in both directions of the stroke. All saw blades should be kept sharp, clean, and lightly oiled. The efficiency of the saw can be increased by periodically spraying light oil on the surface of the material being cut. This oil will reduce friction between the material and the saw blade.

POWER SAWS

Power saws are available in various designs, depending upon their purpose. The operator should know the limitations of each type of power saw. When a saw (or any tool) is pushed beyond the limits of its design and purpose, two things may occur: tool failure (including breakage) and/or injury to the operator. The use of power saws in flammable atmospheres must be prohibited. The saw's motor or sparks from the cutting operation can ignite a fire or explosion.

Circular Saw

A circular saw, also known as a rotary rescue saw, is well suited for cutting a variety of materials because it can be equipped with different blades, each designed for a specific material (Figure 6.7). As with handsaws, a large-toothed blade produces a faster, less precise cut than a fine-toothed blade. Carbide-tipped blades are superior to standard blades because they are less prone to dulling with heavy use. The circular saw must be used with extreme care.

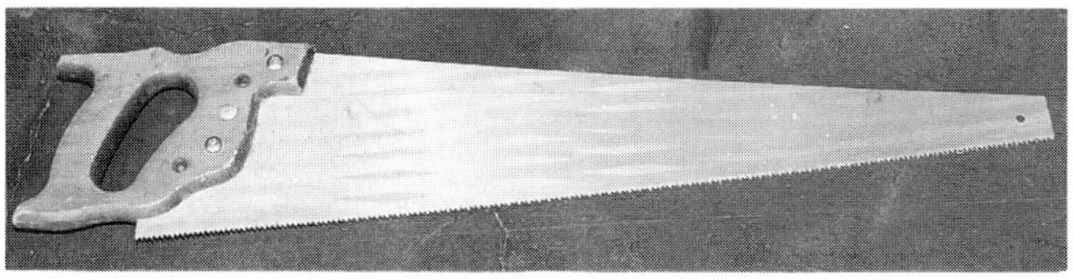

Figure 6.6 A handsaw.

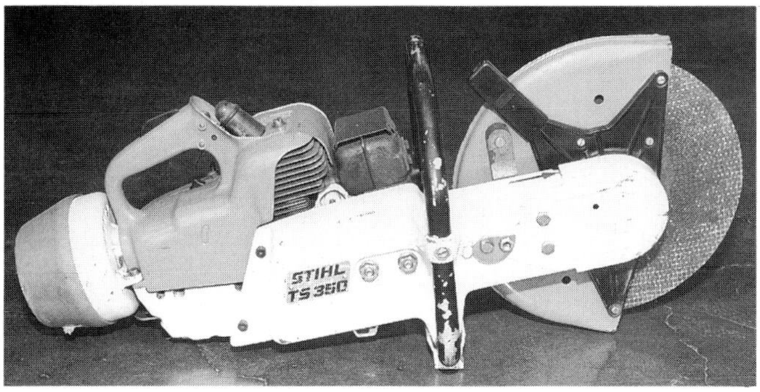

Figure 6.7 A circular saw, which is also known as a rotary rescue saw.

If necessary, cool the blade with a fine water mist from a hoseline. Make sure that the water mist is on the blade *before* cutting any metal. Applying water on a hot blade can result in blade disintegration. Wear full protective clothing with face and eye protection when using circular saws because molten metal is discharged from the saw (Figure 6.8). Do not use one manufacturer's blade on another manufacturer's saw. Blades from different manufacturers may look alike, but they should not be used interchangeably. Always store blades in a clean, dry environment. Blades should not be stored in any compartment where gasoline fumes accumulate (for example, where spare saw fuel is kept) as the hydrocarbons will attack the bonding material in the blades and make them subject to sudden disintegration during use.

Reciprocating Saw

The reciprocating saw is a highly controllable saw that is well suited for cutting wood or metal. A reciprocating saw may be required when overhead work must be done in an area where space is tight (Figure 6.9). This saw has a short, straight blade that moves forward and backward with an action similar to that of a handsaw.

Chain Saws

Chain saws are gaining popularity for forcible entry and ventilation work. The best type of saw is one powerful enough to penetrate dense material yet lightweight enough to be easily handled in awkward positions (Figure 6.10). Chain saws equipped with carbide-tipped chains are capable of penetrating a large variety of materials, including light sheet metal. Although carbide-tipped chains cost almost four times as much as standard chains, they last twelve times longer.

Figure 6.8 When using a power saw, always wear full protective equipment with eye and face protection.

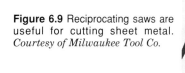

Figure 6.9 Reciprocating saws are useful for cutting sheet metal. *Courtesy of Milwaukee Tool Co.*

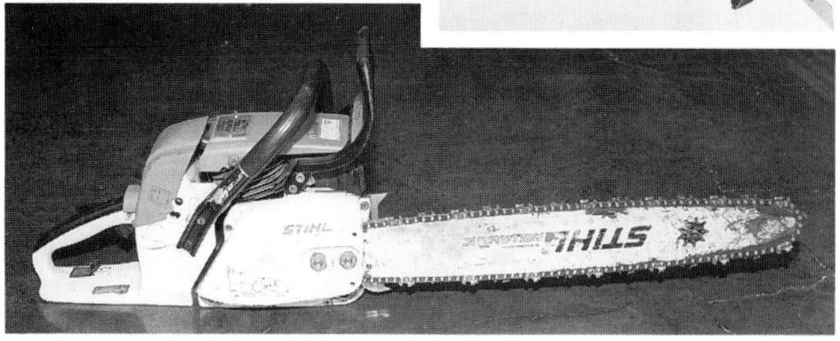

Figure 6.10 A chain saw.

Power Saw Safety Rules

Following a few simple safety rules when using power saws will prevent most typical accidents:

- Match the saw to the task and the material to be cut. Never push a saw beyond its design limitations.

- Always wear proper protective equipment, including gloves and eye protection.
- Do not use any power saw when working in a flammable atmosphere or near flammable liquids.
- Keep unprotected and nonessential people out of the work area.
- Follow manufacturer's guidelines for proper saw operation.
- Keep blades and chains well sharpened (Figure 6.11). A dull saw is more likely to cause an accident than a sharp one.

Figure 6.11 Keep the chains well sharpened on a chain saw.

METAL CUTTERS AND CUTTING TORCHES

Bolt cutters are used as a forcible entry tool to cut iron bars, bolts, cables, padlock hasps, and other objects. Wire cutters are used to cut fences and wire other than electrical.

The oxyacetylene cutting torch is a common forcible entry tool that cuts by burning (Figure 6.12). It is useful for penetrating heavy metal enclosures that are resistant to more conventional forcible entry equipment. The torch preheats metal to its ignition temperature and then burns a path in the metal with an extremely hot cone of flame caused by the introduction of pure oxygen into the flame.

Because a cutting torch operates with a highly flammable gas and generates a flame, it should be used with extreme caution. Do not use a cutting torch if there is any reason to believe that the atmosphere is flammable. As a precautionary measure, it is advisable to have charged hoselines in place before beginning cutting torch operations (Figure 6.13). Cutting torch operators should be experienced and efficient in using the tool in all situations. Anyone who uses a cutting torch should train regularly in exercises that present a variety of cutting problems such as cutting metal objects of various shapes and thicknesses.

Another hazard associated with a cutting torch is the storage of oxygen and acetylene. Always keep oxygen

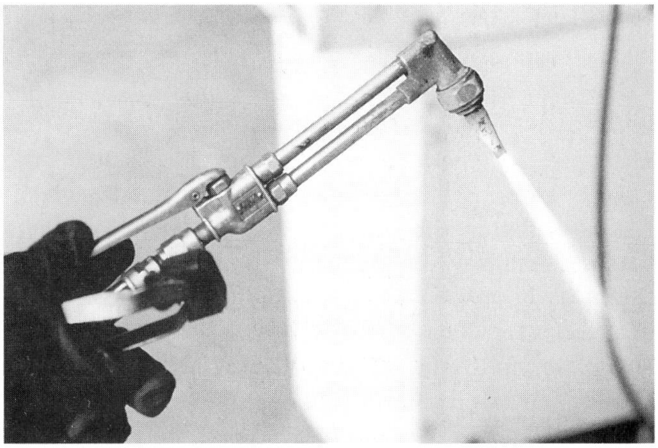

Figure 6.12 Oxyacetylene torches are effective for cutting metal that is too thick to be easily cut by saws.

Figure 6.13 Charged hoselines should be present when forcing entry into a potential fire area.

and acetylene cylinders in an upright position, whether they are in use or in storage (Figure 6.14). Acetylene is an unstable gas that is pressure and shock sensitive. Acetylene storage cylinders, however, are designed to keep the gas stable and safe to use. The cylinders contain a porous filler of calcium silicate, which prevents accumulations of free acetylene within the cylinder bottle. They also contain liquid acetone to dissolve the acetylene. When an acetylene cylinder's valve is opened, the acetylene gas leaves the mixture as it travels through the torch hoseline assembly.

WARNING

Keep acetylene cylinders in an upright position to prevent acetone, a flammable liquid, from flowing through the cylinder valve and pooling in the work area.

The cutting torch generates an extremely hot flame. For preheating metal, the flame temperature in air is approximately 4,200°F (2 316°C). When pure oxygen is added through the torch-assembly handle, a flame of over 5,700°F (3 149°C) is created. This is hot enough to burn through iron and steel with ease.

Cutting Torch Safety Rules

Observe the following safety rules when using oxy-acetylene cutting equipment:

- Store and use acetylene cylinders in an upright position to prevent loss of acetone. Remember, when an acetylene cylinder is "empty" of acetylene, it is always full of acetone. Never place empty cylinders on their sides.
- Handle cylinders carefully to prevent damage to the cylinder and to the filler. A dent in the cylinder indicates that the filler may be damaged. If the filler is damaged, voids are created where free acetylene can pool and decompose, creating a dangerous condition. Dropping a cylinder may cause fuse plugs to leak, also creating a dangerous condition. Mark and return dented cylinders to the supplier.
- Avoid exposing cylinders to excessive heat. This means that an ambient air temperature exceeding 130°F (54°C) is undesirable for storing or using acetylene cylinders.
- Do not place cylinders on wet or damp surfaces. Cylinder bottoms rust as protective paint is worn away.
- Store acetylene cylinders in an area separate from oxygen cylinders and other oxidizing gas cylinders. Separate full acetylene cylinders from empty or partially full cylinders. Design storage areas to prevent cylinders from falling when they are bumped (Figure 6.15).

Figure 6.14 Keep cylinders in an upright position at all times.

Figure 6.15 Stored cylinders should be chained in the upright position to keep them from falling over.

- Perform a soap test (applying a solution of soap and water on fittings) to detect leaks at regulator, torch, hose, and cylinder connections. Slow leaks in confined areas permit acetylene to quickly accumulate in concentrations above the lower flammability limit. Acetylene has a wide flammability range: 2.5 percent to 81 percent by volume in air. Remove leaking cylinders to an open area immediately. Do not attempt to stop a fuse plug leak.
- Open acetylene cylinder valves no more than a three-quarter turn. Do not use wrenches on cylinders that have handle valves. If the valve is faulty, do not force it. Take the cylinder out of service immediately and return it to the supplier for service.
- Do not use acetylene at pressures greater than 15 psi (103 kPa). Acetylene decomposes rapidly at high pressures and may explode as decomposition occurs.

- Do not exceed the withdrawal rate of one-seventh of the cylinder capacity per hour.
- Keep valves closed when cylinders are not in use and when they are empty. After the valves are closed, bleed off the pressure in the regulator and in the torch assembly. Keep unconnected cylinders capped, whether they are full or empty, to prevent damage to fittings.

Prying Tools

Hand prying tools use leverage to provide a mechanical advantage. This means that a person can generate more force to an object using a tool than they could without one. These tools are very effective in breaking locks, opening doors, forcing windows, and prying up objects. They can even be used to lift automobiles and other heavy objects. The pry-axe, Halligan (Hooligan) tool, crowbar, claw tool, pry bar, Kelly tool, spanner wrench, and Quic-Bar are examples of hand prying tools (Figure 6.16). Crowbars and other prying tools are excellent for widening a small opening for larger power tools to fit into.

When used correctly, prying tools are safer than striking tools. As with other tools, using prying tools incorrectly creates a safety hazard. For example, it is not acceptable to use a "cheater bar" or to strike the handle of a pry bar with other tools (Figure 6.17). A cheater bar is a piece of pipe added to a prying tool to lengthen the handle, thus providing additional leverage. Use of a cheater bar can put forces on the tool that are greater than the tool was designed to handle. This can cause serious injury and destroy the tool. If a job cannot be done with one tool, use another. Do not use a prying tool as a striking tool unless it has been designed for that purpose.

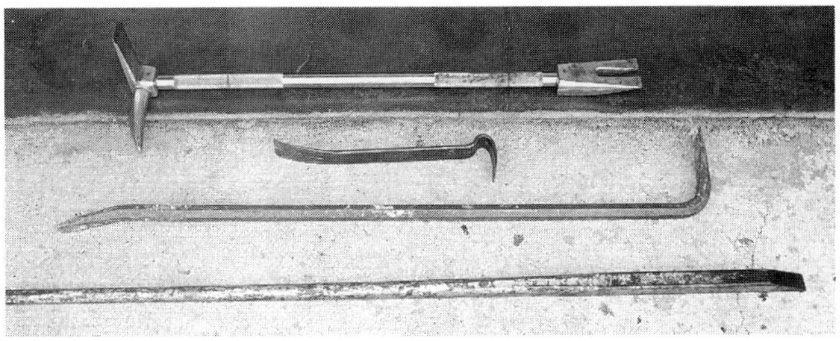

Figure 6.16 Several types of prying tools. From top to bottom they are a Halligan tool, a small pry bar/nail puller, a crow bar, and a large pry bar.

Figure 6.17 Never use a cheater bar to extend a prying tool.

POWERED HYDRAULIC SPREADER

This rescue tool, which one person can operate, has proven very effective in extrication rescues and is also useful in forcible entry situations. This tool is useful for a variety of different operations involving either prying, pushing, or pulling (Figure 6.18). The spreader is powered by hydraulic fluid that is supplied from either a gas, electric, power take-off, or hand pump. Depending on the brand, this tool can produce up to 22,000 psi (154 000 kPa) of force at the tips. The tips of the tool may spread as much as 32 inches (813 mm) apart. Its ability to exert pressure in both spreading and pulling modes makes it a versatile device (Figure 6.19).

RABBIT TOOL

The rabbit tool is a prying tool that is specifically designed for prying open doors and windows (Figure 6.20). The working end of this tool is a pair of intermeshed spreading tips that can be inserted into a very narrow opening such as that between a door and its casing. A few strokes to the hand-pump compressor are usually sufficient to force open any door that swings inward.

Pushing/Pulling Tools

Highlighted in the following sections are some of the pushing and pulling tools that have applications for forcible entry operations.

PIKE POLES AND PLASTER HOOKS

The pike pole is useful as a pushing and pulling tool. This tool can be used to open windows, ceilings, and partitions. The plaster hook has two knifelike wings that

Figure 6.18 A powered hydraulic spreader is effective in extrication rescues and forcible entry situations.

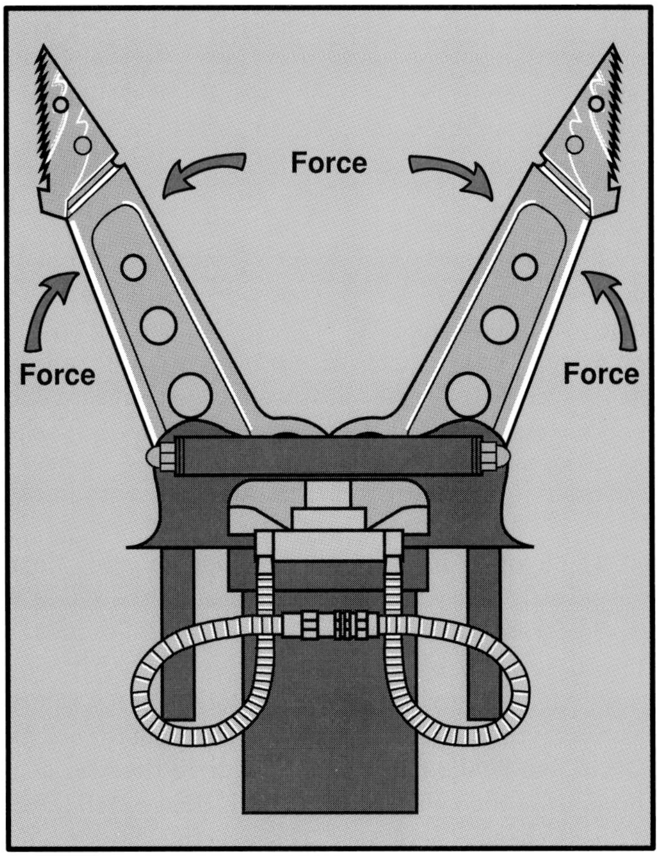

Figure 6.19 Powered hydraulic spreaders produce force in both opening and closing directions.

Figure 6.20 The rabbit tool is used to open doors that swing away from the firefighter.

depress as the head is driven through an obstruction and reopen or spread outward under the pressure of self-contained springs (Figures 6.21 a and b). A pike pole and plaster hook should be carried with the sharp ends toward the front and lowered.

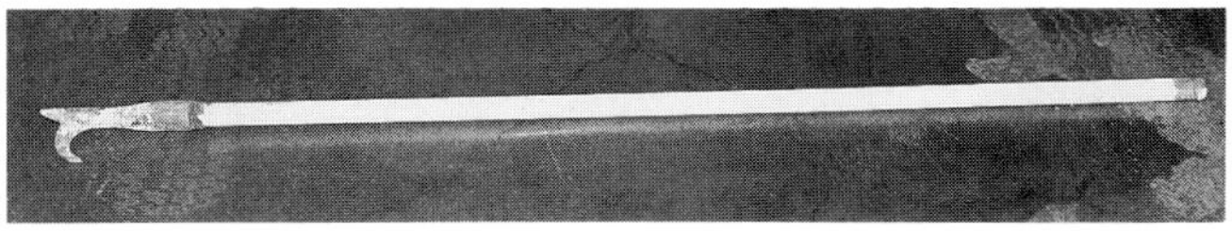

Figure 6.21a A pike pole is used for pushing and pulling operations such as opening windows, ceilings, and partitions.

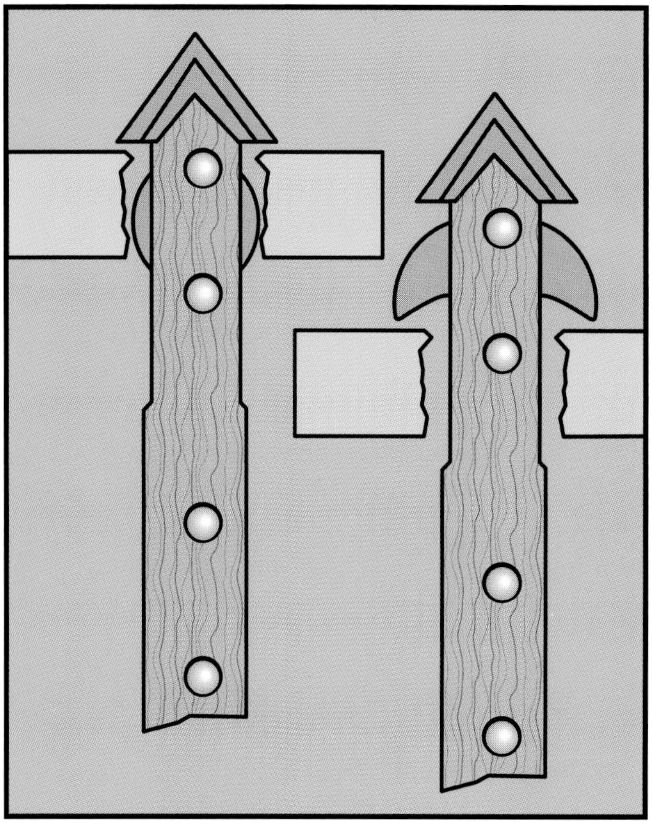

Figure 6.21b This illustrates the operation of a plaster hook.

POWERED HYDRAULIC EXTENSION RAMS

Powered hydraulic extension rams were designed primarily for extrication operations; however, they can be useful in some forcible entry operations. One use is to place the ram in between either side of the door frame to spread the frame apart far enough to allow the door to swing open (Figure 6.22 on next page).

Powered hydraulic extension rams are designed primarily for straight pushing operations, although they are effective at pulling as well. They are especially valuable when it is necessary to push objects farther than the maximum opening distance of the hydraulic spreaders. The largest of these extension rams can extend from a closed length of 36 inches (914 mm) to an

Figure 6.22 The hydraulic ram can be used to spring the door open.

extended length of nearly 63 inches (1 600 mm). They open with a pushing force of about 15,000 psi (103 425 kPa). The closing force is about one-half that of the opening force.

Striking Tools

Striking tools, characterized by large, weighted heads on handles, are the most common and basic hand tools. This category of tools includes axes, battering rams, ram bars, punches, mallets, hammers, sledgehammers or mauls, chisels, automatic center punches, and picks (Figures 6.23 a and b).

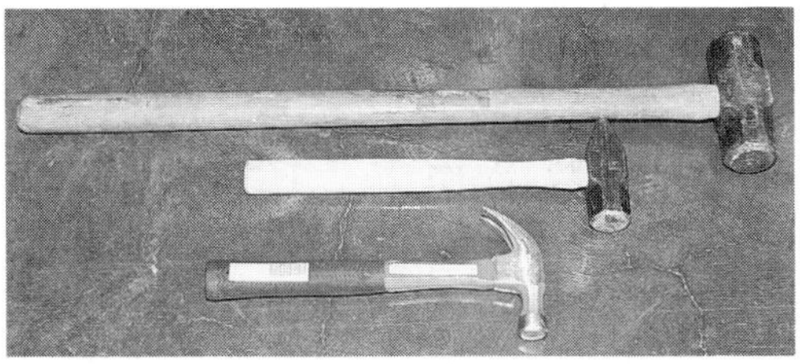

Figure 6.23a Various striking tools. From top to bottom they are a sledge hammer, a mallet, and a claw or carpenter's hammer.

Figure 6.23b Battering rams are forms of striking tools. *Courtesy of Detroit (MI) Police Department, Narcotics Division.*

Striking tools can be dangerous and when used improperly may crush or sever fingers, feet, or other parts of the body. Striking tools can also send chips and splinters into the air, piercing skin and eyes.

CARE AND MAINTENANCE OF FORCIBLE ENTRY TOOLS

It is very important to properly clean, dry, and repair forcible entry tools. The proper care of forcible entry tools increases their span of service depending on the intensity of their use. Important maintenance requirements specific to axe heads as well as general care and maintenance tips for all types of tools are covered in the following sections.

Care Of Axe Heads

The manner in which the axe head is maintained will directly affect how well it works. If the blade is extremely sharp and the body of the blade is ground too thin, pieces of the blade may be broken out when cutting gravel roofs or striking nails and other materials in flooring. If the body is too thick, regardless of the sharpness of the blade, it is difficult to drive the axe head through ordinary objects. Medium thickness of the body of the axe head is necessary. The body thickness should be about $\frac{1}{4}$-inch (6 mm) at $\frac{3}{4}$-inch (19 mm) from the edge, $\frac{3}{8}$-inch (10 mm) at $1\frac{1}{4}$ inches (32 mm) from the edge, and $\frac{1}{2}$-inch (13 mm) at 2 inches (50 mm) from the edge. The measurements are to be taken at the center of the blade (Figure 6.24). The temper should be such that the blade will not bend easily or break off when nails, gravel, or other foreign matter are struck. Temper is a quality introduced into metal by heating it and then cooling it in oil. Care must be used when filing the blade in order to prevent heating and the consequent softening of the steel. The axe should be filed to preserve the body thickness and not merely to sharpen the blade. The effectiveness of an axe depends on its proper use.

General Care And Maintenance Principles For Tools

In order for tools to be ready for action at any given moment, they must be properly maintained. Failure to properly maintain tools could result in tool failure at a critical time. Some procedures for forcible entry tool care are addressed in the following sections.

Wooden Handles

- Check for cracks, blisters, or splinters in the wood.
- Sand wooden handles to minimize hand injuries.
- Clean wooden handles with soapy water, rinse, and dry after use (Figure 6.25).

- Apply a coat of boiled linseed oil to the handle to prevent roughness and warping (Figure 6.26). Paint or varnish on the handles tends to blister when exposed to heat, and they also hide cracks. A ½-inch (13 mm) stripe painted around the handle for identification purposes is suggested. Some departments prefer to mix ground walnut shells or similar material in the oil that is applied to the area near the end of the handle to increase grip and decrease the possibility of the handle slipping out of the operator's hand and possibly injuring someone nearby.
- Check to ensure that the head is on tight.

Fiberglass Handles

- Wash with warm, soapy water.
- Dry with a soft, dry cloth.
- Check to ensure that the head is on tight.

Cutting Edges

- Check to ensure that the cutting edge is free of nicks or tears.
- Replace cutting edge of bolt cutters when needed.
- File the edges by hand. Grinding takes the temper out of the metal (Figure 6.27).

Plated Surfaces

- Inspect for damage.
- Wipe plated surfaces clean or wash with soap and water.
- **Do Not Paint Axe Heads**. Painting makes it difficult to inspect for cracks and other wear on the head.

Unprotected Metal Surfaces

- Keep clean of rust.
- Keep oiled when not used. Any light machine oil will work.
- Do not completely paint; it hides cracks.
- Check to see that the metal surfaces are free of burred or sharp edges. File off when found.

Power Equipment

- Check to see whether the equipment starts manually.
- Check blades and equipment for completeness and readiness.
- Check electric tool cords for cuts and frays.
- Make sure that the appropriate guards are in place.

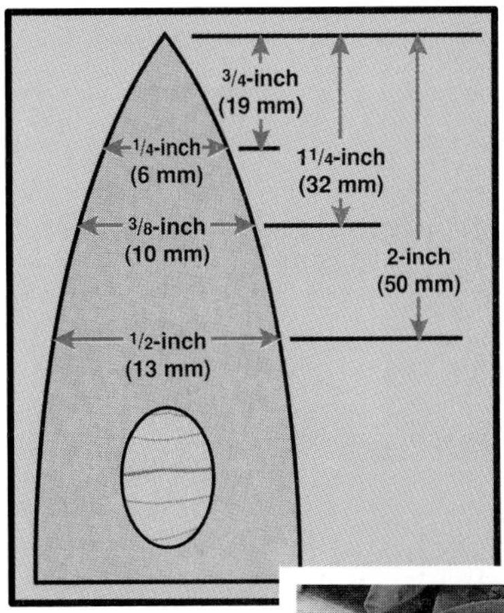

Figure 6.24 The dimensions for an axe head.

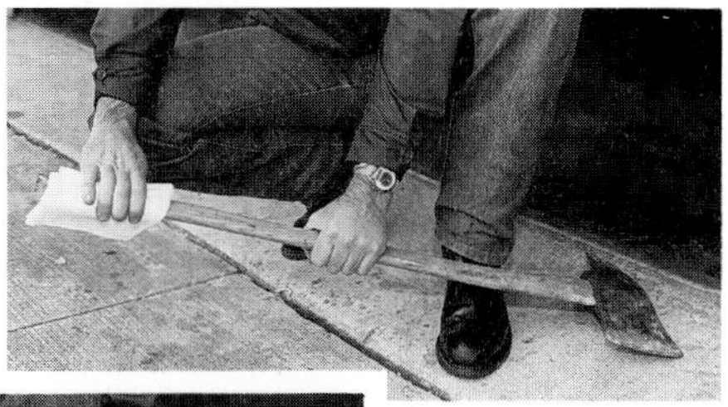

Figure 6.25 Clean wooden handles with soapy water, rinse, and dry after use.

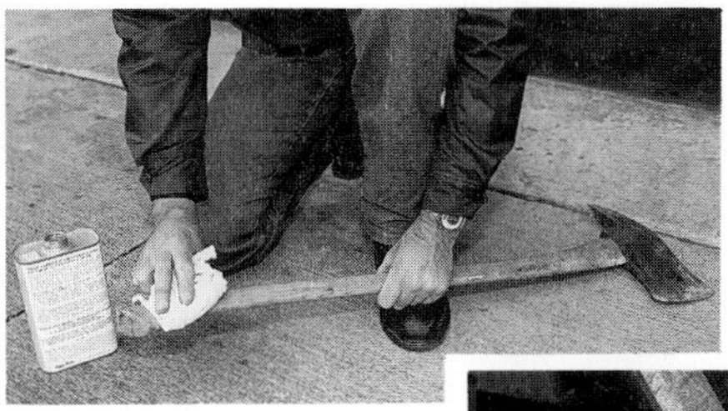

Figure 6.26 Apply a coat of boiled linseed oil to the handle to prevent roughness and warping.

Figure 6.27 File the edges by hand.

OPENING DOORS

Locked doors are obstacles to firefighters when they endeavor to reach various areas of a building. From a firefighter's standpoint of forcible entry, doors may be classified as either swinging, revolving, sliding, or overhead. Regardless of the type of door, firefighters should try the door to make sure that it is locked before force is used (Figure 6.28 on next page). Remember, **"Try Before You Pry!"** If the door is locked, examine it to determine which way it swings and which method of forcible entry will prove most effective. The construction features of some doors render them practically impossible to force, and entry may be more easily made by some other means such as breaching the wall next to the door or making entry through a nearby window.

The following sections on opening locked doors discuss each type of door on the basis of its construction features and the opening techniques that have been

Figure 6.28 Remember to always try the door before you pry.

proved through practice. No attempt will be made to designate one product as better than another, and the illustrations used herein have been selected for clarity of pattern rather than comparison of design.

Swinging Doors

Door locks and fasteners for swinging doors consist of a bolt or bar, called a tang, that protrudes from the door into a metal keeper that is mortised into the doorjamb. This bolt or bar may be part of the lock assembly or it may be entirely separate. In either case, to open the door the jamb must be sprung enough to permit the bolt to pass the keeper. Some special instal-

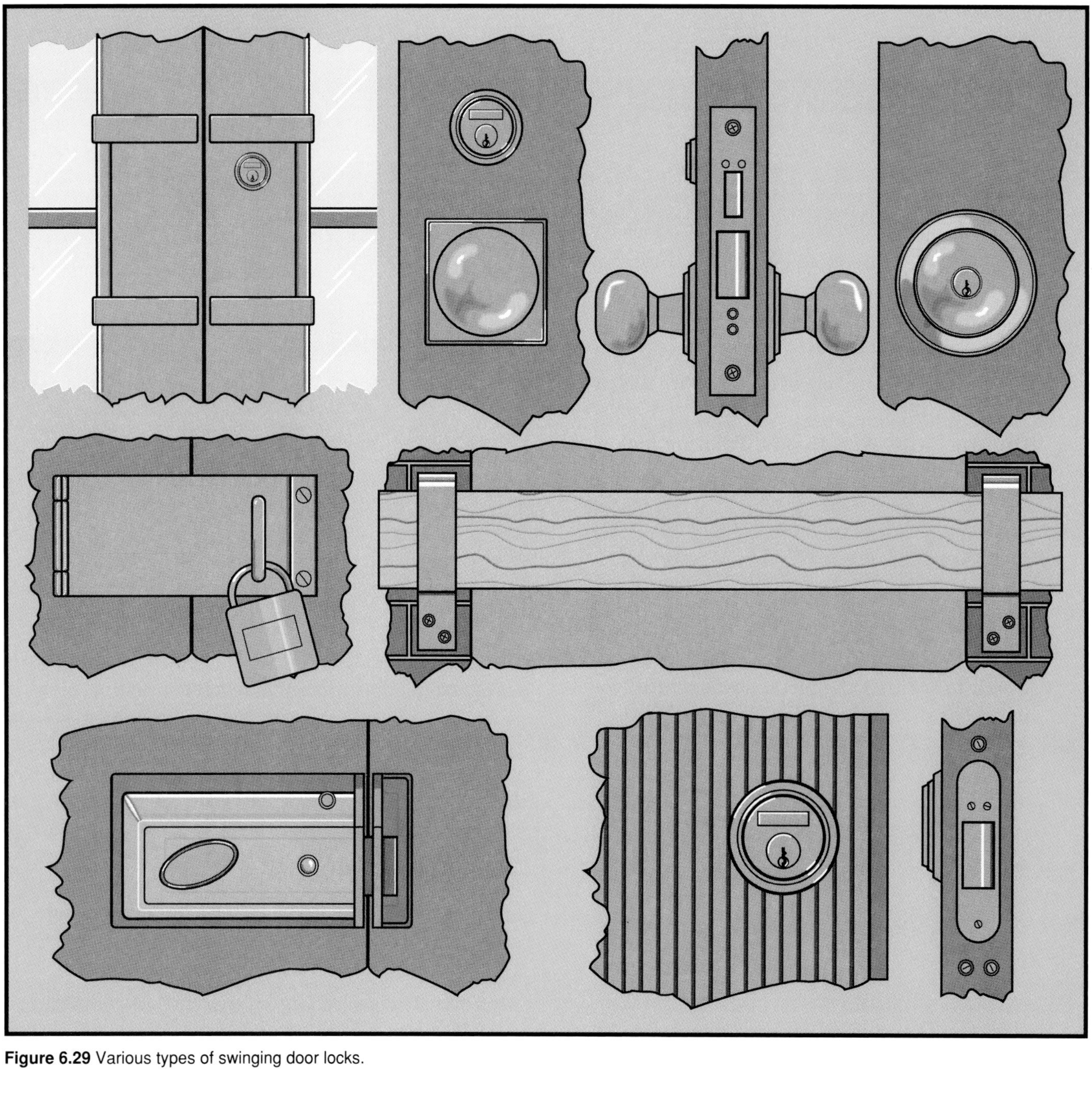

Figure 6.29 Various types of swinging door locks.

lations have two bars, one at the top of the door and one at the bottom, and such door locks are exceedingly difficult to force. A record of the type of door and how it is locked can be valuable to firefighters, if such information is collected during inspections and pre-incident planning surveys. Some of the more common types of locks for swinging doors are shown in Figure 6.29.

WOODEN DOOR CONSTRUCTION

Three general types of wooden swinging doors are panel, slab, and ledge. Front doors in residences may be either panel or slab, often with glass panels in the upper area. Some panels of glass are held in place with moldings, which may be pried off to remove the panel. Residence doors generally open inward; in this way they differ from outside doors in public buildings that should open outward for easy exiting during emergencies.

Slab doors are very popular, and they may be constructed as either hollow core or solid core (Figure 6.30). The term "hollow core" implies that the entire core of the door is hollow, without fillers, which is not correct. Instead, the core is made up of an assembly of wood strips formed into a grid or mesh. These strips are glued within the frame, forming a rigid and strong core. Several layers of plywood veneer paneling are then glued over this framework and grid. The purpose of the hollow-core grid is to decrease the weight and cost of the door. Most exterior slab doors found on newly constructed residences will be hollow core, but the exterior slab doors on older homes may be solid core.

The term "solid core" means that the entire core of the door is constructed of solid material. Some solid-core slab doors have a core built up of tongue-and-groove blocks or boards, which are glued within the frame. Other solid-core doors may be filled with a compressed mineral substance that is fire resistant. In either case, the door is solid with a plywood veneer covering. This type of construction adds considerably to the weight of the door. Some exterior slab doors on industrial and mercantile buildings are solid-core slab doors.

Ledge doors, sometimes called batten doors, are found on warehouses, storerooms, barns, etc. (Figure 6.31) They are made of built-up material and are locked with either surface locks, hasps and padlocks, bolts, or bars. Hinges on ledge doors, generally of the surface type with stationary pins, are fastened to the door and facing with screws or bolts.

Doorjambs are the sides of the doorway's opening. Doorjambs for wood swinging doors may be rabbeted jambs or stopped jambs. The rabbeted jamb is one into which a shoulder has been milled to permit the door to close against the provided shoulder (Figure 6.32). Stopped

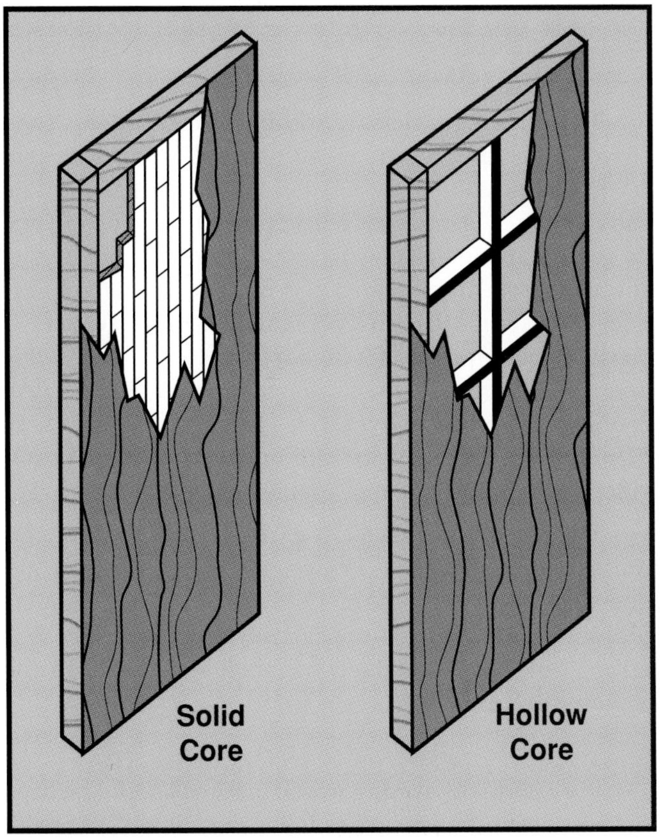

Figure 6.30 Doors may be of the solid or hollow core types.

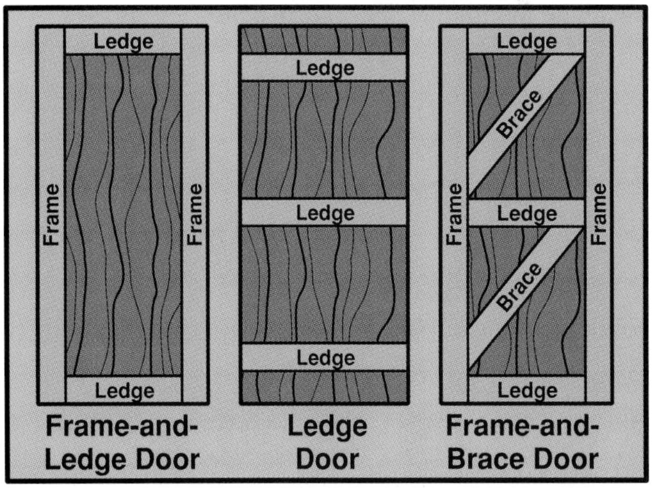

Figure 6.31 Types of ledge doors.

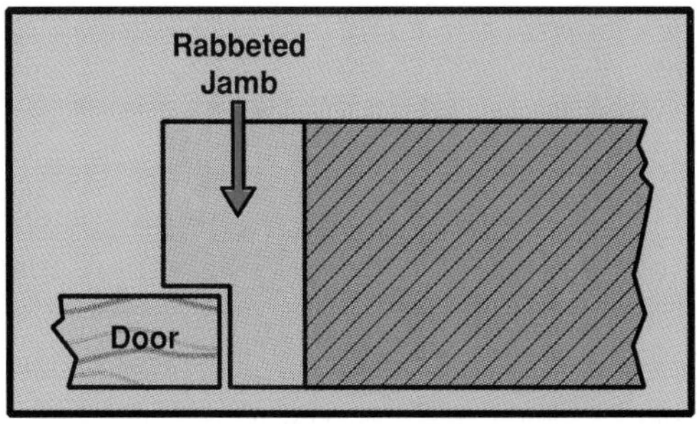

Figure 6.32 This illustration highlights the position of the rabbeted jamb and the door.

doorjambs are provided with a wooden strip or doorstop that is nailed inside the jamb against which the door closes (Figure 6.33). This stop may be easily removed with most prying tools.

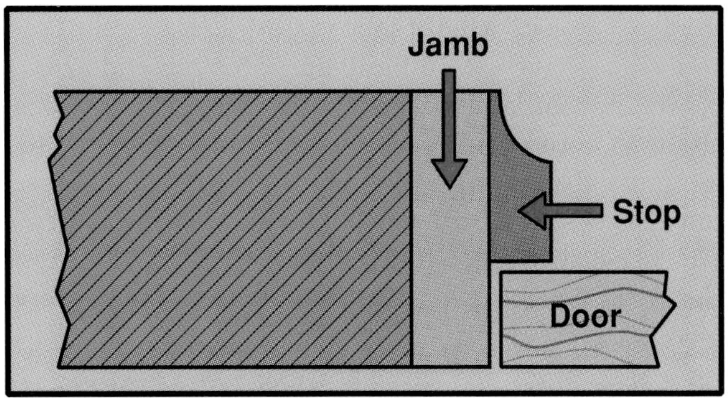

Figure 6.33 A typical stopped jamb.

METAL DOOR CONSTRUCTION

Metal swinging doors may be classified as hollow metal, metal covered, and tubular (Figure 6.34). Metal swinging doors are generally more difficult to force due to the manner in which the door and doorjamb are constructed. This difficulty is also more acute when the jamb is set in masonry. It is generally considered impractical to force metal doors. The framework of hollow metal doors is constructed entirely of metal. The jambs are hollow and are fastened to the walls by specially designed metal anchors (Figure 6.35).

Figure 6.35 Most metal or glass and metal doors will have metal door frames.

Figure 6.34 A typical metal swinging door.

Metal-covered doors are constructed essentially as hollow metal doors except that there may be a wooden core or metal ribs over which the metal covering is placed. The paneling sometimes consists of metal-covered asbestos (Figure 6.36).

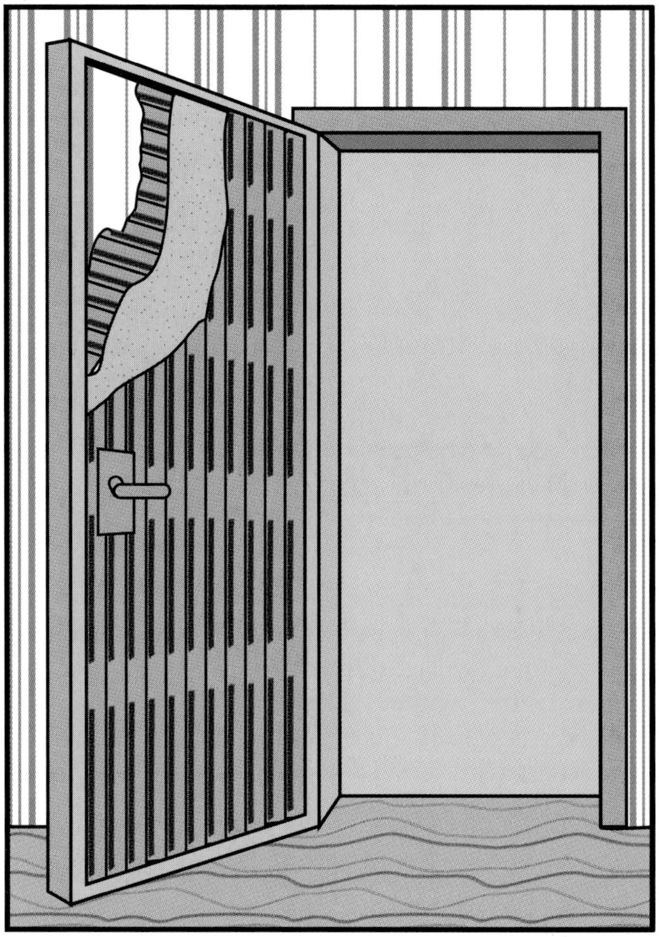

Figure 6.36 Metal doors usually have some type of filler material or insulation inside them.

The structural design of tubular metal doors is of seamless rectangular tube sections. A groove is provided in the rectangular tube for glass or metal panels. The tube sections form a door with unbroken lines all in one piece and are sometimes found on exterior openings of the more recently constructed buildings. The tubular doors are hung with conventional hardware except that the balance principle of hanging is sometimes used. The operating hardware consists of an upper and lower arm connected by a concealed pivot. The arms and pivots are visible from the exterior side only. From the interior side, the balanced door resembles any other door.

Tubular aluminum doors with narrow stiles are also quite commonly used. The panels of these doors are generally glass but some metal panels are used. Tubular aluminum doors are comparatively light in weight, are strong, and are not subject to much spring within the aluminum frame.

Opening Techniques For Swinging Doors

In order to determine which method should be used to force a swinging door, the following should be considered in the order given.

- What type of door it is
- How the door is hung
- How it is locked

Again, before attempting to force any door, check to see whether the door is locked and whether or not the hinge pins can be removed. The conditions of the building should be observed, and hoselines should be made available for use if there is any suspicion that fire conditions may be present on the other side of the door (Figure 6.37). The following sections highlight some of the techniques that can be used to force swinging doors.

Figure 6.37 Have a hoseline in place prior to forcing a door.

BREAKING GLASS

In some cases, less damage may be done by breaking a small glass near the lock, through which the door can be opened from the inside. Note that this glass does not always have to be in the door but may be next to the door (Figure 6.38). Because glass will shatter into fragments with keen cutting edges, the act of breaking glass must be done in a manner to ensure the safety of the firefighter. Some of the principal safety features for breaking glass are as follows:

- Stand to the windward side of the glass pane to be broken if possible (Figure 6.39).
- Strike the tool at the top of the pane (Figure 6.40).
- Keep the hands above the point of impact (Figure 6.41 on next page).
- Wear full protective clothing, including hand and eye protection. If making entry into a fire building, SCBA should be worn and charged hoselines should be in place.

Figure 6.38 In some cases, it would be quicker and less expensive to the owner to break a small glass next to the door and reach through to unlock the door.

Figure 6.39 Stand to the windward side when breaking a window.

Figure 6.40 Break the glass near the top of the pane.

Figure 6.41 The firefighters' hands should be higher than the point of impact to prevent shards of glass from falling onto them.

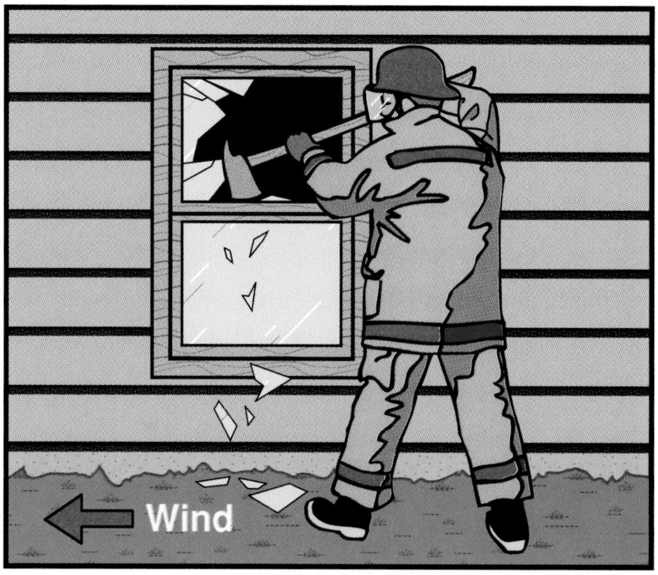

Figure 6.42 Clean all the glass from the window opening.

This procedure permits the broken particles of glass to fall downward and away from the hands and to the side of where the firefighter stands. The glass may be broken with an axe or other appropriate tools. Never break glass with the hands or helmet. After the glass is broken, all jagged pieces should be removed from the sash. Glass removal may be done with the same tool that was used to break the glass. Removing all pieces of glass prevents anyone who goes through or reaches through the opening from being cut and will prevent damage to hose, ropes, or other materials that may be passed through the opening (Figure 6.42).

Figure 6.43 The components of the K-tool are kept in a small pouch.

BREAKING THE LOCK

A lock may sometimes be removed or destroyed with less damage and expense than might occur if the door is forced. An A-tool or K-tool may be used to pull or force a cylinder lock (Figure 6.43). These devices require little force and accomplish a fast, neat job of forcible entry. Both tools are used in conjunction with a flat-head axe or other suitable striking tools. A key tool, which has a straight and bent end, is used to move the bolt or plunger from the keeper.

Figure 6.44 The flat side of an axe may be used to drive the K-tool down over the lock cylinder.

Both the A-tool and the K-tool are used to physically pull the cylinder out of the door. The K-tool is useful in pulling all types of lock cylinders (rim, mortise, or tubular). Used with a Halligan or other prying tool, the K-tool is forced behind the ring and face of the cylinder until the wedging blades take a bite into the cylinder (Figure 6.44). Light blows of a hammer help this forcing. The K-tool is positioned correctly when the blades are engaged into the body of the cylinder. This eliminates the likelihood of breaking off the ring and face of a cylinder made of soft white metal. The front metal loop of the tool acts as a fulcrum for leverage and holds the fork end of the prying tool.

When a cylinder is found close to the threshold or jamb, the narrow blade side of the tool will usually fit behind the ring. A close-clearance situation will often be found on a glass sliding door, but only a half-inch clearance is needed. Once the cylinder is removed, a key tool can be used in the hole to move the locking bolt to the open position.

The A-tool is a different tool which accomplishes the same job as the K-tool (Figure 6.45). The A-tool is slightly more damaging to the door than a K-tool, but it will rapidly pull the cylinder. Sometimes the fire situation will necessitate quick entry at whatever the damage to a door. The jaws of the A-tool are forced around and behind the protruding rim of a cylinder. The curved head and long handle are then used to provide the leverage for pulling the cylinder. The chisel head, on the other end of the tool, is used when necessary to gouge out the wood around the cylinder for a better bite of the working head. Insert either the straight or bent end of the key tool into the hub of the lock and turn (Figure 6.46). If for any reason this fails and it is a rim lock, insert the straight end of the A-tool through the hole, and drive the lock off the door with an axe.

Figure 6.45 A lock can be removed by prying it loose with an A-tool.

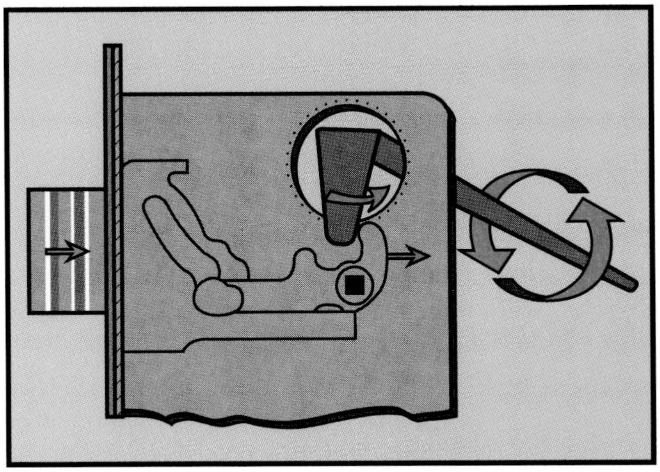

Figure 6.46 Use the key tool to manipulate the lock.

FORCING DOORS THAT OPEN TOWARD THE FIREFIGHTER

The direction in which the door swings may be determined by examining the door hinges and the jamb. If the door opens toward the firefighter, the hinges will be on the same side as the firefighter. Several forcible entry tools may be used for this operation, including the Halligan tool, Kelly tool, or pry bar to name a few. The techniques of their use are similar, and the steps are as follows:

Step 1: Insert the blade of the tool between the door and jamb near the lock (Figure 6.47).

Step 2: Force the blade in and against the rabbet or stop by working and pushing on the tool (the tool may be hammered with another object) (Figure 6.48).

Figure 6.47 Step 1: Insert the blade of the tool between the door and jamb near the lock.

Figure 6.48 Step 2: Force the blade in and against the rabbet or stop by working and pushing on the tool (the tool may be hammered with another object).

Step 3: Pry the tool bar away from the door to move the door and jamb apart.

Step 4: Pull the door open or pry open with another tool when the lock has cleared the keeper.

The use of different tools for this operation is shown in Figure 6.49. These methods may work very well on ordinary and inexpensive doors, but variations may need to be made to force more sturdy doors. Two tools may be used together to open a door. If using two tools, insert both tools between the door and jamb, one just above the lock and the other just below the lock. By alternately prying with one and getting more purchase with the other, more force can be applied.

Figure 6.49 If using two tools, insert both tools between the door and jamb, one just above the lock and the other just below the lock.

FORCING DOORS THAT OPEN AWAY FROM THE FIREFIGHTER

Swinging doors that open away from the firefighter present greater difficulties than do doors that open toward the firefighter. If the door is in a stopped frame, the blade of the tool may be inserted between the stop and the jamb, the stop lifted, and the tool inserted between the door and jamb near the lock. By separating the door from the jamb, the operator may spring it sufficiently to permit the bolt to pass the keeper. Under certain conditions, it may be better to remove the stop completely. The steps in forcing a door that opens away from the operator when stops are used on the jamb are as follows (**NOTE:** As shown in the illustrations, the stop has not been completely removed from the jamb.):

Step 1: Bump the cutting edge of the tool against the stop to break the paint or varnish so that the blade can be inserted.

Step 2: Loosen the stop at the lock or remove the stop completely.

Step 3: Start the blade between the door and the jamb.

Step 4: Make the initial pry only after the blade is halfway in, to permit the blade to be worked and pushed.

Step 5: With a full bite behind the door, pry the door away from the jamb until the bolt passes the keeper (Figure 6.50).

Figure 6.50 With a full bite behind the door, pry the door away from the jamb until the bolt passes the keeper.

If the door opens away from the operator and is in a rabbeted jamb, it can more easily be forced by using two tools. Prying against the door with one tool should open a crack between the door and the rabbet, into which the blade of the second tool can be inserted. After the blade of the second tool has been forced well into the opening between the door and jamb, the door may be pried sufficiently to permit the bolt to pass the keeper. Even with two tools, forcing this type of door construction may be quite difficult. The steps in working the blades of the tools into the crack between the door and the rabbeted jamb are as follows:

Step 1: Lay the blade of the tool flat against the door, and insert the blade between the rabbet and the door (Figure 6.51).

Step 2: Make short pries with the first tool to spread the jamb.

Step 3: Work the blade of the second tool between the door and the jamb, hammering the blade well into the opening (Figure 6.52).

Step 4: With a full bite behind the door, pry the door away from the jamb until the bolt passes the keeper (Figure 6.53).

Figure 6.51 Step 1: Lay the blade of the tool flat against the door, and insert the blade between the rabbet and the door.

Figure 6.52 Step 3: Work the blade of the second tool between the door and the jamb, hammering the blade well into the opening.

Figure 6.53 Step 4: With a full bite behind the door, pry the door away from the jamb until the bolt passes the keeper.

A hydraulically powered multiuse tool known as the "rabbit tool" is specifically designed to pry inward-swinging doors mounted in metal frames. However, it has many other applications in forcible entry work. The steps for prying a door with the rabbit tool are as follows:

Step 1: Insert the rabbit tool jaws next to the lock between the stop and the door (Figure 6.54). This can usually be done by driving the jaws in with the palm of the hand; for a tight-fitting door, use a striking tool. If there is more than one lock, place the jaws between the locks.

Step 2: Place one foot on the leg of the pump while holding the jaws in place with one hand, and grasp the pump handle with the free hand (Figure 6.55).

Step 3: Pump the portable hydraulic pump while maintaining a grip on the back of the jaws (Figure 6.56). Hold the jaws in place throughout the entire operation to prevent them from slipping out of position as they spread. Be prepared for the door to open suddenly as the latch clears its strike.

Figure 6.54 Step 1: Insert the rabbit tool jaws next to the lock between the stop and the door.

Figure 6.55 Step 2: Place one foot on the leg of the pump while holding the jaws in place with one hand, and grasp the pump handle with your free hand.

Figure 6.56 Step 3: Pump the portable hydraulic pump.

DOUBLE SWINGING DOORS

Double swinging doors may be forced with most pry tools by prying the two doors sufficiently apart at the lock to permit the lock bolt to pass the keeper (Figure 6.57). Sometimes a wooden molding is fastened to one or two wooden doors where they come together at the center. The purpose of this molding is to cover the crack between the doors when they are closed. This molding must be removed before the blade of the tool can be inserted. Some double doors have a permanently mounted center stile. These types of doors may need to be opened by forcing between the stile and door.

Double swinging doors are sometimes secured with a bar on the inside wall. If the opening between the doors

Figure 6.57 With double swinging doors, one door is sometimes locked in placed.

is sufficient to permit the insertion of a flat tool or object, a bar can sometimes be lifted or bumped from the stirrups (Figure 6.58). When this process is not possible, doors secured with a bar may need to be battered in if forcing is necessary. It may be possible, however, to cut a hole in a door panel or break glass and remove the bar.

TEMPERED PLATE GLASS DOORS

Tempered plate glass doors are commonly found in commercial, institutional, and light industrial occupancies (Figure 6.59). The breaking characteristics of tempered plate glass are quite different from those of ordinary plate glass. This difference is due to the heat treatment given to the glass during tempering. The results of heat tempering plate glass produce high tension stresses in the center of the glass and high compression stresses on the exterior surfaces. These tension and compression stresses balance each other and may be visualized when the glass is subjected to polarized light.

Figure 6.58 The brackets on this door indicate that a burglar bar is normally present here.

Figure 6.59 Tempered glass doors typically are set in metal frames.

The heat treatment given to tempered plate glass increases its strength and flexibility. Its resistance to shock, pressure, impacts, and temperatures is also increased. Plate glass is approximately four times stronger after tempering. It will withstand, without breaking, a temperature of 650°F (343°C) on one side while the other side is exposed to ordinary atmospheric temperature. When broken, the sheet of glass suddenly disintegrates into relatively small, cubical pieces.

Although a tempered plate glass door may be locked at the center, top, or bottom, its resistance to shock and its rigid characteristics make it almost impossible to spring with forcible entry tools. Tests conducted warrant the basic conclusion that every other available means of forcible entry should be employed before deciding to gain entrance through a tempered plate glass door. Tempered plate glass door panels are considerably more expensive than other glass-paneled doors of similar size. Each door, in a sense, is custom-built and the cost of installation varies. The time necessary to prepare and install a replacement may be considerably longer than for other type doors.

When it becomes necessary to break through a tempered plate glass door, the glass should be shattered at a bottom corner. To break the glass, use the pick end of a pick-head axe. The firefighter should wear a suitable faceshield to protect against eye injury, or turn away from the door as the glass is being broken. Some departments place a shield, made from a salvage cover, as close to the glass as possible, and the blow is struck through the cover (Figure 6.60). The remaining glass should then be removed from the frame.

Figure 6.60 Prior to breaking the glass, cover the door with a salvage cover to protect the firefighters.

Revolving Doors And Forcing Techniques

Revolving doors consist of quadrants that revolve around a center shaft (Figure 6.61). The revolving wings turn within a metal or glass housing that is open on each side and through which pedestrians may travel as the door is turned. The mechanism of the revolving door is usually collapsible and panic-proof, and each of the four revolving wings is held in position when the hangers are collapsed. Some revolving doors will collapse automatically when forces are exerted in opposite directions on any two wings. All revolving doors do not collapse in the same manner, and it is a good policy to collect such information when fire department inspection surveys are made. There are three basic types of revolving doors. The following sections briefly describe each type and how they are collapsed.

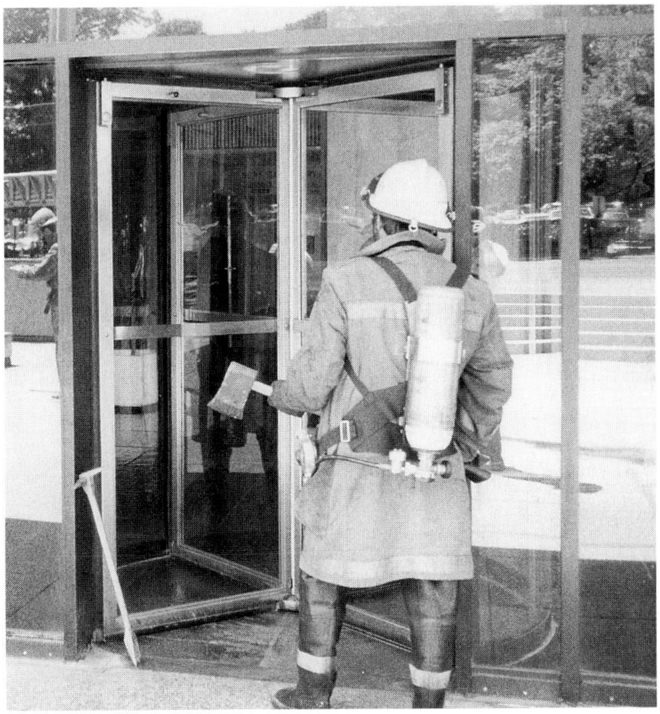

Figure 6.61 It is important for firefighters to know how revolving doors collapse.

PANIC-PROOF TYPE

The panic-proof, collapsible mechanism has a 1/4-inch (6 mm) cable holding the wings apart. To collapse the mechanism, push or press the doors or wings in opposite directions (Figure 6.62).

DROP-ARM TYPE

The drop-arm mechanism has a solid arm passing through one of the doors. A pawl will be found on the door through which the arm passes. To collapse the mechanism, press this pawl to disengage it from the arm and then push the wing to one side.

METAL-BRACED TYPE

The metal-braced mechanism is held in position by arms that resemble a gate hook with an eye. To collapse the mechanism, it is only necessary to lift a hook and fasten it back against the fixed door or wing. The hooks are located on both sides of the doors. The pivots are, in

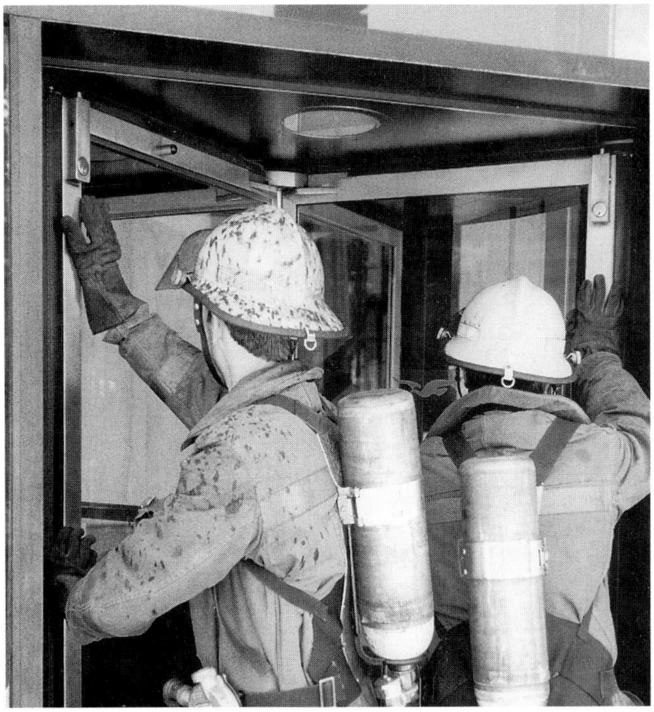

Figure 6.62 Firefighters are collapsing a panic-proof revolving door.

most cases, cast iron and can be broken by forcing the door with a bar at the pivots.

Revolving doors may be locked in various ways, and in general, they are considered difficult to force when locked. Swinging doors are usually found on either side of a revolving door, and large transom lights may be located over the doors.

Sliding Doors

Sliding doors are generally considered to be those doors that can travel either to the right or left of their opening and in the same plane (Figure 6.63 on next page). Sliding doors are usually supported upon a metal track, and their side movement is made easier by small rollers or guide wheels. The ordinary sliding door travels into a partition or wall when it is pushed open. This type of installation is more common for interior openings. These doors may be forced similarly to the swinging door except that they must be pried straight backward from the lock. These units consist of heavy-duty, full-panel glass set into a metal or wood frame. These glass panels are sometimes doubled (thermopane) or tempered, which adds to their value.

Patio sliding doors usually slide past stationary glass panels instead of disappearing into a wall. Sliding patio doors can be forced open by inserting a wedge tool between the jamb and door near the lock and prying the door away from the frame (Figure 6.64 on next page). Patio sliding doors may sometimes be barred or blocked

Figure 6.63 Many residential occupancies have sliding doors that exit to a patio.

Figure 6.64 To force open sliding patio doors, insert a wedge tool between the jamb and door near the bottom and pry the door away from the frame.

by a metal rod or a special device. These devices are commonly called "burglar blocks." This feature can easily be seen from the outside, and it practically eliminates any possibility of forcing without causing excessive damage. If it is necessary to enter, the glass will have to be broken using the techniques described previously for plate glass windows.

Overhead Doors

Overhead doors may be constructed of metal, fiberglass, or wood. There are three common types of overhead doors: sectional or folding, rolling steel, and slab (Figures 6.65 a through c). From a forcible entry point of view, the sectional door does not present a serious problem unless it is motor driven or remote controlled. The latch is usually in the center of the door, and it controls the locks on each side of the door. The lock and latch may also be located on only one side. These latches and locks are illustrated in Figure 6.66.

Overhead doors may be forced by prying upward at the bottom of the door with a good prying tool, but less damage will be done and time will be saved if a panel is knocked out and the latch is turned from the inside (Figure 6.67). Some overhead doors may be locked with a padlock through a hole at either end of the bar, or the padlock may even be in the track. These systems of locking may make it necessary to cut a hole in the door to gain access and remove the padlock. If the door is motor driven, it will be necessary to pull the manual release chain or rope on the inside of the door. This release is generally hanging on either side of the door near the roller track.

Pivoting or overhead slab doors, sometimes referred to as awning type, are locked similarly to the previously described sectional or folding door. Sometimes it is possible to pry outward with a bar at each side near the bottom (Figure 6.68 on the following page). This action will tend to bend the lock bar enough to pass the keeper.

Rolling steel doors are among the toughest forcible entry challenges faced by firefighters. They are best accessed by cutting a triangle-shaped opening large enough for firefighters to crawl through. This can be done with a rotary rescue saw or a cutting torch. All

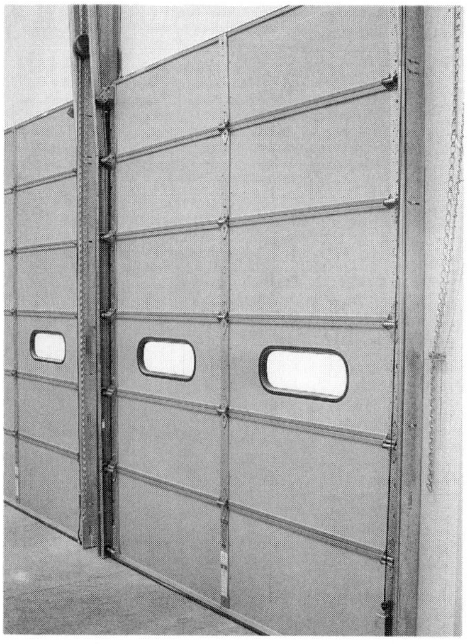

Figure 6.65a Usually the horizontal lines across the face of the door will indicate that the door is a folding type.

Figure 6.65b A rolling steel door.

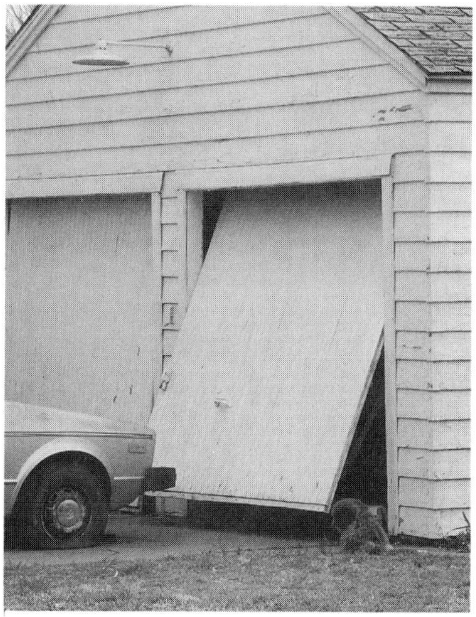

Figure 6.65c A slab-type garage door. Note that during forcible entry operations this door might not be able to be opened all the way unless the car in front of it is moved.

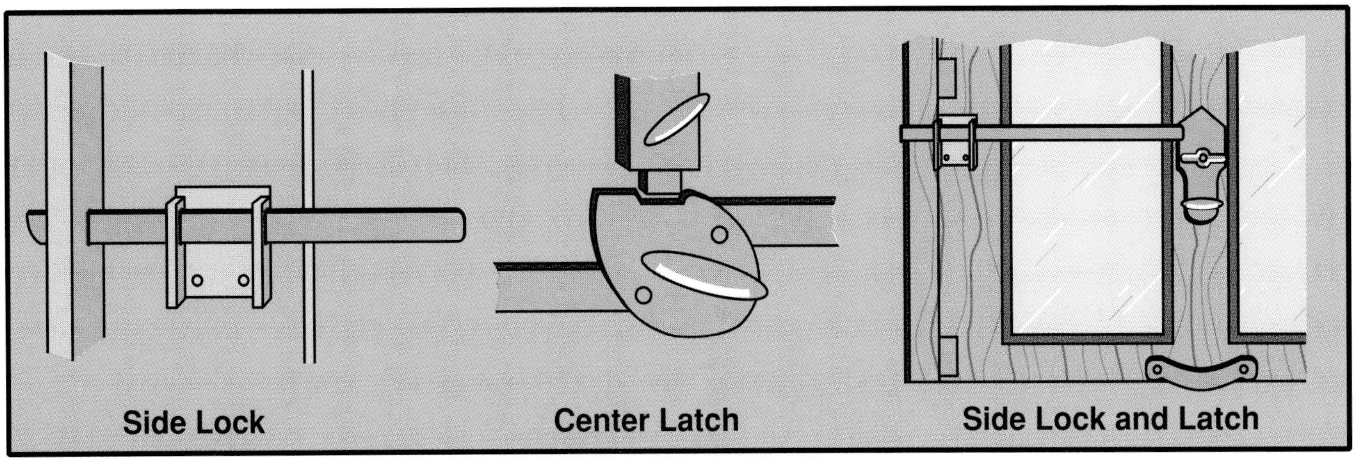

Figure 6.66 Types of overhead door latches.

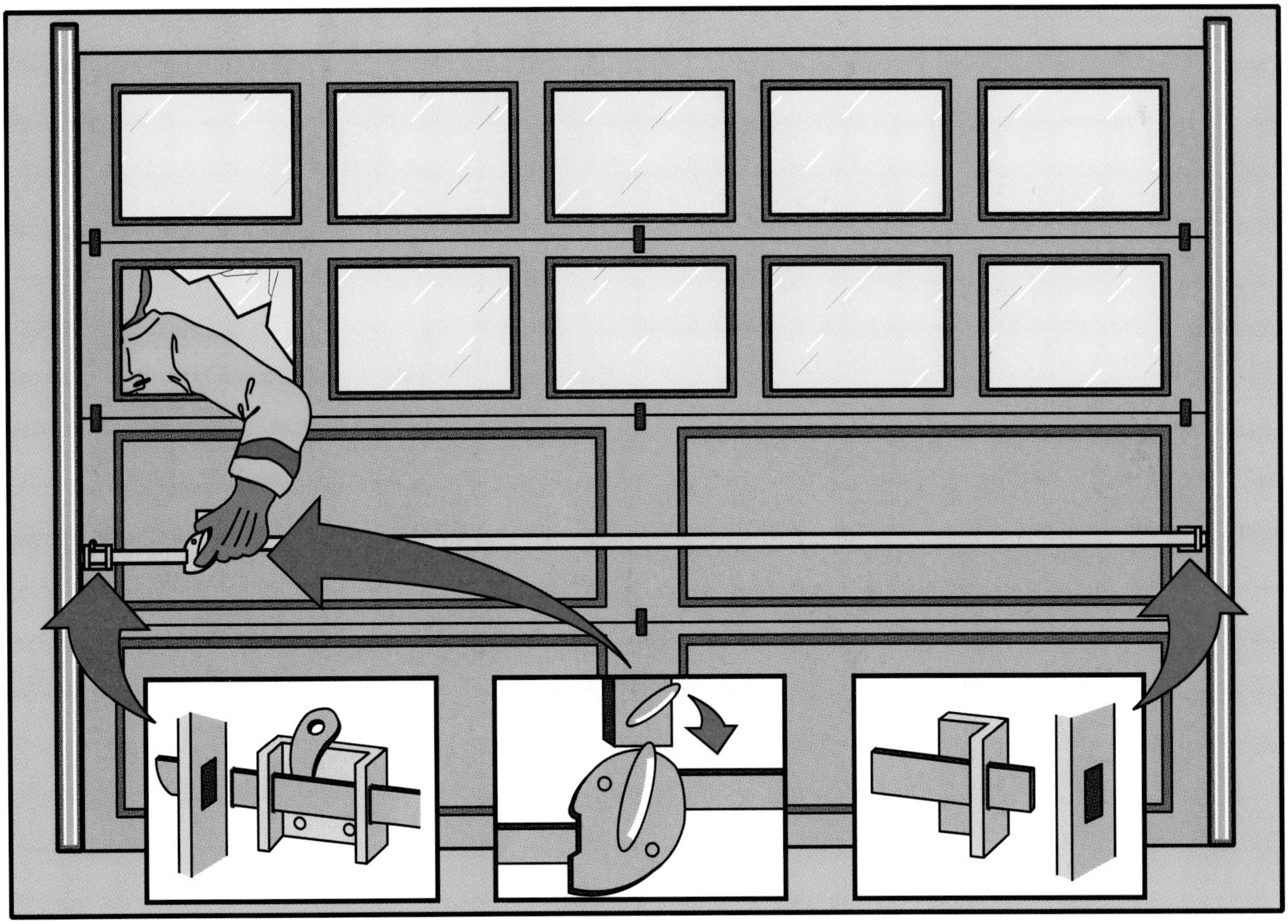

Figure 6.67 After knocking out a panel or pane of glass, reach through to open the lock.

overhead doors should be blocked open (up position) to prevent injury to firefighters should the control device fail.

Nondestructive Rapid Entry Method

The problem of gaining rapid entry without destruction has confronted fire departments for as long as locks have existed. In trying to find a solution, many departments have attempted to keep an inventory of keys to all the buildings in their area. While this procedure does reduce damage from forcible entry, it also presents a problem of maintaining an inventory of keys and gaining quick access to the right key at the right time. The problems presented by locked doors can be eliminated

Figure 6.68 Pry open a slab door by using a pry bar at each corner.

through the use of a rapid entry key box system (Figure 6.69). All necessary keys to the building, storage areas, gates, and elevators are kept in a key box mounted at a high-visibility location on the building's exterior. Only the fire department carries a master key that will open all boxes in its jurisdiction.

Proper mounting is the responsibility of the property owner. The fire department should indicate the desired location for mounting, inspect the completed installation, place the building keys inside, and lock the box with the department's master key. Unauthorized duplication of the master key is prevented because key blanks are not available to locksmiths and cannot be duplicated with conventional equipment.

Figure 6.69 Knox boxes contain keys for the building.

Fire Doors

Fire doors are used primarily to protect openings in division walls and walls of vertical shafts. However, for certain exposures, fire doors may be found on door openings (Figure 6.70). Openings for fire doors are classified as to the character and location of the wall in which they may be situated. Although classifications exist from "A" through "F," only Classes A and B will be discussed here. Class A openings are located in walls separating buildings or in walls within a building that is separated into distinct fire areas (Figure 6.71). Class B openings are vertical enclosures, such as elevators, stairways, or dumbwaiters, that may allow fire to spread within a building (Figure 6.72).

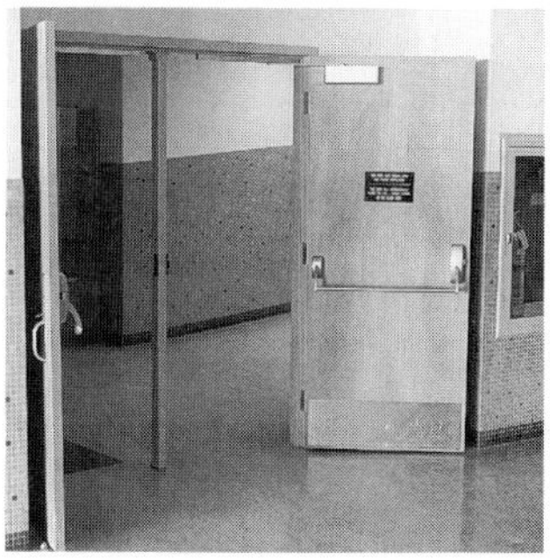

Figure 6.70 Double-swinging fire doors, common in the corridors of commercial buildings, automatically close when alarm systems are triggered.

Figure 6.71 Rolling steel fire doors can be used to protect Class A openings.

Figure 6.72 A common vertical shaftway fire door.

Types of standard fire doors include horizontal and vertical sliding, single and double swinging, and overhead rolling. They may or may not be counterbalanced. Counterbalanced doors are generally employed on openings to freight elevators, and they are mounted on the face of the wall inside the shaft (Figure 6.73). There are two standard means by which fire doors operate: self-closing, which when opened return to the closed position; and automatic closing, which normally remain open but close when heat actuates the closing device (Figure 6.74).

Figure 6.73 A counterbalance fire door.

Figure 6.74 An automatic fire door closer.

Swinging fire doors are generally used on stair enclosures and in other areas where they must be opened and closed frequently in normal service (Figure 6.75). Vertical sliding fire doors are normally open and so arranged as to close automatically. They are employed where horizontal sliding or swinging fire doors cannot be used. Overhead rolling fire doors may be installed where space limitations prevent the installation of other types (Figure 6.76). Like vertical sliding doors, overhead rolling doors are arranged to close automatically.

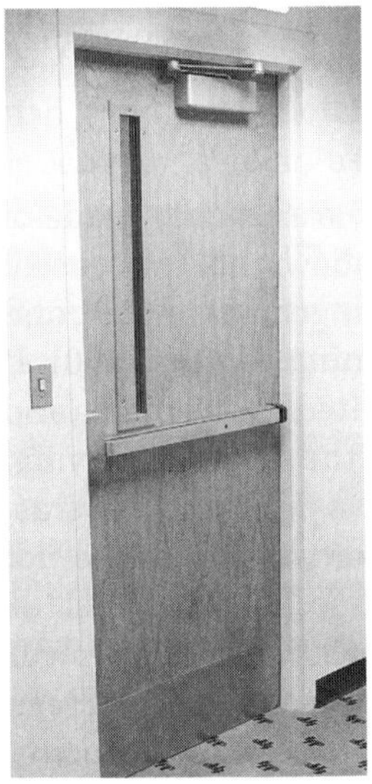

Figure 6.75 A typical swinging fire door.

Figure 6.76 A rolling steel door is made of interlocking steel plates or slats.

Fire doors that slide horizontally are preferable to other types when floor space is limited. They operate on overhead tracks that are mounted in such a way that when a fusible link releases the door, a counterweight causes the door to move across the opening. Vertical sliding fire doors are normally used where horizontal sliding doors or swinging doors cannot be used. Some models utilize telescoping sections that slide into position vertically on side-mounted tracks; the sections are operated by counterweights. Both horizontal and vertical fire doors close automatically.

Fire doors may be mechanically, manually, or electrically operated. The barrel on which the door is wound is usually turned by a set of gears located near the top of the door on the inside of the building. This feature makes the door exceptionally difficult to force. Therefore, whenever possible, entrance to the building should be gained at some other point and the door operated from the inside.

Most interior fire doors do not lock when they close; therefore, they can be opened without using forcible entry techniques. Fire doors that are used on exterior openings may be locked; therefore, the lock must be forced. A precautionary measure that firefighters should take when passing through an opening protected by a fire door is to block the door open to prevent its closing and trapping them. Fire doors have also been known to close behind firefighters and cut off the water supply in a hoseline.

FENCES

Fences of wood, metal, masonry, and woven wire often pose an entrance problem (Figure 6.77). The gates on these fences are usually locked with padlock and hasps. These locks may be pulled apart by a claw tool or cut by some available cutting tool such as a large bolt cutter. The staple of the lock can sometimes be pried or twisted off by using a pry tool or the point of the claw tool. Another quick method, which applies to inexpensive locks, is to brace the bow of the lock against the hasp or staple and then strike the bottom of the lock with the

Figure 6.77 Fences are common barriers to fireground access.

is fastened over the sheathing. When opening a wood frame wall, it is extremely important to watch for electrical wiring and pipes. The need for opening a wooden wall is often to gain access to the area involved by fire between studs. The procedure for opening a wood frame wall is the same as for roofs and floors except that the opening will be on a vertical instead of horizontal plane (Figure 6.98). Remove the siding, sound the wall for stud supports, and cut along a stud. In some areas, the siding will be plywood nailed directly to the studs. When this is removed, firefighters need only to break through the interior side of the wall.

Figure 6.98 Open plaster walls to search for fire extension.

Partition Walls

This section will address the general types of partition construction such as hollow clay tile, covered wood or metal studding, and solid concrete blocks (Figure 6.99). Solid concrete blocks should be opened in the same manner as described for the exterior masonry wall, and the same precautions should be taken. Unless previous inspections have been conducted on the building involved, it is very difficult to determine the construction of a partition during fire fighting procedures. If an opening needs to be made in a partition, the following procedures can be followed:

Step 1: Select the location of the opening, and before attempting to open the partition, check for electric wall plugs and switches.

Step 2: Have sufficient tools available such as power tools, picks, fire axes, sledges, and pry bars.

Step 3: If the studding in the partition is wood or metal, locate the studs by sounding.

Step 4: Cut along the studs with the fire axe or other cutting tool.

If the partition is hollow clay tile or concrete block, crush one or two blocks with a sledgehammer or a jackhammer, and the other blocks or tiles will remove more easily.

Partitions should be opened if there is any indication of fire in the wall. Four ways to determine whether the partitions contain fire are:

- Feeling the wall for hot spots
- Looking for discolored wallpaper or blistered paint
- Listening for the sound of burning
- Using electronic sensors

If opening the partition is part of the overhaul operation, open the partition on the side where the least damage will result. If possible, place a salvage cover on the floor before opening the partition. When making an opening in a partition, cut the wall covering along the studs, and remove it instead of ripping it off in a haphazard manner (Figure 6.100).

Figure 6.99 Dry wall and plaster are used to make partitions between rooms.

Figure 6.100 When making an opening in a partition, cut the wall covering along the studs.

7

Ventilation

This chapter provides information that addresses performance objectives described in NFPA 1001, *Standard for Fire Fighter Professional Qualifications* (1992), particularly those referenced in the following sections:

3-9 Ventilation

3-9.1

3-9.2

3-9.3

3-9.4

3-9.5

3-9.6

3-9.7

3-9.8

3-9.9

3-9.10

3-9.11

3-9.12

3-9.13

4-9 Ventilation

4-9.1

4-9.2

4-9.3

4-9.4

4-9.5

Chapter 7
Ventilation

Ventilation is the systematic removal and replacement of heated air, smoke, and gases from a structure with cooler air. The cooler air facilitates entry by firefighters for rescue and other fire fighting operations. The importance of ventilation cannot be overlooked. It increases visibility for quicker location of the seat of the fire. It decreases the danger to trapped occupants by channeling away hot, toxic gases. Ventilation also reduces the chance of flashover or backdraft.

Modern technology requires the emphasis on ventilation. As a result of the increased use of plastics and other synthetic materials, the fuel load in all occupancies has increased dramatically. The products of combustion produced during fires are becoming more dangerous and in larger quantities than ever before. Prompt ventilation for the saving of lives, suppression of fire, and reduction of damage becomes more important every day.

Modern energy conservation practices using increased insulation may be creating additional ventilation problems. A well-insulated structure will retain heat much better than a less-insulated, older structure. This means that the heat from a fire will be retained better, and flashover may occur much faster than it would in a less-insulated structure. Insulation installed over roof coverings of fire-rated roof construction will effectively retain heat and may reduce the fire rating drastically, causing premature roof failure. Therefore, the need for ventilation is increased and must be accomplished much sooner than has been practiced in the past.

Pre-incident inspections should note roof construction. The inspection should also note areas where extra insulation has been added to existing roofs and attic areas. This information will alert firefighters to possible problems while performing ventilation.

When a fire officer determines the need for ventilation, he or she must consider the precautions necessary to control the fire and assure the safety of firefighters performing the ventilation. Firefighters must wear full protective clothing, including SCBA. A charged hoseline must be in place at the ventilation location (Figure 7.1). Before, during, and following the ventilation work being performed, it is important to consider the possibility of fire spreading throughout a building and the danger of exposure fires.

This chapter covers the basics of ventilation operations. The advantages of proper ventilation are discussed, and considerations are given for deciding if and where to ventilate. Also, vertical (roof or topside ventilation), horizontal (using wall openings such as doors or windows), and forced ventilation procedures are covered.

Figure 7.1 A charged line must always be present when cutting a vent hole.

ADVANTAGES OF VENTILATION

Ventilation during fire fighting aids in meeting fire fighting objectives. There are certain advantages that result from proper ventilation. The following sections describe some of the advantages that proper ventilation holds for various parts of the overall fire fighting operation.

Rescue Operations

Proper ventilation simplifies and expedites rescue by removing smoke and gases that endanger trapped or unconscious occupants. The replacement of heat, smoke, and gases with cooler, fresh air helps victims breathe better (Figure 7.2). Proper ventilation also makes conditions safer for firefighters and improves visibility so that unconscious victims may be located more easily.

Fire Attack And Extinguishment

When a ventilation opening is made in the upper portion of a building, a chimney effect, drawing air currents from throughout the building in the direction of the opening, is created (Figure 7.3). For example, if this opening is made directly over the fire, it will tend to localize the fire. If it is made elsewhere, it may contribute to the spread of the fire (Figure 7.4). The channel

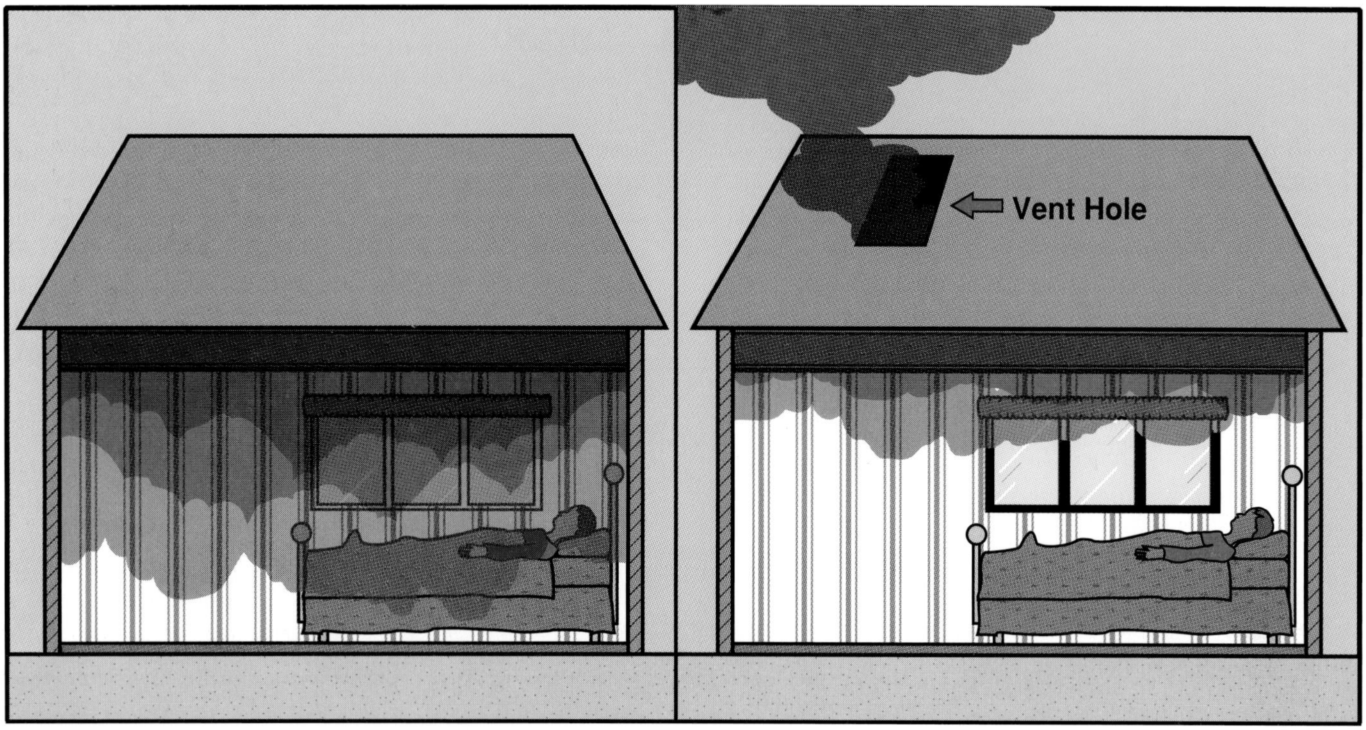

Figure 7.2 Ventilation increases the firefighter's ability to see in the structure and helps lift the smoke and toxic gases off the victims.

Figure 7.3 This illustrates the chimney effect in the building.

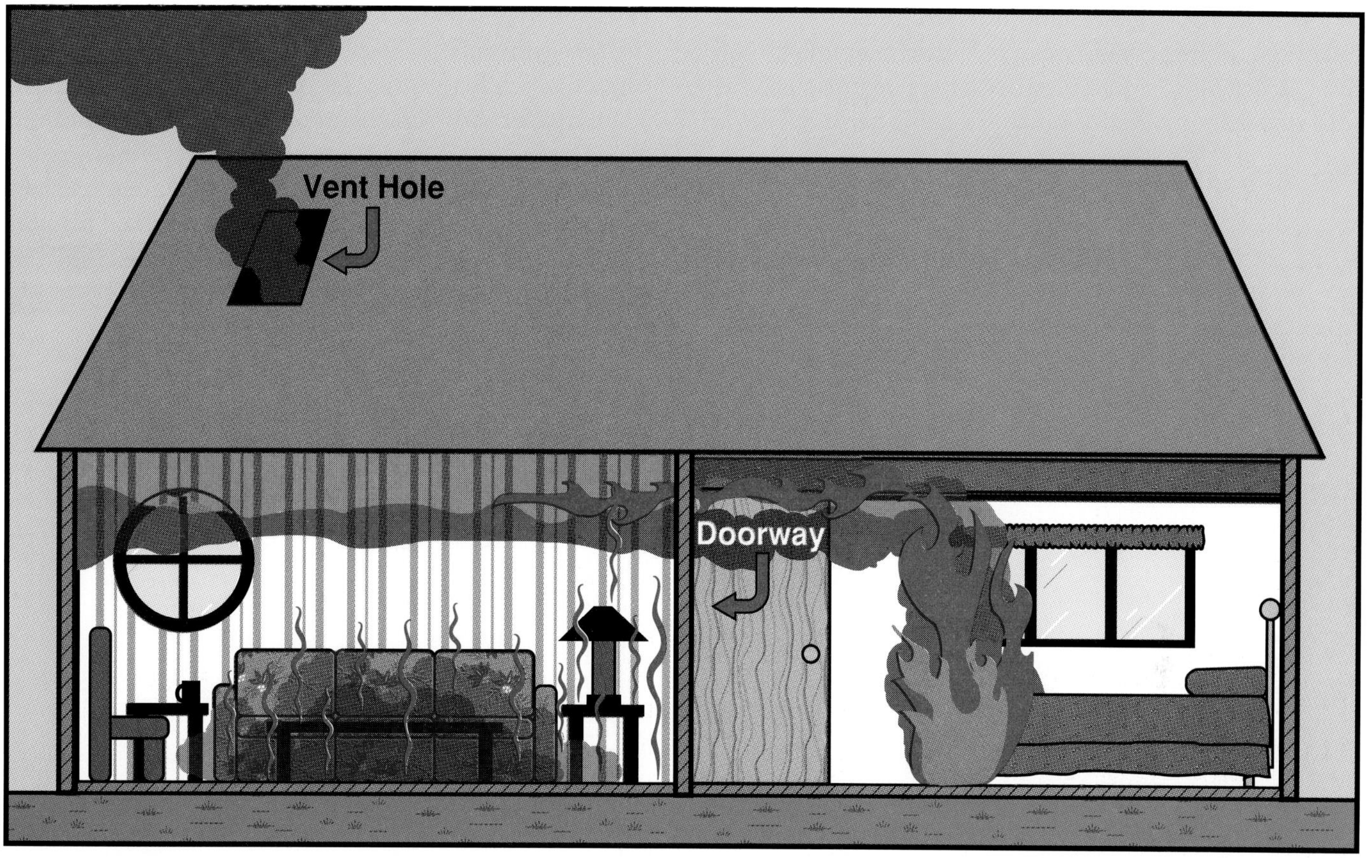

Figure 7.4 If the vent hole is not placed directly over the fire, the fire will be pulled through the structure toward the hole, thus increasing the fire damage to the structure.

effect from a properly placed hole will aid in the removal of smoke, gases, and heat from a building, which in turn permits firefighters to more rapidly locate the fire and proceed with extinguishment. It also reduces the chance of firefighters receiving steam burns when the water converts to steam. Proper ventilation reduces obstacles, such as limited visibility and excessive heat, that hinder firefighters while they perform fire extinguishment, salvage, rescue, and overhaul procedures.

Property Conservation

Rapid extinguishment of a fire reduces water, heat, and smoke damage. Proper ventilation assists in making this damage reduction possible. One method of ventilation that may prove advantageous is applying water to the heated area in the form of water fog or spray. The gases and smoke may be dissipated, absorbed, or expelled by the rapid expansion of the water when it is converted to steam. In addition to removing gases, smoke, and heat, this method also reduces the amount of water that may be required to extinguish the fire.

Smoke may be removed from burning buildings by controlling heat currents, by dissipating smoke through the expansion of water as it turns into steam, or by mechanical processes. Mechanical processes include blowers and smoke ejectors (Figure 7.5). Regardless of the method used, ventilation reduces smoke damage because fuel vapors and carbon particles are removed.

Figure 7.5 Mechanical blowers or ejectors help speed the ventilation process.

When smoke, gases, and heat are removed from a burning building, the fire can be quickly confined to an area. This will permit effective salvage operations to be initiated even while fire control is being accomplished.

Fire Spread Control

Convection causes heat, smoke, and fire gases to travel upward to the highest point in an area until they are trapped by a roof or ceiling. As the heat, smoke, and fire gases are trapped and begin to accumulate, they bank down and spread laterally to involve other areas of the structure (Figure 7.6). This process is generally termed *mushrooming*.

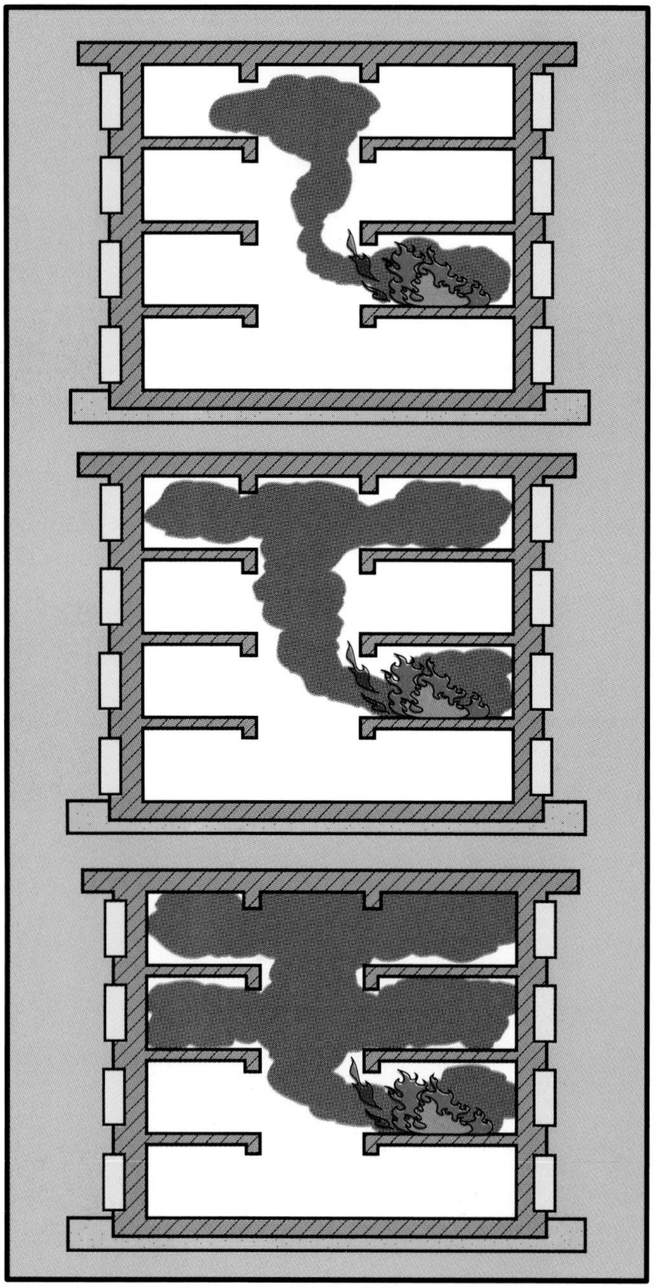

Figure 7.6 Mushrooming will occur in unventilated structures.

Proper ventilation of a building during a fire reduces the possibility of mushrooming. This in turn reduces the rate at which fire will spread over an area by providing an escape for the rising heated gases, at least for a short time. However, even with proper ventilation, if the fire is not extinguished soon after ventilation is accomplished, the increased supply of fresh air will feed the fire and eventually allow it to grow. Ventilation should occur when the hoseline crews are ready to move in and attack the fire (Figure 7.7).

Figure 7.7 A hoseline must be ready and in place when ventilation is being performed. *Courtesy of Bob Esposito.*

Reduction Of Flashover Potential

Flashover is a condition that may occur due to an excessive amount of heat. As the fire continues to burn, combustibles in the room are heated to their ignition temperatures. Once their ignition temperatures are reached, the entire room will be involved in flames with dire consequences to anyone in the room. Ventilation helps to alleviate this condition because the heat is removed before it reaches the necessary levels for mass ignition.

Reduction Of Backdraft Potential

When sufficient heat is confined in an area, the temperatures of combustible materials rise to their ignition points. These materials will not ignite, however, unless sufficient oxygen is available to support combustion. In this situation, a very dangerous condition exists because the admittance of an air supply (which provides the necessary oxygen) is all that is needed to explosively change the superheated area into an instant inferno. This sudden ignition is referred to as a backdraft. In order to prevent this critical situation from occurring, top ventilation must be provided to release superheated fire gases and smoke (Figure 7.8).

Firefighters must be aware of this explosion potential and must proceed cautiously in areas where excessive amounts of heat have accumulated. As previously explained in Chapter 1, the firefighters should be observant for the signs of possible backdraft conditions (Figure 7.9). These signs include the following:

- Smoke-stained windows
- Smoke puffing at intervals from the building
- Pressurized smoke coming from small cracks

Figure 7.8 Vertical ventilation will reduce the chance of a backdraft occurring.

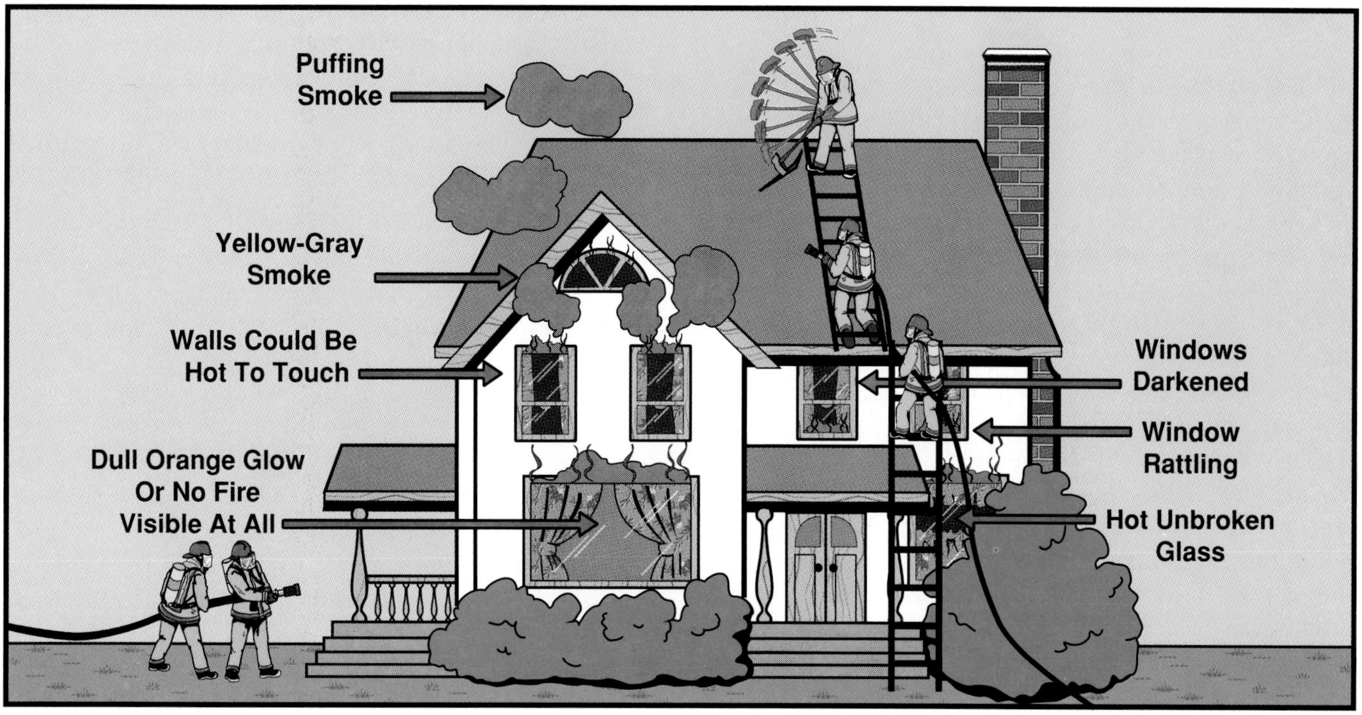

Figure 7.9 Some of the indicators of a potential backdraft situation.

- Not much visible flame from the exterior of the building

If any signs of backdraft are present, firefighters should stay away from doors and windows until vertical ventilation has had the chance to reduce the severity of the situation.

CONSIDERATIONS AFFECTING THE DECISION TO VENTILATE

The requirements for a plan of attack must be considered before a fire officer directs or orders ventilation to be started. A series of decisions should first be made that pertain to ventilation needs. These decisions will, by the nature of fire situations, fall into the following order:

- *Is there a need for ventilation at this time?* The need must be based upon the heat, smoke, and gas conditions within the structure, structural conditions, and the life hazard.
- *Where is ventilation needed?* This involves construction features of the building, contents, exposures, wind direction, extent of the fire, location of the fire, top or vertical openings, and cross or horizontal openings.
- *What type of ventilation should be used?* Horizontal (natural or mechanical)? Vertical (natural or mechanical)?

To answer these questions, firefighters will have to evaluate several pieces of information and take into account numerous factors. These are detailed in the following sections.

Life Safety Hazards

Dealing with the danger to human life is of utmost importance. The first consideration is the safety of firefighters and occupants. The life hazards are generally reduced in an occupied building involved by fire if the occupants are awake. If, however, the occupants were asleep when the fire developed and are still in the building, either of two situations may be expected. First, the occupants may have been overcome by smoke and gases. Second, they might have become lost in the building and are probably panicking. In either case, proper ventilation will be needed in conjunction with rescue operations. Depending on fire conditions, ventilation may need to be performed before rescue operations begin, or if conditions warrant, spreading flames may need to be attacked first; sometimes both must be performed simultaneously.

In addition to the hazards that endanger occupants, there are potential hazards to firefighters and rescue workers. The type of structure involved, whether natural openings are adequate, and the need to cut through roofs, walls, or floors (combined with other factors) add more problems requiring consideration to the decision process.

The hazards that can be expected from the accumulation of smoke and gases in a building include:

- Obscurity caused by dense smoke
- Presence of poisonous gases
- Lack of oxygen
- Presence of flammable gases

Another consideration facing firefighters is flammability of the materials and contents within a building. While these combustible materials may be heated above their ignition temperatures, they may not burn because of a lack of oxygen. The hazard lies in the fact that preheated combustibles will burst into a free-burning fire when provided with a supply of fresh air.

Visible Smoke Conditions

When first arriving at the scene of a fire, firefighters can make some ventilation decisions, as well as other tactical decisions, based on visible smoke conditions. Smoke accompanies most ordinary forms of combustion, and it differs greatly with the substance of the materials being burned. The density and color of the smoke is in direct ratio to the amount of suspended particles. Smoke conditions will vary according to how burning has progressed. A free-burning fire must be treated differently than one in the smoldering phase. A fire that is just beginning and is consuming wood, cloth, and other ordinary furnishings will ordinarily give off gray-white or blue-white smoke of no great density (Figure 7.10). As the burning progresses, the density may increase; the smoke may become darker because of the presence of large quantities of carbon particles (Figure 7.11).

Black smoke is usually the result of burning rubber, tar, or other flammable liquids (Figure 7.12). Brown smoke may indicate the oxides of nitrogen fumes, and gray-yellow smoke is a danger signal of approaching backdraft. A firefighter should remember that the materials contained in smoke can only be determined by chemical analysis. Although the color of the smoke may be of some value in determining what is burning, it is not always a reliable indicator.

The Building Involved

Knowledge of the building involved is a great asset when making decisions concerning ventilation. Building type and design are the initial factors to consider in determining whether to use horizontal or vertical ventilation. Other determining factors include:

Figure 7.10 In the early stages of a house fire, the smoke will be white or gray in color. *Courtesy of Joel Woods, University of Maryland Fire & Rescue Institute.*

Figure 7.11 During the latter stages of a house fire, the smoke will become markedly darker. *Courtesy of Bob Norman, Elkton (MD) V.F.D.*

- The number and size of wall openings
- The number of stories, staircases, shafts, dumbwaiters, ducts, and roof openings
- The availability and involvement of exterior fire escapes and exposures (Figure 7.13)

Building permits that are issued within the fire department's jurisdiction may enable the department to know when buildings are altered or subdivided. Checking these permits will also often reveal information concerning heating, ventilating, and air-conditioning systems and avenues of escape for smoke, heat, and fire gases. The extent to which a building is connected to adjoining structures also has a bearing on the decision to ventilate. In-service company inspection and pre-incident planning may provide more valuable and detailed information.

Figure 7.12 Rubber or petroleum fires will produce a thick, black smoke. *Courtesy of Ron Jeffers.*

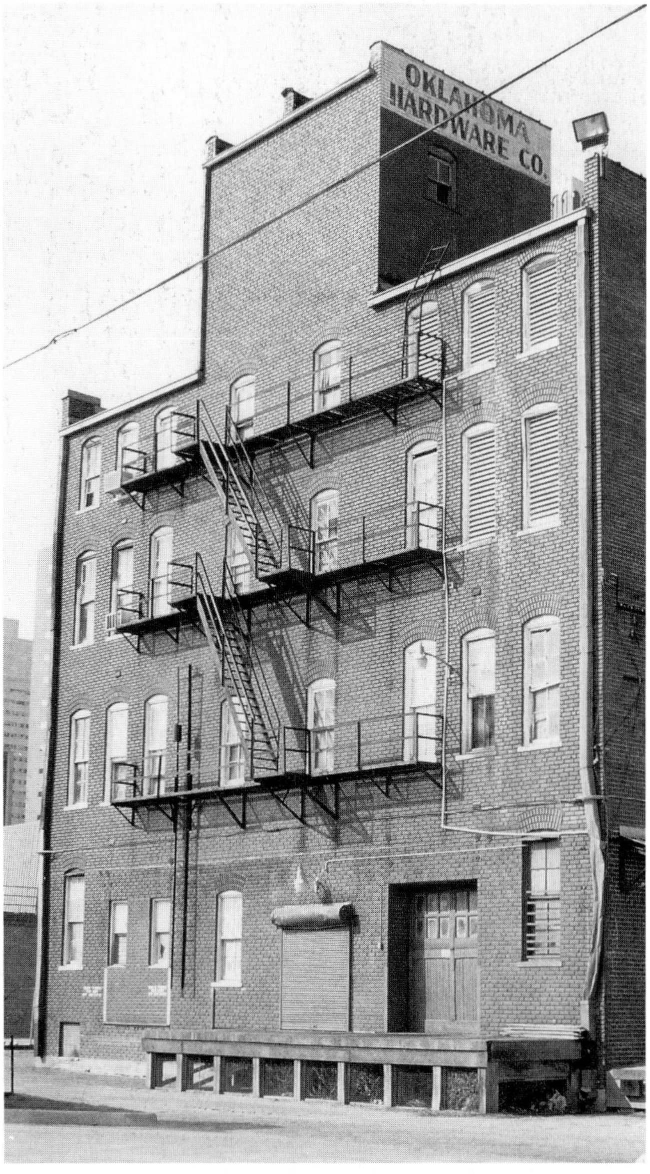

Figure 7.13 Note the availability and condition of fire escapes.

HIGH-RISE BUILDINGS

The danger to occupants from heat and smoke is a major consideration in high-rise buildings. High-rise buildings are normally occupied by hospitals, hotels, apartments, and business offices. In any case, a great number of people may be exposed to danger.

Fire and smoke may spread rapidly through pipe shafts, stairways, elevator shafts, air-handling systems, and other vertical openings. These openings contribute to a "stack effect," creating an upward draft and interfering with evacuation and ventilation (Figure 7.14).

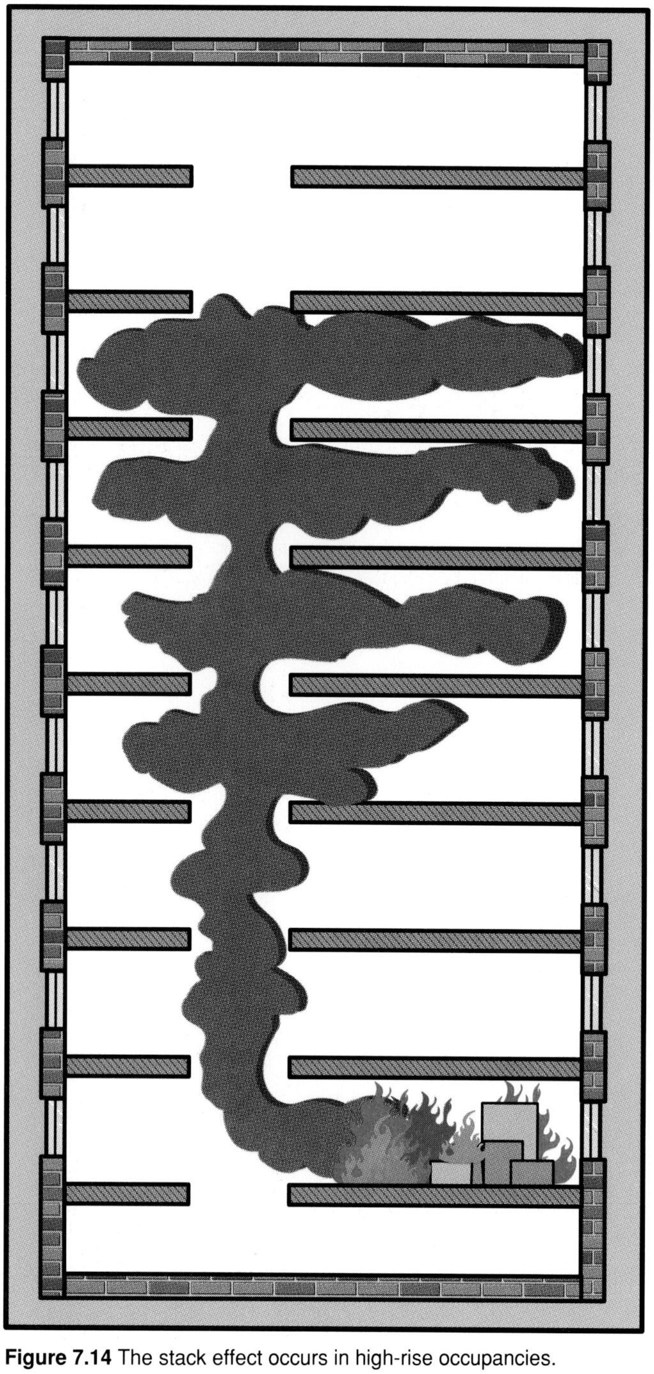

Figure 7.14 The stack effect occurs in high-rise occupancies.

The creation of layers of smoke and fire gases on floors below the top floor of unvented multistory buildings is possible. Smoke and fire gases travel through a building until their temperatures are reduced to the temperature of the surrounding air. When this stabilization of temperature occurs, the smoke and fire gases form layers or clouds within the building. The mushrooming effect, which is usually expected on top floors, does not occur in tall buildings until sufficient heat is built up to move the stratified, or layered, smoke and fire gas clouds that have gathered on lower floors in an upward direction (Figure 7.15). Pre-incident planning should include tactics and strategy that can cope with the ventilation and life hazard problems inherent in stratified smoke.

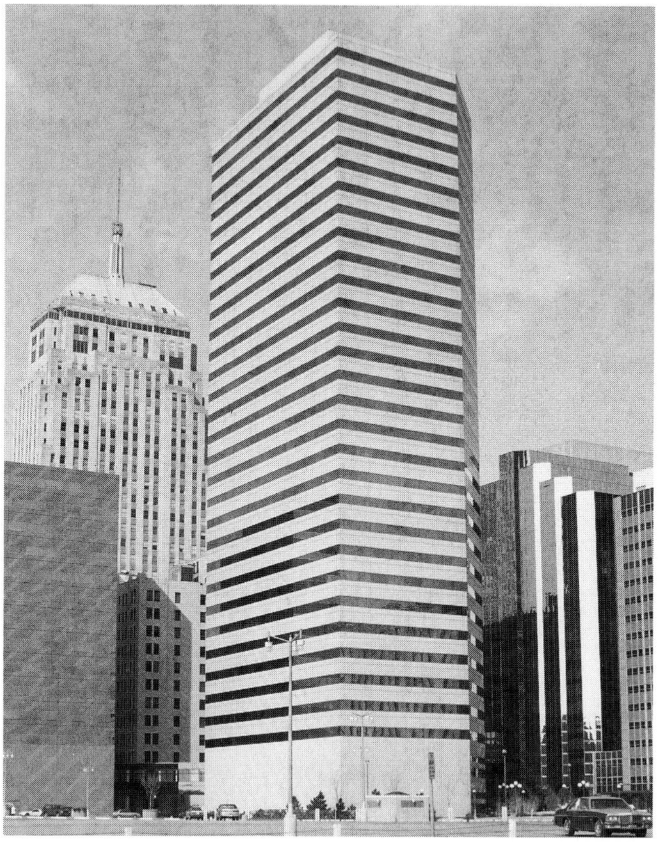

Figure 7.15 High-rise structures pose many challenges to the ventilation process.

Ventilation in a high-rise building must be carefully coordinated to ensure the effective use of personnel, equipment, and extinguishing agents. The personnel demand for this type of building is approximately four to six times as great as required for an average residential fire. In many instances, ventilation must be accomplished horizontally with the use of mechanical ventilation devices. Protective breathing equipment will be in great demand, and the ability to provide large quantities of fresh SCBA cylinders must be addressed. The problems of communication and coordination among the various attack and ventilating teams become more involved as the number of participants increases.

Top ventilation in high-rise buildings must be considered during pre-incident planning. In many buildings, only one stairwell pierces the roof. This vertical "chimney" must be used to ventilate smoke, heat, and fire gases from various floors. Before the doorways on the fire floors are opened and the stair shaft is ventilated, the door leading to the roof must be blocked open or removed from its hinges (Figure 7.16). Removal of the door at the top of the shaft ensures that it cannot close and allow the shaft to become filled with superheated gases after ventilation tactics are started. Many elevator shafts penetrate the roof line and may be used for ventilation.

Figure 7.16 Roof doors should be blocked open to facilitate ventilation.

BASEMENTS AND WINDOWLESS BUILDINGS

Basement fires are among the most challenging that firefighters will face. Access into the basement is difficult because the firefighters have to descend through the worst heat and smoke to get to the seat of the fire. Access to the basement may be via interior or exterior stairs, exterior windows, or hoistways. Many outside entrances to basements may be blocked or secured by iron gratings, steel shutters, wooden doors, or combinations of these for protection against weather and burglars (Figure 7.17). All of these features serve to impede attempts at natural ventilation.

Many buildings, especially in business areas, have windowless wall areas. While windows may not be the most desirable means for escape from burning buildings, they are an important consideration for ventilation. Windowless building designs create an adverse effect on fire fighting and ventilation operations (Figure 7.18).

Figure 7.17 Exterior basement doors are commonly made of steel and are difficult to force. *Courtesy of Bob Esposito.*

Figure 7.18 Buildings with few or no windows are difficult to ventilate.

The ventilation of a windowless building may be delayed for a considerable time, allowing the fire to gain headway or to create backdraft conditions.

Problems inherent in ventilating this type of building are many and varied depending on the size, occupancy, configuration, and type of material from which the building is constructed. Windowless buildings usually require mechanical ventilation for the removal of smoke. Most buildings of this type are automatically cooled and heated through ducts (Figure 7.19). Mechanical ventilation equipment can sometimes effectively clear the area of smoke by itself.

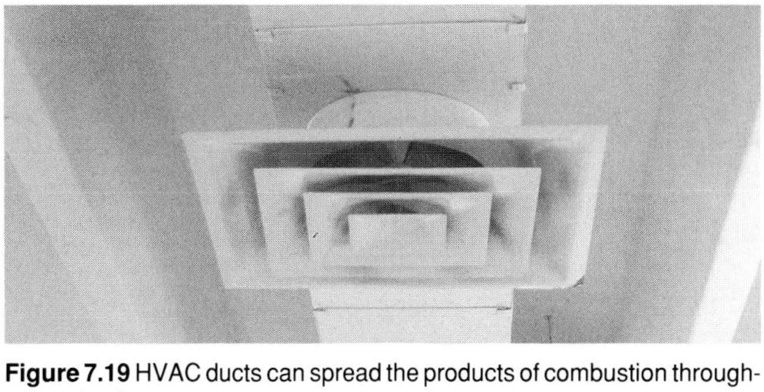

Figure 7.19 HVAC ducts can spread the products of combustion throughout a structure.

Location And Extent Of The Fire

The fire may have traveled some distance throughout the structure by the time fire fighting forces arrive, and consideration must be given to the extent of the fire, as well as to its location. Opening for ventilation purposes before the fire is located may spread the fire throughout areas of the building that otherwise would not have been affected. The severity and extent of the fire usually depend upon the type of fuel and the amount of time it has been burning, installed fire protection devices, and the degree of confinement of the fire. The phase to which the fire has progressed is a primary consideration in determining ventilation procedures. Some of the ways vertical extension occurs are as follows:

- Through stairwells, elevators, and shafts by direct flame contact or by convected air currents (Figure 7.20)
- Through partitions and walls and upward between the walls by flame contact and convected air currents (Figure 7.21)
- Through windows or other outside openings where flame extends to other exterior openings and enters upper floors (commonly called "lapping") (Figure 7.22)
- Through ceilings and floors by conduction of heat through beams, pipes, or other objects that extend from floor to floor (Figure 7.23)
- Through floor and ceiling openings where sparks and burning material fall through to lower floors
- By the collapse of floors and roofs

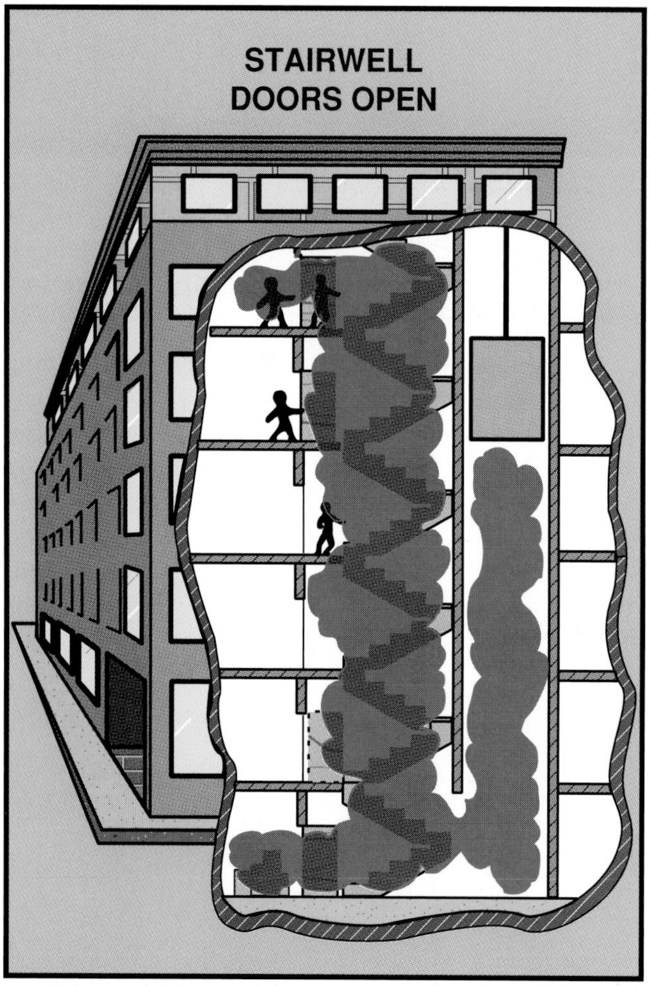

Figure 7.20 Fire and smoke can spread throughout the stairwell system if doors are left open.

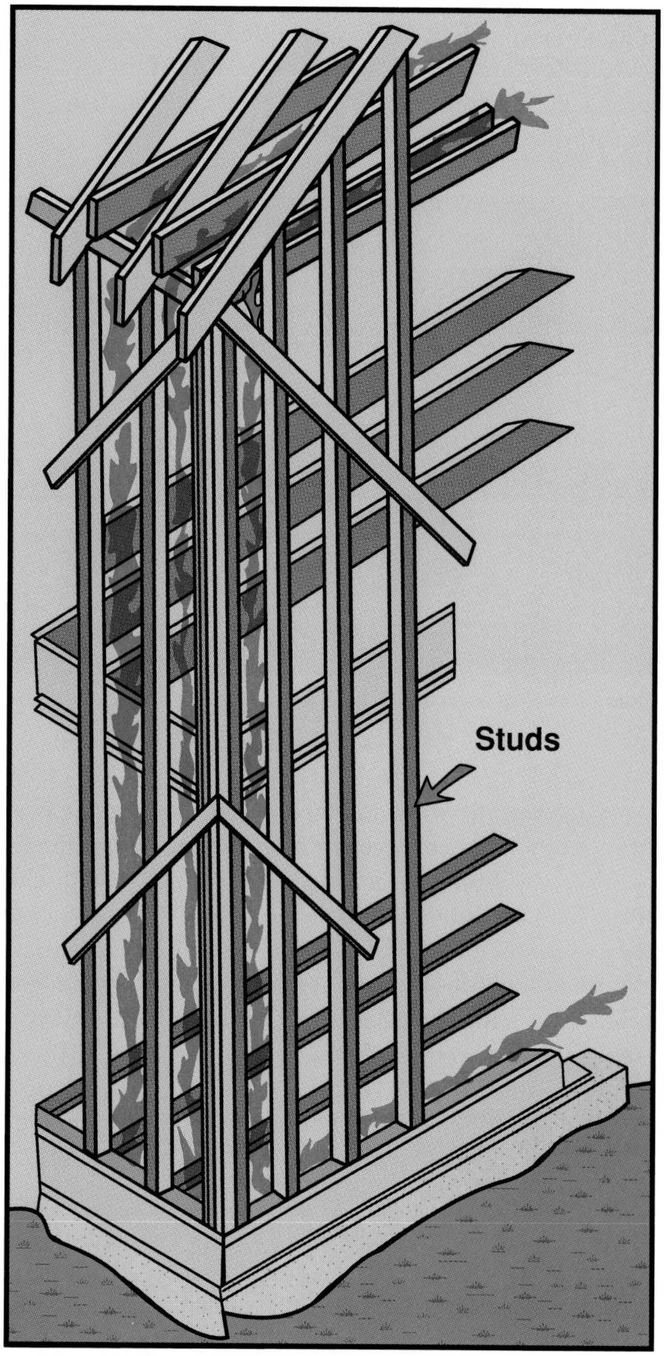

Figure 7.21 Fire will extend through open spaces within the wall assembly.

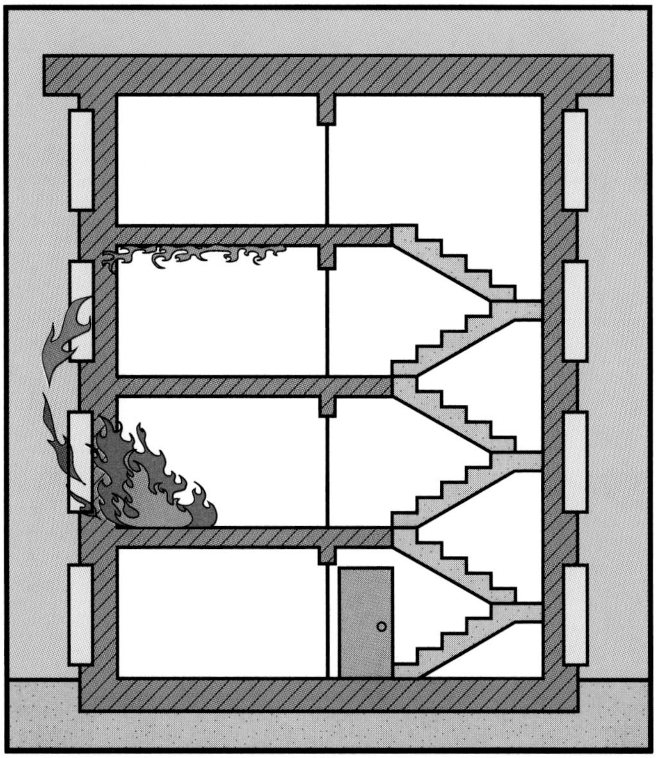

Figure 7.22 Fire can spread vertically by lapping from window to window.

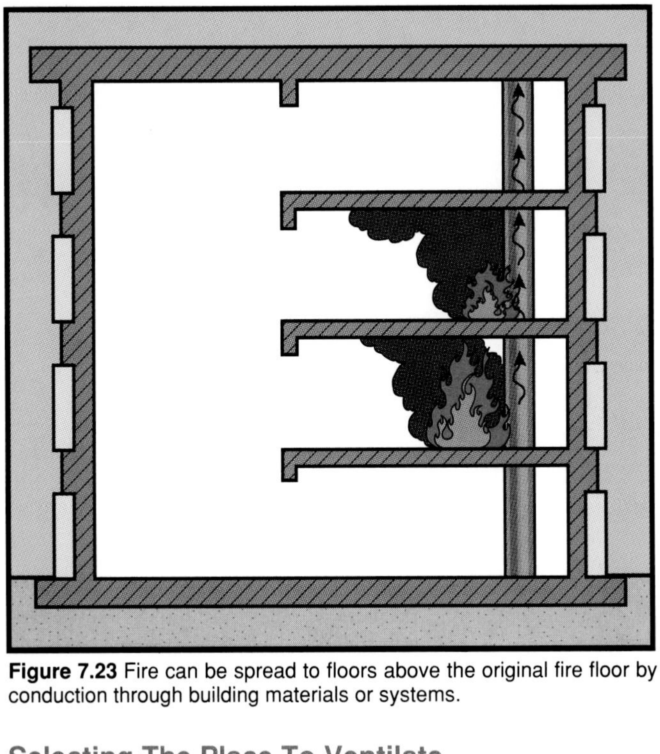

Figure 7.23 Fire can be spread to floors above the original fire floor by conduction through building materials or systems.

Selecting The Place To Ventilate

The ideal situation in selecting a place to ventilate is one in which firefighters have prior knowledge of the building and its contents. There is no rule of thumb in selecting the exact point at which to open a roof except "as directly over the fire as possible." Many factors will have a bearing on where to ventilate. Some of these factors include:

- The availability of natural openings such as skylights, ventilator shafts, monitors, and hatches (Figure 7.24)
- Location of the fire and the direction in which the incident commander wishes it to be drawn
- Type of construction
- Wind direction (Figure 7.25)

Figure 7.24 Existing roof openings include (from top to bottom) scuttle hatches, skylights or monitors, chimneys and vent pipes, and stairwell openings.

Figure 7.25 The direction of the wind will influence the manner in which a structure is ventilated.

- The extent of progress of the fire and the condition of the building and its contents
- Bubbles or melting of roof tar
- Indications of lessening structural integrity of the roof
- The effect that ventilation will have on the fire
- The consequences its release will have on exposures
- The attack crew's state of readiness
- The ability to protect exposures prior to actually opening the building (Figure 7.26)

Before ventilating a building, adequate personnel and fire control equipment must be ready, because the fire may immediately increase in intensity when the building is opened. These resources should be provided for both the building involved and other exposed buildings. As soon as the building has been opened to permit hot gases and smoke to escape, an effort to reach the seat of the fire for extinguishment should be made at once, if conditions permit this to be done safely. If wind direction permits, entrance should be made into the building as near the fire as possible (Figure 7.27). It is at this opening that charged hoselines should be positioned in case of violent burning or an explosion (Figure 7.28). Charged lines should also be in place at critical points of exposure to prevent the fire from spreading.

Figure 7.26 Use caution when venting next to an exposure that is taller than the fire building.

Figure 7.27 It is generally safer to enter the building after ventilation has been accomplished.

Figure 7.28 Stretch a hoseline to the ventilation area.

VERTICAL VENTILATION

Vertical ventilation generally means the opening of the roof or existing roof openings for the purpose of allowing heated gases and smoke to escape to the atmosphere. In order to properly ventilate a roof, the firefighter must understand the basic types and designs of roofs. Many designs are used, and their names vary with the locality.

A roof covering is the exposed part of the roof. Its primary purpose is to afford protection against the weather. Roof coverings may be wood shingles, composition shingles, composition roofing paper, tile, slate, synthetic membrane, or a built-up tar and gravel surface. The type of roof covering is important from a fire protection standpoint because it may be subjected to sparks and blazing brands.

A study of local common roof types and the manner in which their construction affects opening procedures is necessary to develop effective vertical ventilation policies and procedures. The firefighter is concerned with three prevalent types of roof construction: flat, pitched, and arched (Figure 7.29). Buildings may be constructed with a combination of roof designs. Some of the more common styles are the flat, gable, gambrel, shed, hip, mansard, dome, lantern, and butterfly (Figure 7.30).

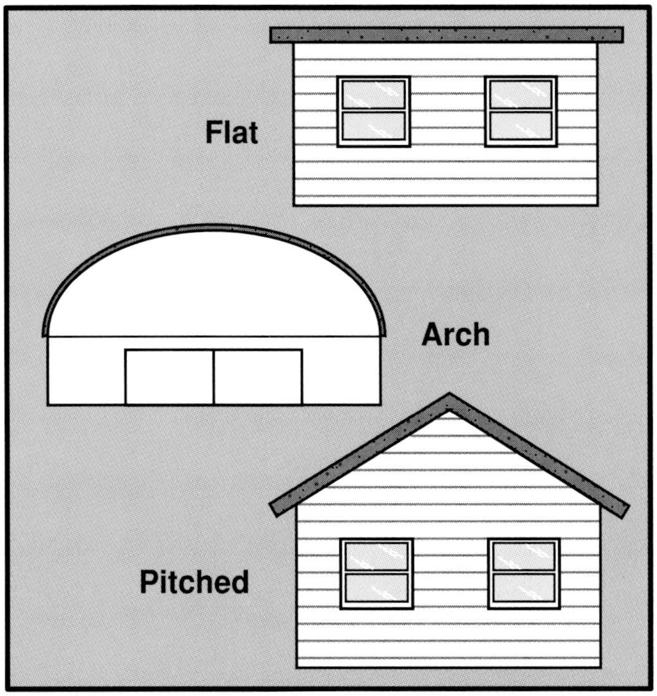

Figure 7.29 The three basic shapes of roofs are flat, arched, and pitched.

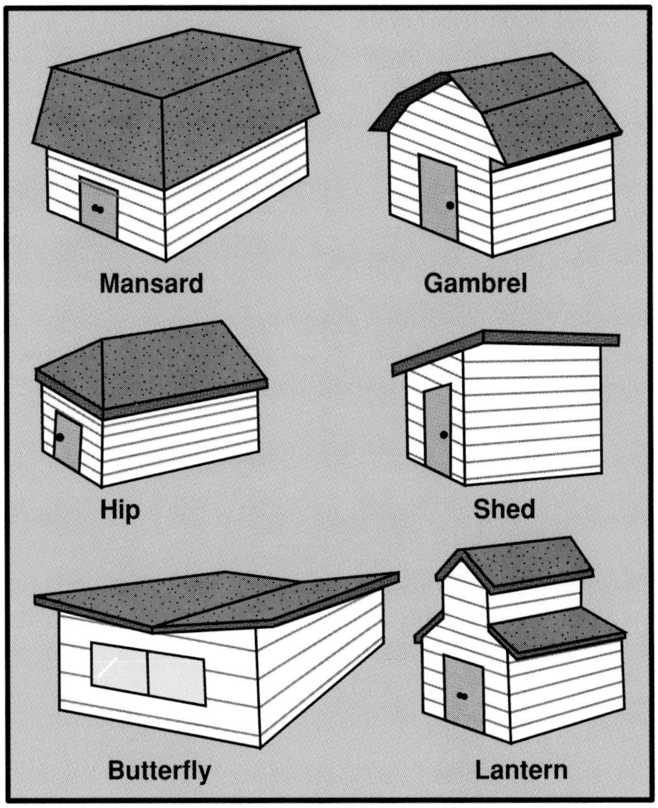

Figure 7.30 Common roof styles.

Safety Precautions

Vertical ventilation can be undertaken after the fire officer has completed the following:

- Considered the type of building involved
- Considered the location and extent of the fire
- Moved personnel and tools to the roof
- Observed safety precautions
- Selected the place to ventilate

The ranking firefighter on top of the roof should be in constant communication with the operations chief or the incident commander. Portable radios are most adaptable to this type of communication (Figure 7.31). Responsibilities of the leader on the roof include:

- Ensuring that only the required openings are made
- Coordinating the crew's efforts with those of the firefighters inside the building
- Ensuring the safety of all personnel who are assisting in the opening of the building

Figure 7.31 By using a portable radio, the leader on the roof can have constant communication with the incident commander.

Some of the safety precautions that should be practiced include the following:

- Observe the wind direction with relation to exposures.
- Work with the wind at your back or side to provide protection while cutting the roof opening.
- Note the existence of obstructions or excess weight on the roof. These may make operations more difficult or reduce the amount of time before a roof fails.

- Provide a secondary means of escape for crews on the roof (Figure 7.32).
- Exercise care in making the opening so that main structural supports are not cut.
- Guard the opening to prevent personnel from falling into the building.
- Evacuate the roof when ventilation work is complete.
- Use lifelines, roof ladders, or other means to protect personnel from sliding and falling off the roof (Figure 7.33).
- Exercise caution in working around electric wires and guy wires.

Figure 7.32 Provide two means of escape from the roof. They should be well apart from each other, but close to the work area.

Figure 7.33 Firefighters using lifelines to keep from falling off the roof.

- Ensure that all personnel on the roof are wearing full personal protective equipment, including SCBA (Figure 7.34).
- Keep other firefighters out of range of the axe.
- Caution axe users to beware of overhead obstructions within the range of their axe.
- Start power tools on the ground to ensure operation; however, it is important that the tool be shut off before hoisting or carrying the tool to the roof.

Figure 7.34 Personnel working on a roof must wear full protective equipment, including their SCBA.

- Make sure that the angle of the cut is not toward your body.
- Extend ladders at least three rungs above the roof line. When possible, 6 feet (2 m) or more is desirable (Figure 7.35). When using elevating platforms, the floor of the platform should be even with or slightly above roof level.
- "Sound" the roof for structural integrity before stepping on it; do not jump without checking it first (Figure 7.36).
- Use supporting members of the structure (for example, beam or parapet) for travel; use no diagonal travel.
- Use pre-incident planning and inspections to identify buildings that have roofs supported by lightweight or wooden trusses. Realize that these roofs may fail early into a fire and are extremely dangerous to be operating on or below.
- When using a roof ladder, make sure that it is firmly secured over the peak of the roof before operating from it (Figure 7.37).

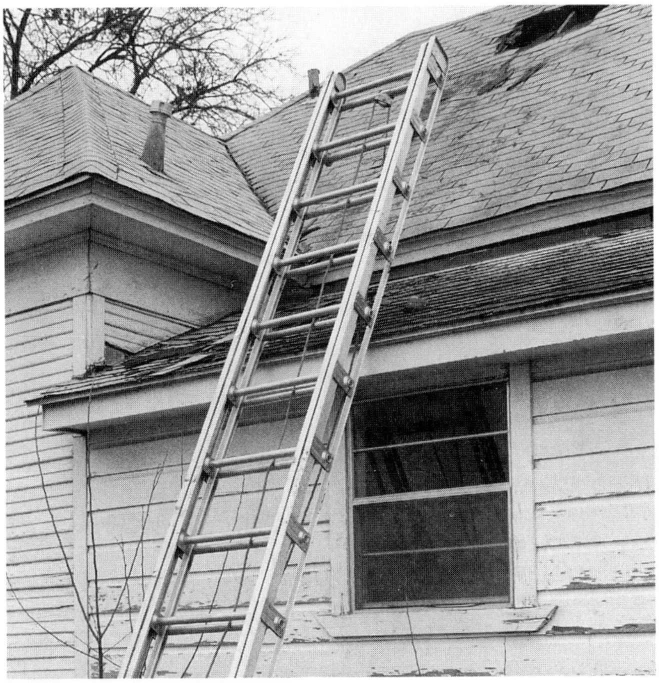

Figure 7.35 Extend ladders at least three rungs above the roof line.

Figure 7.36 Always "sound" the roof before stepping on it.

Figure 7.37 The roof ladder must be secured over the peak of the roof before working from it.

- Be aware of the following warning signs of an unsafe roof condition (Figure 7.38):
  - — melting asphalt
  - — "spongy" roof, a normally solid roof that springs back when walked upon
  - — smoke coming from the roof
  - — fire coming from the roof
- Work in groups of at least two, with no more people than absolutely necessary to get the job done.

Figure 7.38 Fire will often come from the vent hole as superheated fire gases get the oxygen they need to burn. *Courtesy of Harvey Eisner.*

Existing Roof Openings

Existing roof openings, such as scuttle hatches, skylights, monitors, ventilating shafts, and stairway doors, may be found on various types of roofs (Figure 7.39). Almost every roof opening will be locked or secured

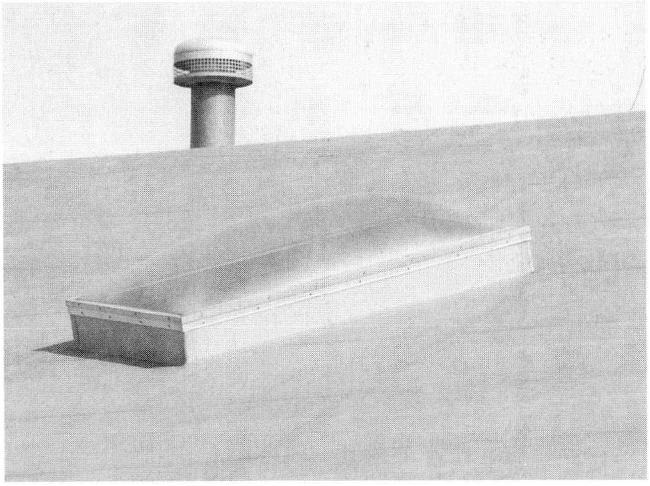

Figure 7.39 Skylights are one type of existing roof opening.

in some manner. Scuttle hatches are normally square and large enough to permit a person to climb onto the roof (Figure 7.40). A scuttle hatch may be metal or wood, and generally speaking, it does not provide an adequate opening for ventilation purposes. If skylights contain ordinary shatter-type glass, they may be conveniently opened. If they contain wired glass, Plexiglas®, or Lexan®, they are very difficult to shatter and are more easily opened by removing the frame. The sides of a monitor may contain glass (which is easily removed) or louvers made of wood or metal (Figure 7.41). The sides, which are hinged, are easily forced at the top. If the top of the monitor is not removable, at least two sides should be opened to create the required draft. Stairway doors may be forced open in the same manner as other doors of the same type.

Existing openings should be used for vertical ventilation purposes whenever possible. Typically, it is quicker to open one of these man-made openings than it is to cut a hole in the roof. However, firefighters must realize that these openings are rarely in the best location or large enough for adequate ventilation. Most often they will simply supplement holes that have to be cut.

Figure 7.40 A scuttle hatch.

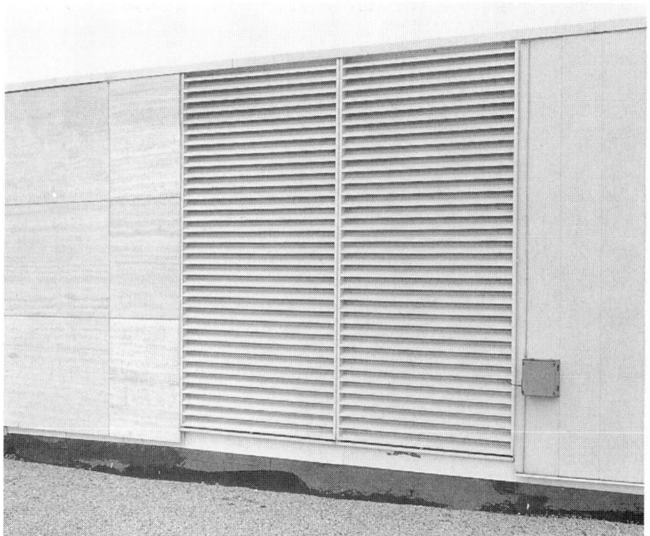

Figure 7.41 Louvers on a roof monitor.

Flat Roofs

Flat roofs are most commonly found on commercial, industrial, and apartment buildings. This type of roof may or may not have a slight slope to facilitate water drainage. The flat roof is frequently pierced by chimneys, vent pipes, shafts, scuttles, and skylights (Figure 7.42). The roof may be surrounded and/or divided by parapets, and it may support water tanks, air-conditioning equipment, antennas, and other obstructions that may interfere with ventilation operations.

Figure 7.42 The flat roof is frequently pierced by vent pipes.

The structural part of the flat roof is generally similar to the construction of a floor that consists of wooden, concrete, or metal joists covered with sheathing. The sheathing is covered with a layer of waterproofing material and an insulating material (Figure 7.43). Instead of joists and sheathing construction, flat roofs are sometimes poured reinforced concrete or lightweight concrete, precast gypsum, or concrete slabs set within metal joists.

The best way for fire departments to determine the material from which roofs are constructed is through inspection and pre-incident planning surveys. The materials used in flat roof construction determine what equipment will be necessary to open holes in it. When cutting through a roof, the firefighter should make the opening rectangular or square to facilitate repairs to the roof. One large opening, at least 4 x 4 feet (1.2 m by 1.2 m), is much better than several small ones (Figure 7.44). A procedure for opening a wood joist or rafter roof with

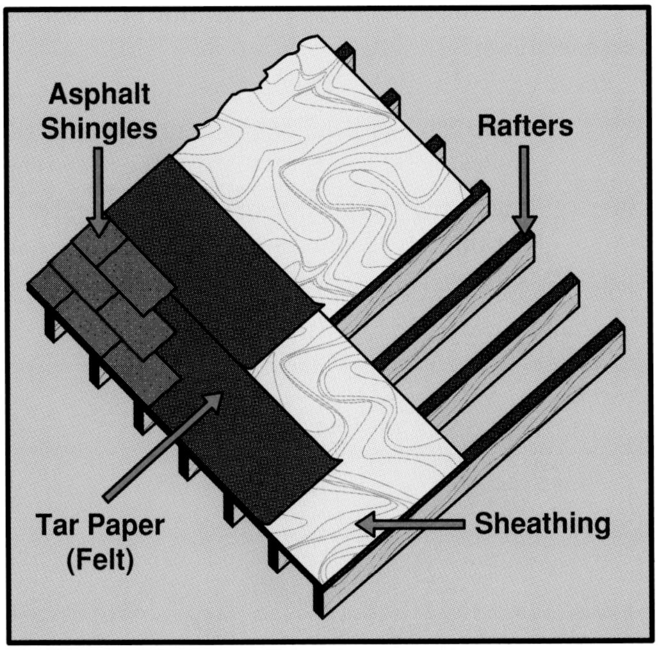

Figure 7.43 This illustration shows the construction of a typical composite roof assembly.

Figure 7.44 One large hole is better than several small holes.

an axe is suggested in the following sequence. Remember that this procedure should always be carried out by at least two firefighters.

Step 1: Use the following factors to determine the location for the opening:

— Location of seat of fire

— Direction of wind

— Existing exposures

— Extent of fire

— Obstructions

Step 2: Locate roof supports by sounding with the axe. Between the rafters or joists it will sound hollow and the axe will bounce. When near or on top of a support, it will sound dull and solid.

Step 3: Mark the location for the opening by scratching a line on the roof surface with the pick head.

Step 4: Remove the built-up roof material or metal by cutting the material and using the pick head to pull the material out of the way.

Step 5: Cut the wood decking diagonally alongside the joist toward the hole. (**NOTE:** The joist should never be cut. Cutting alongside a joist permits the blows to have a solid base to avoid bouncing.)

Step 6: Pry up the roof boards with the pick end of the axe. After the opening has been cut in the roof, push the blunt end of a pike pole, plaster hook, or some other suitable tool through the roof opening to open the ceiling below.

Power equipment for opening roofs is most useful and often provides a means by which ventilation procedures may be accelerated. Rotary rescue saws or carbide-tipped chain saws are excellent for roof-cutting operations. Use the saw to make the same cut as previously described for the axe. Care should always be taken to ensure that the saw operator has good footing and does not operate the saw in a manner that might allow it to accidentally come in contact with any parts of the body. Always shut the saw off when it is being transported to or from the point of operation.

Pitched Roofs

The pitched roof is elevated in the center and thus forms a pitch to the edges (Figure 7.45). Pitched roof construction involves timber rafters or metal trusses that run from the ridge to a wall plate on top of the outer wall at the eaves level. The rafters or trusses that carry the sloping roof can be made of various materials. Over these rafters, the sheathing boards are applied either squarely or diagonally. These sheathing boards are sometimes applied solidly over the entire roof. Pitched roofs sometimes have a covering of roofing paper applied before shingles are laid. Shingles may be wood, metal, composition, asbestos, slate, or tile.

Figure 7.45 Most single-family dwellings have pitched roofs.

Pitched roofs on barns, churches, supermarkets, and industrial buildings may have roll felt applied over the sheathing and then mopped with asphalt roofing tar. Gypsum slabs, approximately 2 inches (50 mm) thick, may be laid between the metal trusses of a pitched roof instead of wood sheathing. These conditions can only be determined by inspection surveys conducted by fire department personnel.

Pitched roofs have a more pronounced downward incline than those of flat roofs. This incline may be gradual or steep. The procedures for opening pitched roofs are quite similar to those for flat roofs except that additional precautions must be taken to prevent slipping (Figure 7.46). Suggested steps for opening pitched roofs are as follows:

Step 1: Locate the position where the opening is to be made. The location should usually be at the highest point of the roof, above the fire area.

Step 2: Place a roof ladder on the roof so that personnel working off it will be upwind from the hole (Figure 7.47).

Figure 7.46 Shingles can be cleared away prior to making the roof opening.

Figure 7.47 Always work upwind of the vent hole.

Step 3: Bounce an axe or other tool on the roof to sound for solid supports or rafters (Figure 7.48). Mark their location by scratching with the pick of an axe.

Step 4: Rip off the shingles or roofing felt sufficiently to permit the initial cut to be made (Figure 7.49). (**NOTE:** In some cases, it is best to first

Figure 7.48 Bounce an axe or other tool on the roof to sound for solid supports or rafters.

Figure 7.50 Cut the sheathing alongside a rafter the distance required for the opening. The opening should be square or rectangular.

Figure 7.49 Rip off the shingles or roofing felt sufficiently to permit the initial cut to be made.

Figure 7.51 Remove sheathing boards with the pick of the axe or some other suitable tool.

Figure 7.52 Stick the blunt end of a pike pole through the vent hole to break a hole in the plaster of the ceiling below to further facilitate the venting of the structure below the attic.

remove all shingles or roofing felt from the entire area where the hole is to be made.) The removal of these coverings is most important when the hole is going to be made with an axe.

Step 5: Cut the sheathing alongside a rafter the distance required for the opening (Figure 7.50). The opposite side of the opening may then be cut in a like manner. (**NOTE:** The opening should be square or rectangular.) Use care not to cut the rafters.

Step 6: Remove sheathing boards with the pick of the axe or some other suitable tool (Figure 7.51).

Step 7: Push the blunt end of a pike pole or other long-handled tool through the hole to open the ceiling (Figure 7.52).

Other types of pitched roofs may require different opening techniques. For example, some slate and tile roofs may require no cutting. Slate and tile roofs can be opened by using a large sledgehammer to smash the slate or tile and the thin lath strips or the 1- x 4-inch (25 mm by 100 mm) boards that support the tile or slate. Tin roofs can be sliced open and peeled back with tin snips or a large device similar to a can opener. Fire departments should formulate plans for dealing with types of roof construction specific to their jurisdiction.

Arched Roofs

Arched roof construction has many desirable qualities for certain types of buildings. One form of arched roof construction uses the bow-string truss for supporting members. The lower chord of the truss may be covered with a ceiling to form an enclosed cockloft or roof space (Figure 7.53). Such concealed, unvented spaces are definitely ventilation problems and contribute to the spread of fire and early failure of the roof.

Trussless arched roofs are made up of relatively short timbers of uniform length. These timbers are beveled and bored at the ends where they are bolted together at an angle to form a network of structural timbers. This network forms an arch of mutually braced and stiffened timbers (Figure 7.54). Being an arch rather than a truss, the roof exerts a horizontal reaction in addition to the vertical reaction on supporting structural components. Trussless arch construction enables all parts of the roof to be visible to firefighters. A hole of considerable size may be cut or burned through the network sheathing and roofing any place without causing collapse of the roof structure, instead the loads are distributed to less damaged timbers around the opening.

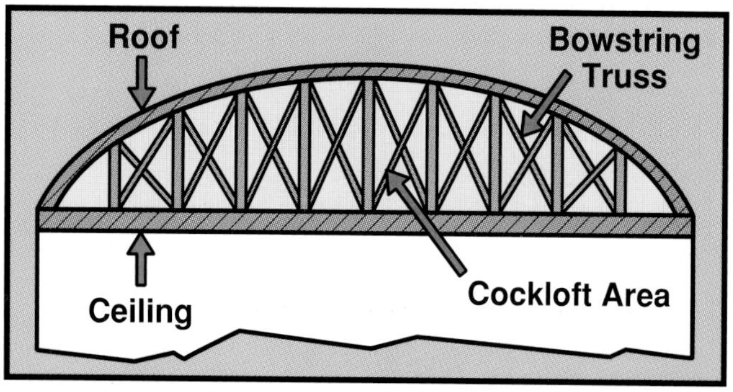

Figure 7.53 Notice the large open areas within the truss roof structure.

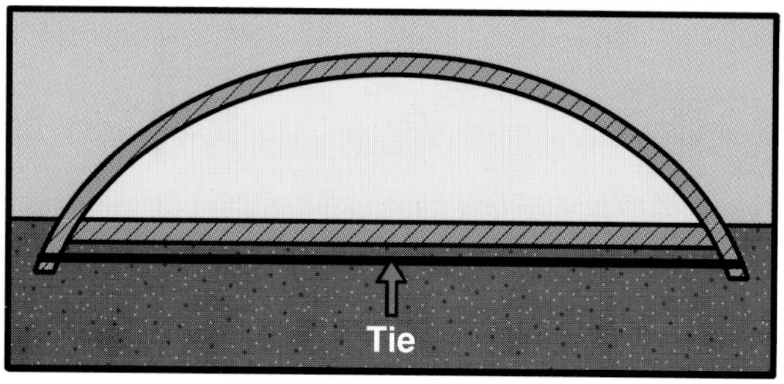

Figure 7.54 A trussless arched roof.

Cutting procedures for opening arched roofs are the same as for flat or pitched roofs except that it is doubtful that a roof ladder can always be used on an arched roof. Regardless of the method used to support the firefighter, the procedure is difficult and dangerous because of the curvature of the roof. Because of the potential for sudden collapse of this type of roof under fire conditions, only allow firefighters to work from an aerial ladder or platform extended to the roof (Figure 7.55).

Figure 7.55 Always work from an aerial device when ventilating a trussed arch roof.

Concrete Roofs

The use of precast concrete is very popular with certain types of construction. Precast roof slabs are available in many shapes, sizes, and designs. These precast slabs are hauled to the construction site, ready for use. Other builders form and pour the concrete on the job. Roofs of either precast or reinforced concrete are extremely difficult to break through, and opening them should be avoided whenever possible (Figure 7.56). Natural roof openings and horizontal openings should be used on buildings with heavy concrete roofs.

A popular lightweight material made of gypsum plaster and portland cement mixed with aggregates, such as perlite, vermiculite, or sand, provides a lightweight floor and roof assembly. This material is sometimes referred to as lightweight concrete. Lightweight precast planks are manufactured from this material, and the slabs are reinforced with steel mesh or rods. Lightweight concrete roofs are usually finished with roofing felt and a mopping of hot tar to make them watertight.

Lightweight concrete roof decks are also poured in place over permanent formboards, steel roof decking, paper-backed mesh, or metal rib lath. These lightweight concrete slabs are relatively easy to penetrate. Some

Figure 7.56 Most concrete roof structures are of the pre-fabricated type.

types of lightweight concrete can be penetrated with a hammer-head pick, power saw with concrete blade, jackhammer, or any other penetrating tool.

Metal Roofs

Metal roof coverings are made from several different kinds of metal and constructed in many styles. Light-gauge steel roof decks can either be supported on steel framework or they can span across wider spaces (Figure 7.57). Other types of corrugated roofing sheets are made from light-gauge cold-formed steel, galvanized sheet metal, and aluminum. The light-gauge cold-formed steel sheets are used primarily for the roofs of industrial buildings. Corrugated galvanized sheet metal and aluminum are seldom covered with a roof material, and the sheets can usually be pried from their supports (Figure 7.58). Metal cutting tools or power saws with metal cutting blades must be employed to open metal roofs. Metal roofs on industrial buildings are usually provided with adequate roof openings, skylights, or hatches. Older buildings may have roofs that are made of large, fairly thin sheets of tin laid over lath strips. These can be opened by cutting with a power saw, axe, or a large sheet-metal cutter similar to a can opener.

Figure 7.57 Some sheet metal roofs will have a slight texture to them.

Figure 7.58 A corrugated sheet metal roof.

Trench Or Strip Ventilation

Trench ventilation (also referred to as strip ventilation) is used in a slightly different way than the standard vertical ventilation techniques previously described. Standard vertical ventilation is used simply to remove heated smoke and gases from the structure and is best done directly above the fire. Trench ventilation is used to stop the spread of fire in a long, narrow structure. Trench ventilation is performed by cutting a large hole, or trench, that is at least 4 feet (1.3 m) wide and extends from one exterior wall to the opposite exterior wall (Figure 7.59). This hole is usually cut well ahead of the advancing fire with the purpose of setting up a defensive line at which the fire's progress will be halted. This technique is most commonly used in low, expansive occupancies such as schools, shopping malls, and motels.

In some jurisdictions, the term strip ventilation is used to describe a slightly different ventilation technique.

Figure 7.59 A trench cut covers the entire width of the roof.

Strip ventilation, in this sense, refers to the opening of large holes on roofs, but not completely side to side as described for trench ventilation. Large strips are cut in the roof and then folded back to form a louver. This is particularly useful on large structures (Figure 7.60).

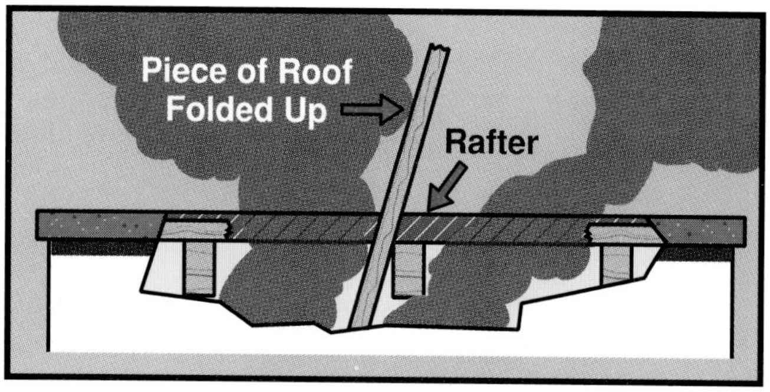

Figure 7.60 Bend the cut piece of the roof around the rafter to form a louver of sorts.

Basement Fires

The importance of ventilation when attacking basement fires cannot be overemphasized. In the absence of built-in vents from the basement, heat and smoke from basement fires will quickly spread upward into the building (Figure 7.61). Especially in buildings of balloon-frame construction, the first extension of a basement fire will commonly be into the attic. The likelihood of vertical extension of the fire may be reduced by direct ventilation of the basement during fire attack. After the basement fire is confirmed to be out, the attic may be vented to remove residual smoke.

Direct ventilation of a basement can be accomplished several ways. If the basement has ground-level windows or even below-ground-level windows in wells, horizontal ventilation can be employed effectively (Figure 7.62). If these windows are not available, interior vertical ventilation must be performed. Natural paths from the basement, such as stairwells and hoistway shafts, can be used to evacuate heat and smoke provided there is a means to expel the heat and smoke to the atmosphere without placing other portions of the building in danger (Figure 7.63). As a last resort, a hole may be cut in the floor

Figure 7.61 Products of combustion from a basement fire will quickly collect in the upper reaches of a structure.

Figure 7.62 Some basement windows will actually be at ground level outside the building.

Figure 7.63 Stairways can be used to vent a basement.

near a ground-level door or window, and the heat and smoke can be forced from the hole through the exterior opening using smoke ejectors or blowers (Figure 7.64).

Figure 7.64 A hole may be cut in the floor, if needed, to vent a basement.

Precautions Against Upsetting Established Vertical Ventilation

When vertical ventilation is accomplished, the natural convection of the heated gases creates upward currents that draw the fire and heat in the direction of the upper opening. This natural, vertical heat and smoke movement throughout a building is called the stack effect. Fire fighting teams take advantage of the improved visibility and less contaminated atmosphere to attack the fire at its lower point. If the stack effect is interrupted, the heat, smoke, and steam back up, hampering extinguishment efforts. The magnitude of the stack effect is determined by the differences in temperature and the densities between the air inside and outside the building. Some common factors that can destroy the effectiveness of ventilation are:

- Improper use of forced ventilation
- Excess breakage of glass
- Fire streams directed into ventilation holes
- Breakage of skylights
- Explosions
- A burn-through
- Additional openings between the attack team and the upper opening

Usually, these problems are avoided by having well-trained firefighters on the ground who are working a well-coordinated attack.

Elevated streams are frequently used to lessen sparks and flying brands from a burning building or to reduce the thermal column of heat over a building. However, when elevated or handline streams are projected downward through a ventilation opening or used to reduce the thermal column to a point where ventilation is hindered, these streams either destroy or upset the orderly movement of fire gases from the building. This can force superheated air and gases back down on firefighters, causing serious injury or death. At the least, it will contribute to the spread of fire throughout the structure. Streams that are being operated just above ventilated openings should be projected slightly above the horizontal plane (Figure 7.65). In this position, they will help cool the thermal column and extinguish sparks. The movement of the stream may even increase the rate of ventilation.

WARNING:

Never operate any type of fire stream through a ventilation hole while firefighters are still inside the building. This stops the ventilation process and places interior crews in serious danger.

Figure 7.65 A stream directed above the vent opening can aid the ventilation process and reduce the chance of secondary flying brand fires.

Vertical ventilation cannot be the solution to all ventilation problems because there may be many instances where its application would be impractical or impossible. In these cases, other strategies, such as the use of strictly horizontal ventilation, must be employed.

HORIZONTAL VENTILATION

Horizontal ventilation is the venting of heat, smoke, and gases through wall openings such as windows and doors. Structures that lend themselves to the application of horizontal ventilation include:

- Residential-type buildings in which the fire has not involved the attic area
- The involved floors of multistoried structures below the top floor, or the top floor if the attic is uninvolved
- Buildings with large, unsupported open spaces under the roof in which the structure has been weakened by the effects of burning

Many of the aspects of vertical ventilation also apply to horizontal ventilation. However, a different procedure must be followed in ventilating a room, floor, cockloft, attic, or basement. The procedure to be followed will be influenced by the location and extent of the fire. Some of the ways by which horizontal extension occurs are as follows:

- Through wall openings by direct flame contact or by convected air
- Through corridors, halls, or passageways by convected air currents, radiation, and flame contact (Figure 7.66)

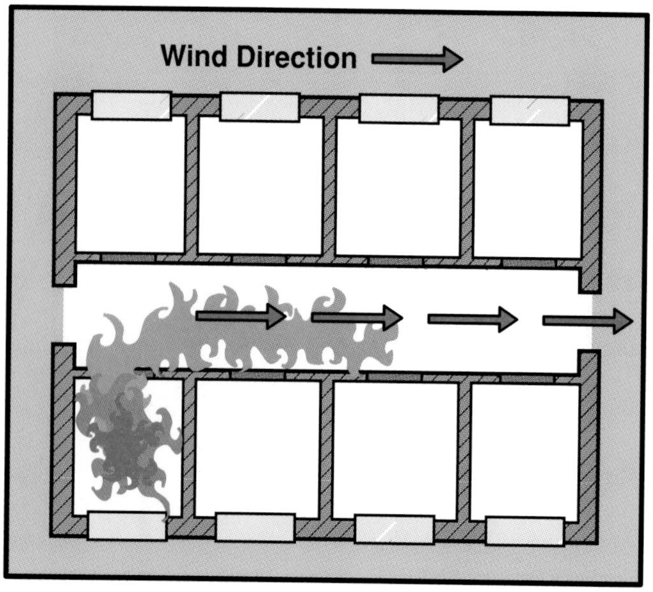

Figure 7.66 Fire can spread down a hallway, especially when the wind is aiding the process.

- Through open space by radiated heat or by convected air currents (Figure 7.67)
- In all directions by explosion or flash burning of fire gases, flammable vapors, or dust
- Through walls and interior partitions by direct flame contact
- Through walls by conduction of heat through beams, pipes, or other objects that extend through walls

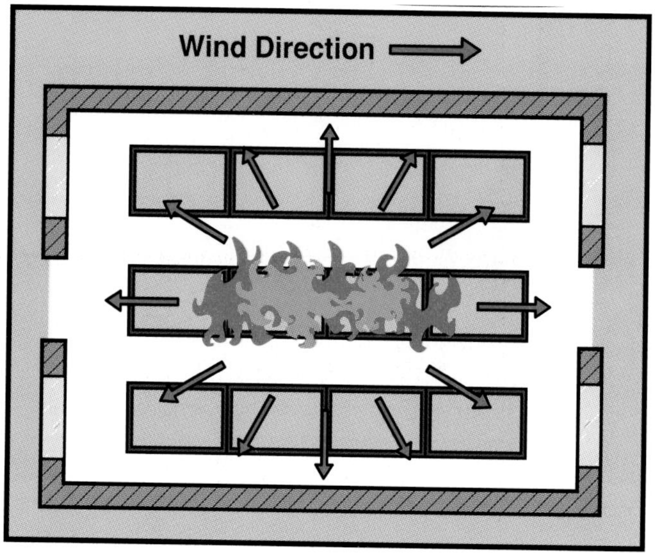

Figure 7.67 Fire can spread rapidly in all directions in a large area.

Weather Conditions

Weather conditions are always a consideration in determining the proper ventilation procedure. The wind plays an important role in ventilation. Its direction may be designated as windward or leeward. The side of the building the wind is striking is the windward, the opposite is leeward (Figure 7.68). Under certain conditions, when there is no wind, natural horizontal ventilation is less effective because the force to remove the smoke is absent. In other instances, natural horizontal ventilation cannot be accomplished due to the danger of wind blowing toward an exposure or feeding oxygen to the fire.

Exposures

Because horizontal ventilation does not normally release heat and smoke directly above the fire, some routing is necessary. Firefighters should be aware of internal exposures as well as external exposures. The routes by which the smoke and heated gases would travel to the exit may be the same corridors and passageways that occupants will be using for evacuation. Therefore, the practice of horizontal ventilation, without first considering occupants and rescue procedures, may

Figure 7.68 Fresh air enters the windward side, and smoke, heat, and gases exit the leeward side.

block the escape of occupants. The theory of horizontal ventilation is basically the same as that of vertical ventilation inasmuch as the release of the smoke and heat is an aid in fighting the fire and reducing damage.

Because horizontal ventilation is not accomplished at the highest point of a building, there is the constant danger that when the rising heated gases are released, they will ignite higher portions of the fire building. They may ignite eaves of adjacent structures or be drawn into windows above their liberation point (Figure 7.69). Unless for the specific purpose of aiding in rescue, a building should not be opened until charged lines are in place at the attack entrance point, at the intermediate point where fire might be expected to spread, and in positions to protect other exposures (Figure 7.70 on the next page).

Precautions Against Upsetting Horizontal Ventilation

The opening of a door or window on the windward side of the structure prior to first opening a door on the leeward side may pressurize the building and upset the normal process of thermal layering (Figure 7.71 on the next page). Opening doors and windows between the advancing fire fighting crews and the established

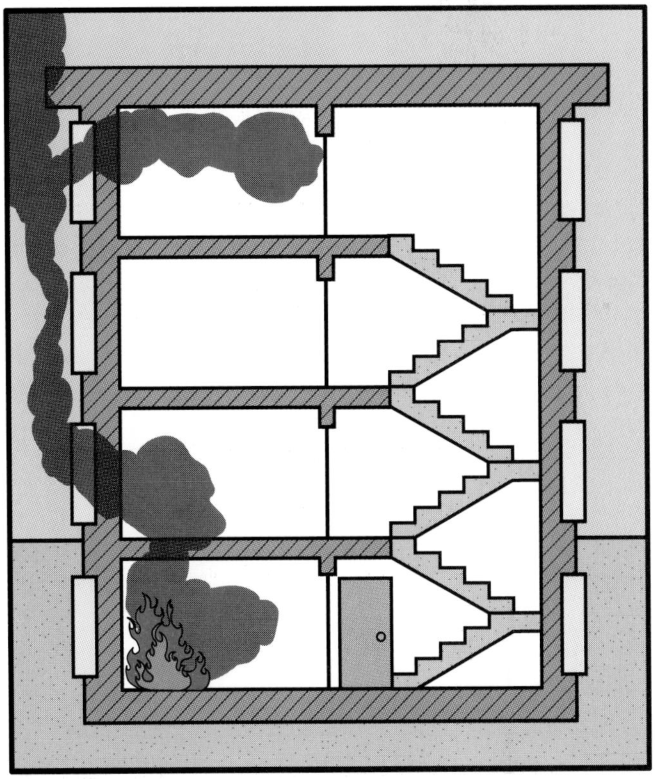

Figure 7.69 If care is not taken, smoke removed from lower levels can be drawn back in at higher parts of the structure.

Figure 7.70 Be ready to enter the structure as soon as ventilation is accomplished.

Figure 7.71 Opening extra doors or windows can disrupt the ventilation process, especially when positive-pressure ventilation is used.

ventilation exit point reduces the intake of fresh air from the opening behind the firefighters. Firefighters following established cross ventilation currents are illustrated in Figure 7.72. The smoke and heat will intensify if the established current is interrupted by a firefighter or other obstruction in the doorway (Figure 7.73).

Figure 7.72 Proper ventilation allows for more effective fire attacks.

Figure 7.73 A person or object blocking the ventilation opening can disrupt the whole process.

FORCED VENTILATION

Up to this point, ventilation has been considered from the standpoint of the natural flow of air currents and the currents created by fire. Forced ventilation is accomplished mechanically (with blowers or ejectors) or hydraulically (with fog streams). The fact that forced ventilation is effective for heat and smoke removal when other methods are not adequate proves its value and importance.

It is difficult to classify forced ventilation equipment by any particular type. The principle applied is that of moving large quantities of air and smoke. These portable ejectors and blowers are powered by electric motors, gasoline-driven engines, or water pressure from hoselines. Portable ejectors and several methods of using them to ventilate are shown in Figure 7.74.

Forced air blowers should always be equipped with explosion-proof motors and power cable connections when used in a flammable atmosphere. Forced air blowers should be shut down when they are moved, and they should be moved by the handles provided for this purpose. Before starting forced air blowers, be sure that no one is near the blades and that clothing, curtains, or draperies are not in a position to be drawn into the fan. The discharge stream of air should be avoided because of particles that may be picked up and blown by the venting equipment.

Advantages Of Forced Ventilation

Even when fire may not be a factor, contaminated atmospheres must be rapidly and thoroughly cleared. Forced ventilation, if not the only means of clearing a contaminated atmosphere, is always a welcome addition to normal ventilation. Some of the reasons for employing mechanical or forced ventilation include the following:

- It ensures more positive control.
- It supplements natural ventilation.

Figure 7.74 There are many ways a smoke ejector can be positioned in a doorway or window opening.

- It speeds the removal of contaminants, facilitating a more rapid rescue under safer conditions.
- It reduces smoke damage.
- It promotes good public relations.

Disadvantages Of Forced Ventilation

If mechanical or forced ventilation is misapplied or improperly controlled, it can cause a great deal of harm. Forced ventilation requires supervision because of the mechanical force behind it. Some of the disadvantages of forced ventilation include:

- The introduction of air in such great volumes can cause the fire to intensify and spread.
- It is dependent upon a power source.
- It requires special equipment.

Negative-Pressure Ventilation

The term negative-pressure ventilation is used to describe the oldest of mechanical forced ventilation techniques. This technique uses smoke ejectors to develop artificial circulation and pull smoke out of a structure. Smoke ejectors are placed in windows, doors, or roof vent holes, and they pull the smoke, heat, and gases from inside the building and eject them to the exterior (Figure 7.75).

The ejector should be placed to exhaust in the same direction as the natural wind. This will aid the ejection process by supplying fresh air to replace that which is being drawn from the building. If the natural wind is too light to be effective, smoke ejectors on one side of the structure can be turned to blow air into the building while the ejectors on the other side are turned so that they exhaust the smoke and other combustion by-products from the building.

Recirculating air in the opening around the smoke ejector can be a problem. When air is allowed to recirculate around the sides of the smoke ejector and in and out of nearby openings, it causes a churning action that reduces efficiency (Figure 7.76 on the next page). If the area surrounding the fan is left open, atmospheric pressure pushes the air through the bottom of the doorway and pulls the smoke back into the room. To prevent churning air, cover the area around the unit with salvage covers or other material.

Establish the desired draft path and keep the airflow in as straight a line as possible. Every corner causes turbulence and decreases efficiency. Avoid opening windows or doors near the exhausting smoke ejector unless opening them definitely increases circulation. Remove all obstacles to the airflow. Even a window

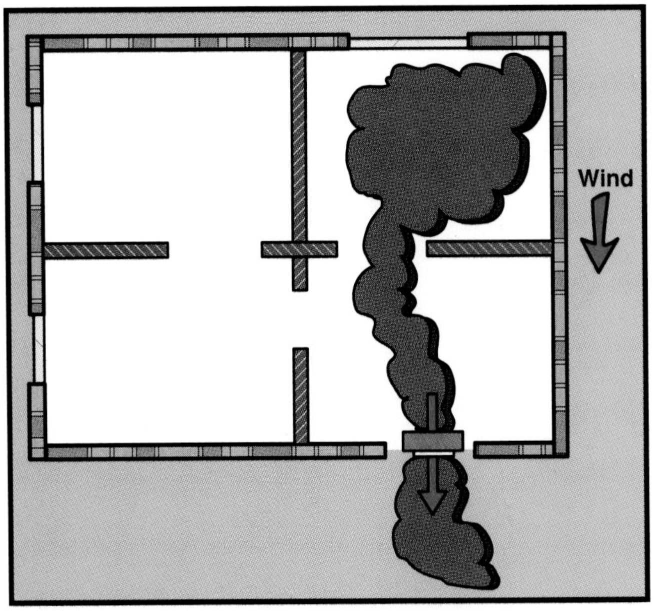

Figure 7.75 Negative-pressure ventilation is accomplished by using a smoke ejector to draw the smoke out of the building.

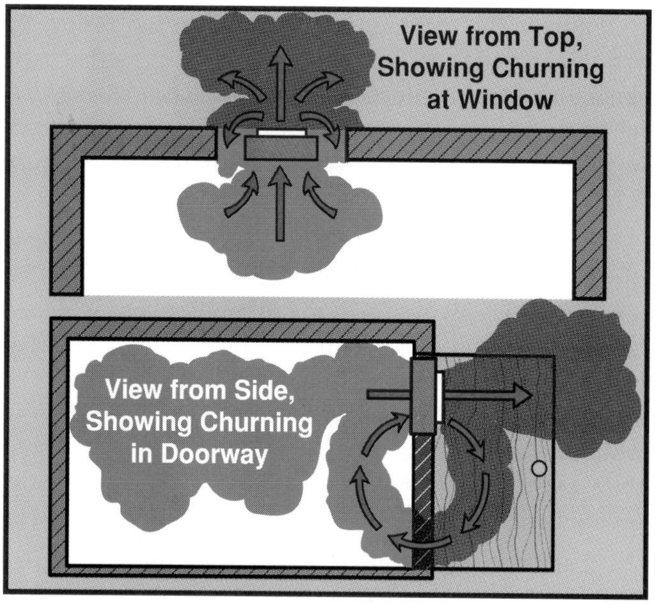

Figure 7.76 Avoid churning situations when using smoke ejectors.

screen will cut effective exhaust by half. Avoid blockage of the intake side of the smoke ejector by debris, curtains, drapes, or anything that can decrease the amount of intake air.

Positive-Pressure Ventilation

Positive-pressure ventilation is a forced ventilation technique that uses the principle of creating pressure differentials. By using high volume blowers, a higher pressure is created inside a building than that of the outside environment (Figure 7.77). As long as the pressure is higher inside the building, the smoke within the building will seek an outlet to a lower pressure zone through openings controlled by firefighters.

Figure 7.77 A positive-pressure ventilation blower.

The advantages of positive-pressure ventilation include:

- Firefighters can set up forced ventilation procedures without entering the smoke-filled environment.
- Because positive-pressure ventilation merely supplements natural ventilation currents, it is equally effective with horizontal or vertical ventilation.
- It allows more efficient removal of smoke and heat from the structure or vessel.
- The velocity of air currents within the building are minimal and have little, if any, effects that disturb the building contents or smoldering debris. Yet, the total exchange of air within the building is faster than using negative-pressure ventilation.
- The placement of blowers does not interfere with ingress or egress (Figure 7.78).
- The cleaning and maintenance of blowers is greatly reduced compared to that of smoke ejectors.
- This system is applicable to all types of structures or vessels and is particularly effective at removing smoke from large, high-ceiling areas where negative-pressure ventilation is ineffective.
- Heat and smoke may be directed away from unburned areas or paths of exit.

The positive pressure is created by using the high volume blowers to blow fresh air into the structure. The location where this is done, usually an exterior doorway,

Figure 7.78 The positive-pressure blower is set back from the door or opening.

is called the point of entry. The blower is placed several feet outside the door so that the cone of air from the blower completely covers the door opening (Figure 7.79). The smoke is then ejected from another opening on the opposite side of the structure. For maximum efficiency, the exit opening should be from 75 to 150 percent as large as the entry opening. It is important that no other exterior openings be opened while the positive-pressure operation is in use.

Figure 7.79 The cone of air must cover the entire opening.

By closing doors within the structure and pressurizing one room or area at a time, the process of removing the smoke is speeded up because the velocity of the air movement is increased. The process can also be speeded up by placing additional blowers at the entry point. If none of the doors inside the structure were opened and closed systematically, the process would still work, but it would take more time.

Positive-pressure ventilation works equally well in aboveground fire fighting operations. The same principles apply as far as creating a higher pressure inside the building than outside the building. However, greater coordination is required to control interior fire doors so that the positive pressure is directed to the appropriate floor. When using positive pressure to remove smoke from multiple floors of a building, it is generally best to apply the positive pressure at the lowest point (Figure 7.80). Smoke can then be systematically removed one floor at a time starting with the floor most heavily charged with smoke. Note that positive pressure is applied to the building at ground level through the use of one or more blowers. The positive pressure is then directed throughout the building by opening and closing doors until the building is totally evacuated of smoke through any opening selected by firefighters (Figure 7.81). This is accomplished by either cross ventilating fire floors or directing smoke up a stairwell and out the stairshaft rooftop opening.

Figure 7.80 Place the blower at a low point in the structure.

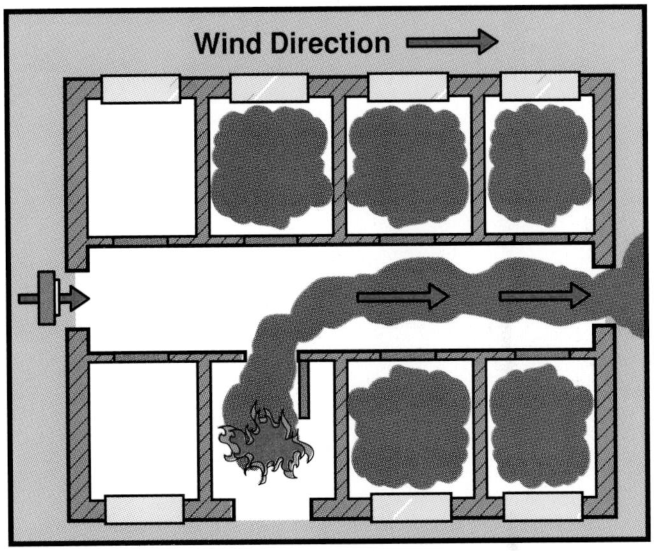

Figure 7.81 Use doors to control the ventilation effort. This could even allow the structure to be cleared one room at a time.

The main problem in using positive-pressure ventilation in aboveground operations is coordinating the opening and closing of the doors in the stairwell being used to ventilate the building. Curious tenants will often stand with the door to the stairshaft or the door to their room open and thereby redirect the positive pressure away from the fire floor. To control openings or pressure leaks, place one person in charge of the pressurizing process. It is helpful to use portable radios and to have firefighters patrol the stairwell and hallways.

In order to ensure an effective positive-pressure ventilation operation, the following points should be taken into consideration:

- Take advantage of existing wind conditions.
- Make certain that the cone of air from the blower covers the entire entry opening.
- To speed up the process, reduce the size of the area being pressurized by systematically opening and closing doors or by increasing the number of blowers.
- The size of the exit opening must be kept in proportion to the entry opening.

Hydraulic Ventilation

Hydraulic ventilation may be used in situations where other types of forced ventilation are not being used (Figure 7.82). Hydraulic ventilation is performed by hose teams making an interior attack on the fire. Typically, this technique is used to clear the room or building of smoke, heat, steam, and gases following the initial knockdown of the fire. This technique takes advantage of the air that is drawn into a fog stream to help push the products of combustion out of the structure.

Figure 7.82 Hydraulic ventilation in progress.

To perform hydraulic ventilation, a fog stream is set on a wide fog pattern that will cover 85 to 90 percent of the window or door opening from which the smoke will be pushed out. The nozzle tip should be at least 2 feet (0.6 m) back from the opening (Figure 7.83). The larger the opening, the faster the ventilation process will go.

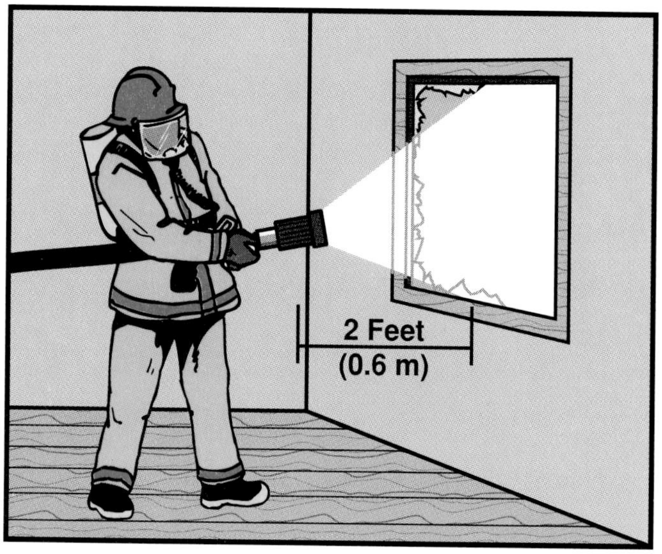

Figure 7.83 The nozzle should be 2 feet (0.6 m) back from the window.

There are drawbacks to the use of fog streams in forced ventilation. These drawbacks include the following:

- There may be an increase in the amount of water damage within the structure.
- There will be a drain on the available water supply. This is particularly crucial in rural fire fighting operations where water shuttles are being used.
- In climates subject to freezing temperatures, there will be an increase in the problem of ice in the area surrounding the building.
- The firefighters operating the nozzle must remain in the heated, contaminated atmosphere throughout the operation.
- The operation may have to be interrupted when the nozzle team goes for fresh SCBA tanks.

THE EFFECT OF BUILDING VENTILATION SYSTEMS IN FIRE SITUATIONS

Most buildings, other than sheds and small storage facilities, have heating, ventilation, and air-conditioning (HVAC) systems. These systems have several fire fighting implications that firefighters should be aware of. If they are not functioning properly or fail to shut down when they are supposed to, these systems can significantly contribute to the spread of smoke and fire throughout a structure. Pre-incident planning should include information on the design capabilities of the HVAC system. Also included should be diagrams of the duct system throughout the building and information on fire protection systems (sprinkler, smoke, or heat detection) within the HVAC ductwork. Fire personnel should be familiar with the location and operation of controls that will manually shut down the system when so desired.

Because the system may draw heat and smoke into the duct before it is shut down, firefighters should always check around the ductwork for fire extension during overhaul operations. Also, personnel should be familiar with the best ways to rid the system of smoke before reactivating it. Because of the variety and complexity of these systems, firefighters should not attempt to operate these systems under fire conditions. Building engineers should be called to the scene to operate the system under the fire department's direction.

8

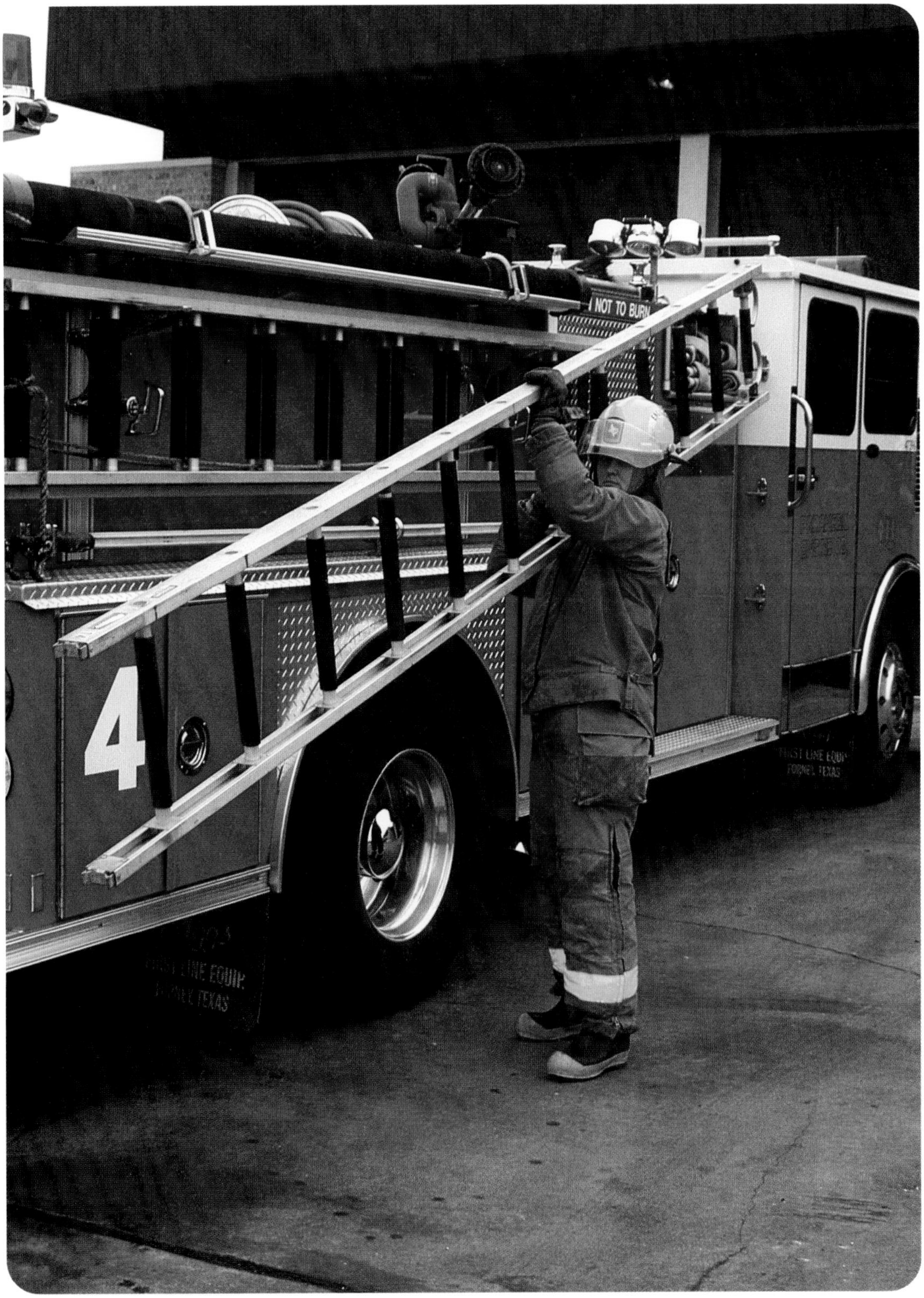

Ladders

This chapter provides information that addresses performance objectives described in NFPA 1001, *Standard for Fire Fighter Professional Qualifications* (1992), particularly those referenced in the following sections:

3-11 Ladders

3-11.1

3-11.2

3-11.3

3-11.4

3-11.5

4-11 Ladders

4-11.1

4-11.2

4-11.3

4-11.4

4-11.5

Chapter 8
Ladders

Fire service ladders are essential in the performance of many fireground and rescue scene functions. From both the tactical and safety standpoints, it is crucial that firefighters be knowledgeable of the characteristics and proper uses of ground ladders. Fire service ladders are similar to any other ladder in shape and design; however, they tend to be built more rigidly and are capable of withstanding heavier loads than commercial ladders. Their use under adverse conditions requires that they provide a margin of safety not usually expected of commercial ladders. NFPA 1931, *Standard on Design of and Design Verification Tests for Fire Department Ground Ladders*, provides recommended specifications of fire department ground ladders.

This chapter first introduces the reader to basic ladder parts and terms that are common to most ladders. Then, the various types of ladders in use by the fire service are reviewed. The chapter also details the proper care, carrying, deployment, and use of fire service ground ladders. For more information on the operation and use of fire department aerial apparatus, see the IFSTA validated **Fire Department Aerial Apparatus** manual.

BASIC PARTS OF A LADDER

In order to successfully continue this discussion of fire service ladders, the firefighter must have a basic understanding of the various parts of the ladder. Many of these terms apply to all types of ladders; others may be specific to a certain type of ladder. The firefighter needs to be familiar with all of them.

Aerial Apparatus — A fire apparatus equipped with a powered extension ladder, telescoping aerial platform, or articulating aerial platform.

Base Section (Also called bed section or main section) — The bottom section of an extension ladder (Figure 8.1).

Beam — The side rail of a ladder (Figure 8.2).

Beam Bolts — Bolts that pass through both rails at the truss block of a wooden ladder to tie the two truss rails together.

Butt (Also called heel) — The bottom end of the ladder that will be placed on the ground or other supporting surface when the ladder is raised.

Butt Spurs — Metal safety plates or spikes attached to the butt end of ground ladder beams (Figure 8.3).

Figure 8.1 The base section of an extension ladder.

Figure 8.3 A butt spur.

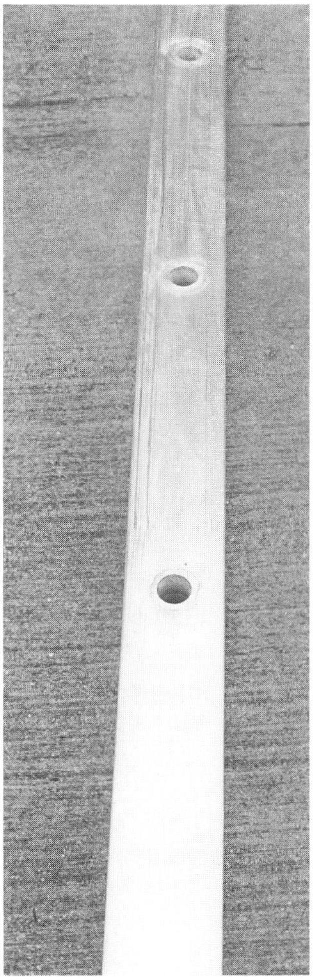

Figure 8.2 The beam of a ladder.

Dogs — See Pawls

Extension Ladder — A ladder with a base section and one or more telescoping sections.

Fly — The upper section or top sections of an extension ladder or aerial device (Figure 8.4).

Ground Ladder — A ladder that is manually carried to the desired position and manually raised, positioned, and lowered.

Guides — Wood or metal strips, sometimes in the form of slots or channels, on an extension ladder that guide the fly section while being raised.

Halyard — A rope or cable used for hoisting and lowering the fly sections of a ground ladder.

Heat Sensor Label — A label affixed to the ladder beam near the tip to provide a warning that the ladder has been subjected to excessive heat (Figure 8.5).

Figure 8.5 Heat sensor labels indicate when a ladder should be service tested due to extreme heat exposure.

Figure 8.4 The fly section of an extension ladder.

Hooks — A pair of sharp curved devices at the top of a roof ladder that fold outward from each beam (Figure 8.6).

Pawls (Also called dogs or locks) — Devices attached to the inside of the beams on fly sections used to hold the fly section in place after it has been extended (Figure 8.7).

Protection Plates — Plates fastened to a ladder to prevent wear at points where it comes in contact with mounting brackets.

Pulley — A small, grooved wheel through which the halyard is drawn on an extension ladder.

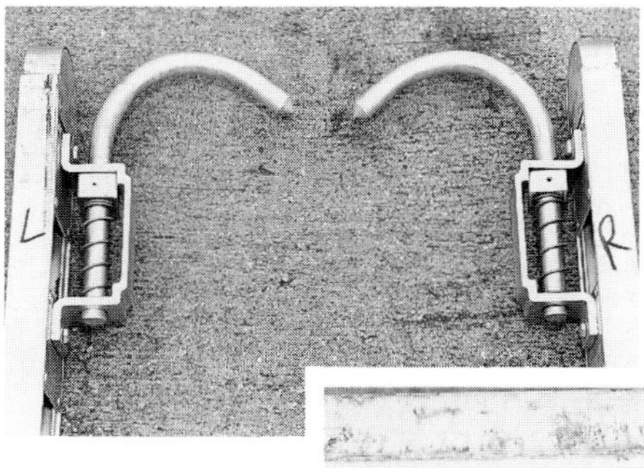

Figure 8.6 The hooks of a roof ladder.

Figure 8.7 Pawls hold the fly section in place after it has been extended.

Rails — The two lengthwise members of a trussed ladder beam that are separated by truss or separation blocks (Figure 8.8).

Rungs — Cross members (usually round or oval) between the beams on which the climber steps.

Safety Shoes — Rubber or neoprene foot plates, usually of the swivel type, attached to the butt end of the beams of a ground ladder (Figure 8.9).

Figure 8.8 The rails of a trussed ladder beam.

Figure 8.9 Safety shoes are attached to the butt end of the beams of a ground ladder.

Single Ladder (Also called straight ladder) — A ladder consisting of one section.

Spurs — Metal points at the lower end of staypoles.

Staypoles (Also called tormentor poles) — The poles attached to long extension ladders to assist in raising and steadying the ladder. Some poles are permanently attached, and some are removable (Figure 8.10).

Stops — Wood or metal pieces that prevent the fly section from being extended too far.

Tie Rods — Metal rods running from one beam to the other.

Toggle — A hinge device by which a staypole is attached to a ladder.

Top or Tip — The extreme top of a ladder.

Truss Block — Separation pieces between the rails of a trussed ladder. Sometimes used to support rungs (Figure 8.11).

Figure 8.11 A truss block.

LADDER TYPES

All of the various types of fire service ladders have a purpose. Many of them, however, are more adaptable to a specific function than they are to general use. Their identifying name is often significant regarding their use, and firefighters frequently make reference to them by association. The following descriptions more clearly identify fire service ladders.

Single Ladders

A single ladder is nonadjustable in length and consists of only one section (Figure 8.12 on the next page). Its size is designated by the overall length of the side rails. The single ladder, sometimes called a wall ladder, is used for quick access to windows and roofs on one- and two-story buildings. Single ladders must be constructed to have a maximum strength and minimum weight. These ladders may be of the trussed type in order to reduce their weight. Single ladders are generally used in lengths of 12, 14, 16, 18, and 24 feet (4 m, 4.3 m, 5 m, 5.5 m, and 8 m), but some longer single ladders do exist.

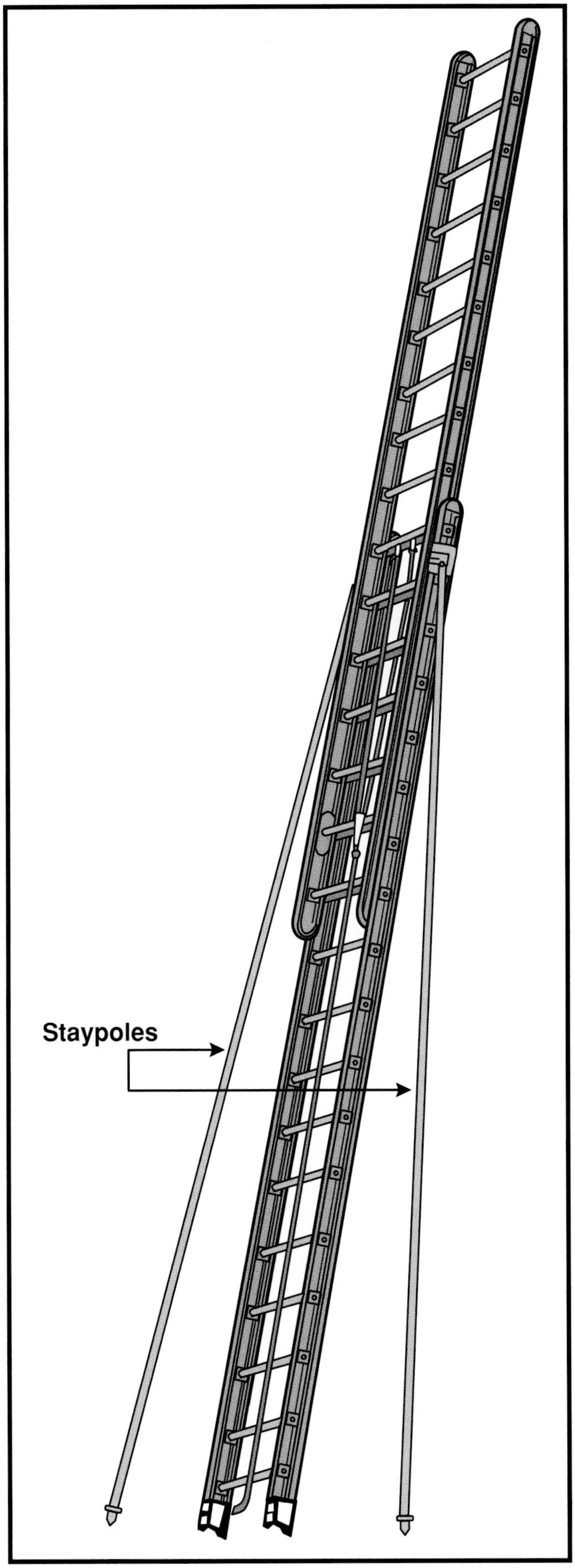

Figure 8.10 Staypoles are attached to the base section of the ladder.

Figure 8.12 A single or wall ladder.

NFPA 1932, *Standard on Use, Maintenance, and Service Testing of Fire Department Ground Ladders*, requires single ladders to be rated for a total weight load of at least 750 pounds (340 kg).

Roof Ladders

Roof ladders are single ladders equipped at the tip with folding hooks that provide a means of anchoring the ladder over the roof ridge or other roof part (Figure 8.13). Roof ladders are generally required to lie flat on the roof surface so that the firefighter may stand on the ladder for roof work (Figure 8.14). The ladder distributes the firefighter's weight and helps prevent slipping. Roof ladders may also be used as single wall ladders. Their lengths range from 12 to 20 feet (4 m to 6 m). NFPA 1932 requires roof ladders to be rated for a total weight load of not less than 750 pounds (340 kg).

Figure 8.13 A roof ladder.

Figure 8.14 The roof ladder helps to distribute the firefighter's weight and prevents slipping.

Folding Ladders

Folding ladders are single ladders that have hinged rungs allowing them to be folded so that one beam rests against the other (Figures 8.15 a and b). This allows them to be carried in narrow passageways and used in attic scuttle holes and small rooms or closets. Folding ladders are commonly found in 10-foot (3 m) lengths because they are only required to reach a short distance. All folding ladders should be equipped with safety shoes to prevent the ladder from slipping (Figure 8.16). NFPA 1932 requires folding ladders to be rated for a total weight load of not less than 300 pounds (136 kg).

Extension Ladders

An extension ladder is adjustable in length. It consists of a base section and one or more fly sections that travel in guides or brackets to permit length adjustment (Figure 8.17). Its size is designated by the length of the sections and measured along the side rails when fully extended. An extension ladder provides access to windows and roofs within the limits of its length. Most of the longer extension ladders are trussed. Extension ladders are heavier than single ladders, and more personnel are needed to safely handle them. Extension ladders gener-

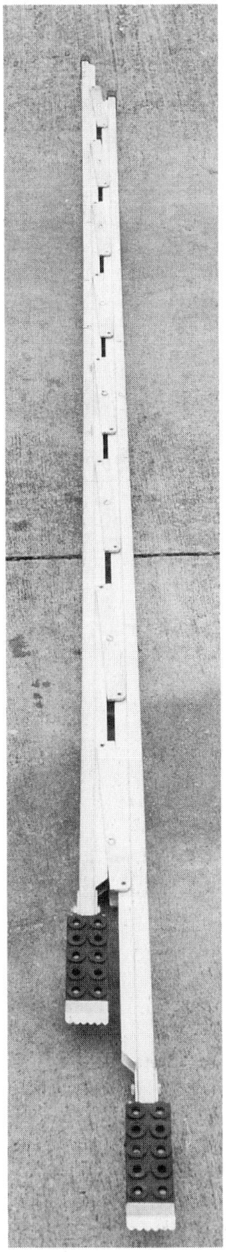

Figure 8.15a A closed folding ladder.

Figure 8.15b An open folding ladder.

Figure 8.16 All folding ladders should be equipped with safety shoes.

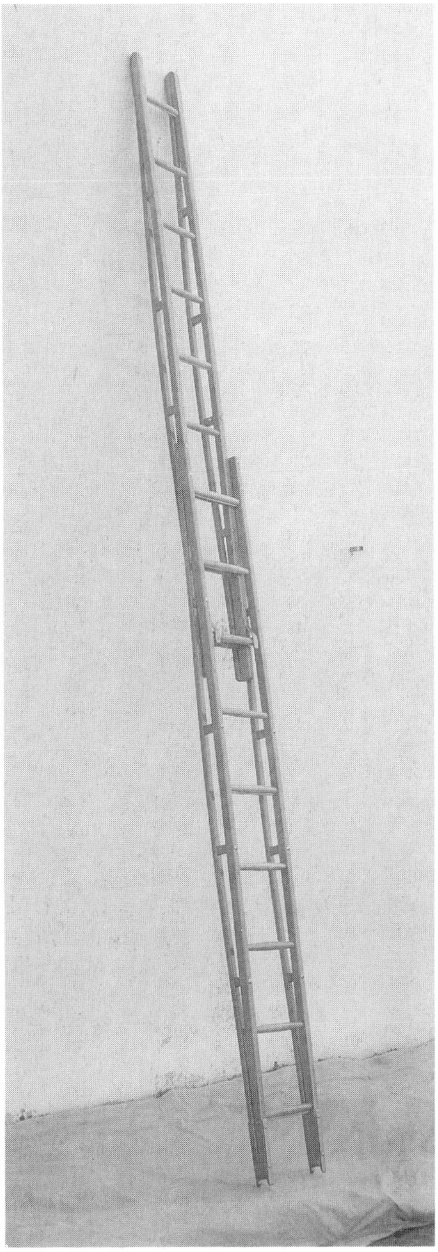

Figure 8.17 Wooden ground ladders are used with the fly in toward the building. *Courtesy of ALACO Ladder Co.*

ally range in length from 24 to 39 feet (7 m to 12 m). Extension ladders manufactured before 1984 have a maximum load limit of 500 pounds (227 kg) distributed over the length of the ladder or 400 pounds (181 kg) concentrated at one spot. Extension ladders manufactured after 1984 must be rated for a total weight of not less than 750 pounds (340 kg).

Pole Ladders

Pole ladders are extension ladders that have staypoles for added leverage and stability when raising the ladder (Figure 8.18 on the next page). NFPA 1931, *Standard on Design of and Design Verification Tests for Fire Department Ground Ladders*, requires all extension ladders that are 40 feet (12.1 m) or longer to be equipped with staypoles. Some pole ladders up to 65 feet (20 m) may still be in service; however, most do not exceed 50 feet (15 m). They are of truss construction and have one to three fly sections (Figure 8.19). They are subject to the same weight load requirements as other extension ladders. NFPA aerial apparatus standards no longer require aerial apparatus to carry pole ladders, and NFPA 1001 no longer requires firefighters to be required to raise them; therefore, their deployment and use are not covered in this manual. For more information on using pole ladders, see the IFSTA validated **Fire Service Ground Ladders** manual.

Figure 8.18 A pole ladder.

Figure 8.19 Most pole ladders are of the truss type.

Combination Ladders

Combination ladders are designed so that they may be used as either a single, extension, or A-frame ladder (Figures 8.20 a and b). Lengths range from 8 to 14 feet (2 m to 4.3 m), the latter being the maximum allowed by NFPA 1931. The ladder must be equipped with positive locking devices to hold the ladder in the open position. Combination ladders must be able to withstand a total weight load of not less than 750 pounds (340 kg).

Pompier Ladders

The pompier ladder, sometimes referred to as a scaling ladder, is a single-beam ladder with rungs projecting from both sides. It has a large metal "gooseneck" projecting at the top for inserting into windows or other openings (Figure 8.21). It is used to climb from floor to floor, via exterior windows, on a multistory building. NFPA 1931 specifies that these ladders should not exceed 16 feet (5 m). NFPA 1932 specifies that pompier ladders should be rated for a total weight load of not less than 300 pounds (136 kg).

Figure 8.20a A combination ladder may be used as a short straight ladder. *Courtesy of ALACO Ladder Co.*

Figure 8.20b The combination ladder used as an A-frame ladder. *Courtesy of ALACO Ladder Co.*

AERIAL APPARATUS

Fire apparatus are classified according to the function for which they were designed. The NFPA has divided aerial fire apparatus into two specific groups: aerial ladders and elevating platforms. Both of these groups are covered in NFPA 1904, *Standard for Aerial Ladder and Elevating Platform Fire Apparatus*. Other types of apparatus are covered in separate standards. IFSTA places aerial apparatus into four specific categories: aerial ladders, aerial ladder platforms, telescoping aerial platforms, and articulating aerial platforms.

Aerial Ladder Apparatus

An aerial ladder is a power-operated ladder mounted on a special truck chassis (Figure 8.22). The fully ex-

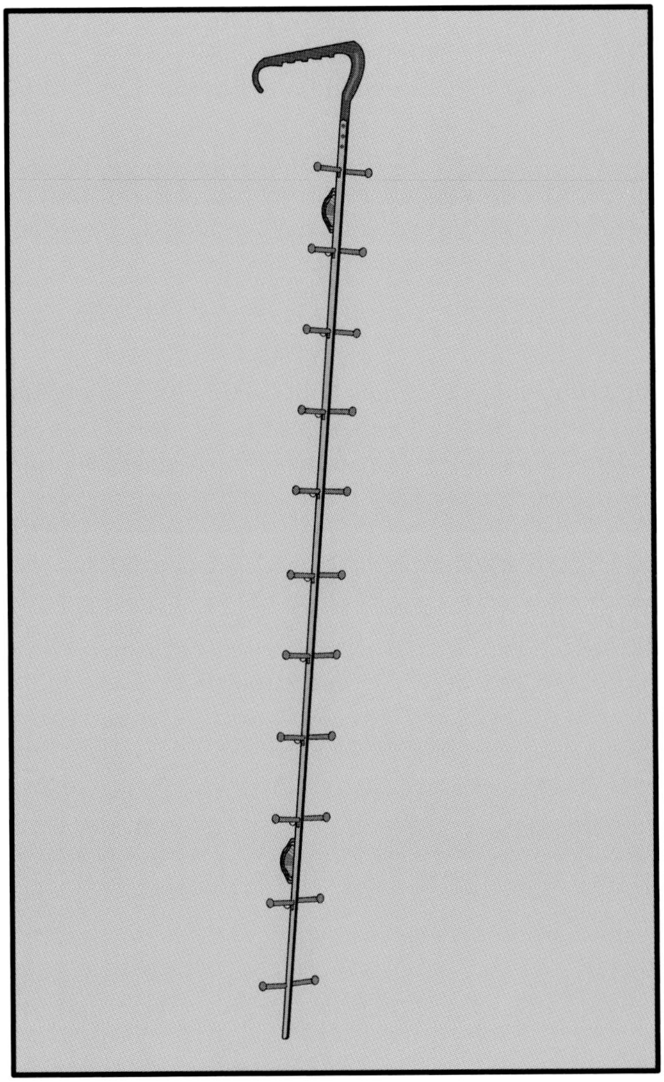

Figure 8.21 A pompier ladder.

Figure 8.22 A typical aerial ladder apparatus.

tended length, also referred to as the working height, of North American-made aerial ladders is 50 to 135 feet (15 m to 41 m). The main uses of aerial ladders are rescue, ventilation, elevated master stream application, and gaining access to upper levels. Most aerial ladders are powered by hydraulic pumps, cylinders, and motors. The aerial ladder may be mounted on either a two- or three-axle, single-chassis vehicle or on a three-axle tractor-trailer vehicle.

Aerial Ladder Platform Apparatus

Aerial ladder platform apparatus are similar to aerial ladder apparatus except that a work platform (sometimes referred to as basket or bucket) is attached to the end of the aerial ladder (Figure 8.23). These vehicles are always single-chassis and are usually of the three-axle design. The aerial ladder platform combines the safe work area of a platform with a safe, climbable aerial ladder. NFPA 1904 requires that platforms be constructed of metal; they are usually made of steel or aluminum alloy. The working height of all types of elevating platforms is measured from the ground to the top surface of the highest platform handrail with the aerial device at maximum extension and elevation. The working height of aerial ladder platforms range from 85 to 110 feet (26 m to 34 m).

Figure 8.23 An aerial ladder platform apparatus. *Courtesy of Joel Woods, University of Maryland Fire & Rescue Institute.*

Telescoping Aerial Platform Apparatus

NFPA places aerial ladder platforms and telescoping aerial platforms under the same definition. However, each type of apparatus has different capabilities on the fireground. Aerial ladder platforms are designed with a large ladder that allows firefighters to routinely climb back and forth from the platform. During mass evacuations, victims may also exit from the platform down the ladder. Telescoping aerial platforms are not intended for this purpose. They are generally equipped with a small ladder attached to the boom (Figure 8.24 on the next page). This ladder is designed primarily as an escape ladder for firefighters in the platform to use during an emergency. Telescoping aerial platforms are equipped with built-in piping and nozzles for providing

Figure 8.24 Telescoping aerial apparatus are not designed to be regularly climbed. *Courtesy of Joel Woods, University of Maryland Fire & Rescue Institute.*

elevated streams. The working height of telescoping aerial platforms range from 50 to 100 feet (15 m to 30 m).

Articulating Aerial Platform Apparatus

Articulating aerial platform apparatus are similar to the telescoping aerial platform apparatus. The primary difference is in the operation of the aerial device. Instead of telescoping into each other, the boom sections are connected by a hinge and fold like an elbow (Figure 8.25). The booms are constructed in basically the same manner as telescoping platforms. These units are often used along with or in place of aerial ladder apparatus and perform many of the same functions. These functions include rescue, ventilation, master stream application, and gaining access to upper floors. NFPA 1904 requires that a permanent turret nozzle and supply system be provided on this and all other platform apparatus. These units range in size from 55 to 102 feet (17 m to 31 m).

Figure 8.25 An articulating aerial platform. *Courtesy of Joel Woods, University of Maryland Fire & Rescue Institute.*

CONSTRUCTION AND MAINTENANCE

Fire service ladders must be able to take considerable abuse such as sudden overloading, exposure to temperature extremes, and falling debris.

Because of the importance of eliminating, or at least reducing, any structural defects and design weaknesses, ladder specifications should be written to meet NFPA 1931, *Standard on Design of and Design Verification Tests for Fire Department Ground Ladders*. All ladders meeting NFPA 1931 are required to have a certification label affixed to the ladder by the manufacturer indicating that the ladder meets the standard.

Construction Features

The major components of ladder construction are the beams and the rungs. Solid-beam and truss-beam construction are the two designs that are used (Figures 8.26 a and b). However, both types must meet the same requirements in NFPA 1931 and 1932.

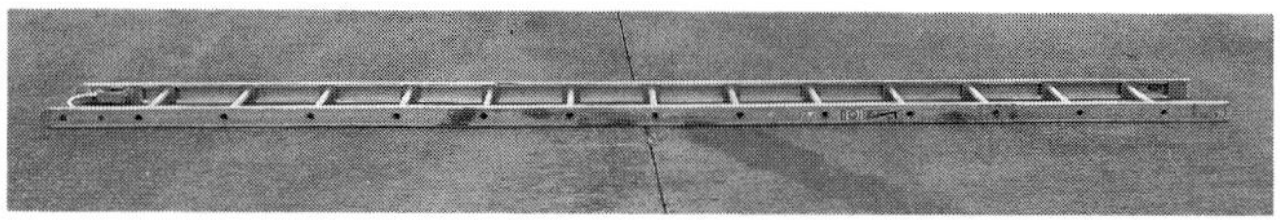

Figure 8.26a This ladder is made by solid-beam construction.

Figure 8.26b A trussed-beam ladder.

For lengths of 24 feet (8 m) or less, solid-beam construction provides a lighter weight ladder. For intermediate lengths of 25 to 35 feet (8 m to 11 m), solid- beam metal ladders will usually still be lightest in weight. However, with wooden ladders, the truss construction will be lighter than solid beams. For longer ladders of lengths over 35 feet (11 m), truss construction provides the lightest weight ladder regardless of whether the ladder is made of wood or metal.

Ladder rungs must not be less than 1¼ inches (30 mm) in diameter and spaced on 14-inch (360 mm) centers (Figure 8.27). When metal rungs are used, they must be constructed of heavy-duty corrugated, serrated, knurled, or dimpled material. As an alternative, the rungs may be coated with a skid-resistant material.

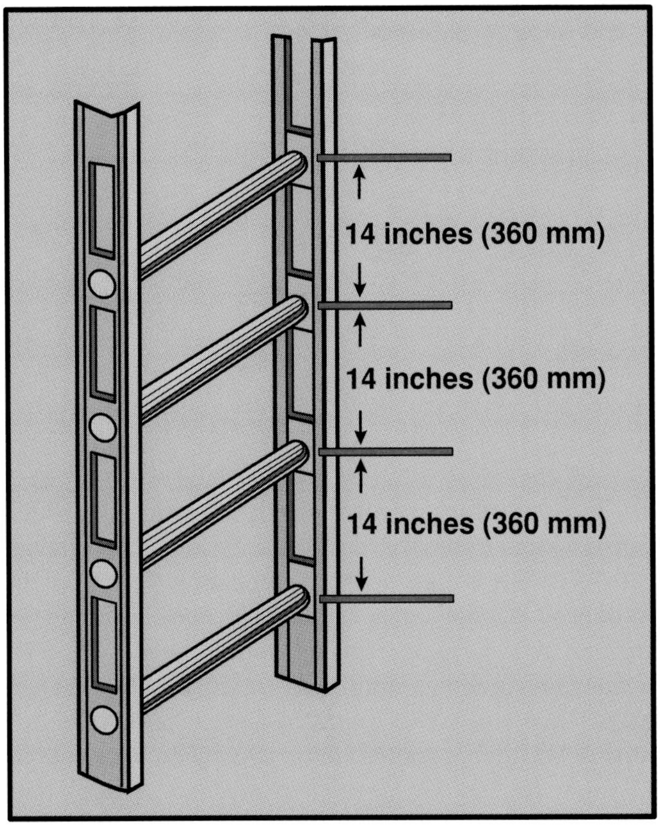

Figure 8.27 Rungs should be on 14-inch (360 mm) centers.

Fire service ladders may be constructed of metal, wood, or fiberglass. Regardless of the type of material used in ladder construction, it must be of high quality. The variety of materials used is probably a result of material availability and a constant effort to improve the construction of ladders. The following sections highlight the characteristics of each of these materials.

METAL LADDER CONSTRUCTION

Metal fire service ladders are usually built with heat-treated aluminum alloy. Aluminum is used because it is light, has adequate strength, and permits a reliable visual inspection of all ladder parts.

Aluminum ladders usually require less maintenance than wood because wood requires an artificial coating to protect its exterior. Also, it is sometimes difficult for a person to perform a visual maintenance inspection of wooden ladders, because pitch pockets, knots, nodes, and other defects may sometimes be concealed. It is a relatively simple process to keep an aluminum ladder clean, and other maintenance requirements are seldom necessary.

There are, however, some disadvantages of using aluminum for fire department ladders. They are good conductors of electricity, so caution must be exercised whenever metal ladders are used near electrical power sources (Figure 8.28). Another disadvantage is the increased possibility of freezing to the ladder in extremely cold weather. Aluminum can become very cold in winter and noticeably hot in summer because of its good conductive qualities.

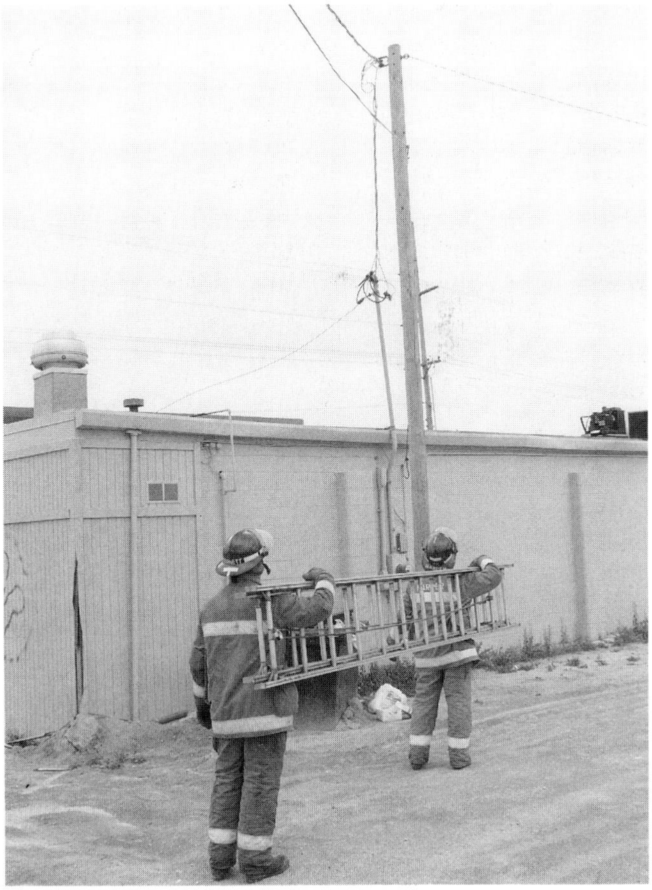

Figure 8.28 Use caution when raising ladders around electrical wires.

WOODEN LADDER CONSTRUCTION

Although there are relatively few wooden fire department ladders manufactured today, many are still in service. Clear, straight-grained, coast Douglas fir has long been a favored wood for ladder beams because it is relatively free of knots and pitch pockets. White ash or hickory is used for rungs, with oak added to some critical wear spots. A two-year drying period is usually required before ladder stock can qualify with a moisture content between 9 and 12 percent. Ladders constructed from wood with excessive moisture content are likely to shrink, resulting in warping or loose rungs. Wood shrinkage is particularly noticeable after the ladder has been subjected to low humidity and artificial heat. This slow drying process is a primary cause of loose rungs and cracked rails in wooden ladders.

FIBERGLASS LADDER CONSTRUCTION

Ladders made of fiberglass are becoming increasingly popular in the fire service. These ladders are not totally constructed of fiberglass but a combination of fiberglass beams and metal rungs. The major advantage of fiberglass is that it is a nonconductor of electricity and a poor conductor of heat. However, in order to meet the specifications of NFPA 1931, *Standard on Design of and Design Verification Tests for Fire Department Ground Ladders*, these types of ladders become relatively heavy. This is a result of the dense qualities of fiberglass and the amount of material needed to meet the strength requirements of the standard. All fiberglass ladders are of the solid-beam type (Figure 8.29). Another disadvantage of fiberglass is that it tends to chip and crack when it is struck.

Figure 8.29 Fiberglass ladder beams are configured in a C-shape. *Courtesy of Phoenix (AZ) Fire Department.*

Maintenance

In order for ladders to remain safe and reliable, they must be properly maintained. This includes regularly cleaning and inspecting all parts. The following sections highlight these tasks.

CLEANING LADDERS

Regular and proper cleaning of ladders is more than a matter of appearance. Unremoved dirt or debris from a fire may collect and harden to the point where ladder sections are no longer operable. They should be cleaned after every use.

Remove dirt with a brush and running water. Use solvent cleaners, approved by the manufacturer, to remove any oily or greasy residues. After rinsing, or anytime a ladder is wet, wipe it dry (Figure 8.30). During each cleaning period, firefighters should look for defects in the ladder. Any defects identified should be handled through local fire department procedures.

Figure 8.30 After a ladder is wiped dry, look for any defects in the ladder.

INSPECTING LADDERS

It is possible for all ladders, regardless of their construction, to have or develop defects and to deteriorate because of improper maintenance. Damaged or weakened portions of a ladder can best be found by regular and systematic inspections (Figure 8.31). The inspection should cover all parts of a ladder, and when a part shows excessive wear, the cause should be determined. Fire department ladders should be inspected at

Figure 8.31 Inspect a ladder for damaged or weakened portions.

regular intervals, inspected after each use or major repair, and tested once each year. Appropriate records should be kept for each ladder.

Thoroughly inspect ladders before they are subjected to any physical test. Rungs should be checked for tightness; bolts, rivets, and welds for looseness; and beams, trusses, and truss blocks for evidence of compression failure. Compression failure in wood is not easily seen, but it can sometimes be detected by a wavy condition of the wood grain. Exposure of metal ladders to temperatures over 300°F (149°C) should be cause for testing. An indication of high temperature would be if water boils when sprayed on the ladder or the manufacturer's heat indicator label has changed colors.

During the inspection of ladders, the inspector should mark all defects with chalk or other suitable marker (Figure 8.32). Legible marks permit repairs to be made without the chance of missing some defect that was previously found. Major repairs that must be made on metal ladders may often require special tools that are usually not available in most fire departments. Repairs to ladders should be made according to the manufacturer's recommendations.

Figure 8.32 Use a suitable marker to mark any defects on the ladder.

Wooden ladders require a much closer inspection than metal ladders, because their trusses and beams are subject to cracks and splinters. Wooden rungs are susceptible to damage at the point where they come in contact with the locks. Wooden staypoles have similar damage characteristics.

Varnish may be used to preserve wooden ladders, because it seals in the natural oils and resins and keeps the moisture out. Varnish also prevents dry rot and attack from fungus growth. When the varnish finish becomes worn or scratched, the ladder should be revarnished without delay. Paint is not recommended on fire department ladders except to identify the ladder ends, balance point, or length. Paint used as an outer cover on a wooden ladder makes it practically impossible during inspection to detect fungus growth, dry rot, or cracks.

Metal ladders are not subject to some of the problems, such as moisture and climate conditions, that affect wooden ladders. However, all braces, slides, stops, locks, halyards, rivets, pulleys, and other movable parts should be examined. The movable parts should be lubricated at least every six months with a waterproof grease. Old grease should be removed with a solvent.

Fiberglass ladders should be checked for cracks, chips, or other flaws. Rungs should be checked for tightness. Any flaws found should be repaired as specified by the manufacturer.

SERVICE TESTING GROUND LADDERS

Because fire service ladders are subject to harsh conditions and physical abuse, it is important that they be service tested to ensure they are fit for use. NFPA 1932, *Standard on Use, Maintenance, and Service Testing of Fire Department Ground Ladders*, should serve as the guideline for ground ladder service testing. This standard recommends that only the tests specified be conducted either by the fire department or an approved testing organization. NFPA 1932 further recommends that caution be used when performing service tests on ground ladders to prevent damage to the ladder or injury to personnel.

CAUTION: There is a possibility of sudden failure of the ground ladder during service testing. All possible safety precautions should be taken to avoid injury.

Service tests of ground ladders may require the purchase of measuring instruments. Testing should be performed only by personnel trained in service test procedures and operation of the service test equipment.

NFPA 1932 contains the following requirements pertaining to frequency of service testing:

- At least annually.
- Any time a ladder is suspected of being unsafe.
- After the ladder has been subjected to overloading.
- After the ladder has been subjected to impact loading or unusual conditions of use.

- After heat exposure. (**NOTE:** Metal ground ladders being tested because of exposure to heat may be subjected to either the Strength Service Test or the Hardness Service Test; both are not required. However, if the Hardness Service Test is used and the ladder fails, a Strength Service Test shall be conducted.)
- After any deficiencies have been repaired, unless the only repair was replacing the halyard.

Any signs of failure during service testing shall be sufficient cause for the ground ladder to be removed from service. It should either be repaired and retested or be destroyed. The following sections highlight the service tests that should be performed on ground ladders.

Strength Tests For All Ladders, Except Pompier And Folding Ladders

The following tests are designed to ensure that the ladder has adequate strength to be used safely.

HORIZONTAL BENDING TEST

All ladders should be subjected to the horizontal bending test. The following procedure may be used:

Step 1: The ladder should rest on 1-inch (25 mm) cylindrical supports that are placed 6 inches (150 mm) from each end (Figure 8.33). These supports should be high enough off the ground to allow for normal ladder deflection (sagging). If the ladder is an extension or combination ladder, it should be fully extended.

Step 2: The test load area should be 32 inches (800 mm) long and centered over the actual center of the ladder. It should be some type of board or plate on which the free weights can be set (Figure 8.34). Remember to include the weight of the board or plate as part of the overall test weight.

Step 3: Add the preload test weights slowly until 350 pounds (159 kg) is in place (Figure 8.35). (**NOTE:** For ladders made before 1984, use 300 pounds [136 kg].) Allow the load to sit for one minute, and then remove the weights.

Step 4: Measure and record the distance from the bottom of each side rail to the ground (Figure 8.36).

Step 5: Apply the test load of 500 pounds (227 kg) to the ladder (Figure 8.37 on the following page). (**NOTE:** For ladders made before 1984, use 400 pounds [181 kg].) Allow the load to sit for five minutes, and then remove the weight.

Step 6a: **For Metal and Fiberglass Ladders** — Five minutes after the load has been removed, again measure the distance from the bottom of each side rail to the ground. The difference in distances should be less than those listed in Table

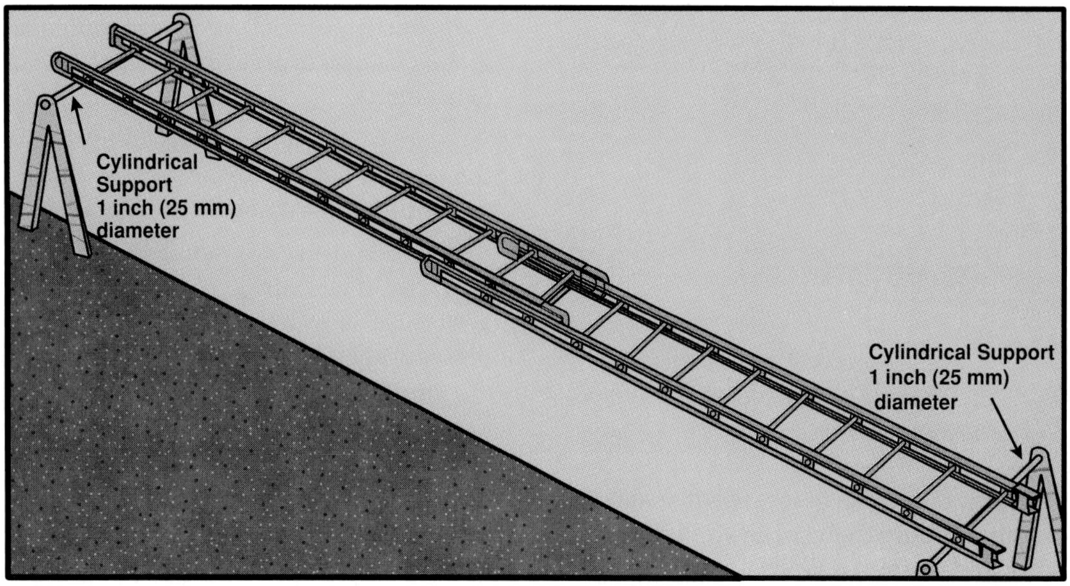

Figure 8.33 Support the ladder on 1-inch (25 mm) rods.

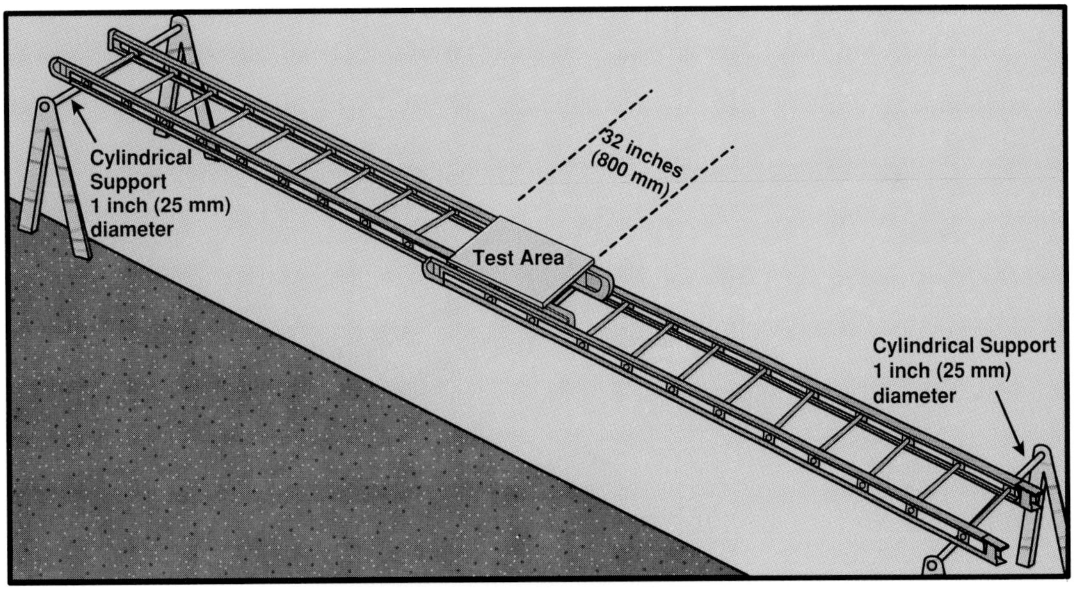

Figure 8.34 The test area is in the center of the ladder.

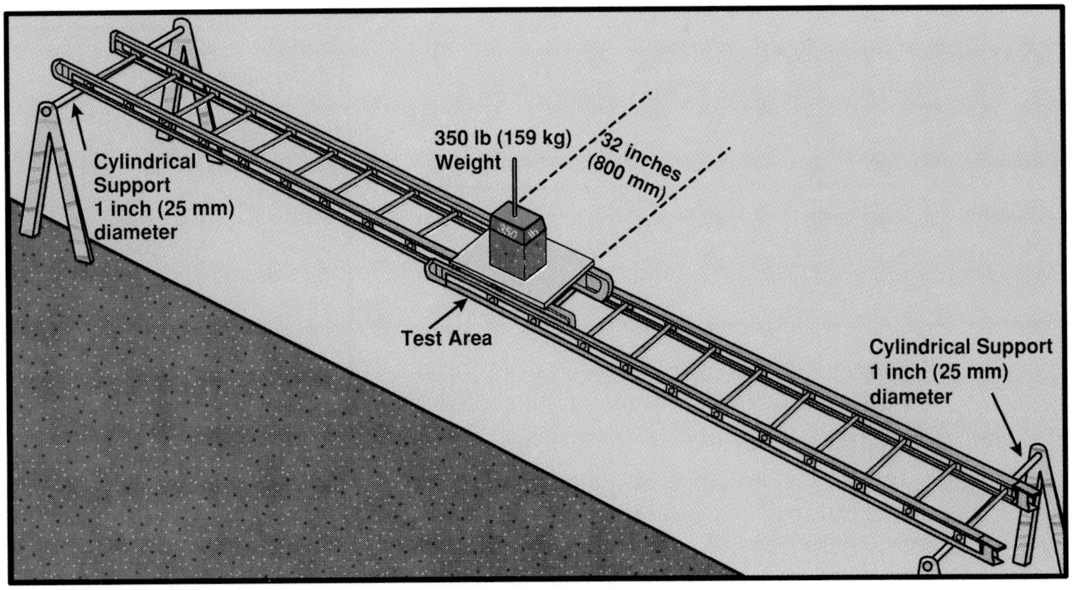

Figure 8.35 The 350 pounds (159 kg) is slowly added to the test area.

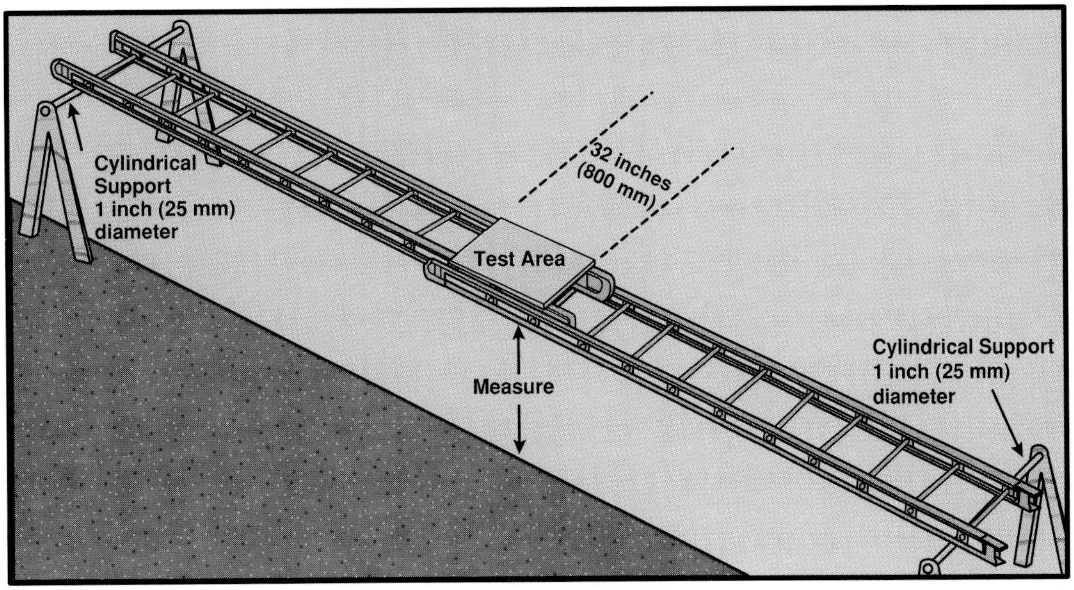

Figure 8.36 Measure the distance from the ladder to the ground.

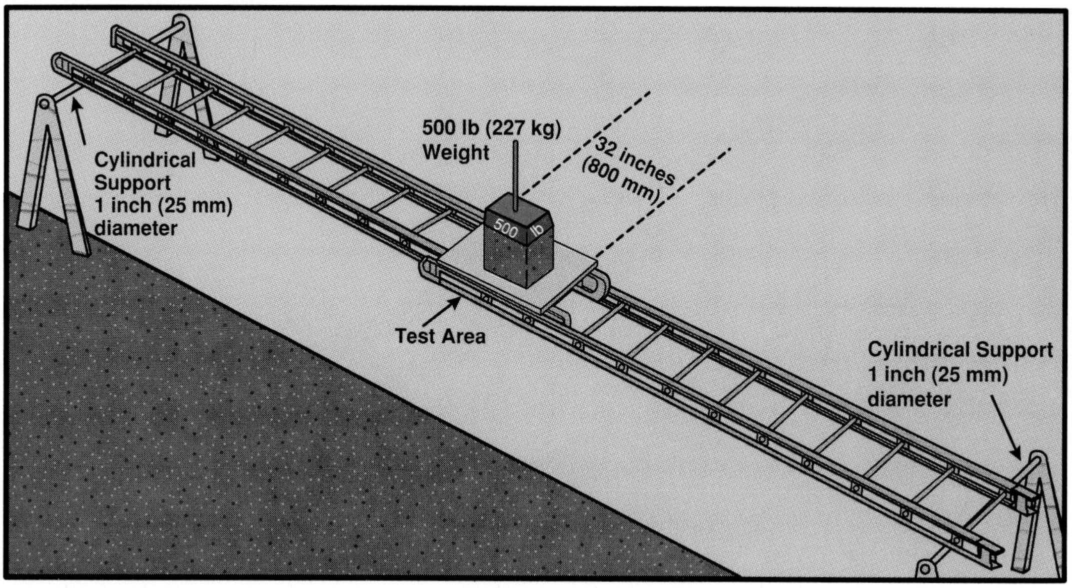

Figure 8.37 Add 500 pounds (227 kg) to the ladder.

TABLE 8.1
HORIZONTAL BENDING TEST DISTANCES

Length of Ladder	*Difference in Measurements*
25 feet or less (7.6 m)	½ inch (13 mm)
26 feet to 34 feet (7.7 m to 10.4 m)	1 inch (25 mm)
35 feet or over (10.5 m or over)	1½ inch (38 mm)

8.1; any ladder that does not fall within these measurements should be removed from service.

Step 6b: **For Wooden Ladders** — No measurements are taken. The ladder must simply not show signs of failure during the testing.

ROOF LADDER HOOK TEST

The following procedure should be used to ensure that the hooks on a roof ladder are structurally sound and capable of holding the weight of firefighters who work on the ladder.

Step 1: Position the ladder over the edge of a wall or platform so that it is hanging solely by the points of its hooks (Figure 8.38). The wall or platform must be high enough for the ladder to hang straight with the test weights suspended from it. A safety restraint should be used to fasten the ladder to the wall so that it will not fly away and injure people should a failure occur.

Step 2: Using webbing of appropriate size and strength, suspend the test weight of 1,000 pounds (454 kg) from the ladder for at least one minute (Figure 8.39).

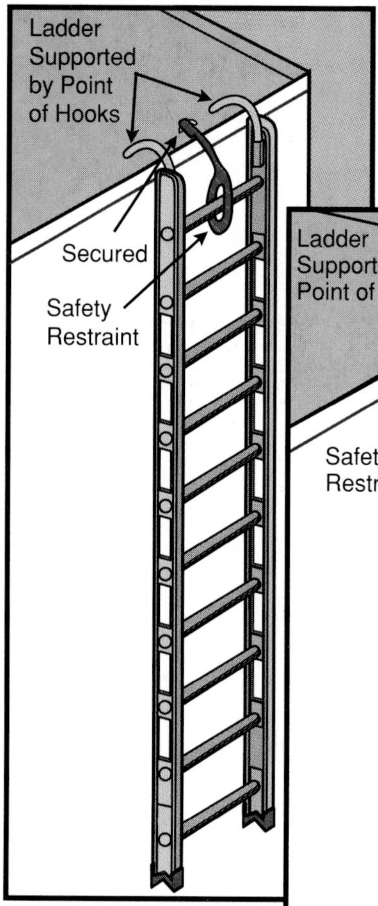

Figure 8.38 Hook the ladder over a sturdy wall.

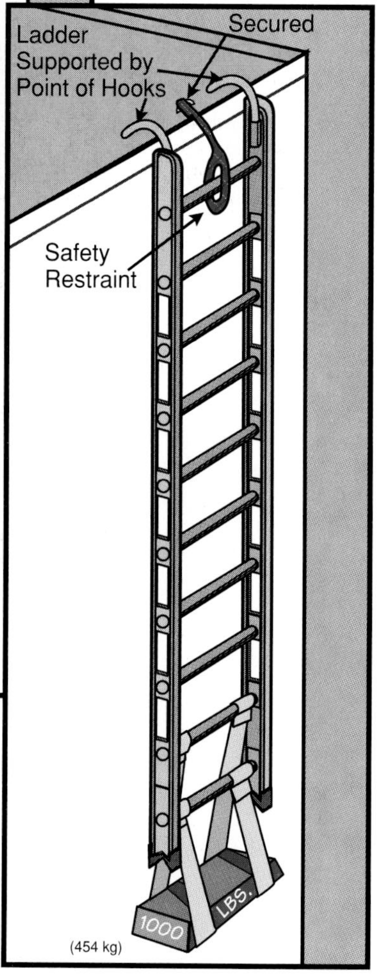

Figure 8.39 Add the weight to the ladder slowly.

Step 3: Remove the test weight and inspect the hook for damage (Figure 8.40). The ladder shall pass the test if no permanent deformation can be found.

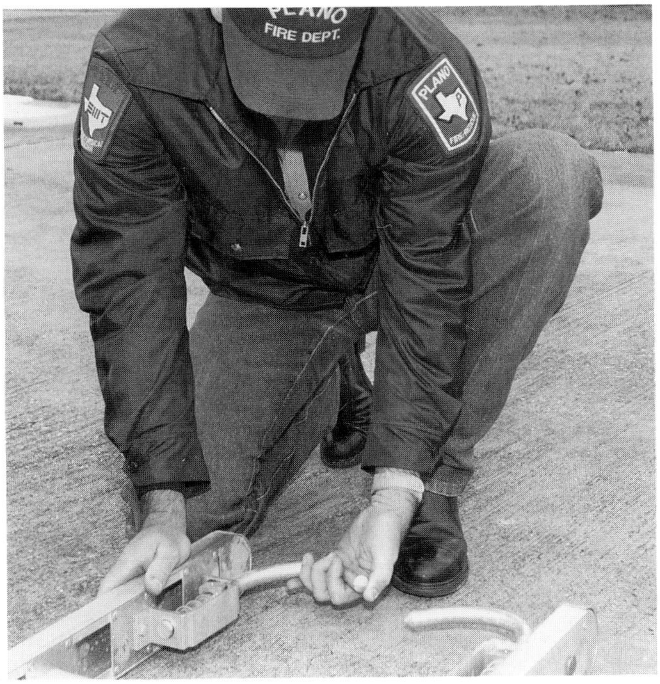

Figure 8.40 Step 3: Remove the test weight and inspect the hooks for damage.

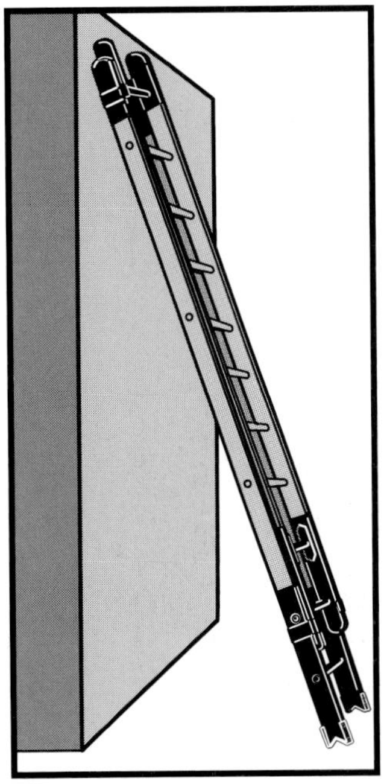

Figure 8.41 Place the ladder against the wall.

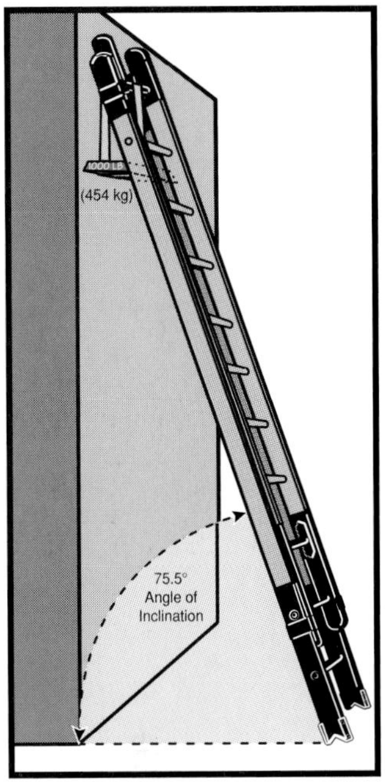

Figure 8.42 Add 1,000 pounds (454 kg) to the ladder.

EXTENSION LADDER HARDWARE TEST

The following test is used to ensure that ladder pawls and other components of extension ladders are strong enough to hold the weight of firefighters who will work on them.

Step 1: Position the ladder against a wall so that it is at a 75½-degree angle. The ladder should be extended a minimum of one rung's length (Figure 8.41).

Step 2: Using webbing of appropriate size and strength, suspend the test weight of 1,000 pounds (454 kg) from the ladder for at least one minute (Figure 8.42).

Step 3: Remove the test weight and inspect the ladder hardware for damage (Figure 8.43). The ladder shall pass the test if no permanent deformation can be found.

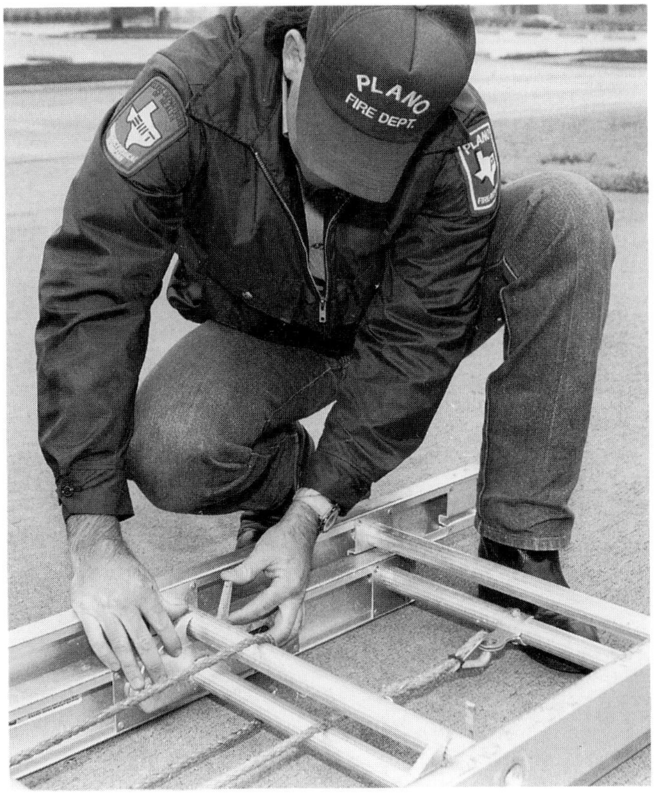

Figure 8.43 Step 3: Remove the test weight and inspect the ladder hardware for damage.

Strength Tests For Pompier Ladders

Use the following procedure to check the integrity of pompier ladders.

Step 1: Position the ladder over the edge of a wall or platform so that it is hanging solely by its hook. The wall or platform must be high enough for the ladder to hang straight with the test weights suspended from it. A safety restraint should be used to fasten the ladder to the wall so that it will not fly away and injure people should a failure occur.

Step 2: Using webbing of appropriate size and strength, suspend the test weight of 1,000 pounds (454 kg) from the ladder for at least one minute.

Step 3: Remove the test weight and inspect the hook for damage. The ladder shall pass the test if it does not ultimately fail (break).

Strength Test For Folding Ladders

The following horizontal bending test should used for determining the strength of folding ladders.

Step 1: The ladder should rest on 1-inch (25 mm) cylindrical supports that are placed 6 inches (150 mm) from each end (Figure 8.44). These supports should be high enough off the ground to allow for normal ladder deflection (sagging).

Step 2: The test load area should be 16 inches (400 mm) long and centered over the actual center of the ladder. It should be some type of board or plate on which the free weights can be set (Figure 8.45). Remember to include the weight of the board or plate as part of the overall test weight.

NOTE: For wooden folding ladders, proceed to Step 5.

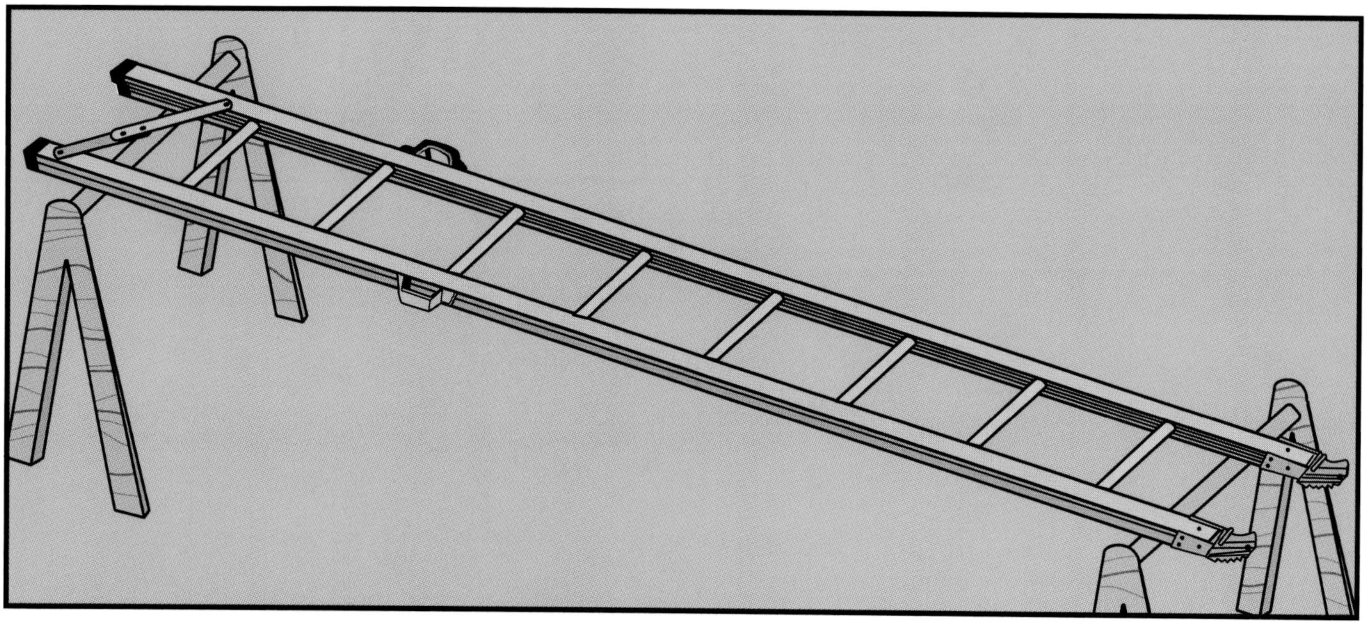

Figure 8.44 Support the folding ladder on 1-inch (25 mm) cylinders.

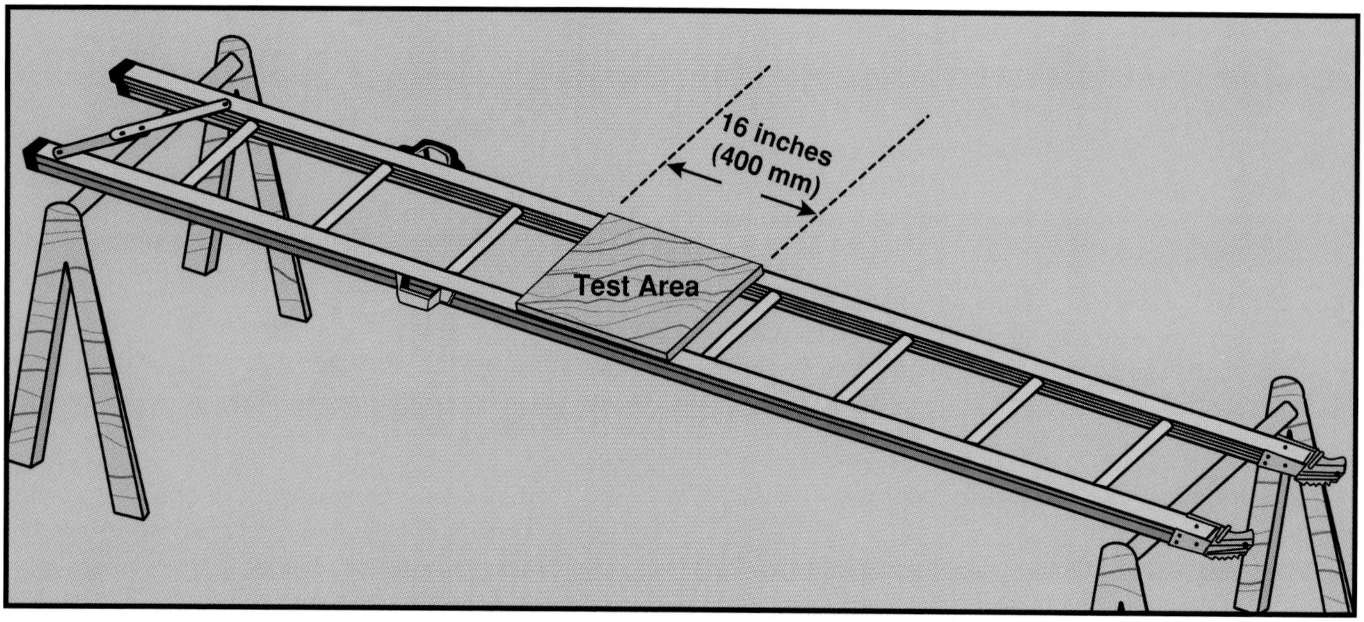

Figure 8.45 The test area is in the center of the ladder.

Step 3: Add the preload test weights slowly until 160 pounds (73 kg) is in place (Figure 8.46). Allow the load to sit for one minute, and then remove the weights.

Step 4: Measure and record the distance from the bottom of each side rail to the ground (Figure 8.47).

Step 5: Apply the test load of 225 (102 kg) pounds to the ladder. Allow the load to sit for five minutes, and then remove the weight (Figure 8.48 on the next page).

Step 6a: **For Metal and Fiberglass Ladders** — Five minutes after the load has been removed, again measure the distance from the bottom of each side rail to the ground. The difference in distances should be less than ½-inch (13 mm). Any ladder that does not fall within these measurements should be removed from service.

Step 6b: **For Wooden Ladders** — No measurements are taken. The ladder must simply not show signs of failure during testing.

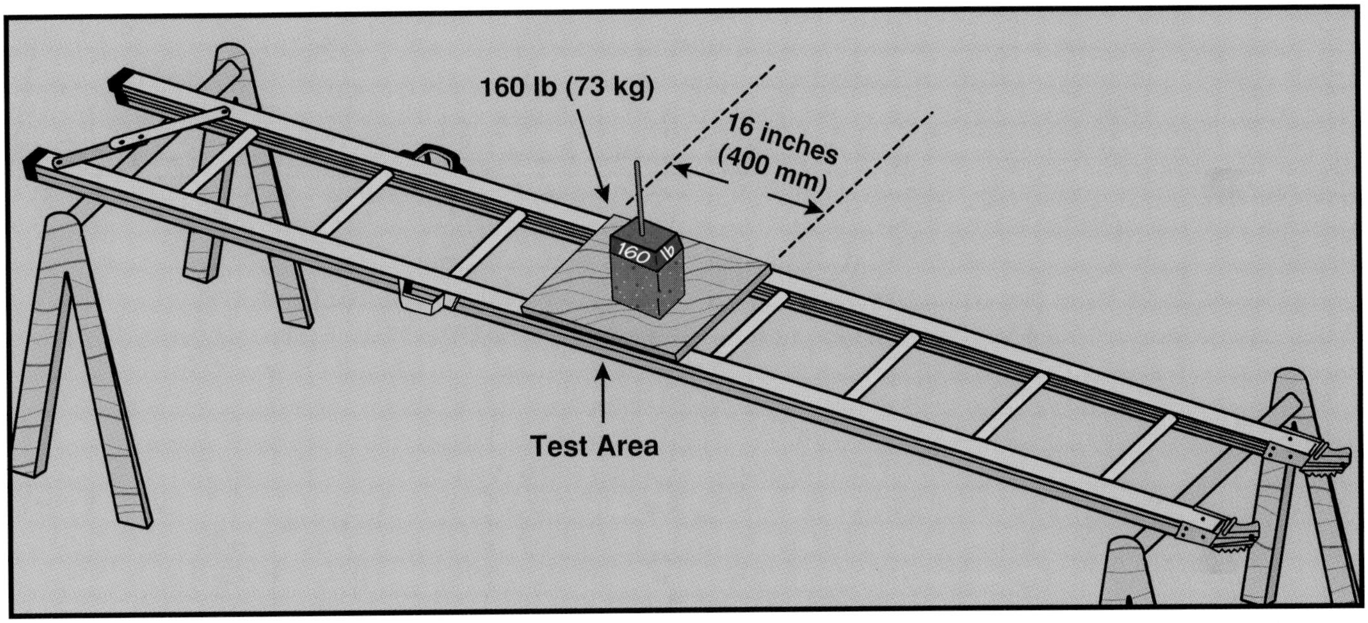

Figure 8.46 Add 160 pounds (73 kg) to the test area.

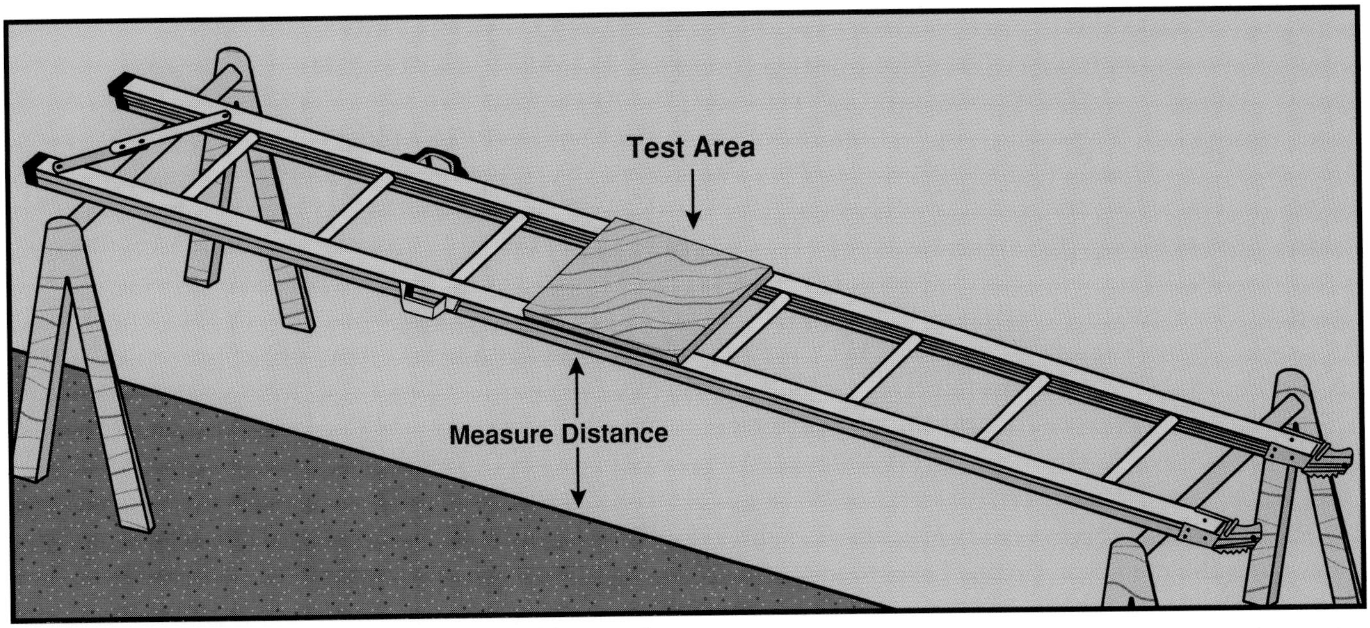

Figure 8.47 Measure the distance between the ladder and the ground.

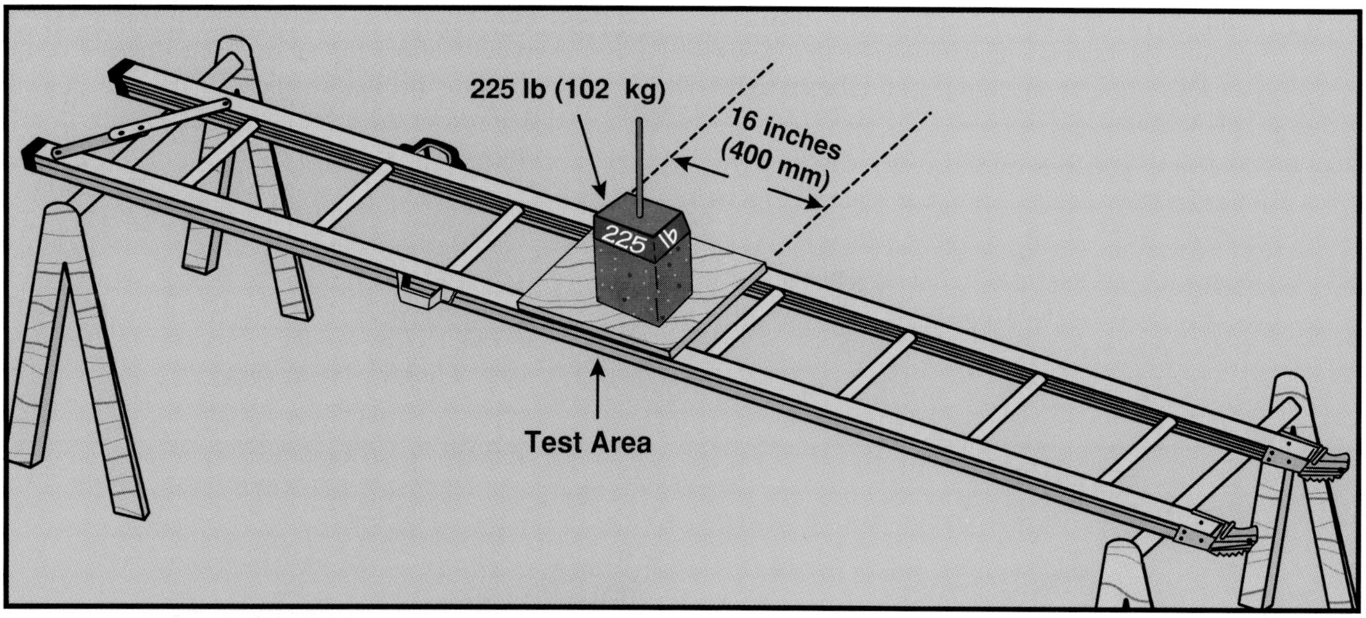

Figure 8.48 Add 225 pounds (102 kg) to the ladder slowly.

HANDLING LADDERS

Before raising ground ladders, the firefighter(s) must first select the proper ladder for the given job and then carry it to its location for use.

Selecting The Ladder For The Job

Selecting a ladder to do a specific job requires that the firefighter be a good judge of distance, stress, and strength. Roughly speaking, a residential story will average 8 to 10 feet (2 m to 3 m) from floor to floor, with a 3-foot (1 m) distance from the floor to windowsill. Stories of commercial buildings will average 12 feet (4 m) from floor to floor, with a 4-foot (1.3 m) distance from the floor to windowsill. In general, Table 8.2 can be used in selecting ladders.

Working rules for ladder length include the following:

- Ladders should extend a few feet (preferably five rungs) beyond the roof edge to provide a footing and handhold for persons stepping on and off the ladder (Figure 8.49).
- When rescue from a window opening is to be performed, the tip of the ladder should be placed just below the windowsill (Figure 8.50).
- When used for access from the side of a window or for ventilation, the tip should be three or four rungs above the windowsill (Figure 8.51).

Finally, to select the proper ladder, the firefighter must understand how the length of the ladder is measured. The ladder length that will be marked on the side of the ladder gives the actual length of the ladder when

TABLE 8.2
LADDER SELECTION GUIDE

First story roof	16 to 20 feet (4.9 m to 6.0 m)
Second story window	20 to 28 feet (6.0 m to 8.5 m)
Second story roof	28 to 35 feet (8.5 m to 10.7 m)
Third story window or roof	40 to 50 feet (12.2 m to 15.2 m)
Fourth story roof	over 50 feet (15.2 m)

Figure 8.49 At least five rungs should be above the roof level.

Figure 8.50 For rescue purposes, the tip of the ladder should be at windowsill level.

Figure 8.51 For side access or ventilation purposes, place the tip of the ladder three or four rungs above windowsill level.

it is fully extended, not the reach of the ladder. The reach of the ladder will be less than the length because of having to set the ladder at an angle. For ladders of 35 feet (11 m) or less, the reach will be about 1 foot (0.3 m) less than the length. For ladders over 35 feet (11 m), the reach will be about 2 feet (0.6 m) less than the length.

Methods Of Mounting Ground Ladders On Apparatus

The method used to mount ground ladders on fire apparatus varies depending on the departmental requirements, the type of apparatus, the type of ladder, and the manufacturer's policies. There are no established standards for mounting ground ladders on fire apparatus, and a training sequence for ladder removal is a local procedure.

NFPA 1901, *Standard for Pumper Fire Apparatus*, requires fire department pumpers to be equipped with an extension ladder at least 24 feet (8 m) long, a roof ladder at least 14 feet (4.3 m) long, and a folding ladder at least 10 feet (3 m) long. These ladders are usually mounted on edge at or above shoulder level on one side of the apparatus (Figure 8.52). There are many types of brackets provided and methods used to hold the ladder.

NFPA 1904, *Standard for Aerial Ladder and Elevating Platform Fire Apparatus*, requires ladder apparatus to carry the following complement of ground ladders:

- One 10-foot (3 m) folding ladder
- Two 16-foot (5 m) roof ladders with folding hooks

Figure 8.52 Most pumpers have their ladder mounted on the side.

- One 14-foot (4.3 m) combination ladder
- One 24-foot (8 m) extension ladder
- One 35-foot (11 m) extension ladder

Ladders may be mounted on aerial apparatus in several different ways:

- *Loaded from the rear, lying flat in tiers usually two rows wide and two rows high* (Figure 8.53). This type mounting requires the ladders to fit into runners or troughs. The ladders are locked into position at each tier by a bar that is controlled either by a manually-operated lever or a power-operated bar activated by electric solenoid. When ladders are mounted in this manner, room must be left at the rear of the apparatus to allow for removal and loading.

Figure 8.53 Ladders mounted in this manner must be unloaded from the rear.

- *Nested vertically (on edge) or in flat tiers on the sides of the apparatus* (Figure 8.54). This arrangement eliminates the problem of another piece of apparatus stopping too close.
- *Nested vertically (on edge) on each side* (Figure 8.55). These ladders are arranged so that they may be loaded and unloaded from the rear by sliding them in and out of the troughs. This requires that room be left at the rear of the apparatus for removal and loading.

Figure 8.54 Ladders may be mounted on edge or in flat tiers on the sides of the apparatus. *Courtesy of Joel Woods, University of Maryland Fire & Rescue Institute.*

Figure 8.55 Ladders mounted vertically on the side of the aerial apparatus may be removed over the side.

When ladders must extend beyond the rear of the apparatus to the extent that they create an accident hazard, some type of guard should be placed over the heel plates (Figure 8.56). Short pieces of fire hose or brightly painted protectors are sometimes used for this

Figure 8.56 A cushion or some type of guard should be placed over heel plates to prevent an accident hazard.

purpose. Before firefighters are drilled in removing ladders from the apparatus, each firefighter should be able to answer the following questions concerning the equipment.

- Are the ladders mounted on the right or left side of the pumper?
- Are the butt plates toward the front or the rear of the apparatus?
- Can the roof ladder be removed, leaving the extension ladder securely in place on the pumper?
- Is the fly section of the extension ladder on the inside or outside when the ladder is on the pumper?
- How are the ladders secured in place?
- Are the ladders provided with protection plates?
- Where two or more ladders are mounted flat, one inside the other, will all the ladders need to be removed from the tier to get only one ladder?

Removing Ladders From Apparatus

Ladders carried on pumpers are sometimes mounted so that it is necessary to remove the complete assembly of roof and extension ladders before they are separated. Unless the roof and extension ladders have individual brackets and holding devices, one ladder may fall from the apparatus when the other is removed. For safety reasons, consideration should be given to providing a separate holding device for each ladder. Unused ladders should be put back on apparatus or kept in a safe and available location (Figure 8.57). Each department should develop a procedure for placing unused ladders in such a manner that personnel will not trip or vehicles will not run over them.

If two firefighters remove the ladder or ladders from a pumper, they should position themselves at each locking device. After the locks have been released, each person should grasp a convenient group of rungs on each side of the locks and lift the entire assembly from the apparatus (Figure 8.58). If one firefighter is to remove a ladder, he or she takes a position at the center of the ladder near the balance point after releasing the locks (Figure 8.59). The ladder can then be lifted over the retaining brackets. Once removed from the pumper, the ladder can be prepared for the appropriate carry.

Figure 8.57 If the roof ladder is not going to be used, it should be placed back on the apparatus after the extension ladder has been removed.

Figure 8.58 When possible, two firefighters should be used to remove the extension ladder from the apparatus.

Figure 8.59 One firefighter can easily handle the roof ladder.

Some departments mount a three-section, 35-foot (11 m) extension ladder on the side of the pumper. It can be removed by two or three firefighters, following the same procedure as for the two-firefighter method. The firefighters would then stay in the same position on one side of the ladder when making the desired carry. Larger ladders may require up to six people to remove them.

When ladders are carried flat, as with overhead racking on some pumpers and on aerial apparatus, the procedure begins with the release of the locking device. After the ladder locks have been released, the ladder should be pulled straight back from the ladder bed on a level plane (Figure 8.60). This can be accomplished by two to six firefighters, depending on the size of the ladder and the number of firefighters available to remove and carry the ladder. As the ladder is pulled from the bed, the firefighters should evenly space themselves along the

Figure 8.60 First, pull the ladder straight back.

length of the ladder as they grasp the beam of the ladder. The person closest to the apparatus should give a signal to stop when the ladder is about ready to clear the ladder bed (Figure 8.61). At this point, all firefighters lift the ladder from the apparatus for the desired carry (Figure 8.62).

Figure 8.61 The firefighter(s) on the ground should begin to lower their end of the ladder as the other end nears the end of the rack.

Figure 8.62 All firefighters must control the ladder after it clears the bed.

Proper Lifting And Lowering Methods

When lifting any object from the ground, the lifting force should come from the legs and NOT the back (Figure 8.63). To lift a ladder from the ground, use the following procedure:

- Obtain adequate personnel for the job.
- Place firefighters parallel to the ladder at the ends and in the middle if necessary (Figure 8.64).
- Bend your knees, keeping your back as straight as possible. On command, lift with your legs, not with your back or arms. (**NOTE:** The command to lift should be given by one of the firefighters at the rear who can see the operation.) Lifting should be done in unison and as a team (Figure 8.65). If one firefighter is not ready to lift, he or she should make it known to the rest of the team. Examples of commands might include the following:
  "Prepare to beam ladder."
  "Beam ladder."
  "Prepare to raise ladder."
  "Raise ladder."
  "Advance."
- When lowering a ladder, be sure to lower it with your leg muscles and not your back muscles. Also, be sure to keep your body and feet parallel to the ladder so that when the ladder is set down, it does not injure your toes. The knee closest to the ladder should be the knee to contact the ground (Figure 8.66).

Ladder Carries

Once the ladder has been removed from its mounting, there are numerous ways it can be transported to its

Figure 8.63 To prevent back injury, lift with the legs and not the back.

Figure 8.64 Place firefighters parallel to the ladder at the ends and in the middle.

Figure 8.65 After the command to lift is given, lift in unison.

Figure 8.66 The knee closest to the ladder should be the knee to contact the ground.

point of use. The following sections highlight these carries.

ONE-FIREFIGHTER LOW-SHOULDER METHOD

Short, light ladders may be carried on the shoulder by one person. When taking the ladder from an apparatus mount, use the following procedure:

Step 1: Select a balance point near the center of the ladder (**NOTE:** Some departments mark the center for convenience.)

Step 2: Face toward the butt and insert one arm between the beams. Rest the upper beam on your shoulder, and steady the ladder with both hands (Figure 8.67).

Step 3: Lower the butt (front) end slightly for better balance and vision during the carry (Figure 8.68).

To pick up the ladder for the low-shoulder carry when the ladder is on the ground, use the following procedure:

Step 1: Face toward the tip, crouch beside the ladder, and grasp the middle rung with your hand nearest the ladder (Figure 8.69).

Step 2: Stand, using your leg muscles and keeping your back straight and vertical. As the ladder is brought up, pivot into the ladder, and insert

Figure 8.67 Select a balance point near the center of the ladder. Face toward the butt and insert one arm between the beams. Rest the upper beam on your shoulder, and steady the ladder with both hands.

Figure 8.68 Step 3: Lower the butt (front) end slightly for better balance and vision during the carry.

Figure 8.69 Step 1: Face toward the tip, crouch beside the ladder, and grasp a middle rung with your hand nearest the ladder.

your other arm through the rungs so that the upper beam rests on your shoulder (Figure 8.70).

Step 3: Face toward the butt, steady the ladder, and lower the butt (front) end slightly for better balance and visibility (Figure 8.71).

Figure 8.70 Step 2: Stand, using your leg muscles and keeping your back straight and vertical. As the ladder is brought up, pivot into the ladder, and insert your other arm through the rungs so that the upper beam rests on your shoulder.

Figure 8.71 Step 3: Face toward the butt, steady the ladder, and lower the butt (front) end slightly for better balance and visibility.

ONE-FIREFIGHTER HIGH-SHOULDER METHOD

The high-shoulder carry is good to use when the ladder will be raised on the beam.

Step 1: Face the butt, and at the balance point, place the palm of one hand under the bottom beam and the palm of your other hand on the top beam (Figure 8.72).

Step 2: Lift the ladder, and rest the lower beam on your shoulder with the butt end lowered slightly. Hold the upper beam with one hand, palm down (Figure 8.73).

CARRYING ROOF LADDERS

Normally, the roof ladder is carried to its place of use with the hooks closed. If the hooks are to be used, they should be opened before taking the ladder onto the roof. When the roof ladder is carried up another ladder and its hooks are to be placed on a sloping roof, the roof ladder should be carried with the tip forward (Figure 8.74).

Figure 8.72 Face the butt with the palms of your hands on the top and bottom of the ladder.

Figure 8.73 Lower the butt before moving out.

Figure 8.74 Some departments prefer their firefighters to carry a roof ladder with the tip forward.

Another method is to carry it butt-forward in the high-shoulder carry and, upon reaching the ladder to be climbed, to raise the roof ladder on the beam and lean it against the other ladder (Figure 8.75). The firefighter then climbs the ladder until he or she is even with the top of the roof ladder. Once the hooks are open, the firefighter steps down one rung and slips the near arm through the roof ladder (one or two rungs from the tip) and then proceeds up the ladder to the edge of the roof (Figure 8.76).

Figure 8.75 Another method for carrying a roof ladder is to carry it with the butt forward. This makes it easy to place the butt at the desired position.

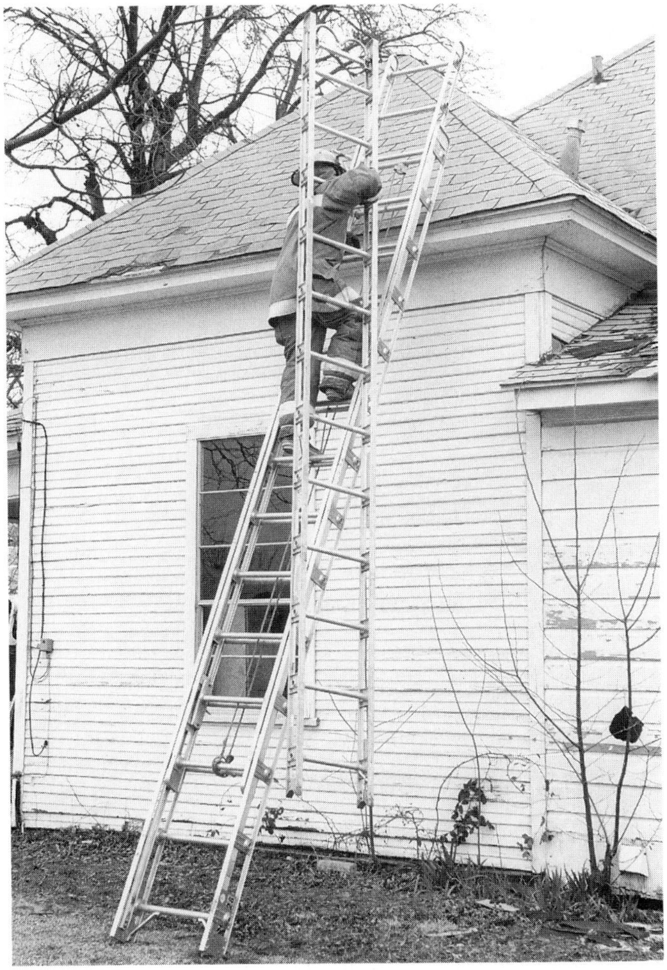

Figure 8.76 To place the roof ladder, the firefighter steps down one rung and slips the near arm through the roof ladder and then proceeds up the ladder.

TWO-FIREFIGHTER LOW-SHOULDER METHOD (FROM VERTICAL RACKING)

Use the following method to perform a two-firefighter low-shoulder carry from a vertical rack.

Step 1: The two firefighters stand facing the ladder. One firefighter is positioned near the tip end, and the other is positioned near the butt end (Figure 8.77).

Step 2: Each firefighter uses both hands to grasp the ladder to remove it from the rack (Figure 8.78).

Figure 8.77 Step 1: The two firefighters stand facing the ladder. One firefighter is positioned near the tip end, and the other is positioned near the butt end.

Figure 8.78 Step 2: Each firefighter uses both hands to grasp the ladder to remove it from the rack.

Step 3: As soon as the ladder clears the rack, the firefighters continue to grasp the ladder with the hand nearest the butt end while they place the other arm between two rungs, pivot, and bring the upper beam onto the shoulder (Figure 8.79).

Step 4: The forward firefighter is in a position to use one hand to push people out of the way to prevent them from being struck by the butt spur (Figure 8.80).

Figure 8.79 Step 3: As soon as the ladder clears the rack, the firefighters continue to grasp the ladder with the hand nearest the butt end while they place the other arm between two rungs, pivot, and bring the upper beam onto the shoulder.

Figure 8.80 Step 4: The forward firefighter is in a position to use one hand to push people out of the way to prevent them from being struck by the butt spur.

TWO-FIREFIGHTER LOW-SHOULDER METHOD (FROM FLAT RACKING)

The two-firefighter low-shoulder carry may be performed from the flat racking position using the following procedure.

Step 1: As the ladder clears the rack, each firefighter grasps two rungs holding the ladder flat (Figure 8.81).

Step 2: For ladders stored at or above shoulder level, the outside beam is lowered and the inside beam raised to the shoulder simultaneously with each firefighter pivoting and placing the arm farthest from the butt end between two rungs (Figure 8.82). If the ladder is stored below shoulder level, raise the outside beam and lower the inside one.

Step 3: The upper beam of the ladder should now be on the firefighters' shoulders with the firefighters facing the butt end (Figure 8.83).

Figure 8.81 Step 1: As the ladder clears the rack, each firefighter grasps two rungs, holding the ladder flat.

Figure 8.82 Step 2: The firefighters simultaneously lower the outside beam and raise the inside beam to the shoulder, while they pivot and place the arm farthest from the butt end between two rungs.

Figure 8.83 Step 3: The upper beam of the ladder should now be on the firefighters' shoulders with the firefighters facing the butt end.

TWO-FIREFIGHTER LOW-SHOULDER METHOD (FROM THE GROUND)

To pick up the ladder when it is on the ground, the following procedure is used:

Step 1: The two firefighters position themselves on the same side of the ladder — one near the butt end, the other near the tip end (Figure 8.84).

Step 2: Both firefighters kneel next to the ladder facing the tip end and grasp a convenient rung with their near hand, palm forward (Figure 8.85).

Step 3: The firefighter at the heel gives the command to "shoulder the ladder." Both firefighters should stand up, using their leg muscles to lift the ladder (Figure 8.86).

Figure 8.84 Step 1: The two firefighters position themselves on the same side of the ladder — one near the butt end, the other near the tip end.

Figure 8.85 Step 2: Both firefighters kneel next to the ladder facing the tip end and grasp a convenient rung with their near hand, palm forward.

Step 4: As the ladder and the firefighters rise, the far beam is tilted upward, and the firefighters pivot and place their free arm between two rungs. The upper beam is placed on the shoulders, with the firefighters facing the butt end (Figure 8.87).

Figure 8.86 Step 3: Both firefighters should stand up, using their leg muscles to lift the ladder.

Figure 8.87 Step 4: The upper beam is placed on the shoulders, with the firefighters facing the butt end.

OTHER TWO-FIREFIGHTER CARRIES

There are other two-firefighter carries that may be used for ground ladders. Two commonly used methods are the hip or underarm method and the arm's length method.

THREE-FIREFIGHTER FLAT-SHOULDER METHOD (FROM FLAT RACKING)

The three-firefighter flat-shoulder method may be performed from the flat racking position by using the following procedure:

Step 1: As the ladder is pulled from the rack, two firefighters position themselves on one side of the ladder near each end, and the third firefighter takes a position at midpoint on the opposite side (Figure 8.88 on the next page).

Step 2: All three firefighters grasp the rungs and beam, lift the ladder clear of the apparatus, and prepare to pivot toward the butt end (Figure 8.89 on the next page).

Step 3: The firefighters raise the ladder to shoulder height while pivoting and place the beam upon their shoulders (Figure 8.90).

Figure 8.88 Step 1: Two firefighters position themselves on one side of the ladder near each end, and the third firefighter takes a position at midpoint on the opposite side.

Figure 8.89 Step 2: All three firefighters grasp the rungs and beam, lift the ladder clear of the apparatus, and prepare to pivot toward the butt end.

Figure 8.90 Step 3: The firefighters raise the ladder to shoulder height while pivoting, and place the beam upon their shoulders.

THREE-FIREFIGHTER FLAT-SHOULDER METHOD (FROM THE GROUND)

To raise the ladder from the ground using the flat-shoulder carry use the following procedure:

Step 1: Two firefighters position themselves on one side of the ladder near each end, and the third firefighter takes a position at midpoint on the opposite side (Figure 8.91).

Step 2: Facing the tip end, each firefighter kneels so that his or her back is straight and the knee closest to the ladder is the one touching the ground. Each grasps the beam with the hand closest to the ladder (Figure 8.92).

Step 3: The firefighter at the heel first gives the command to "prepare to shoulder the ladder," followed by the command to "shoulder the ladder," when everybody is ready. All firefighters stand up, using their leg muscles to lift the ladder, and pivot toward the butt end (Figure 8.93).

Step 4: The beam is placed on the shoulders with the firefighters facing toward the butt end (Figure 8.94).

Figure 8.91 Step 1: Two firefighters position themselves on one side of the ladder near each end, and the third firefighter takes a position at midpoint on the opposite side.

Figure 8.92 Step 2: Facing the tip end, each firefighter grasps the beam with the hand closest to the ladder.

Figure 8.93 Step 3: To shoulder the ladder, all firefighters stand up and pivot toward the butt end.

Figure 8.94 Step 4: The beam is placed on the shoulders with the firefighters facing toward the butt end.

THREE-FIREFIGHTER FLAT ARM'S LENGTH METHOD

Use the following procedure to perform the three-firefighter flat arm's length carry:

Step 1: Two firefighters position themselves on one side of the ladder near each end. The third firefighter takes a position at midpoint on the opposite side (Figure 8.95). All three firefighters face the butt of the ladder.

Step 2: All firefighters kneel beside the ladder with the knee closest to the ladder touching the ground. The firefighters grasp the beam (Figure 8.96).

Step 3: The firefighter at the heel gives the command to "pick up the ladder," and all stand, using their leg muscles to lift the ladder to the arm's length position (Figure 8.97).

Figure 8.95 Step 1: All three firefighters face the butt end of the ladder.

Figure 8.96 Step 2: The firefighters grasp the beam.

Figure 8.97 Step 3: The firefighter at the heel gives the command to "pick up the ladder," and all stand.

FOUR-FIREFIGHTER CARRIES

By making a slight change in the positioning of the firefighters, the same methods may be used by four firefighters as those used by three firefighters. For the flat-shoulder and flat arm's length methods, two firefighters are positioned at each end of the ladder opposite one another (Figures 8.98 a and b). When using the low-shoulder or arm's length on-edge methods, firefighters are equally spaced along one side of the ladder (Figures 8.99 a and b).

Figure 8.98a A four-person flat-shoulder carry.

Figure 8.98b A four-person flat arm's length carry.

Figure 8.99a A four-person low-shoulder carry.

Figure 8.99b A four-person arm's length on-edge carry.

Proper Ladder Placement

The placement of ladders is determined by the intended use of the ladder and the angle of the ladder for safe and easy climbing. Ladder placement for ventilation, rescue, and fire attack should be as follows:

- If the ladder is to be used to effect ventilation from a window, it should be positioned alongside the window to the windward side with the tip about even with the top of the window (Figure 8.100). The same position is used when firefighters desire to climb into or out of narrow windows.
- If the ladder is to be used for rescue from a window, consideration must be given to the size of the window. Normally, the ladder tip is placed even with or slightly below the sill (Figure 8.101). If the sill projects out from the wall, the tip of the ladder can be wedged up under the sill for additional stability (Figure 8.102). Where the window opening is wide enough to permit placing the ladder inside the window opening and still leave room beside it to facilitate the rescue, it should be placed so that two or three rungs extend above the sill (Figure 8.103). The same position can be used for firefighters to climb in or out of wide window openings.
- When a ladder is used as a vantage point from which to direct a hose stream into a window opening and no entry is to be made, it is raised directly in front of the window with the tip on the wall above the window opening (Figure 8.104).
- Ladders placed to the roof should be extended five rungs above the roof edge to aid in climbing on and off the ladder (Figure 8.105). This placement also allows firefighters on the roof to find the ladder.

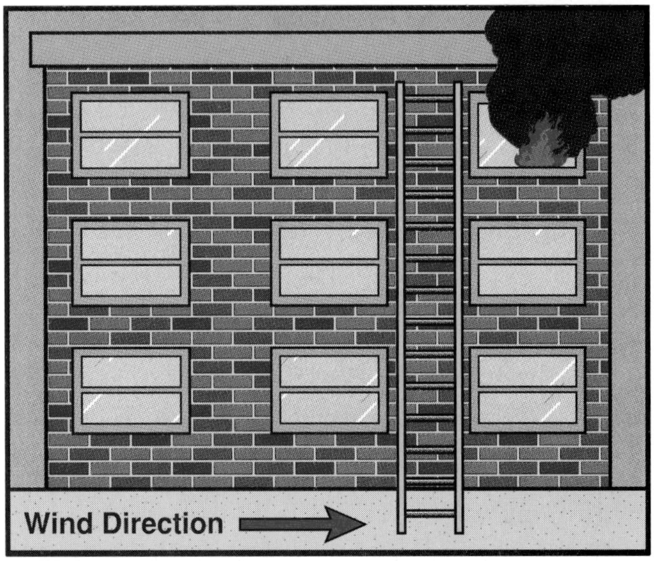

Figure 8.100 Place the ladder to the upwind side of the window being ventilated with the tip of the ladder at the top of the window.

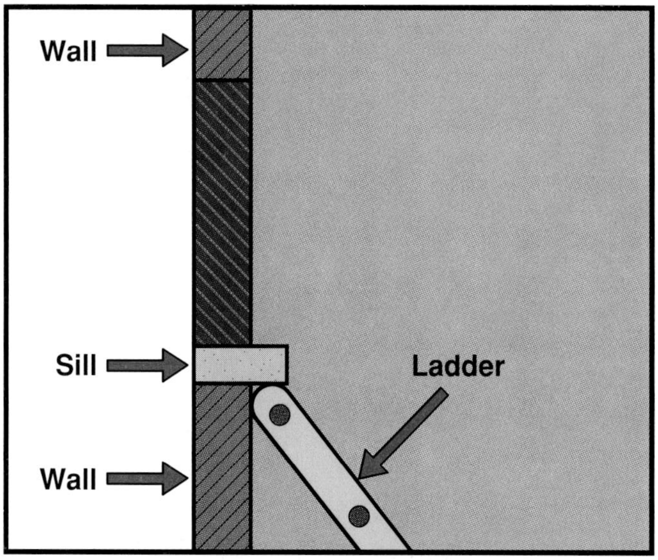

Figure 8.102 Wedge the ladder under the sill when possible.

Figure 8.101 Place the tip of the ladder at sill level.

Figure 8.103 The ladder may be projected into the opening on wide windows.

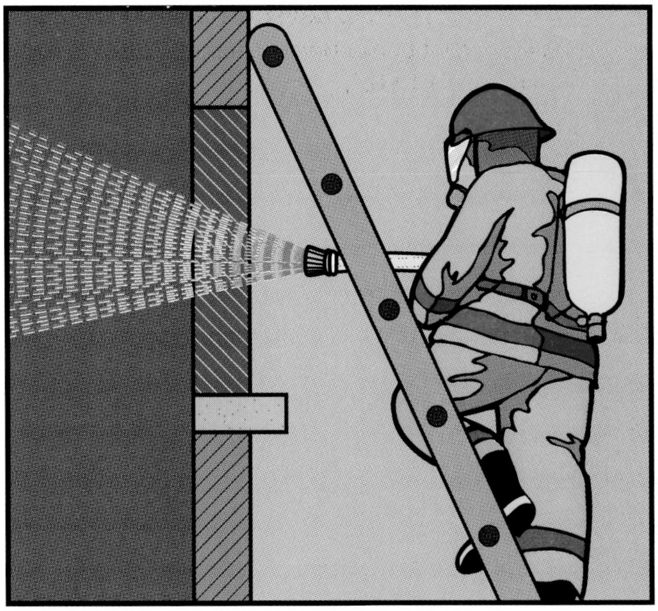

Figure 8.104 Ladders may be used to provide a base for hose stream operations.

Figure 8.105 At least five rungs of the ladder should be above roof level.

Other factors that affect ladder placement include:

- Overhead obstructions such as wires, tree limbs, signs, cornices, and building overhangs
- Uneven terrain and soft spots
- Obstructions on the ground such as bushes and parked cars
- Main paths of travel that firefighters or evacuees may use (Figure 8.106)

Any of these factors may require the ladder to be relocated or placed at a less optimum climbing angle.

When ladders are raised into place, they should be at an angle that is safe and easy to climb. The distance of the butt end from the building establishes the angle formed by the ladder and the ground. If the butt is too close to the building, the ladder's stability is reduced, and when it is climbed, the tip of the ladder may be pulled away from the building. If the butt is too far from the building, the load-carrying capacity of the ladder is reduced, and the ladder may also slip away from the building. An angle of 75½ degrees gives adequate stability, ensures safe stress on the ladder, and is easy to climb. With a ladder at this angle, a climber can stand upright on the rungs and climb at arm's length from the rungs. The proper angle and the position of a person climbing is shown in Figure 8.107.

Figure 8.107 Check for a proper climbing angle before ascending the ladder.

Figure 8.106 Do not place a ladder in front of an exit.

An easy way to determine the proper distance between the heel of the ladder and the building is to divide the used length of the ladder by four. For example, if 20 feet (6 m) of a 24-foot (8 m) ladder is to be used, the butt end should be 5 feet (1.5 m) (20 feet ÷ 4 [6 m ÷ 4]) from the building (Figure 8.108 on the next page). Exact measurements are unnecessary on the fire scene. Firefighters develop the experience to judge visually the proper positioning for the ladder. The proper angle can also be checked by standing on the bottom rung and reaching out for the rung in front of the firefighter. The firefighter should be able to grab the rung, while standing straight up, with arms extended straight out in front (Figure 8.109). Newer ladders are equipped with an inclination marking on the outside of the beam whose lines become perfectly vertical and horizontal when the ladder is properly set (Figure 8.110).

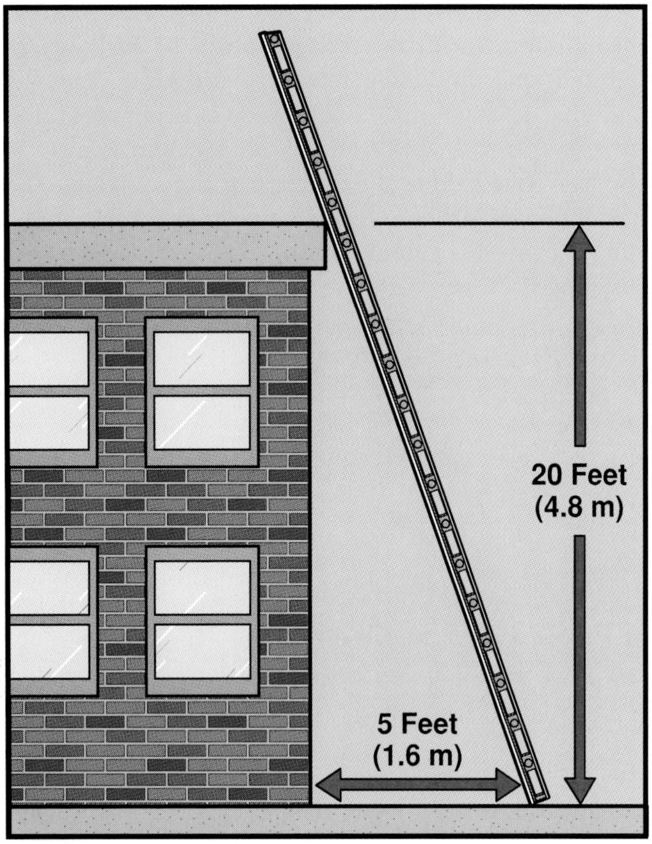

Figure 8.108 The base of the ladder should be one-fourth the distance away from the building as the height it is being raised.

Figure 8.109 To check the proper angle, the firefighter should be able to grab the rung with arms extended straight out in front.

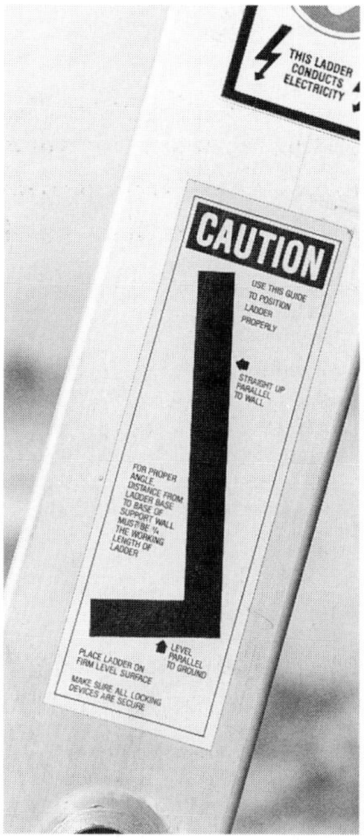

Figure 8.110 Newer ladders are equipped with an inclination marking on the outside of the beam.

GENERAL PROCEDURES FOR RAISING AND CLIMBING LADDERS

A well-positioned ladder becomes a means by which important fire fighting operations can be performed. If speed and accuracy are to be developed, teamwork, smoothness, and rhythm are necessary when raising and lowering fire department ladders. However, before learning the technique of raising ladders, firefighters should be aware of certain general procedures that affect the raising of ladders.

Transition From Carry To Raise

With the exception of pole ladders, it is not necessary to place the ladder flat on the ground before raising it; only the butt end needs to be placed on the ground. The transition from carry to raise should be a smooth, continuous series of motions.

Electrical Hazards

A major concern when raising ladders is contact with live electrical wires or equipment either by the ladder or by the person climbing the ladder. The danger of metal ladders in this respect has been previously stressed; however, many firefighters do not realize that wet wooden or fiberglass ladders present the same hazard. Care must be taken before beginning a raise to be sure that this hazard is avoided.

Regardless of the material from which the ladder is made, firefighters risk electrocution if they contact energized wires or equipment while climbing the ladder. This must be considered when placing the ladder.

Position Of The Fly Section On Extension Ladders

The question of whether the fly on an extension ladder should be in (next to the building) or out (away from the building) must be settled before starting the discussion on individual raises. This has been a matter of controversy for many years.

The position of the fly depends on the manufacturer's recommendation for that specific ladder. However, presently all North American manufacturers of metal and fiberglass ground ladders require that their ladders be used in the *fly out* (fly away from the building) position (Figure 8.111). This requirement is based on strength tests conducted on ground ladders by the National Bureau of Standards (now called the National Institute for Science and Technology [NIST]) that showed metal and fiberglass extension ladders were stronger with the fly out.

The exception to the fly out rule is when the wooden ladder is used. Historically, most wooden ground ladders were designed to be stronger in the *fly in* (fly toward the building) position (Figure 8.112). This is true of all wooden ground ladders being manufactured today.

Figure 8.111 Metal and fiberglass extension ladders should be positioned with the fly out.

Figure 8.112 Wooden extension ladders should be positioned with the fly in toward the building. *Courtesy of ALACO Ladder Co.*

Fire departments should consult the manufacturer of their ladders for specific information. Failure to use the ladder in the recommended position limits the manufacturer's liability and increases the user's liability should the ladder fail.

NOTE: Various types of ladders are shown in this manual. Each will be positioned according to the manufacturer's recommendations.

Pivoting Ladders

It may be necessary to pivot a ladder when it has been raised to the vertical position because of a beam raise, a flat raise parallel to a building, or the need to reposition the ladder so that the fly section is facing out. Whenever possible, pivoting should be done before extending the ladder. The procedure for pivoting a ladder is as follows:

Step 1: Two firefighters face each other through the ladder and grasp the beams with both hands. Each firefighter places his or her foot against the beam that the ladder will pivot on (Figure 8.113). (**NOTE:** Another method is for one firefighter to place a foot on the bottom rung.)

Step 2: The firefighters tilt the opposite beam of the ladder until it clears the ground (Figure 8.114).

Figure 8.113 Step 1: Two firefighters face each other through the ladder and grasp the beams with both hands. Each firefighter places his or her foot against the beam on which the ladder will pivot.

Figure 8.114 Step 2: The firefighters tilt the opposite beam of the ladder until it clears the ground.

Step 4: Grasp the halyard and extend the fly section with a hand-over-hand motion (Figure 8.125). When the tip is at the desired elevation, make sure that the ladder locks are in place.

Step 5: To lower the ladder, place at least one foot against a butt spur or on the bottom rung while grasping the beams, and lower the ladder gently into the building (Figure 8.126).

Step 6: In order to position the ladder so that the fly section is facing out, the ladder must be rolled over (Figure 8.127).

Figure 8.125 Step 4: Grasp the halyard and extend the fly section with a hand-over-hand motion.

Figure 8.126 Step 5: To lower the ladder, place at least one foot against a butt spur or on the bottom rung while grasping the beams.

Figure 8.127 Step 6: To position the ladder so that the fly section is facing out, the ladder must be rolled over.

The major difference in using the one-firefighter raise from the low-shoulder carry is the placement of the butt. In this instance, a building is used to heel the ladder to prevent the ladder butt from slipping while the ladder is brought to the vertical position. When raising the ladder from the low-shoulder carry, the following procedure is used.

Step 1: Place the butt end of the ladder on the ground with the butt spurs against the wall of the building (Figure 8.128).

Step 2: With your free hand, grasp a rung in front of your shoulder while removing your opposite arm from between the rungs (Figure 8.129).

Step 3: Step beneath the ladder and grasp a convenient rung with your other hand (Figure 8.130).

CAUTION: The area overhead should be visually checked for obstructions before bringing the ladder to a vertical position. The terrain in front of the firefighter should also be visually checked before stepping forward.

Step 4: Advance hand-over-hand down the rungs toward the butt until the ladder is in a vertical position (Figure 8.131).

Step 5: To extend the ladder, pull the halyard until the ladder has been raised to the desired level and

Figure 8.128 Step 1: Place the butt end of the ladder on the ground with the butt spurs against the wall of the building.

Figure 8.129 Step 2: With your free hand, grasp a rung in front of your shoulder while removing your opposite arm from between the rungs.

the pawls are engaged (Figure 8.132). Care must be taken to pull straight down on the halyard so that the ladder is not pulled over.

Step 6: To position the ladder for climbing, push against an upper rung to keep the ladder against the building. Grasp a lower rung with your other hand and carefully move the ladder butt out from the building to the desired location (Figure 8.133). If necessary, roll the ladder to bring the fly to the out position.

Figure 8.130 Step 3: Step beneath the ladder and grasp a convenient rung with your other hand.

Figure 8.131 Step 4: Advance hand-over-hand down the rungs toward the butt until the ladder is in a vertical position.

Figure 8.132 Step 5: To extend the ladder, pull the halyard until the ladder has been raised to the desired level and the pawls are engaged.

Figure 8.133 Step 6: To position the ladder for climbing, push against an upper rung to keep the ladder against the building.

Two-Firefighter Raises

Space permitting, it makes little difference if a ladder is raised parallel with or perpendicular to a building. If raised parallel with the building, the ladder can always be pivoted after it is in the vertical position. Whenever two or more firefighters are involved in raising a ladder, the firefighter at the butt end, called the heeler, is responsible for placing it at the desired distance from the building and determining whether the ladder will be raised parallel with or perpendicular to the building. There are two basic ways for two firefighters to raise a ladder: the flat raise and the beam raise. The following are step-by-step procedures for doing each:

FLAT RAISE

Step 1: When the desired location for the raise has been reached, the heeler places the ladder butt end on the ground while the firefighter at the tip rests the ladder beam on a shoulder (Figure 8.134).

Step 2: The heeler heels the ladder by standing on the bottom rung, crouches down to grasp a convenient rung or the beams with both hands, and leans back. The firefighter at the tip steps beneath the ladder and grasps a convenient rung with both hands (Figure 8.135 on the next page).

CAUTION: Visually check the area overhead for obstructions before bringing the ladder to a vertical position. Before stepping forward, visually check the terrain.

Figure 8.134 Step 1: The heeler places the ladder butt end on the ground while the firefighter at the tip rests the ladder beam on a shoulder.

Step 3: The firefighter at the tip advances hand-over-hand down the rungs toward the butt end until the ladder is in a vertical position (Figure 8.136). As the ladder comes to a vertical position, the heeler grasps successively higher rungs or higher on the beams until he or she is standing upright.

Figure 8.135 Step 2: The heeler heels the ladder by standing on the bottom rung, crouches down to grasp a convenient rung or the beams with both hands, and leans back. The firefighter at the tip steps beneath the ladder and grasps a convenient rung with both hands.

Figure 8.136 Step 3: The firefighter at the tip advances hand-over-hand down the rungs toward the butt end until the ladder is in a vertical position.

Step 4: Both firefighters face each other and heel the ladder by placing their toes against the same beam. When raising extension ladders, pivot the ladder to position the fly away from the building (fly in for wooden ladders), if it is not already in that position. The firefighter on the halyard side of the ladder grasps the halyard and extends the fly section with a hand-over-hand motion (Figure 8.137). When the tip is at the desired elevation, make sure the ladder locks are in place.

Step 5: To lower the ladder, the firefighter on the outside of the ladder places one foot against a butt spur or on the bottom rung and grasps the beams. Both firefighters gently lower the ladder into the building (Figure 8.138). If the ladder has not yet been turned to position the fly in the out position, it can be done at this time.

Figure 8.137 Step 4: The firefighter on the halyard side of the ladder grasps the halyard and extends the fly section with a hand-over-hand motion.

Figure 8.138 Step 5: To lower the ladder, the firefighter on the outside of the ladder places one foot against a butt spur or on the bottom rung and grasps the beams.

BEAM RAISE

Step 1: When the desired location for the raise has been reached, the heeler places the ladder beam on the ground. The firefighter at the tip rests the beam on one shoulder while the heeler places one foot on the lower beam at the butt spur (Figure 8.139).

Step 2: The heeler then grasps the upper beam with hands apart and the other foot extended back to act as a counterbalance (Figure 8.140).

An alternate method of heeling the ladder is to stand parallel to the ladder at the butt. Place one foot against the butt spur and position the other forward toward the tip of the ladder (Figure 8.141).

Step 3: The firefighter at the tip advances hand-over-hand down the beam toward the butt until the ladder is in a vertical position (Figure 8.142).

CAUTION: Visually check the area overhead for obstructions before bringing the ladder to a vertical position. Before stepping forward, visually check the terrain.

Step 4: Pivot the ladder to position the fly away from the building (fly in for wooden ladders), if it is not already in that position. The firefighter on the halyard side of the ladder grasps the halyard and extends the fly section with a hand-over-hand motion (Figure 8.143). When the tip is at the desired elevation, make sure the ladder locks are in place.

Step 5: To lower the ladder, the firefighter on the outside of the ladder places one foot against a butt spur or on the bottom rung and grasps the rung or beams. Both firefighters gently lower the ladder into the building (Figure 8.144).

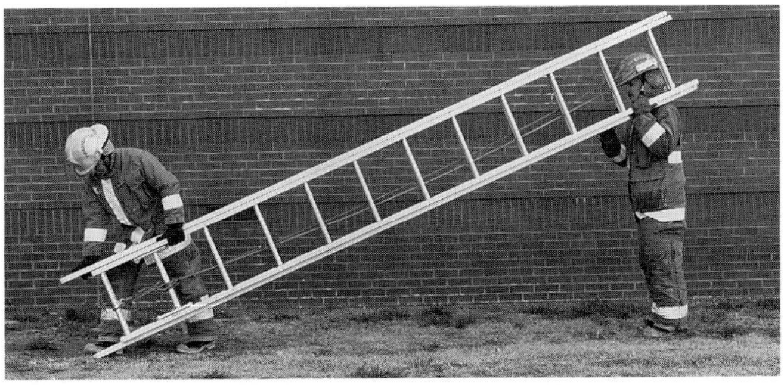

Figure 8.139 Step 1: The firefighter at the tip rests the beam on one shoulder while the heeler places one foot on the lower beam at the butt spur.

Figure 8.140 Step 2: The heeler then grasps the upper beam with hands apart and the other foot extended back to act as a counterbalance.

Figure 8.141 An alternate method of heeling the ladder is to stand parallel to the ladder at the butt. Place one foot against the butt spur and position the other forward toward the tip of the ladder.

Figure 8.142 Step 3: The firefighter at the tip advances hand-over-hand down the beam toward the butt until the ladder is in a vertical position.

Figure 8.143 Step 4: Pivot the ladder to position the fly away from the building. The firefighter on the halyard side of the ladder grasps the halyard and extends the fly section with a hand-over-hand motion.

Figure 8.144 Step 5: To lower the ladder, the firefighter on the outside of the ladder places one foot against a butt spur or on the bottom rung and grasps the rung or beams.

Three-Firefighter Raise

As the length of the ladder increases, the weight also increases. This requires more personnel for raising the larger extension ladders. Typically, ladders of 35 feet (11 m) or larger should be raised by at least three firefighters. The procedure for raising ladders with three firefighters is as follows:

Step 1: When the desired location for the raise has been reached, the heeler places the ladder butt end on the ground while the firefighters at the tip rest the ladder flat on their shoulders (Figure 8.145).

Step 2: The heeler heels the ladder by standing on the bottom rung, crouches down to grasp a convenient rung with both hands, and leans back (Figure 8.146).

CAUTION: Visually check the area overhead for obstructions before bringing the ladder to a vertical position. Before stepping forward, visually check the terrain.

Figure 8.145 Step 1: When the desired location for the raise has been reached, the heeler places the ladder butt end on the ground while the firefighters at the tip rest the ladder flat on their shoulders.

Figure 8.146 Step 2: The heeler heels the ladder by standing on the bottom rung, crouches down to grasp a convenient rung with both hands, and leans back.

Step 3: The firefighters at the tip advance in unison, with their outside hands on the beams and inside hands on the rungs, until the ladder is in a vertical position (Figure 8.147).

Step 4: If necessary, the firefighters pivot the ladder to position the fly section away from the building, using the procedure for three firefighters described earlier. If using a wooden ladder, the fly should be in toward the building.

Step 5: To extend the ladder, one firefighter grasps the halyard and extends the fly section with a hand-over-hand motion while the other two steady the ladder (Figure 8.148). When the tip is at the desired elevation, make sure that the ladder locks are in place.

Step 6: To lower the ladder, the firefighters on the outside of the ladder each places one foot against a butt spur or on the bottom rung and grasps the beam or a convenient rung. Either method is acceptable as long as both do it the same way. The third firefighter steadies the ladder from the inside position. All firefighters gently lower the ladder into the building (Figure 8.149).

To raise a ladder using the beam method with three firefighters, follow the same procedures for the two-firefighter raise. The only difference is that the third firefighter is positioned along the beam (Figure 8.150). Once the ladder has been raised to a vertical position, follow the procedures described for the flat raise.

Figure 8.147 Step 3: The firefighters at the tip advance in unison, with their outside hands on the beams and inside hands on the rungs, until the ladder is in a vertical position.

Figure 8.148 Step 5: To extend the ladder, one firefighter grasps the halyard and extends the fly section with a hand-over-hand motion while the other two steady the ladder.

Figure 8.149 Step 6: To lower the ladder, the firefighters on the outside of the ladder each place one foot against a butt spur or on the bottom rung and grasp the beams or a convenient rung. The third firefighter steadies the ladder from the inside position.

Four-Firefighter Raise

When personnel are available, four firefighters can be used to better handle the larger and heavier extension ladders. A flat raise is normally used, and the procedures for raising the ladder are similar to the

Figure 8.150 Using the beam method with three firefighters requires that the third firefighter be positioned along the beam.

three-firefighter raise except for the placement of personnel. A firefighter at the butt is responsible for placing the butt at the desired distance from the building and determining whether the ladder will be raised parallel with or perpendicular to the building. The procedure for raising ladders with four firefighters is as follows:

Step 1: When the desired location for the raise has been reached, the heelers place the ladder butt on the ground while the firefighters at the tip rest the ladder flat on their shoulders (Figure 8.151).

Step 2: The heelers heel the ladder by placing their inside feet on the bottom rung and their outside feet on the ground outside the beam. They each then grasp a convenient rung with the inside hand and the beam with the other hand and pull back (Figure 8.152).

CAUTION: Visually check the area overhead for obstructions before bringing the ladder to a vertical position. Before stepping forward, visually check the terrain.

Figure 8.151 Step 1: The heelers place the ladder butt on the ground while the firefighters at the tip rest the ladder flat on their shoulders.

Step 3: The firefighters at the tip advance in unison, with their outside hands on the beams and inside hands on the rungs, until the ladder is in a vertical position (Figure 8.153).

Step 4: If necessary, pivot the ladder to position the fly section away from the building. Wooden ladders should be positioned with the fly section in toward the building.

Step 5: To extend the ladder, either one or both of the firefighters on the halyard side of the ladder grasp the halyard and extend the fly section with a hand-over-hand motion (Figure 8.154). If both do this, they must coordinate their actions so as not to drop the fly section accidentally. When the tip is at the desired elevation, make sure that the ladder locks are in place.

Step 6: To lower the ladder, the firefighters on the outside of the ladder place their inside feet against the butt spur or bottom rung and grasp the beams (Figure 8.155). All firefighters gently lower the ladder into the building.

Figure 8.152 Step 2: The heelers heel the ladder by placing their inside feet on the bottom rung and their outside feet on the ground outside the beam. They each then grasp a convenient rung with the inside hand, the beam with the other hand, and pull back.

Figure 8.153 Step 3: The firefighters at the tip advance in unison, with their outside hands on the beams and inside hands on the rungs, until the ladder is in a vertical position.

Figure 8.154 Step 5: To extend the ladder, either one or both of the firefighters on the halyard side of the ladder grasp the halyard and extend the fly section with a hand-over-hand motion.

Figure 8.155 Step 6: To lower the ladder, the firefighters on the outside of the ladder place their inside feet against the butt spur or bottom rung, and grasp the beams.

Securing The Ladder

A ladder should be kept from slipping whenever firefighters are climbing, especially if the ladder is at a lower-than-desirable angle, if there are strong winds, or if the ground is icy or unstable. Added stability might also be necessary when operating hoselines from the ladder or when using the ladder for rescue. The ladder must first be secured by tying the halyard off to make sure the ladder does not retract unexpectedly. Then, the ladder must be prevented from falling away from the building. This can be accomplished by manually heeling the ladder or tying it off to a nearby object or anchor point.

TYING THE HALYARD

Once the extension ladder is resting against the building and before it is climbed, the excess halyard should be tied to the ladder to prevent the fly from slipping or to keep anyone from tripping over the rope. The same tie can be used for either a closed- or open-ended halyard. The procedure for tying the halyard is described as follows:

Step 1: Wrap the excess halyard around two convenient rungs, and pull it taut (Figure 8.156).

Step 2: Hold the halyard between your thumb and forefinger with your palm down (Figure 8.157).

Step 3: Turn your hand palm up and push the halyard underneath and back over the top of the rung (Figure 8.158).

Step 4: Grasp the halyard with your thumb and fingers, and pull it through the loop, making a clove hitch (Figure 8.159). Finish the tie by making the half hitch or overhand safety on top of the clove hitch (Figure 8.160).

Figure 8.156 Step 1: Wrap the excess halyard around two convenient rungs, and pull it taut.

Figure 8.157 Step 2: Hold the halyard between your thumb and forefinger with your palm down.

Figure 8.158 Step 3: Turn your hand palm up and push the halyard underneath and back over the top of the rung.

Figure 8.159 Step 4: Grasp the halyard with your thumb and fingers, and pull it through the loop, making a clove hitch.

Figure 8.160 Finish the tie by making the overhand safety on top of the clove hitch.

HEELING

There are several methods of properly heeling a ladder. One method is for the firefighter to stand underneath the ladder. The firefighter stands with feet about shoulder-width apart, grasps the ladder beams at about eye level, and pulls backward to press the ladder against the building (Figure 8.161). When using this method, the firefighter must wear head and eye protection and must not look up when there is someone climbing the ladder. The firefighter must be sure to grasp the beams and not the rungs.

Another method of heeling a ladder is for the firefighter to stand on the outside of the ladder and chock the butt end with his or her feet (Figures 8.162 a and b). With this method, either the firefighter's toes are placed against the butt spur or one foot is placed on the bottom rung. The hands grasp the beams, and the ladder is pressed against the building. The firefighter heeling the ladder must stay alert for descending firefighters.

Figure 8.161 One method for properly heeling a ladder.

Figure 8.162a Another method used to heel a ladder is for the firefighter to stand on the outside of the ladder and place his or her toes against the butt spur.

Figure 8.162b The firefighter is heeling the ladder with one foot placed on the bottom rung.

TYING IN

Whenever possible, a ladder should be tied securely to a fixed object. Tying a ladder in is simple, can be done quickly, and is strongly recommended to prevent the ladder from slipping or pulling away from the building. Tying in also frees personnel who would otherwise be holding the ladder in place. A rope hose tool or safety strap can be used between the ladder and a fixed object (Figures 8.163 a and b).

Figure 8.163a Secure a ladder to the building by tying it securely to a fixed object.

Figure 8.163b The ladder can also be secured by tying the lower section to a fixed object.

CLIMBING LADDERS

Ladder climbing should be done smoothly and with rhythm. The climber should ascend the ladder so that there is the least possible amount of bounce and sway. This smoothness is accomplished if the climber's knee is bent to ease the weight onto each rung. Balance on the ladder will come naturally if the ladder is properly spaced from the building, for the body will be perpendicular to the ground.

The climb may be started after the climbing angle has been checked and the ladder is properly secured. The eyes of the climber should be focused forward, with an occasional glance at the tip of the ladder. The climber's arms should be kept straight during the climb, which will keep the body away from the ladder and permit free knee movement during the climb (Figure 8.164). The hands should grasp the rungs with the palm down and the thumb beneath the rung. Some people find it natural to grasp every rung with alternate hands while climbing; others prefer to grasp alternate rungs. An option for hand placement when climbing ground ladders is to climb with both hands sliding up behind the beams to maintain constant contact, as when climbing and carrying equipment. A person should try both methods and select the one that is most natural or required by departmental SOP.

If the feet should slip, the arms and hands are in a position to stop the fall. All of the upward progress should be done by the leg muscles, not the arm muscles. The arms and hands should not reach upward during the climb because reaching upward will bring the body too close to the ladder.

Figure 8.164 The firefighter is using proper form for climbing a ladder.

Practice climbing should be done slowly to develop form rather than speed. Speed will be developed as the proper technique is mastered. Too much speed results in lack of body control, and quick movements cause the ladder to bounce and sway.

Many times during fire fighting, a firefighter is required to carry equipment up and down a ladder. This procedure interrupts the natural climb either by the added weight on the shoulder or the use of one hand to hold the item. If the item is to be carried in one hand, it is desirable to slide the free hand under the beam while making the climb (Figure 8.165). This method permits constant contact with the ladder. If the firefighter's hands are large enough, the hand with the tool may also be slid along the beam (Figure 8.166). Whenever possible, a handline rope should be used to hoist tools and equipment rather than carrying them up the ladder.

The technique used for climbing aerial ladders is basically the same as for ground ladders. The only difference is that the firefighter has the option of grasping the handrails as well as the rungs when climbing (Figure 8.167).

Figure 8.165 When carrying an object up a ladder, slide your free hand under the beam.

Figure 8.166 Slide tools, such as an axe, up the beam.

Figure 8.167 Firefighter using handrails while climbing an aerial ladder.

Working On A Ladder

Firefighters must sometimes work while standing on a ground ladder and must have both hands free. A Class I life safety harness (ladder belt) or a leg lock can be used to safely secure the firefighter to a ladder while working from it.

> **WARNING**
>
> **Never use a leg lock on an aerial ladder.**

When using a life safety harness, it should be strapped tightly around the waist. The hook may be moved to one side, out of the way, while the firefighter is climbing the ladder. After reaching the desired height, the firefighter returns the hook to the center and attaches it to a rung (Figure 8.168 on the next page). All life safety harnesses should meet the requirements set forth in NFPA 1983, *Standard on Life Safety Rope, Harness, and Hardware.*

The following steps should be used for taking a leg lock on a ground ladder:

Step 1: Climb to the desired height.

Step 2: Advance one rung higher (Figure 8.169 on the next page).

Step 3: Slide the leg on the opposite side from the working side over and behind the rung that you will lock onto (Figure 8.170 on the next page).

Step 4: Hook your foot either on the rung or on the beam (Figure 8.171 on the next page).

Figure 8.168 Firefighter's life safety harness hook attached to the appropriate rung.

Figure 8.169 Step 2: Advance one rung higher.

Figure 8.170 Step 3: Slide the leg on the opposite side from the working side over and behind the rung that you will lock onto.

Figure 8.171 Step 4: Hook your foot either on the rung or on the beam.

Figure 8.172 Step 6: Step down with the opposite leg.

Step 5: Rest on your thigh.

Step 6: Step down with the opposite leg (Figure 8.172).

Placing A Roof Ladder

There are a number of ways to get a roof ladder in place on a sloped roof, using either one or two firefighters. A roof ladder can be carried conveniently by one firefighter, using the shoulder method. The hooks should be closed while the ladder is being carried. The following procedure shows one method of placing a roof ladder in position.

Step 1: After carrying the roof ladder to the ladder that is to be ascended, set the roof ladder down and open the hooks (Figure 8.173).

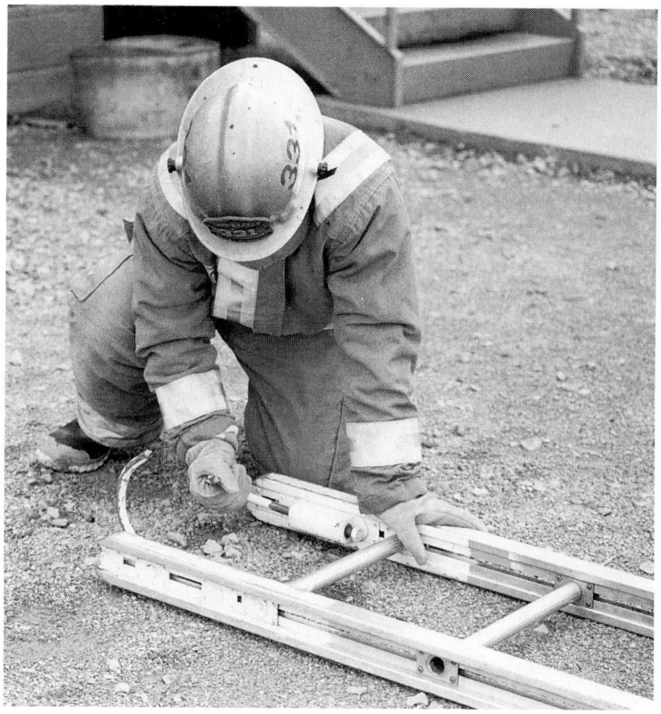

Figure 8.173 Open the hooks before raising the ladder.

Step 2: Tilt the roof ladder up so that it rests against the other ladder, with the hooks facing outward (Figure 8.174).

Step 3: Climb the main ladder until your shoulder is about two rungs above the midpoint of the roof ladder (Figure 8.175).

Step 4: Reach through the rungs of the roof ladder and hoist it onto your shoulder (Figure 8.176).

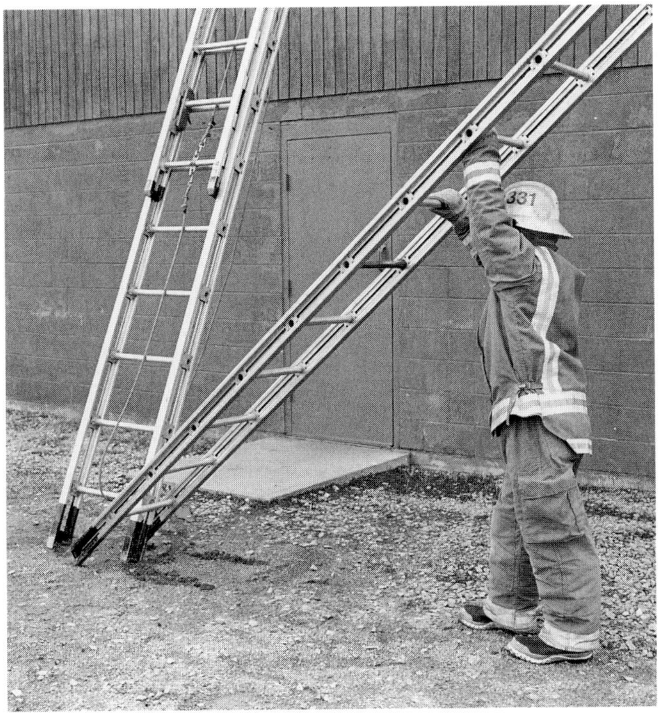

Figure 8.174 Raise the roof ladder and lean it against the extension ladder.

Step 5: Climb to the top of the ladder, and use a leg lock or life safety harness to lock into the ladder (Figure 8.177).

Step 6: Once locked in, take the roof ladder off your shoulder, and use a hand-over-hand method to push the roof ladder onto the roof (Figure 8.178). The ladder should be pushed onto the roof so that the hooks are in the down position.

Step 7: Push the roof ladder up the roof until the hooks go over the edge of the peak and catch solidly (Figure 8.179).

Step 8: To remove the roof ladder, reverse the process.

This evolution may also be accomplished with two firefighters if so desired. In this case, the second firefighter allows the climbing firefighter to start up the ladder and then passes the roof ladder up.

Figure 8.175 Shoulder the roof ladder while on the extension ladder.

Figure 8.176 Carry the roof ladder up to the roofline.

Figure 8.177 Lock into the ladder with a lifebelt or leg lock.

Figure 8.178 Raise the roof ladder hand over hand until it rests on the roof and then slide it up the roof.

Figure 8.179 Make sure the hooks are securely grabbing the peak of the roof.

ASSISTING A VICTIM DOWN A LADDER

When it is known in advance the ladder will be used for a window rescue, the ladder tip is raised only to the sill. This gives the victim easier access to the ladder. All other loads and activity should be removed from the ladder, which should be securely anchored at both the top and bottom if possible.

Use three or four firefighters on a ladder rescue. For ground ladder rescues, at least two firefighters should be in the building and one on the ladder. If the victim is conscious, the firefighters in the building should lower the victim feet first from the building to the ladder (Figure 8.180). The rescuer on the ladder supports the victim and descends the ladder. The rescuer descends first, keeping both arms around the victim under the armpits, with hands on the rungs in front of the victim, who is facing the ladder, for support in case the victim slips or passes out (Figure 8.181). When descending aerial ladders, it may not be possible to place the arms around the victim as the side rails may be in the way. The rescuer must reassure the victim constantly while descending because the victim will probably be nervous and panicky.

> **WARNING**
>
> **Firefighters must exercise care not to exceed the rated capacity of the ladder when performing rescues. Failure to observe the ladder load capacities may result in dramatic failure of the ladder.**

Figure 8.180 Assist a conscious victim from the building onto the ladder.

Figure 8.181 The rescuer descends first, keeping both arms around the victim under the armpits.

There are several ways to rescue an unconscious victim. The first method is to hold on to the unconscious victim in the same way as a conscious victim, except that the body rests on the rescuer's supporting knee. The firefighter places the victim's feet outside the rails to prevent entanglement (Figure 8.182). Another way to prevent entanglement of the victim's limbs in the rungs is to turn the victim around so that he or she faces the rescuer (Figure 8.183). A third way is to have the victim facing the rescuer, but with the victim's knees over the shoulder of the rescuer (Figure 8.184). This method works best if the ladder is placed at a slightly steeper than normal angle. The victim's armpits are supported by the rescuer's forearms. The rescuer descends the ladder rung by rung, while sliding his or her hands down the beams. In order to increase the control over the victim, the rescuer can lean in toward the ladder to slow the process. Smaller-sized adults and children can be brought down a ladder by cradling them across the arms (Figure 8.185).

LADDER SAFETY

A firefighter's safety and well-being while on a ladder depend on commonsense precautions. Firefighters should check important items at every opportunity. Points to ensure a safe climb are:

- Check the ladder for the proper angle.
- Make sure that the ladder is secure at the top or the bottom (preferably both) before climbing.
- Check the locks to be sure that they are seated over the rungs (Figure 8.186).

Figure 8.182 An unconscious victim may be brought down facing the ladder on the rescuer's knee.

Figure 8.183 An alternative is to have the victim facing the rescuer.

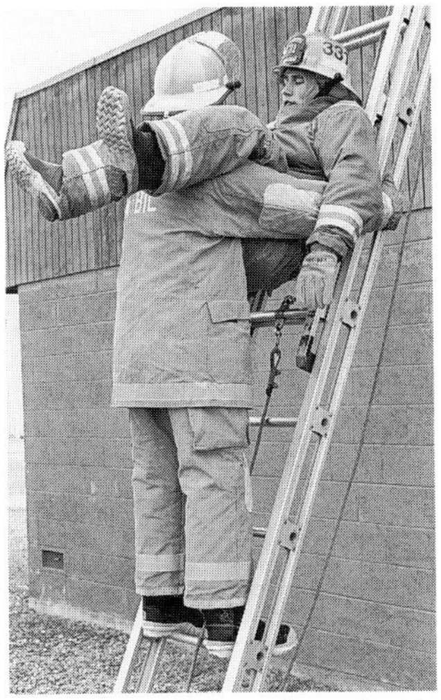

Figure 8.184 This cradle method is preferred by some firefighters.

Figure 8.185 Carry smaller-sized adults and children down a ladder by cradling them across your arms.

- Climb smoothly and rhythmically.
- Always tie in to ground ladders with a leg lock or ladder belt when working from the ladder. Use only a ladder belt and not a leg lock when working from an aerial ladder, tower ladder, or articulating platform.
- Do not overload the ladder.

Figure 8.186 Make sure that the ladder locks are properly set.

- Always wear protective gear, including gloves, when working with ladders.
- Choose the proper ladder for the job.
- Use the proper number of firefighters for each raise.
- Use leg muscles, not back or arm muscles, when lifting ladders below the waist.
- Make sure that ladders are not raised into electrical wires.
- Inspect ladders for damage and wear after each use.

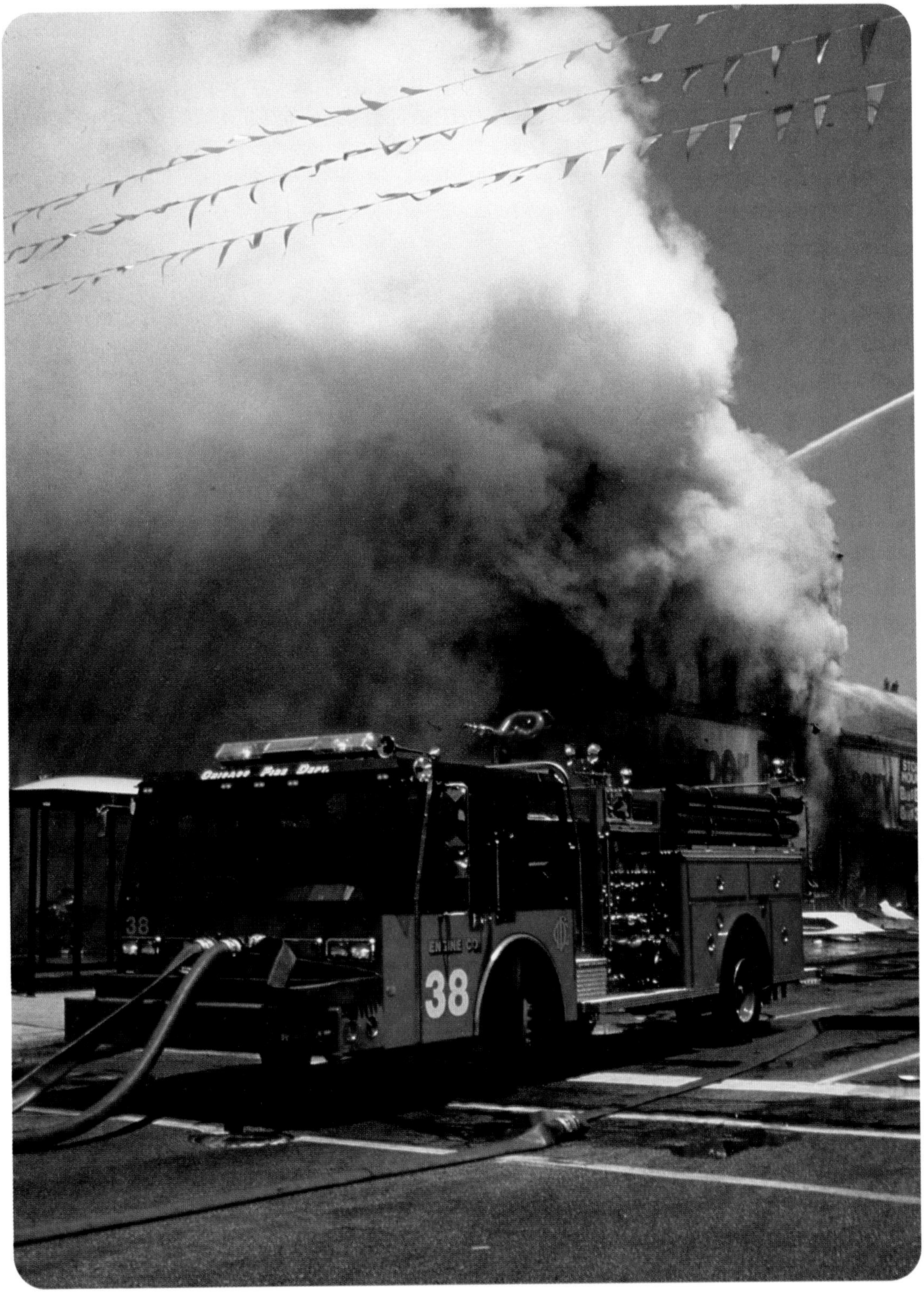

9

Water Supply

This chapter provides information that addresses performance objectives described in NFPA 1001, *Standard for Fire Fighter Professional Qualifications* (1992), particularly those referenced in the following sections:

3-19 Water Supplies

3-19.1

3-19.2

3-19.3

3-19.4

3-19.5

3-19.6

4-19 Water Supplies

4-19.1

4-19.2

4-19.3

4-19.4

4-19.5

4-19.6

4-19.7

4-19.8

4-19.9

4-19.10

4-19.11

Chapter 9
Water Supply

Technology keeps advancing new methods and materials for extinguishing fires. However, water still remains the primary extinguishing agent because of its universal abundance and ability to absorb heat. Two primary advantages of water are that it can be conveyed long distances and that it can be easily stored. These are the fundamental principles of a water supply system. Because water remains the primary extinguishing agent used by firefighters, it is important that they have a good working knowledge of water supply systems.

PRINCIPLES OF MUNICIPAL WATER SUPPLY SYSTEMS

Public and/or private water systems provide the methods for supplying water to the more populated areas. As rural areas increase in population, rural communities seek to improve efficient water distribution systems from a reliable source.

The water department may be a separate city-operated utility or a regional or private water authority. Its principal function is to provide water safe for human use. Water department officials should be considered the experts in water supply problems. The fire department must work with the water department in planning fire protection. Water department officials should realize that fire departments are vitally concerned with water supply and should work with them on supply needs, location, and types of fire hydrants.

The intricate working parts of a water system are many and varied. Basically, the system can be described by the following fundamental components (Figure 9.1):

- The source of supply
- The mechanical or other means of moving water
- The processing or treatment facilities
- The distribution system, including storage

The following sections explore these fundamental components in more detail.

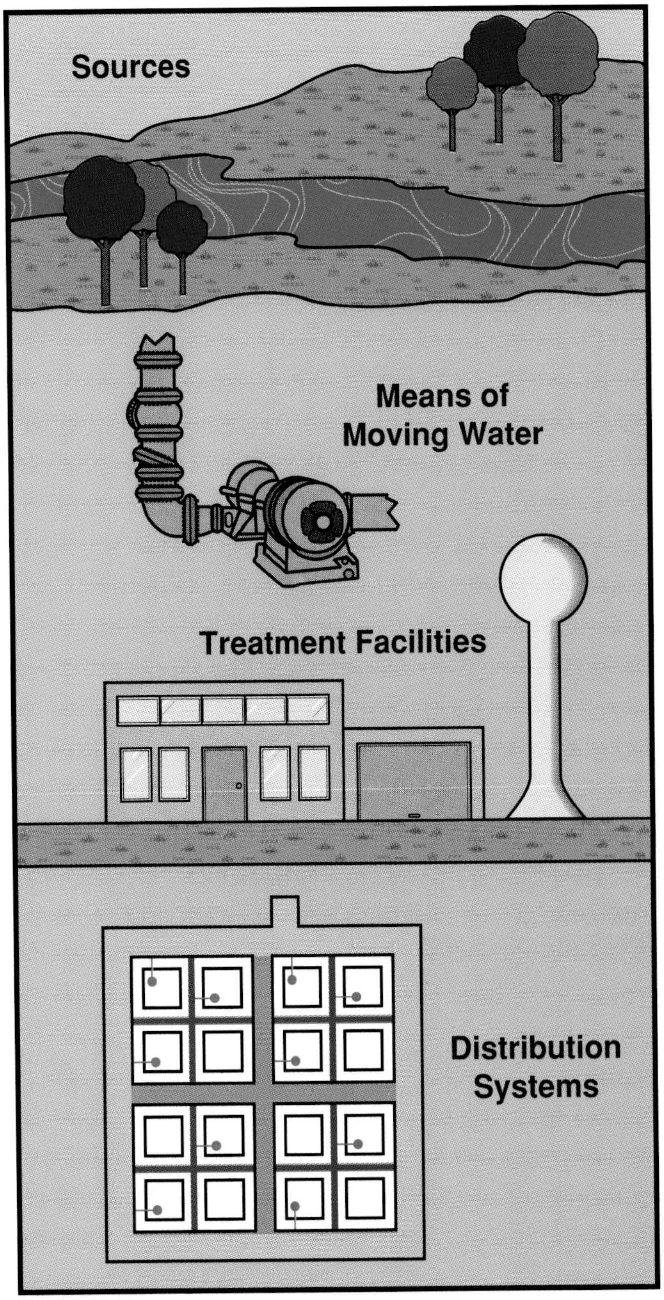

Figure 9.1 There are four components of any municipal water supply system.

Sources Of Supply And Means Of Movement

The primary water supply can be obtained from either surface water or groundwater. Although most water systems are supplied from only one source, there are instances where both sources are used. Two examples of surface water supply are river supply and lake supply. Groundwater supply can be water wells or water-producing springs (Figures 9.2 a and b).

Figure 9.2a Open ground reservoirs are generally located at the water treatment plant and may be used as a water source for fire fighting operations.

Figure 9.2b Ground level storage tanks can supply large amounts of water for fire fighting operations.

The amount of water that a community may need can be determined by an engineering estimate. This estimate is the total amount of water supply needed to furnish domestic and industrial use and for fire fighting use. In cities, the domestic/industrial requirement will far exceed that required for fire protection. In small towns, the requirements for fire protection exceed other requirements.

There are three methods of moving water in a system:

- Direct pumping system
- Gravity system
- Combination system

DIRECT PUMPING SYSTEM

Direct pumping systems use one or more pumps that take the water from the primary source and discharge it through the filtration and treatment processes (Figure 9.3). From here, a series of pumps force the water into the distribution system. If purification of the water is not needed, the water can be pumped directly into the distribution system. Failures in supply lines and pumps can usually be overcome by duplicating these units and providing a secondary power source.

Figure 9.3 Municipal water supply pumps.

GRAVITY SYSTEM

This is a primary water source located at a higher elevation than the distribution system. The gravity flow from the higher elevation provides the pressure (Figure 9.4). This is usually only sufficient when the primary water source is located at least several hundred feet higher than the highest point in the water distribution system. The most common scenarios would be a mountain reservoir that supplies water to a city below or a system of elevated tanks.

COMBINATION SYSTEM

Most communities use the combination system. This system uses a combination of the direct pumping and gravity systems. In most cases, the gravity flow is supplied by elevated storage tanks (Figure 9.5). These tanks serve as emergency storage and can provide adequate pressure through the use of gravity. When the system pressure is high during periods of low consumption, automatic valves open and allow the elevated storage tanks to fill. When the pressure drops during periods of

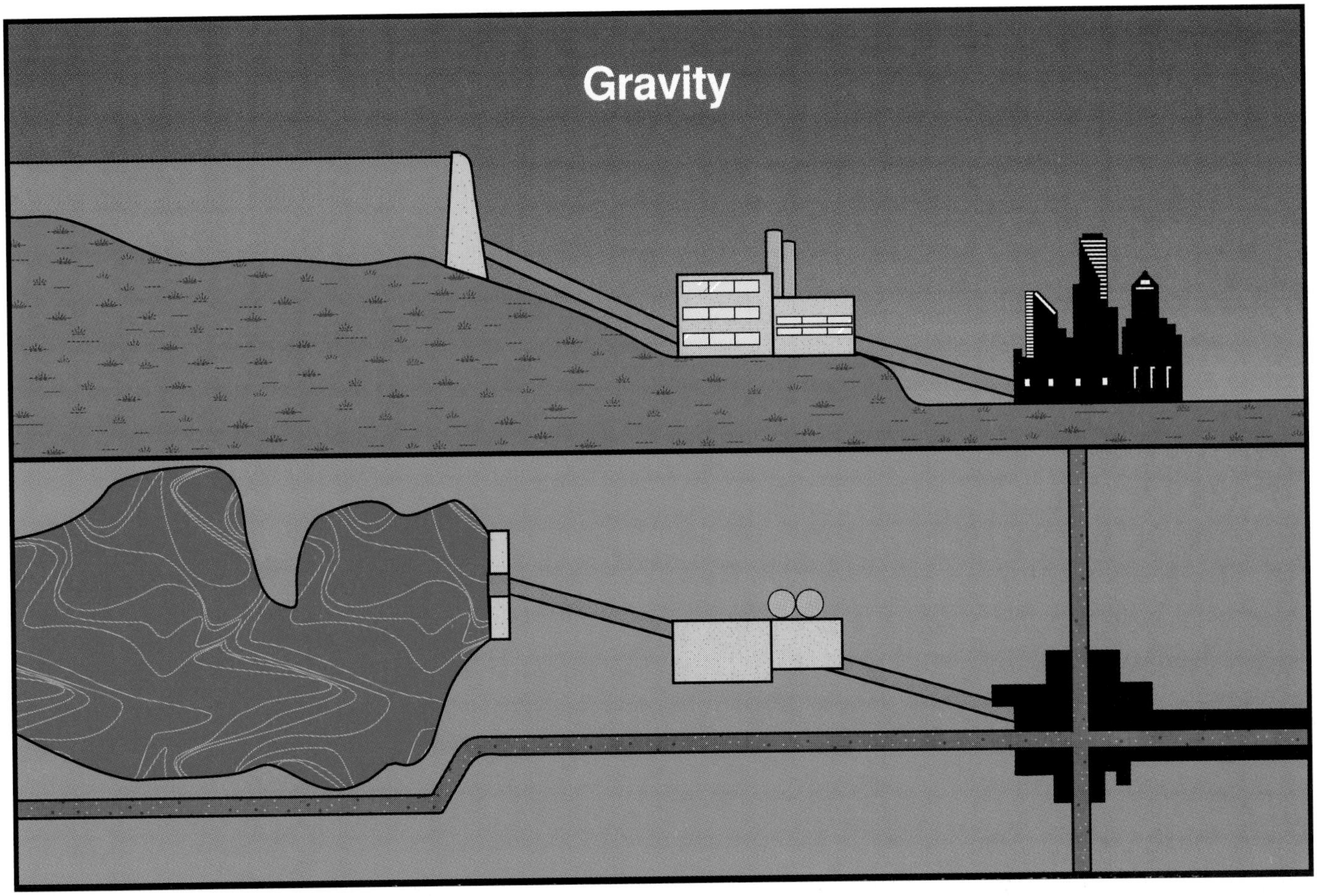

Figure 9.4 Gravity systems rely on Mother Nature to move water within the system.

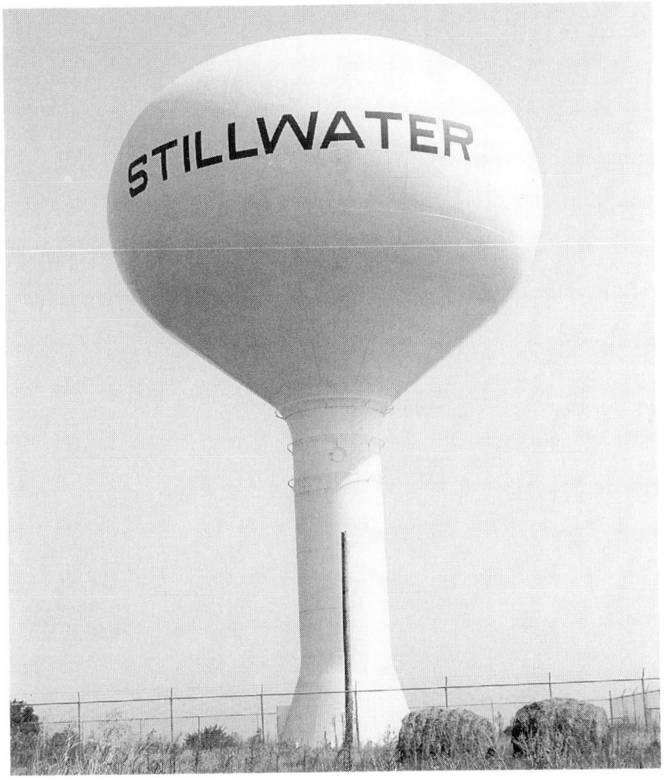

Figure 9.5 An elevated storage tank.

heavy consumption, the storage containers provide extra water by feeding back into the distribution system. Providing a good combination system involves reliable and duplicated equipment and proper-sized storage containers strategically located.

The storage of water in elevated reservoirs can also ensure water supply when the system becomes otherwise inoperative. According to the American Insurance Association (AIA), storage should be sufficient to provide domestic and industrial demands (based on any five-day period of maximum consumption) plus fire flow for 2 to 10 hours. Such storage should be sufficient to permit making most of the repairs, alterations, or additions to the system. Location of the storage and the capacity of the mains leading from this storage are also important factors.

Many industries provide their own private systems, such as elevated storage tanks, which are available to the fire department (Figure 9.6 on the next page). Water for fire protection may be available to some communities from storage systems, such as cisterns, which are considered a part of the distribution system. The fire department pumper removes this water by draft and provides pressure by the pump.

Figure 9.6 An industry's private water supply.

Processing Or Treatment Facilities

The treatment of water for the water supply system is a vital process. The water is treated to remove contaminants that may be detrimental to the health of those who use it. Water may be treated by coagulation, sedimentation, filtration, or the addition of chemicals, bacteria, or other organisms. In addition to removing things from the water, some things may be added, such as fluoride or oxygen.

The fire department's main concerns regarding treatment facilities are that a maintenance error, natural disaster, loss of power supply, or fire could knock out the pumping station(s) or severely hamper the purification process. This would drastically reduce the volume and pressure of water available for fire fighting operations. Another problem would be the inability of the treatment system to process water fast enough to meet the demand. In either case, fire officials must have a plan to deal with these potential shortfalls.

The Distribution System

The distribution system of the overall water supply system is that part which receives the water from the pumping station and delivers it throughout the area to be served (Figure 9.7). The ability of a water system to deliver an adequate quantity of water relies upon the carrying capacity of the system's network of pipes. When water flows through pipes, its movement causes friction which results in reduction of pressure. There is much less pressure loss in a water distribution system when fire hydrants are supplied from two or more directions. A fire hydrant that receives water from only one direction is known as a dead-end hydrant (Figure 9.8). When a fire hydrant receives water from two or more directions, it is said to have circulating feed (Figure 9.9). A distribution system that provides circulating feed from several mains constitutes a grid system (Figure 9.10). A grid system should consist of the following:

- *Primary feeders.* Large pipes (mains), with relatively widespread spacing, that convey large quantities of water to various points of the system for local distribution to the smaller mains.
- *Secondary feeders.* A network of intermediate-sized pipes that reinforce the grid within the various loops of the primary feeder system and aid the concentration of the required fire flow at any point.
- *Distributors.* A grid arrangement of smaller mains serving the individual fire hydrants and blocks of consumers.

To ensure sufficient water, two or more primary feeders should run by separate routes from the source of supply to the high-risk and industrial districts of the community. Similarly, secondary feeders should be arranged as far as possible in loops to give two directions of supply to any point. This practice increases the capacity of the supply at any given point and ensures that a break in a feeder main will not completely cut off the supply.

In residential areas, fire hydrant supply mains should be at least 6 inches (150 mm) in diameter. These should be closely gridironed by 8-inch (200 mm) cross-connecting mains at intervals of not more than 600 feet (180 m) (Figure 9.11).

Figure 9.7 A typical distribution system arrangement.

Figure 9.8 Dead-end hydrants receive water from only one direction.

Figure 9.9 A circulating-feed hydrant receives water from more than one direction.

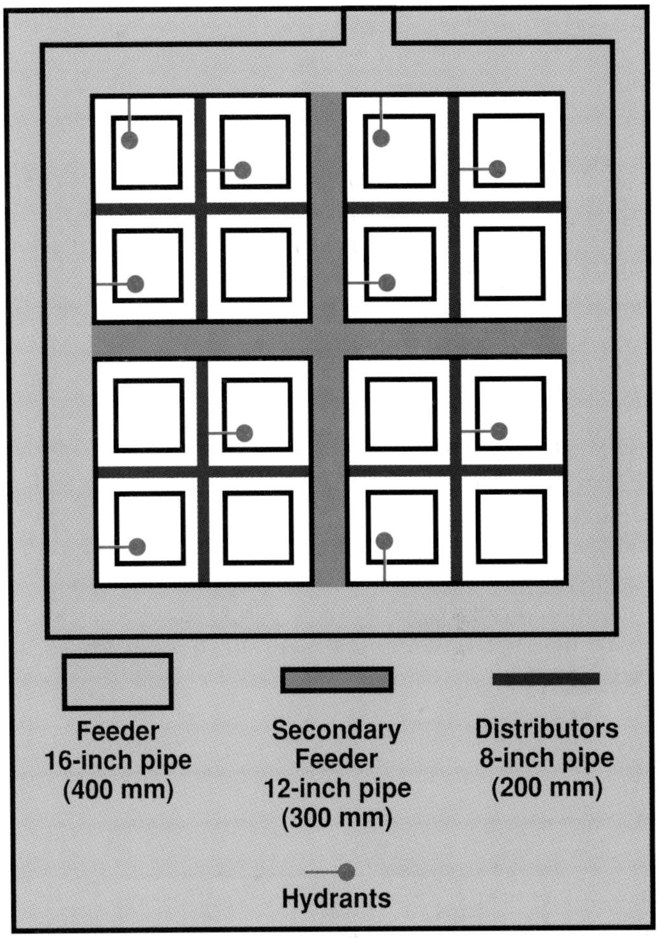

Figure 9.10 Grid system of water mains.

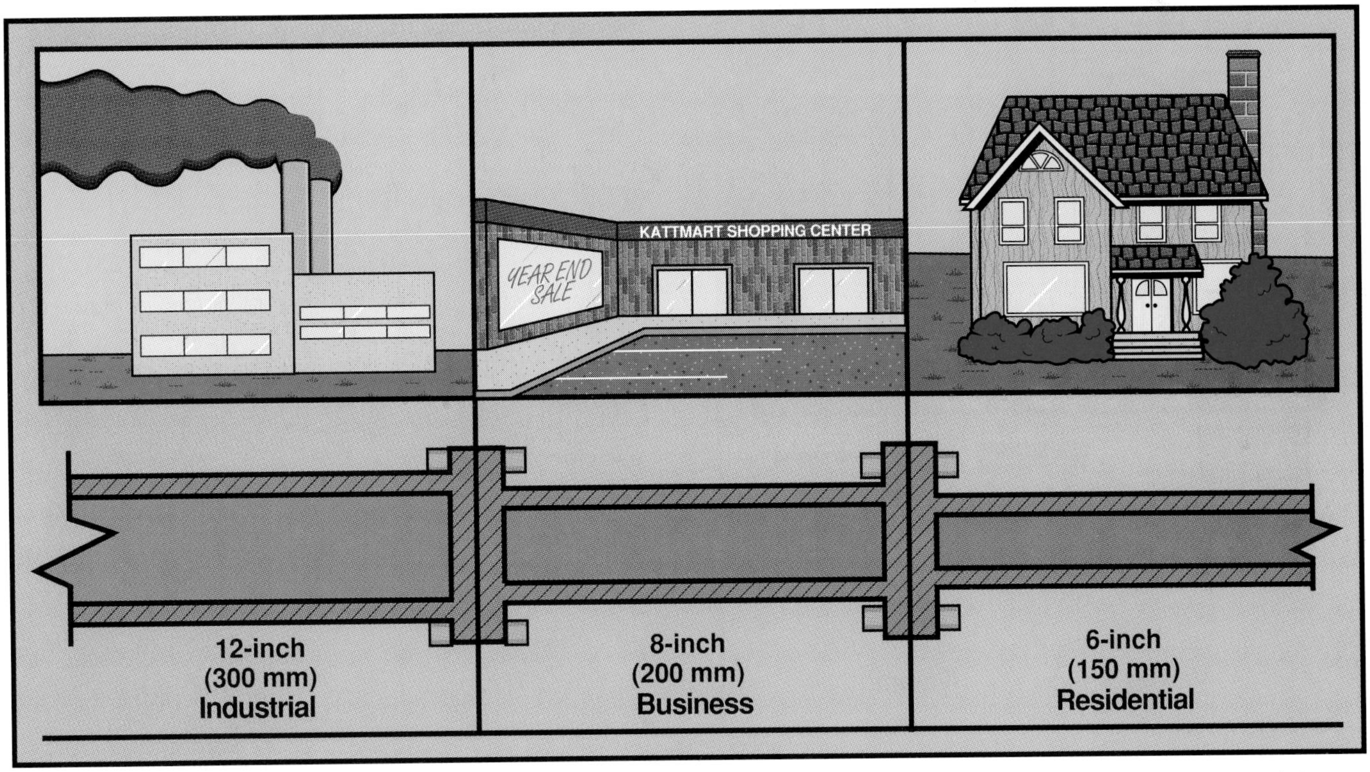

Figure 9.11 The minimum recommended sizes of water mains for various occupancies.

In the business and industrial districts, the minimum size is an 8-inch (200 mm) main with cross-connecting mains every 600 feet (180 m). Twelve-inch (300 mm) mains should be used on all principal streets and for long mains not cross-connected at frequent intervals.

WATER MAIN VALVES

The function of a valve in a water distribution system is to provide a means for controlling the flow of water through the distribution piping. Valves should be located at frequent intervals in the grid system so that only small districts will need to be cut off when necessary to stop the flow at specified points (Figure 9.12). Valves should be operated at least once a year to keep them in good condition. The actual need for valve operation in a water system rarely occurs, sometimes not for many years. Valve spacing should be such that a minimum length of pipe will be out of service at one time. The maximum lengths should be 500 feet (150 m) in high-value districts and 800 feet (240 m) in other areas, as recommended by Commercial Risk Services (formerly ISO) engineers.

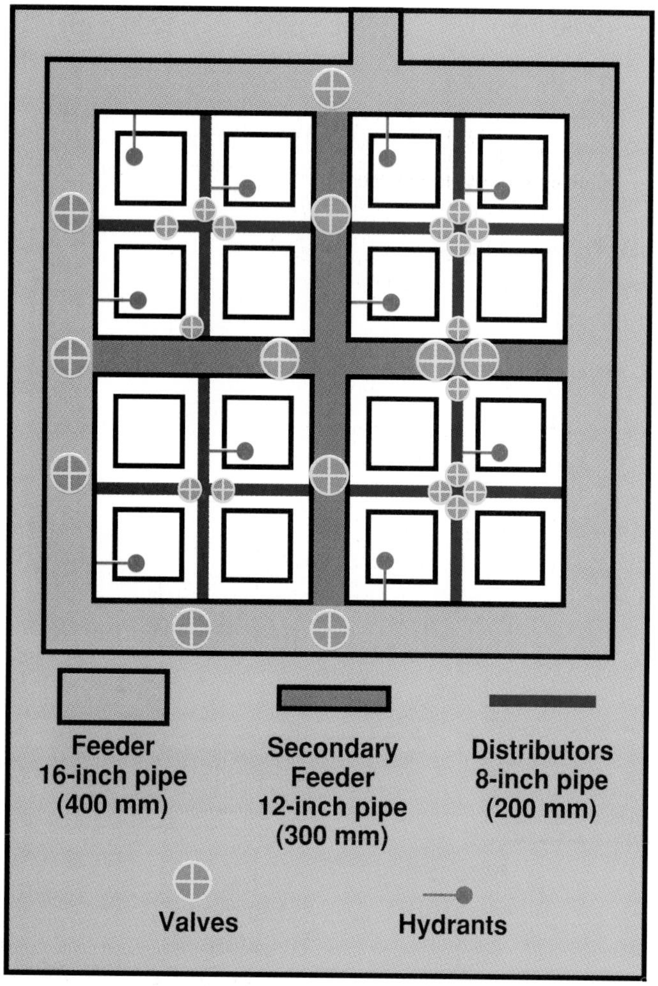

Figure 9.12 Valves located at frequent intervals in the system make it easier to isolate problems without large-scale disruptions of service.

One of the most important factors in a water supply system is the water department's ability to promptly operate the valves during an emergency or breakdown of equipment. A well-run water utility has records of the location of all valves. All valves should be inspected and operated at least once a year. If each fire department company is informed of the location of valves in the distribution system, their condition and accessibility can be noted during fire hydrant inspection and the water department informed of needed attention.

Valves for water systems are broadly divided into indicating and nonindicating types. An indicating valve visually shows whether the gate or valve seat is open, closed, or partially closed. Valves in private fire protection systems are usually of the indicating type. Two common indicator valves are the post indicator valve (PIV) and the outside screw and yoke valve (OS&Y). The post indicator valve is a hollow metal post that is attached to the valve housing. The valve stem is inside of this post with the words OPEN and SHUT printed on the stem so that the position of the valve is shown (Figure 9.13). The OS&Y valve has a yoke on the outside with a threaded stem that controls the gate's opening or closing (Figure 9.14). The threaded portion of the stem is out of the yoke when the valve is open and inside the yoke when the valve is closed.

Nonindicating valves in a water distribution system are normally buried or installed in manholes. If a buried valve is properly installed, the valve can be operated aboveground through a valve box (Figure 9.15). A special socket wrench on the end of a reach rod is available to operate the valve (Figure 9.16). Control valves in water distribution systems may be either gate valves or butterfly valves. Gate valves are usually the nonrising stem type, and as the valve nut is turned by the valve key (wrench), the gate either rises or lowers to control the water flow (Figure 9.17). Gate valves should be marked with a number indicating the number of turns necessary to

Figure 9.13 Post indicator valves clearly show whether the water supply is turned on or off.

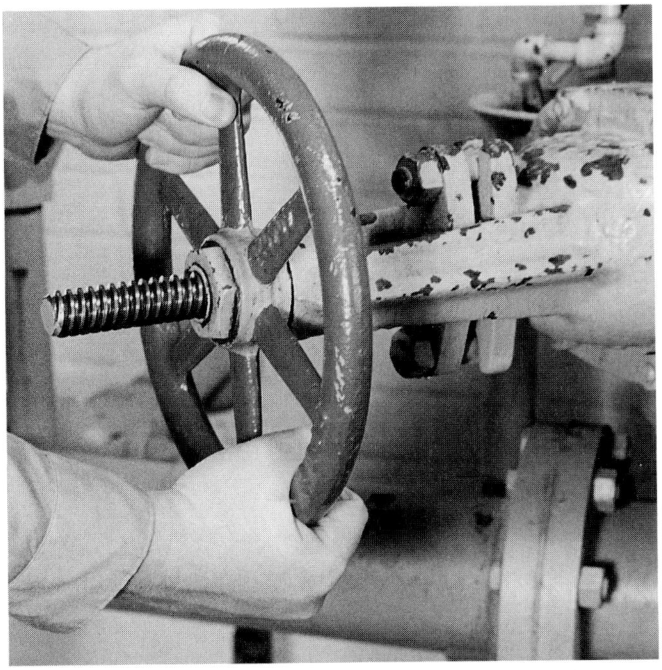

Figure 9.14 An O.S.& Y. valve.

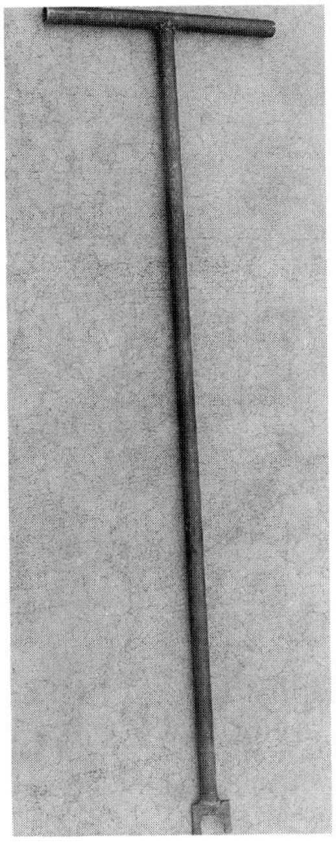

Figure 9.16 A water valve key.

Figure 9.17 A gate valve.

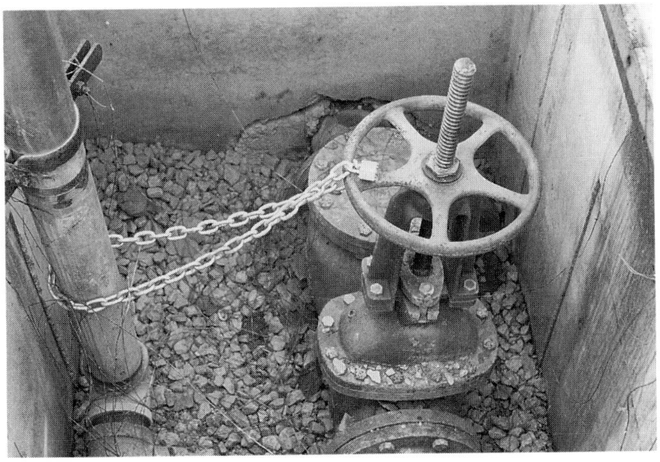

Figure 9.15 Some valve boxes are below grade.

completely close the valve. If the valve resists turning after fewer than the indicated number of turns, it usually means there is debris or another obstruction in the valve. Butterfly valves are tight closing, and they usually have rubber or a rubber composition seat that is bonded to the valve body. The valve disk rotates 90 degrees from the fully open to the tight-shut position (Figure 9.18). The nonindicating butterfly type also requires a valve key. Its principle of operation provides satisfactory water control after long periods of inactivity.

The advantages of proper valving in a distribution system are readily apparent. If valves are installed according to established standards, it will normally be necessary to close off only one or perhaps two fire hydrants from service while a single break is being repaired. The advantage of proper valving will, however, be reduced if all valves are not properly maintained and kept fully open. High friction loss is caused by valves that are only partially open. When valves are closed or partially closed, the condition may not be noticeable

Figure 9.18 A butterfly valve.

during ordinary domestic flows of water. As a result, the impairment will not be known until a fire occurs or at a time when a detailed inspection and a fire flow test is made. A fire department will experience difficulty in obtaining water in areas where there are closed or partially closed valves in the distribution system.

WATER PIPE

Water pipe, which is used underground, is generally made of either cast iron, ductile iron, asbestos cement, steel, plastic, or concrete. Whenever pipe is installed, it should be of the proper type for the soil conditions and pressures to which it will be subjected. When water mains are installed in unstable or corrosive soils or in difficult access areas, steel or reinforced concrete pipe may be used to give the strength needed. Some conditions that may require extra protection include beneath railroad tracks and highways, areas close to heavy industrial machinery, earthquake areas, or in rugged terrain.

The internal surface of the pipe, regardless of the material from which it is made, offers resistance to water flow. Some materials, however, have considerably less resistance to water flow than others. The engineering division of the water department should determine the type of pipe best suited for the conditions at hand.

The amount of water able to flow through a pipe and the amount of friction loss created can also be affected by other factors. Frequently, friction loss will be increased by incrustation of minerals on the interior surfaces of the pipe. Another problem can be sedimentation that drops out of the water. Both of these conditions result in a restriction of the pipe size, increased friction loss, and a proportionate reduction in the amount of water able to be drawn from the system.

KINDS OF PRESSURE

The term *pressure*, in connection with fluids, has a very broad meaning. Technically, pressure is defined as *force per unit of area*. In fire service terms, we most commonly think of pressure as the velocity of water in a conduit (either pipe or hose) of a certain size. Pressure, in the fire service sense, is measured in pounds per square inch (psi) (kilopascals [kPa]). It is essential to understand the following terms that identify the kinds of pressure with which the fire service is concerned.

- Static pressure
- Normal operating pressure
- Residual pressure
- Flow pressure

Static Pressure

If the water is not moving, the pressure exerted is static (Figure 9.19). Static pressure is stored potential energy that is available to force water through pipe, fittings, fire hose, and adapters. Because true static pressure is rarely found in a water supply system, there is a different use for the term in water supply system application. In these cases, static pressure is defined as the normal pressure existing on a system before the flow hydrant is opened.

Normal Operating Pressure

The flow of water through a distribution system fluctuates during the day and night. Normal operating pressure is that pressure which is found on a water distribution system during normal consumption demands.

Figure 9.19 Static pressure on hydrant.

Residual Pressure

The term residual pressure represents the pressure that is left in a distribution system at a specific location when a quantity of water is flowing (Figure 9.20). Residual pressure is that part of the total available pressure that is not used to overcome friction or gravity while forcing water through pipe, fittings, fire hose, and adapters.

Flow Pressure

The forward velocity of a water stream exerts a pressure that can be read on a pitot tube and gauge (Figure 9.21). Flow pressure is that forward velocity pressure at a discharge opening, either at a hydrant discharge or a nozzle discharge orifice, while water is flowing.

Figure 9.20 The residual pressure at a hydrant is measured while water is being discharged from another hydrant nearby.

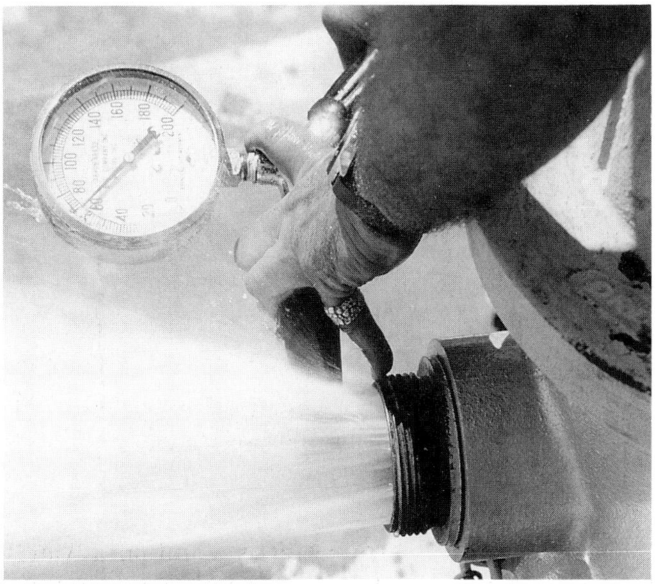

Figure 9.21 A pitot tube in use.

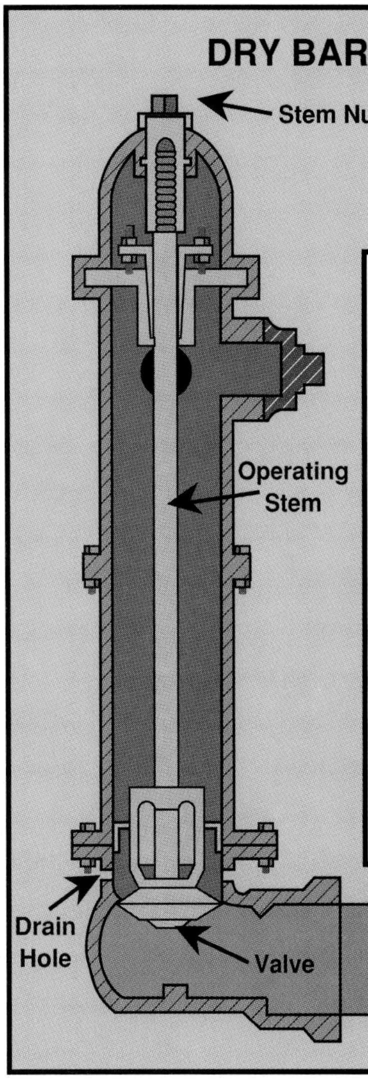

Figure 9.22a A dry barrel hydrant schematic.

Figure 9.22b A typical dry barrel hydrant.

FIRE HYDRANTS

The two main types of fire hydrants are dry barrel and wet barrel. The dry barrel hydrant, used in climates where freezing is expected, is usually classified as a compression, gate, or knuckle-joint type that opens either with pressure or against pressure (Figures 9.22 a and b). The actual valve holding back the water is well below ground, below the anticipated frost line for that geographic location. When the hydrant is closed, the barrel from the top of the hydrant down to the valve should be empty. Any water that remains in a closed, dry barrel hydrant will drain through a small valve at the bottom of the hydrant near the valve. This draining feature of a dry barrel hydrant is very important in determining hydrant usability. The drain on the dry barrel hydrant is open when the hydrant is not flowing water and shut when the hydrant is operating. If the hydrant is not completely open, the drain is partly open and the resulting flow out of the hydrant will contribute to ground erosion. This explains the old adage that a hydrant must be either completely open or completely shut — there is no halfway point.

The hydrant's ability to drain may be tested in the following manner. After allowing the hydrant to flow some water, shut it, and cap all discharges except one. Place your hand over the discharge (Figure 9.23 on the next page). At this time, you should feel a slight vacuum pulling your palm toward the discharge. If you do not feel this vacuum, notify the waterworks authority and have them inspect it.

Wet barrel hydrants may only be used in areas that do not have freezing weather. Wet barrel hydrants

Figure 9.23 Check to see if the hydrant is draining.

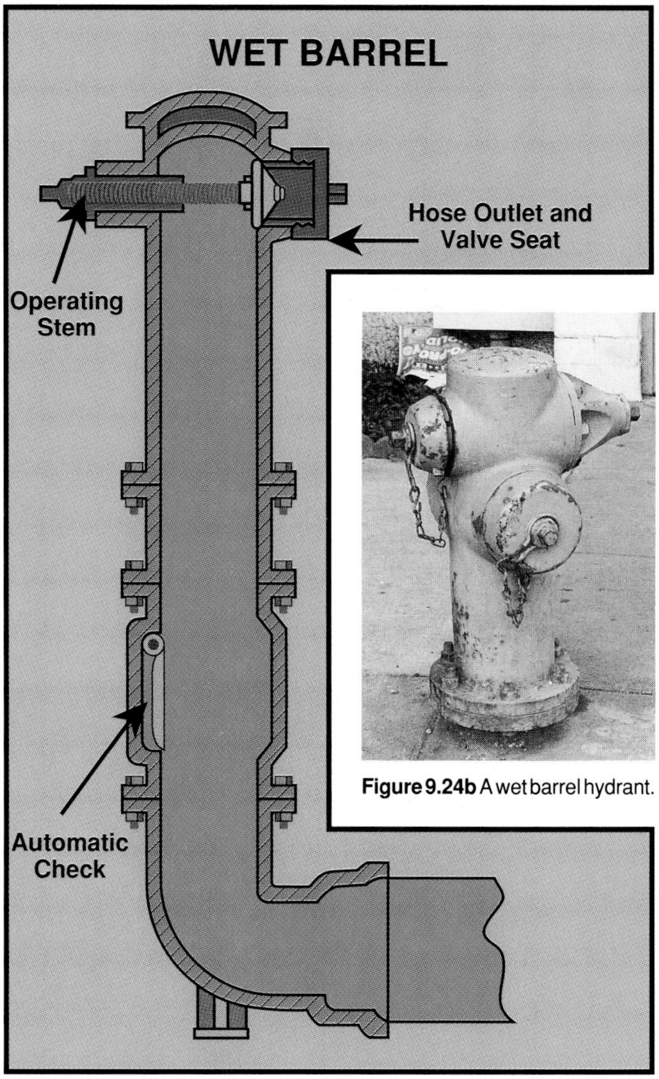

Figure 9.24b A wet barrel hydrant.

Figure 9.24a A wet barrel hydrant schematic.

usually have a compression-type valve at each outlet, or they may have only one valve located in the bonnet that controls the flow of water to all outlets (Figures 9.24 a and b). The entire hydrant is always filled with water to the valves near the discharges.

In general, all hydrant bonnets, barrels, and foot pieces are made of cast iron. The important working parts are usually made of bronze, but valve facings may be made of rubber, leather, or composition materials.

The flow of a hydrant will vary for several reasons. First, and most obviously, the proximity of feeder mains and the size of the mains to which the hydrant is connected will have a major impact on amount of flow. Sedimentation and deposits within the distribution system may increase the resistance to water flow. These problems may occur over a period of time; therefore, older water systems may experience a decline in the available flow. Firefighters can make better decisions affecting a fire attack if they at least know the relative available water flow of different hydrants in the vicinity. To aid them, a system of coloring hydrants to indicate a range of water flow was developed by NFPA. With this system, hydrants are classified as shown in Table 9.1. Local variation of coloring may be found, but the main intent of any color scheme is simplicity.

TABLE 9.1
HYDRANT COLOR CODES

Hydrant Class	*Color*	*Flow*
Class AA	Light Blue	1,500 gpm (5 678 L/min) or greater
Class A	Green	1,000-1,499 gpm (3 785 L/min-5 677 L/min)
Class B	Orange	500-999 gpm (1 893 L/min-3 784 L/min)
Class C	Red	less than 500 gpm (1 893 L/min)

Location Of Fire Hydrants

Although the installation of fire hydrants is usually performed by water department personnel, the location, spacing, and distribution of fire hydrants should be the responsibility of the fire chief or fire marshal. In general, fire hydrants should not be spaced more than 300 feet (90 m) apart in high-value districts. A basic rule to follow is to place one hydrant near each street intersection and to place intermediate hydrants where distances between intersections exceed 350 to 400 feet (105 m to 120 m). This basic rule represents a minimum requirement and should be regarded only as a guide for spacing hydrants. Other factors more pertinent to the particular locale include types of construction, types of occupancy, congestion, the sizes of water mains, fire flows, and pumping capacities.

Fire Hydrant Maintenance

In most cities, repair and maintenance of fire hydrants are the responsibility of the water department, because this department is in a better position to do this work than any other agency. However, in many cases, the fire department performs water supply testing and hydrant inspections. Therefore, firefighters should look for the following potential problems when checking fire hydrants:

- Have obstructions, such as sign posts, utility poles, or fences, been erected too near the hydrant?
- Do the outlets face the proper direction, and is there sufficient clearance from them to the ground?
- Has the hydrant been damaged by traffic (Figure 9.25)?
- Is the hydrant rusting or corroded?
- Is the operating stem easily turned?
- Are the hydrant caps stuck in place with paint?

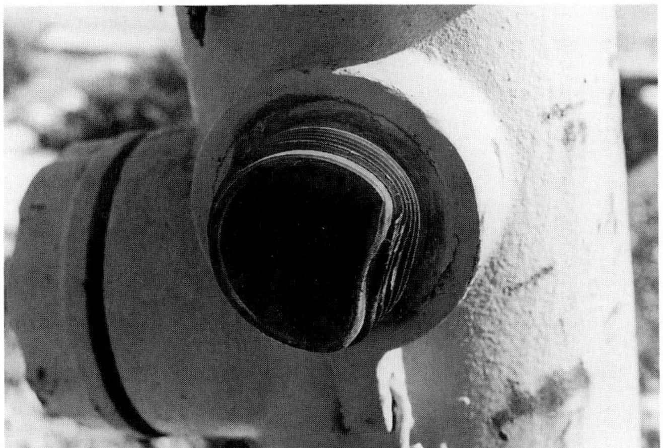

Figure 9.25 Check hydrants for signs of damage.

Using A Pitot Tube

Firefighters who assist in hydrant testing and inspections will be required to use a pitot tube to measure the flow pressure coming from a hydrant. There are two methods of holding the pitot tube in the proper position. The first is to grasp the pitot tube just behind the blade with the first two fingers and thumb of the left hand while the right hand holds the air chamber. The little finger of the left hand rests upon the hydrant outlet or nozzle to steady the instrument (Figure 9.26). Another method of holding the pitot tube is to have the fingers of the left hand split around the gauge outlet and the left side of the fist placed on the edge of the hydrant orifice or outlet (Figure 9.27). The blade can then be sliced into the stream in a counterclockwise direction. The right hand once again steadies the air chamber.

Figure 9.26 The little finger is used to help steady the pitot tube.

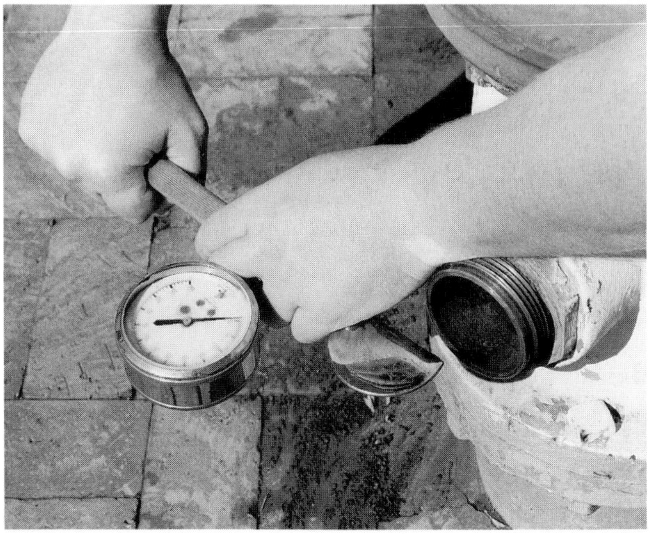

Figure 9.27 Steady the pitot tube by holding the left side of the fist against the discharge outlet. Then, slice the blade into the stream.

The procedure for using the pitot tube and gauge is as follows:

Step 1: Open the petcock on the pitot tube, make certain the air chamber is drained, and then close the petcock (Figure 9.28).

Step 2: Edge the blade into the stream, with the small opening or point being centered in the stream and held away from the orifice at a distance approximately half the diameter of the orifice (Figure 9.29). For a 2½-inch (65 mm) hydrant butt, this distance would be 1¼ inches (32 mm).

Step 3: The pitot tube should now be parallel to the outlet opening, with the air chamber kept above the horizontal plane passing through the center of the stream (Figure 9.30). This will increase the efficiency of the air chamber and help avoid needle fluctuations.

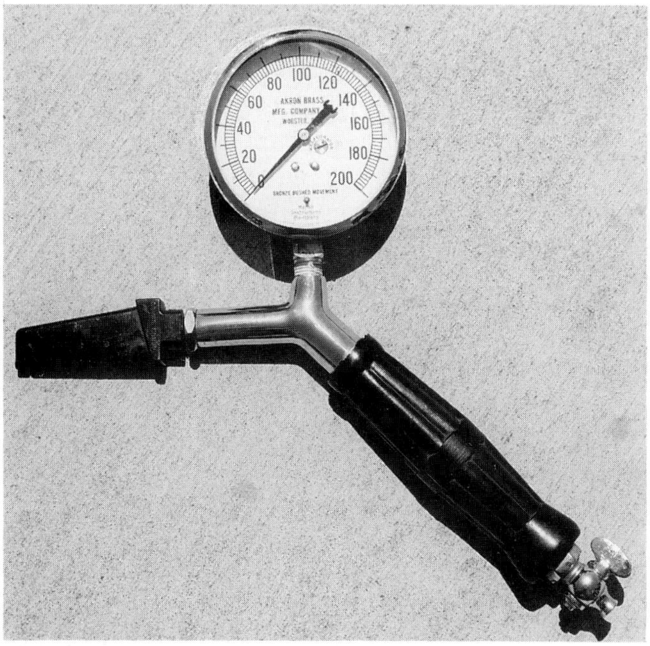

Figure 9.28 The petcock is located on the bottom of the pitot tube. When properly drained and ready, the needle should read zero.

Figure 9.29 Blade placed in stream.

Step 4: Take and record the velocity pressure reading from the gauge. If the needle is fluctuating, read and record the value located between the high and low extremes (Figure 9.31).

For more information on testing fire hydrants, see the IFSTA validated **Water Supplies for Fire Protection** or **Fire Inspection and Code Enforcement** manuals.

Figure 9.30 The air chamber should be kept above the horizontal plane.

Figure 9.31 In cases where a poor stream of water is discharging from the hydrant, a nozzle may be added to the discharge to boost velocity pressure.

ALTERNATIVE WATER SUPPLIES

Fire departments should not limit their study of water supplies to the piped public distribution system. Areas outside the public system should be studied for available water. Even areas with water systems should be surveyed for alternative supplies in the event that the water system fails or a fire occurs that requires more water than the system can supply. Water supply can be provided where an industry has its own private water system. With today's modern pumpers, water can be drawn from many natural sources such as the ocean, lakes, ponds, and rivers (Figure 9.32). Water can sometimes also be found in farm stock tanks and swimming pools. A good method of providing water for fire protection is to construct storage tanks at strategic locations.

Figure 9.32 A drafting operation is a good way to provide water for fire protection.

Almost any static source of water can be used if it is sufficient in quantity and is not contaminated to the point of creating a health hazard. Fire departments should make every attempt to identify, mark, and record alternative water supply sources in pre-incident planning. Consideration should be given to the effect that weather will have on the amount of water available and access to the water source.

RURAL WATER SUPPLY OPERATIONS

Rural water supply operations consist principally of tanker (tender) shuttles and relay pumping. For either type of operation to succeed, it requires pre-incident planning and practice. Adequate resources must be dispatched promptly, and an organizational structure is necessary for control and coordination. The following sections briefly highlight each of these operations.

The keys to efficient tanker shuttles are fast-fill and fast-dump times. Water supply officers should be positioned at both the dump and fill sites. As manpower is available, consideration should be given to assigning people to traffic control, hydrant operations, hookups, and tank venting. If possible, the tanker drivers should remain in their vehicles.

Water Shuttles

Water shuttling is the hauling of water from a supply source to the portable tanks from which water may be drawn to fight the fire (Figure 9.33). Water shuttles are recommended for distances greater than one-half mile or greater than the fire department's capability of laying supply hoselines.

There are three key elements to water shuttle operations:

- Attack apparatus at the fire (dump site)
- Fill apparatus at the fill site

Figure 9.33 Portable tanks are necessary for efficient shuttle operations. *Courtesy of Bob Esposito.*

- Mobile water supply apparatus (tankers/tenders to haul the water from the fill site to the dump site (Figures 9.34 a and b)

It is critical to have an adequate number of tankers/tenders for the needed fire flow.

The dump site is generally located very near the actual fire or incident. The dump site consists of one or more portable water tanks into which water-hauling apparatus dump their water before returning to the fill site. Apparatus attacking the fire may draft directly from the portable tanks, or other apparatus may draft from the tanks and supply the attack apparatus

Figure 9.34a A common, single axle elliptical tanker (tender) apparatus. *Courtesy of Bob Esposito.*

Figure 9.34b Some tankers (tenders) have large pumps and are equipped similar to a standard engine company. *Courtesy of Joel Woods, University of Maryland Fire & Rescue Institute.*

(Figure 9.35). Low-level intake devices, commercial or homemade, permit use of most of the water in the portable reservoir (Figure 9.36).

When large flows must be maintained, multiple portable tanks will be required. Capacities of portable tanks range from 1,000 gallons (4 000 L) upward. When multiple portable tanks are used, a jet siphon will maintain the water level in one tank for the pumper, while tankers dump into the others. Plain siphons or commercial tank connecting devices are also sometimes used for this purpose, although they are not generally as efficient as a jet siphon.

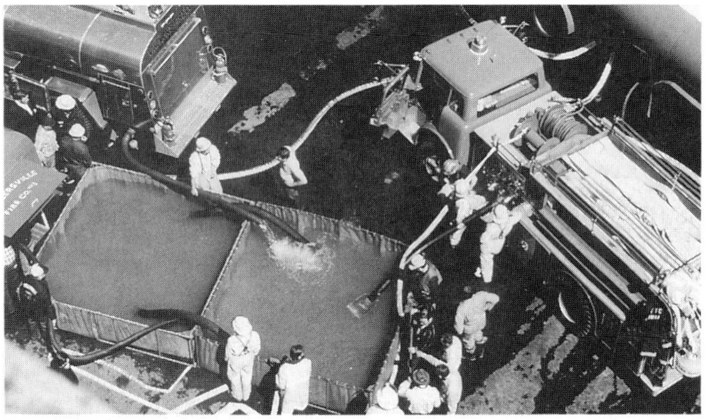

Figure 9.35 Drafting from a folding tank.

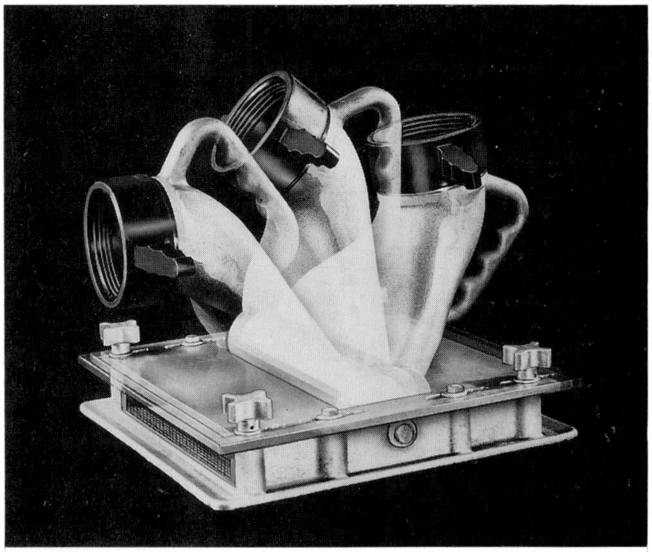

Figure 9.36 A low level strainer designed to be used in a portable tank. *Courtesy of Ziamatic Corporation.*

There are varying methods of construction for portable reservoirs. The most common is the collapsible or folding style, using a square metal frame and a synthetic or canvas duck liner (Figure 9.37). Another style is a round, synthetic tank with a floating collar that rises as the tank is filled, making it self-supporting. These portable reservoirs should be mounted for easy removal from the apparatus.

Figure 9.37 Folding portable tanks are put into service by simply pulling on each side of the tank. *Courtesy of Bob Esposito.*

Before opening a portable tank, a heavy tarp should be spread out on the ground to help protect the liner once water is dumped into it. Portable tanks should, if possible, be positioned in a location that allows easy access from multiple directions but does not inhibit access of other apparatus to the fire scene. If more than one tanker has arrived, empty one completely, and send it for another load before emptying the second one. This will better sequence the tankers at both the dump and fill sites.

There are four basic methods by which tankers unload water:

- Gravity dumping through large (10- or 12-inch [250 mm or 300 mm]) dumps (Figure 9.38)
- Jet dumps (Figure 9.39)
- Pumping off the water (Figure 9.40)
- A combination of these methods

NFPA 1903, *Standard for Mobile Water Supply Fire Apparatus*, requires apparatus to be capable of dump or fill rates of at least 1,000 gpm (4 000 L/min). This necessitates adequate tank venting and openings in tank baffles. Pumping the water off the truck will need to be done by a trained apparatus driver/operator. However, jet dumps or gravity dumps may be activated by a firefighter. This relieves the driver/operator from having to exit the cab and saves time in the overall process. Most dumps are activated by a lever near the outlet.

In order to fill tankers quickly, use the best fill site or hydrant available, large hoselines, multiple hoselines, and if necessary, a pumper for an adequate flow. In some situations multiple portable pumps may be necessary. Both fill sites and dump sites should be arranged so that a minimum of backing (or jockeying) of apparatus is required.

Figure 9.38 Dump valves allow for maximum unloading capacity. They may be located on any side of the apparatus but are most commonly found on the rear.

Figure 9.39 Jet dumps increase the flow rate of any dump valve.

Figure 9.40 Stream shapers are used to ensure that all water is directed toward the portable tank. They are most commonly used when water is being pumped off the apparatus.

Relay Pumping

Sometimes a water source is close enough to the fire scene that relay pumping can be used. Some departments use variations of a combination tanker shuttle and relay pumping to minimize congestion at the fire scene (Figure 9.41). There are two important factors to be considered when contemplating the establishment of a relay operation:

- The water supply must be capable of maintaining the desired volume of water required for the duration of the incident.
- The relay must be able to be established quickly enough to be worthwhile.

Figure 9.41 Some tankers have the capability of pumping their load.

The number of pumpers needed and the distance between pumpers is determined by several factors such as volume of water needed, distance, hose size, amount of hose available, and pumper capacities. The apparatus with the greatest pumping capacity should be located at the water source. Large diameter hose or multiple hoselines will increase the distance and volume that a relay can supply because of reduced friction loss. A water supply officer must be appointed to determine the distance between pumpers and to coordinate water supplies.

In order to determine the distance between pumpers, a quick calculation must be made by the water supply officer. It is important to know the friction loss at particular flows for the size hose being used. This can be made into a chart for quick reference on the pumper. The best way to prepare for relay operations is to plan them in advance and to try them in practical training exercises.

10

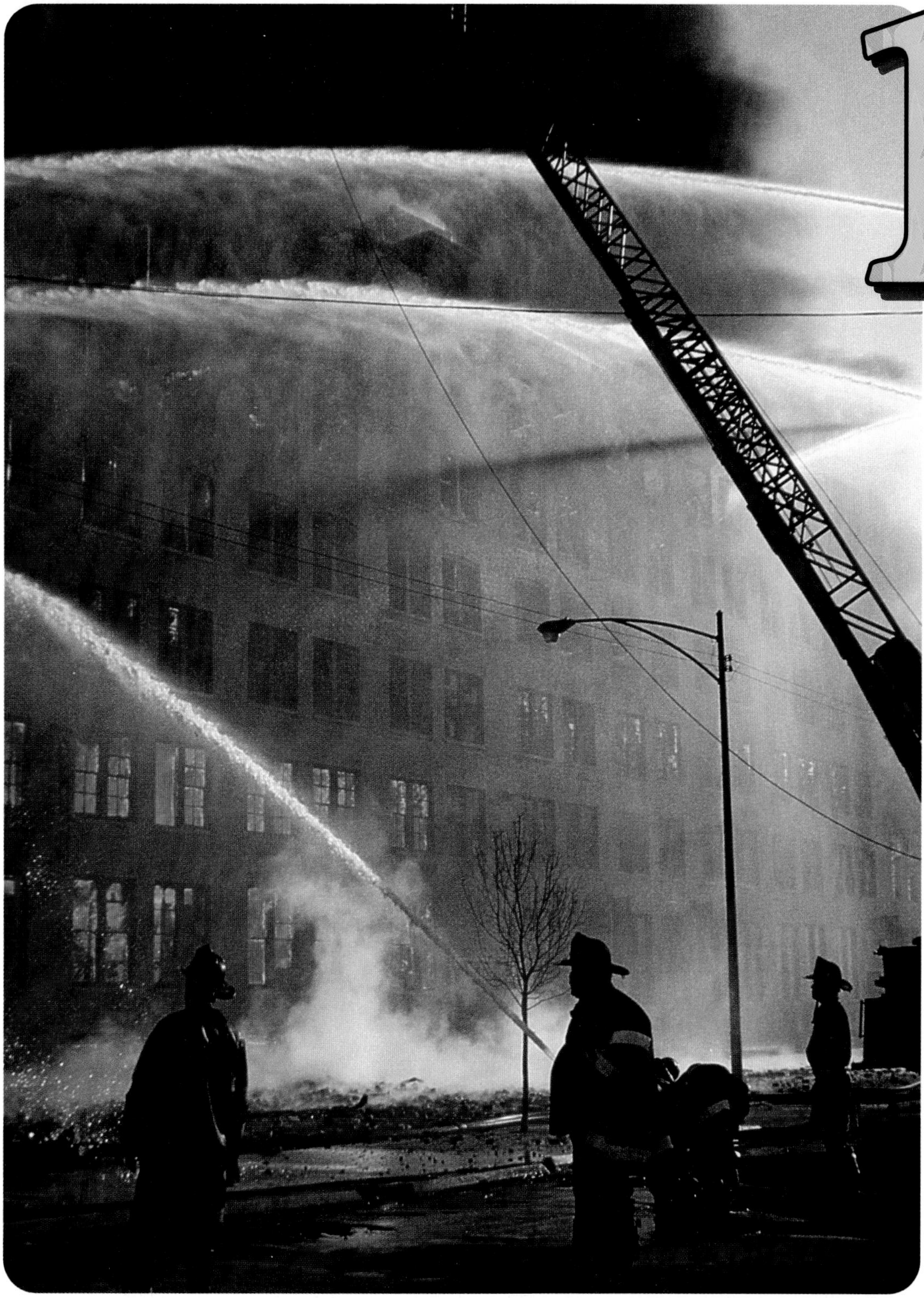

Fire Streams

This chapter provides information that addresses performance objectives described in NFPA 1001, *Standard for Fire Fighter Professional Qualifications* (1992), particularly those referenced in the following sections:

3-12 Fire Hose, Appliances and Streams

3-12.8

3-12.9

3-12.10

3-12.11

3-13 Foam Fire Streams

3-13.1

3-13.2

4-12 Fire Hose, Appliances and Streams

4-12.5

4-12.7

4-13 Foam Fire Streams

4-13.1

4-13.2

4-13.3

4-13.4

4-13.5

Chapter 10
Fire Streams

A fire stream can be defined as a stream of water, or other extinguishing agent, after it leaves the fire hose and nozzle until it reaches the desired point. The perfect fire stream can no longer be sharply defined because individual desires and extinguishing requirements vary. During the time a stream of water passes through space, it is influenced by its velocity, gravity, wind, and friction with the air. The condition of the stream when it leaves the nozzle is influenced by operating pressures, nozzle design, nozzle adjustment, and the condition of the nozzle orifice.

Fire streams are intended to reduce temperatures and provide protection by one of the following methods:

- Applying water directly to the burning material
- Reducing high atmospheric temperature
- Dispersing hot smoke and fire gases from a heated area
- Reducing the temperature over an open fire, thus permitting a closer proximity with handlines to effect extinguishment
- Protecting firefighters and property from heat through the use of fire streams as a water curtain

EXTINGUISHING PROPERTIES OF WATER

Water has the ability to extinguish fire in several ways. The primary way is by cooling, which removes the heat from the fire. Another way water extinguishes fire is by smothering. Water has the ability to absorb large quantities of heat. Amounts of heat transfer are measured in terms of British thermal units (Btu) or in joules (1 Btu = 1 055 J or 1.055 kJ). A British thermal unit is the amount of heat needed to raise the temperature of one pound of water one degree Fahrenheit. The joule, also a unit of work, has taken the place of the calorie in the International System of Units for heat measurement (1 calorie = 4.19 J).

Water is a compound formed of hydrogen and oxygen when two parts of hydrogen combine with one part oxygen (H_20). At normal temperatures (32° to 212°F [0°C to 100°C]), water exists in a liquid state. Below 32°F (0°C) (the freezing point of water), it converts to a solid state in the form of ice. Above 212°F (100°C) (the boiling point of water), it converts into a gas called water vapor or steam. Water cannot be seen in this vapor form. When steam starts to cool, however, its visible form is called condensed steam (Figure 10.1 on the next page).

For all practical purposes, water is considered to be incompressible, and its weight varies at different temperatures. Water's density, or its weight per unit of volume, is measured in pounds per cubic foot (kg/L). Water is heaviest (approximately 62.4 lb/ft^3 [1 kg/L]) close to its freezing point. Water is lightest (approximately 60 lb/ft^3 [0.96 kg/L]) close to its boiling point. For general purposes, ordinary fresh water can be considered to weigh 62.5 lb/ft^3 (1 kg/L).

Complete vaporization does not happen the instant the water reaches the boiling point, because each pound of water requires approximately 970 Btu's of additional heat to completely turn into steam. When a fire stream is broken into small particles, it absorbs heat and converts into steam more rapidly than it would in a compact form because more of the water surface is exposed to the heat. For example, a one-square-inch cube of ice dropped into a glass of water takes some time for it to absorb its capacity of heat. This is because a surface area of only 6 square inches (3 870 mm^2) of the ice is exposed to the water. However, if that cube of ice is divided into ⅛-inch (3 mm) cubes and dropped into the water, a surface area of 48 square inches (30 967 mm^2) of the ice is exposed to the water. The finely divided particles of ice absorb heat more rapidly. This same principle applies to water in the liquid state.

Another characteristic of water that is sometimes an aid to fire fighting is its expansion when converted into steam. This expansion helps cool the fire area by driving heat and smoke from the area. This steam, however, can

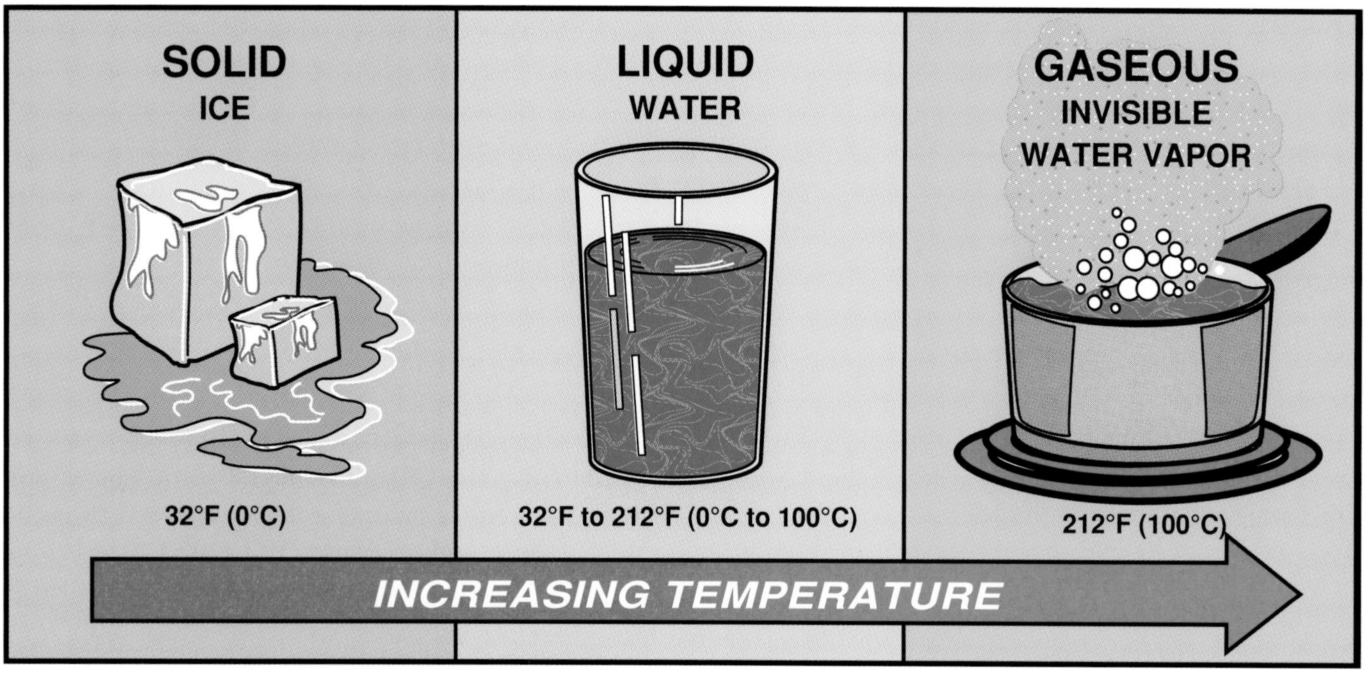

Figure 10.1 Water is found in the solid, liquid, and gaseous states.

cause serious burn injuries to firefighters and occupants. The amount of expansion varies with the temperatures of the fire area. At 212°F (100°C), a cubic foot of water expands approximately 1,700 times its original volume (Figure 10.2). The greater the temperature, the higher the amount of expansion. For example, at 500°F (260°C), the expansion ratio is approximately 2,400 times, and at 1,200°F (649°C), the ratio is approximately 4,200 times.

Visualize a nozzle discharging 150 gallons (600 L) of water fog every minute into a heated area of approximately 500°F (260°C) and the water fog being converted into steam. During one minute of operation, 20 cubic feet (0.57 m^3) of water will have been discharged and vaporized. The 20 cubic feet (0.57 m^3) of water will have expanded to approximately 48,000 cubic feet (1359 m^3) of steam. This is enough steam to fill a room approximately 10 feet high (3 m), 50 feet (15 m) wide, and 96 feet (30 m) long (Figure 10.3). In extremely hot atmospheres, steam expands to even greater volumes.

Steam expansion is not gradual, but rapid. If the room is already full of smoke and gases, the steam that is generated will displace these gases if adequate ventilation openings are provided (Figure 10.4). As the room cools, the steam condenses and allows the room to refill with cooler air. The steam produced can also be an aid in fire extinguishment by smothering when certain types of materials burn. Smothering is accomplished when the expansion of steam reduces oxygen in a confined space. The use of a fog stream in a direct or combined attack

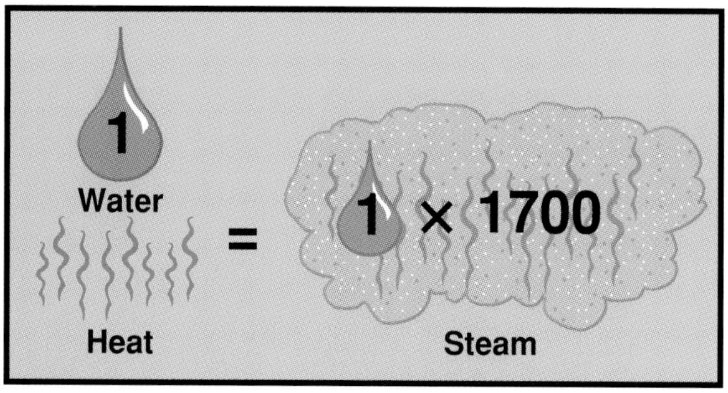

Figure 10.2 Water expands to 1,700 times its original volume when it converts to steam.

requires that adequate ventilation be provided ahead of the hoseline. Otherwise, there is a high possibility of steam or even fire rolling back over and around the hose team, and the potential for injury is great.

Water may be used as a smothering agent when it floats on liquids that are heavier than water such as carbon disulfide. It can be used to smother fires in liquids that are lighter than water or with which water is usually miscible if a foam concentrate is added to the water (Figure 10.5). In some cases, water can be used to dilute a miscible liquid enough that it will be below its fire point. This can only be done if the container holding the liquid is large enough to hold all of the fuel and water.

Some of the observable results of the proper application of a fire stream are: the fire is extinguished or reduced in size, visibility may be maintained, and the room temperature will be reduced.

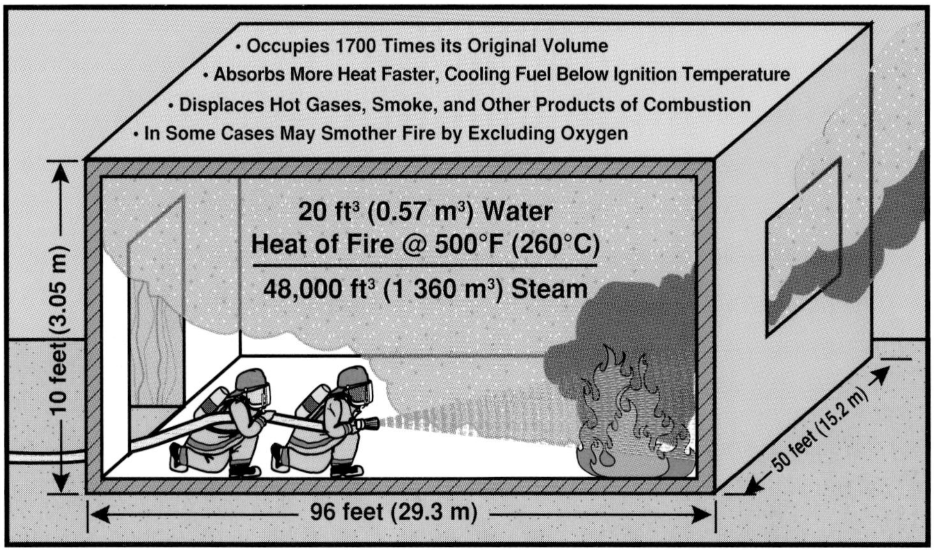
• Occupies 1700 Times its Original Volume
• Absorbs More Heat Faster, Cooling Fuel Below Ignition Temperature
• Displaces Hot Gases, Smoke, and Other Products of Combustion
• In Some Cases May Smother Fire by Excluding Oxygen
20 ft³ (0.57 m³) Water
Heat of Fire @ 500°F (260°C)
48,000 ft³ (1 360 m³) Steam
10 feet (3.05 m)
96 feet (29.3 m)
50 feet (15.2 m)

Figure 10.3 Water's expansion rate makes it very effective for fire extinguishment.

Figure 10.4 Water can be used to disperse the products of combustion from an enclosed space.

Figure 10.5 Foam is often required to extinguish Class B fires. *Courtesy of Bob Esposito.*

Several characteristics of water that are extremely valuable for fire extinguishment are as follows:

- Water has a greater heat-absorbing capacity than other common extinguishing agents.
- A relatively large amount of heat is required to change water into steam.
- The greater the surface area of the water exposed, the more rapidly heat will be absorbed.

FRICTION LOSS

A fire stream definition of friction loss is as follows: *Friction loss is that part of total pressure that is lost while forcing water through pipe, fittings, fire hose, and adapters.* The difference in pressure on a hoseline between the nozzle and pumper, excluding pressure lost due to a change in elevation between the two, is a good example of friction loss. Friction loss can be measured by inserting in-line gauges at different points in a hoseline. The difference in the pressures between gauges when water is flowing through the hose will be the friction loss for the length of hose between the gauges for that rate of flow.

One point that should be considered in applying pressure to water in a hoseline is that there is a limit to the velocity or speed at which the stream can travel. If the velocity is increased beyond these limits, the friction becomes so great that the entire stream is agitated by resistance. This agitation causes a point of turbulence, called *critical velocity*. Beyond this point, it becomes necessary to parallel or siamese hoselines in order to increase the flow and reduce friction.

Certain characteristics of hose layouts affect friction loss. In order to reduce pressure loss due to friction, consider the following:

- Check for rough linings in old hose.
- Replace damaged couplings.
- Eliminate sharp bends.
- Use adapters only when necessary.
- Keep nozzles and valves fully open when possible.
- Use proper size gaskets.
- Use short lines as much as possible.
- When flow must be increased, use larger hose or multiple lines.

WATER HAMMER

When the flow of water through fire hose or pipe is suddenly stopped, the resulting surge is referred to as *water hammer*. Water hammer can often be heard as a distinct sharp clank, very much like a hammer striking a pipe. This sudden stoppage results in a change in the direction of energy, and this energy is instantaneously multiplied many times. These excessive pressures can cause considerable damage to water mains, plumbing, fire hose, and fire pumps. Nozzle controls, hydrants, valves, and hose clamps should be operated slowly to prevent water hammer (Figure 10.6).

FIRE STREAMS

A fire stream is identified by the type and size of stream. The type of fire stream indicates a specific pattern of water needed for a specific job. There are three major types of fire streams:

- Solid
- Fog
- Broken

The rate of discharge of a fire stream is measured in gallons per minute (gpm) or liters per minute (L/min). Fire streams are classified into one of three sizes: low volume streams, handline streams, and master streams. A low volume stream is one that discharges less than 40 gpm (150 L/min) including those fed by booster hoselines. Handline streams are supplied by 1½- to 3-inch (38 mm to 77 mm) hose. These handline streams will flow from 40 to 350 gpm (120 L/min to 1 400 L/min). Nozzles with flows in excess of 350 gpm (1 400 L/min) are not recommended for handlines. A master stream is a large-volume stream which discharges more than 350 gpm (1 400 L/min) and should be fed by multiple 2½- or 3-inch (65 mm or 77 mm) hoselines or large diameter hoselines connected to a master stream nozzle. In any classification, the stream may be either solid, fog, or broken.

Regardless of the type of fire stream, several items are needed to produce an effective fire stream. All fire streams must have a pressuring device, hose, an agent, and the nozzle (Figure 10.7). The following sections more closely examine the different types of streams.

Solid Streams

A solid stream is a fire stream produced from a fixed orifice, smoothbore nozzle (Figure 10.8). The solid stream nozzle is designed to produce a stream as compact as possible with little shower or spray. It has the ability to reach areas that might not be reached by other streams and also minimizes the chance of steam burns. This has reestablished the solid stream as an important fire service tool. When solid stream nozzles are used on handlines, they should be operated at 50 psi (350 kPa) nozzle pressure. A solid stream master stream device should be operated at 80 psi (560 kPa). Flow from a solid stream nozzle is determined by the velocity of the stream (nozzle pressure) and the size of the discharge opening.

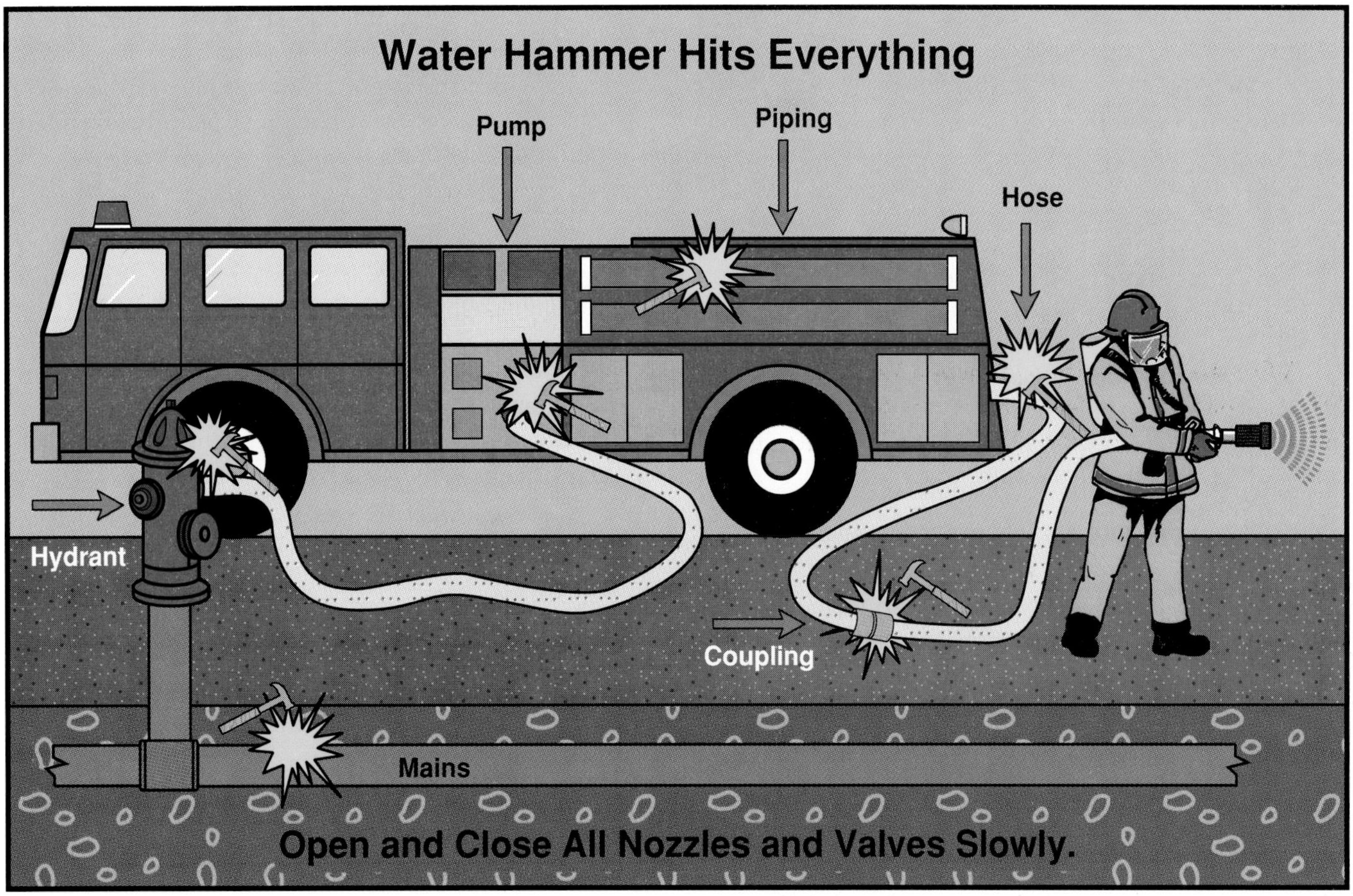

Figure 10.6 Water hammer can cause damage to all parts of the water system and to fire equipment.

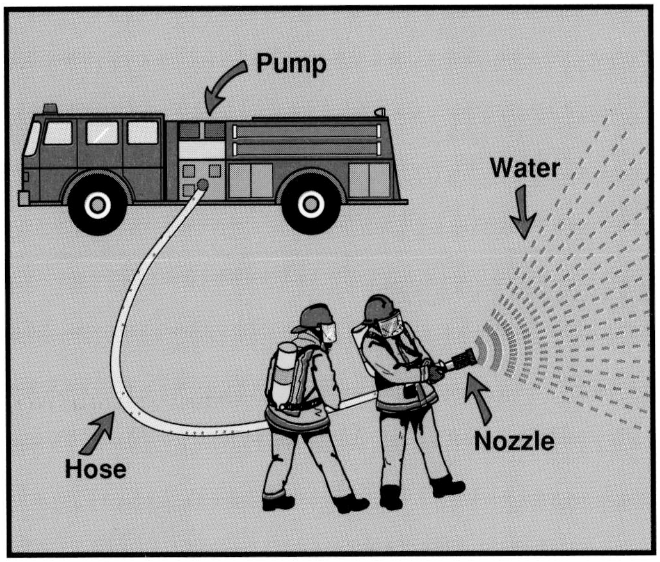

Figure 10.7 The four elements that make a fire stream are a pump, hose, nozzle, and water.

Figure 10.8 A smoothbore nozzle produces a solid fire stream.

The extreme limit at which a solid stream of water can be classified as a good stream cannot be sharply defined and is, to a considerable extent, a matter of judgment. It is difficult to say just exactly where the stream ceases to be good. Observations and tests, covering the effective range of fire streams, classify effective streams as follows:

- A stream that at the point of breakover has not lost continuity by breaking into showers of spray.

- A stream that would probably be stiff enough to attain, in fair condition, the height required even though a fresh breeze was blowing.

Water from a solid stream nozzle produces a compact jet to the breakover point, but beyond this point, it tends to be in the form of a heavy rain, which is easily carried away. These characteristics and the point of breakover are illustrated in Figure 10.9.

ADVANTAGES AND DISADVANTAGES OF SOLID STREAMS

The following is a synopsis of the advantages and disadvantages of solid streams:

Advantages

- Greater reach than other types of streams
- Greater penetration power than other types of streams
- Less likely to disturb normal thermal layering of heat and gases during interior structural attacks

Disadvantages

- Unable to select different stream patterns when desired
- May not be used for foam application
- Less heat absorption per gallon delivered

Fog Streams

The term fog stream is commonly used in the fire service to describe a patterned stream composed of water droplets. The design of most fog nozzles permits adjustment of the fog tip to produce different stream patterns from the nozzle (Figure 10.10). Water droplets, in a shower or spray, are formed to expose the maximum water surface for heat absorption. Fog nozzles permit settings of wide-angle fog, narrow-angle fog, and straight stream. It should be understood that a straight stream is a pattern of the adjustable fog nozzle, whereas a solid stream is discharged from a smoothbore nozzle. When fog nozzles are used, they should be operated at 100 psi (700 kPa) nozzle pressure.

The design of the fog nozzle controls certain factors that influence the fog stream. A particular fog nozzle may be considered to have a perfect design and meet all requirements. However, without the required 100 psi (700 kPa) pressure at the nozzle, it may prove to be inadequate. As pressures are increased, factors appear that influence the development of the fog stream pattern. The velocity of the jet, the size of the water particles, and the volume of water discharged are more important than any excess in pressure that may be necessary to produce a stream that will reach a desired objective.

Although nozzle designs differ, the water pattern that is produced by the nozzle setting may affect the ease with which a particular nozzle may be handled. Fire stream nozzles, in general, are not considered to be

Figure 10.10 A fog stream is a patterned stream composed of water droplets.

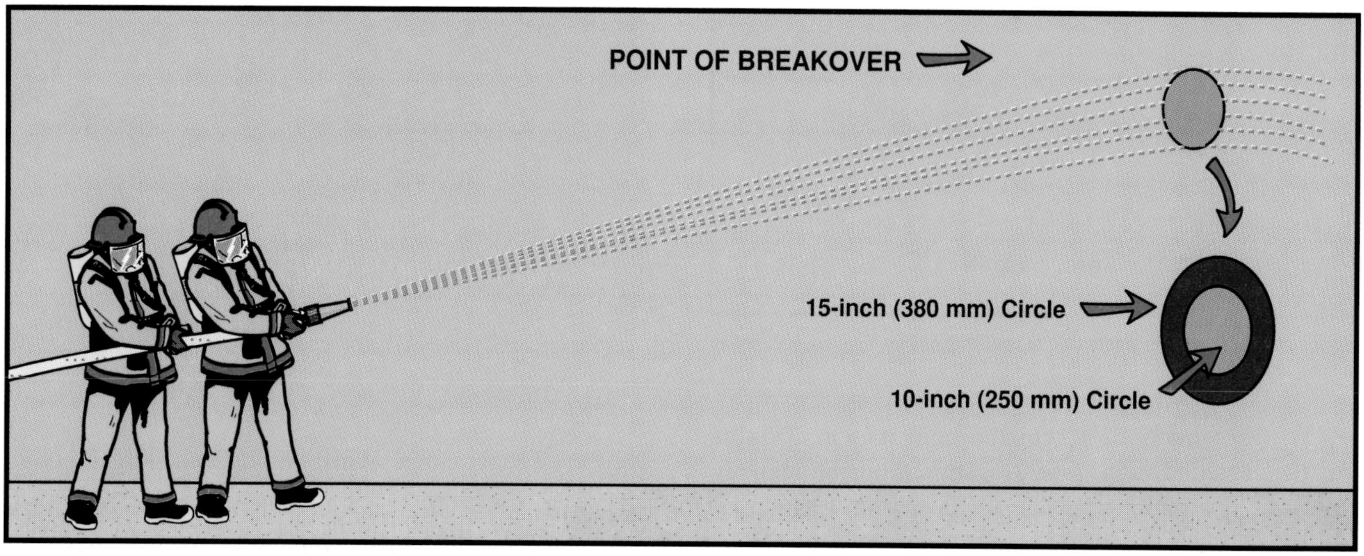

Figure 10.9 The breakover point is that point at which the solid stream begins to lose its forward velocity.

easily handled. This difficulty is due to the fact that when water is projected directly away from the nozzle, the reaction is equally strong in the opposite direction. Thus, the reaction caused by the velocity and mass of the stream, which acts against the nozzle and the curves in the hose, makes the nozzle hard to handle. If water travels at angles to the direct line of discharge, the reaction forces may be made to more or less counterbalance each other and reduce the nozzle reaction. This balancing of forces is the reason why a wide-angle fog pattern can be handled more easily than a straight-stream pattern. The desired performance of fog stream nozzles is judged by the amount of heat that a fog stream absorbs and the rate by which the water is converted into steam or vapor.

Figure 10.11 A wide fog pattern has less reach and forward velocity than a narrow fog pattern.

Figure 10.12 An adjustment of the fog head produces different stream patterns such as this wide pattern.

VELOCITY OF THE JET

Velocity is the rate of motion of a particle in a given direction. Direction and rate of motion are both essential parts of velocity. As soon as the water is discharged from a nozzle, it is carried forward by its momentum and downward by the force of gravity, and its motion is retarded by friction with the air. Obviously, the faster the water is traveling forward, the farther it will travel before it is pulled to earth by gravity. Forward velocity, therefore, is one factor which governs the reach of a fog stream. The reach varies in proportion to the pressure applied. There is a maximum reach, however, and once the nozzle pressure is sufficient to produce a satisfactory reach, further increases in the nozzle pressure will only cause the stream to break up and be difficult to handle.

The effect that nozzle adjustment has upon a fog stream may be illustrated by operating two identical adjustable nozzles at identical pressures. Because the pressures are equal at both nozzles, it can be assumed that the water is being discharged from both at the same velocity. However, the difference in the adjustment of these nozzles demonstrates a marked difference in the two streams. The fog stream in Figure 10.11 has considerably more reach and forward velocity than the one in Figure 10.12. The design of the nozzle permits an adjustment of the fog head to produce different stream patterns from the two nozzles. This adjustment changes the angle of the discharge orifice and increases the forward velocity of the stream. Fog streams should be evenly divided into a fine spray with a uniform discharge pattern around the "cone" regardless of the nozzle setting.

SIZE OF THE WATER PARTICLES

Any jet of water is retarded in its path through the air because of the friction between the outer surface of the jet and the air. A fog stream is composed of droplets of water surrounded by air. The air is entrained in the stream and becomes part of the stream. A fog stream is retarded by the friction between the outer surface of the fog cone and the surrounding air, turbulence within the fog stream, and the process of entraining air. If two fog streams are of the same diameter and the water droplets have the same initial velocity, the one containing the most water has the greatest reach because it has a greater momentum. The size of the droplets has some effect on the reach of the fog stream because of the amount of entrained air. Because fog streams have a larger diameter in comparison to straight streams, there is more area subject to friction with the air, and its forward velocity is retarded more rapidly.

Water may be divided so finely that the air appears to be saturated with a fine mist, and the small water particles appear to be suspended in the air. Water particles in this form are so fine that they may be carried away from the fire by air currents and never penetrate the fire area. The result is that they do not have an opportunity to absorb sufficient heat to be effective. Water droplets must be large enough to maintain momentum and obtain desired reach and penetration. Fog patterns should also be sufficiently heavy to work into a moderate wind without being destroyed.

VOLUME OF WATER DISCHARGED

Another factor that is determined by the design of the nozzle and the pressure at the nozzle is the volume of water discharged. Fog streams must have sufficient volume to penetrate the heated area. If a low volume nozzle producing finely divided particles is used where the heat is generated faster than it is absorbed, extinguishment will not be accomplished until the fuel is completely consumed or is shut off. It is essential for a fog stream to deliver a volume of water sufficient to absorb heat more rapidly than it is generated.

ADVANTAGES AND DISADVANTAGES OF FOG STREAMS

The following is a synopsis of the advantages and disadvantages of fog streams:

Advantages

- The discharge pattern may be adjusted to suit the situation.
- Some fog stream nozzles have adjustable gallonage settings to control the amount of water being used (Figure 10.13).
- Fog streams may be used in close proximity to energized electrical equipment with a reduced chance of electrocution to the firefighter.

Disadvantages

- Fog streams do not have the reach or penetrating power of solid streams.
- Fog streams are more susceptible to wind currents than are solid streams.
- When improperly used during interior attacks, they can spread fire, create heat inversion, and cause steam burns to firefighters.

Broken Streams

A broken fire stream is a stream of water that has been broken into coarsely divided drops (Figure 10.14). A solid stream may become a broken stream because of reacting forces, and it may be desirable that a stream break near the fire to sprinkle the burning material. Other means may be employed to produce broken streams such as a rotary distributor nozzle, water curtain nozzle, or by directing two solid streams together in midair (Figure 10.15). The droplets of a broken stream are larger than those of a fog stream and have greater

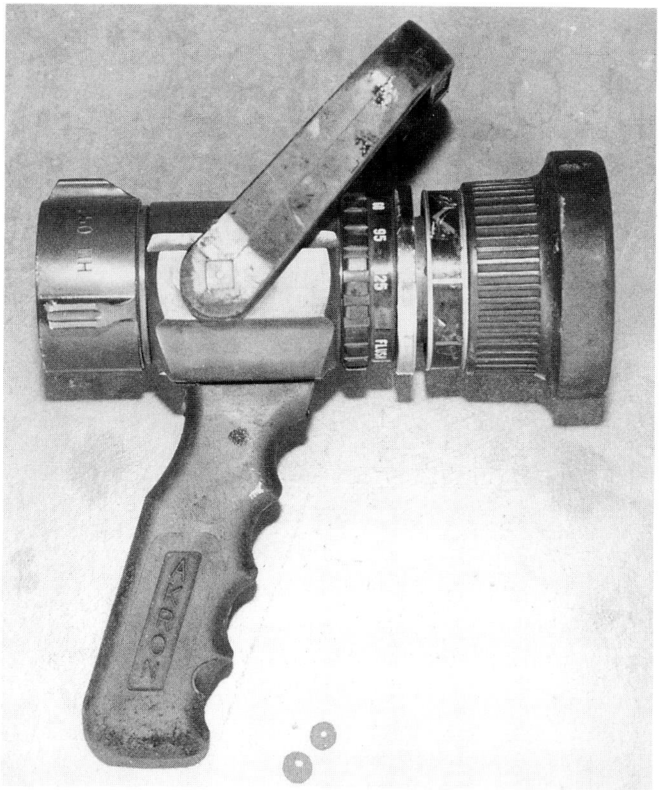

Figure 10.13 This fog stream nozzle has an adjustable gallonage setting.

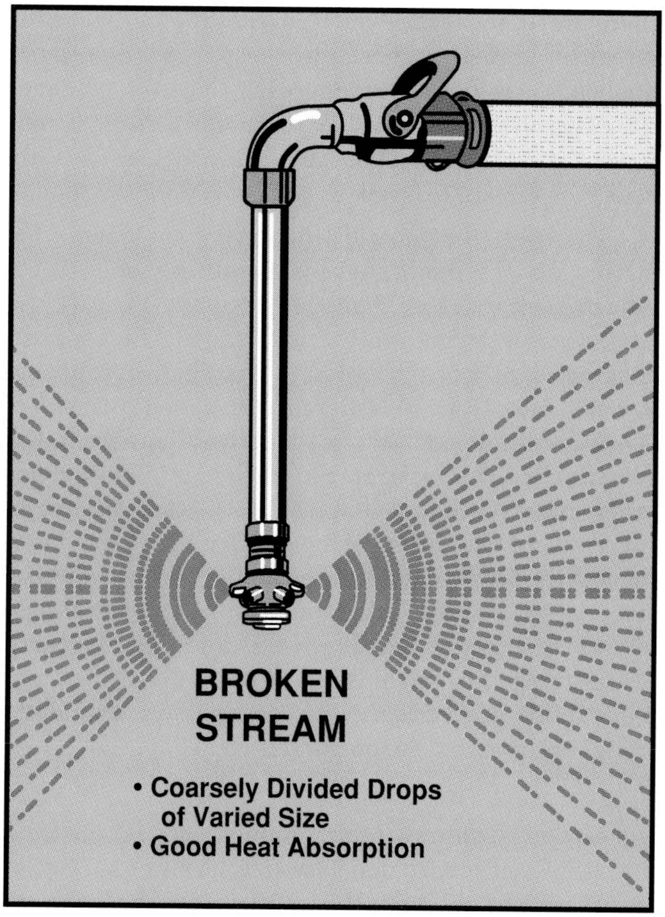

Figure 10.14 A broken stream produces large droplets.

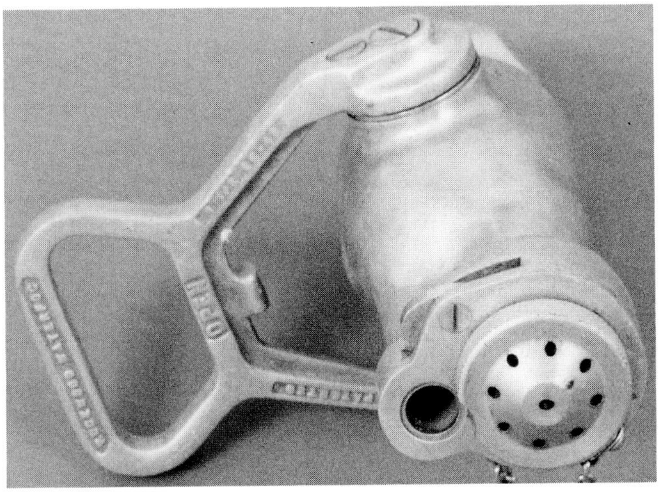

Figure 10.15 A broken stream nozzle. *Courtesy of Fred Myers.*

penetration. Therefore, the broken stream can be useful where neither a fog stream nor a solid stream would be as effective. For more information on broken streams, see the IFSTA validated **Fire Stream Practices** manual.

FIRE HOSE NOZZLES

Fire hose nozzles are grouped according to the type and volume of the stream (Figures 10.16 a through c). The most common types of nozzles are:

- Solid stream — handline and master
- Fog stream — handline and master
- Broken stream

Figure 10.16a A solid stream nozzle.

Figure 10.16b A fog stream nozzle.

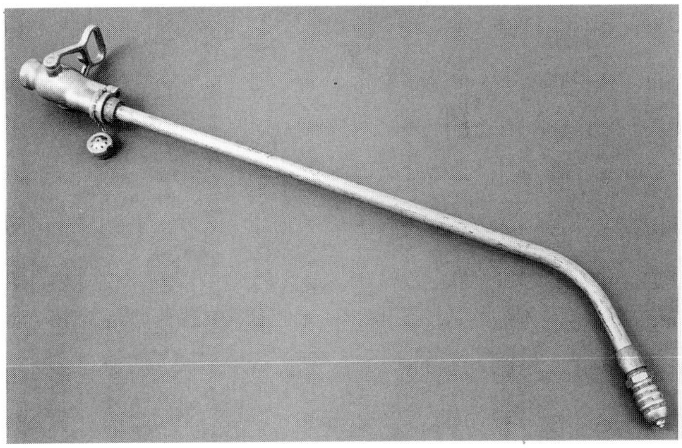

Figure 10.16c A broken stream nozzle with an applicator pipe. *Courtesy of Fred Myers.*

Solid Stream Nozzles

Many years ago, experiments were conducted to compare streams from many different kinds, shapes, and sizes of nozzles. These experiments have become known as the Freeman experiments, named after the leader of the project. The results of these comparisons were surprising. The individuals conducting the tests expected a polished stream of water with a clear, glassy surface and little spray. Not one of the many nozzles tested produced such a stream. After the experiments, Freeman designed a standard tip for ordinary use. Most manufacturers still use this standard tip as a basis for design (Figure 10.17).

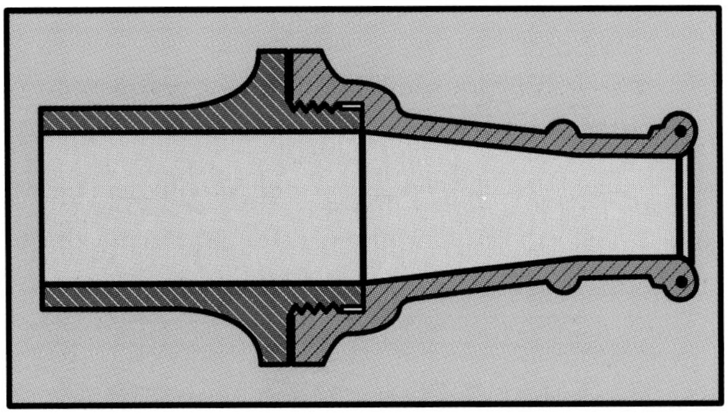

Figure 10.17 The basic design for a solid stream nozzle.

At present, solid stream nozzles are designed so that the shape of the water in the nozzle is gradually reduced until it reaches a point a short distance from the outlet. At this point, the nozzle becomes a cylindrical bore whose length is from one to one and one-half times its diameter. The purpose of this short, truly cylindrical bore is to round out and give the water its shape before discharge. A smooth-finished waterway contributes to both the shape and reach of the stream. Alteration or damage to the nozzle can significantly alter stream shape and performance.

Although solid stream nozzles differ in shape and size, they all conform to the required physical characteristics for solid stream nozzles. Both ends of the shutoff valve should be threaded with 1½-inch (38 mm) National Standard threads (Figure 10.18). The discharge orifice of the nozzle should be no greater than one-half the size of the hoseline to which the nozzle is attached.

Figure 10.18 Most stack tip nozzles will have a 1½-inch (38 mm) thread orifice ahead of the tip to allow the line to be broken down to a smaller size.

Fog Stream Nozzles

A fog stream may be produced by deflection at the periphery, by impinging jets of water, or by a combination of the two. Periphery-deflected streams are produced by deflecting water from the periphery of an inside, circular stem in a periphery-deflected fog nozzle (Figure 10.19). This water is again deflected by the exterior barrel. The relative positions of the deflecting stem and the exterior barrel determine the shape of the fog stream.

The impinging-stream nozzle drives several jets of water together at a set angle to break the water into finely divided particles (Figure 10.20). Impinging-stream nozzles usually produce a wide-angle fog pattern, but a narrow pattern is possible. Impinging jets are also used on periphery-deflected fog stream nozzles.

Fog stream nozzles may be of the set gallonage, adjustable gallonage, or automatic gallonage type. Set gallonage nozzles are designed to be used for only one flow rate at a given discharge pressure. Adjustable gallonage nozzles have a rotating ring on them that allows any one of several preset gallon settings to be selected. Automatic nozzles will discharge a wide range of flows depending on the pressure being supplied to the nozzle (Figure 10.21).

Broken Stream Nozzles

Broken stream nozzles are basically limited to special applications. Piercing nozzles are one type of broken stream nozzle. They are commonly used to apply water to areas which may otherwise be inaccessible. This nozzle is generally a 3- to 6-foot (1 m to 2 m), 1½-inch (38 mm) diameter hollow steel rod. The discharge end of the piercing nozzle is usually a hardened steel point that is suitable for being driven through concrete block or other types of wall or partition assemblies (Figure 10.22). Built into the point is an impinging-jet nozzle that is generally capable of delivering about 100 gpm (379 L/min) of water. These nozzles may also be used to deliver AFFF or high-expansion foam to a confined area. Oppo-

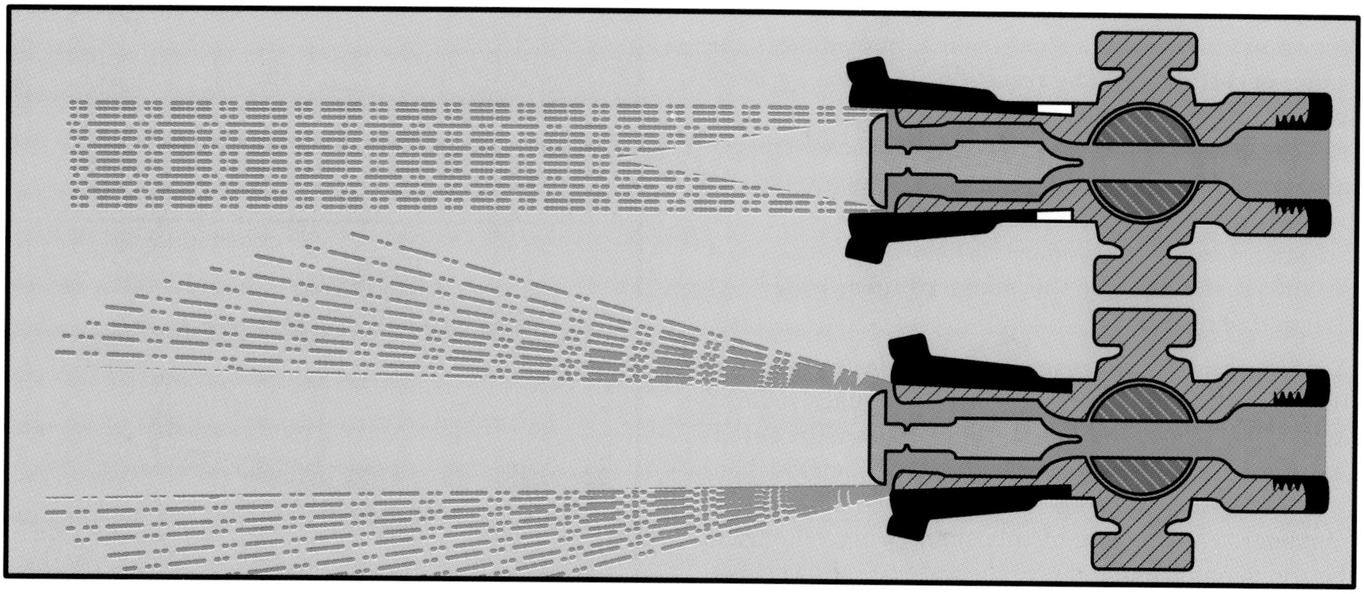

Figure 10.19 A periphery fog stream.

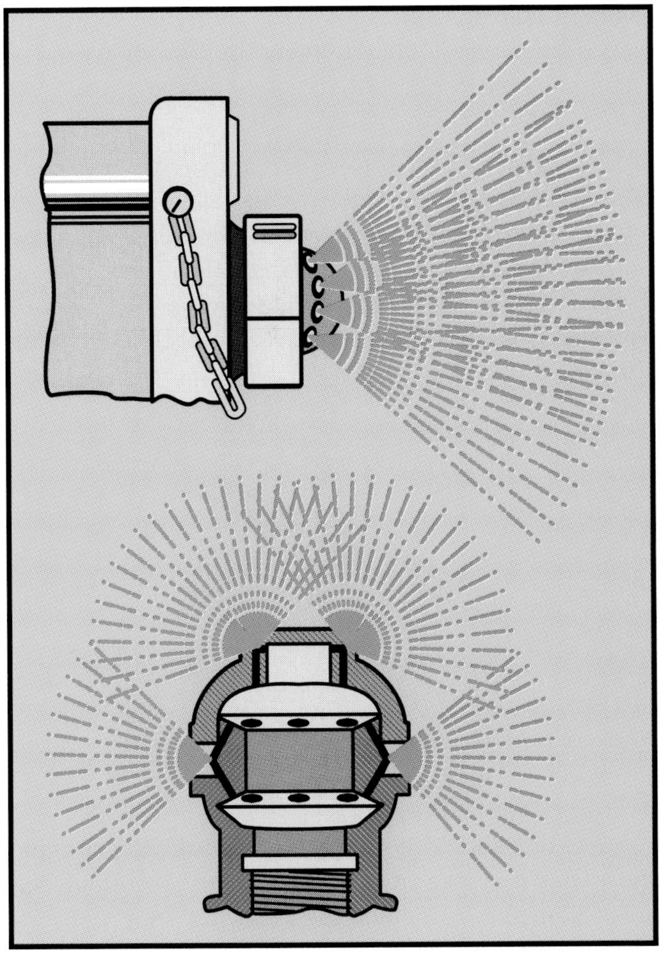

Figure 10.20 An impinging jet fog stream.

Figure 10.21 An automatic nozzle.

site the pointed nozzle end of the piercing nozzle is the driving end. This end of the nozzle is driven with a sledgehammer to force the point through an obstruction. At this end is also the inlet for a 1½-inch (38 mm) hoseline to be connected to supply the nozzle. This nozzle is most commonly used to extinguish fires in walls and in the trunk or motor areas of automobiles. There are also air-driven penetrating nozzles used for aircraft fire fighting, which bore through the skin of an aircraft to reach the seat of the fire.

Figure 10.22 Piercing nozzles can be driven through masonry walls.

In the past, water curtain broken stream nozzles were commonly used for exposure protection (Figure 10.23). However, research has indicated that these nozzles are only effective if the water is sprayed directly against the exposure being protected.

A special nozzle is available for use in fighting chimney flue fires. It is similar to homemade nozzles that have been in use since the 1940s. The chimney nozzle is designed to be placed on the end of a booster hose. The nozzle is a solid piece of brass or steel with numerous, very small holes in it (Figures 10.24 a and b on the next page). At a nozzle pressure of 100 psi (700 kPa), it will generally produce only 6 to 12 quarts of water per minute (6 L/min to 12 L/min) in a very fine mist fog cone. The hose and nozzle are dropped down the entire length of the chimney and then quickly pulled back out (Figure 10.25 on the next page). The mist from the nozzle immediately turns to steam and chokes the flue fire, as well as loosens the soot on the inside of the chimney. Because such a small amount of water is used and it converts to steam so quickly, no damage due to

Figure 10.23 A water curtain nozzle in use.

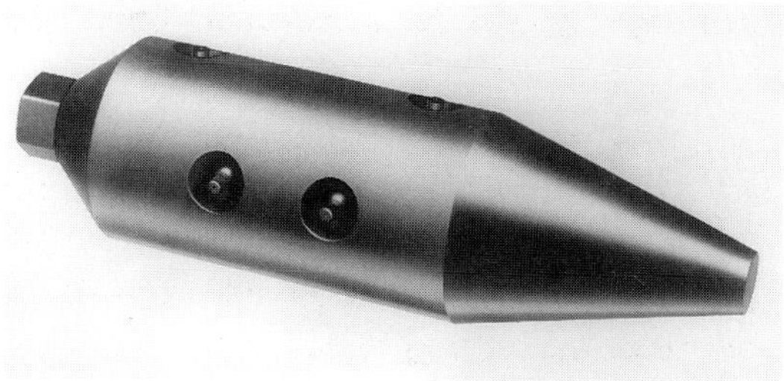

Figure 10.24a A commercially produced chimney nozzle. *Courtesy of Jaffery Fire Protection Co.*

Figure 10.24b The stream from a homemade chimney nozzle. *Courtesy of Eugene Knause, Pennsburg (PA) Fire Co. No. 1.*

Figure 10.25 The chimney nozzle is lowered down the flue to extinguish the fire.

cooling-induced expansion of the flue liner occurs. Because this process may damage the booster hose, it is recommended that an old section be placed on the end of the regular section whenever this nozzle is used.

Another type of broken stream nozzle sometimes still in use today is the cellar, or distributor, nozzle. This nozzle is designed to be raised or lowered through holes in floors or ceilings to attack fires on a level above or below that of the firefighters. They usually have a rotating head that sprays water in all directions (Figure 10.26).

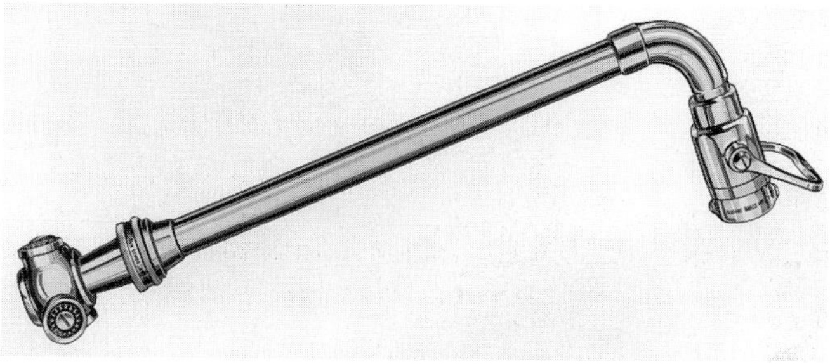

Figure 10.26 A typical cellar nozzle. *Courtesy of Elkhart Brass Mfg. Co.*

Maintenance Of Nozzles

Nozzles should be inspected periodically to make sure that they are in proper working condition. This inspection should include the following:

- Check the gasket.
- Check for external damage.
- Check for internal damage and debris.
- Check the ease of operation.

When necessary, the nozzles should be thoroughly cleaned with soap and water, using a soft bristle brush (Figure 10.27). If the gasket is worn or missing, it should be replaced. Any moving parts that appear to be sticking should be cleaned and lubricated appropriately according to manufacturer's recommendations.

FOAMS

Water alone is not always effective as an extinguishing agent. Under certain circumstances, foam is needed. Fire fighting foam (especially the low-expansion form) is especially effective on the two basic categories of flammable liquids: hydrocarbon fuels and polar solvents.

Hydrocarbon fuels, such as crude oil, fuel oil, gasoline, benzene, naphtha, jet fuel, and kerosene, are petroleum based and float on water. Fire fighting foam is effective as an extinguishing agent and vapor suppressant because it can float on the surface of these fuels.

Figure 10.27 Clean nozzles with soap and water when necessary.

Polar solvents, such as alcohols, acetone, lacquer thinner, ketones, esters, and acids, are flammable liquids that have an attraction for water, much like a positive magnetic pole attracts a negative pole. Fire fighting foam is effective on these fuels, but only in special alcohol-resistant formulations (referred to as Alcohol Type Concentrate [ATC]).

Fire fighting foam is also used for acid spills, pesticide fires, confined- or enclosed-space fires, and deep-seated Class A fires. In addition to fire fighting foams, there are special low-expansion foams designed solely for use on unignited spills of hazardous liquids. These foams are necessary because unignited chemicals have a tendency to change the pH of or remove the water from fire fighting foams, thereby rendering them ineffective.

Foam extinguishes and/or prevents fire in several ways:

- *Smothering:* preventing air and flammable vapors from combining
- *Separating:* intervening between the fuel and the fire
- *Cooling:* lowering the temperature of the fuel and adjacent surfaces
- *Suppressing:* preventing the release of flammable vapors (Figure 10.28)

In general, foam works by forming a blanket on the burning fuel. The foam blanket excludes oxygen and stops the burning process. Hazardous materials mitigation foam blankets the potential fuel and prevents vapors from escaping and being ignited.

Figure 10.28 Foam cools, smothers, separates, and suppresses vapors.

How Foam Is Generated

Foams in use today are of the mechanical type, that is, they must be proportioned and aerated (mixed with air) before they can be used. Before discussing types of foams and the foam-making process, it is important to understand the following terms:

Foam Concentrate — The raw foam liquid as it sits in its storage container, usually a 5-gallon (20 L) pail, 55-gallon (220 L) drum, or an apparatus storage tank (Figure 10.29).

Foam Proportioner — The device that injects the correct amount of foam concentrate into the water stream to make the foam solution (Figure 10.30 on the next page).

Figure 10.29 Foam concentrate is commonly carried in 5-gallon (20 L) containers, although in this case it is also carried in the apparatus booster tank. *Courtesy of Harvey Eisner.*

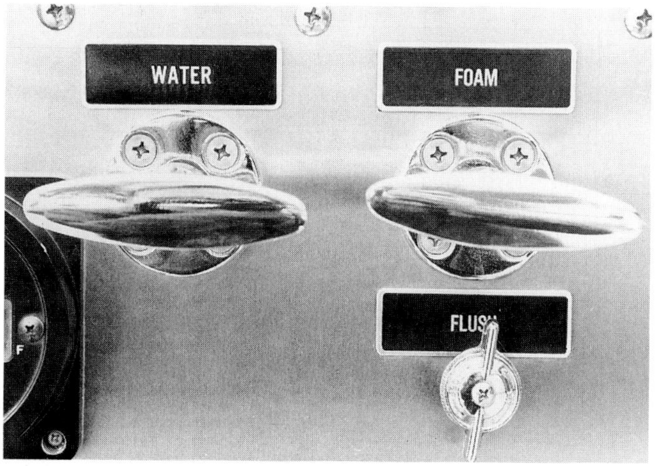

Figure 10.30 The foam proportioner injects the correct amount of foam concentrate into the water stream. Some foam proportioners are mounted directly in the fire pump.

Foam Solution — The mixture of foam concentrate and water that is discharged from the proportioner and passed through the hoseline.

Finished Foam — The completed product after the foam solution reaches the nozzle and air is introduced into the solution (aeration).

Four elements are necessary to produce high-quality fire fighting foam: foam concentrate, water, air, and mechanical agitation (aeration) (Figure 10.31). All these elements must be present and blended in the correct ratios. Removing any element results in either no foam or a poor quality foam.

There are two stages in the formation of foam. First, water is mixed with foam liquid concentrate to form a *foam solution*. This is the proportioning stage of foam production. Second, the foam solution passes through the hoseline to the foam maker, which is usually part of the foam nozzle (Figure 10.32). The foam maker aerates the foam solution to form *finished foam*.

Aeration produces an adequate amount of foam bubbles to form an effective foam blanket. Proper aeration also produces uniform-sized bubbles, which form a longer lasting blanket. A good foam blanket is required to maintain an effective cover over the flammable or toxic liquid.

Proportioners and foam nozzles (also called foam makers) are engineered to work together. Using a foam proportioner that is not matched to the foam maker (even if the two are made by the same manufacturer) can result in unsatisfactory foam or no foam at all. There are numerous appliances for making and applying finished foam. A number of these are discussed later in this chapter.

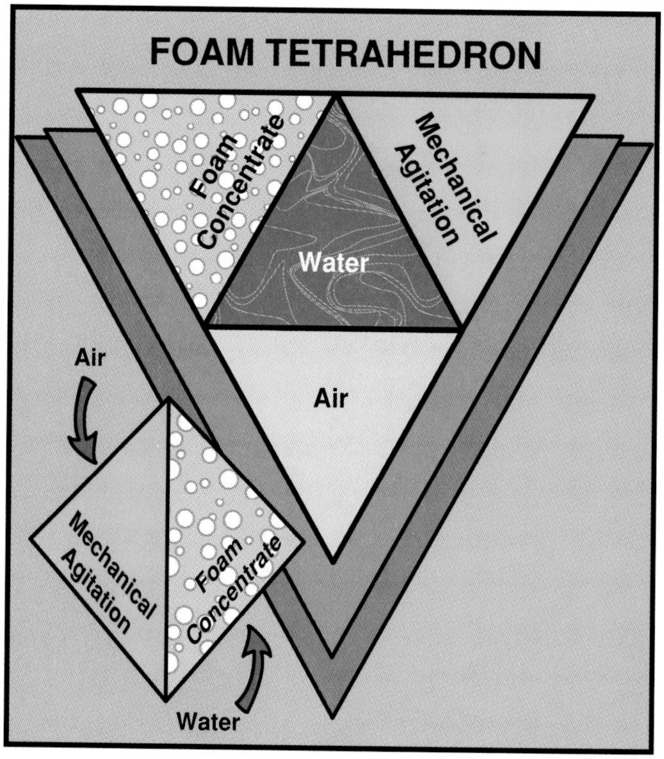

Figure 10.31 The foam tetrahedron.

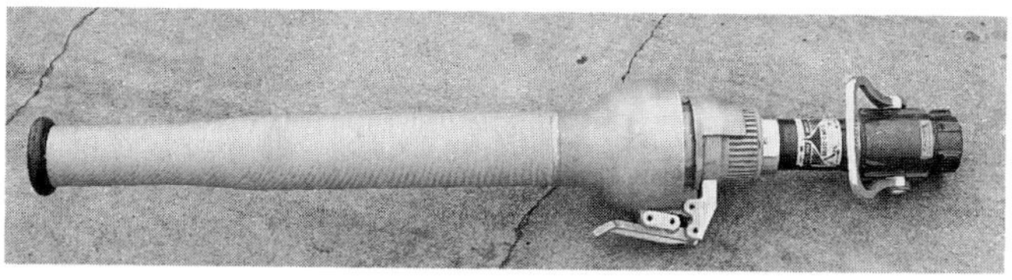

Figure 10.32 A standard fog nozzle may be turned into a foam nozzle with the addition of an aeration device.

Fire fighting foam is 90 to 99 percent water. Foams in use today are designed to be used at 1, 3, or 6 percent concentrations, although very strong polar solvents may require foam concentrations of up to 10 percent. In general, foams designed for hydrocarbon fires are used at 1 to 6 percent concentrations. Polar solvent fuels require 6 to 10 percent concentrates. Medium- and high-expansion foams are typically used at 1½, 2, or 3 percent concentrations.

NOTE: To ensure maximum effectiveness, use foam concentrates *only* at the specific percentage for which they are intended to be proportioned. This percentage rate is clearly marked on the outside of every foam container.

To be effective, foam concentrates must also match the fuel to which they are applied. Foams designed for hydrocarbon fires will not extinguish polar solvent fires regardless of the concentration at which they are used. Many foams that are designed for polar solvents may be used on hydrocarbon fires, but this should not be attempted unless the particular concentrate being used specifically says this can be done. This is why it is extremely important to identify the type of fuel involved before beginning to apply foam.

Depending on their purpose, foams are designed for low, medium, or high expansion. Low-expansion foams, such as AFFF or FFFP foams, have a small air/solution ratio generally in the area of 5:1 to 10:1. They are used primarily to extinguish fires in the two basic categories of liquid fuels: hydrocarbon fuels and polar solvent fuels. They are also used for vapor suppression on unignited spills. Low-expansion foams are effective when the temperature of the fuel liquid does not exceed 250°F (121°C).

Medium-expansion foams typically have expansion ratios between 100:1 and 300:1. High-expansion foams generally have expansion ratios of 500:1 to about 1,200:1. Both foams are especially useful as space-filling agents in such hard-to-reach spaces as basements, mine shafts, and other subterranean areas. Steam dilution caused by the vaporization of the foam in heated areas displaces gas and smoke, thus cooling the environment and extinguishing confined-space fires.

Application Rates

All AFFFs and FFFPs require an application rate of 0.10 gpm foam solution per square foot (4.1 L/min/square meter) for *ignited* hydrocarbon fuel. Fires involving polar solvent fuels generally require a higher rate of application and an alcohol-resistant foam concentrate. As a general rule, apply at least 0.24 gpm per square foot (9.8 L/min/square meter) to polar solvent fires. Apply protein and fluoroprotein foams at a rate of 0.16 gpm per square foot (6.5 L/min/square meter).

Unignited spills do not require the same application rates as ignited spills because radiant heat, open flame, and thermal drafts do not attack the finished foam as they would under fire conditions. For unignited spills, a lower application rate may be satisfactory. In case the spill does ignite, however, firefighters should be prepared to flow at least the minimum application rate for fire conditions.

Specific Foam Concentrates

Fire fighting foam concentrate is manufactured with either a synthetic or protein base. Protein-based foams are derived from plant or animal matter. Detergent-based foam is made from a mixture of detergents. Some foams are a combination of the two.

PROTEIN FOAMS

Protein foams are chemically broken down (hydrolyzed) protein solids. The end product of this chemical digestion is protein liquid concentrate. The characteristics of protein foam are as follows:

- Excellent water-retention capabilities
- High heat resistance
- Performance not affected by freezing and thawing
- Can be stored at temperatures from 20° to 120°F (-7°C to 49°C)
- Can be compounded for freeze protection using a nonflammable antifreeze solution

When applying regular protein foam to burning liquids, DO NOT allow the foam to plunge into the fuel. Apply it gently in a finished foam spray, or apply it indirectly by directing the stream against a stationary object. Applying foam indirectly allows it to spread over the fuel with as little agitation as possible (Figure 10.33).

Figure 10.33 Protein foam requires an indirect application technique.

WARNING

Take great care when applying protein foam because this application requires a close approach by firefighters. Allowing protein foam to plunge into the fuel allows burning to continue and may splash the fuel, increasing the fire hazard.

Regular protein foam can be made alcohol resistant by adding heavy metal salts suspended in organic solvents. Alcohol-resistant protein foam solution loses its resistance to solvent attack very quickly; the finished foam must be applied within seconds of proportioning.

FLUOROPROTEIN FOAMS

Like regular protein foams, fluoroprotein foams are based on hydrolized protein solids. Fluoroprotein foams are fortified with fluoronated surfactants. These surfactants enable the foam to shed, or separate from, hydrocarbon fuels. This ability makes them ideally suited for direct foam application using a plunge technique. This means that firefighters can apply fluoroprotein foam streams from a distance. Fluoroprotein foams can also be injected at the base of a burning storage tank and allowed to surface and extinguish the fire. This method is called subsurface injection (Figure 10.34). Fluoroprotein foams also provide a strong "security blanket" for long-term vapor suppression. Vapor suppression is especially critical with unignited spills.

The characteristics of fluoroprotein foams are as follows:

- They may be used with fresh or salt water.
- They have good water retention and extreme resistance to heat.
- They can be stored at temperatures from 20° to 120°F (-7°C to 49°C).
- Their performance is not affected by freezing and thawing.
- The concentrate can be freeze-protected with a nonflammable antifreeze solution.
- They can be stored premixed for a short period of time.
- They maintain rather low viscosity at low storage and use temperatures.
- They are compatible with simultaneous application of dry chemical extinguishing agents.

Fluoroprotein foam can be formulated to be alcohol resistant by adding ammonia salts suspended in organic solvents. Alcohol-resistant fluoroprotein foams maintain their alcohol-resistive properties in solutions much longer — 15 minutes as compared with seconds for alcohol-resistant protein foam.

FILM FORMING FLUOROPROTEIN FOAM (FFFP)

Film forming fluoroprotein foam (FFFP) concentrate is based on fluoroprotein foam technology with aqueous film forming foam (AFFF) capabilities (**NOTE:** AFFF is discussed in the next section). These film forming fluoroprotein foams incorporate the benefits of AFFF for fast fire knockdown and the benefits of fluoroprotein foam for long-lasting heat resistance. The characteristics of FFFP are as follows:

- They can be stored at temperatures from 10° to 120°F (-12°C to 49°C) with good low-temperature viscosity.
- They can be stored premixed, that is, already proportioned with water for use in portable fire extinguishers and fire apparatus water tanks.
- They are compatible with simultaneous application of dry chemical fire fighting agents.
- Their performance is not affected by freezing and thawing.

As with other fluoroprotein-based fire fighting foams, FFFP foam is available in an alcohol-resistant formulation. Alcohol-resistant FFFP has all the fire fighting

Figure 10.34 Fluoroprotein foam is used for subsurface injection applications.

capabilities of regular FFFP, plus some additional advantages:

- Is multipurpose — can be used on polar solvent fuels at 6 percent and on hydrocarbon fuels at 3 percent
- Can be stored at temperatures from 25° to 120°F (-4°C to 49°C)

AQUEOUS FILM FORMING FOAM (AFFF)

Since detergent foam's entry into the fire service some years earlier, the U.S. Navy was interested in using it as a flammable liquid fire fighting foam. They discovered that when a fluoronated surfactant was added to detergent foam, the water that drained from the foam blanket actually floated on hydrocarbon fuel spills — "light water." This film is known as an aqueous film (Figure 10.35). The characteristics of AFFF are as follows:

- Is detergent based and is premixable in portable fire extinguishers and apparatus water tanks (Figure 10.36).

Figure 10.35 An AFFF container.

Figure 10.36 An AFFF extinguisher.

- Can be used through nonaerating nozzles, which gives AFFF penetrating capabilities in baled storage fuels or high surface tension fuels such as treated wood (Figure 10.37).
- Can be stored at temperatures from 25° to 120°F (-5°C to 49°C). (**NOTE:** Some AFFFs are adversely affected by freezing and thawing, so consult the manufacturer.)
- Can be freeze-protected with a nonflammable antifreeze solution.
- Good low-temperature viscosity.

When AFFF is applied to a hydrocarbon fire, three things take place:

- The foam blanket immediately begins to drain water, which floats on the fuel (Figure 10.38). In this way, the light water sends an air-excluding film ahead of the foam blanket. Excluding air leads to a dramatic fire knockdown.
- The rather fast-moving foam blanket then moves across the spill, adding further insulation.
- As the aerated (7-10:1) foam blanket continues to drain its water, more film is released. This gives AFFF the ability to "heal" over areas where the foam blanket is disturbed.

Today, AFFF is available in concentrations from 1 to 6 percent for use with fresh or salt water. AFFF is ideal for crash rescue fire fighting involving spills or anywhere transportation-related fires occur.

Alcohol-resistant AFFFs are available from most foam manufacturers; like other AFFFs, these are also

Figure 10.37 Foam can be useful when extinguishing deep-seated fires such as a hay barn fire.

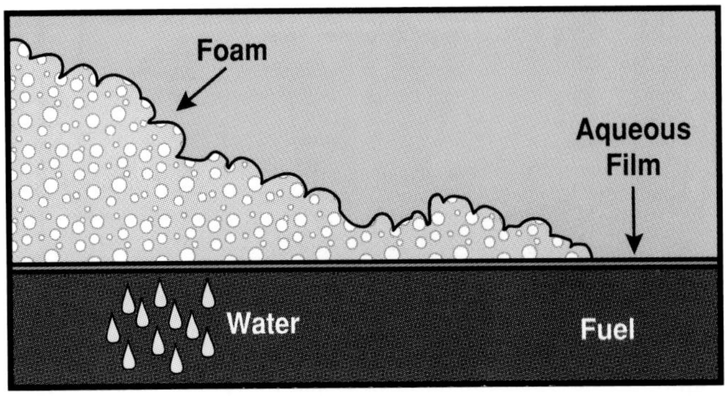

Figure 10.38 The film of AFFF will actually float ahead of the foam blanket.

submerged in the foam concentrate draws concentrate into the water stream, creating a foam water solution.

Balanced-pressure proportioners and around-the-pump proportioners are both systems that are built into the apparatus fire pump. Apparatus equipped with a balanced-pressure proportioner have a foam concentrate line connected to each discharge outlet (Figure 10.44). This line is supplied by a foam concentrate pump separate from the main fire pump. The foam concentrate pump draws the concentrate from a fixed tank on the apparatus. This pump is designed to supply foam concentrate to a desired outlet at the same pressure at which the fire pump is supplying water to that discharge.

The around-the-pump proportioning system consists of a pickup line from the discharge side of the pump back to the intake side of the pump (Figure 10.45). An inline eductor is positioned on the pump bypass. These units are rated for a specific flow and should be used at this rate, although they do have some flexibility. For example, a unit designed to flow 500 gpm (2 000 L/min) at a 6 percent concentration will flow 1,000 gpm (4 000 L/min) at a 3 percent rate.

Foam Makers

Nozzles designed to discharge foam are called foam makers. There are many types that can be used, including standard water stream nozzles. The following sections highlight some of the more common foam makers.

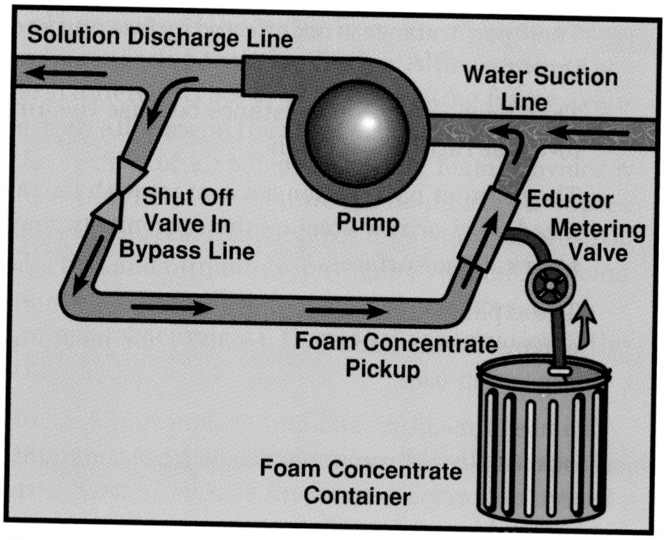

Figure 10.45 An around-the-pump proportioning system.

AIR-ASPIRATING FOAM NOZZLE

The most effective appliance for the generation of low-expansion foam is the air-aspirating foam nozzle. The foam nozzle is especially designed to provide the aeration required to make the highest quality foam possible. These makers may be hand held (30 to 250 gpm [125 L/min to 1 000 L/min]) or monitor-mounted (250 gpm [1 000 L/min] or more) units (Figure 10.46).

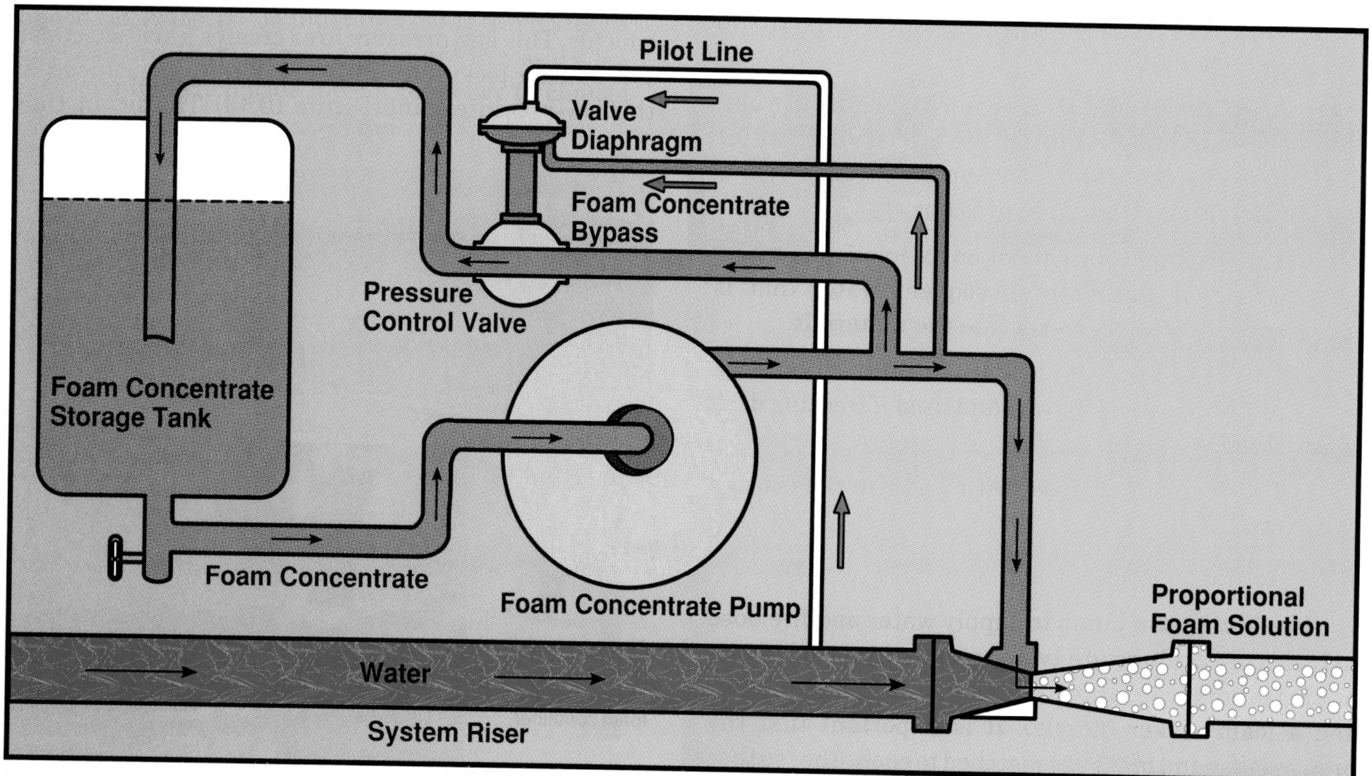

Figure 10.44 A balanced pressure proportioning system.

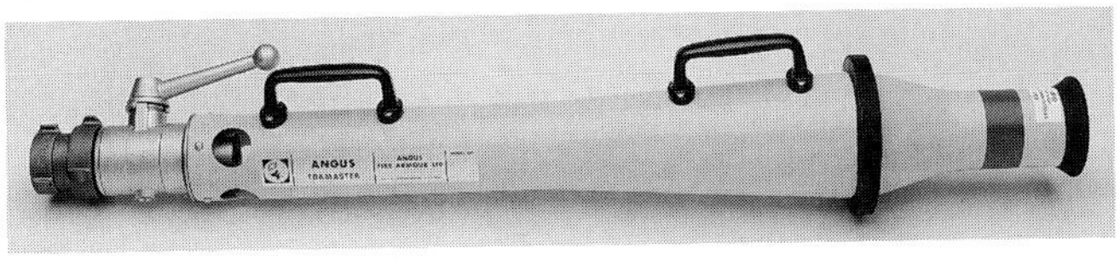

Figure 10.46 An air-aspirating foam nozzle. *Courtesy of Angus Fire Armour, Ltd.*

STANDARD FIXED-FLOW FOG NOZZLE

The fixed-flow, variable-pattern fog nozzle is used with foam solution to produce a low-quality, short-lasting foam. This nozzle breaks the foam solution into tiny droplets and uses the agitation of water droplets moving through air to achieve its foaming action (Figure 10.47). Its best application is when coupled with AFFF, which due to its filming characteristic does not require a high-quality foam to be effective. These nozzles cannot be used with protein or fluoroprotein foams. Water fog nozzles have found a growing acceptance because of their ability to be used as both a standard fire fighting nozzle and as a foam nozzle.

Figure 10.47 A standard fixed-flow fog nozzle can be used to deliver a low quality foam blanket.

AUTOMATIC NOZZLES

An automatic nozzle operates with an eductor in the same way that a fixed-flow fog nozzle operates, providing that the eductor is operated at the inlet pressure for which it was designed and provided that the nozzle is fully open. An automatic nozzle may cause problems if the eductor is operated at a lower pressure than that recommended by the manufacturer or if the nozzle is gated down.

HIGH-EXPANSION FOAM GENERATORS

High-expansion foam generators produce a high-air-content, stable foam. The air content ranges from 100 parts air to one part foam solution (100:1) to 1,000 parts air to one part foam solution (1,000:1). There are two basic types of high-expansion foam generators: the mechanical blower and the water aspirating. The water aspirating type is very similar to the other foam-producing nozzles except it is much larger and longer. The back of the nozzle is open to allow airflow. The foam solution is pumped through the nozzle in a fine spray that mixes with air to form a moderate-expansion foam. The end of the nozzle has a screen, or series of screens, that breaks the foam up and further mixes it with air (Figure 10.48). These nozzles typically produce a lower-air-volume foam than do mechanical blower generators.

Mechanical blower generators resemble smoke ejectors in appearance (Figures 10.49 a and b). They operate along the same principle as the water-aspirating nozzle except the air is forced through the foam spray instead of being pulled through by the water movement. These devices produce a higher-air-content foam and are typically associated with total-flooding applications.

Figure 10.48 High-expansion foam may be delivered from a special nozzle.

Figure 10.49a A large-capacity high-expansion foam blower. *Courtesy of Walter Kidde, Inc.*

Figure 10.49b A smaller high-expansion foam blower. *Courtesy of College Park (MD) Fire Department.*

Assembling A Foam Fire Stream System

To provide a foam fire stream, the firefighter or apparatus driver/operator must be able to correctly assemble the components of the system. The following procedure describes the steps for placing a foam line in service using an in-line proportioner.

Step 1: Select the proper foam concentrate for the burning fuel involved.

Step 2: Check the eductor and nozzle to make sure that they are hydraulically compatible (rated for same flow).

Step 3: Check to see that the foam concentration listed on the foam container matches the eductor percentage rating. If the eductor is adjustable, set it to the proper concentration setting.

Step 4: Attach the eductor to a hose capable of efficiently flowing the rated capacity of the eductor and the nozzle.

— Avoid kinks in the hose.

— If the eductor is attached directly to a pump discharge outlet, make sure that the ball valve gates are completely open. In addition, avoid connections to discharge elbows. This is important because any condition that causes water turbulence will adversely affect the operation of the eductor.

Step 5: Attach the attack hoseline and desired nozzle to the discharge end of the eductor. The length of the hose should not exceed the manufacturer's recommendations.

Step 6: Open enough buckets of foam concentrate to handle the task. Place them at the eductor so that the operation can be carried out without any interruption in the flow of concentrate.

Step 7: Place the eductor suction hose into the concentrate. Make sure that the bottom of the concentrate is no more than 6 feet (2 m) below the eductor (Figure 10.50).

Step 8: Increase the water supply pressure to that required for the eductor. Be sure to consult the manufacturer's recommendations for your specific eductor. Foam should now be flowing.

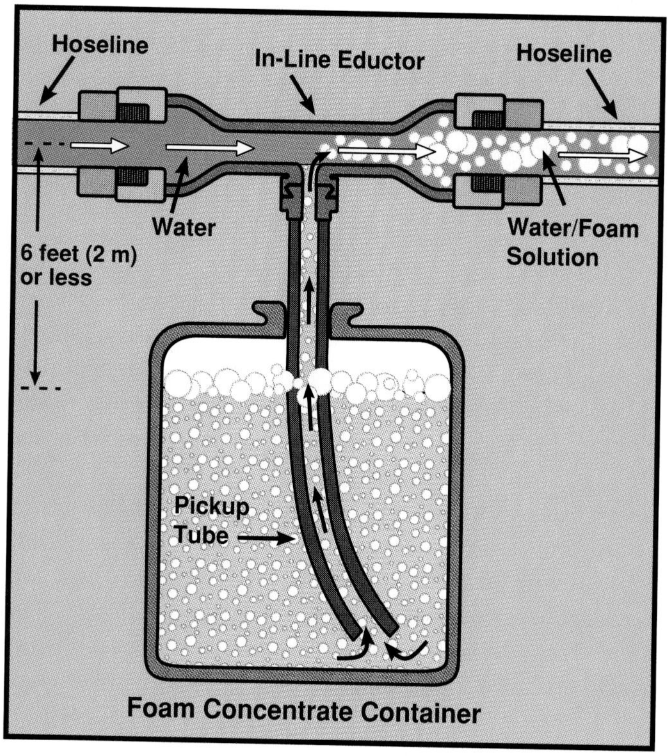

Figure 10.50 The eductor cannot be more than 6 feet (2 m) above the foam concentrate.

Troubleshooting Foam Operations

There are a number of reasons for failure to generate foam or for generating poor quality foam. The most common reasons for failure are as follows:

- Failure to match eductor and nozzle flow, resulting in no pickup of foam concentrate
- Air leaks at fittings that cause loss of suction
- Improper cleaning of proportioning equipment that results in clogged foam passages
- Partially closed nozzle control that results in a higher nozzle pressure
- Too long a hose lay on the discharge side of the eductor
- Kinked hose
- Nozzle too far above eductor (results in excessive elevation pressure)
- Mixing different types of foam concentrate in the same tank, which can result in a mixture too viscous to pass through the eductor

When using other types of foam proportioning equipment, such as balanced pressure proportioners or around-the-pump proportioners, the apparatus operator should be well trained in the proper operation of that particular equipment. Consult the equipment manufacturer for specific operating instructions.

11

Hose

This chapter provides information that addresses performance objectives described in NFPA 1001, *Standard for Fire Fighter Professional Qualifications* (1992), particularly those referenced in the following sections:

3-12 Fire Hose, Appliances and Streams

3-12.1

3-12.2

3-12.3

3-12.4

3-12.5

3-12.6

3-12.7

4-12 Fire Hose, Appliances and Streams

4-12.1

4-12.2

4-12.3

4-12.4

Chapter 11
Hose

The term *fire hose* identifies a type of flexible tube used by firefighters to carry water under pressure from the source of supply to a point where it is discharged. In order to be reliable, fire hose should be constructed of the best materials, and it should not be used for purposes other than fire fighting. Fire hose is the most used item in the fire service. It must be flexible, be watertight, have a smooth lining, and have a durable covering.

CONSTRUCTION OF HOSE

Fire hose is classified by the material from which it is constructed and its size or internal diameter. Present-day fire hose is made of materials that may be susceptible to deterioration and wear. Fire hose materials can be made in several grades and degrees of quality. Fire hose is manufactured in four basic construction methods: braided, wrapped, woven jacket, and rubber covered (Table 11.1). Some of the major fibers or materials used in the construction of the outer jacket of fire hose are cotton, nylon, rayon vinyl, rubber blends, or polyester fibers. Fire hose must be able to withstand relatively high pressures, must be able to transport water with a minimum loss in pressure, and must be sufficiently flexible to load into a hose compartment without occupying excessive space (Figure 11.1).

TABLE 11.1
USE CLASSIFICATION/TYPE OF CONSTRUCTION

Woven-Jacket	Rubber-Covered	Braided	Wrapped
Attack	Attack	Booster	Hard Suction
MDH Relay-Supply	MDH Relay-Supply	Extinguisher	
LDH Relay-Supply	LDH Relay-Supply		
Soft Sleeve	Soft Sleeve		

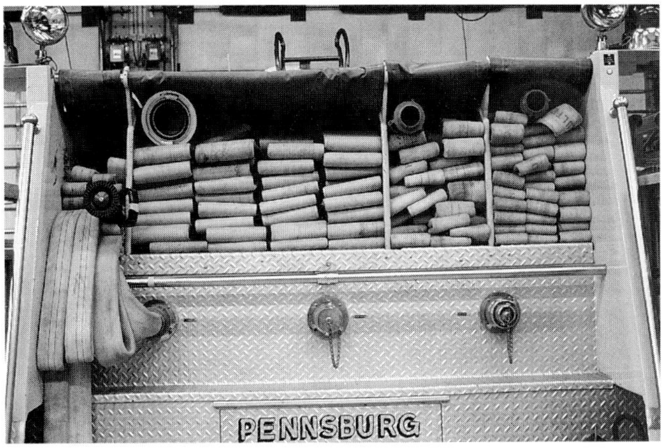

Figure 11.1 Hose loaded in an efficient manner.

HOSE SIZE AND USES

Each size of fire hose is designed for a specific purpose. Reference made to the diameter of the fire hose refers to the dimensions of the inside diameter of the hose. Fire hose is most commonly cut and coupled into lengths of 50 or 100 feet (15 m or 30 m) for convenience of handling and replacement, but other lengths may be obtained. These lengths are also referred to as sections, and they must be coupled together to produce a continuous hoseline. NFPA 1961, *Standard on Fire Hose*, lists specifications for fire hose; NFPA 1963, *Standard for Screw Threads and Gaskets for Fire Hose Connections*, lists specifications for fire hose couplings and screw threads. Table 11.2 on the next page shows the common sizes of fire hose. NFPA 1901, *Standard for Pumper Fire Apparatus*, requires pumpers to carry at least 15 feet (5 m) of large intake hose, 1,200 feet (360 m) of 2½-inch (65 mm) or larger supply hose, and 400 feet (124 m) of 1½-, 1¾-, or 2-inch (38 mm, 45 mm, or 50 mm) attack hose. These lengths and sizes may be increased, depending on the needs of the department.

CAUSES AND PREVENTION OF HOSE DAMAGE

Fire hose is a tool used during fire fighting that is subjected to many potential sources for damage. Usu-

TABLE 11.2
HOSE SIZES

The following table outlines the common hose sizes in the United States, the metric sizes commonly used in Canada, the metric sizes as written in NFPA 1961, and IFSTA's metric equivalent for those not otherwise covered.

U.S. (Inches)	Canada (mm)	NFPA (mm)	IFSTA (mm)
¾ booster	20	—	—
1 booster or forestry	25	—	—
1½	38	38	—
1¾	45	44	—
2	50	51	—
2½	65	64	—
2¾	—	—	70
3	77	76	—
3½	90	89	—
4	100	102	—
4½	115	114	—
5	125	127	—
6	150	152	—

ally, little can be done at fires to provide safe usage and to protect the hose from injury. The most important factor relating to the life of fire hose is the care it gets after fires, in storage, and on the fire apparatus. Fire hose should be selected with caution to ensure its lasting qualities. Regardless of being constructed of quality materials, it cannot stand up under mechanical injury, heat, mildew and mold, and chemical contacts. The life of fire hose is, however, considerably dependent upon how well the hose is protected against these destructive causes.

Mechanical Damage

Fire hose may be damaged in a variety of ways while being used at fires. Some common mechanical injuries are worn places, rips, abrasions, crushed or damaged couplings, and cracked interlinings (Figure 11.2). To prevent these damages, the following practices are recommended:

- Avoid laying or pulling hose over rough, sharp corners (Figure 11.3).
- Provide warning devices (traffic cones, barricades, etc.), and use hose bridges in traffic lanes (Figure 11.4).
- Prevent vehicles from running over fire hose.
- Close the nozzle slowly to prevent a water hammer.
- Change position of bends in hose when reloading.
- Provide chafing blocks to prevent abrasion to hose when it vibrates near the pumper (Figure 11.5).
- Avoid excessive pump pressure on hoselines.

Figure 11.2 Remove burrs and abrasions on the thread surface with a three-cornered file.

Figure 11.3 To prevent damage to hose, avoid pulling hose over sharp corners.

Figure 11.4 To prevent damage to hose, use traffic cones as a warning device in the hose area. *Courtesy of George Braun, Gainesville (FL) Fire-Rescue.*

Figure 11.5 Chafing blocks help prevent hose from being damaged by apparatus vibrations and rubbing on the pavement.

Thermal Damage

The exposure of hose to excessive heat or its contact with fire will char, melt, or weaken the fabric and dry the rubber lining. A similar drying effect may occur to interlinings when hose is hung in a drying tower for a longer period of time than is necessary or when dried in intense sunlight (Figure 11.6). To prevent thermal damage, firefighters should conform to the following recommended practices.

Figure 11.6 Hose towers are designed for hanging and properly drying hose.

Figure 11.7 Laying hose close to the apparatus exhaust pipe can cause thermal damage to the hose.

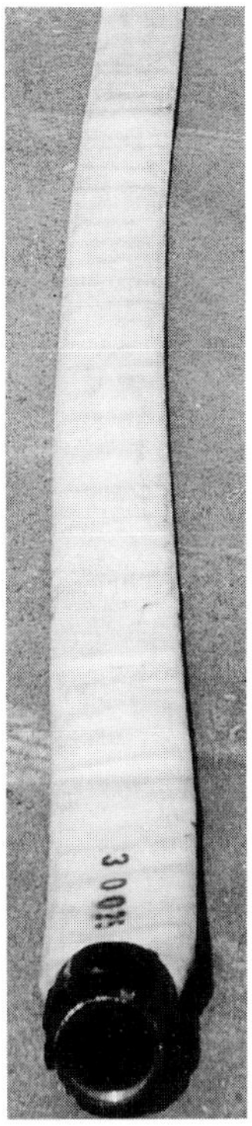

Figure 11.8 Hose should not be dried by laying it on the hot pavement.

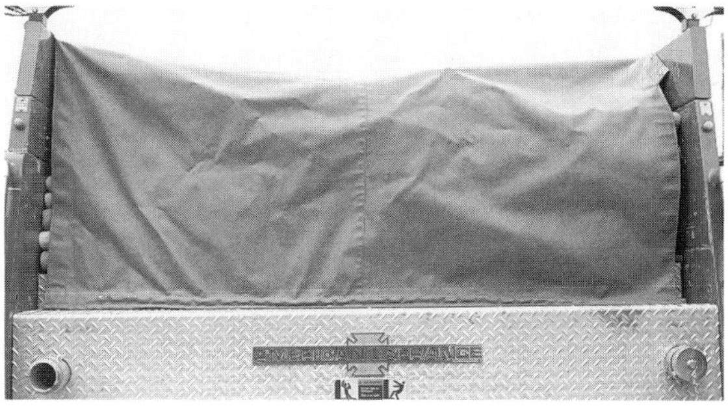

Figure 11.9a Hose can be covered with a tarp to protect it from the sun.

Figure 11.9b Metal hose bed covers have been installed on some apparatus. *Courtesy of Emergency One, Inc.*

- Protect hose from excessive heat or fire when possible (Figure 11.7).
- Do not allow hose to remain in any heated area after it is dry.
- Use moderate temperature for drying. A current of warm air is much better than hot air.
- Keep the outside woven jacket of the hose dry.
- Run water through hose that has not been used for some time to prolong its life.
- Avoid drying fire hose on hot pavement (Figure 11.8).
- Prevent hose from coming in contact with, or being in close proximity to, vehicle exhaust systems.
- Use hose bed covers on apparatus to shield the hose from the sun (Figures 11.9 a through c).

Hose can also be damaged by freezing temperatures. Hose, wet or dry, should not be subjected to freezing conditions for prolonged periods of time.

Figure 11.9c View of the hose beneath a metal hose bed cover. *Courtesy of Santa Rosa (CA) Fire Department.*

Mildew And Mold Damage

Mildew and mold may occur on woven-jacket hose when moisture remains on the outer surfaces (Figures 11.10 a and b). Mildew and mold cause decay and the consequent deterioration of the hose. Rubber-jacket hose

Figure 11.10a Mold and mildew on the outer surface of woven-jacket hose.

Figure 11.10b Remove wet woven-jacket hose from the apparatus to prevent hose from mildewing.

is not subject to mold and mildew damage. Some methods of preventing mildew and mold on woven-jacket hose are as follows:

- All wet hose should be removed from the apparatus after a fire and replaced with dry hose.
- If hose has not been unloaded from the apparatus during a period of 30 days, it should be removed, inspected, swept, and reloaded.
- Some fire hose has been chemically treated to resist mildew and mold but such treatment is not always 100 percent effective. Regardless of this, hose should be exercised every 30 days, and water should be run through it every 90 days to prevent drying and cracking of the rubber lining.

Chemical Damage

Chemicals and chemical vapors will damage the rubber lining and often cause the lining and jacket to separate. When hose is exposed to petroleum products, paints, acids, or alkalies, it may be weakened to the point of bursting. Run-off water from the fire may also carry foreign materials that can damage fire hose. After being exposed to chemicals or chemical vapors, hose should be cleaned as soon as practical. Some recommended good practices are as follows:

- Thoroughly scrub and brush all traces of acid contacts with a solution of baking soda and water. The baking soda neutralizes the acids.
- Periodically remove hose from the apparatus, wash it with plain water, and dry thoroughly.
- Properly test hose if there is the least suspicion of damage.
- Avoid laying hose in the gutter or where automobiles have been parked next to the curb, because parked cars drop oil from the mechanical components and drop acid from the battery (Figure 11.11).
- Properly dispose of hose that has been exposed to hazardous materials and cannot be decontaminated.

Figure 11.11 Avoid laying hose in the gutter where it would be subjected to debris and runoff.

GENERAL CARE AND MAINTENANCE OF HOSE

The life span of fire hose can be appreciably extended if the hose is properly cared for. The techniques of washing and drying and the provisions for storage are very important functions in the care of fire hose. The following sections highlight the proper care of fire hose.

Washing Hose

The method used to wash fire hose depends on the type of hose. Hard-rubber booster hose, hard suction hose, and rubber-jacket collapsible hose require little more than rinsing with clear water, though a mild soap may be used if appearance is important (Figure 11.12).

Most woven-jacket fire hose requires a little more care than the previously mentioned ones. After woven-jacket hose has been in use, the usual accumulation of dust and dirt should be thoroughly brushed from it (Figure 11.13). If the dirt cannot be removed by brushing, the hose should be washed and scrubbed with clear water. When fire hose has been exposed to oil, it should be washed with a mild soap or detergent making sure that all oil is completely removed. The hose should then be properly rinsed. If a commercial hose washing machine is not available, common scrub brushes or brooms can be used with streams of water from a hoseline and nozzle (Figure 11.14). As another alternative method, several hoses may be laid together and scrubbed with a floor buffer.

A hose washing machine is a very important appliance in the care and maintenance of fire hose (Figure 11.15). The most common type washes almost any size of fire hose up to three inches (77 mm). The flow of water into this device can be adjusted as desired, and the movement of the water assists in propelling the hose through the device. The hoseline that supplies the washer with water can be connected to a pumper or used directly from a hydrant. It is obvious that higher pressure gives better results.

Figure 11.12 Hose can be cleaned by rinsing it with water.

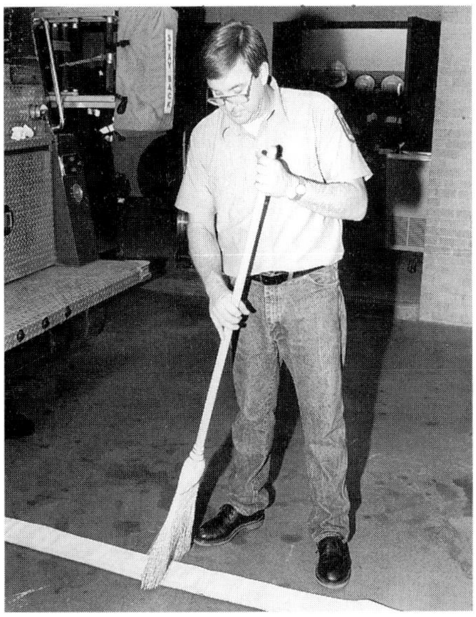

Figure 11.13 Using a broom, brush dirt from woven-jacket hose.

Figure 11.14 Use a scrub brush and water to remove excess dirt and stains.

Figure 11.15 A jet spray washer cleans the hose jacket with a high-pressure water stream that surrounds the hose.

A cabinet-type machine that washes, rinses, and drains fire hose is designed to be used in the station (Figure 11.16). This type of machine can be operated by one person, is self-propelled, and can be used with or without detergents.

Hose washing machines will not clean hose couplings sufficiently when the coupling swivel becomes stiff or sluggish from dirt or other foreign matter. The swivel part should be submerged in a container of warm soapy water and worked forward and backward to thoroughly clean the swivel. The male threads should be cleaned with a suitable brush, and it may be necessary to use a wire brush if clogged by tar, asphalt, or other foreign matter (Figure 11.17).

Drying Hose

The methods used to dry hose depend on the type of hose. Hard-rubber booster hose, hard suction hose, and rubber-jacket collapsible hose may be placed back on the apparatus wet with no ill effects. Woven-jacket hose requires thorough drying before being reloaded on the apparatus.

Once a woven-jacket hose has been thoroughly washed, it should be hung to dry in a hose tower, placed on an inclined drying rack, or placed in a cabinet-type hose dryer (Figures 11.18 a through c on the next page). Where hose drying towers are available, they are usually built as a part of the fire station. Tower drying has proved to be quite successful in most installations.

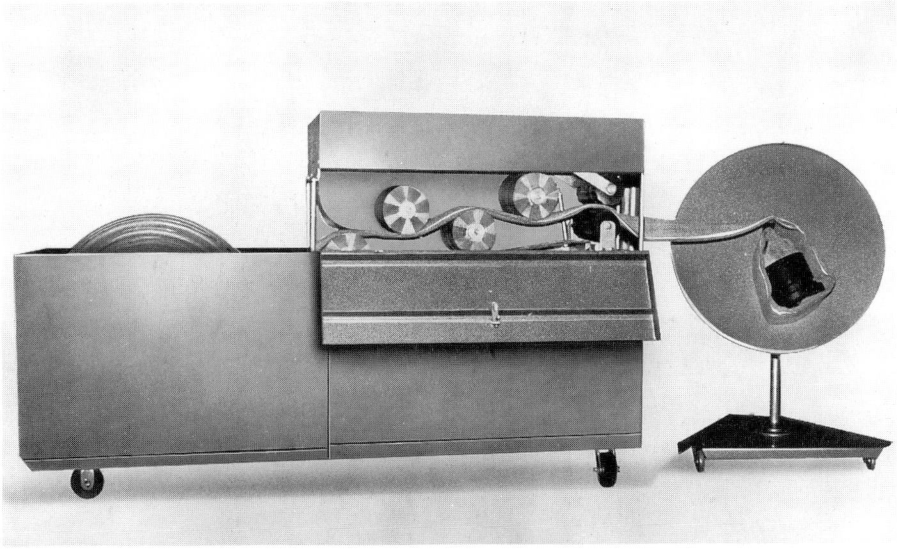

Figure 11.16 A commercially produced hose washing machine used within the fire station. *Courtesy of Circul-Air Corporation.*

Figure 11.17 Use a wire brush to clean male threads that are clogged with foreign materials.

Figure 11.18a Hang washed hose in a hose tower to dry.

Figure 11.18b An indoor hose drying rack.

Figure 11.18c A mechanical hose dryer. *Courtesy of Circul-Air Corp.*

However, the tower must be adequately ventilated, and the hose must not be exposed to excessive temperatures or direct sunlight.

Most authorities feel that it is best for the inside of the rubber tube to remain damp or moist while in storage or on the fire apparatus. Because tower drying provides complete draining and often drying of the rubber tube, it may be desirable to dampen the inner tube after the outer jacket has dried.

Hose drying racks can be constructed with a wood or metal framework and covered with wood slats. Drying racks are more adaptable to long, spacious interior walls that are free from window openings. Wall openings may permit the sun's rays to shine upon the outstretched fire hose. Each deck should be approximately 52 feet (16 m) long, and an overall space of 60 feet (18 m) should be provided to permit adequate working space at each end. If hose sections that are longer than 50 feet (15 m) are used, the racks must be designed so that the entire section can be put on the rack without having to place a bend in the middle that will prevent drainage from the hose. In areas where the climate is agreeable, outdoor hose drying racks may be constructed. However, these outdoor racks should be shielded from direct sunlight, as direct sunlight can be damaging to the hose in prolonged exposures (Figure 11.19).

Mechanical cabinet dryers may be purchased commercially, or they may be built into the structure of a building. Built-in dryers should be provided with adequate fans for forced air circulation within the drying space. Most mechanical cabinet dryers are equipped with thermostats to control temperature and fans to circulate the warm air. When possible, provisions to vent the damp air to the outside should also be made (Figure 11.20). Racks, upon which the hose is placed, are usually removable so that the sections of hose can be arranged for complete drying.

Figure 11.19 Outdoor hose drying rack.

Figure 11.20 Hose dryers should be vented to the outside of the station.

Storing Hose

After fire hose has been adequately brushed, washed, dried, and rolled, it should be stored in suitable racks. Hose racks should be located in a clean, well-ventilated room in or close to the apparatus room for easy access. Racks can be mounted permanently on the wall or can free-stand on the floor (Figure 11.21). Mobile hose racks can be used to both store hose and move hose from storage rooms to the apparatus for loading.

Figure 11.21 Clean, dry hose should be rolled and stored on racks.

HOSE COUPLINGS

Fire hose couplings are made of durable materials and designed so that it is possible to couple and uncouple them in a short time with little effort. The materials used for fire hose couplings are generally alloys in varied percentages of brass, aluminum, or magnesium. These alloys make the coupling more durable and easier to attach to the hose. Much of the efficiency of the fire hose operation depends upon the condition and maintenance of its couplings. Firefighters should be knowledgeable of the type of couplings with which they work.

Types Of Hose Couplings

There are several types of hose couplings used in the fire service. The most commonly used couplings are the threaded and Storz types (Figures 11.22 a and b). Other types of couplings used with less frequency are the quarter turn, oilfield rocker lug, and snap (sometimes referred to as the Jones snap) (Figure 11.23). Couplings

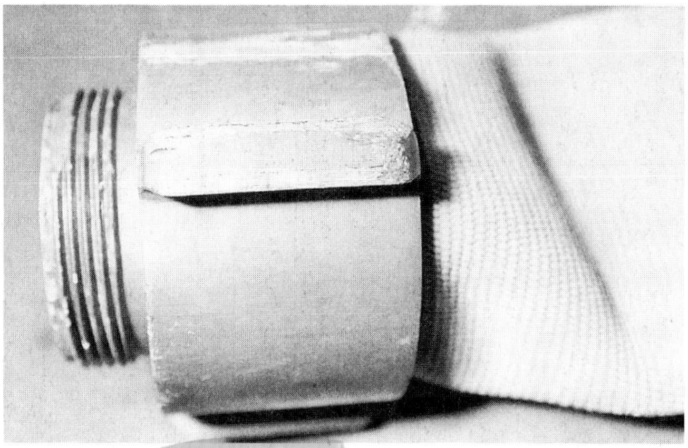

Figure 11.22a A threaded coupling.

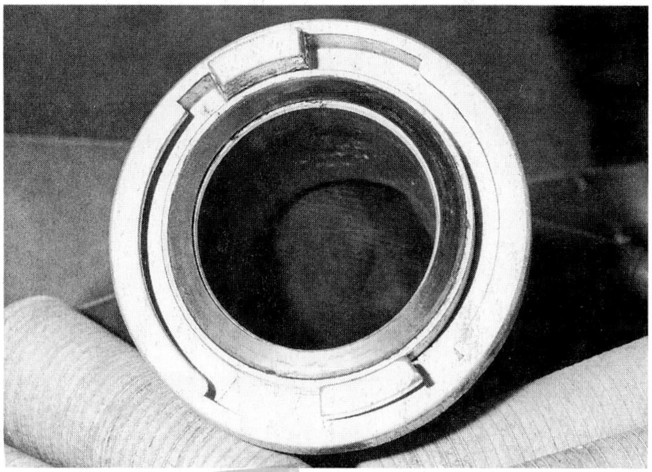

Figure 11.22b A Storz coupling.

Figure 11.23 Snap couplings interlock when the two spring-loaded hooks on the female coupling engage a ring on the shank of the male coupling.

constructed of metals, such as brass, aluminum alloy, and aluminum alloy with a hard coating, are not subject to rusting. The couplings are made by forging, extruding, or casting. Drop-forged couplings are the hardest type and stand up well to normal use. Extruded couplings tend to be somewhat weaker than drop-forged couplings, although they are fine for fire fighting operations. Cast couplings are the weakest and are rarely used on modern fire hose.

Threaded Couplings

Threaded couplings are either three piece or five piece (Figure 11.24). Five-piece types are reducing couplings that are used when the needed coupling size is smaller than the hose it is attached to. This is done so that hoses of different sizes can be connected without using adapter fittings. The portion that serves as a point of attachment to hose is called the shank (also called the tailpiece, bowl, or shell). For the purpose of this manual, it will be referred to as the shank. The male side of a connected coupling can be distinguished from the female side by noting the rocker lugs or pins on the shank. Only male shanks have rocker lugs or pins (Figure 11.25) .

An added feature that may be obtained with screw-type couplings is the Higbee cut and indicator. The Higbee cut is a special type of thread design in which the beginning of the thread is "cut" to provide a positive connection between the first thread of opposing couplings, which tends to eliminate cross-threading (Figure 11.26).

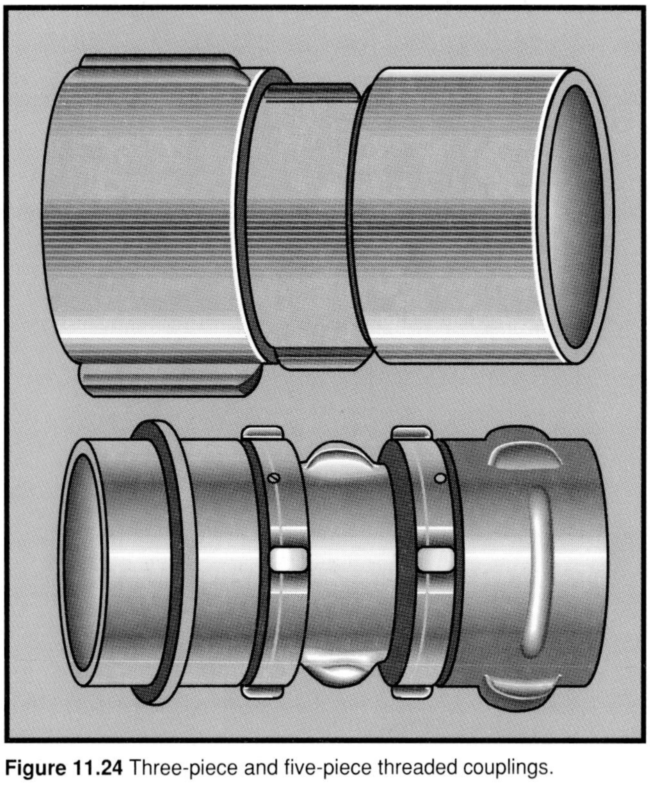

Figure 11.24 Three-piece and five-piece threaded couplings.

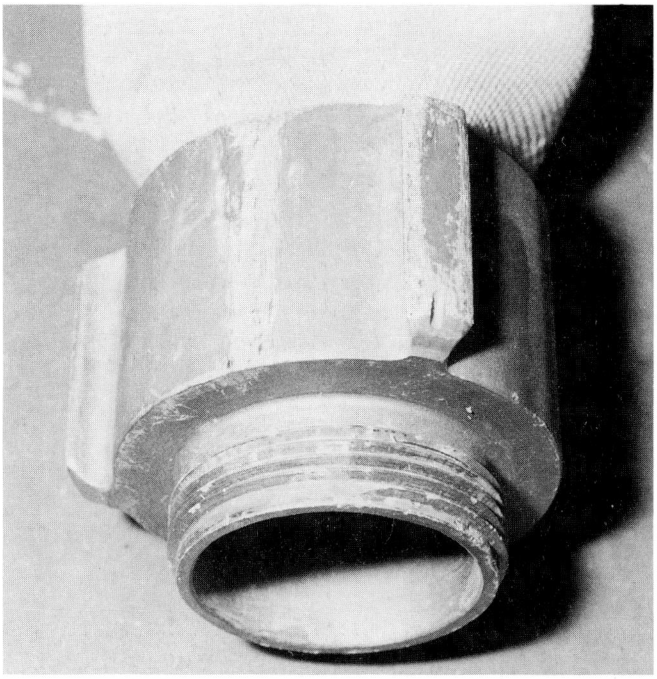

Figure 11.25 The male side of a connected coupling can be distinguished from the female side by noting the rocker lugs or pins on the shank.

Figure 11.26 This shows the location of the Higbee cut.

Each threaded coupling is manufactured with lugs to aid in tightening and loosening connections. They also aid in grasping the coupling when making and breaking coupling connections. Connections may be made by hand or with spanners, which are special tools that fit against the lugs. There are three types of lugs: pin, rocker, and recessed (Figures 11.27 a through c). Although still available, pin-lug couplings are not commonly ordered with new hose because of their tendency to hang up when hose is dragged over objects. Booster hose normally has couplings with recessed lugs, which are simply shallow holes drilled into the coupling. This lug design prevents abrasion that would occur if the hose had protruding lugs and was wound onto reels. These holes are designed to accept a special spanner wrench that can be used to

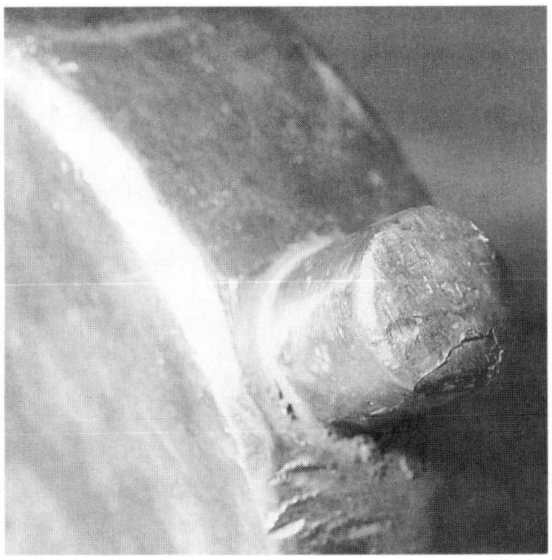

Figure 11.27a Pin-lug couplings are not commonly ordered with new hose because of their tendency to hang up when hose is dragged over objects.

Figure 11.27b Recessed lugs are simply shallow holes drilled into the coupling.

Figure 11.27c Most of the hose purchased today comes equipped with rocker lugs, which helps the coupling slide over obstructions when the hose is moved on the ground or around objects.

couple or uncouple the hose (Figure 11.28). Modern threaded couplings have rounded rocker lugs. Most hose purchased today comes equipped with rocker lugs to help the coupling slide over obstructions when the hose is moved on the ground or around objects. Hose couplings may be obtained with either two or three rocker lugs. On the couplings, one of the rocker lugs on the swivel is scalloped with a shallow indention, called the Higbee indicator, to mark where the Higbee cut begins. This indicator aids in matching the male coupling thread to the female coupling thread, which is not readily visible.

Figure 11.28 This special spanner wrench is used to couple or uncouple hose with recessed lug couplings.

Hose couplings for the various sizes of intake hose are equipped with extended lugs which afford convenient handles for attaching intake hose to a hydrant or pump intake (Figure 11.29). The general three-piece construction of fire hose couplings is also used for intake hose couplings.

Storz-Type Couplings

Storz-type couplings are sometimes referred to as "sexless" couplings. This term means that there are no distinct male and female components; both couplings are identical and may be connected to each other. These couplings are designed to be connected and disconnected

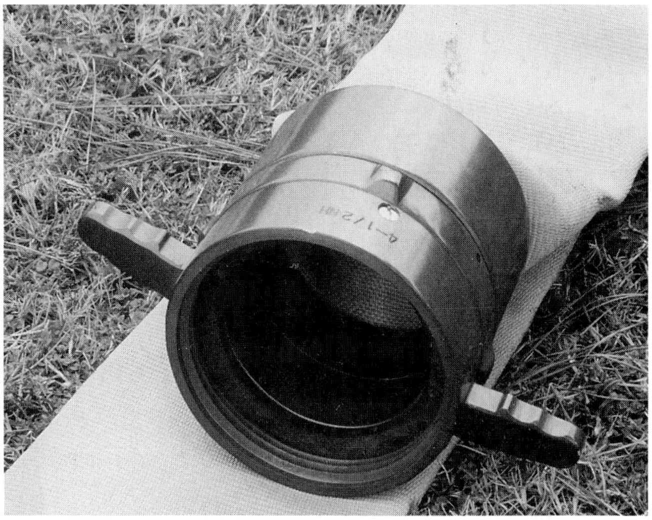

Figure 11.29 An intake hose coupling.

with only one-third of a turn. The locking components are grooved lugs and inset rings built into the swivels of each coupling (Figure 11.30 on the next page). When mated, the lugs of each coupling fit into the recesses in the opposing coupling ring, and then slide into locking position with a one-third turn.

For more information on how both types of couplings are fastened to the fire hose, see IFSTA's **Hose Practices** manual.

Care Of Fire Hose Couplings

All parts of the fire hose coupling are susceptible to damage. On threaded couplings, the male threads are exposed when not connected and are subject to damage. The female threads are not exposed, but the swivel is subject to bending and damage. When coupling parts are connected, there is less danger of damage to either part during common usage; however, they can be bent or crushed if they are run over by heavy vehicles. This is

Figure 11.30 A view of the Storz locking components.

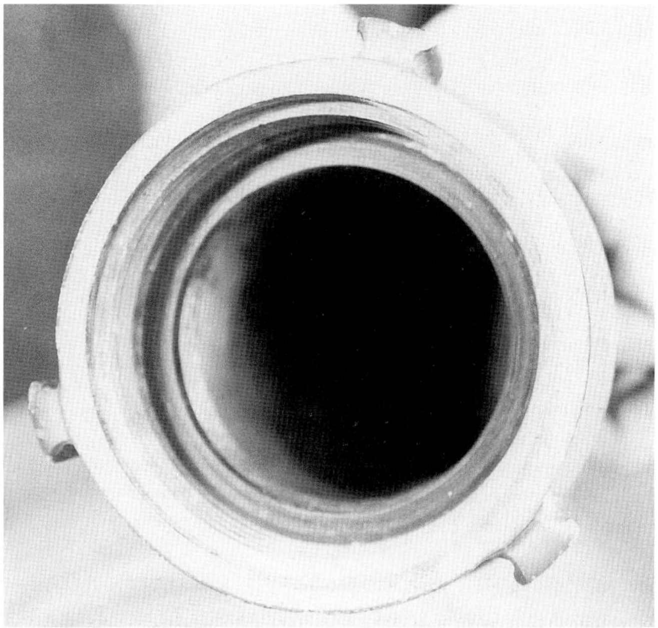

Figure 11.31 A view of a gasket in position.

reason enough to prohibit vehicles from running over fire hose. Some simple rules for the care of fire hose couplings are as follows:

- Avoid dropping and/or dragging couplings.
- Do not permit vehicles to run over fire hose.
- Examine couplings when hose is washed and dried.
- Remove the gasket and twist the swivel in warm, soapy water.
- Clean threads to remove tar, dirt, gravel, and oil.
- Inspect gasket, and replace if cracked or creased.

The swivel gasket and the expansion-ring gasket are two types of gaskets used with fire hose couplings. The swivel gasket is used to make the connection watertight when a female and male end are connected (Figure 11.31). The expansion-ring gasket is used at the end of the hose where it is expanded into the shank of the coupling. These two gaskets are not interchangeable. The difference lies between their thickness and width. Swivel gaskets should occasionally be removed from the coupling and checked for cracks, creases, and general elastic deterioration. The gasket inspection can be made by simply pinching the gasket together between the thumb and index finger (Figure 11.32). This method usually discloses any defects and demonstrates the inability of the gasket to return to normal shape. When replacing the gasket, use the following technique:

Step 1: Hold the gasket between your middle finger and thumb with your index finger resting on the inside rim of the gasket (Figure 11.33).

Figure 11.32 To inspect the gasket, simply pinch the gasket together between the thumb and index finger.

Figure 11.33 Step 1: Hold the gasket between your middle finger and thumb.

Step 2: Fold the outer rim of the gasket upward by pulling with your index finger, and place the gasket into the swivel by permitting the large loop of the gasket to enter into the coupling swivel at the place provided for the gasket (Figure 11.34).

Step 3: Allow the small loop to fall into place by releasing your grip on the gasket (Figure 11.35).

Step 4: If a swivel has become stiff or sluggish, place the swivel into a container of warm, soapy water and work the swivel forward and backward in the solution.

When a coupling has been severely bent (into an egg shape), the usual solution is to replace the coupling.

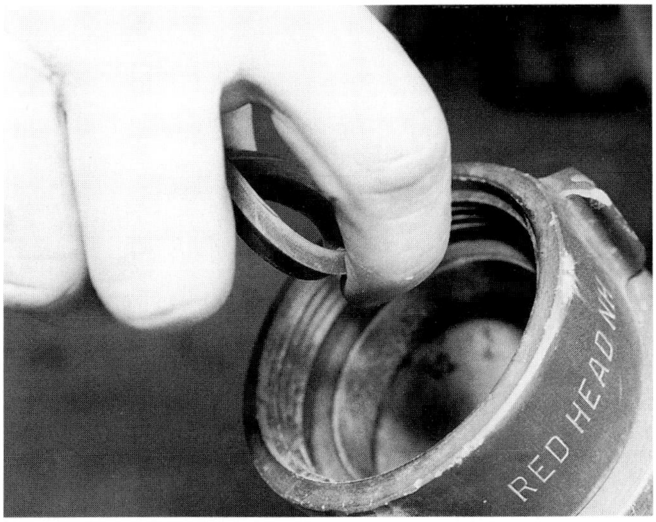

Figure 11.34 Step 2: Fold the outer rim of the gasket upward by pulling with your index finger, and place the gasket into the swivel by permitting the large loop of the gasket to enter into the coupling swivel at the place provided for the gasket.

Figure 11.35 Step 3: Allow the small loop to fall into place by releasing your grip on the gasket.

There are cases, however, when slightly bent or egg-shaped couplings can sometimes be straightened by using the expander tool to help round the shape of the coupling. Tapping the coupling with a small hammer while under slight expansion sometimes helps. After a coupling has been straightened, the threads should be rerun by a tap or die threading tool. Male threads can sometimes be repaired by using a small three-cornered file.

HOSE APPLIANCES AND TOOLS

A complete hose layout for fire fighting purposes includes one end of the hose attached to a source of water supply and the other to a nozzle or similar discharge device. There are various devices other than hose couplings and nozzles used with fire hose to complete such an arrangement. These devices are usually grouped into two categories: hose appliances and hose tools. A simple way to remember the difference between hose appliances and hose tools is that appliances have water flowing through them and tools do not. Appliances include valves; valve devices, such as wyes, siameses, and water thieves; and fittings, which include adapters. Examples of hose tools include hose rollers, spanner wrenches, rope hose tools, and hose clamps. The following section highlights some of the more common hose appliances and hose tools.

Valves

The flow of water is controlled by various valve devices in hoselines, at hydrants, and at pumpers. These devices include the following valve types:

- Ball valves — Open when the handle is in line with the hose and closed when it is at a right angle to the hose. Also used in fire pump piping systems (Figure 11.36).

Figure 11.36 A partially open (or closed) ball valve.

- Gate valves — Used to control the flow from a hydrant. Gate valves have a baffle that is moved by a handle and screw arrangement (Figure 11.37).
- Butterfly valves — Used on large pump intakes. A butterfly valve uses a flat baffle operated by a quarter-turn handle. The baffle is in the center of the waterway when the valve is open (Figure 11.38).

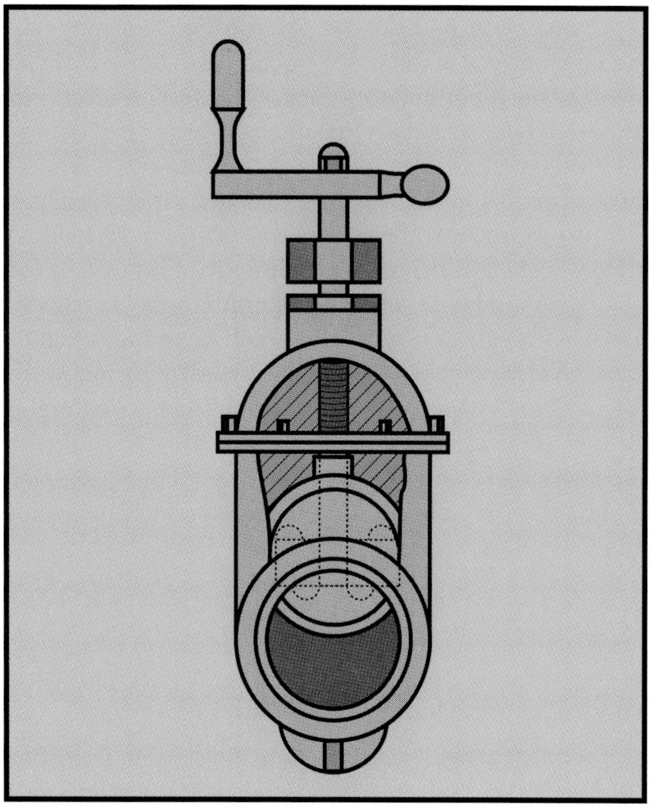

Figure 11.37 A gate valve.

- Clapper valves — Used in siamese appliances to allow only one intake hose to be connected and charged before the addition of more hoses. The clapper is a flat disk that is hinged on one side and swings in a door-like manner (Figure 11.39).

Figure 11.39 This cutaway shows the position of the clapper valve.

Valve Devices

WYE APPLIANCES

Certain situations make it desirable to divide a line of hose into two or more lines. Various types of wye connections are used for this purpose. The most common wye has a 2½-inch (65 mm) inlet to two 1½-inch (38 mm) outlets, though there are many other combinations commonly found in use (Figure 11.40). The 2 ½-inch (65 mm) wye is also used to divide one 2½-inch (65 mm) or larger hoseline into two 2½-inch (65 mm) lines (Figure 11.41). Wye appliances are often gated so that water being fed into the hoselines may be controlled at the gate.

Figure 11.38 A butterfly valve is commonly used on large fire pump intakes.

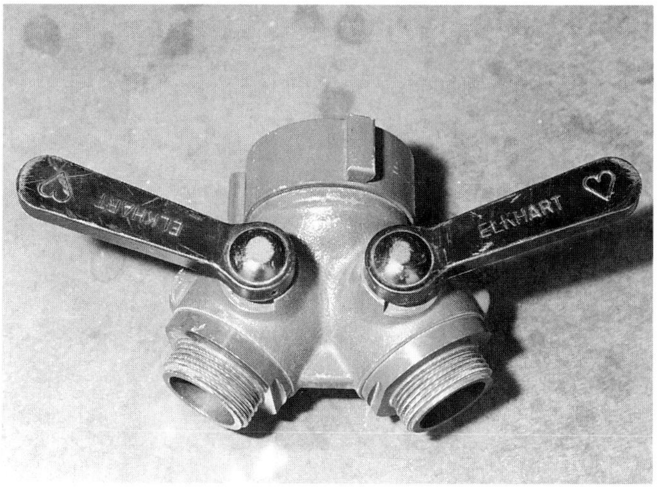

Figure 11.40 This common wye has a 2½-inch (65 mm) inlet to two 1½-inch (38 mm) outlets. It is often referred to as a "leader-line" wye.

Figure 11.41 A 2½-inch (65 mm) wye is used to divide one 2½-inch (65 mm) or larger hoseline into two 2½-inch (65 mm) lines.

SIAMESE APPLIANCES

The siamese and wye adapters are often confused because of their close resemblance. Siamese fire hose layouts consist of two or more hoselines that are brought into one hoseline or device. The typical siamese will have two or three female connections coming into the appliance and one male discharge exiting the appliance (Figure 11.42). Siamese appliances may be equipped with or without clapper valves. The clapper valves allow the siamese to be used with only one incoming supply hoseline attached to it. Siamese appliances are commonly used to overcome the problems encountered due to friction loss in hose lays which need to carry a large flow or cover a long distance. They are also used quite frequently when supplying ladder pipes that are not equipped with a permanent waterway. Two or three lines will be used to supply the one line that is actually going up the ladder. With the increased popularity of large diameter hose, siamese appliances are being used to feed a large diameter hoseline when multiple smaller hoselines have to be used in the same relay as larger diameter hose.

Figure 11.42 A siamese.

WATER THIEF APPLIANCES

The water thief is a variation of the wye appliance. The most common water thief consists of one 2½-inch (65 mm) inlet with one 2½-inch (65 mm) and two 1½-inch (38 mm) discharge outlets, though other versions, such as a 1½-inch (38 mm) to 1-inch (25 mm) model, are in use (Figures 11.43 a and b). Quarter-turn gate valves control the outlets. The water thief is intended to be used on a 2½-inch (65 mm) or larger hoseline, usually near the nozzle, so that 2½-inch (65 mm) and 1½-inch (38 mm) hoselines may be used as desired from the same layout.

Figure 11.43a A standard water thief.

Figure 11.43b This forestry water thief is designed to split a 1-inch (25 mm) hose off the main 1½-inch hose.

fire fighting conditions are such that it is not possible to shut down the hoseline and replace the bad section, a hose jacket can be installed on the hose at the point of rupture. The hose jacket encloses the hose so effectively that it can continue to operate at full pressure. A hose jacket can also be used to connect hose with mismatched or damaged screw-thread couplings.

A hose jacket consists of a two-piece metal cylinder that hinges open and closed (Figure 11.50). Rubber gaskets at each end of the cylinder seal against the hose to prevent leakage. A clamp device locks the cylinder closed when in use. Hose jackets are made in two sizes: 2½-inch and 3-inch (65 mm and 77 mm).

Figure 11.50 A hose jacket in use.

HOSE CLAMPS

A hose clamp is used to stop the flow of water in a hoseline:

- To prevent charging the hose bed during hose layout operations
- To allow replacement of a burst section without shutting down the water supply
- To allow extension of a hoseline without shutting down the water supply

Based on the method by which they work, there are three types of hose clamps: screw-down, press-down, and hydraulic press (Figure 11.51). It is important to remember that a hose clamp can cause injury to firefighters or damage hose if not used correctly. Some general rules that apply to all hose clamps are as follows:

- Apply the hose clamp at least 20 feet (6 m) behind the apparatus.
- Apply the hose clamp no farther than 5 feet (2 m) from the coupling on the incoming water side (Figure 11.52).
- Stand to one side when applying or releasing the press-down type of hose clamp (the operating handle is prone to suddenly snapping open) (Figure 11.53).

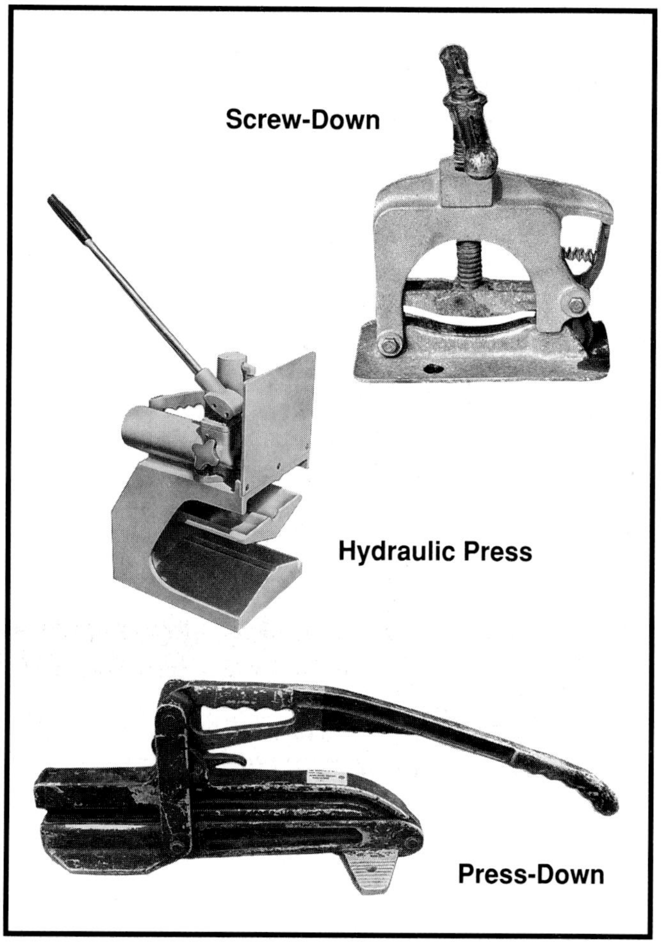

Figure 11.51 Various types of hose clamps.

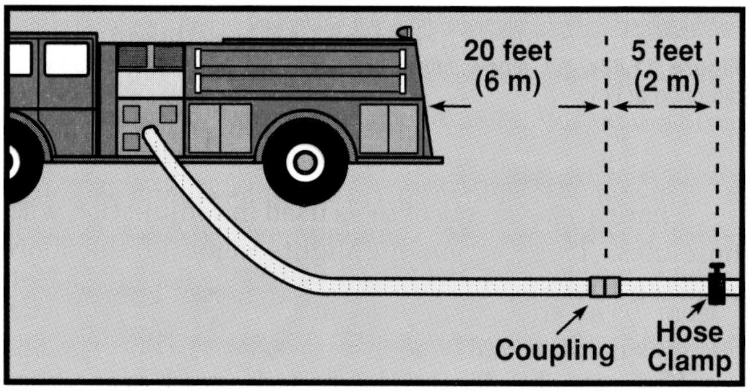

Figure 11.52 Place the hose clamp at least 20 feet (6 m) behind the apparatus and no farther than 5 feet (2 m) behind the coupling.

CAUTION: Never stand over the handle of the hose clamp when applying or releasing it. The handle may swing upward in a violent motion and injure the firefighter attempting to operate the handle.

- Center the hose evenly in the jaws to avoid pinching.
- Close and open the hose clamp slowly to avoid water hammer.

Figure 11.53 Always stand to one side when applying or releasing the press-down type of hose clamp.

SPANNER AND HYDRANT WRENCHES AND RUBBER MALLETS

The primary purpose of a spanner wrench, or spanner, is to tighten and loosen hose couplings (Figure 11.54). A number of other features have been built into some spanner wrenches:

- A wedge for prying
- An opening to fit gas valves
- A slot for pulling nails
- A flat surface for hammering

Hydrant wrenches are primarily used to remove caps from fire hydrant outlets and to open fire hydrant valves (Figure 11.55). The hydrant wrench is usually equipped with a pentagon opening in its head that will fit most standard fire hydrant operating nuts. The lever

Figure 11.54 A spanner wrench.

handle may be threaded into the operating head to make it adjustable, or the head and handle may be of the ratchet type. The head may also be equipped with a spanner to help make and break coupling connections.

The rubber mallet is used to tighten and loosen intake hose couplings. It is sometimes difficult to get a completely airtight connection with intake hose couplings even though these couplings may be equipped with long operating lugs. Thus, the rubber mallet is used to strike the lugs and further tighten the connection (Figure 11.56).

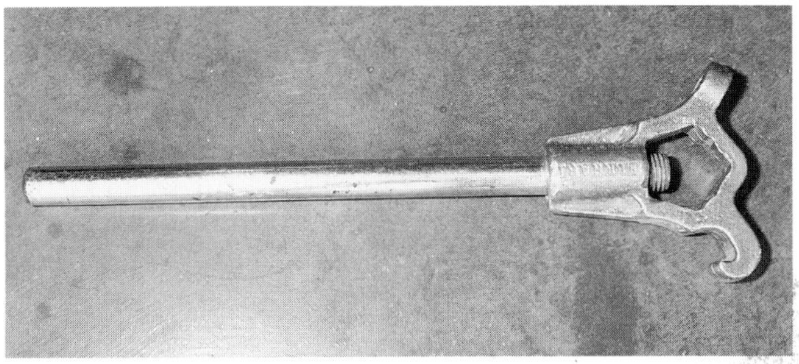

Figure 11.55 A hydrant wrench.

Figure 11.56 A rubber mallet is used to strike the lugs and further tighten the connection.

HOSE RAMPS OR BRIDGES

Hose bridges (also called hose ramps) help prevent injury to hose when vehicles cross it (Figure 11.57). They

Figure 11.57 Hose bridges in use.

should be used whenever a hoseline crosses a street or other area where vehicular traffic cannot be diverted. Hose ramps can also be used as chafing blocks. Some ramps can also be positioned over small spills to keep hoselines out of potentially damaging liquids.

CHAFING BLOCKS

Chafing blocks are devices that are made to fit around fire hose where the hose is to be subjected to rubbing from vibrations (Figure 11.58). Chafing blocks are particularly useful where intake hose comes in contact with pavement or curb steps. At these points, wear on intake hose is most likely because pumper vibrations may be keeping the intake hose in constant motion. Chafing blocks may be made of wood, leather, or sections from old truck tires.

Figure 11.58 Chafing blocks prevent damage to intake hose.

HOSE STRAP, ROPE, AND CHAIN TOOLS

One of the most useful tools to aid in handling a charged hoseline is a hose strap. Similar to the hose strap are the hose rope and hose chain. These devices can be used to carry and pull fire hose, but their primary value is to provide a more secure means to handle pressurized hose when applying water. Another important use of these tools is to secure hose to ladders and other fixed objects (Figures 11.59 a and b).

HOSE ROLLS

There are a number of different methods for rolling hose, depending on its intended use. In all methods, care must be taken to protect all couplings. Some of the various hose rolls will be discussed in the following sections.

Straight Roll

The basic hose roll consists of starting at one end, usually at the male coupling, and rolling the hose toward the other end to complete the roll. When the roll is completed, the female end is exposed, and the male end is protected in the center of the roll. The straight roll is commonly used for hose that is:

- Placed in storage (especially rack storage)
- Returned to quarters for washing (Figure 11.60)
- Loaded back on the apparatus at the fire scene.

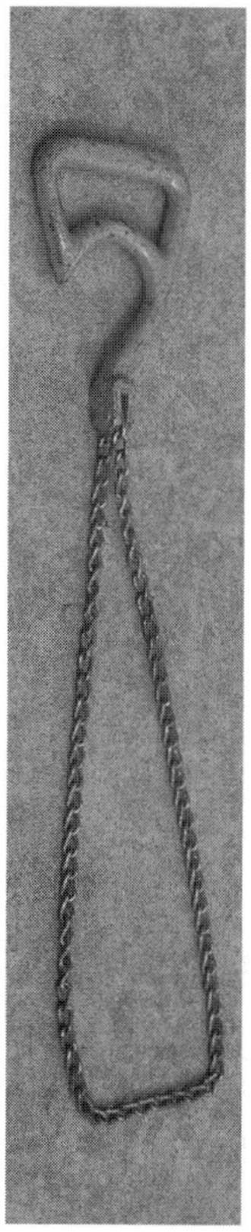

Figure 11.59a A rope hose tool.

Figure 11.59b Attach a hose strap or rope tool so that the nozzle is within easy reach.

Figure 11.60 Straight rolls can be carried back to the station for washing.

The procedure for making the straight roll is as follows:

Step 1: Lay the hose out straight and flat on a clean surface (Figure 11.61).

Step 2: Roll the male coupling over onto the hose to start the roll. Form a coil that is open enough to allow your fingers to be inserted (Figure 11.62).

Step 3: Continue to roll the coupling over onto the hose, forming an even roll. As the roll increases in size, keep its edges aligned on the remaining hose to make a uniform roll (Figure 11.63).

Step 4: Lay the completed roll on the ground. With your foot, tamp any protruding coils down into the roll (Figure 11.64).

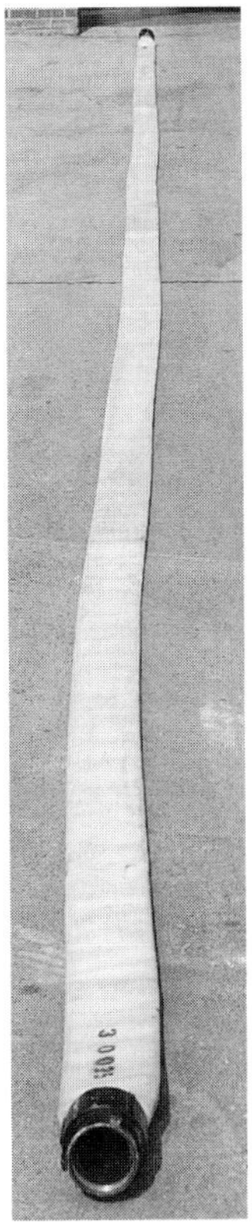

Figure 11.61 Step 1: Lay the hose out straight and flat on a clean surface.

Figure 11.62 Step 2: Roll the male coupling over onto the hose to start the roll.

Figure 11.63 Step 3: Continue to roll the coupling over onto the hose, forming an even roll. As the roll increases in size, keep its edges aligned on the remaining hose to make a uniform roll.

Figure 11.64 Step 4: Lay the completed roll on the ground. With your foot, tamp any protruding coils down into the roll.

A variation of this method is to begin the roll at the female coupling so that when the roll is completed, the male coupling is exposed. This is often done to denote a damaged coupling or piece of hose. A tag is usually attached to the male coupling indicating the type and location of the damage. This is also done when the hose is going to be reloaded on the apparatus for a straight or forward lay.

Donut Roll

The donut roll is commonly used in situations where hose is going to be deployed for use directly from a roll. The donut roll has certain advantages that the straight roll does not possess. Three main advantages are that both ends are available on the outside of the roll, the hose may be quickly unrolled and placed into service, and the hose is less likely to spiral or kink when unrolled. When a section of fire hose needs to be rolled into a donut roll, one or two firefighters may perform the task. The following describes two methods used to make the donut roll.

DONUT ROLL (METHOD ONE)

The first procedure for making the donut roll is as follows:

Step 1: Lay the section of hose flat and in a straight line. Start the roll from a point 5 or 6 feet (2 m) off center toward the male coupling, and roll toward the female end. Leave sufficient space at the center loop to insert your hand for carrying (Figure 11.65 on the next page).

Step 2: Near the completion of the roll, the male coupling will be enclosed within the roll as the hose is rolled over it. Use the short length of hose at the female end to protect the male threads (Figure 11.66 on the next page).

Step 3: When the roll is complete, the male coupling will be inside the roll. The female coupling will be about 3 feet (1 m) ahead of the male coupling (Figure 11.67 on the next page).

DONUT ROLL (METHOD TWO)

A second procedure for making the one-person donut roll is as follows:

Step 1: Grasp either coupling end, carry it to the opposite end, and cause the looped section to lie flat, straight, and without any twists (Figure 11.68 on the next page).

Step 2: Face the coupling ends, start the roll on the male coupling side about 2½ feet (0.7 m) from the bend (1½ feet [0.4 m] for 1½-inch [38 mm] hose), and roll toward the male coupling (Figure 11.69 on the next page).

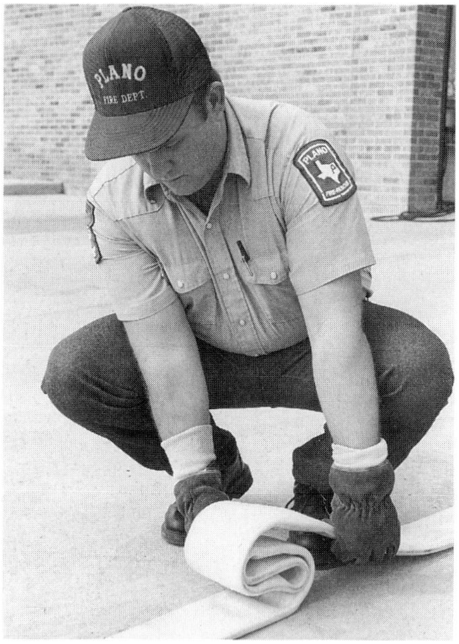

Figure 11.65 Step 1: Lay the section of hose flat and in a straight line. Start the roll from a point 5 or 6 feet (2 m) off center toward the male coupling, and roll toward the female end.

Figure 11.66 Step 2: Near the completion of the roll, the male coupling will be enclosed within the roll as the hose is rolled over it.

Figure 11.67 The completed roll.

Step 3: If the hose behind the roll becomes tight during the roll, pull the female side back a short distance to relieve the tension (Figure 11.70).

Step 4: As the roll approaches the male coupling, lay the roll flat on the ground and draw the female coupling end around the male coupling to complete the roll (Figure 11.71).

Twin Donut Roll

The twin donut roll is more adaptable to 1½-inch (38 mm) and 1¾-inch (45 mm) hose, although 2-, 2½-, or 3-inch (50 mm, 65 mm, or 77 mm) hose can be used. Its purpose is to arrange a compact roll which may be transported and carried for special applications such as high-rise operations. The following steps describe how the twin donut roll can be made.

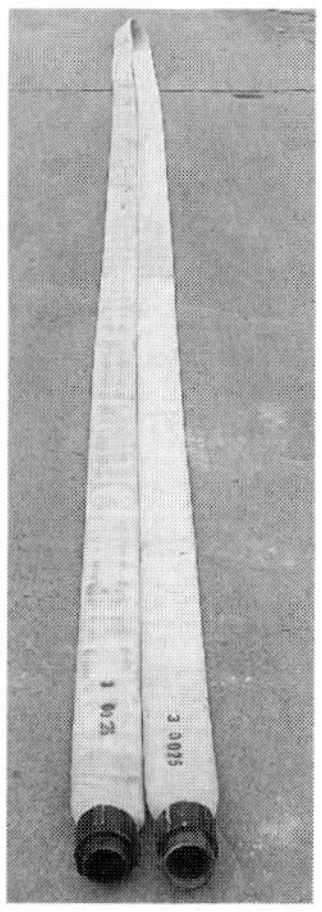

Figure 11.68 Step 1: Grasp either coupling end, carry it to the opposite end, and cause the looped section to lie flat, straight, and without any twists.

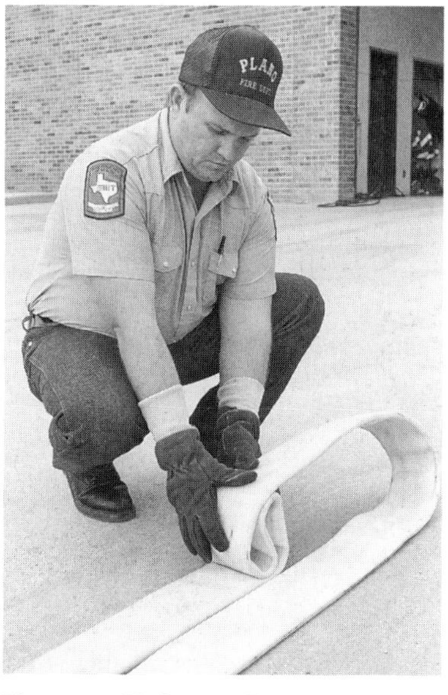

Figure 11.69 Step 2: Face the coupling ends, start the roll on the male coupling side about 2½ feet (0.7 m) from the bend (1½ feet [0.4 m] for 1½-inch [38 mm] hose), and roll toward the male coupling.

Figure 11.70 Step 3: If the hose behind the roll becomes tight during the roll, pull the female side back a short distance to relieve the tension.

Figure 11.71 Step 4: As the roll approaches the male coupling, lay the roll flat on the ground and draw the female coupling end around the male coupling to complete the roll.

Step 1: Place the male and female couplings together. Lay the hose flat, without twisting, to form two parallel lines from the loop end to the couplings (Figure 11.72).

Step 2: Fold the loop end over and upon the two lines to start the roll (Figure 11.73).

Step 3: Continue to roll both lines simultaneously toward the coupling ends, forming a twin roll with a decreased diameter (Figure 11.74).

Step 4: Carry the twin donut roll in the same manner as the standard donut roll. Or, loop a short piece of strap or rope through the roll and tie it with a quick-releasing hitch for fireground operations or storage on fire apparatus (Figure 11.75).

Alternative Method: If the couplings are offset by about one foot (0.3 m) at the beginning, they can be coupled together after the roll is tied or strapped. This forms a convenient loop which can be slung over one shoulder for carrying while leaving the hands free (Figure 11.76). By offsetting the couplings at the beginning they do not dig into your shoulder, but are still readily accessible when you need to place the length in service.

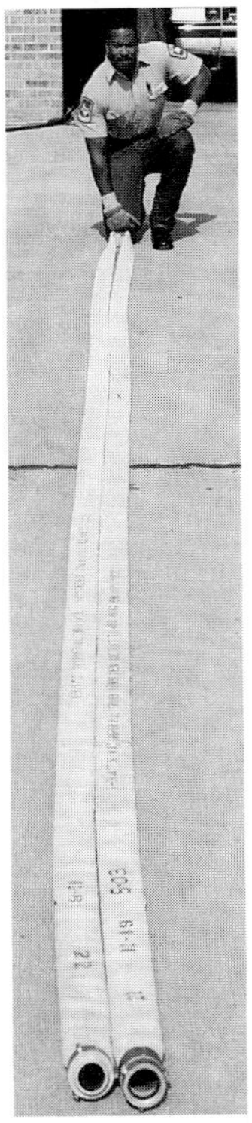

Figure 11.72 Step 1: Stretch the hose out and place the couplings together.

Figure 11.73 Step 2: Fold the loop end over and upon the two lines to start the roll.

Figure 11.74 Step 3: Continue to roll both lines simultaneously toward the coupling ends, forming a twin roll with a decreased diameter.

Figure 11.75 Step 4: One method of tying a roll is to loop a short piece of strap or rope through the roll, and tie it with a quick-releasing hitch.

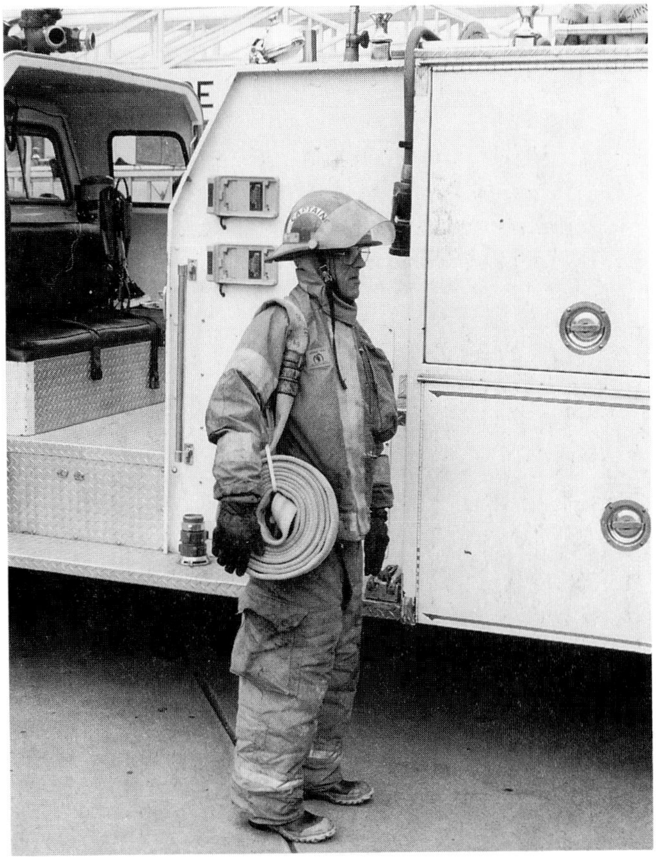

Figure 11.76 An alternative donut roll.

Self-Locking Twin Donut Roll

The self-locking twin donut roll is a twin donut roll that has a built-in carrying strap formed from the hose itself. This strap locks over the couplings to keep the roll intact for carrying. The length of the carrying strap may be adjusted to accommodate the height of the person carrying the hose. The following steps describe how to make the roll:

Step 1: Place the male and female couplings together and lay the hose flat, without twisting, to form parallel lines from the loop end to the couplings (Figure 11.77).

Step 2: Move one side of the hose up and over 2½ to 3 feet (0.8 m to 1 m) to the opposite side without turning. This lay-over method prevents a twist in the hose at the big loop. The size of this loop, known as a butterfly loop, determines the length of the shoulder loop for carrying (Figure 11.78).

Step 3: Face the coupling ends, bring the back side of the loop forward toward the couplings, and place it on top of where the hose crosses. This action forms a loop on each side without twist (Figure 11.79).

Step 4: Start rolling toward the coupling ends, forming two rolls side by side (Figure 11.80).

Step 5: When the rolls are completed, allow the couplings to lie across the top of each roll, and adjust the loops, one short and one long, by pulling only one side of the loop through (Figure 11.81).

Step 6: Place the long loop through the short loop, just behind the couplings, and tighten snugly. The loop forms a shoulder sling (Figure 11.82).

Step 7: Carry the coupling ends in front or to the rear (Figure 11.83).

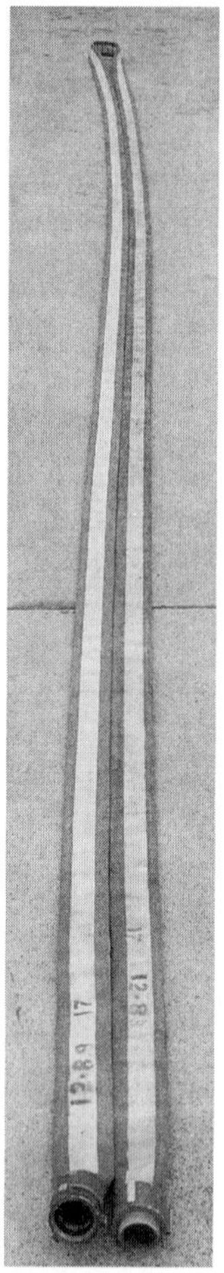

Figure 11.77 Step 1: Place the male and female couplings together and lay the hose flat, without twisting, to form parallel lines from the loop end to the couplings.

Figure 11.78 Step 2: Move one side of the hose up and over 2½ to 3 feet (0.8 m to 1 m) to the opposite side without turning.

Figure 11.79 Step 3: Face the coupling ends, bring the back side of the loop forward toward the couplings, and place it on top of where the hose crosses.

Figure 11.80 Step 4: Start rolling toward the coupling ends, forming two rolls side by side.

Figure 11.81 Step 5: When the rolls are completed, allow the couplings to lie across the top of each roll, and adjust the loops, one short and one long, by pulling only one side of the loop through.

Figure 11.82 Step 6: Place the long loop through the short loop, just behind the couplings, and tighten snugly. The loop forms a shoulder sling.

Figure 11.83 Step 7: Carry the coupling ends in front or to the rear.

DIRECTION OF HOSE LAYS

Threaded-coupling hose must be arranged in the hose bed so that when hose is laid, the end with the female coupling is toward the water source, and the end with the male coupling is toward the fire. When hose is laid in this manner, several options are available. At the water source, hose can be connected to the male threads of a pumper discharge valve or to the male threads of a hydrant. At the fire end, it can be connected to the auxiliary intake valve of a pumper, or it can be connected directly to nozzles and appliances, all of which have female threads. There are three basic hose lays for supply hose: the forward lay, the reverse lay, and the split lay (sometimes called the combination lay).

Figure 11.84 The female coupling comes off first on a forward or straight lay.

Forward Lay

With the forward lay, hose is laid from the water source to the fire. Hose beds set up for forward lays should be loaded so that the first coupling to come off the hose bed is female (Figure 11.84). This method is often used when the water source is a hydrant and the pumper must stay at the fire location (Figure 11.85). The primary advantage with this lay is that a pumper can remain at the incident scene so that its hose, equipment, and tools can be quickly obtained if needed. The pump operator also has visual contact with the fire fighting crew and can better react to changes in the fire operation than if the pumper were at the hydrant. A disadvantage with the lay, however, is that if a long length of medium diameter (2½- or 3-inch [65 mm or 77 mm]) hose is laid, it may be necessary for a second pumper to boost the pressure in the line at the hydrant. This requires the use of a four-way hydrant valve so that the transition from hydrant pressure to pump pressure can be made without interrupting the flow of water in the supply hose. Another disadvantage is that one member of the crew is temporarily unavailable for a fire fighting assignment because that person must stay at the hydrant long enough to make the connection and open the hydrant.

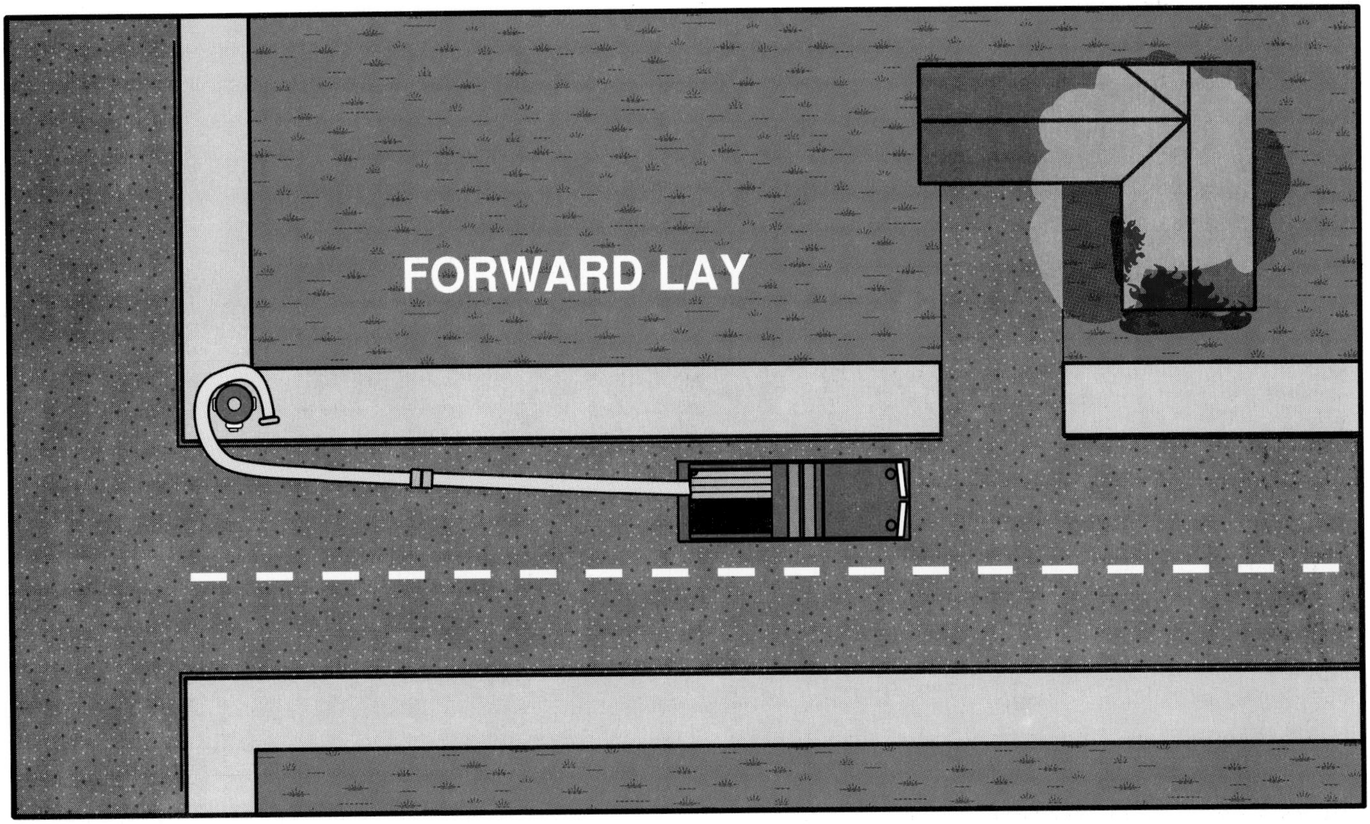

Figure 11.85 The forward lay proceeds from the water source to the scene.

Reverse Lay

With the reverse lay, hose is laid from the fire to the water source. Hose beds set up for reverse lays should be loaded so that the first coupling to come off the hose bed is male (Figure 11.86). This method is used when a pumper must first go to the fire location so that a size-up can be made before laying a supply line (Figure 11.87). It is also the most expedient way to lay hose if the apparatus that lays the hose must stay at the water source, as when drafting or boosting hydrant pressure to the supply line. A disadvantage with the reverse lay, however, is that essential fire fighting equipment, including attack hose, must be removed and placed at the fire location before the pumper can proceed to the water source. This causes some delay in the initial attack. The reverse lay also obligates one person, the pump operator, to stay with the pumper at the water source, thus preventing that person from performing other essential fire location activities.

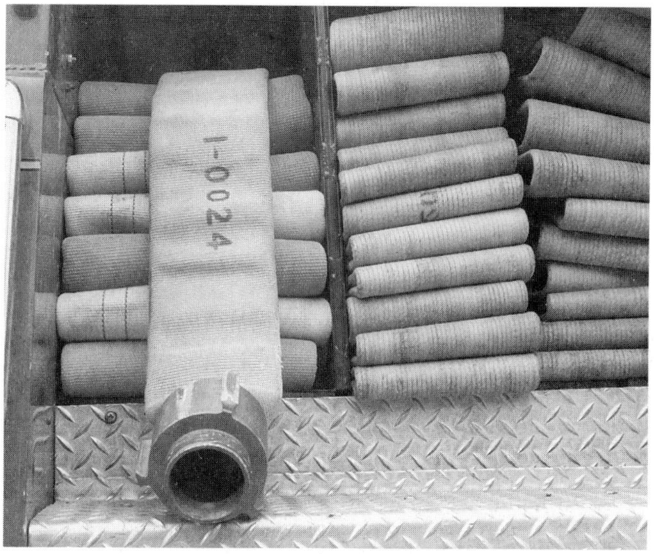

Figure 11.86 Set up hose beds for reverse lays so that the first coupling to come off the hose bed is male.

Split Lay

The split lay is a hoseline laid in part as a forward lay and in part as a reverse lay. This can be accomplished by one pumper making a forward lay from an intersection or driveway entrance toward the fire. A second pumper can then make a reverse lay to the water supply source from the point where the initial line was laid (Figure 11.88). Care must be taken to avoid making the lay too long for the pump, hose size, and required gallon per minute delivery.

It must be noted that when using hose equipped with Storz (sexless) couplings, the direction of lay is a moot point. The hose may be laid in either direction with the same result. The only thing that the firefighter and

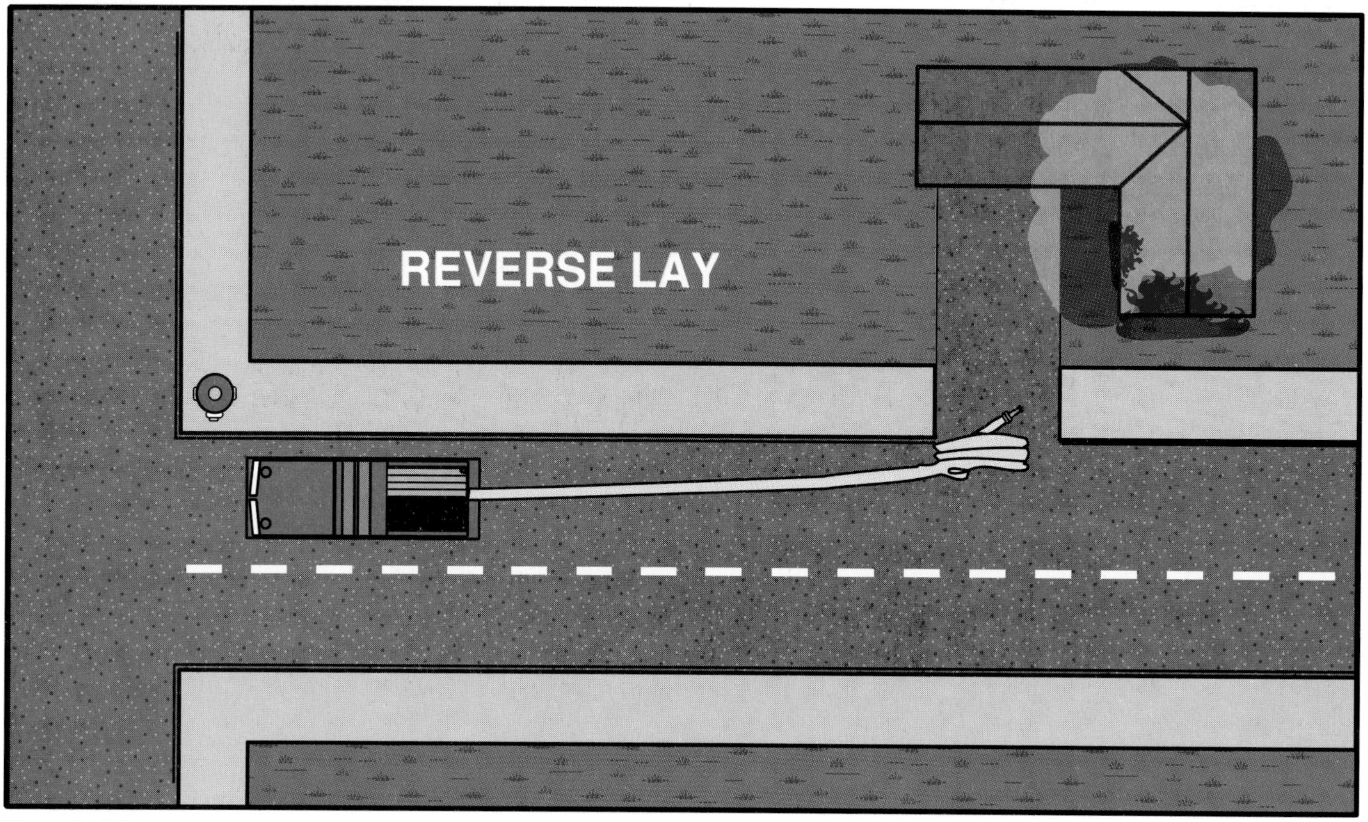

Figure 11.87 The reverse lay proceeds from the scene to the water source.

Figure 11.88 A split lay is composed of both a forward and a reverse lay.

driver/operator must be concerned with is making sure that the proper adapters are present at each end of the lay to make the appropriate connections.

BASIC HOSE LOADS

The most common terminology used to describe a fire hose compartment is "hose bed." Hose beds vary in size and shape, and they are sometimes built for specific needs. In this manual, the front of the hose bed is designated as that part of the compartment toward the front of the apparatus, and the rear of the hose bed is designated as that part of the compartment toward the rear of the apparatus. Most hose beds have open slats in the bottom that enable air to circulate throughout the hose load. Without this feature, the jacketed hose could mildew and rot in a very short time.

A hose bed may be divided or separated at some point for the compartment to hold two or more separate loads of hose (Figure 11.89). The divider (or separator) is usually made of sheet metal. A split bed allows the apparatus to have both forward and reverse lays if desired. Hose in a split bed should be stored so that both beds may be connected together when a long lay is required. The following sections highlight the three most common loads for supply hoselines.

The Accordion Load

The accordion load derives its name from the manner in which the hose appears after loading. The hose is laid progressively on edge in folds that lie adjacent to each other (accordionlike). An advantage of this load is its ease of loading: its simple design requires only two or three people (although four people are best), and it can be loaded in a matter of minutes. Another advantage is

Figure 11.89 Hose beds have dividers to separate the various loads.

that hose for shoulder carries can be easily taken from the load by simply picking up a number of folds and placing them on the shoulder.

The first coupling placed in the bed should be located to the rear of the bed. It can be put to either side if the bed is not split. The following procedure is for loading an accordion load into a split hose bed for a reverse lay:

Step 1: Lay the first length of hose in the bed on edge against the partition. Allow the female coupling to hang below the hose bed so that it can later be placed on top of the hose in the adjacent bed. At the front of the hose bed, fold the hose back on itself, and lay it back to the rear next to the first length (Figure 11.90).

Step 2: At the rear of the hose bed, fold the hose so that the bend is even with the rear edge of the bed, and then lay the hose back to the front (Figure 11.91).

Step 3: Continue laying the hose in folds across the hose bed. At the rear edge of the bed, stagger the folds so that every other bend is approximately 2 inches (50 mm) shorter than the edge of the bed. This stagger may also be done at the front of the bed, if desired. Angle the hose upward to start the next tier (Figure 11.92).

Step 4: Make the first fold of the second tier directly over the last fold of the first tier at the rear of the bed (Figure 11.93).

Step 5: Continue with the second tier in the same manner as the first, progressively laying the hose in folds across the hose bed. Stagger the folds as before so that every other bend is approximately 2 inches (50 mm) inside adjacent bends (Figure 11.94).

Step 6: Make the third and succeeding tiers in the same manner as the first two tiers. Move to the opposite bed, and load the hose in the same manner as the first side. Start with the first female coupling against the front wall of the hose bed so that it will be pulled straight from the bed when this section of hose is pulled (Figure 11.95)

Step 7: Load the second side in the same manner as the first side. When the load is completed, connect the last coupling on top with the female coupling from the first side. Lay the connected couplings on top of the hose load and pull out the slack so that the crossover loop lies tightly against the hose load (Figure 11.96).

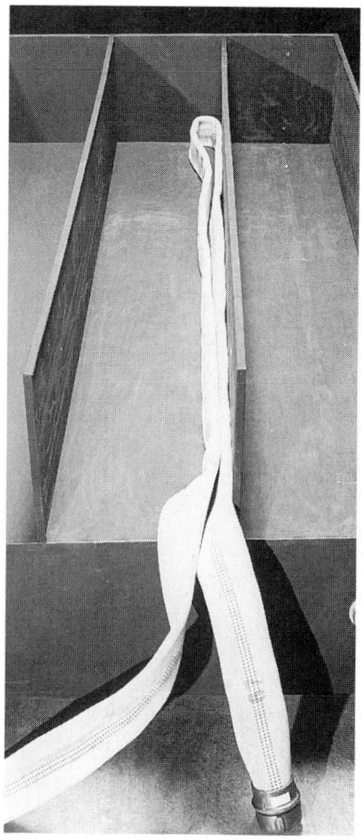

Figure 11.90 Step 1: Lay the first length of hose into the hose bed with the female coupling hanging below the hose bed. At the front of the bed, fold the hose back onto itself, and lay it back to the rear adjacent to the first length.

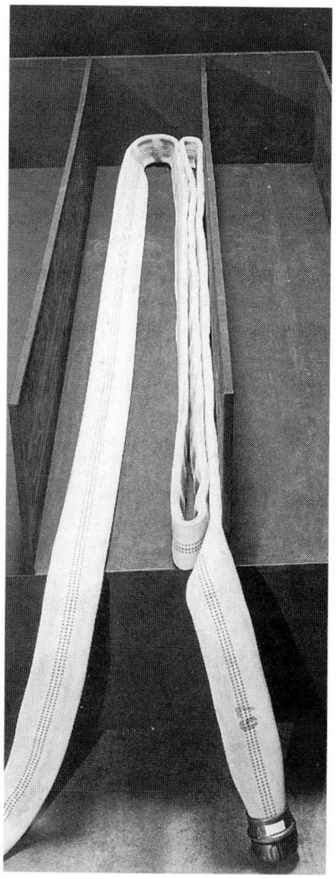

Figure 11.91 Step 2: Fold the hose even with the rear edge of the bed, and lay it back to the front.

Figure 11.92 Step 3: Continue laying the hose in folds across the hose bed. Stagger the folds so that every other bend is approximately 2 inches (50 mm) shorter than the edge of the bed.

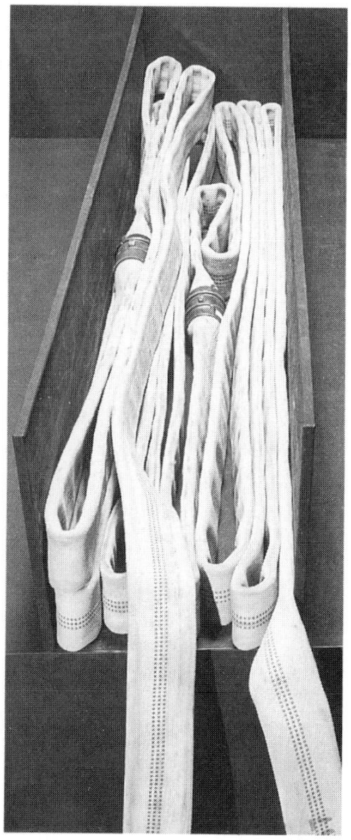

Figure 11.93 Step 4: Place the first fold of the second tier directly over the last fold at the rear of the bed.

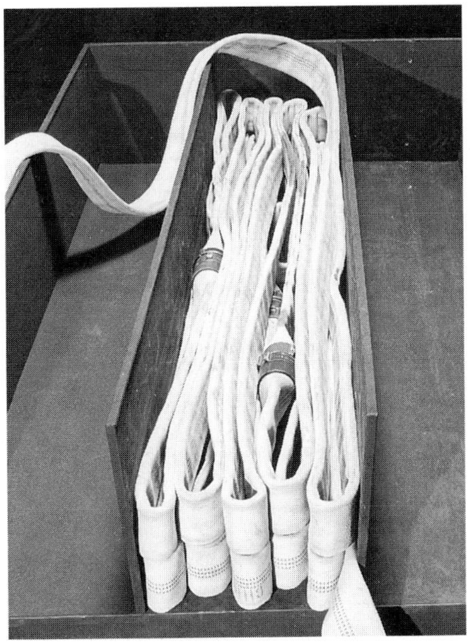

Figure 11.94 Step 5: Continue with the second tier in the same manner as the first.

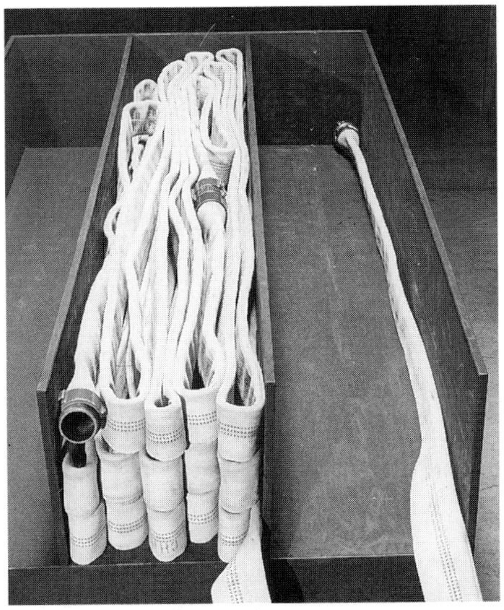

Figure 11.95 Step 6: Load the third and succeeding tiers in the same manner as the first two tiers. Start loading the opposite bed by placing the female coupling against the front wall of the hose bed.

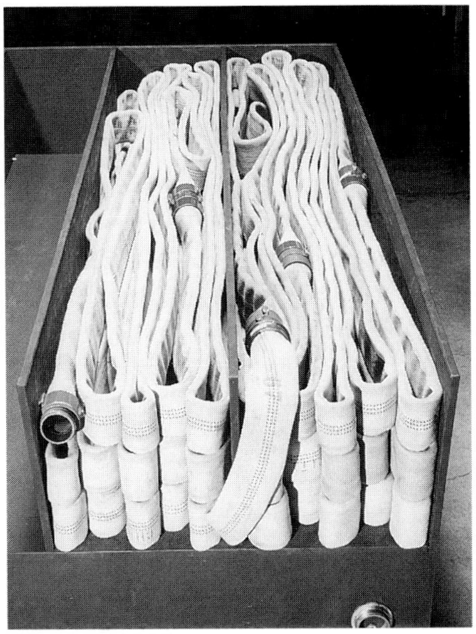

Figure 11.96 Step 7: When the second side is completed, interconnect the beds and lay the connected couplings on top of the hose bed.

Horseshoe Load

The horseshoe load is also named for the way it appears after loading — it resembles a horseshoe. Like the accordion load, it is loaded on edge, but in this case the hose is laid around the perimeter of the hose bed in a U-shaped configuration. Each length is progressively laid from the outside of the bed toward the inside so that the last length is at the center of the horseshoe. The primary advantage of the horseshoe load is that it has fewer sharp bends in the hose than the accordion or flat load. A disadvantage of the horseshoe load occurs most often in wide hose beds. The hose sometimes comes out in a wavy, or snakelike, lay in the street as the hose is pulled alternately from one side of a bed and then the other. Another disadvantage with the horseshoe load is that folds for a shoulder carry cannot be obtained as easily as with an accordion load. In this case, two people are required to make the shoulder folds for the carry. As with the previous load, the hose is loaded on edge, which can promote wear on the hose edges. The horseshoe load is not recommended for large diameter hose because as the hose pays off, the hose remaining in the bed tends to fall over, which can cause the hose to become entangled.

In a single hose bed, the horseshoe load may be started on either side. In a split hose bed, lay the first length against the partition with the coupling hanging an appropriate distance below the hose bed. Gauge this distance on the anticipated height of the completed hose load so that the coupling can be connected to the last coupling of the load on the opposite side (crossover) and laid on top of the load. This allows easy disconnection of the couplings when the load must be split to lay dual lines. With a combination load (one side loaded for a reverse lay, the other side for a forward lay), use an adapter to connect identical couplings.

The procedure that follows is for a single-bed horseshoe load set up for a reverse lay:

Step 1: Place the female coupling in a front corner of the hose bed. Lay the first length of hose on edge against the wall. Make the first fold at the rear even with the edge of the hose bed (Figure 11.97).

Step 2: Lay the hose to the front and then around the perimeter of the bed so that it comes back to the rear along the opposite side (Figure 11.98).

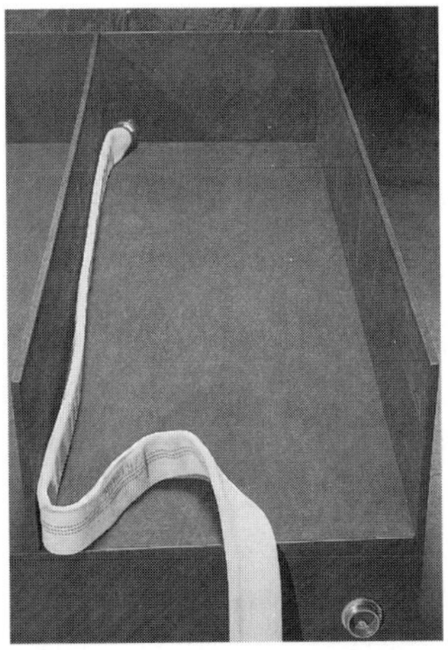

Figure 11.97 Step 1: Lay the first length of hose on edge against the wall with the female coupling in the front corner. Make the first fold at the rear even with the edge of the hose bed.

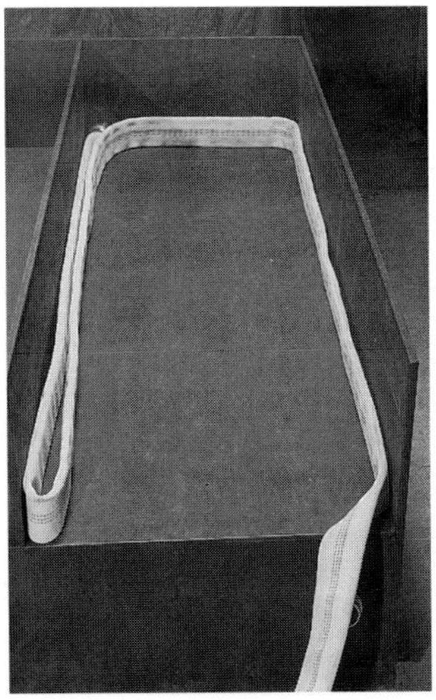

Figure 11.98 Step 2: Lay the hose around the perimeter of the bed so that it comes back to the rear along the opposite side.

Step 3: Make a fold at the rear in the same manner as before, and then lay the hose back around the perimeter of the hose bed inside the first length of hose (Figure 11.99).

Step 4: Lay succeeding lengths progressively inward toward the center until the entire space is filled. If desired, stagger the folds so that every other bend is approximately 2 inches (50 mm) inside adjacent bends (Figure 11.100). Start the second tier by extending the hose from the last fold directly over to a front corner of the bed, laying it flat on the hose of the first tier.

Step 5: Make the second and succeeding tiers in the same manner as the first. Lay the crossover length flat on the second tier, but lay it to the opposite corner from that in the first tier (Figure 11.101). Make crossovers in succeeding tiers to alternate corners.

Flat Load

Of the three supply hose loads, the flat load is the easiest to load. It is suited for any size of supply hose, and it is the best way to load large diameter hose. As the name implies, the hose is laid so that its folds lie flat rather than on edge (Figure 11.102). Hose loaded in this manner is less subject to wear from apparatus vibration during travel. As with the accordion load, a disadvantage of this load is that the hose folds contain sharp bends at both ends, which requires that the hose be reloaded periodically to relocate bends within each length to prevent damage to the lining.

In a single hose bed, the flat load may be started on either side. In a split hose bed, lay the first length against the partition with the coupling hanging an appropriate distance below the hose bed. Gauge this distance on the anticipated height of the hose bed so that the coupling can be connected to the last coupling of the load on the opposite side (crossover) and laid on top of the load. This allows easy disconnection of the couplings

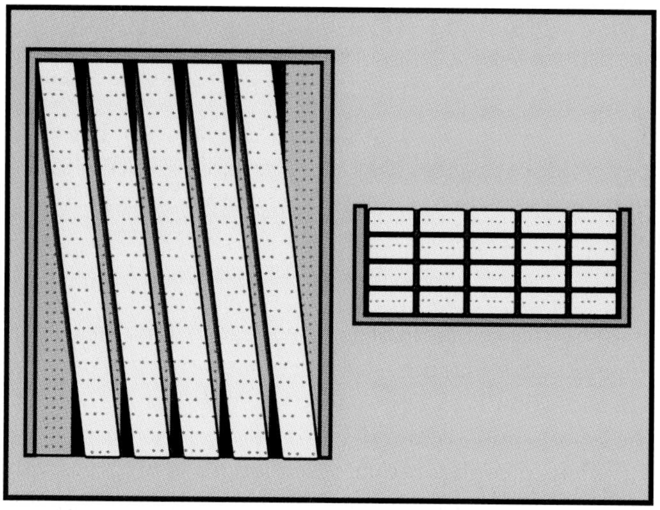

Figure 11.102 The hose in the flat load is laid flat across the bed, rather than on edge.

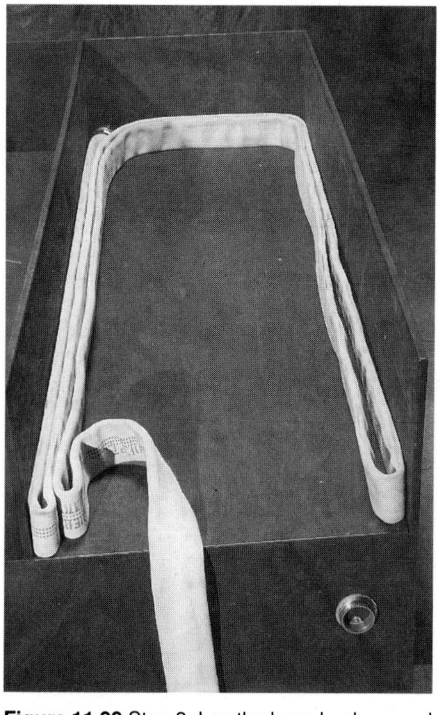

Figure 11.99 Step 3: Lay the hose back around the perimeter of the hose bed inside the first length of hose.

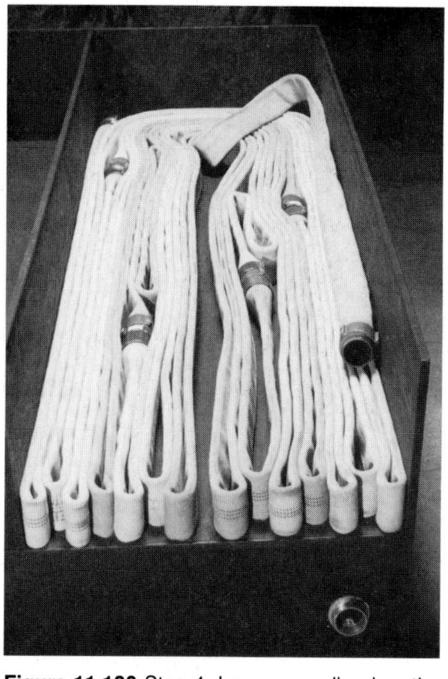

Figure 11.100 Step 4: Lay succeeding lengths progressively inward toward the center until the entire space is filled. Lay the hose from the last fold flat over to a front corner of the bed to start the second tier.

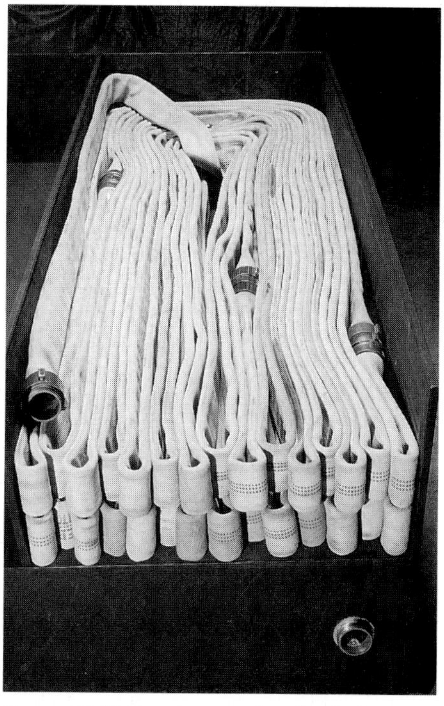

Figure 11.101 Load succeeding tiers in the same manner to complete the load. Make crossovers in succeeding tiers to alternate corners.

when the load must be split to lay dual lines. With a combination load, use an adapter to connect identical couplings.

The procedure that follows is for a split-bed flat load setup for a reverse lay:

Step 1: Lay the first length of hose flat in the bed against the partition with the female coupling hanging below the hose bed (to be connected later to hose in the adjacent bed) (Figure 11.103).

Step 2: Fold the hose back on itself at the front of the hose bed, and lay it back to the rear on top of the previous length. At the rear of the hose bed, fold the hose so that the bend is even with the rear edge of the bed (Figure 11.104).

Step 3: Lay the hose back to the front of the bed, angling it to make the front fold adjacent to the previous fold (Figure 11.105).

Step 4: Continue to lay the hose in folds progressively across the bed to complete the first tier (Figure 11.106).

Step 5: Continue with the second tier in the same manner as the first, laying the hose in folds progressively across the hose bed. If desired, make the folds of the second tier approximately 2 inches (50 mm) shorter than the folds of the first tier (Figure 11.107 on the next page).

Step 6: Make the third and succeeding tiers in the same manner as the first and second tiers. Align the bends of the third tier even with those of the first tier, the bends of the fourth tier with those of the second tier, and so on until the load is completed (Figure 11.108 on the next page).

Step 7: Move to the opposite bed and load the hose in the same manner as the first side. Start the first male coupling against the front wall of the hose bed so that it will be pulled straight from the bed when this last section of hose is pulled.

Step 8: When the opposite side is loaded, connect the last coupling on top with the female coupling from the first side using a double male. Lay the connected couplings on top of the hose load, and pull out the slack so that the crossover loop lies tightly against the hose load (Figure 11.109 on the next page).

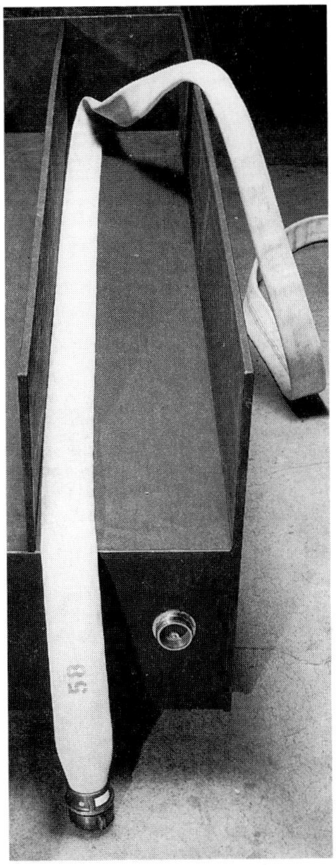

Figure 11.103 Step 1: Lay the hose against the partition with the female coupling hanging below the hose bed.

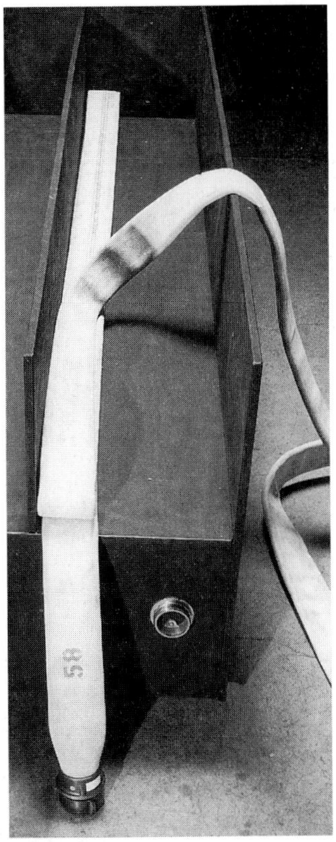

Figure 11.104 Step 2: Make a bend in the hose at the front, and lay the hose back to the rear on top of the previous length. Fold the hose even with the rear edge of the bed.

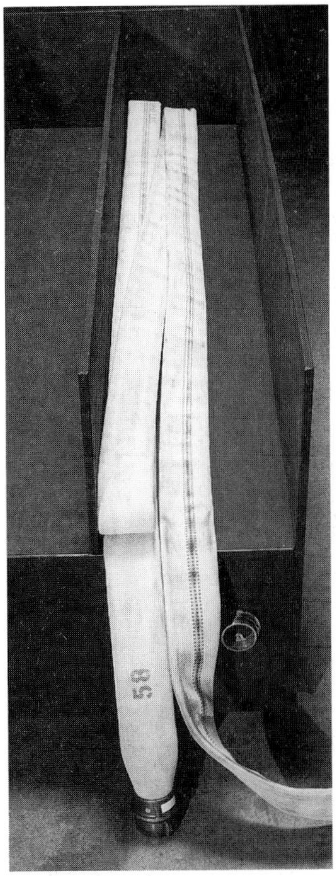

Figure 11.105 Step 3: Lay the hose back to the front of the bed, angling it so that the front fold is made adjacent to the previous fold.

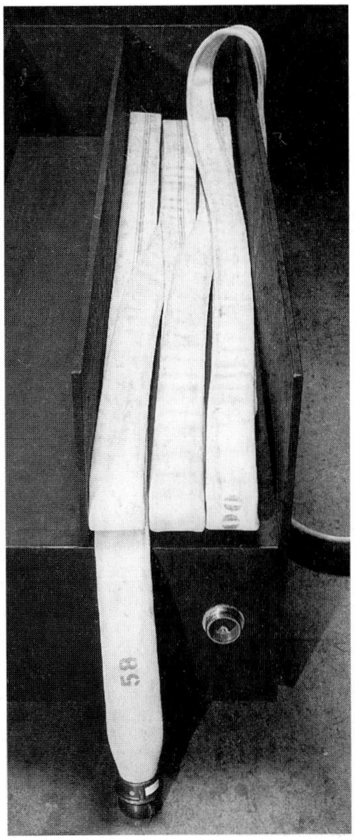

Figure 11.106 Step 4: Continue with the second tier in the same manner as the first, laying the hose in folds progressively across the hose bed.

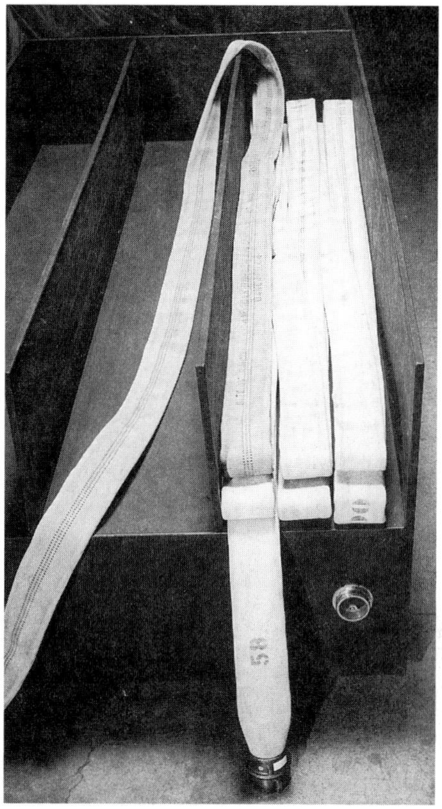

Figure 11.107 The completed first tier.

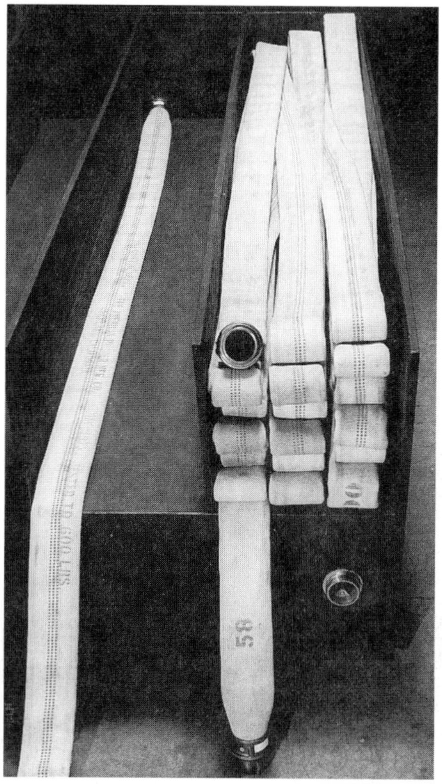

Figure 11.108 Once the first side is complete, the second bed may be started.

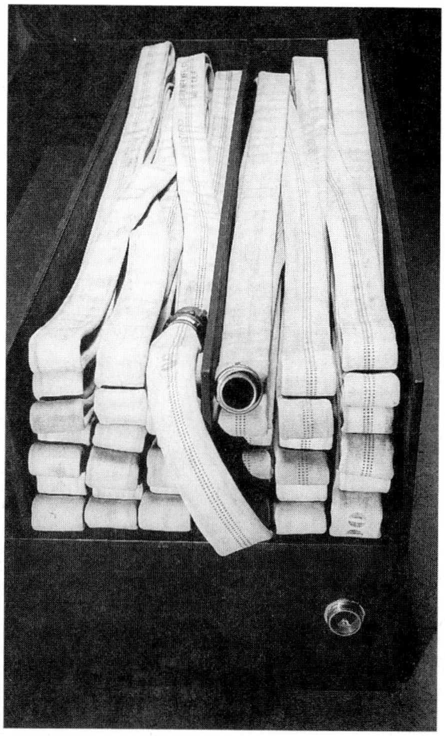

Figure 11.109 Step 8: Connect the two loads, and lay the couplings on top of the load. Pull out the slack so that the crossover loop lies tightly against the hose load.

This method can be adapted for loading large diameter hose. Large diameter hose can be loaded directly from the street, after an incident, by straddling the hose with the pumper and driving slowly backwards as the hose is progressively loaded into the bed (Figure 11.110). A hose wringer or roller can be used to expel the air and water from the hose as it is placed in the hose bed (Figure 11.111). This is done to have a neat and space-efficient load of large diameter hose. The hose lay should be started 12 to 18 inches (300 mm to 450 mm) from the front of the hose bed. This extra space should be reserved for couplings, and all couplings should be laid in a manner that will allow them to pay out without having to flip over (Figure 11.112). To do this it may be necessary to make a short fold or reverse bend in the hose (Figure

Figure 11.110 Large diameter hose can be loaded directly by straddling the hose with the pumper and driving backward slowly as the hose is progressively loaded into the bed.

Figure 11.111 A hose wringer can be used to rid the hose of excess water and air. The reduction of excess water and air makes the hose lie flatter in the bed.

11.113). This practice is commonly referred to as a "dutchman." The dutchman serves two purposes: to change the direction of a coupling and to change the location of a coupling.

Figure 11.112 Some departments prefer to load large diameter hose with all the couplings near the front of the hose bed.

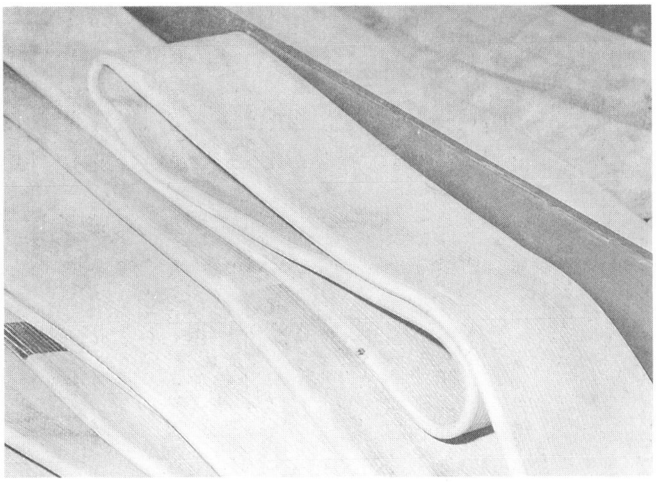

Figure 11.113 An LDH "dutchman."

Hose Load Finishes

Another way to arrange hose is to "finish" a hose load with additional hose that can be quickly pulled at the beginning of a forward or reverse lay. Finishes are arrangements of hose that are usually placed on top of a hose load and are connected to the end of the load.

Finishes fall into two categories: those for forward lays and those for reverse lays. A finish for a reverse lay expedites making a full or partial strip of equipment for fire fighting. Finishes for forward lays are usually designed to speed the pulling of hose when making a hydrant and are not as elaborate as finishes for reverse lays.

Hose load finishes are added to the basic hose load to increase the versatility of the load. Finishes are normally loaded to provide enough hose to make a hydrant connection and to provide a working line at the fire scene.

STRAIGHT FINISH

A straight finish consists of the last length or two of hose flaked loosely back and forth across the top of the hose load. This finish is normally associated with forward lay operation. A hydrant wrench, gate valve, and any necessary adapters should be strapped at or near the female coupling (Figure 11.114).

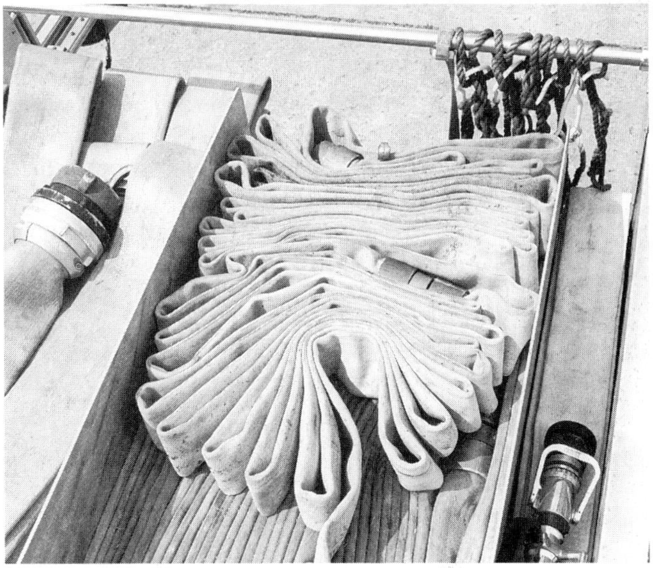

Figure 11.114 A straight finish.

REVERSE HORSESHOE FINISH

This finish is similar to the horseshoe load, except that the bottom of the "U" portion of the horseshoe is at the rear of the hose bed. It is made of one or two 100-foot (30 m) lengths of hose, each connected to one side of a wye. Any size of attack hose can be used, 1½, 1¾, or 2½ inch (38 mm, 45 mm, or 65 mm). The smaller sizes require a 2½- x 1½-inch (65 mm by 38 mm) gated reducing wye; the 2½-inch (65 mm) hose requires a 2½- x 2½-inch (65 mm by 65 mm) gated wye. Two nozzles of the appropriate size are also needed. The procedure that follows is for making a reverse horseshoe finish with 1½-inch (38 mm) hose:

Step 1: Connect the wye to the end (male) coupling of the hose load at the rear of the bed, and then place the wye in the center of the hose load with the two male openings toward the rear of the bed (Figure 11.115 on the next page).

Step 2: Connect one 1½-inch (38 mm) hose to the wye, and then lay the hose on edge to the front of the bed and make a fold (Figure 11.116 on the next page).

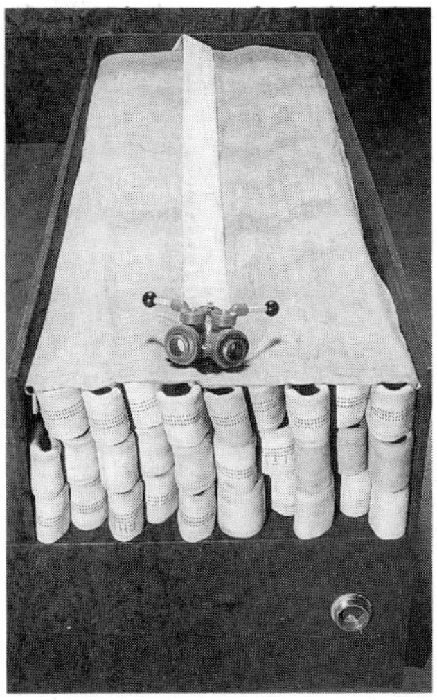

Figure 11.115 Step 1: Connect the wye, and place it in the center of the hose load.

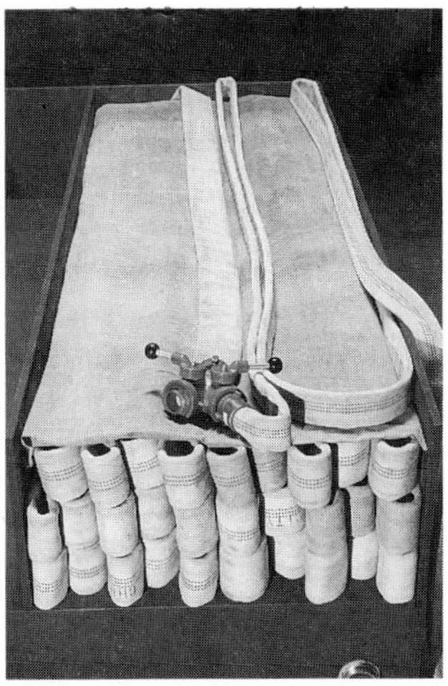

Figure 11.116 Step 2: Connect one 1½-inch (38 mm) hose to the wye, and then lay the hose on edge to the front of the bed and make a fold.

Step 3: Lay the hose back to the rear alongside the first length. Form a "U" at the edge of the bed, and then return the hose to the front and make a fold.

Step 4: Lay the hose back inside the previously laid length in the same manner as before, and continue until the entire length has been loaded (Figure 11.117).

Step 5: Wrap the male end of the hose once around the horseshoe loops (Figure 11.118).

Step 6: Form a small loop by bringing the end back under the center of the loops and then over the top (Figure 11.119).

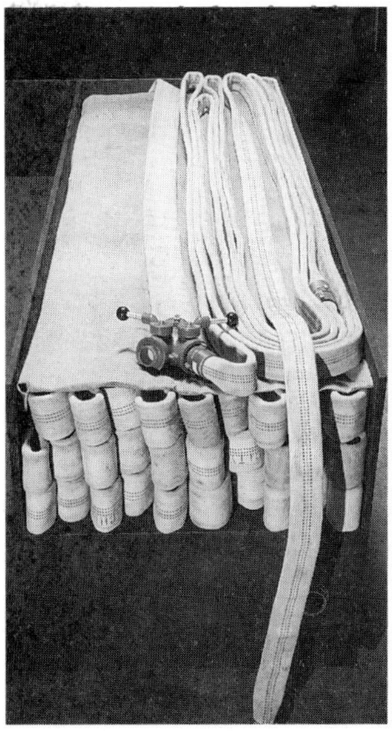

Figure 11.117 Step 4: Lay the hose back inside the previously laid length in the same manner, and continue until the entire length has been loaded.

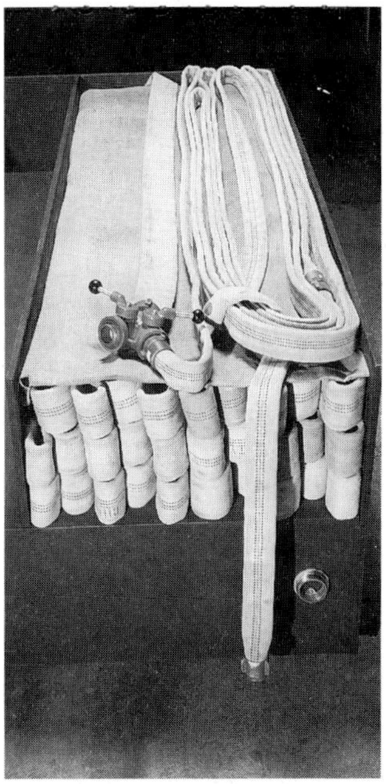

Figure 11.118 Step 5: Wrap the male end of the hose once around the horseshoe loops.

Step 7: Attach the nozzle, and place it inside the small loop. Pull the remaining slack hose back into the center of the horseshoe to tighten the loop against the nozzle (Figure 11.120).

Step 8: Load the second length of hose in the same manner on the opposite side of the bed if desired (Figure 11.121).

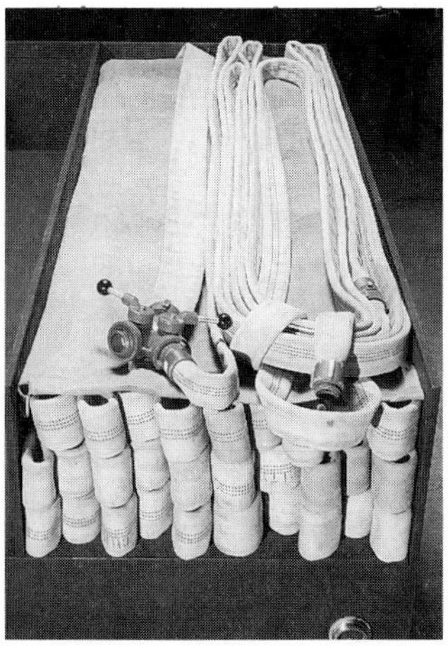

Figure 11.119 Step 6: Form a small loop by bringing the end back under the center of the loops, and then over the top.

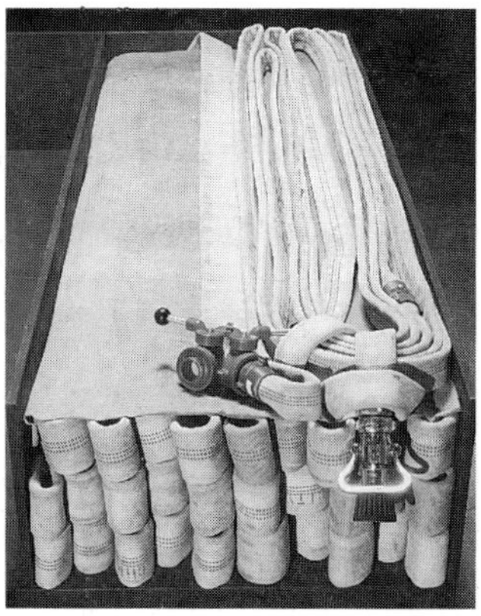

Figure 11.120 Step 7: Attach the nozzle, and place it inside the small loop. Pull the remaining slack hose back into the center of the horseshoe to tighten the loop against the nozzle.

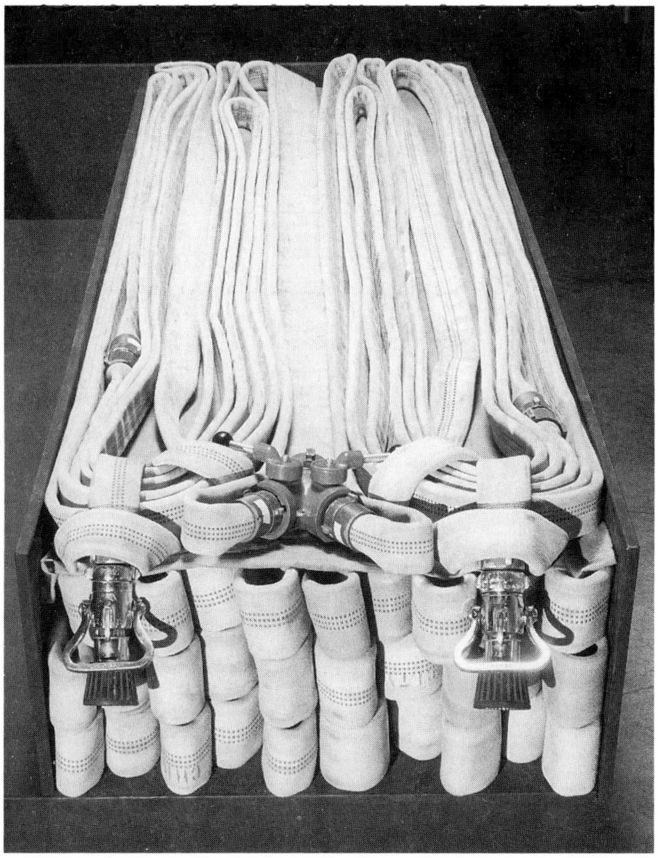

Figure 11.121 Step 8: Load the second length of hose in the same manner on the opposite side of the bed.

The reverse horseshoe load can also be used for a preconnected line and loaded in two or three layers. With the nozzle extending to the rear, it can be placed over a shoulder and the opposite arm extended through the loops of the layers, pulling the hose from the bed for an arm carry. A second preconnect can be bedded below when there is sufficient depth.

HOSE LOADING GUIDELINES

Although the loading of fire hose on fire apparatus is not an emergency operation, it is a very vital operation that must be done correctly. When fire hose is needed at a fire, the proper hose load permits efficient and effective operations to be carried out. The following are some general guidelines that should be followed, regardless of the type of hose load that is going to be used:

- Before connecting any coupling, check gaskets and swivel.
- When two sections of hose are connected, keep the flat sides of the hose in the same plane (Figure 11.122). The alignment of the lugs on the couplings is not important.

Figure 11.122 When two sections of hose are connected, keep the flat sides of the hose in the same plane.

- When two sections of hose are connected, tighten the couplings hand-tight. Do not use wrenches or undue force.
- When fire hose must be bent to form a loop in the hose bed, remove all wrinkles by pressing with the fingers so that the inside of the bend is smoothly folded.
- During the loading process, a coupling will frequently come in position so that it must turn around to be pulled out. To avoid this situation, make a short fold or reverse bend in the hose (Figure 11.123). This practice is commonly referred to as a "dutchman."
- Load large diameter hose (3½-inch [90 mm] or larger) with all couplings to the front of the bed. This saves space and allows the hose to lie flat. Couplings should again be laid in a manner that does not require them to flip over as the hose pays out of the bed.

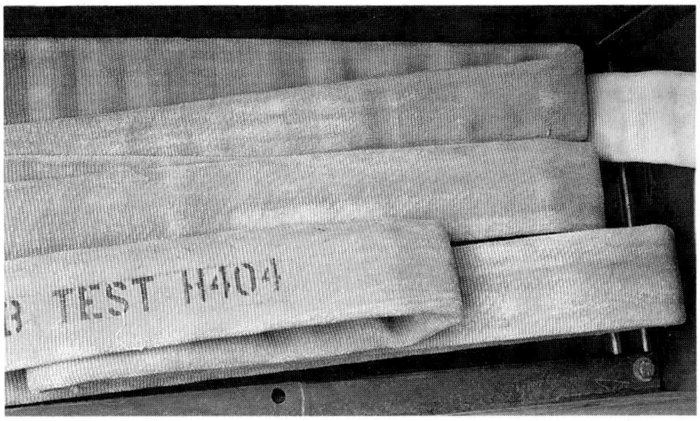

Figure 11.123 A short fold or reverse bend in the hose is commonly referred to as a "dutchman."

- Do not pack hose too tightly. This puts excess pressure on the folds of the hose, and it causes couplings to get hung up when the hose pays out of the bed. A rule of thumb is that the hose should be loose enough to allow a hand to be easily inserted between the folds (Figure 11.124).

Figure 11.124 If hose is loaded properly, a firefighter should be able to slide his or her hand between the folds.

COUPLING AND UNCOUPLING HOSE

The process of coupling hose is, for the most part, a simple process of fastening together the male and female hose couplings. The need for speed and accuracy under emergency conditions requires that specific techniques for coupling hose be developed. When coupling two sections of hose, keep the flat sides of the hose in the same plane. This practice makes it easier to handle and load the hose. Nozzles may be attached to the hose by using the same methods as when coupling two sections of hose together.

Coupling Hose

The following techniques describe the coupling and uncoupling of threaded couplings; however, the same techniques can be applied to Storz (sexless) couplings.

FOOT-TILT METHOD (ONE-FIREFIGHTER METHOD)

Step 1: Stand facing the two couplings so that one foot is near the male end.

Step 2: Place your foot on the hose directly behind the male coupling, and apply pressure to tilt it upward. Position your feet well apart for balance (Figure 11.125).

Step 3: Grasp the female end by placing one hand behind the coupling and your other hand on the coupling swivel.

Step 4: Bring the two couplings together, and turn the swivel clockwise with your thumb to make the connection.

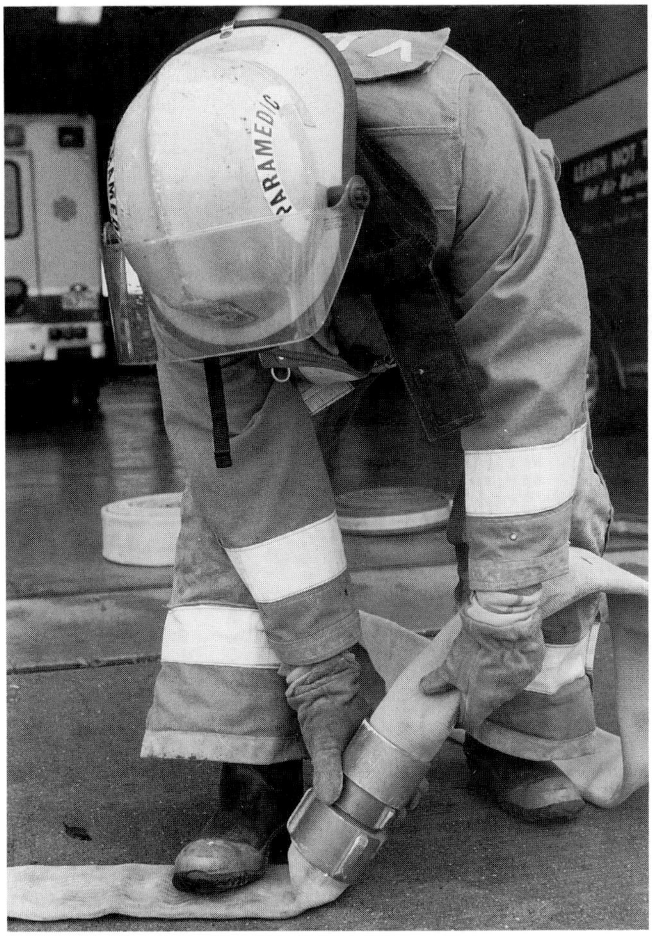

Figure 11.125 Place your foot on the hose directly behind the male coupling, and apply pressure to tilt it upward. Position your feet well apart for balance.

TWO-FIREFIGHTER METHOD

Step 1: Two firefighters holding opposite couplings stand facing each other. The firefighter holding the male coupling grasps the coupling with both hands, bends the hose directly behind the coupling, and holds the coupling and hose tightly against the upper thigh or mid-section with the male threads pointed outward (Figure 11.126).

Step 2: The alignment of the hose must be done by the firefighter with the female coupling. It may help for the firefighter with the male coupling to look in another direction in order to prevent trying to help align the couplings.

Step 3: The firefighter with the female coupling grasps the coupling with both hands, brings the two couplings together, and aligns their positions. The Higbee indicator, discussed earlier in this chapter, can be used to align the couplings. When the couplings are placed together, the firefighter with the female coupling turns it counterclockwise until a click is heard. This indicates that the threads are aligned. Then turn the female swivel clockwise to complete the connection.

Figure 11.126 Step 1: Two firefighters holding opposite couplings stand facing each other. The firefighter holding the male coupling grasps the coupling with both hands, bends the hose directly behind the coupling, and holds the coupling and hose tightly against the upper thigh or mid-section with the male threads pointed outward.

Breaking A Tight Coupling

It may sometimes be necessary to break a tight coupling when spanner wrenches are not available. Following are two methods by which one or two firefighters may accomplish this task.

KNEE-PRESS METHOD (ONE-FIREFIGHTER METHOD)

Step 1: Grasp the hose behind the female coupling, and stand the connection on end with the male coupling below. Best results will be obtained if the hose is bent close to the male coupling.

Step 2: Set your feet well apart for balance, and place one knee upon the hose and shank of the female coupling (Figure 11.127). Keep your thigh in a vertical plane with the couplings, and apply body weight to the connected coupling.

Step 3: As this weight is applied, quickly snap the swivel in a counterclockwise direction to loosen the connection.

Figure 11.127 Set your feet well apart for balance, and place one knee upon the hose and shank of the female coupling.

STIFF-ARM METHOD (TWO-FIREFIGHTER METHOD)

Step 1: Two firefighters face each other with the hose couplings between them.

Step 2: The firefighters take a firm two-handed grip on their respective coupling and press the couplings toward each other, thereby compressing the gasket in the coupling (Figure 11.128).

Step 3: Keeping their arms stiff, both firefighters use the weight of their bodies to turn their respective hose coupling counterclockwise, thus loosening the connection.

Figure 11.128 The firefighters take a firm two-handed grip on their respective coupling and press the couplings toward each other, thereby compressing the gasket in the coupling.

PRECONNECTED HOSE LOADS FOR ATTACK LINES

Another way to carry attack hose is to preconnect it to a discharge valve and place it in an area other than the main hose bed. Preconnected hoselines are the primary lines used for fire attack by most fire departments. Preconnected hoselines generally range from 50 to 250 feet (15 m to 75 m) in length. There are several places in which preconnected attack lines can be carried:

- Longitudinal beds (Figure 11.129)
- Raised trays

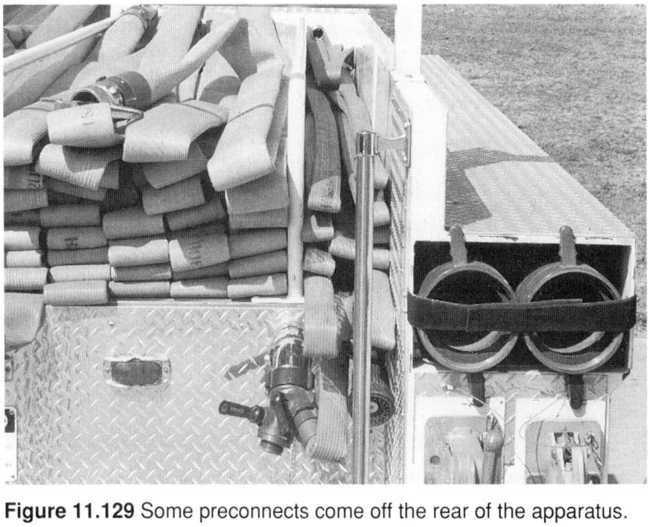

Figure 11.129 Some preconnects come off the rear of the apparatus.

- Transverse beds (Figure 11.130)
- Tailboard compartments
- Side compartments or bins
- Front bumper wells (Figure 11.131)
- Reels

There are several different loads that can be used for preconnected lines. The following sections detail some of the more common ones. Special loads to meet local requirements may be developed based on individual experiences and apparatus limitations.

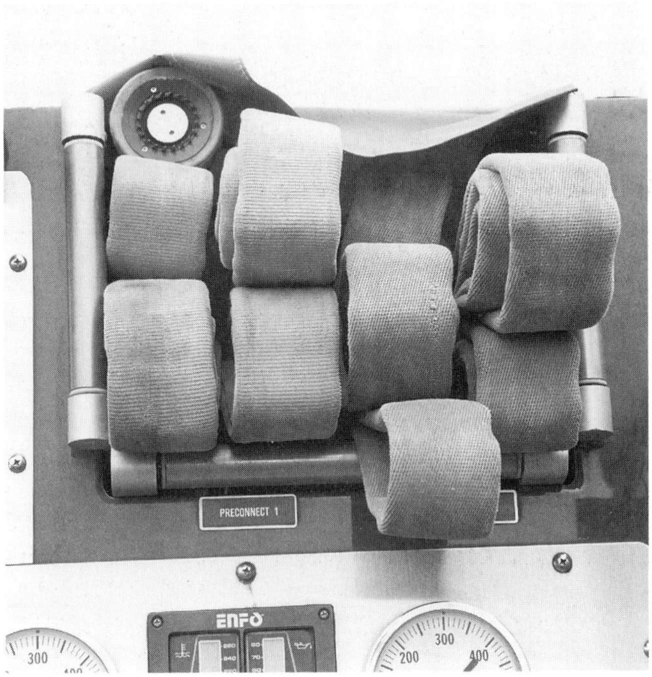

Figure 11.130 Mattydale or cross-lay preconnects are located above the pump panel.

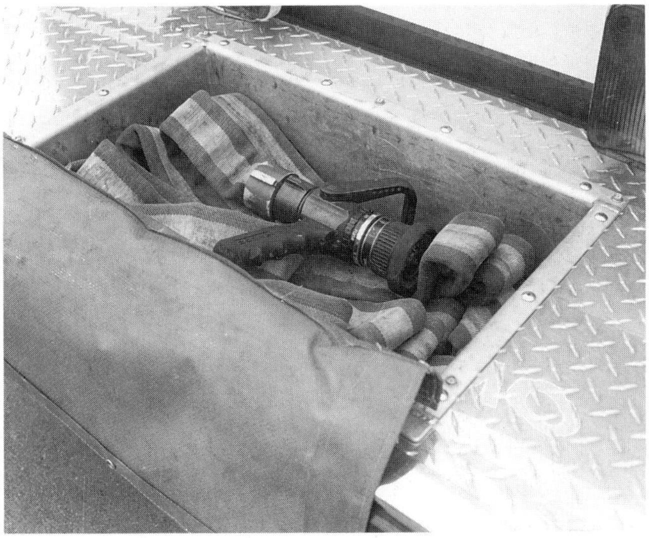
Figure 11.131 Preconnected attack lines can be carried in front bumper wells.

Preconnected Flat Load

The flat load is adaptable for varying widths of hose beds and is often used in transverse beds. This load is similar to the flat load for larger supply hose with two exceptions: it is preconnected and loops are provided to aid in pulling the load from the bed. The pull loops should be placed at regular intervals within the load so that equal portions of the load are pulled from the bed. The number of loops and the intervals at which they are placed are dependent upon the size and total length of the hose. The procedure shown can be adapted for any type of hose bed.

Step 1: Attach the female coupling to the discharge. Lay the first length of hose flat in the bed against the side wall (Figure 11.132).

Step 2: Angle the hose to lay the next fold adjacent to the first fold. Continue building the first tier in this manner (Figure 11.133).

Step 3: At a point approximately one-third of the total length of the load, make a fold that extends approximately 8 inches (203 mm) beyond the load (Figure 11.134). This loop will later serve as a pull handle.

Step 4: Continue laying the hose in the same manner, building each tier with folds laid progressively across the bed. At a point approximately two-thirds of the total length of the load, make a fold that extends approximately 14 inches (356 mm) beyond the load (Figure 11.135). This loop will also serve as a pull handle.

Step 5: Complete the load, and then attach the nozzle and lay it on top of the load (Figure 11.136).

Figure 11.132 Step 1: Attach the female coupling to the discharge. Lay the first length of hose flat in the bed against the side wall.

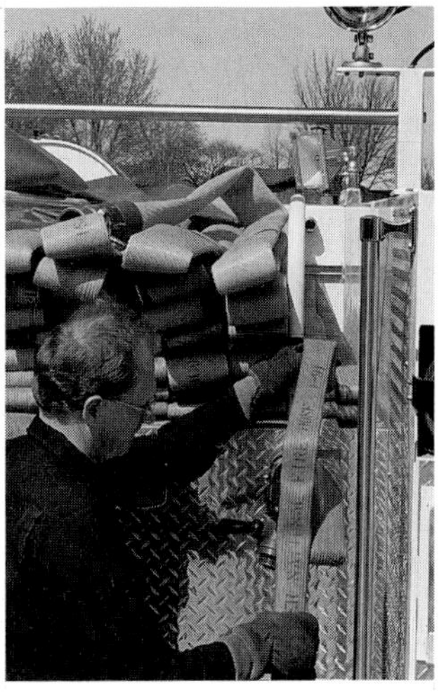
Figure 11.133 Step 2: Angle the hose to lay the next fold adjacent to the first fold. Continue building the first tier in this manner.

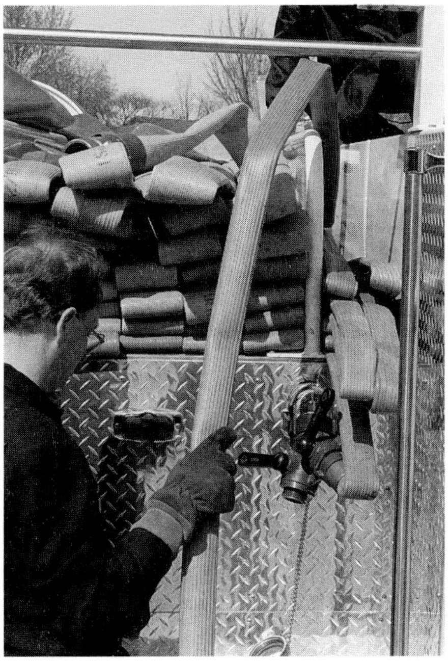

Figure 11.134 Short loops are formed about one-third of the way through the load.

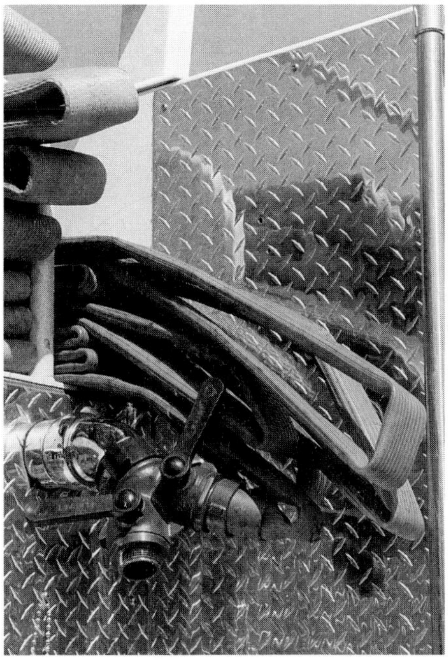

Figure 11.135 Long loops are formed about two-thirds of the way through the load.

Figure 11.136 The completed load has the nozzle attached with it laying on top of the hose.

Triple Layer Load

The triple layer load gets its name because the load begins with hose folded in three layers. The three folds are then laid into the bed in an S-shaped fashion. The load is designed to be pulled by one person. A disadvantage with the triple layer load is that the three layers, which may be as long as 50 feet (15 m), must be completely removed from the bed before leading in the nozzle end of the hose. This could be a problem if other apparatus are parked directly behind the hose bed. While this hose can be used for all size attack lines, it is often preferred for larger (2- and 2½-inch [50 and 65 mm]) lines that may be too cumbersome for shoulder carries.

Start with the sections of hose connected and the nozzle attached. The procedure shown here is for 200 feet (60 m) of 1½- or 1¾-inch (38 mm or 45 mm) hose.

Step 1: Connect the female coupling to the discharge, and extend the hose in a straight line to the rear (Figure 11.137).

Step 2: Pick up the hose at a point two-thirds of the distance from the tailboard to the nozzle, and carry it to the tailboard. This will form three layers of hose stacked one on the other with a fold at each end (Figure 11.138).

Step 3: Use several people to pick up the entire length of the three layers, and begin laying the hose into the bed by folding over the three layers into the hose bed (Figure 11.139 on the next page).

Step 4: Fold the layers over at the front of the bed, and lay them back to the rear on top of the previously laid hose (if the hose compartment is wider than one hose width, alternate folds on each side of the bed). Make all folds at the rear even with the edge of the hose bed (Figure 11.140 on the next page).

Step 5: Continue to lay the hose into the bed in an S-shaped configuration until the entire length is loaded (Figure 11.141 on the next page). If desired, use a rope or strap to secure the nozzle to the first set of loops. Some departments like to pull the loop at the end through the nozzle bale. This can be a problem if the line is charged before removing the loop from the bale. Once charged, it may not be possible to pull the loop through the bale.

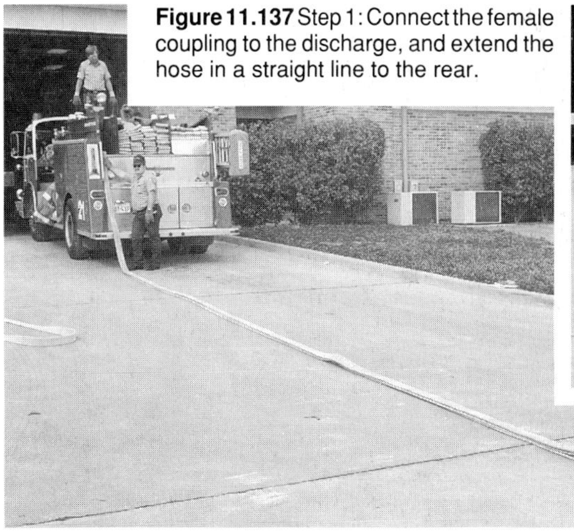

Figure 11.137 Step 1: Connect the female coupling to the discharge, and extend the hose in a straight line to the rear.

Figure 11.138 Step 2: Pick up the hose at a point two-thirds of the distance from the tailboard to the nozzle, and carry it to the tailboard. This will form three layers of hose stacked one on the other with a fold at each end.

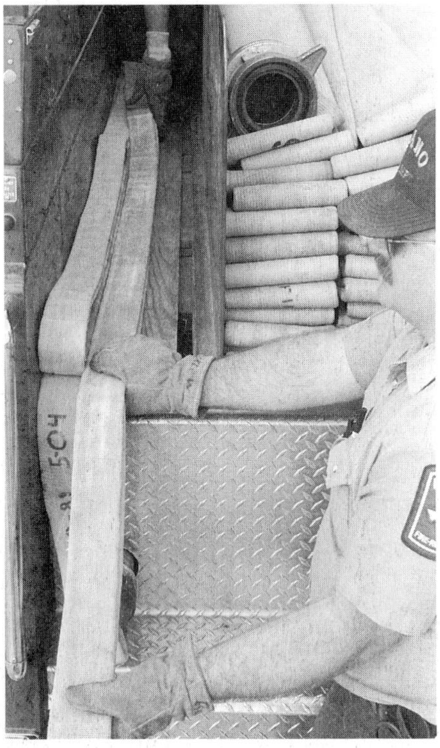

Figure 11.139 Step 3: Use several people to pick up the entire length of the three layers, and begin laying the hose into the bed by folding over the three layers into the hose bed.

Figure 11.140 Step 4: Fold the layers over at the front of the bed, and lay them back to the rear on top of the previously laid hose (if the hose compartment is wider than one hose width, alternate folds on each side of the bed). Make all folds at the rear even with the edge of the hose bed.

Figure 11.141 Step 5: Continue to lay the hose into the bed in an S-shaped configuration until the entire length is loaded.

Minuteman Load

The minuteman load is designed to be pulled and advanced by one person. The primary advantage with this load is that it is carried on the shoulder, completely clear of the ground, so that it will not snag on obstacles. The load pays off the shoulder as the firefighter advances toward the fire. The load is also particularly well suited for a narrow bed. A disadvantage with the load is that it can be awkward to carry when wearing an SCBA. If the load is in a single stack, it may also collapse on the shoulder if not held tightly in place.

The following procedure is for 150 feet (45 m) of 1½- or 1¾-inch (38 mm or 45 mm) hose loaded in a double stack:

Step 1: Connect the first section of hose to the discharge. Do not connect it to the other lengths of hose. Lay the hose flat in the bed to the front, and then lay the remaining hose out the front of the bed to be loaded later (Figure 11.142). (**NOTE:** If the discharge is at the front of the bed, lay the hose to the rear of the bed and then back to the front before it is set aside (Figure 11.143). This provides slack hose for pulling the load clear of the bed.)

Step 2: Couple the remaining hose sections together, and attach a nozzle to the male end. Place the

Figure 11.142 Step 1: Connect the first section of hose to the discharge. Do not connect it to the other lengths of hose. Lay the hose flat in the bed to the front, and then lay the remaining hose out the front of the bed to be loaded later.

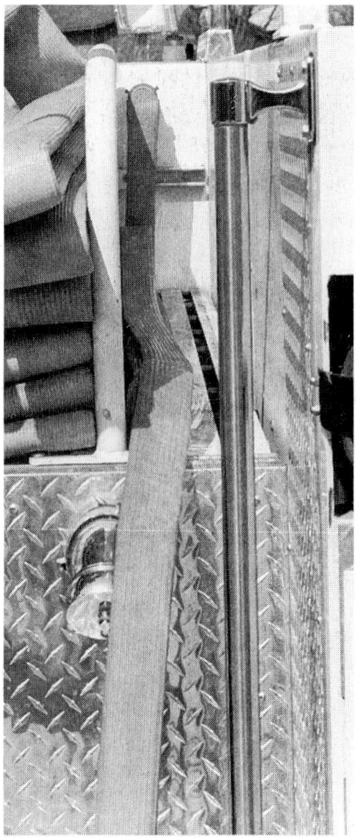

Figure 11.143 If front discharge is used, lay the hose from the front of the bed to the rear and back out the front before continuing the load. This will allow the entire load to be pulled out of the bed and onto the firefighter's shoulder.

nozzle on top of the first length at the rear, and then angle the hose to the opposite side of the bed and make a fold. Lay the hose back to the rear (Figure 11.144).

Step 3: Make a fold at the rear of the bed, and then angle the hose back to the other side and make a fold at the front. The first fold or two may be longer than the other to facilitate the pulling of the hose from the bed (Figure 11.145). Continue loading the hose to alternating sides of the bed in the same manner until the complete length is loaded (Figure 11.146).

Step 4: Connect the male coupling of the first section to the female coupling of the last section (Figure 11.147), and lay the remainder of the first section in the bed in the same manner as before (Figure 11.148).

Figure 11.144 Couple the remaining hose sections together, and attach a nozzle to the male end. Place the nozzle on top of the first length at the rear, and then angle the hose to the opposite side of the bed and make a fold. Lay the hose back to the rear.

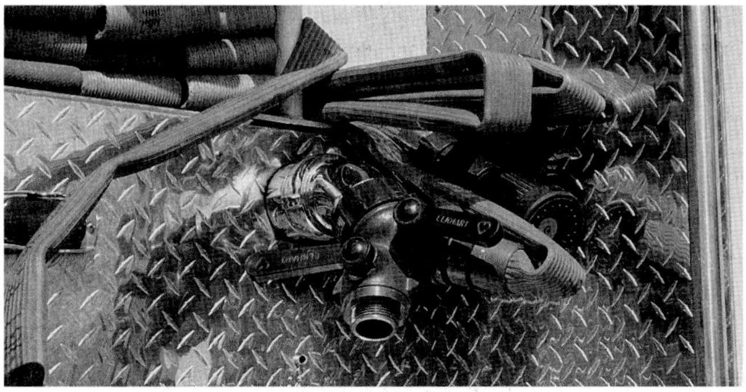

Figure 11.145 Make two loops directly above the nozzle.

Figure 11.146 Continue to lay the hose in the bed.

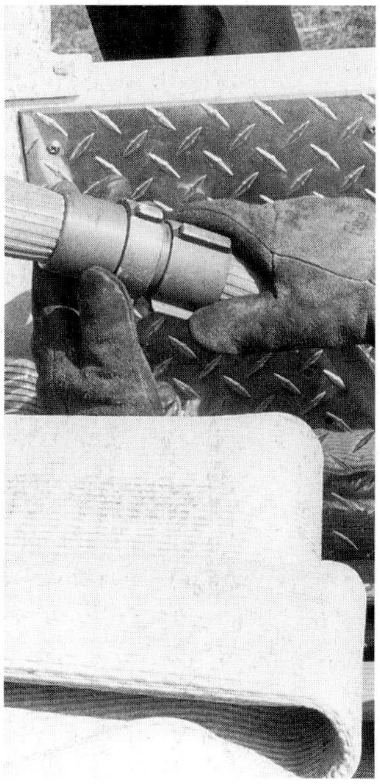

Figure 11.147 Connect the couplings.

Figure 11.148 A completed minuteman.

Booster Hose Reels

Booster hoselines are preconnected hose that are usually carried coiled upon reels. These booster line reels may be mounted several places upon the fire apparatus according to specified needs and the design of the apparatus. Some booster reels are mounted above the fire pump and behind the apparatus cab (Figure 11.149 on the next page). This arrangement provides booster hose which can be unrolled from either side of the apparatus, but its advancement above ground level

Figure 11.149 A top-mount booster reel.

is limited to its length. Other booster reels are mounted on the front bumper of the apparatus or in rear compartments (Figures 11.150 a and b). Hand- and power-operated reels are available. Noncollapsible hose should be loaded one layer at a time, in an even manner (Figure 11.151). This will allow the maximum amount to be loaded and will provide for the easiest removal from the reel.

Figure 11.150a A booster reel located on the front bumper of the apparatus.

Figure 11.150b A booster reel located in a rear compartment of the apparatus. *Courtesy of Phoenix (AZ) Fire Department.*

Figure 11.151 Load noncollapsible hose one layer at a time in an even manner.

HOSE LAY PROCEDURES

Hose lay procedures vary from department to department, but the basic methods of laying hose remain the same. Hose is either laid forward from a water source to the incident scene, reverse from the incident scene to a water source, or split so that the hose can be laid to and from the junction to the water source and the incident scene. These basic methods are presented to provide the foundation for developing hose lays that more specifically suit individual department needs.

When laying hose, a few basic guidelines should be followed:

- Personnel shall not be riding in a standing position any time the apparatus is moving.
- Drive at a speed no greater than that which allows the couplings to clear the tailboard as the hose leaves the bed — generally between 5 and 10 mph (8 km/h and 16 km/h).
- Lay the hose to one side of the roadway so that other apparatus are not forced to drive over it.

Forward Lay

The first layout practices, in the days before the installation of pumps on motor fire apparatus, were from the hydrant (or source of water supply) to the fire and became known as a forward, or straight, lay. The operation consists of stopping the apparatus at the water supply source and permitting the hydrant person to safely leave the apparatus and secure the hose, after which the apparatus proceeds to the fire laying either single or dual hoselines. There are two primary skills that the firefighter who is going to make the hydrant connection — also known as "catching the plug" or "making the hydrant" — must know: the proper procedures for wrapping and connecting to the hydrant and the operation of the hydrant valve, if one is used.

MAKING THE HYDRANT CONNECTION

The first task to be accomplished when laying a hoseline is to manually remove a small amount of hose from the hose bed to start the lay. As a general rule, it is best to pull about 30 feet (10 m) of hose from the apparatus to get things started. This will vary with the distance the hydrant or other anchoring object is from the apparatus. When pulling the hose from the apparatus, it is important that the firefighter also have the other tools necessary to perform the job. This is especially important when making a forward lay from a hydrant that will not be connected to by another pumper to support the line. The person "catching" the hydrant should have a spanner wrench, hydrant wrench, and if not preconnected to the supply line, a hydrant valve (4-Way, Humat, etc.). These tools should be conveniently located so that the person catching the plug can grab them in a hurry. Many departments choose to put all of these tools in a jump kit that is kept on the rear step of the apparatus and easy to grab when needed. It is also desirable that this person have a portable radio so that when the attack engine is ready, water may immediately be sent to the attack engine. However, most departments are not fortunate enough to have a sufficient quantity of radios to do this. In these cases, visual or audible signals are used to tell the firefighter at the hydrant when to start the flow of water. The use of audible warning devices to signal when to start the water can be a problem with other apparatus responding to the scene. The hydrant person might mistakenly charge the line before the driver/operator is ready to accept the water. This can result in a charged hose bed or a loose, flowing hose coupling.

When laying a hoseline from a moving apparatus, it is necessary to anchor the hose at the location from which the lay is being made in order to ensure the end of the hose will remain at the desired location. The best way to do this is to wrap the end of the hose around a stationary object. This object would most likely be a fire hydrant. However when making a split lay from a location where there is no hydrant, a service pole, sturdy sign post, mail box, or parked vehicle can be used as an anchor.

When making hydrant connections from a forward lay, the following procedures should be used:

Step 1: The driver/operator stops the apparatus approximately 10 feet (3 m) beyond the hydrant. Grasp a sufficient amount of hose to reach the hydrant. Step down from the tailboard facing the hydrant with all of the equipment necessary to make the hydrant connection (Figure 11.152).

Step 2: Approach the hydrant and loop the hose around the hydrant. Place your foot on the hose, and signal the driver to proceed (Figure 11.153). An alternate method of wrapping the hydrant is to tie a piece of rope or nylon in the form of a loop to the hose so that it may be dropped over the hydrant.

Step 3: Place the hydrant wrench on the valve stem operating nut. Remove the caps from the hydrant, and place a gate valve on the outlet away from the fire, if department policy calls for this. Remove the hose loop from the hydrant, and

Figure 11.152 The firefighter should leave the apparatus with all equipment necessary to make the hydrant connection.

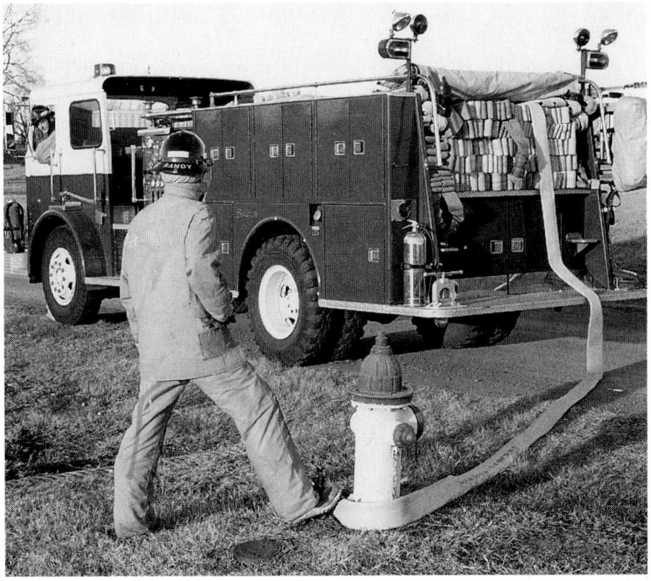

Figure 11.153 Wrap the hose around the hydrant.

connect it to the outlet nearest the fire (Figure 11.154). When using large diameter hose, a threaded-to-quick-coupling adapter must be placed on the hydrant before the hose can be connected to it.

Step 4: When the apparatus has completed the lay, apply the hose clamp, uncouple the hose from the bed (allowing enough hose to reach the pump inlet), and connect the hose to the pump. Give the signal to charge the line. (**NOTE:** Signaling to charge the line can be accomplished by using hand signals, a hand light, a radio or by using the bell, siren, or air horn.) An alternate method is to place a hose clamp on the supply line 20 feet (6 m) behind the apparatus (this allows room to remove fire fighting lines), and signal for the line to be charged. The hose can then be uncoupled from the bed and connected to the pump while the line is being charged. When the pump connection is complete, release the hose clamp.

Step 5: When the appropriate order or signal is given, open the hydrant fully (Figure 11.155). When returning to the apparatus, tighten leaking couplings and push hose toward the curb.

NOTE: If multiple lines are laid, follow the same procedures as for a single line.

Figure 11.154 Connect the hose to the hydrant.

Figure 11.155 Open the hydrant fully.

USING FOUR-WAY HYDRANT VALVES

A four-way hydrant valve allows a forward-laid supply line to be immediately charged and allows a later-arriving pumper to connect to the hydrant. The second pumper can then supply additional supply lines and/or boost the pressure to the original line. Typically, the four-way hydrant valve is preconnected to the end of the supply line. This allows the firefighter who is catching the plug to hook the valve and the hose to the hydrant in one action. There are several manufacturers providing four-way valves that have the same basic operating principles. The following steps describe the typical application of a four-way hydrant valve.

Step 1: The hydrant person on the first pumper wraps the hydrant as previously described for forward lays.

Step 2: Once the hose can be unwrapped from the hydrant and the cap on the steamer connection has been removed, connect the four-way valve to the hydrant (Figure 11.156).

Step 3: When the pumper at the scene is ready for water, turn the hydrant completely on.

Step 4: The second pumper stops at the hydrant and connects its intake sleeve to the large connection on the four-way hydrant valve and opens the valve to permit water flow into the pump (Figure 11.157).

Step 5: The second pumper connects a discharge line to the four-way valve inlet. The discharge line is then brought to the proper pressure to support the first pumper through the original supply line (Figure 11.158). The four-way valve is then switched from hydrant to pumper supply on the valve where necessary. Other supply lines may be charged as well.

Figure 11.156 Initially only the attack apparatus is connected to the hydrant via the hydrant valve. *Courtesy of George Braun, Gainesville (FL) Fire-Rescue.*

Figure 11.157 The second pumper connects directly to the hydrant valve. *Courtesy of George Braun, Gainesville (FL) Fire-Rescue.*

Figure 11.158 When the connections are complete, the second pumper can boost the pressure in the original supply line. *Courtesy of George Braun, Gainesville (FL) Fire-Rescue.*

Reverse Lay

Laying hose from the incident scene back to the water source has become a standard method for setting up a relay pumping operation when using medium diameter hose as a supply line. In most cases with medium diameter hose, it is necessary to place a pumper at the hydrant to supplement hydrant pressure to the supply hose. It is, of course, always necessary to place a pumper at the water source when drafting. The reverse lay is the most direct way to accomplish this.

A common operation involving two pumpers — an attack pumper and a water supply pumper — calls for the first-arriving pumper to go directly to the scene to start an initial attack on the fire using water from its tank, while the second-arriving pumper lays the supply line from the attack pumper back to the water source. This is a relatively simple operation because the second pumper needs only to connect its just-laid hose to a discharge, connect a suction hose, and begin pumping.

When reverse-laying a supply hose, it is not necessary to use a four-way hydrant valve. One can be used, however, if it is expected that the pumper will later disconnect from the supply hose and leave the hose connected to the hydrant. This may be desirable when the demand for water diminishes to the point that the second pumper can be made available for response to other incidents. As with a forward lay, using the four-way valve in a reverse lay provides the means to switch from pump pressure to hydrant pressure without interrupting the flow.

The reverse lay is also used when the first-in pumper arrives at a fire and must work alone for an extended period of time. In this case, the hose laid in reverse becomes an attack line. It is often connected to a reducing wye so that two smaller hoses can be used to make a two-directional attack on the fire (Figure 11.159 on the next page).

The reverse lay procedure described is for the second-in pumper to lay line from an attack pumper to a hydrant (Figure 11.160 on the next page). It can be modified to accommodate most types of apparatus, hose, and equipment.

Step 1: Stop the apparatus so that its tailboard is slightly beyond the intake valve of the attack pumper, and have a firefighter pull sufficient hose to reach the intake valve.

Step 2: When signaled by the firefighter anchoring the hose, lay out the hose to the water source.

Step 3: The firefighter at the attack pumper then applies a hose clamp to the hose.

Step 4: Stop the apparatus at the hydrant, and make an intake hose connection. If laying to a static source, make preparations for drafting.

Step 5: Pull the remaining length of the last section of hose coming from the hose bed, disconnect the couplings, and return the male coupling to the hose bed.

Step 6: Connect the supply hose to a discharge valve, and charge the hose.

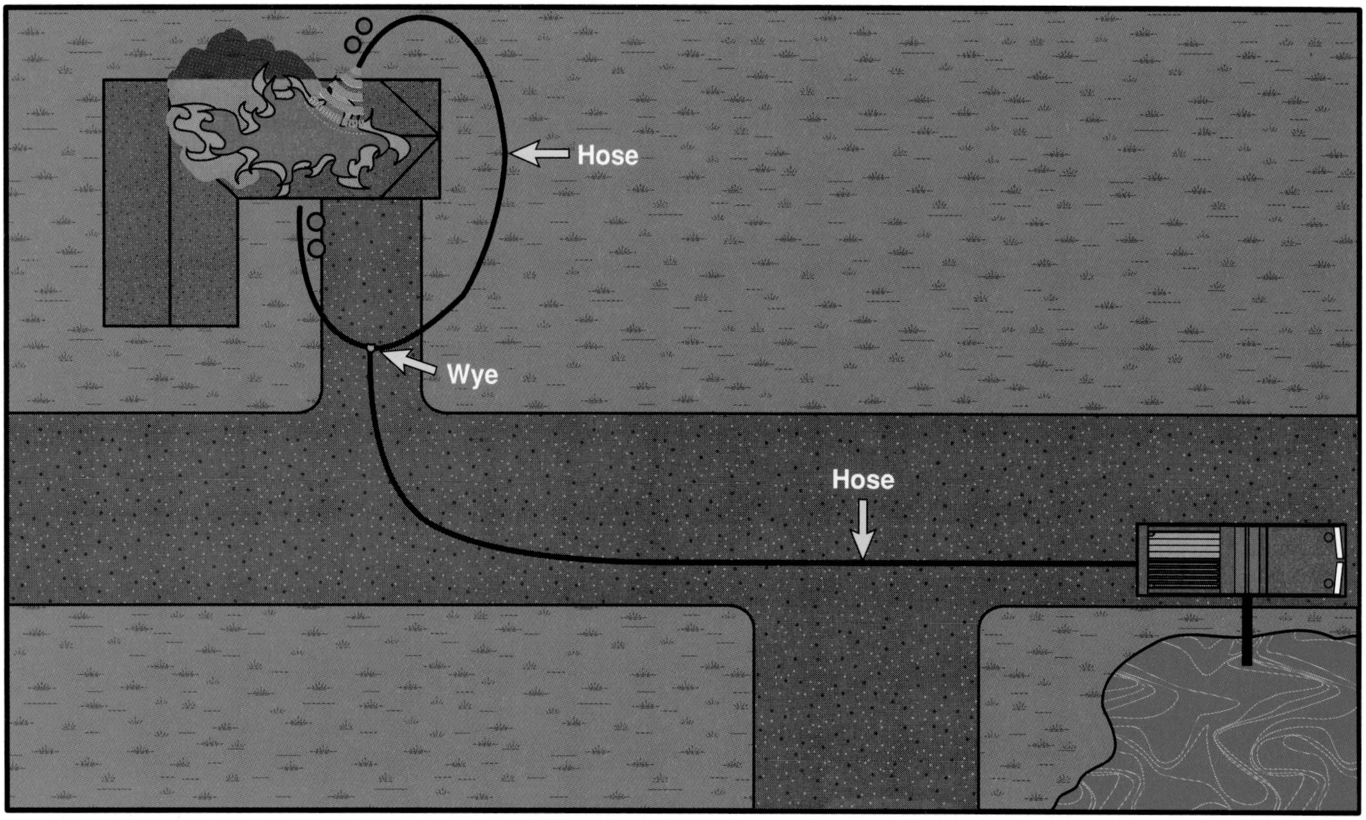

Figure 11.159 A typical reverse lay using wyed hoselines in operation.

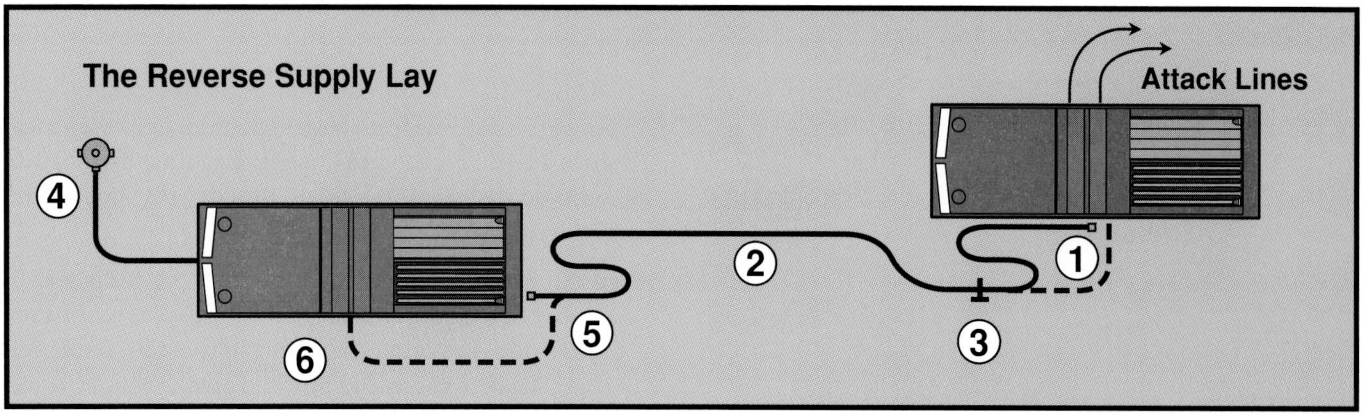

Figure 11.160 The Reverse Lay. Step 1: Pull sufficient hose to reach the intake valve. Step 2: Lay hose to the water source. Step 3: Apply a hose clamp to the hose. Step 4: Make a suction hose connection. Step 5: Disconnect the last section of hose coming from the hose bed. Step 6: Connect the supply hose to the discharge valve and charge the hose.

MAKING PUMPER-TO-HYDRANT CONNECTIONS

Frequently, firefighters will assist pumper driver/operators in making hydrant connections following a reverse lay. Either soft sleeve or hard suction intake hose may be used to connect to hydrants. Hard suction hose must be used when drafting from a static water supply source. Connecting a pumper to a fire hydrant involves coordination and teamwork. More people are needed to connect a hard suction hose than are needed to connect soft sleeve hose. The following steps illustrate the procedure for making a soft sleeve connection to a hydrant:

Step 1: The pumper is positioned so that the pump intake is a few feet ahead or short of the hydrant connection. This allows for a slight bend in the hose, yet avoids kinking (Figure 11.161).

Step 2: Remove the soft sleeve hose, hydrant wrench, and gate valve from the pumper, make pumper connection (if the intake hose is not preconnected), and unroll the intake hose (Figure 11.162). Place the hydrant wrench on the hydrant valve stem operating nut with the handle pointing away from the outlet, if it is a dry barrel hydrant.

Figure 11.161 Stop the pumper a few feet short of the hydrant.

Figure 11.162 First connect the hose to the pumper.

Step 3: Remove the hydrant cap and make the hydrant connection, using any adapters that may be necessary for this operation. Should an adapter be needed, it is usually used at the hydrant connection. Two full twists should be placed in the hose (Figure 11.163). This prevents kinking when the hose is charged.

Figure 11.163 Put two twists in the hose when making the hydrant connection.

Step 4: Open the hydrant slowly and tighten any connection that leaks (Figure 11.164). Chafing blocks may be added to the hose where it contacts the ground to prevent rub-induced damage by water and apparatus vibrations.

Figure 11.164 Open the hydrant completely. Use a chafing block if the hose is rubbing on the street or sidewalk.

Two 2½- or 3-inch (65 or 77 mm) hoselines are sometimes used for hydrants equipped with two 2½-inch (65 mm) outlets (Figure 11.165). These smaller intake hoses can be connected to a siamese on the pump. It is more efficient to connect a 4½-inch (115 mm) or larger intake hose to a hydrant with only 2½-inch (65 mm) outlets. Such a connection is made by using a 4½-inch (115 mm), or whatever size intake hose coupling that is used, to a 2½-inch (65 mm) reducer coupling.

Figure 11.165 Two smaller supply lines may be used to connect the pumper to a hydrant that lacks a large steamer connection.

Making hydrant connections with a hard suction hose is considerably more difficult than with a soft intake hose. The first aspect that is important is the positioning of the pumper from the hydrant. No definite rule can be given to determine this distance because not all hydrants are at the same distance from the curb, and the hydrant outlet may not directly face the street. Another determining factor is that while most apparatus have the pump intakes on both sides, others may also have one at the front or rear. It is considered a good policy to stop the apparatus with the intake of choice just short of the hydrant outlet. Depending on local preferences, the hard suction hose may be hooked to the apparatus or the hydrant first. The procedure for making the hard sleeve connection, either way, is as follows:

Step 1: Spot the pumper at a convenient angle to the hydrant and within the limits of the length of the intake hose. The driver/operator should check to see whether the booster tank valve is closed and should remove the pump intake cap. The other firefighter secures the hydrant wrench and adapter, removes the hydrant outlet cap, and places the hydrant wrench on the hydrant valve stem operating nut with the handle pointing away from the outlet (Figure 11.166).

Figure 11.166 Stop the apparatus a few feet short of the hydrant, but within reach of the suction hose.

Step 2: Place the adapter on the 4½-inch (115 mm) outlet if necessary.

Step 3: Remove the hard sleeve intake hose from the apparatus, and connect it to the hydrant or the apparatus, depending on local preference (Figure 11.167).

Step 4: Connect the opposite end to either the pump or hydrant. It may be necessary to move the apparatus slightly to accomplish this connection. It will also probably be necessary to put at least a slight bend in the hose (Figure 11.168).

Step 5: Open the hydrant. The driver/operator may ready the pump for operation.

Figure 11.167 In this case, the hose is attached to the apparatus first.

Figure 11.168 Connect the hose to the hydrant.

The Split Lay

The term "split lay" can refer to any one of a number of ways to lay multiple supply hoses. Dividing a hose bed into two or more separate sections provides the most options for laying multiple lines. Depending upon whether the beds are set up for forward or reverse lays, lines can be laid in the following ways (assume for now that the hose is in two beds and is of the same diameter):

- Two lines laid forward
- Two lines laid reverse
- Forward lay followed by a reverse lay
- Reverse lay followed by a forward lay
- Two lines laid forward followed by one or two lines laid reverse
- Two lines laid reverse followed by one or two lines laid forward

Clearly, there are many other split lay options when the hose bed is divided. One of the most versatile arrangements is one in which one section of the hose bed contains large diameter hose (LDH), and the other sections contain small diameter hose that can be used for either supply or attack. A pumper set up in this manner can lay LDH when the fire situation requires the pumper to lay its own supply line and work alone (laying it in forward so the pumper stays at the incident scene). It can use its small diameter hose as a supply line at fires with less demanding waterflow requirements, as well as for attack lines on large fires. A split hose bed, therefore, gives the officer the greatest number of choices when determining the best way to use limited resources.

The term split lay may also be used to denote an operation where two different pumpers complete the same lay. For example, an attack pumper lays supply lines up a driveway from the main thoroughfare, while the second pumper attaches to those lines and continues the lay to a water supply source.

HANDLING HOSELINES

To effectively attack and extinguish a fire, hoselines must be removed from the apparatus and advanced to the location of the fire. The techniques used to advance hoselines depend on how the hose is loaded. Hoselines may be loaded preconnected to a discharge outlet or simply placed in the hose bed and not connected.

Preconnected Hoselines

The method used to pull preconnected hoselines varies with the type of hose load that is used. The following sections describe the methods used to pull the preconnected hose loads described earlier in this chapter.

FLAT LOAD

Advancing the preconnected flat load involves pulling the hose from the compartment and walking toward the fire. This procedure is illustrated in the following steps:

Step 1: Put one arm through the longer loop and grasp the shorter pull loop with the same hand. Grasp the nozzle with the opposite hand (Figure 11.169).

Step 2: Using the pull loops, pull the load from the bed (Figure 11.170).

Step 3: Walk toward the fire. As the hose pulls taut in the hand, release the hand loop (Figure 11.171).

Step 4: Continue to lead in the hose. As the shoulder loop becomes taut, drop it to the ground and proceed until the hose is fully extended (Figure 11.172).

Figure 11.169 Grab the nozzle and the loops.

Figure 11.170 Pull the hose from the apparatus.

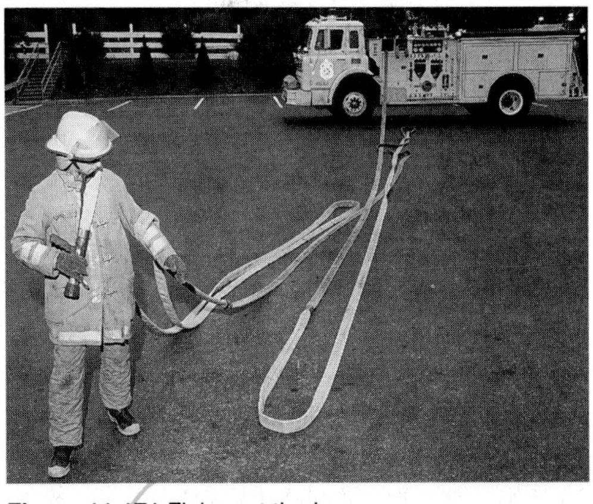

Figure 11.171 Flake out the hose.

Figure 11.172 Drop the hose when most of it has been flaked out.

MINUTEMAN LOAD

The minuteman load is intended to be deployed without dragging any of the hose on the ground. The hose is flaked off the top of the shoulder as the firefighter advances toward the fire. This procedure is illustrated in the following steps:

Step 1: Grasp the nozzle and bottom loops, if provided, and pull the load approximately one-third to one-half of the way out of the hose bed (Figure 11.173).

Step 2: Face away from the apparatus, and place the hose load on your shoulder with the nozzle against your stomach (Figure 11.174).

Step 3: Walk away from the apparatus, pulling the hose out of the bed by the bottom loop (Figure 11.175).

Step 4: Allow the load to pay off from the top of the pile as you advance toward the fire (Figure 11.176).

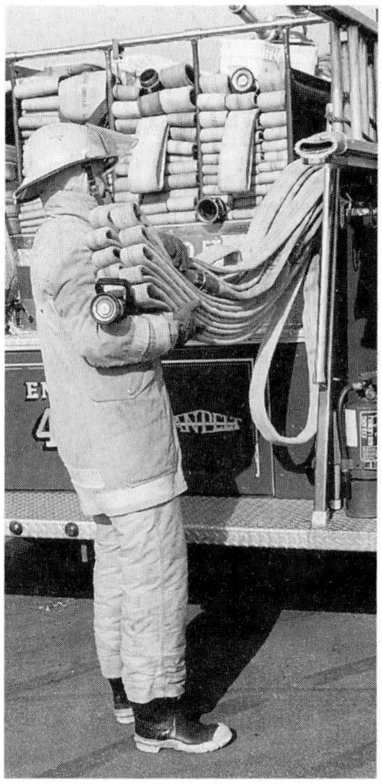

Figure 11.173 Pull the load about halfway out of the bed.

Figure 11.174 Load the hose onto the shoulder.

Figure 11.175 The hose feeds off the top of your shoulder as you advance.

Figure 11.176 Drop the rest of the hose when the end is reached.

TRIPLE LAYER LOAD

Advancing the triple layer load involves placing the nozzle and the fold of the first tier on the shoulder and walking away from the apparatus. This procedure is described and illustrated in the following steps:

Step 1: Place the nozzle and fold of the first tier over your shoulder while facing in the direction of travel (Figure 11.177).

Step 2: Walk away from the apparatus, pulling the hose *completely* out of the bed (Figure 11.178).

Step 3: When the hose bed has been cleared, drop the folded end from your shoulder, and advance the nozzle (Figure 11.179). If the direction of travel is going to be changed, you may wish to hold onto the fold and pull all three layers in that direction before dropping the fold and advancing the nozzle.

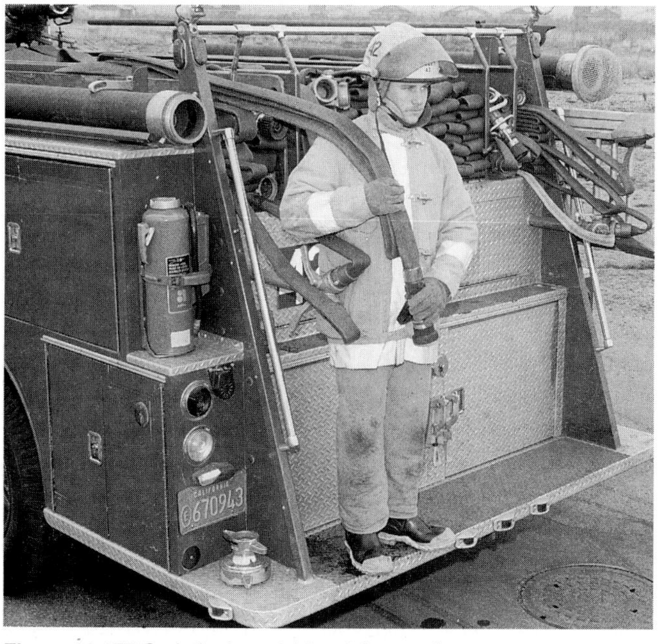

Figure 11.177 Grab the hose fold and the nozzle.

Figure 11.178 Walk straight back from the apparatus until all of the hose has cleared the bed.

Figure 11.179 Grab the nozzle and head in the desired direction.

Nonpreconnected Hoselines

The following procedures are used for handling hose that is not preconnected. This hose is usually 2½ inch (65 mm) or larger.

WYED LINES

The reverse horseshoe and other wyed lines are normally used in connection with a reverse layout because the wye connection is fastened to the 2½- or 3-inch (65 mm or 77 mm) hose. The unloading process involves two operations which can be done consecutively by one person. The following steps for unloading and advancing are as follows:

Step 1: Grasp the nozzle and small loop of one bundle and pull the bundle from the bed. When the end of the bundle clears the tailboard, lay the bundle on the ground (Figure 11.180).

Step 2: Pull the opposite bundle in the same way (Figure 11.181).

Figure 11.180 Pull one bundle clear of the apparatus.

Figure 11.181 Pull the second bundle clear of the apparatus.

Step 3: Pull the wye and attached hose from the bed, and lay the wye between the bundles near the ties (Figure 11.182).

Figure 11.182 Pull the wye back to the bundles of hose.

Step 4: When you are ready to reverse lay to the hydrant, pick up the wye and signal the driver to proceed. Anchor the hose so that it lays out from the bed as the apparatus proceeds toward the water source (Figure 11.183).

Step 5: After the apparatus makes the lay, place one arm through the horseshoe loops of one bundle and lay off the loops one at a time to lead in the hose (Figure 11.184). Lay out the second hose in the same manner.

Step 6: Open the wye when ready for water (Figure 11.185).

Figure 11.183 Anchor the hose as the apparatus lays away.

Figure 11.184 Stretch the bundles to their desired position.

Figure 11.185 Open the wye when the lines are ready to be charged.

SHOULDER LOADS (FROM FLAT OR HORSESHOE LOADS)

Due to the way flat and horseshoe loads are arranged in the hose bed, it becomes necessary to load one section of hose at a time onto the shoulder. The following steps are for shoulder loading and advancing hose from a flat or horseshoe load:

Step 1: If desired, attach the nozzle to the end of the hose. Position a carrier at the tailboard facing in the direction of travel. Place the initial fold of hose over the carrier's shoulder so that the nozzle can be held at chest height. Bring the hose from behind back over the shoulder so that the rear fold ends at the back of the knee (Figure 11.186).

Step 2: Make a fold in front that ends at knee height and bring the hose back over the shoulder. Continue to make knee-high folds until an appropriate amount of hose is loaded. Have the carrier hold the hose to prevent it from slipping off the shoulder (Figure 11.187).

Step 3: Have the carrier move forward approximately 15 feet (5 m) and position another person at the tailboard. Load hose onto the shoulder in the same manner as before, making knee-high folds until an appropriate amount of hose is loaded (Figure 11.188).

Step 4: Repeat the loading process with each carrier until the desired length of hose is loaded. Then uncouple the hose from the hose bed, and hand the coupling to the last carrier (Figure 11.189).

Figure 11.186 Begin to load hose onto the firefighter's shoulder.

Figure 11.187 Load only that hose which the firefighter can easily carry.

Figure 11.188 The second firefighter is loaded with hose.

Figure 11.189 As many firefighters can be used as required by the length being covered.

SHOULDER LOADS (FROM THE ACCORDION OR FLAT LOAD)

Because all of the folds in an accordion and flat load are near the same length, they can be loaded on the shoulder by taking several folds at a time directly from the hose bed. The following steps are for shoulder loading and advancing hose that is loaded in an accordion or a flat load.

Step 1: Face the hose bed, grasp the nozzle or coupling and with both hands grasp the number of folds needed to make up that portion of the shoulder load (Figure 11.190).

Step 2: Pull the folds about one-third of the way out of the bed, twist the folds into an upright position (accordion load only), turn and pivot into the folds placing them on top of your shoulder. Make sure that the hose is flat on your shoulder with the nozzle or coupling in front of your body (Figure 11.191).

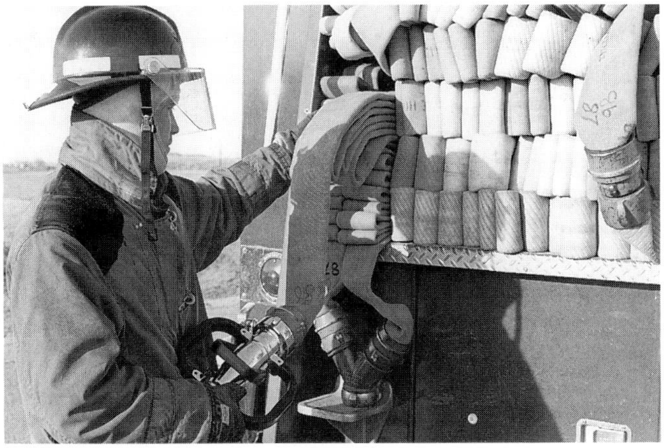

Figure 11.190 Grasp the nozzle and hose.

Figure 11.191 Pull the hose from the bed.

Step 3: Grasp the bundle tightly with both hands, and step away from the apparatus, pulling the shoulder load completely out of the bed (Figure 11.192).

NOTE: Additional firefighters may remove shoulder loads in the same manner.

Figure 11.192 Walk away from the apparatus.

WORKING LINE DRAG

The working line drag is one of the quickest and easiest ways to move fire hose at ground level. Its use is limited by available personnel, but when adapted to certain situations, it proves to be an acceptable method.

The first method is to stand alongside a single hoseline at a coupling or nozzle and face the direction of travel. Place the hose over your shoulder with a coupling in front and resting on your chest. Hold the coupling in place and pull with your shoulder (Figure 11.193). Additional firefighters are needed at each coupling, and about one-third of the hose section should form a loop on the ground between each firefighter (Figure 11.194).

Figure 11.193 One firefighter carries the hose with the nozzle resting on his or her chest.

Figure 11.194 The second firefighter grabs the hose at the next coupling.

ADVANCING HOSELINES

Once hoselines have been laid out and connected for fire fighting, they must be advanced into final position for applying water on the fire. The methods of deploying hose described up to this point work fine if the firefighter is simply advancing hose over flat ground with no obstacles. The advancement of hoselines becomes considerably more difficult when other factors come into play. Having to advance hose up stairways, up ladders, and into buildings are all examples of tasks that require the firefighter to know special techniques. These tasks are more easily accomplished before the hose is charged because water adds considerable weight and makes the lines less maneuverable. However, sometimes it becomes necessary to perform these tasks with charged hoselines, and where appropriate, methods for handling both dry and charged lines are discussed.

Advancing Hose Into A Structure

For maximum firefighter safety, it is necessary that the firefighters advancing hose into a structure be alert to the potential dangers of backdraft, flashover, and structural collapse, among other things. The following are general safety guidelines that should be observed when advancing a hoseline into a burning structure:

- The firefighter on the nozzle and the backup firefighter(s) should be on the same side of the line (Figure 11.195).
- Feel the door using the back of the hand with the glove off. This gives an indication of whether there is an extreme amount of heat built up behind the door and alerts the firefighters to the possibility of a backdraft or flashover.
- Bleed the air out of the line before entry into the fire area.
- Stay low and avoid blocking ventilation openings such as doorways or windows.

Figure 11.195 Both firefighters should be on the same side of the hoseline.

Advancing Hose Up A Stairway

Hose is difficult to drag in an open space and is exceedingly difficult to drag around the obstructions found in a stairway. When safely possible, hose should be advanced up stairways before it is charged with water. If the line has already been charged, clamp it off before advancing up the stairs.

The shoulder carry and the underarm carry are adaptable to stairway advancement, because the hose is carried into position and is fed out as it is needed. The minuteman load and carry is also excellent for use on stairways. During the advancing process, lay the hose on the stairs against the outside wall to avoid sharp bends and kinks (Figure 11.196). Excess hose should be flaked up the stairs, because as hose is advanced into the fire floor, the hose will be much easier to advance.

Figure 11.196 Lay the hose against the outside walls.

Advancing Hose Down Inside And Outside Stairways

Advancing a charged hoseline down a stairway will be difficult because of the awkwardness of the fire hose and because it will be getting very hot as you near the fire floor. It will be necessary to have all available hose when reaching the fire floor, because the advance must be made quickly due to the heat. Firefighters must be stationed at critical points — corners and obstructions — to help feed the hose and to keep it on the outside of the staircase.

The advancement of an uncharged hoseline down a flight of stairs will be considerably easier than with a charged hose. However, because advancing a hoseline down a stairway often subjects the firefighters to intense heat, in most cases the hoseline should be charged when advancing down a stairway. Advancing an uncharged line down stairs is recommended only when you are sure that there is no fire present or it is very minor.

Advancing Hose From A Standpipe

Fighting fires in tall buildings presents the problem of getting hose to upper floors. While hoselines may be pulled from the apparatus and extended to the fire area, it is not considered good practice. It is more practical to have some hose rolled or folded on the apparatus ready for standpipe use.

The manner in which standpipe hose is arranged is a matter of personal preference. It may be in the form of folds or bundles that are easily carried on the shoulder or in specially designed hose packs complete with nozzles, fittings, and tools (Figure 11.197).

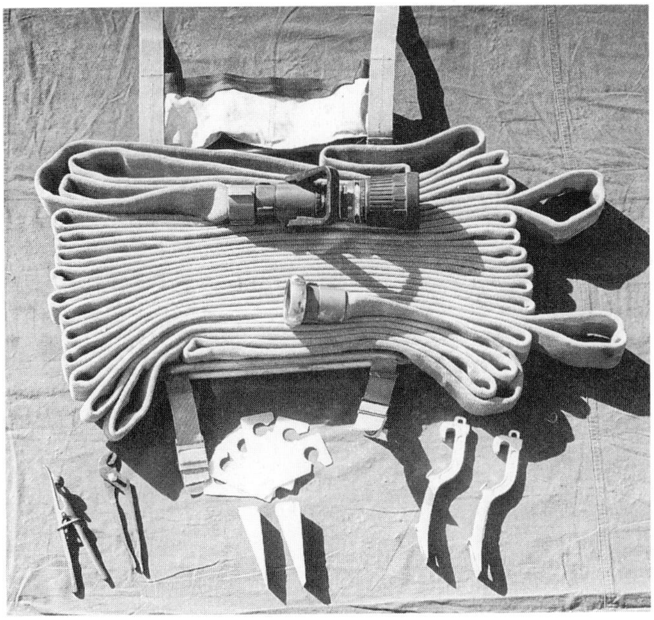

Figure 11.197 The components of a typical hose bundle.

Hose should be brought to the fire floor by taking the stairway. Fire crews should stop one floor below the fire floor and make the connection to the standpipe. The standpipe connection is usually in the stairwell or just outside the stairwell door. Firefighters can also use the floor below to get a general idea of the layout of the fire floor.

Upon reaching the standpipe, detach the building hoseline or outlet cap (whichever is present), check the connection for the correct adapters (if needed) and foreign objects in the discharge, and connect the fire department hose to the standpipe (Figures 11.198 a and b on the next page). If 1½-, 1¾-, or 2-inch (38 mm, 45 mm, or 50 mm) hose is used, it is a good practice to place a gated wye either on the standpipe or at the end of a short piece of

Figure 11.198a Check the discharge for debris.

Figure 11.198b Connect to the discharge.

2½-inch (65 mm) hose connected to the standpipe. A 2½-inch (65 mm) attack line may also be used depending on the size and nature of the fire. Once the standpipe connection is completed, any extra hose should be flaked up the stairs toward the floor above the fire (Figure 11.199). Charged hoselines as well as dry lines may be advanced in this manner.

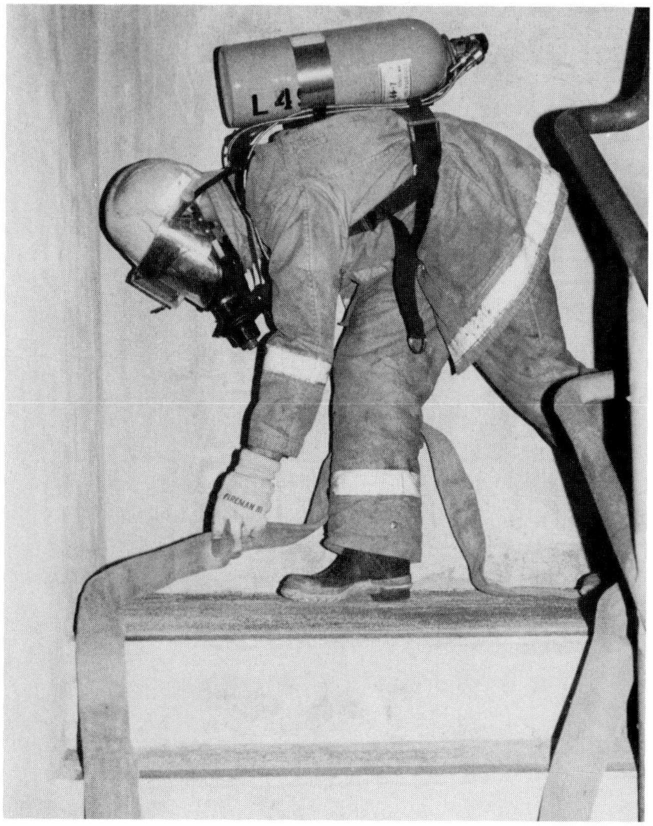

Figure 11.199 Flake extra hose up the stairs.

During pickup operations, the water contained in the hoselines should be carefully drained to prevent unnecessary water damage. This can be accomplished by draining the hose out a window, down a stairway, or by some other suitable drain.

Advancing Hose Up A Ladder

One of the safest ways to get hose to an elevated position is to carry it up in a bundle and drop the end down to be connected to a source; another safe method is to hoist it up using a rope. However, sometimes these methods cannot be used, and it becomes necessary to advance the hose up a ladder. Advancing fire hose up a ladder can be best achieved with a line that is not charged. If the hose is already charged with water, it will be safer, quicker, and easier to relieve the pressure and drain the hose before the advancement is made.

The best way to advance hose up a ladder is to have the lead firefighter drape the nozzle or end coupling over the shoulder on the side on which the hose is being carried. This firefighter then advances up the ladder until the first fly section is reached. At that point, a second firefighter drapes a large loop of hose over his or her shoulder and starts up the ladder. If the ladder is a three-section ladder, a third firefighter may continue the process once the second firefighter reaches the first fly section (Figure 11.200). As a rule of thumb, to avoid overloading the ladder, only one person should be allowed on each section of the ladder at the same time. Rope hose tools or utility straps can also be used for this advancement (Figure 11.201). Whenever possible, it is best to have one firefighter at the base of the ladder to help feed the hose to the carriers and to have one firefighter heel the ladder during the advancement. The hose can be charged once it has reached its point of intended use.

When advancing a charged line up a ladder, firefighters should position themselves on the ladder within reach of each other. Each firefighter should be attached to the ladder via a leg lock or ladder belt, because both hands will be required to move the charged line. The hose is then pushed upward from firefighter to firefighter. The firefighter on the nozzle takes the line into the window, and the other firefighters continue to hoist additional hose as necessary.

WARNING:

Caution must be exercised to ensure that the rated capacity of the ladder is not exceeded. If the hose cannot be passed up the ladder without exceeding the load limit, another method of advancing the hoseline, such as by hoisting, should be used.

Figure 11.200 The hose may be carried up the ladder on the firefighters' shoulders.

Figure 11.201 Rope hose tools may be used to carry hose up the ladder.

Figure 11.202 Secure the hose to the ladder.

Figure 11.203 A completed fire stream operation from a ground ladder.

Sometimes it will be necessary to operate the hoseline from the ladder. When this becomes necessary, the hoseline is passed up the ladder as previously stated. The hose should be secured to the ladder with a hose strap at a point several rungs below the one the firefighter is standing on (Figure 11.202). The firefighter on the nozzle projects the nozzle through the ladder and holds it with a rope hose tool or similar aid. All firefighters on the ladder must use a leg lock or ladder belt to be secured to the ladder. When the line and firefighter are properly secured, the nozzle can be opened (Figure 11.203).

Hoisting Hose

Hoisting hose is possibly the safest way of getting hoselines to upper levels. As with advancing hose up a ladder, it is easier and safer to hoist a dry hoseline; however, charged lines may also be hoisted. There are some precautions that should be taken to reduce the possibility of damaging the coupling or nozzle as it is being raised. When advancing dry hoselines, it is recommended that the end of the hose and nozzle be folded back upon itself a short distance so that the nozzle will point opposite the direction of travel. The following procedure should be used to hoist a dry hoseline:

Step 1: Lower a rope of appropriate size down from the intended destination of the hoseline.

Step 2: Fold the nozzle end of the hoseline back over the rest of the hose so that an overlap of 4 to 5 feet (1.3 m to 1.6 m) is formed.

Step 3: Tie a clove hitch, with a safety knot, around the tip of the nozzle and the hose it is folded against so that they are lashed together (Figure 11.204).

Step 4: Place a half hitch on the doubled hose about 12 inches (300 mm) from the loop end. With the ties properly placed, the hose will turn on the hose roller so that the coupling and nozzle will be on top as the hose passes over the roller (Figure 11.205).

Figure 11.204 The proper tie for hoisting a dry hoseline.

Figure 11.205 Hose is more easily hoisted when it is not charged.

It is most desirable to bleed the pressure off a charged hoseline before hoisting it. However, if it is not possible to do this, use the following procedure to hoist a charged hoseline:

Step 1: Lower a rope of appropriate size from the intended destination of the hoseline.

Step 2: Tie a clove hitch, with a safety knot, around the hose about 1 foot (0.3 m) below the coupling and nozzle.

Step 3: Tie a half hitch through the nozzle bale and around the nozzle itself in a manner that allows the rope to hold the nozzle shut while it is being hoisted.

Extending A Section Of Hose

Occasionally, it becomes necessary to extend the length of a hoseline with hose of the same size or perhaps even smaller hose. The following procedure may be used to extend hoselines:

Step 1: Bring additional sections of hose as needed to the nozzle end of the hoseline.

Step 2: Crack the nozzle open slightly, and apply the hose clamp approximately 3 feet (1 m) behind the nozzle (Figure 11.206).

NOTE: If the line being extended is equipped with a stacked-tip, solid stream nozzle, the hoseline may be extended without using a hose clamp. With the nozzle shut off, the tips ahead of the nozzle may be removed to reveal appropriate threads for hose connection. Once the new hose and nozzle are added, the nozzle may be turned on to resume the flow of water.

Figure 11.206 Apply the hose clamp behind the nozzle.

Step 3: Remove the nozzle, add the new section of hose, and reattach the nozzle (Figure 11.207).

Step 4: Slowly release the clamp allowing water to flow to the nozzle.

Figure 11.207 The additional section may be added when the clamp has been applied.

Retrieving A Loose Hoseline And Replacing Burst Sections

A loose hoseline is one in which water is flowing through a nozzle, an open butt, or a broken line and is not under control by firefighters. This situation is very dangerous because the loose hoseline may whip back and forth and up and down. Firefighters and bystanders may be seriously injured or killed if they are hit by the uncontrolled whipping end.

Closing a valve to shut off the flow of water is the safest way to control a loose line. These valves may be at the pump or a hydrant. Another method is to position a hose clamp at a stationary point in the hoseline. It may also be possible to put a kink in the hose at a point away from the break, until the appropriate valve may be shut off (Figure 11.208). First, obtain sufficient slack in the line, bend the hose over on itself, and apply body weight to the bends in the hose. It is helpful, during the operation, to place one knee directly upon the bend and apply pressure at this point.

When replacing a burst section of hose, two additional sections of hose should be secured from the apparatus for the replacement of one bad section. This is necessary because hoselines stretch to longer lengths when under pressure; thus, the couplings in the line are invariably farther apart than the length of the replacement sections.

Figure 11.208 The hose may be kinked to stop the flow of water.

OPERATING HOSELINES

In order to successfully attack a fire, the firefighter must know how to operate and control the hoseline. There are many methods that may be used. The method that any one particular firefighter finds most comfortable will vary depending on the size, strength, and personal preference of the firefighter. Some of the more popular techniques are described in the following sections.

Handling Medium-Size Attack Lines

The following methods can be used with medium size attack lines of 1½-, 1¾-, and 2-inch (38 mm, 45 mm, 50 mm) hose.

ONE-FIREFIGHTER METHOD

When one firefighter is required to handle a medium-size hose and nozzle, some means must be provided for bracing and anchoring the hoseline. To accomplish this, the firefighter should hold the nozzle with one hand and hold the hose with the other hand, just behind the nozzle (Figure 11.209 on the next page). The hoseline should be straight for at least 10 feet (3 m) behind the nozzle, and the firefighter should face the direction in which the stream is to be projected. Permit the hose to cradle against the inside of the closest leg, and brace or hold it against the front of the body and hip. Anchor the hose to the ground or floor by placing the foot of the supporting leg upon the hose. If the stream is to be moved or directed at an excessive angle from the center line, close the nozzle, straighten the hose, and resume the operation position.

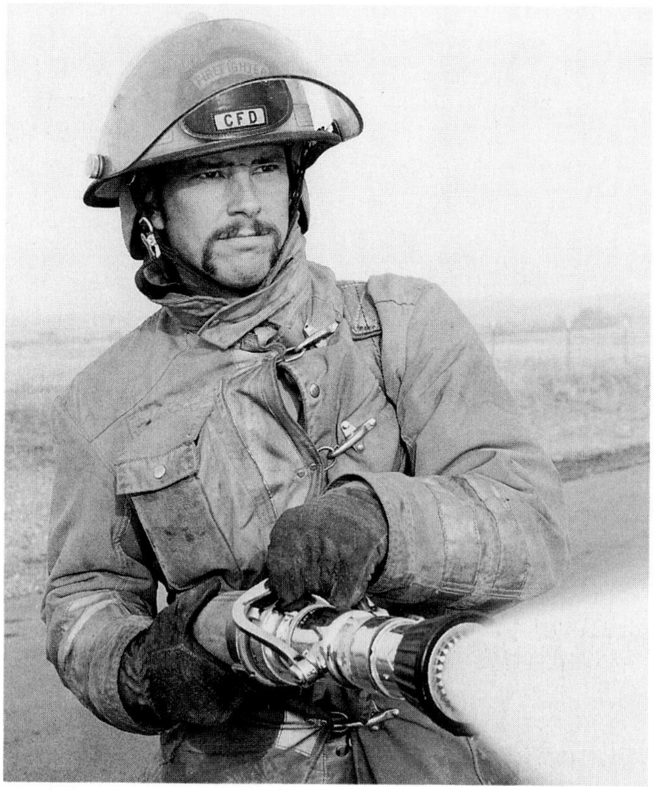

Figure 11.209 One firefighter can operate smaller hoselines.

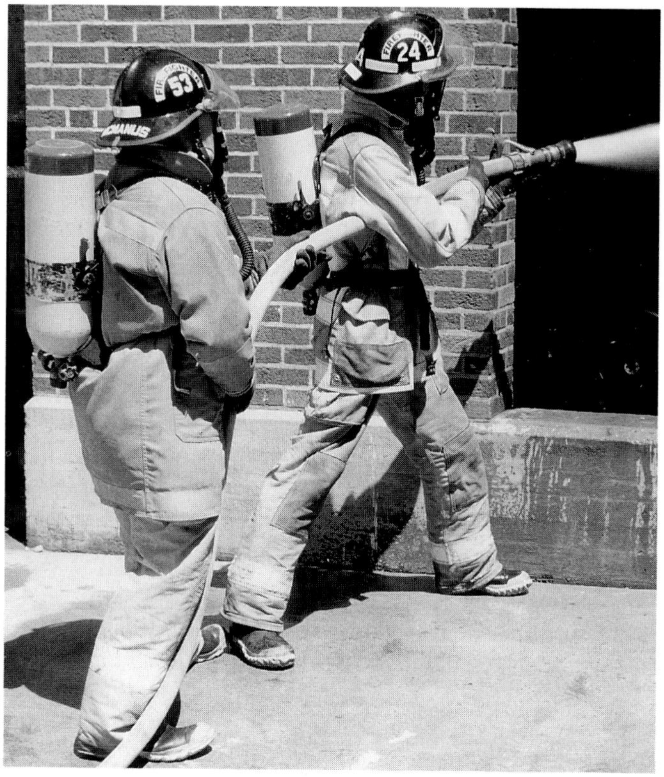

Figure 11.210 When possible, two firefighters should be on the hoseline.

TWO-FIREFIGHTER METHOD

The two-firefighter method of handling a nozzle on a medium-size attack line should be used whenever possible, because it provides a greater degree of safety than the one-firefighter method. The two-firefighter method is usually necessary when the nozzle needs to be advanced. The person at the nozzle holds the nozzle with one hand and holds the hose just behind the nozzle with the other hand. The hoseline is then rested against the waist and across the hip. The backup firefighter takes a position on the same side of the hose about 3 feet (1 m) behind the nozzleman. The second firefighter holds the hose with both hands and rests it against the waist and across the hip or braces it with his or her leg (Figure 11.210). One important function of the backup firefighter is to keep the hose straight behind the person at the nozzle. For extended operation, either or both firefighters may apply a hose strap or utility strap to reduce the effect on the arms of the nozzle reaction.

Handling Large-Size Attack Lines

The following methods can be used with large-size attack lines of $2\frac{1}{2}$-, $2\frac{3}{4}$-, and 3-inch (65 mm, 70 mm, 77 mm) or larger hose.

ONE-FIREFIGHTER METHOD

Whenever a nozzle connected to a large-size attack line is used, a minimum of two firefighters, and preferably three, should be used to handle the line. One firefighter may, however, be found alone with a large charged hoseline that needs to be used. A reasonably safe way by which this task can be performed is illustrated in Figure 11.211. The firefighter secures slack hose from the line, forms a large loop, and crosses the loop over the line about 2 feet (1 m) behind the nozzle. The firefighter then sits where the hose crosses and directs the stream. This method does not permit much maneuvering of the nozzle, but the nozzle can be operated from this point until help is available. If the operation is to be for a long duration and master stream equipment or manpower is not available, tie the hose at the cross to permit ease of operation and greater safety.

Figure 11.211 When looped, a $2\frac{1}{2}$-inch (65 mm) hoseline can be safely operated by one firefighter.

TWO-FIREFIGHTER METHOD

When only two firefighters are available to handle a nozzle on a large hoseline, some means of anchoring the hose must be provided because of the nozzle reaction. The person at the nozzle holds the nozzle with one hand and holds the hose just behind the nozzle with the other hand. The nozzleman rests the hoseline against the waist and across the hip. The backup firefighter must serve as an anchor at a position about 3 feet (1 m) behind the nozzleman. The backup firefighter places the closest knee on the hoseline. In this position, the backup firefighter should be kneeling on one knee with both hands on the hoseline near the other knee. This position prevents the hose from moving back or to either side. Should the hose in front try to come back or up, the backup firefighter is in a position to push it forward.

Another two-firefighter method uses rope hose tools to assist in anchoring the hose. The nozzleman loops a rope hose tool or utility strap around the hose a short distance from the nozzle and places the large loop across his or her back and over the outside shoulder. The nozzle is then held with one hand and the hose just behind of the nozzle with the other hand. The hoseline is rested against the body. Leaning slightly toward the nozzle helps control the nozzle reaction. The backup firefighter again serves as an anchor, and this position is about 3 feet (1 m) back. The backup firefighter also has a rope hose tool around the hose and the shoulder and leans forward to absorb some of the nozzle reaction (Figure 11.212).

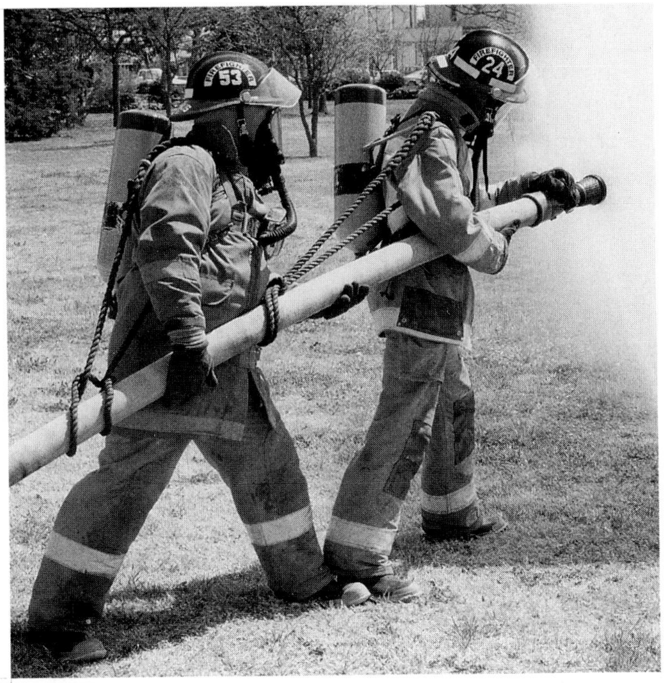

Figure 11.212 When possible, use a rope hose tool to help control the line.

THREE-FIREFIGHTER METHOD

Handling a nozzle on a large-size hoseline can be more easily accomplished by three firefighters. There are several methods for three firefighters to control large hoselines. In all cases, the positioning of the nozzleman is the same as previously described for the two-firefighter method. The only differences will be in the position of the second and third firefighters on the line. Some departments prefer the first backup firefighter to be standing directly behind the firefighter at the nozzle, with the third firefighter kneeling on the hose behind the second firefighter. Another method is for both firefighters to serve as anchors by kneeling on opposite sides of the hoseline. Another technique is for all firefighters to use hose straps and remain in a standing position. The last method is the most mobile (Figure 11.213).

Figure 11.213 Three firefighters make the hoseline more maneuverable.

DEPLOYING MASTER STREAM DEVICES

Master stream devices will see much less use than other types of nozzles. However, when the need does arise for their use, they are generally the only hope left of containing and controlling a fire. Master streams are deployed in situations where the fire is beyond the control of handlines or there is a need for fire streams in a location that is no longer safe for personnel (Figure 11.214). The three main uses for a master stream are:

- For a direct fire attack

Figure 11.214 Master streams are used on large fires where mobility is not crucial. *Courtesy of Joseph J. Marino.*

- To back up handlines that are already attacking the fire
- For exposure protection

The master stream device must be properly positioned to provide an effective stream on the fire. Once the line is in operation, it must be shut down to be moved, which can be a time-consuming process. If the master stream device is being operated into a building, place the device close enough to the window or door so that it is able to hit the base of the fire. This is particularly important when using a fog nozzle, as the fog stream does not have the penetration power of a solid stream.

The second aspect of stream placement is the angle at which the stream is entering the structure. The stream should be aimed so that it is entering the structure at an upward angle, allowing it to be deflected off the ceiling or other overhead objects (Figure 11.215). This makes the stream break up into smaller droplets that will rain down on the base of the fire, providing maximum extinguishing effectiveness. Streams that enter the opening at a perfectly horizontal or less than horizontal angle will not be as effective. Also, operating the stream at too low of an angle might result in a loss of control of the master stream device and hoseline.

Figure 11.215 Deflect the stream off the ceiling.

It is also desirable to place the master stream in a location that will provide maximum coverage on the face of the building. This gives personnel the opportunity to change the direction of the stream and aim it through another opening should the need arise. This is particularly important in situations where there is a large volume of fire and a limited amount of master stream devices.

Supplying The Master Stream

Master stream devices operate at high flow rates, which in turn means large amounts of friction loss. Typically, a master stream device will be expected to flow a minimum of 350 gpm (1 400 L/min). Therefore, it is not practical to supply the master stream appliance with anything less than two 2½-inch (65 mm) hoselines (Figure 11.216). Larger flows will require a third 2½-inch (65 mm) or larger diameter hoses to be used. Some master stream devices are equipped to handle one large diameter (4-inch [100 mm] or larger) supply line. When possible, it is desirable to supply the device with a maximum of 100 feet (30 m) of hose in order to reduce the amount of pressure lost due to friction.

Figure 11.216 A typical master stream device with two intake connections.

Hose for the master stream device will generally be taken from the main supply in the bed of the engine that supplies the device. If the bed is not set up to provide the deployment of two lines at one time, it will be necessary to pull one line to the device, break the connection, and then pull additional lines as required. In some cases, the portable master stream appliance will be preconnected. This allows for faster deployment of the stream.

Manning The Master Stream Device

It will take a minimum of two firefighters to deploy the master stream device and get water to it. If more people are available, it would be desirable to have the extra help. Once the stream is in place, it can be operated by one firefighter (Figure 11.217). When water is flowing, one firefighter should at all times be stationed at the master stream device. This allows the firefighter to change the direction of the stream when required and to prevent it from "crawling away." This movement in master stream devices is caused by the pressure in the hoselines, but it is easily controlled by one firefighter. The exception to this is when the device is being used in a hazardous position such as close to a fire-weakened structure or close to other objects such as an LPG storage tank. These situations may be too dangerous to have personnel that close, so the master stream should be securely anchored, the desired stream put in place, and then personnel should back away. If the device starts to move, the pressure should be decreased at the supply source to curb the movement.

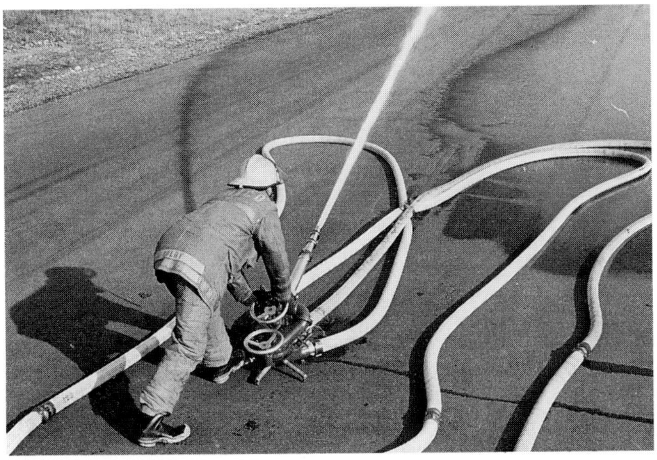

Figure 11.217 One person should operate the master stream device.

SERVICE TESTING FIRE HOSE

Because fire hose is required to be tested annually, firefighters will often be required to assist in the process. There are two types of tests for fire hose: acceptance testing and service testing. At the request of the purchasing agency, coupled hose is acceptance tested by the manufacturer before the hose is shipped. This type of testing is relatively rigorous, and the hose is subjected to extremely high pressures to ensure that it can withstand the most extreme conditions in the field. This type of testing should not be attempted by the fire department. Service testing is performed periodically by the user to ensure that the hose is being maintained in optimum condition. This testing of in-service hose confirms that it is still able to function under maximum pressure during fire fighting or other operations. Guidelines for both types of tests are in NFPA 1962, *Standard for the Care, Use, and Service Testing of Fire Hose Including Couplings and Nozzles.*

Fire department hose should be service tested annually, after being repaired, or after being run over by a vehicle. Unlined standpipe hose should be tested five years from the date of purchase, again at the eighth year, and every two years thereafter.

Before performing a service test, the hose should be examined for jacket defects, coupling damage, and worn or defective gaskets; any defects should be corrected if possible. If damage is not repairable, the hose should be taken out of service.

Test Site Preparation

Hose should be tested in a place that has adequate room to lay out the hose in straight runs, free of kinks or twists. The site should be isolated from traffic. If testing is done at night, the area should be well lighted. The test area should be smooth and free of dirt and debris. A slight grade to facilitate the draining away of water is helpful. A water source sufficient for filling the hose is also necessary.

The following equipment is needed to service test hose:

- A hose testing machine, portable pump, or fire department pumper (Figure 11.218). All such equipment should be equipped with gauges certified as accurate within one year before testing.
- A hose test gate valve.
- A means of recording the hose numbers and test results.
- Tags or other means to identify sections that fail.
- Nozzles with shutoff valves.
- A means of marking each length with the year of the test to easily identify which lengths have been tested and which have not, without having to look it up in the hose records.

Figure 11.218 A typical hose testing machine. *Courtesy of Rice Hydro Equipment Mfg.*

Safety At The Hose Testing Site

As when working with any equipment, exercise care when working with hose, especially when it is under pressure. Pressurized hose is potentially dangerous because of its tendency to whip back and forth if a break occurs such as when a coupling pulls loose. To prevent this, use a specially designed hose test gate valve (Figure 11.219 on the next page). This is a valve with a ¼-inch (6 mm) hole in the gate that permits pressurizing the hose but will not allow water to surge through the hose if the hose fails. Even when using the test valve, stand or walk near the pressurized hose only as necessary. All personnel operating in the area of the pressurized hose should wear their helmets as a safety precaution.

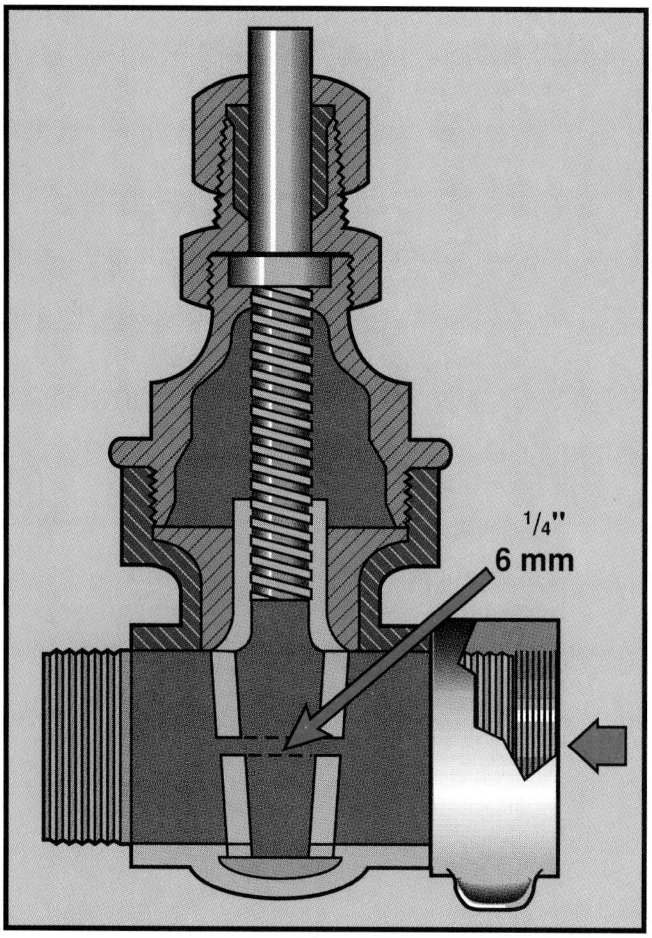

Figure 11.219 A cutaway showing the hole in the valve gate.

When possible, connect the hose to discharges on the side of the apparatus opposite the pump panel. Open and close all valves slowly to prevent water hammer in the hose and pump. Test lengths of hose shall not exceed 300 feet (90 m) in length (longer lengths are more difficult to purge of air).

Laying large diameter hose flat on the ground before charging helps to prevent unnecessary wear at the edges. Stand away from the discharge valve connection when charging because of the hose's tendency to twist when filled with water and pressurized; this could cause the connection to twist loose.

Keep the hose testing area free of water when filling and discharging air from the hoses. During testing, this aids in detecting minor leaks around couplings.

Service Test Procedure

The procedure for service testing lined fire hose and large diameter hose is as follows:

Step 1: Connect a number of hose sections (check the gaskets before connecting) into test lengths of no more than 300 feet (90 m) each. Tighten the connections between the sections with spanner wrenches (Figure 11.220).

Figure 11.220 Connect the hose sections together securely.

Step 2: Connect an open test valve to each discharge valve used (Figure 11.221). Tighten each connection with spanners.

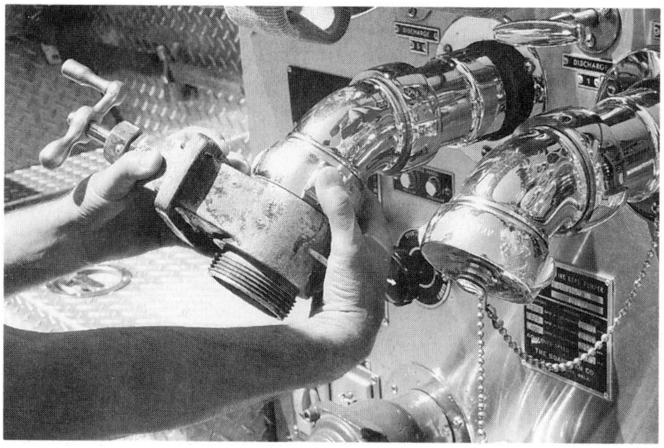
Figure 11.221 Connect the test gate valve to the discharge.

Step 3: Connect a test length to each test valve (Figure 11.222). Tighten each connection with a spanner.

Step 4: Tie a rope, rope hose tool, or hose strap to each test length of hose 10 to 15 inches (254 mm to 380 mm) from the test valve connections. Secure the other end to the discharge pipe or other nearby anchor (Figure 11.223).

Step 5: Attach a shutoff nozzle (or any device that will permit water and air to drain from the hose) to the open end of each test length (Figure 11.224).

Figure 11.222 Connect the hose to the valve.

Figure 11.223 Tie the end of the hose to the apparatus.

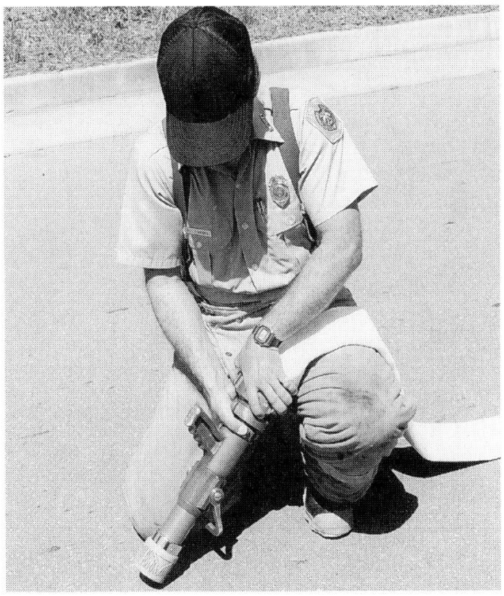

Figure 11.224 Attach a nozzle to the end of the hose.

Step 6: Fill each hoseline with water with a pump pressure of 50 psi (350 kPa) or to hydrant pressure. Open the nozzles as the hoselines are filling. Hold them above the level of the pump discharge to permit all the air in the hose to discharge (Figure 11.225). Discharge the water away from the test area.

Step 7: Close the nozzles after all air has been purged from each test length. Make a chalk or pencil mark on the hose jackets against each coupling (Figure 11.226). Check that all hose is free of kinks and twists and that no couplings are leaking. Any length found to be leaking from BEHIND the coupling should be taken out of service and repaired before being tested. Retighten any couplings that are leaking at the connections. If the leak cannot be stopped by tightening the couplings, depressurize, disconnect the couplings, replace the gasket, and start over at Step 6.

Figure 11.226 Mark the hose at the couplings.

Step 8: Close each hose test gate valve (Figure 11.227).

Step 9: Increase the pump pressure to the required test pressure in NFPA 1962, *Use, Care, and Service Testing of Fire Hose Including Couplings and Nozzles*. Personnel closely monitor the connections for leakage as the pressure increases (Figure 11.228).

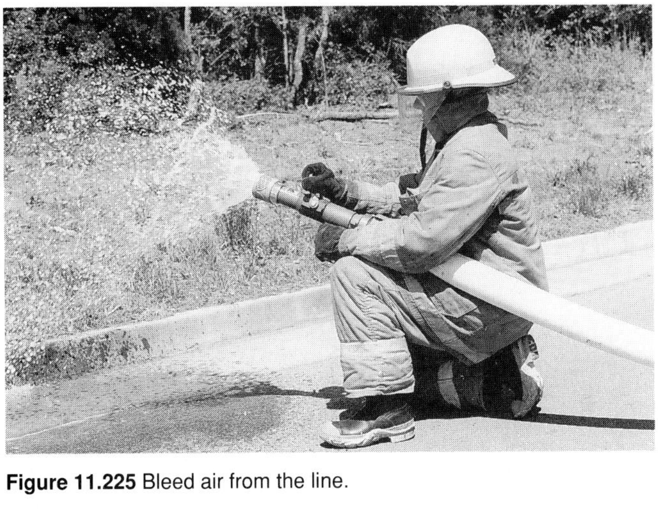

Figure 11.225 Bleed air from the line.

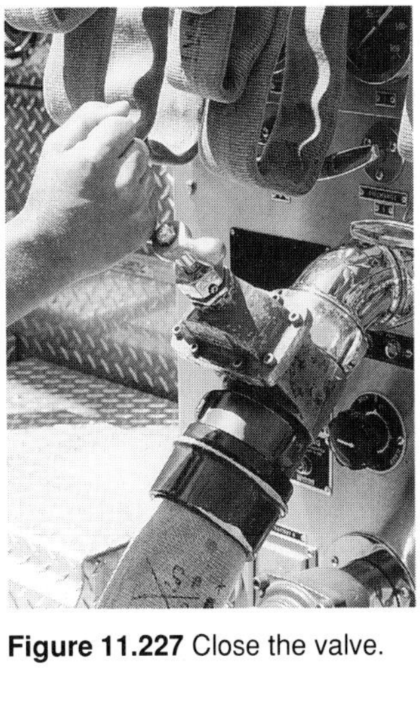

Figure 11.227 Close the valve.

Figure 11.228 Increase the pressure on the hose.

Step 10: Maintain the test pressure for five minutes. Inspect all couplings to check for leakage (weeping) at the point of attachment (Figure 11.229).

Step 11: After five minutes, slowly reduce the pump pressure, close each discharge valve (Figure 11.230), and disengage the pump.

Step 12: Slowly open each nozzle to bleed off pressure in the test lengths. Break all hose connections and drain water away from the test area (Figure 11.231).

Step 13: Observe marks placed on the hose at the couplings. If a coupling has moved during the test, tag the hose section for recoupling. Tag all hose that has leaked or failed in any other way (Figure 11.232).

NOTE: Expect a 1/16- to 1/8-inch (2 mm to 3 mm) uniform movement of the coupling on newly coupled hose. This slippage is normal during initial testing but should not occur during subsequent tests.

Step 14: Record the test results for each section of hose.

Figure 11.229 Watch for couplings to fail.

Figure 11.231 Drain the hose.

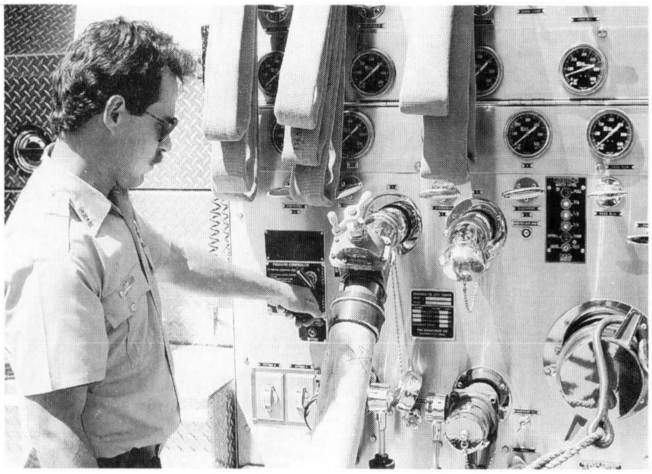

Figure 11.230 Reduce the pressure.

Figure 11.232 Mark sections that fail for repair or disposal.

12

Fire Control

This chapter provides information that addresses performance objectives described in NFPA 1001, *Standard for Fire Fighter Professional Qualifications* (1992), particularly those referenced in the following sections:

3-12 Fire Hose, Appliances and Streams

3-12.12

3-12.13

3-12.14

3-14 Fire Control

3-14.1

3-14.2

4-13 Foam Fire Streams

4-13.6

4-14 Fire Control

4-14.1

Chapter 12
Fire Control

The success or failure of the fire fighting team often depends upon the skill and knowledge of the personnel involved in initial attack operations. A well-trained team of firefighters with an attack plan and an adequate amount of water, properly applied, will contain most fires in their early stages. Failure to make a well-coordinated attack on a fire can permit or allow the fire to gain headway and get out of control. Loss of control of the fire can result in increased damage, as well as further endangerment to firefighters and civilians alike (Figure 12.1).

Figure 12.1 Large fires require firefighters to draw on much of their previous training and experience. *Courtesy of Harvey Eisner.*

Firefighters should be aware of the operations to be performed by each team member at the fire scene. Engine and truck companies typically consist of the following personnel:

- Company Officer — This is the leader of the group. The company officer makes decisions for tactical deployment while assisting and supervising individual members of the company.
- Apparatus Driver/Operator — The driver/operator is responsible for the safe operation of the fire apparatus and the safe transportation of the crew to and from the fire scene. The driver/operator is also responsible for the operation of all associated equipment, such as the fire pump or aerial device, at the emergency scene. The driver/operator maintains the apparatus, and all the equipment on the apparatus, in a state of readiness at all times.
- Firefighters — Firefighters perform all the tasks necessary to control an emergency scene. These tasks may include operating hoselines; deploying ladders; and performing search and rescue, salvage and overhaul, and many other functions.

It is important that all personnel be thoroughly trained in the tactics used by their company and in all equipment they will be expected to operate. The quick and efficient use of tools is further enhanced when the companies that frequently work together train together (Figure 12.2).

Figure 12.2 Companies that frequently work together should also train together.

The need to follow safe procedures and to wear protective clothing cannot be overemphasized. While helmets, gloves, turnout gear, boots, and breathing apparatus protect the firefighter from injury, they also permit the firefighter to apply streams from positions closer to the fire (Figure 12.3). Firefighters should work in pairs whenever they are operating in a hazardous or potentially hazardous location on the emergency scene. Firefighters working alone may overexert themselves or be unable to help themselves when trapped.

This chapter looks at some of the common techniques for fighting different types of fires that firefighters will face. Hazards peculiar to certain situations are discussed. Finally, basic tactics for commonly encountered fire scenarios are covered.

Figure 12.3 The use of protective equipment allows firefighters to make a close fire attack. *Courtesy of Joseph J. Marino.*

SUPPRESSING CLASS A (STRUCTURAL) FIRES

A fire attack must be coordinated to be successful. Firefighters must perform the desired evolutions at the time that the officer wants them performed. Depending on the conditions at the fire scene, the fire officer may choose to perform immediate rescue or to protect exposures rather than attack the fire. Coordination between crews performing different functions is crucial. For example, ventilating a fire before attack lines are in place may result in the unwanted spread of fire due to the increase in air movement through the structure. When properly performed, the ventilation effort will substantially aid the entry and attack of hoseline teams. If the attack is coordinated with ventilation, visibility should improve and entry can be made for rescue, assessment of fire conditions, and suppression.

Teams advancing hoselines should also carry equipment that may be needed to force entry or perform other tasks in addition to operating the fire stream (Figure 12.4). This equipment should include at least a portable light, axe, and prying tool of some type. Before entering the fire area, the person at the nozzle should bleed the air from the line by opening the nozzle slightly. Opening the bale slightly while waiting for water to arrive will hasten this process. The operation of the nozzle should also be checked through the range of the stream or by setting a proper pattern for the attack. Any burning fascia, boxed cornices, or other doorway overhangs should be extinguished before entry.

Firefighters should wait at the entrance, staying low and out of the doorway, until the officer gives the order to advance (Figure 12.5). If the attack is coordinated with ventilation, visibility should improve and a more accurate assessment of fire conditions can be made. When the fire area is reached, firefighters can attack the fire. The fire should always be approached and attacked from the unburned side to keep from spreading it throughout the structure.

When adequate ventilation ahead of the nozzles can be provided, it may be possible to use a 30-degree fog pattern (Figure 12.6). This will give the smoke, heat, and

Figure 12.4 The backup firefighter on the hoseline is carrying an axe and a Halligan tool into the structure.

Figure 12.5 Until the officer gives the order to advance, firefighters should stay low and out of the doorway.

steam a place to go without rolling back over the nozzle and causing injury to the firefighters. It will also help to maintain the normal thermal layering in the area.

If ventilation holes cannot be made large enough or if ventilation will be delayed, then it is important to keep the nozzle on straight stream. By using a straight stream at the base of the fire, the fire may be controlled without upsetting the thermal layering (Figure 12.7). In an unventilated setting, a straight stream does not upset the thermal layering as much as a fog stream does, because a straight stream does not push as much air in front of it as a fog stream does. Some disturbance of the thermal layering may occur when straight or solid streams are used due to the steam production; however, the disturbance will not be as severe as that created by the fog stream.

If a door to the fire area must be opened, position all members of the hose team to one side of the entrance (Figure 12.8 on next page). Remember to stay low before entering a fire area in order to allow fire, smoke, and/or heated gases to remain or exit above you. Unless a protective stream of water is needed, do not open the nozzle until fire is encountered. Discharging water at smoke will decrease visibility and increase water dam-

Figure 12.6 Adequate ventilation must be provided ahead of fog streams that are used for interior attacks.

Figure 12.7 Using a straight stream to attack the seat of the fire reduces the chance of upsetting the thermal layering in the room.

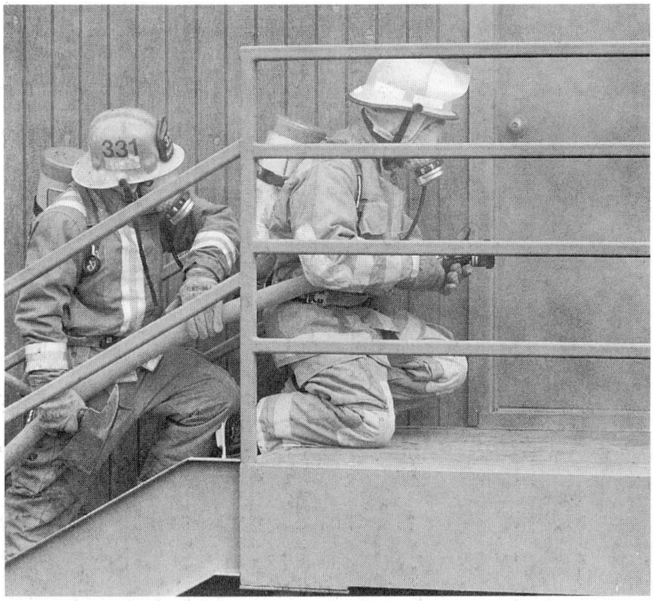

Figure 12.8 If a door to the fire area must be opened, all members of the hose team should be positioned to one side of the entrance.

Figure 12.9 Firefighters must wear SCBA during overhaul operations.

age. If the fire is localized, direct the stream at the base of the fire in short bursts as needed for control. If the area is well involved in fire and sufficiently ventilated, sweeping the ceiling in a side-to-side motion will break the stream up, as will rotating the nozzle clockwise. This nozzle action puts water on the fire and in the upper levels of the room.

The safety of the hose crew is imperative. Remember that if you are required to back out of an area, keep the stream operating until all personnel are out. There are exceptions to this such as an imminent building collapse where everyone must exit immediately.

Once the fire has been contained, it may be necessary to relieve the initial attack crew. Breathing apparatus must still be worn during mop-up and overhaul phases of the operation due to the presence of fire gases (Figure 12.9). Special attention should be directed toward walls, partitions, or overhead loads that may be dislodged by fire fighting activities. Valuables found should be taken immediately to an officer.

Depending on the size of the fire, the type of nozzle being used, ventilation conditions, and other factors, firefighters may choose to use a direct, indirect, or combination method of attacking the fire.

Direct Attack

The most efficient use of water on free-burning fires is made by a direct attack on the base of the fire from a close position with a solid stream or straight stream. The water should be applied in short bursts directly on the burning fuels until the fire "darkens down" (Figure 12.10).

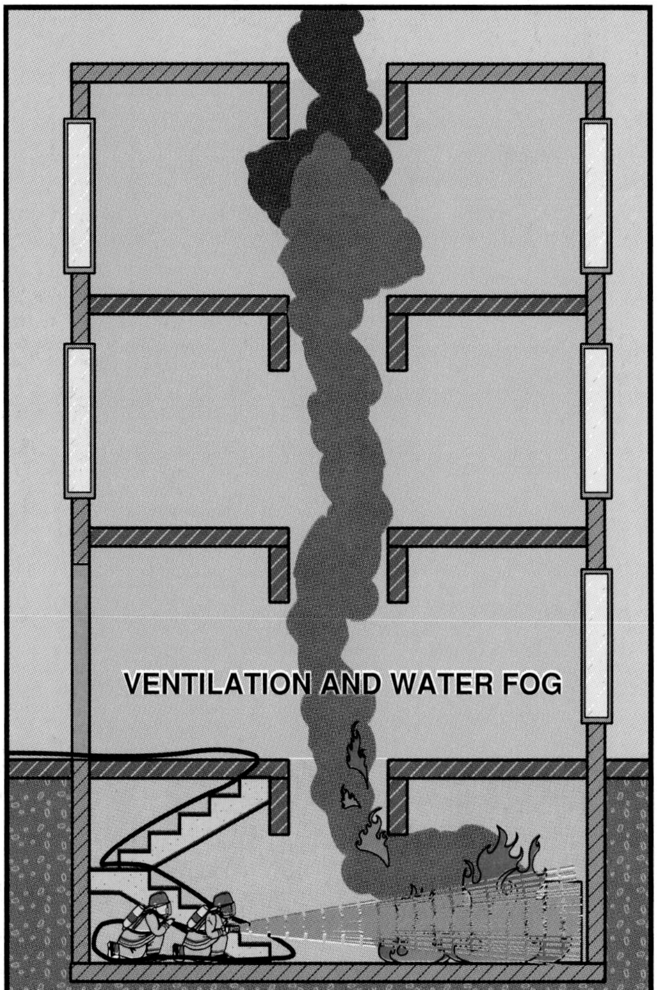

Figure 12.10 A direct attack involves applying water directly to the burning material.

Water should not be applied for too long or the thermal layering (the movement of hot gases toward the ceiling) will be upset. If water is applied for an excessive length of time, the steam produced will begin to condense causing the smoke to drop rapidly to the floor and move sluggishly thereafter.

Indirect Attack

When firefighters are unable to enter the structure or fire area due to intense fire conditions, an indirect attack can be made (Figure 12.11). This attack is not desirable where victims may yet be trapped or where the spread of fire to uninvolved areas cannot be contained. The fire stream, which could be a solid, straight, or narrow fog pattern, should be directed at the ceiling and played back and forth in the superheated gases at the ceiling level. Directing the stream into the superheated atmosphere near the ceiling results in the production of large quantities of steam. One cubic foot (.03 m^3) of water (7.4 gallons [27 L]) completely vaporized will create 1,700 cubic feet (48 m^3) of steam. Once again, the stream should be shut down before it disturbs the thermal layering. Once the fire has been darkened down, and the space has been ventilated, the hoseline can be advanced to extinguish any remaining hot spots with a direct attack.

Combination Attack

The combination method uses the steam-generating technique of ceiling level attack combined with a direct attack on materials burning near the floor level. The nozzle may be moved in a "T," "Z," or "O" pattern starting with a solid, straight, or penetrating fog stream directed into the heated gases at the ceiling level and then dropped down to attack the combustibles burning near the floor level (Figure 12.12 on the next page). The "O" pattern of the combination attack is probably the most familiar method of attack. When performing the "O" pattern, the stream should be directed at the ceiling and rotated with the stream edge reaching the ceiling, wall, floor, and opposite wall. Firefighters should keep in mind that applying water to smoke does not extinguish the fire and only causes unnecessary water damage and disturbance of the thermal layering. Firefighters assisting the person at the nozzle should not bunch up behind the nozzle because this makes manipulation of the nozzle difficult. The assisting team members need to advance hose, as it is needed, to the person at the nozzle. All team members must watch for a number of potential hazardous conditions such as the following:

- Imminent building collapse
- Fire that is behind, below, or above the attack team
- Kinks or obstructions to the hoseline
- Holes, weak stairs, or other fall hazards
- Suspended loads on fire-weakened supports (Figure 12.13 on the next page)
- Hazardous or highly flammable commodities likely to spill
- Backdraft or flashover behavior
- Electrical shock hazards
- Overexertion, confusion, or panic by team members

Figure 12.11 An indirect attack involves aiming the stream at the ceiling above the fire and letting the water rain down on the burning material. Steam conversion by heated air also helps to extinguish the fire.

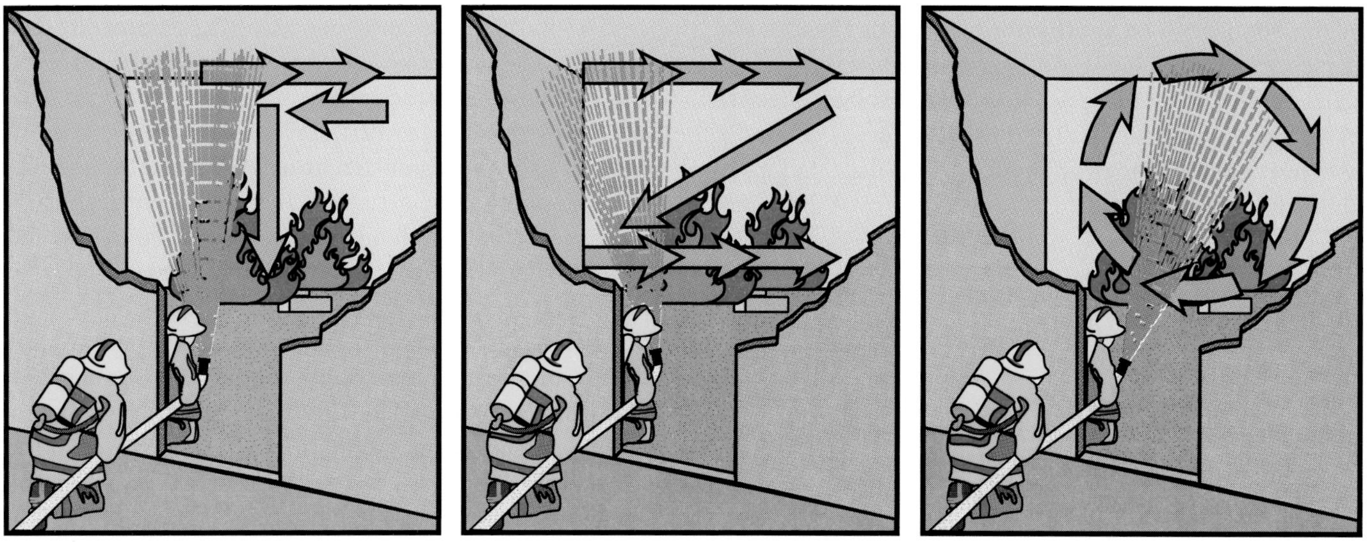

Figure 12.12 T, Z, or O patterns may be used for an indirect attack.

Figure 12.13 Team members must watch for potentially hazardous conditions such as heavy mechanical equipment on the roof.

Stream Selection

As previously mentioned, the technique of water application is only successful if the amount of water applied is sufficient to cool the fuels that are burning. The use of a booster line may not only delay extinguishment, but may be of insufficient volume to protect firefighters from advancing flame fronts. Booster lines should be used only for small, exterior fires such as trash dumpsters and small brush fires. Small interior fires such as those occurring in mattresses, waste baskets, or furniture should be handled by removing the burning object outside and extinguishing it with a booster line or garden hose. *IFSTA recommends that interior fires and larger exterior fires, including automobile fires, be attacked with no less than a 1½-inch (38 mm) hoseline.* Hoseline selection should be dependent upon fire conditions and other factors such as the following:

- Volume of water needed for extinguishment
- Reach needed
- Number of persons available to handle hoseline
- Mobility requirements
- Tactical requirements
- Speed of deployment
- Potential fire spread

Obviously, it would be incorrect to choose a 1½-inch (38 mm) hoseline to attack a fire in a large, well-involved commercial occupancy. The line would have neither the necessary volume nor reach. It would also be overkill to attack a fire involving a single room and contents in a family dwelling with a 2½-inch (65 mm) line discharging 250 gpm (1 000 L). Table 12.1 gives a simple analysis of hose stream characteristics and is not meant to replace the judgment of fire personnel in selecting fire streams.

SUPPRESSING CLASS B FIRES

Class B fires are those that involve flammable and combustible liquids and gases (Figure 12.14 on the following page). Experience has shown that water is effective in extinguishing or controlling many of these fires. Control of Class B fires can be accomplished safely if proper techniques are used. These techniques require a basic understanding of Class B fuel properties and the effects water has on them.

Flammable liquids are those that have flash points less than 100°F (38°C); examples of flammable liquids are gasoline and acetone. Combustible liquids are those that have flash points higher than 100°F (38°C); examples of combustible liquids are kerosene and vegetable oil. Flammable and combustible liquids can be further divided into hydrocarbons and polar solvents. The important thing for firefighters to remember is that hydrocarbon liquids (gasoline, kerosene, and other pe-

TABLE 12.1
HOSE STREAM CHARACTERISTICS

Size	*GPM (L/Min)*	*Reach (Max.)*	*No. of Persons on Nozzle*	*Mobility*	*Control of Damage*	*Control of Direction*	*When Used*	*Estimated Effective Area*
Booster (¾ to 1 inch) (16 mm to 25 mm)	10 to 40 (40 to 60)	25 to 50 feet (8 to 15 m)	1	Excellent	Excellent	Excellent	Small Exterior Fires Chimney Fires Overhaul	Very Small Fires
1½ inch (38 mm)	40 to 125 (160 to 500)	25 to 50 feet (8 to 15 m)	1 or 2	Good	Good	Excellent	Developing fire—still small enough or sufficiently confined to be stopped with relatively limited quantity of water. For quick attack. For rapid relocation of streams. When manpower is limited. When ratio of fuel load to area is relatively light. For exposure protection.	One to Three Rooms
1¾ inch (45 mm)	40 to 175 (160 to 700)	25 to 50 feet (8 to 15 m)	2	Good to Fair	Good	Good		
2 inch (50 mm)	100 to 250 (400 to 1 000)	40 to 70 feet (12 to 21 m)	2 or 3	Fair	Fair	Good	When size and intensity of fire are beyond reach, flow or penetration of 1½-inch (38 mm) line. When both water and manpower are ample. When safety of crew dictates. When larger volumes or greater reach are required for exposure protection.	One Floor or More—Fully Involved
2½ inch (65 mm)	125 to 350 (500 to 1 400)	50 to 100 feet (15 to 30 m)	2 to 4	Fair to Poor	Fair	Good		
Master Stream	350 to 2 000 (1 400 to 8 000)	100 to 200 feet (30 to 60 m)	1	Poor to None (Aerial master streams can be good)	Poor	Good	When size and intensity of fire are beyond reach, flow, or penetration of handlines. When water is ample, but manpower is limited. When safety of personnel dictates. When larger volumes or greater reach are required for exposure protection. When sufficient pumping capability is available. When massive runoff water can be tolerated. When interior attack can no longer be maintained.	Large Structures Fully Involved

Adapted from Joe Batchler, Maryland Fire & Rescue Institute

troleum products) will not mix with water, and polar solvents (alcohols, lacquers, etc.) will mix with water. This fact regarding hydrocarbons and polar solvents affects how fires in each may be extinguished.

Firefighters must exercise caution when attacking fires involving flammable and combustible liquids (Figure 12.15 on the next page). The first precaution to take is to avoid standing in pools of the fuel. Standing in pools of fuel or water runoff containing fuel can result in protective clothing absorbing fuel in a "wicking" action. This wicking action can lead to contact burns of the skin and flaming clothing if an ignition source is present. Even if the wicking does not occur, there exists extreme danger in the event that the pool of liquid ignites.

Figure 12.14 Aircraft emergencies often involve large Class B fires. *Courtesy of Joel Woods, University of Maryland Fire & Rescue Institute.*

Figure 12.15 Firefighters must exercise great care when attacking large Class B fires. *Courtesy of Harvey Eisner.*

Unless the leaking product can be shut off, fires burning around relief valves or piping should not be extinguished (Figure 12.16). Simply try to contain the pooling liquid, if any, until the flow can be stopped. Unburned vapors are usually heavier than air and will form pools or pockets of gas in low spots where they may be ignited. Firefighters must always control all ignition sources in the leak area. Vehicles, smoking materials, electrical fixtures, and sparks from steel tools can all provide an ignition source sufficient to ignite leaking flammable vapors. An increase in the intensity of sound or fire issuing from a relief valve may indicate that rupture of the vessel is imminent. Firefighters should not assume that relief valves are sufficient to safely relieve excess pressures under severe fire conditions. Firefighters have been killed by the rupture of both large and small flammable liquid vessels that have been subjected to flame impingement.

In vessels containing flammable liquids, this condition of the sudden release and consequent vaporization of the liquids is called a BLEVE (Boiling Liquid Expanding Vapor Explosion). A BLEVE is a rupture of the tank that results in the explosive release of vessel pressure, pieces of tank, and a characteristic fireball with radiant heat. BLEVEs most commonly occur when flames contact the vapor space of the flammable liquid vessel and insufficient water is applied to keep the tank cool. When attacking these fires, water should be applied to the upper portions of the tank, preferably from unmanned monitors (Figure 12.17).

Figure 12.16 A coordinated effort is required to effectively attack pressurized fuel fires. *Courtesy of Joel Woods, University of Maryland Fire & Rescue Institute.*

Figure 12.17 Unmanned nozzles should be used to cool tanks.

Using Water As A Cooling Agent

Water can be used as a cooling agent to extinguish fires and to protect exposures. Water without foam additives is not particularly effective on lighter petroleum distillates (such as gasoline or kerosene) or alcohols. However, by applying water in droplet form in sufficient quantities to absorb the heat produced, fires in the heavier oils (such as raw crude) can be extinguished.

Water will be most useful as a cooling agent for protecting exposures. To be effective, water streams need to be applied so that they form a protective water film on the exposed surfaces. This applies to ordinary combustibles and other materials that might weaken or collapse such as metal tanks or support beams. Water applied to burning storage tanks should be directed above the level of the contained liquid to achieve the maximum efficient use of the water.

Using Water As A Mechanical Tool

Water from hoselines can be used to move the fuel, whether it is burning or not, to areas where it can safely burn or where ignition sources are more easily controlled. Fuels must never be flushed down drains or sewers. Firefighters should use wide-angle or penetrating fog patterns for protection from radiant heat and to prevent "plunging" the stream into the liquid. Plunging a stream into burning flammable liquids causes increased production of flammable vapors and greatly increases fire intensity. The stream should be slowly played from side to side and the fuel or fire "swept" to the desired location. Care must be taken to keep the leading edge of the fog pattern in contact with the fuel surface or else the fire may run underneath the stream and flash back around the attack crew. Where small leaks occur, a solid stream may be applied directly to the hole and the escaping liquid forced back. The pressure of the stream must exceed that of the leaking material to perform properly. Care must be taken not to overflow the container. (**NOTE:** The use of foam is the preferred method to control flammable liquid fires.) Water, through the use of fog streams, may also be used to dissipate flammable vapors. Fog streams aid in dilution and dispersion, and these streams control, to a small degree, the movement of the vapors to a desired location.

Using Water As A Substitute Medium

Water can be used to displace fuel from pipes or tanks that are leaking. Fires fed by leaks may be extinguished by pumping water back into the leaking pipe or by filling the tank with water to a point above the level of the leak. This displacement will float the volatile product on top of the water as long as the water application rate equals the leak rate. Due to the large water-to-product ratios, water is seldom used to dilute flammable liquids for fire control. However, this technique may be useful for small fires where the runoff can be contained.

Using Water As A Protective Cover

Hoselines can be used as a protective cover for teams advancing to shut off liquid or gaseous fuel valves (Figure 12.18). Coordination and slow, deliberate movements provide relative safety from flames and heat. While one hoseline can be used as a protective cover, two lines with a backup line are preferred for fire control and safety.

Figure 12.18 Fog streams provide protective cover for firefighters when they advance a fire. *Courtesy of Joel Woods, University of Maryland Fire & Rescue Institute.*

When containers of flammable liquids or gases are exposed to flame impingement, solid streams should be applied from their maximum effective reach until the relief valve closes. This can best be achieved by lobbing the stream along the top of the tank so that the water runs down both sides. This film of water will cool the vapor space of the tank. Steel supports under tanks should also be cooled to prevent their collapse.

Hose streams can then be advanced under progressively widened protective fog patterns to make temporary repairs or shutoffs. A backup line supplied by a separate pump and water source should be provided to protect firefighters in the event other lines fail or additional tank cooling is needed. Approaches to storage vessels should be made at right angles to the tank, never from the ends of the vessels. Rupturing vessels frequently split in two and then become projectiles.

Fires Involving Bulk Transport And Passenger Vehicles

The techniques of extinguishment for fires in vehicles transporting flammable fuels are similar in many ways to fires in storage facilities. The difficulties posed by the amount of fuel to burn, the possibility of vessel failure, and danger to exposures are similar with both. The major differences include the following:

- Increased life safety risks to firefighters from traffic
- Increased life safety risk to passing motorists
- Reduced water supply
- Difficulty in determining the products involved

- Difficulty in containing spills and runoff
- Tanks and piping weakened or damaged by the force of collisions
- Instability of vehicles
- The surroundings of the incident may pose additional concerns (residential neighborhood, schools, etc.)

While a serious accident may bring traffic to a halt, many incidents will be handled with traffic passing the scene at near-normal speeds. A lane of traffic in addition to the incident lane should be closed during initial emergency operations (Figure 12.19). The use of open flame flares should be avoided due to the possibility of their igniting leaking fuels. Fire apparatus should be positioned to take advantage of topography and weather conditions (uphill and upwind) and to protect firefighters from traffic. Firefighters should exit the apparatus and work as much as possible from the curb side away from traffic. In addition, firefighters should avoid working where the apparatus could be pushed into them if it were struck by another vehicle. Where traffic is passing closely, firefighters should be careful not to allow tool handles to extend into the traffic lane where they may be struck. When law enforcement personnel are unavailable, a firefighter should be assigned the role of traffic control officer.

Figure 12.19 Close at least one additional lane of traffic in each direction.

The techniques of approaching and controlling leaks or fires involving vehicles are the same as for storage vessels. Additionally, firefighters should be aware of the failure of vehicle tires that may cause the flammable load to shift suddenly. Crews will need to know the status of their water supply so as not to exceed the limitations of that supply. It may also be necessary to protect trapped victims with hoselines until they can be rescued.

Firefighters must determine, as soon as possible, the exact nature of the cargo from bills of lading, manifests, placards, or the driver of the transport vehicle (Figure 12.20). Unfortunately, cases will exist where these items cannot be found, placards are either wrong or obscured, and drivers are unable to identify their cargo. In these instances, contact should be made with the shipper or manufacturer responsible for the vehicle. Pre-incident plans for transportation emergencies should be followed to reduce life loss, property damage, and environmental pollution.

Figure 12.20 Firefighters can determine from the vehicle's placard what type of cargo it is carrying.

Control Of Gas Utilities

A working knowledge of the hazards and correct procedures for handling incidents related to natural gas and liquefied petroleum gas (LPG) is important to every firefighter. Many houses, mobile homes, and businesses use natural gas or LPG for cooking, heating, or industrial processes. The firefighter familiar with gas distribution and usage will be able to prevent or reduce damage caused by incidents involving these gases.

Natural gas is mostly methane with small quantities of ethane, propane, butane, and pentane added. The gas is lighter than air so that it will tend to rise and diffuse in the open. Natural gas is nontoxic, but it is classified as an asphyxiant because it may displace normal breathing air and lead to asphyxiation. The gas has no odor of its own but a very distinctive odor (mercaptan) is added by the utility. It is distributed from gas wells to its point of usage by a nationwide network of surface and subsurface pipes (Figure 12.21). The pressure in these pipes ranges from ¼ to 1,000 psi (2 kPa to 7 000 kPa). However, the pressure is usually below 50 psi (350 kPa) at the local distribution level. Natural gas is explosive in concentrations between 4 and 14 percent.

The local utility should be contacted when any emergency involving natural gas occurs in its service area. The local utility will provide an emergency response crew equipped with special (nonsparking) tools,

Figure 12.21 Firefighters should be familiar with gas distribution stations and equipment in their area.

Figure 12.22 Many occupancies use compressed fuel gases that are stored in cylinders on their property.

maps of the distribution system, and the training and experience needed to help control the flow of gas. The response time of these crews is usually less than an hour, but the time may be extended in rural areas or in times of great demand. Good relations between the fire department and the utility company are encouraged.

Liquefied petroleum gas (LPG), or bottled gas as it is sometimes known, is a fuel gas stored in a liquid state under pressure. It is used primarily as a fuel gas in campers, mobile homes, agriculture, and rural homes. There is an increased use of LPG as a fuel for motor vehicles. The gas is composed mainly of propane with small quantities added of butane, ethane, ethylene, propylene, isobutane, or butylene. LPG has no natural odor of its own but a very distinctive odor is added. The gas is nontoxic, but it is classified as an asphyxiant because it may displace normal breathing air and lead to asphyxiation. LPG is about one and one-half times as heavy as air, and it will generally seek the lowest point possible. The gas is explosive in concentrations between 1.5 and 10 percent. The gas is shipped from its distribution point to its point of usage in cylinders and in tanks on cargo trucks. It is stored in cylinders and tanks near its point of usage, and then the tank or cylinder is connected by underground piping and copper tubing to the appliances the gas serves (Figure 12.22). All LPG containers are subject to BLEVE's (Boiling Liquid Expanding Vapor Explosions) when exposed to intense heat or open flame.

Incidents involving the distribution system are most often caused by excavation around underground pipes. When these incidents occur, the utility company should be contacted immediately. Even if the gas has not yet ignited, apparatus should approach from and stage on the upwind side. Firefighters should be prepared for the event of an explosion and any accompanying fire. The firefighter's first concerns should be the evacuation of the area around the break and down wind from it, and elimination of ignition sources in that area. The broken main may have damaged service connections near the break so surrounding buildings should be checked for gas buildup. Firefighters should not attempt to operate main valves because incorrect action may worsen the situation or cause unnecessary loss of service to areas unaffected by the break. If the gas is burning, the flame *should not be extinguished*. Exposures can be protected by hose streams if necessary.

WARNING

If the gas is burning, do not extinguish the fire.

Figure 12.23 A natural gas meter.

The most common situation firefighters face involves the service meter. The meter is usually located outside the building and is normally visible from the street (Figure 12.23). The flow of gas into the building may be stopped by turning the cutoff valve to the closed position, which will be at a right angle to the pipe (Figure 12.24).

Firefighters involved in this operation should advance a hoseline on a fog pattern to protect themselves. If the fire is extinguished, the vapors could build up and then reignite with disastrous results. If for some reason the meter cock is inoperable, the pipe may be partially closed by pinching the line with a hydraulic rescue tool. This action may not stop the flow of gas, but it will reduce the flow.

LPG or bottled gas is most often stored in one or more outdoor tanks or cylinders. The supply of gas into a structure may be stopped by shutting a valve on the pipe leading to the building (Figure 12.25). If the valve is inoperable, the gas may be stopped by pinching the copper line leading into the structure with a pair of pliers or, in the case of larger lines, a rescue tool. Unburned gas may be dissipated by a fog stream of at least 100 gpm (400 L/min). If there are any problems with a cylinder or tank, the company responsible for it should be contacted.

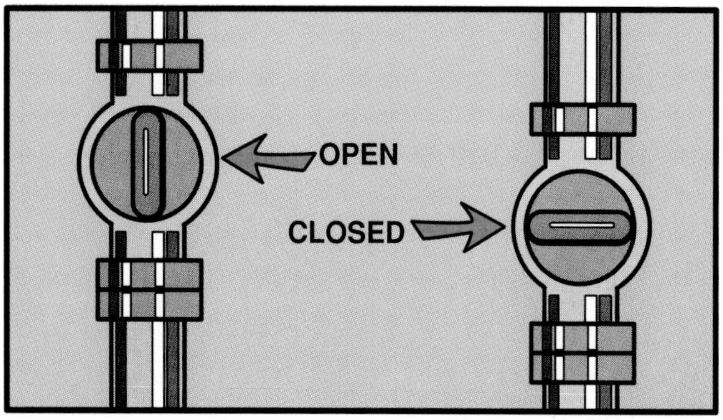

Figure 12.24 This shows the open and closed position for gas line petcocks.

Figure 12.25 Compressed fuel gas cylinders have a shutoff valve that is clearly marked for the direction of operation.

SUPPRESSING CLASS C FIRES

Firefighters must be able to control the flow of electricity into structures where emergency operations are being performed. In order to avoid injury and to protect electrical equipment, firefighters should be familiar with electrical transmission and its hazards. While high-voltage equipment is usually associated with severe shocks, conventional residential current is sufficiently powerful to deliver fatal shocks (Figure 12.26). In addition to reducing the risk of injury or fatal shock, controlling electrical flow reduces the danger of igniting combustibles or accidental equipment start-ups.

The consequences of electrical shock include:

- Cardiac arrest
- Ventricular fibrillation
- Respiratory arrest
- Involuntary muscle contractions
- Paralysis
- Surface or internal burns
- Damage to joints
- Ultraviolet arc burns to the eyes

Figure 12.26 High-voltage equipment pose severe hazards to firefighters.

Factors most affecting the seriousness of electrical shock include:

- The path of electricity through the body
- The degree of skin resistance — wet (low) or dry (high)
- Length of exposure
- Available current — amperage flow
- Available voltage — electromotive force
- Frequency — AC or DC

WARNING

Regardless of the techniques described in this section on handling fires in energized electrical equipment, stop the flow of electricity to the object involved before initiating fire suppression activities.

Fires in electrical equipment occur quite frequently, but once deenergized they can be handled with relative ease.The primary danger of electrical fires is the failure of emergency personnel to recognize the hazard. It is the responsibility of the fire officer to ensure that appropriate power breakers are opened to control power flow into structures. In certain commercial and/or high-rise buildings, power is necessary to operate elevators and/or air handling equipment and thus the entire building should not be unilaterally deenergized. Similarly, a crew member should be assigned to control power at vehicle fires and other emergencies. Once the power has been shut down, these fires may self-extinguish or if they continue to burn they will fall into either class "A" or "B" fires.

When handling fires in delicate electronic or computer equipment, such clean agents as carbon dioxide or Halon should be used to prevent further damage to equipment (Figure 12.27). Multipurpose dry chemical agents present a considerable clean-up problem in addition to being chemically active with some electrical components. The use of water on energized equipment is discouraged unless absolutely necessary because of the inherent shock hazard. If water must be applied, it should be applied from fog streams and from a distance.

Figure 12.27 Halon is often used on Class C fires.

The most common electrical emergencies firefighters will face are emergencies involving bulk electrical transmission lines and equipment (Figure 12.28). When fires occur as a result of transmission lines breaking, an area equal to a span between poles should be cleared on either side of the break. Fires in transformers can present a serious health and environmental risk because of coolant liquids that contain PCBs (polychlorinated biphenyls). These liquids are flammable because of their oil base and are carcinogenic (cancer causing). Transformers at ground level should be extinguished carefully with a dry chemical extinguisher. Transformers aboveground should be permitted to burn until qualified utility personnel can extinguish the fire with a dry chemical extinguisher from an aerial device. Placing a ground ladder against the pole places personnel under risk from both the power source and from the liquid. Applying hose streams to these fires can result in spreading the material onto the ground.

To reduce the risk to life and property in these types of incidents, consultation and cooperation with power company officials is vital. Other unusual electrical hazards can be found in railroad locomotives, telephone relay switching stations, and electrical substations. Procedures for fighting fires in these occupancies should be established in pre-incident plans. For maximum safety on the fireground, live wires should not be cut except by experienced power company personnel with proper equipment.

WARNING:

Consider all wires live until confirmed to be otherwise by the power company.

Figure 12.28 Fires at electrical substations are common electrical emergencies firefighters will face.

Underground Lines

Underground transmission systems consist of cableways and vaults beneath the surface. The most frequent hazard that these systems present is explosions that may blow manhole covers a considerable distance (Figure 12.29). The cause of this occurrence is the accumulation of gases that are ignited by a spark from fuses blowing or the arcing from a short circuit. This is a danger to the public as well as firefighters. If these situations are suspected, firefighters must keep the public clear of the area and must make sure that apparatus is not parked over a manhole.

Firefighters should not enter a manhole except to attempt a rescue. Fire fighting can be accomplished from outside. Firefighters should simply discharge carbon dioxide or dry chemical into the manhole and replace the cover. A wet blanket or salvage cover should be placed over the manhole to exclude the oxygen and assist in extinguishing the fire. Water is not suggested for extinguishment because of the close proximity of electrical equipment. The runoff of water would also create puddles that could become dangerous conductors of electricity.

When circumstances dictate that a firefighter must enter a manhole, it should be done only with approved positive-pressure self-contained breathing apparatus and full protective clothing. A filter-type mask should never be used under any conditions. Any equipment or tools should be nonsparking because the smallest spark might ignite the explosive mixture.

Figure 12.29 Apparatus should never be parked over a manhole because of the chance of an explosion.

Commercial High-Voltage Installations

Many industries, large buildings, and apartment complexes have electrical equipment that uses current in excess of 600 volts. The obvious clue to this condition would be "high-voltage" signs on the doors of vaults or fire-resistive rooms housing high-voltage equipment such as transformers or large electric motors (Figure 12.30). Some transformers use flammable oils as coolants that present a hazard in themselves. Water should not be used in this situation, even in the form of fog, because the hazard of shock is greater and extensive damage may occur to electrical equipment not involved in the fire.

Because of toxic chemicals used in plastic insulation and coolants, smoke becomes a hazard. Enter only when rescue operations require it, wearing positive-pressure self-contained breathing apparatus and using a safety line monitored by someone outside the enclosure. When searching, do so with a clenched fist or the back of the hand to prevent reflex actions of grabbing live equipment which may be contacted (Figures 12.31 a and b). If it is believed that toxic materials were involved in the fire, appropriate decontamination procedures should be followed.

Figure 12.30 Exercise caution when entering rooms that contain high-voltage equipment.

Figure 12.31a Perform search with a clenched fist.

Figure 12.31b Another method used to perform a search is with the back of the hand.

Electrical Power Shutoff

From a safety standpoint, during structural fire fighting operations, power should remain on as long as possible to provide lighting, ventilation, or to run special pumps. When the building becomes damaged to the point that service is interrupted or an electrical hazard exists, power should be shut off by a power company employee if possible. When the fire department must do this, only trained personnel who are aware of the effects should be assigned the task. When a fire involves only one area, it would be pointless to shut down the entire building.

Although it used to be a common practice in residential fires, it is no longer recommended to pull the electrical meter to shut off the electricity. The firefighter should control the power at the panel box by opening the main switch or removing the fuses (Figure 12.32). If further control of the electricity becomes necessary, it should be done only by utility personnel using approved, tested equipment. With some residential and commercial meters, their removal does not discontinue the flow of electricity. Firefighters should be alert for installations with emergency power capabilities such as emergency generators (Figure 12.33). In such cases, pulling the meter or master switch does not shut off the power entirely.

Figure 12.32 The firefighter should control the power at the panel box by opening the main switch.

Figure 12.33 Many occupancies have backup electrical generators that become active when the main power supply is interrupted.

Guidelines For Electrical Emergencies

The following list contains some tips to help deal with electrical emergencies. The list is not totally inclusive but gives some principles which should be considered to maintain a safe working environment for personnel.

- When downed wires are encountered, a danger zone of one span in either direction should be considered for safety (Figure 12.34). This is because other wires may have also been weakened by the short and may fall at a later time.

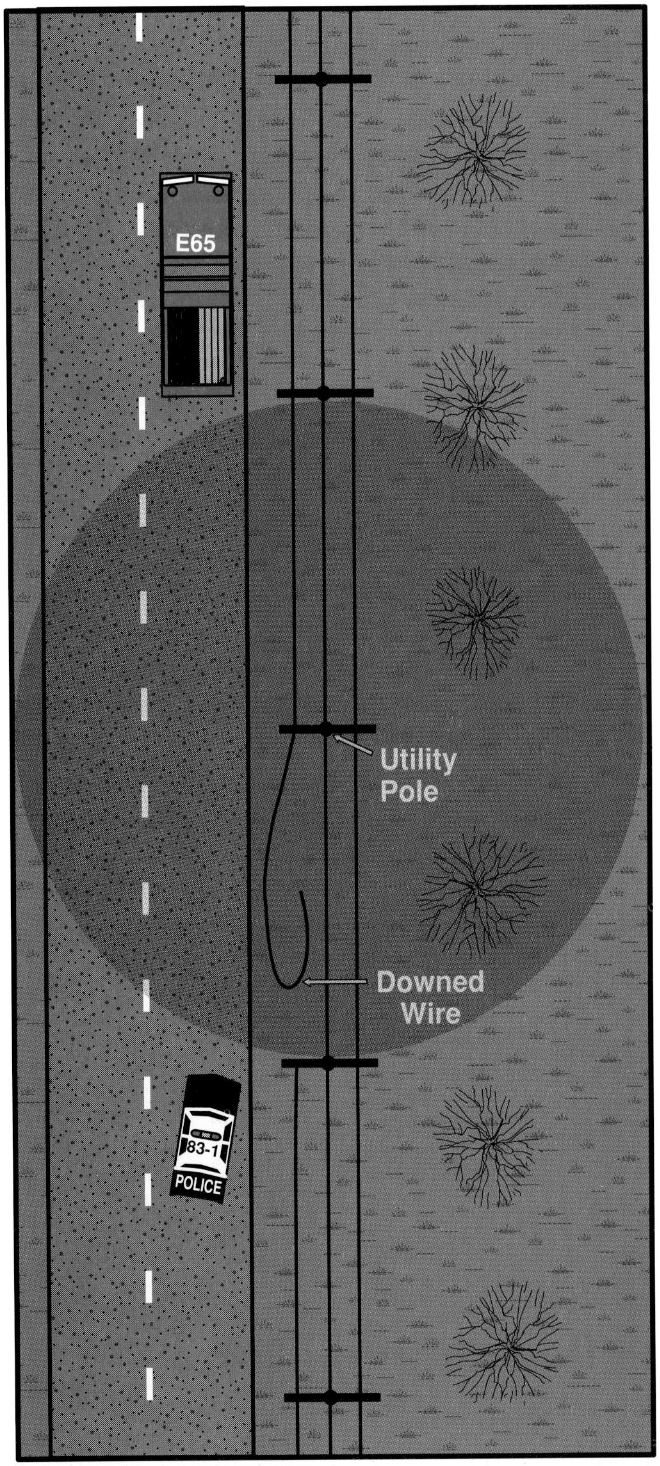

Figure 12.34 A one-span danger zone should be established around the downed wire.

may be turned to controlling the fire, which includes protecting the exposures as well. Lastly, firefighters should make all efforts possible to minimize damage to the structure. This can be accomplished through proper fire fighting tactics and good salvage and overhaul techniques.

SOPs are used to standardize general activities at any emergency scene. Special SOPs can be implemented to make operations at preplanned target hazards more efficient. The use of SOPs reduces confusion and increases efficiency on the fireground. Also, during the process of putting SOPs into effect, the fire officer will have an opportunity to apply other specific pre-incident plans and to develop a fire offense tailored to the needs of the incident. Following are some examples of fire department standard operating procedures for fires in various locations.

Tactics For Fires In Single-Family Dwellings

Fires in single-family dwellings are the most common structural fire scenario firefighters will face. The rescue, exposure protection, ventilation, confinement, and extinguishment functions must be performed in a coordinated way for the operation to be successful. The following information highlights a typical response to a fire in a single-family dwelling.

ROLE OF THE FIRST-DUE ENGINE

The first engine company to arrive at the scene will usually initiate incident command and the fire attack, taking into consideration the present and expected behavior of the fire (Figure 12.40). However, depending on conditions, the first engine may also need to perform search and rescue functions or exposure protection. The first engine company makes a radio report to the dispatch center regarding the exact location, exposures, conditions found at the incident, and, if necessary, additional resources that will be needed. If smoke or fire is visible as the engine company approaches the scene, firefighters should stop and lay a supply line from a hydrant or from the end of the driveway into the scene (Figure 12.41). If a hydrant valve is used on the supply line, the line may be charged as soon as a hose clamp is applied at the scene.

Once the location of the fire is known, assuming that rescue is not needed, the officer of the first company will position the initial attack line to cover the following priorities:

- Placement of lines to intervene between trapped occupants and the fire or to protect rescuers
- Protection of primary means of egress
- Protection of interior exposure (other rooms)

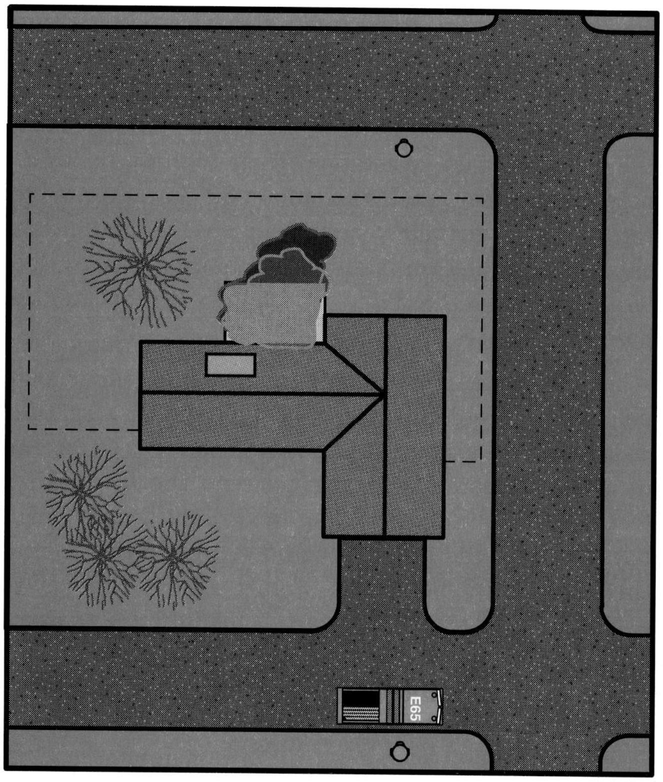

Figure 12.40 More times than not, an engine company will be the first to arrive on the scene.

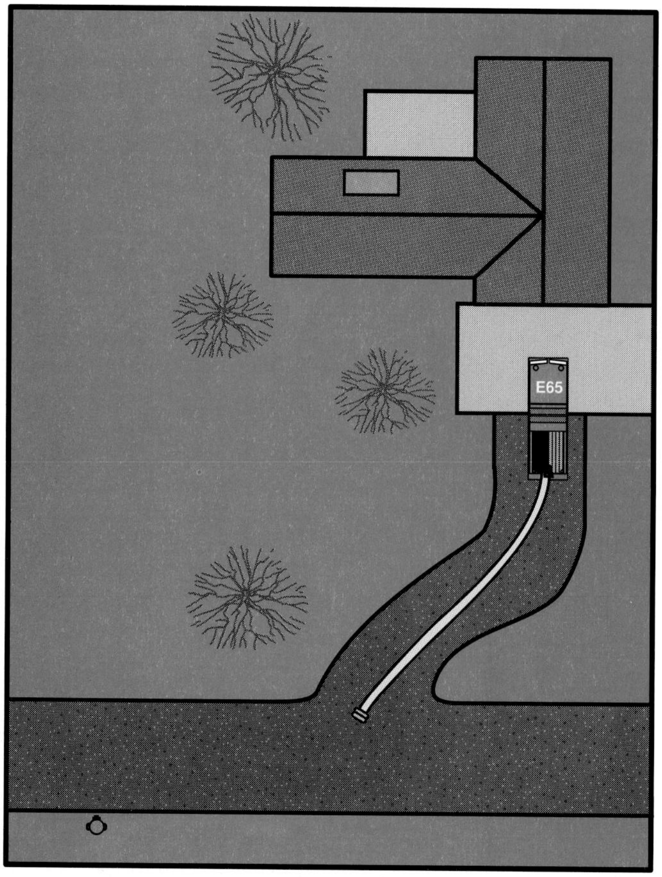

Figure 12.41 If the fire building is up a long driveway, the first engine should lay a supply line up the driveway.

- Protection of exterior exposure (other buildings)
- Initiation of extinguishment from the unburned side
- Operation of master streams

IFSTA recommends that attack lines for structural fire fighting be no less than 1½-inch (38 mm). Local procedures or fire conditions may dictate the use of larger lines; however, lines smaller than 1½-inch (38 mm) should never be used for structural fire attacks. The amount of flow from smaller lines is not sufficient for firefighter protection.

ROLE OF THE SECOND-DUE ENGINE COMPANY

Unless otherwise assigned, the second engine company must first make sure that adequate water supply is established to the fireground. Depending on the situation, it may be necessary to finish a hose lay started by the first engine, lay an additional line, or connect to the hydrant to support the original or additional lines that have been laid (Figures 12.42 a and b). The need to pump the lines from the hydrant will be dependent on local factors that include the size of the hose being used, the distance from the hydrant to the scene, and the residual water pressure available.

Once the water supply has been established, proceed according to the following priorities:

- Back up the initial attack line.
- Protect secondary means of egress.
- Prevent fire extension (confinement).
- Protect the most severe exposure.
- Assist in extinguishment.

In extreme emergencies, it may be necessary for the first-arriving company to begin rescue operations and for the second company to stretch the first fire attack line. Usually, however, the rescue function will be performed by the ladder company under the cover of hose streams from the engine companies.

ROLE OF THE TRUCK COMPANY

The truck company most often arrives with or after the first engine company and is responsible for forcible entry, search and rescue, ventilation, and securing utilities (Figure 12.43 on the next page). Initially, the truck company will observe the outside of the building for signs of victims needing immediate rescue. The truck company can then begin to search for victims either using interior or exterior routes of entry. The truck company will also ventilate as it advances (unless positive-pressure ventilation is being done), staying on the alert for signs of fire spread above the fire floor. The truck

Figure 12.42a Firefighters must understand the operation of hydrant valves used by their department. *Courtesy of George Braun, Gainesville (FL) Fire-Rescue.*

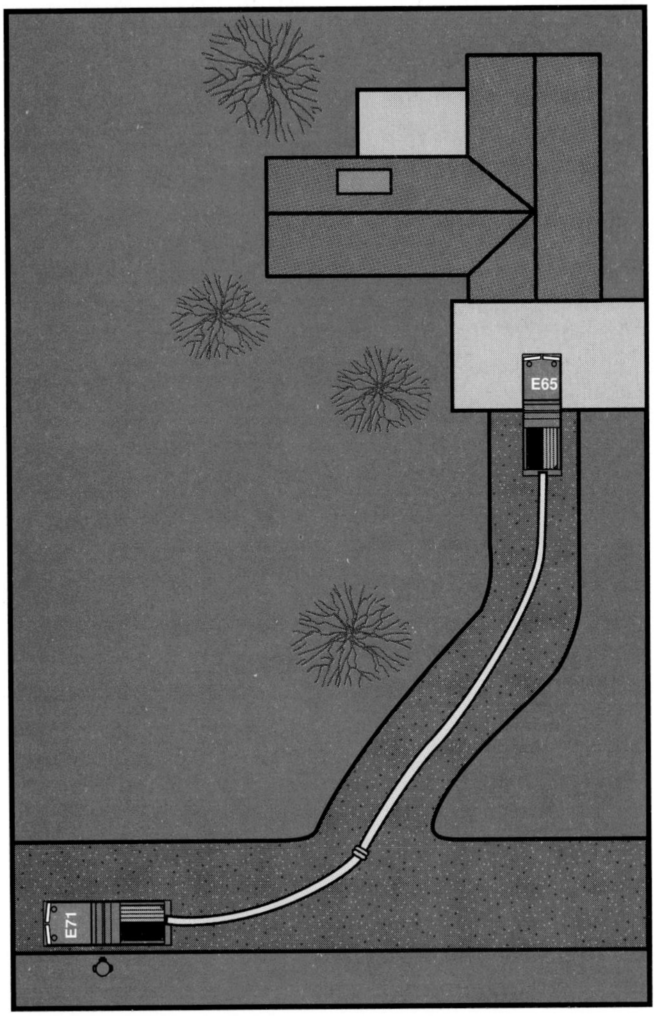

Figure 12.42b A second engine will come along and lay from the end of the driveway to the water supply source.

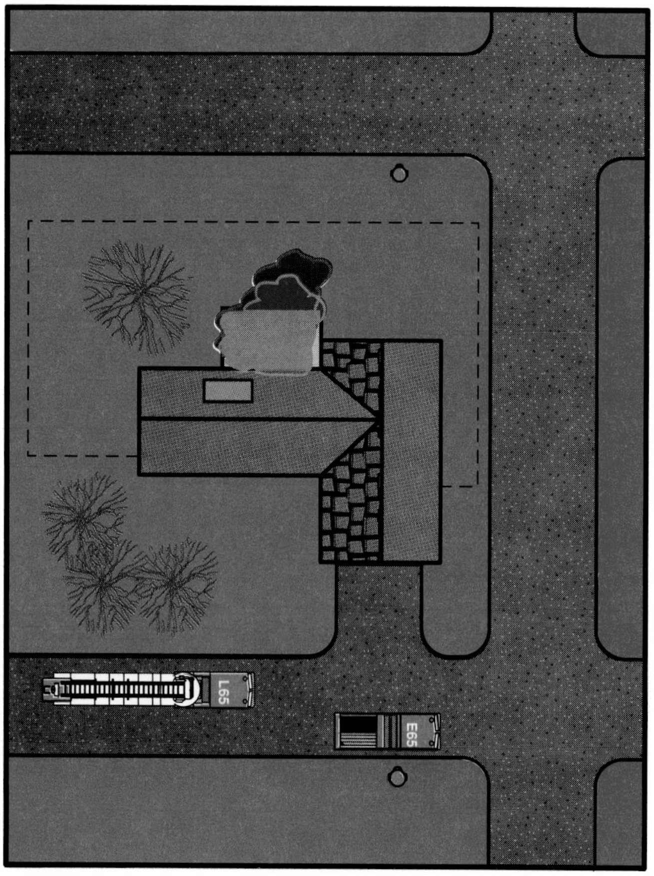

Figure 12.43 Early arriving engines should leave adequate room for aerial apparatus to position properly on the scene.

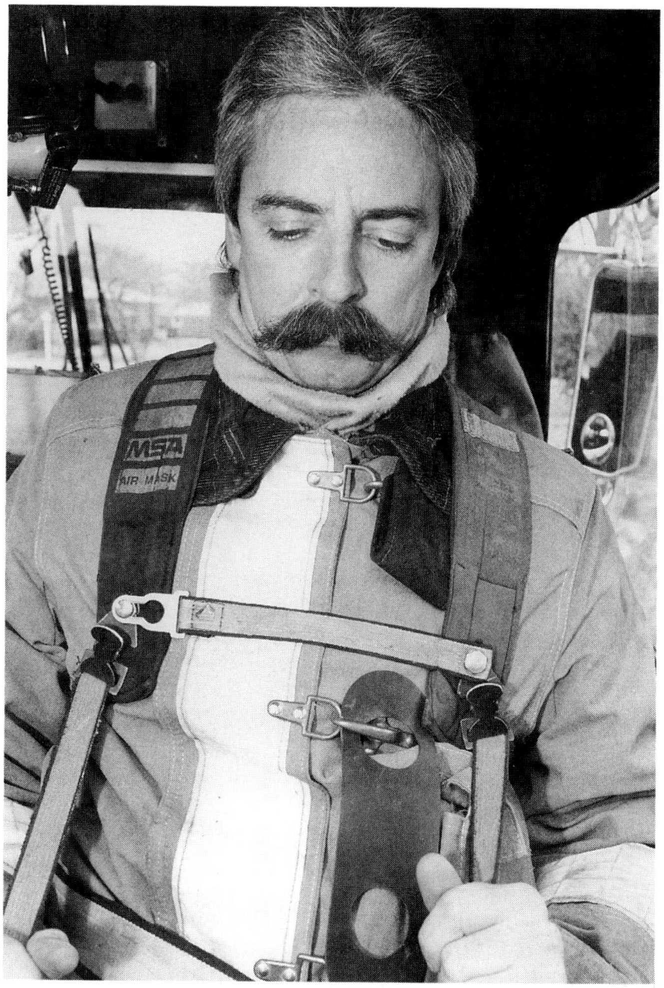

Figure 12.44 A firefighter donning his SCBA en route to the scene.

company must always be prepared to enter the building upon arrival. This means that SCBA should be donned en route to the incident if safely possible (Figure 12.44). Generally, it is desirable to have the truck company personnel assigned as interior or exterior team members. The officer and one or two members with tools enter to begin the interior search before moving up. Simultaneously, the driver with one or two firefighters raises the necessary ladders to enter or ventilate the building from the outside. For instance, a ground ladder may be used to effect a second floor window rescue while the elevating device is raised to the roof for the purpose of ventilation.

Both interior and exterior teams should search first in areas that are most likely to be inhabited according to the situation. As previously described, searches must be conducted systematically to avoid missing areas.

In addition to search and ventilation procedures, it is often necessary for the truck company to assist the engine companies in making the attack. This can be done by the placement of ground ladders as requested by the officer in charge. The truck company can frequently be used to knock down large fires above the first floor with an aerial master stream device (Figure 12.45). This

Figure 12.45 A blitz attack may be accomplished with an elevated master stream device.

action must be coordinated with other operations to both prevent injury to personnel inside and to avoid spreading the fire to uninvolved parts of the building. Interior attack teams have frequently been injured or driven out by poorly directed outside master streams. Elevating devices, ladders, platforms, and ladder towers can also be used as a substitute standpipe from which engine company fire fighting teams can advance hoselines.

ROLE OF THE CHIEF OFFICER

Upon arriving at the scene, the chief officer may choose to assume command from the person who is in command at that time. If the chief officer takes command, he or she will coordinate the overall activities at the scene (Figure 12.46). The situation must be constantly evaluated to be sure that the resources on the scene are being properly allocated. The need for additional resources should be constantly considered. If additional companies are called for, the chief officer should assign them according to the plan as they arrive. The chief officer may also have to coordinate among other entities such as mutual aid units, EMS personnel, utility crews, and members of the media. Depending on the number of fire personnel at hand and the scope of the incident, the chief officer may assign other personnel to act as liaisons to these other organizations operating on the fire scene.

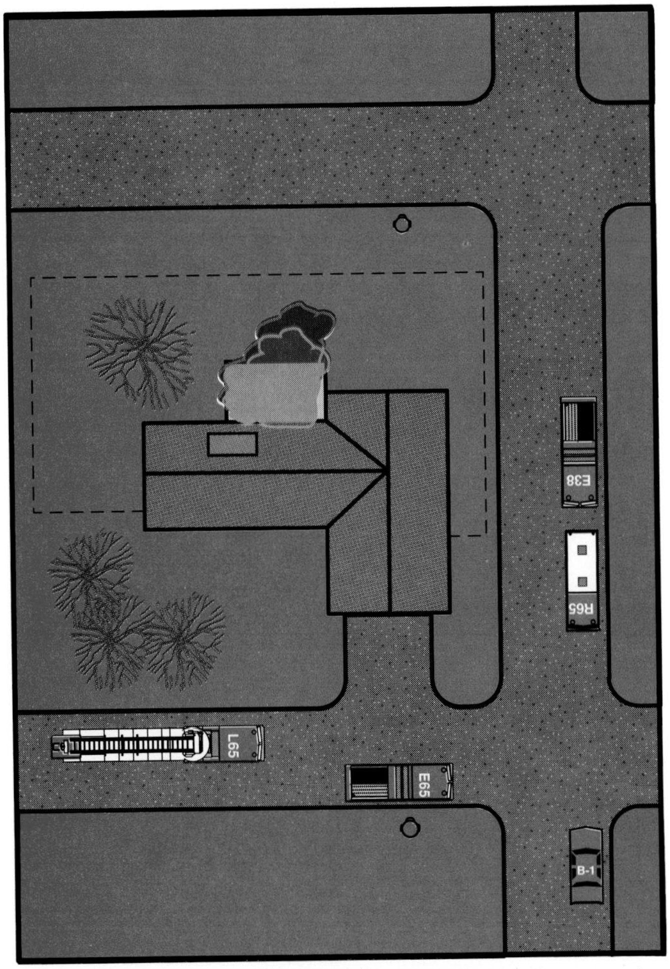

Figure 12.46 The chief officer should be in a position to view the scene but not close enough to get in the way.

Tactics For Fires In Upper Levels Of Structures

Fires in upper levels of structures, such as high-rise buildings, can be very challenging to fire fighting personnel. In order for these fires to be successfully handled, they will usually require a response greater than the two engines and one truck used in the single-family dwelling scenario. This is because extensive manpower may be required to first transport fire fighting equipment to the level of the building where it is needed. This will tire personnel before they get a chance to begin the actual attack on the fire. Elevators should not be relied upon to provide transportation to the fire level, as they can be unreliable and accidentally deposit firefighters directly into the fire area. In most cases equipment must be hand carried up the stairs. If the building has low-rise elevators that end before the level of the fire, they may be used to their top level, and then the equipment must be carried up the rest of the way.

Typically, the fire attack will be initiated from the floor below the fire floor (Figure 12.47). Firefighters may wish to look at the floor below to get a general idea of the layout of the fire floor. Standpipe pack hoselines should be connected to the standpipe connections on the floor below the fire floor. Extra hose should be flaked up the stairs above the fire floor so that it will feed more easily into the fire floor as the line is advanced. In addition to attacking the fire directly, crews should be checking floors above the main fire floor for fire extension and any victims who may have been unable to escape. The staging of extra equipment and personnel is usually established two floors below the fire floor.

Caution must be exercised around the outside perimeter of a high-rise building that is on fire. Falling

Figure 12.47 Organize the fire attack from the floor below the fire.

glass and other debris from many stories above the street level can severely damage equipment, cut hoselines, and injure or kill firefighters. The area should be cordoned off and safe methods of entry to the building identified (Figure 12.48). Conditions will dictate how large of an area has to be cordoned off; however, a minimum of 200 feet (60 m) is recommended.

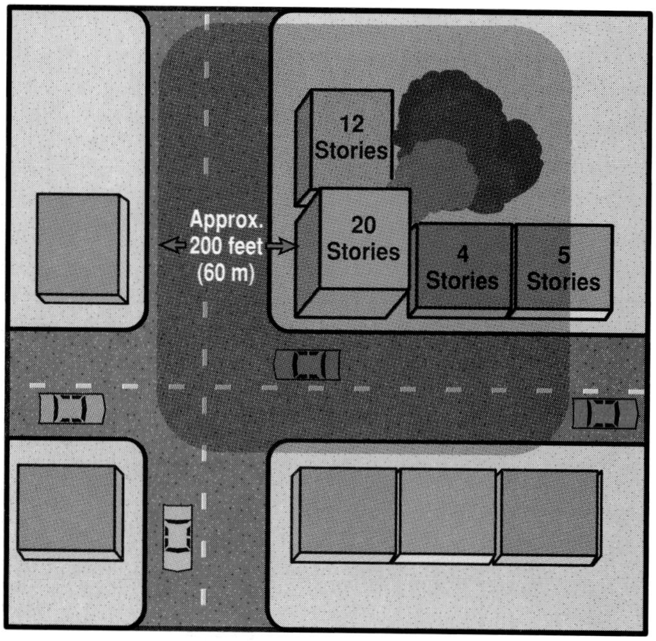

Figure 12.48 A falling debris zone should be established around taller structures that are on fire.

Tactics For Fire Below Grade In Structures

Structural fires that are below ground expose firefighters to extremely punishing conditions. In order to enter a burning basement, firefighters usually have to descend the stairs to that level (Figure 12.49). These stairways are in effect chimneys for the superheated air and fire gases being given off by the fire. Thus, it is imperative that firefighters descend the stairs as quickly as possible, preferably after proper ventilation has been effected. Once the firefighter reaches the basement, the heat conditions will be similar to those in a standard structural fire attack.

Good ventilation techniques are extremely important when fighting fires that are below ground. The ground level floor should be vented to remove as much smoke and heat from the basement as possible. The ventilation point will ideally be at a point far away from the stairs that are going to be descended. The firefighters descending the stairs may put the nozzle on a wide fog pattern to make the trip down the stairs more bearable

Figure 12.49 Descending the stairs into a below ground fire can be very difficult and dangerous.

(Figure 12.50). This is only possible if proper ventilation has been performed away from the stairs. It is also important not to overlook the attic or highest floor of the building. Heat and fire gases accumulate in these areas and could eventually flash over if not vented.

In cases where the basement is inaccessible, high-expansion foam may be used to flood the basement and extinguish the fire. This must be carefully coordinated with ventilation activities to keep from pushing all the heat and fire gases to other areas and causing major fire extension or a flashover. Cellar nozzles can also be put through holes in the floor above the fire to provide water to fires in inaccessible areas (Figure 12.51).

Tactics For Fires At Properties Protected By Fixed Fire Extinguishing Systems

Fire personnel should be familiar with the fixed fire extinguishing systems in buildings protected by their department (Figure 12.52 on the next page). Fire department operating procedures at these occupancies must take into account the necessity of supporting these systems. In particular, the fire department must establish as a high priority the support of fixed fire suppression systems when:

- Initial attack companies must be immediately committed to victim search or evacuation.

Figure 12.51 Cellar nozzles may be used to knock down fires in areas that are otherwise inaccessible.

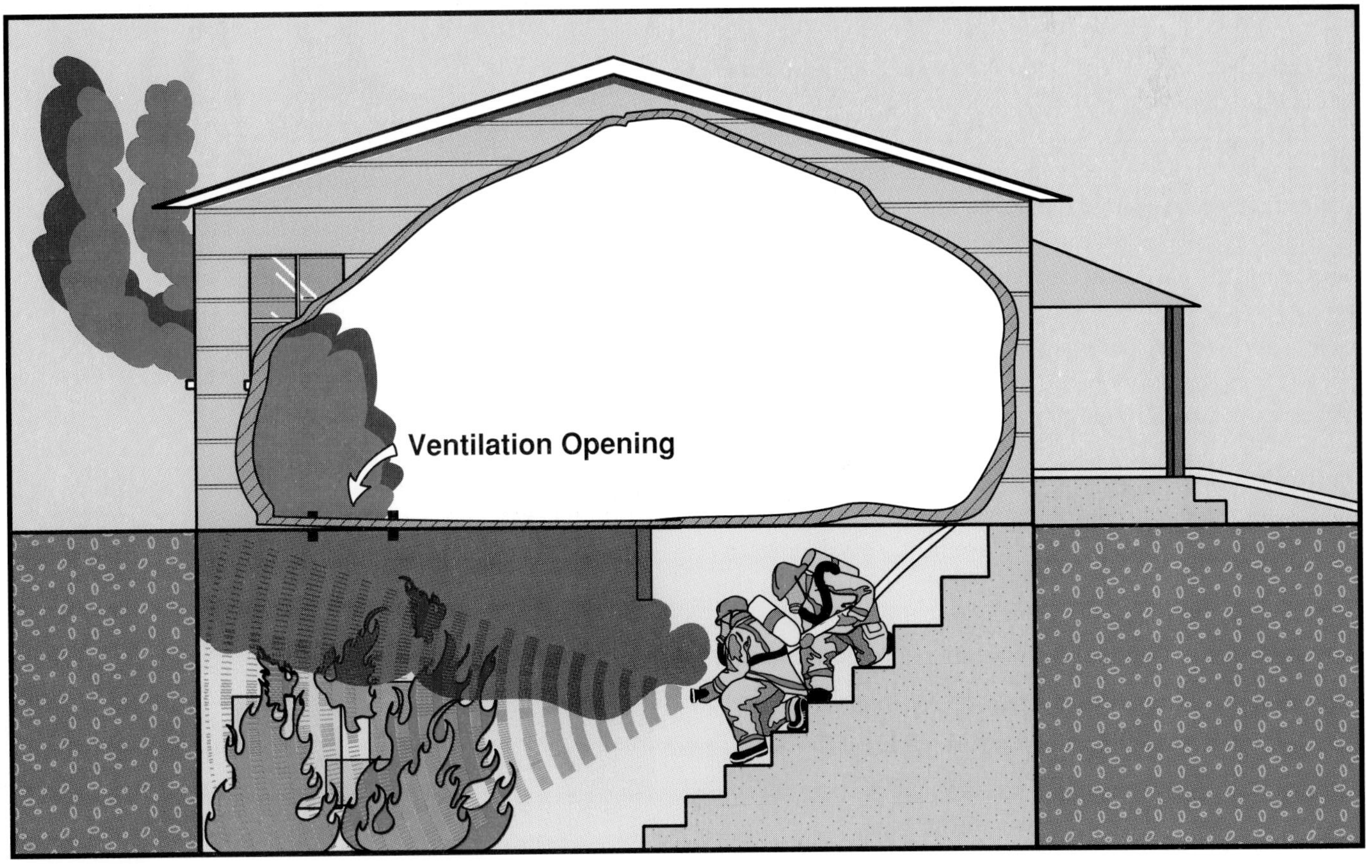

Figure 12.50 Proper ventilation ahead of attack crews can make entering the basement considerably easier.

Figure 12.52 Fire personnel inspecting a fixed fire extinguishing system.

- Streams will be unable to reach areas in large undivided or high-rise occupancies.
- Water is flowing due to sprinklers operating.
- Systems are dependent on fire department support for their ability to function.
- Tactics require the use of standpipes that are not tied in to water supplies.
- Standpipes are supplied by private water sources and pressure is provided by private pumps.

Standard operating procedures used at these occupancies are most likely to be incorporated as a part of the pre-incident plan. This plan includes a detailed account of the construction features, contents, protection systems, and surrounding properties. The pre-incident plan also outlines the procedures to be used by each company according to the conditions that they find. A building map showing water supplies, protection system connections, and company placement should be an integral part of the plan and must be updated to reflect changes affecting fire department operations.

Tactics For Vehicle Fires

Fires in single passenger vehicles are among the most common type of fires that firefighters will encounter (Figure 12.53). These fires should be treated with the same basic care that structural fires are treated with. Firefighters should be in full personal protective clothing, including SCBA. The attack line should be at least a 1½-inch (38 mm) hoseline. The fire should be attacked from the upwind and uphill side when possible.

The basic procedure for attacking a vehicle fire is to extinguish any ground fire around or under the vehicle first and then attack the remaining fire in the vehicle. Firefighters should avoid standing in front of the shock-absorber type bumpers on newer vehicles because they may explode. If the vehicle has combustible metal components that become involved, large amounts of water or Class D extinguishing agents will be needed. Firefighters should use extra caution when water is first applied to these burning parts, because fire intensity will be greatly increased. If a large amount of burning fuel has been spilled, foam may be required for successful extinguishment and continued flame suppression.

Firefighters should not assume that private vehicles or small vans are without extraordinary hazards such as saddle fuel tanks, propane tanks, explosives, or hazardous materials (Figure 12.54). Vans and other small passenger-type vehicles are often used to transport small amounts of radioactive materials for hospital use. Also, large money losses can occur from fires in messenger or courier vehicles. Certainly, firefighters should view any military vehicle as a target hazard. They could be carrying munitions or other hazardous cargoes.

Figure 12.53 Motor vehicle fires pose many hazards to firefighters. *Courtesy of Bob Norman, Elkton (MD) V.F.D.*

Figure 12.54 Some vehicles are powered by compressed gases rather than liquid fuels.

Tactics For Trash And Dumpster Fires

The possibility of exposure to dangerous by-products of combustion should not be overlooked when dealing with trash and dumpster fires. In many cases, the contents of the trash pile or dumpster may include hazardous materials or plastics that give off highly toxic smoke and gases. For this reason, full personal protective equipment and SCBA should be worn when attacking all rubbish fires. Which size hoseline to use depends on the size of the fire and its proximity to exposures. Small piles of trash, garbage cans, and small dumpsters can be handled with a booster line. Larger piles, large dumpsters, and fires that pose exposure problems should be attacked with at least a 1½-inch (38 mm) line. Firefighters should make sure that the fire has not extended into any exposures in the proximity of the fire. Standard overhaul techniques can be used to accomplish this.

Fires And Emergencies In Confined Enclosures

Fire fighting and rescue operations must often be carried out in locations that are below ground or otherwise cut off from natural or forced ventilation. Basements, caves, sewers, storage tanks, and trenches are just a few examples of these types of areas (Figure 12.55). The single most important factor in safely operating at these emergencies is recognition of the inherent hazards of confined enclosures. The atmospheric conditions that may be expected include:

- Oxygen deficiencies
- Flammable gases and vapors
- Toxic gases
- Elevated temperatures

In addition, physical hazards may also be present such as:

- Limited means of entrance and egress
- Cave-ins or unstable support members
- Deep standing water or other liquids
- Utility hazards — electricity, gas, sewage

The importance of wearing personal protective clothing, particularly SCBA, cannot be overemphasized. When firefighters must enter confined spaces separately from the bottle and harness of SCBA, extreme caution must be exercised not to pull the mask off the rescuer. Air supply masks are also available that have long supply hoses without the need for bulky tanks (Figure 12.56).

A lifeline should be tied to each rescuer before entry (Figure 12.57 on the next page). This line must be constantly monitored, and a properly outfitted standby crew equal in number to the rescuers working inside must be available. A system of communication between inside and outside team members should be prearranged as portable radios may prove unreliable. One

Figure 12.55 Most response areas have a variety of potential confined space rescue sites.

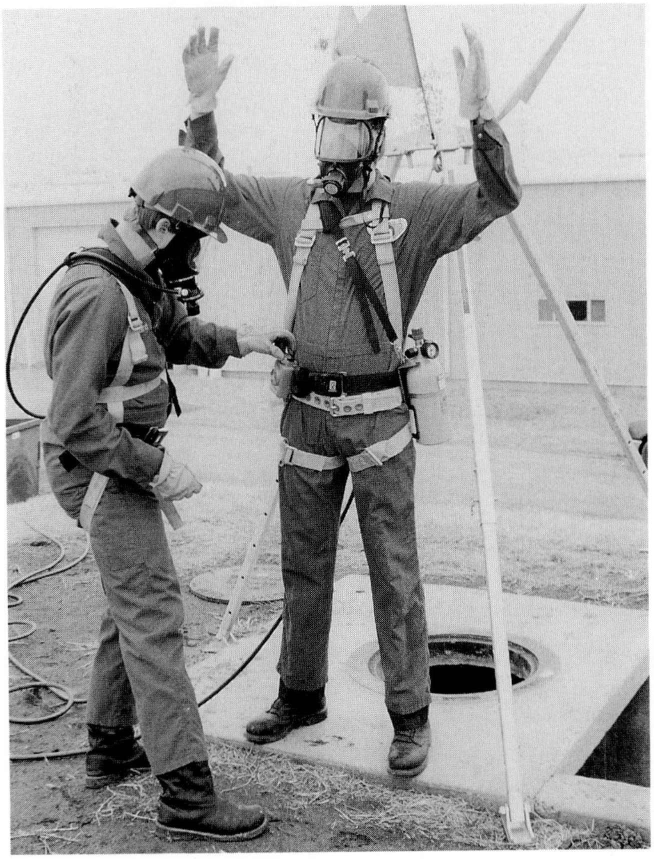

Figure 12.56 The rescuer's partner inspects the rescuer's airline unit to ensure that it has been properly donned.

Figure 12.57 Rescuers entering a confined space must have a lifeline attached to them. *Courtesy of Joel Woods, University of Maryland Fire & Rescue Institute.*

method of signaling is called the O-A-T-H method. The letter O stands for OK, A stands for Advance, T for Take-up, and H for Help. One tug would mean everything is OK, two tugs Advance, three tugs Take-up, and four tugs Help. This system may be used by the firefighter inside the building to communicate to the safety person outside, or for the safety person to communicate with the firefighter.

Any signal given, such as a sharp pull on the line, should be acknowledged by the other party on the line. Another method of safe communication is the use of sound-powered phones that do not require a power source. Also, the rescuer must be able to use the selected communication system without removing the SCBA mask.

Communication with plant or building supervisors or other knowledgeable people at the scene is also important, because they may be able to give valuable information on hazards that are present and the number of victims and their probable location. Likewise, pre-incident plans of existing enclosed spaces in the fire department's jurisdiction reduce guesswork and should be referred to during operations in these locations. The firefighter should be ready to implement prearranged methods of extinguishment or rescue without delay. These plans should include provisions for victim and rescuer protection, control of utilities and other physical hazards, communications, ventilation, and lighting. Power equipment that is used during nonfire rescue operations should be rated for use in explosive atmospheres; this includes flashlights, smoke ejectors, and radios.

Because the entrances to these incidents are generally restricted, the establishment of a command post and staging area are vital to a successful operation. The staging area should be near to, but not obstructing, the entrance and be supplied with the manpower and equipment to be used (Figure 12.58). Firefighters should not enter these enclosures until the incident commander has decided upon a course of action and issued specific orders. A safety officer should be stationed at the entrance to keep track of personnel and equipment entering and leaving the enclosure. The safety officer should check and record each entering member's mission, tank pressure, name, and estimated safe working time. This procedure allows for the accounting of all team members and reduces the possibility of a member being unaccounted for after the safe working time limit has passed.

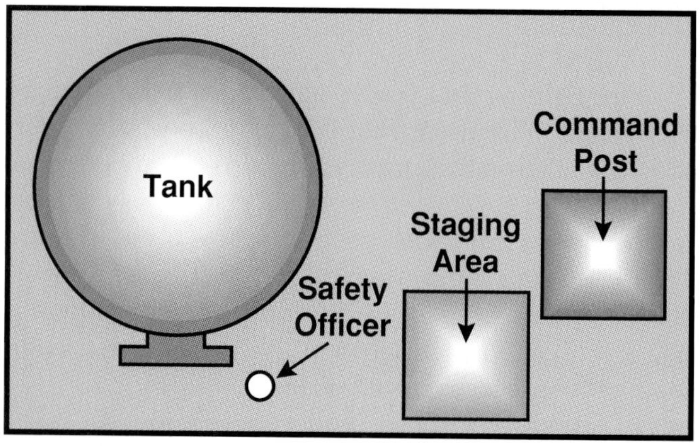

Figure 12.58 The command post and staging areas should be fairly close to the working area. A safety officer should control the entry and exit of rescuers from the hazard area.

When fighting fires in confined spaces such as basements, it is desirable to have other companies locate heavy objects above the fire fighting teams. Fire-weakened supports, such as unprotected steel beams and columns, can be frozen if a hoseline is played on them. Firefighters should be aware that unprotected steel supports yield quickly when exposed to temperatures in excess of 600°F (315°C). The longer supports have been subjected to fire, the more likely they are to fail regardless of their composition. In addition, piled stock will become more unstable as fire damage progresses and water is absorbed by the stock. The application of hose streams must be performed with prudent care because of the difficulty of ventilating generated steam.

Fires in confined areas may also be attacked indirectly with piercing nozzles, cellar nozzles, distributors, or high-expansion foam (Figure 12.59). While the fire is being attacked, special attention should be paid to vertical means of fire spread. Due to the confinement of heat, firefighters may find that they tire more quickly and use their positive-pressure SCBA air supply faster. Firefighters should call for relief before they are exhausted and must pay rigid attention to air conservation techniques and pressure gauges. Firefighters must not advance into confined spaces farther than their air supplies will allow them safe margins for retreating.

Wildland Fire Techniques

Wildland fires include fires in weeds, grass, field crops, brush, forests, and similar vegetation (Figure 12.60). Wildland fires have characteristics of their own that are not comparable to other forms of fire fighting. Local topography, fuel type, and weather present different problems. The local experiences of fire suppression forces will determine the methods and techniques used to control wildland fires.

Once a wildland fire starts, burning is generally rapid and continuous. There are many factors that affect wildland fire behavior, but the three most important factors are fuel, weather, and topography. Any one factor may be dominant in influencing what an individual fire will do, but usually the combined strength of all three factors dictates a fire's behavior.

Figure 12.60 Large brush fires are difficult to extinguish and are taxing on firefighters. *Courtesy of Ron Jeffers.*

FUEL

Fuels are generally classified by grouping together fuels with similar burning characteristics. This method classifies ground cover fuels as ground, surface, and crown fuels (Figure 12.61 on the next page).

- Ground fuels (duff) — Small twigs, leaves, and needles that are decomposing on the ground.

Figure 12.59 High-expansion foam may be used to control basement fires.

- Surface fuels — Living surface vegetation includes grass, brush, and other low vegetation. Nonliving surface vegetation includes downed logs, heavy limbs, etc.
- Crown fuels — Suspended and upright fuels physically separated from the ground fuels to the extent that air can circulate freely around the fuels causing them to burn more readily.

Several factors affect the burning characteristics of fuels, such as:

- Fuel size — Small or light fuels burn faster.
- Compactness — Tightly compacted fuels, such as the ground or surface type, burn slower than the crown type.
- Continuity — When fuels are close together, the fire spreads faster because of the effects of heat transfer. Patchy fuels may spread irregularly or not at all.
- Volume — The amount of fuel present in a given area. The volume will determine the fire's intensity and the amount of water needed to perform extinguishment.

WEATHER

All aspects of the weather have some effects upon the behavior of a wildland fire. Some weather factors that influence wildland fire behavior are:

- Wind — Fans the flames into greater intensity and supplies fresh air that speeds combustion.

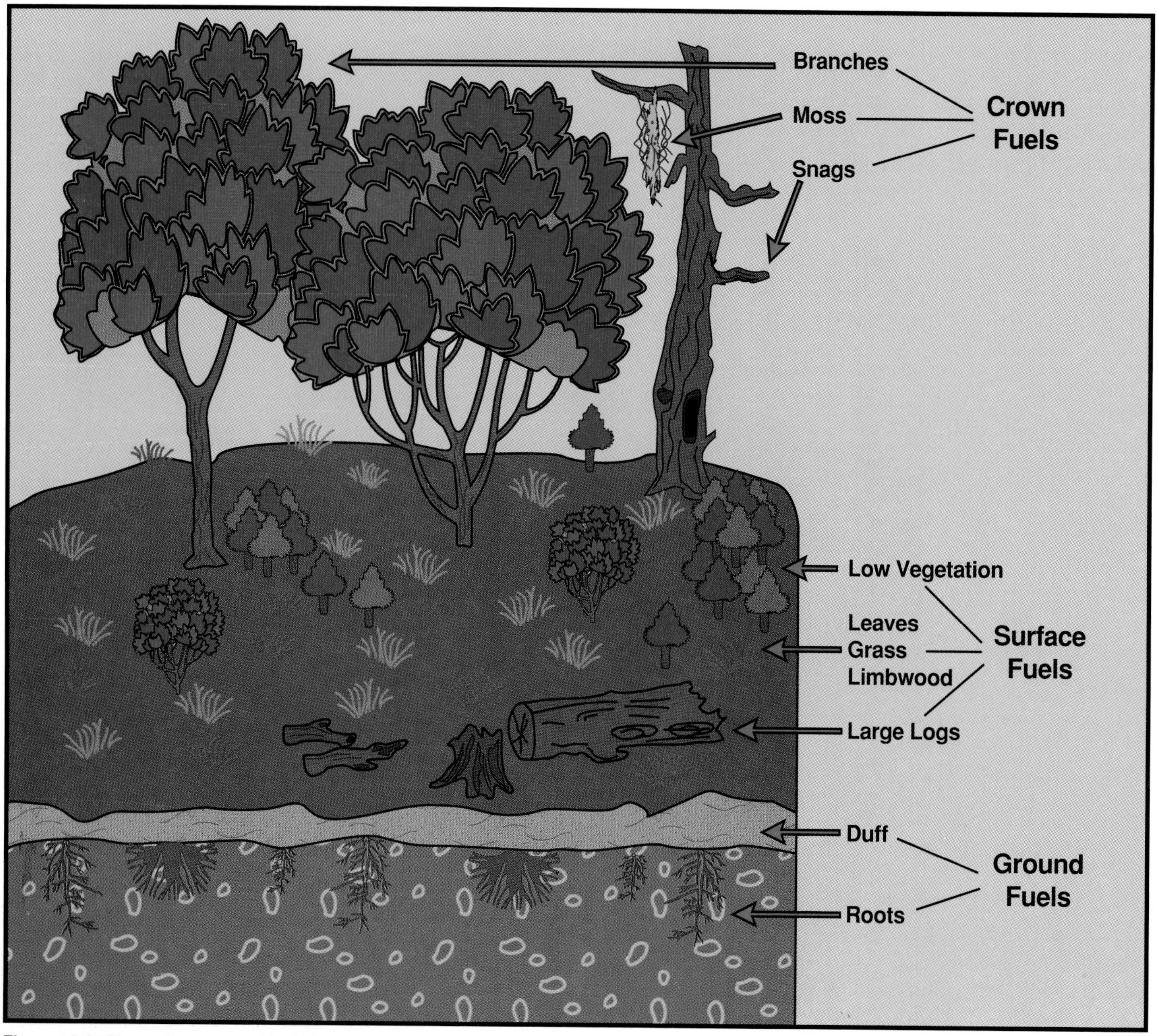

Figure 12.61 The various types of fuels found at brush fires.

Medium- and large-sized fires may create their own winds.

- Temperature — Has effects on wind and is closely related to relative humidity; primarily affects the fuels as a result of long-term drying.
- Relative humidity — The greatest impact is on dead fuels that no longer have any moisture content of their own.
- Precipitation — Largely determines the moisture content of live fuels. Although dead flash fuels may dry quickly, large dead fuels will retain this moisture longer and burn slower.

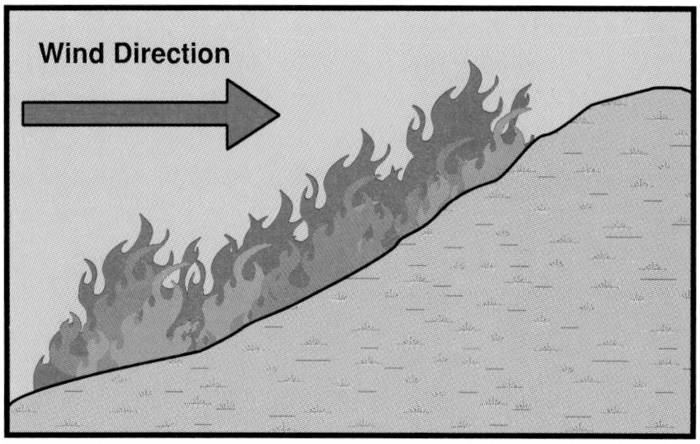

Figure 12.62 Brush fires will move uphill rapidly.

TOPOGRAPHY

Topography refers to the lay of the land and has a decided effect upon fire behavior. The steepness of the slope affects both the rate and direction of the spread. Fires will usually move faster uphill than downhill, and the steeper the slope, the faster the fire will move (Figure 12.62). Other topographical factors influencing wildland fire behavior are:

- Slope aspect — The direction the slope faces. Full southern exposures (north of the equator) receive more of the sun's direct rays and therefore receive more heat. Wildland fires typically burn faster on southern exposures.
- Local terrain features — Directly affect air movements. Obstructions, such as ridges, trees, and even large rock outcroppings, may alter air flow and cause turbulence or eddies resulting in erratic fire behavior.
- Canyons (or other areas with wind flow restrictions) — Result in increased wind velocity. Wind movement can be critical in chutes of steep "V" drainages. These terrain features create turbulent updrafts causing a chimney effect. Fires in these chutes or drainages can spread at an extremely fast rate and are very dangerous.

PARTS OF A WILDLAND FIRE

The typical parts of a wildland fire are shown in Figure 12.63. Any wildland fire will contain at least two of these parts, if not all of them.

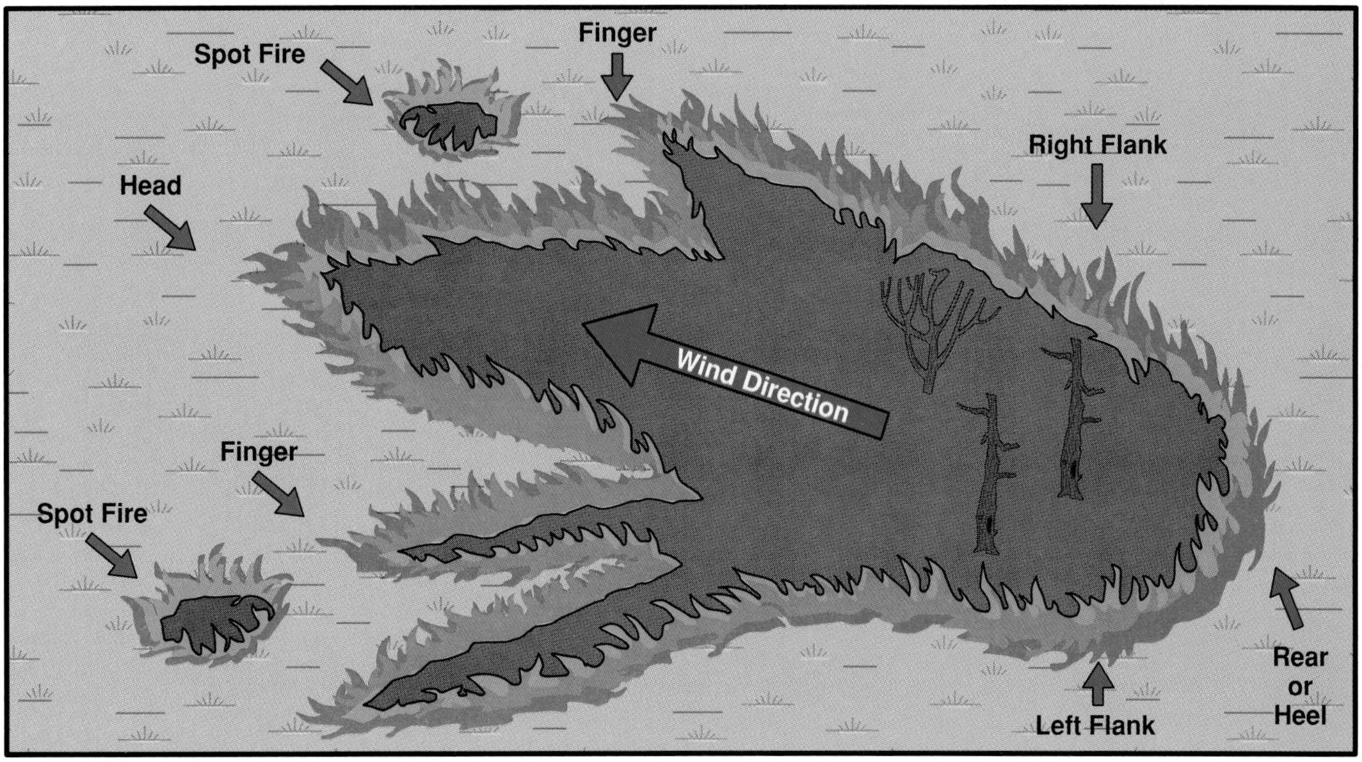

Figure 12.63 The parts of a wildfire.

Head. The head is the part of a wildland fire that travels or spreads most rapidly. The head is usually found on the side of the fire opposite the direction from which the wind is blowing. The head burns intensely and usually does the most damage. Usually, the key to controlling the fire is to control the head and prevent the formation of a new head.

Finger. Fingers are long narrow strips extending out from the main fire. They usually occur when the fire hits an area that has both light and heavy fuels in patches. The light fuel burns faster than the heavy fuel, which gives the fingered effect. When not controlled, these fingers form new heads.

Rear. The rear, or heel of a wildland fire, is the side opposite the head. The rear usually burns slowly and quietly and is easier to control. In most cases, the rear will be found burning downhill or against the wind.

Flanks. The flanks are the sides of a wildland fire. The right and left flank separate the head from the rear. It is from these flanks that fingers are formed, which accounts for the importance of controlling them. A shift in the wind direction can change a flank into a head.

Perimeter. The perimeter of a wildland fire is the boundary of the fire. It is the total length of the outside edge of the burning or burned area. Obviously, the perimeter is ever changing until the fire is suppressed.

Spot Fire. A fire caused by flying sparks or embers landing outside the main fire. Spot fires present a hazard to personnel (and equipment) working on the main fire, because they could become trapped between the two fires. Spot fires must be extinguished quickly or they will form a new head and continue to grow in size.

ATTACKING THE FIRE

The methods used to attack wildland fires revolve around perimeter control. The control line may be established at the burning edge of the fire, next to it, or at a considerable distance away. The objective is to establish fire breaks that completely encircle the fire with all the fuel inside the breaks rendered harmless.

The direct and indirect approaches are the two basic attack methods for attacking wildland fires. The direct method is action taken directly against the flames (Figure 12.64). The indirect attack is used at varying distances from the advancing fire to halt its progress. This method is generally used against fires that are either "too hot," "too fast," or "too big" (Figure 12.65). Because a wildland fire is constantly changing, it is quite possible to begin with one attack method and end with another. Size-up must be continued during the fire so that these adjustments can be made when required.

Fighting wildland fires is a very dangerous occupation. Many firefighters have lost their lives or have been seriously injured while trying to control these fires. Thoroughly think out the situation, then do what will most likely be correct for that situation. Remember: The safety of personnel and equipment always comes first.

Figure 12.64 A direct attack involves trying to extinguish the fire.

Figure 12.65 An indirect attack involves trying to cut the progress of the fire off or removing fuel sources.

13

Automatic Sprinkler Systems

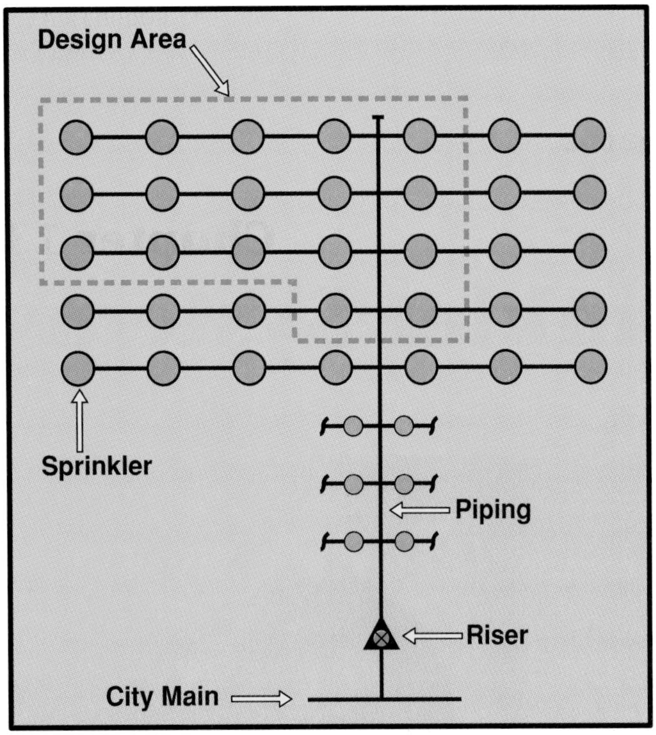

Figure 13.2 Sprinkler systems are designed under the assumption that only a portion of the total number of sprinklers will actually operate.

In general, reports reveal that only in rare instances do automatic sprinkler systems fail to operate. When failures are reported, the reason is rarely because of failure of the actual sprinklers. A sprinkler system may not perform properly due to:

- Partially or completely closed main water control
- Interruption to the municipal water supply
- Damaged or painted-over sprinklers
- Frozen or broken pipes
- Excess debris or sediment in the pipes
- Failure of a secondary water supply

SPRINKLER SYSTEM EFFECTS ON LIFE SAFETY

The life safety of building occupants is enhanced by the presence of a sprinkler system, because it discharges water directly on the fire while it is relatively small. Because the fire is extinguished or controlled in the incipient stage, the combustion products are limited. The few fire fatalities in sprinklered buildings (that have been recorded) have been attributed to asphyxiation from a small fire that did not generate sufficient heat to fuse a sprinkler, or fatal injuries suffered by the victim by the time the sprinkler operated. Sprinklers may also be unable to help sleeping, intoxicated, or handicapped persons whose clothing or bedding ignites early in the fire process. However, in these cases, the sprinkler system will protect the lives of people in other parts of the building.

Sprinkler protection is especially crucial for life safety in high-rise occupancies (Figure 13.3). History has proven that sprinklers are more effective in preventing the spread of upper-story fires than are manual fire suppression operations. Sprinkler systems give people on floors above the fire floor a considerably better chance of survival.

Figure 13.3 Sprinkler systems are crucial in high-rise buildings such as this dormitory.

SPRINKLER SYSTEM FUNDAMENTALS

The principal parts of an automatic sprinkler system are illustrated in Figure 13.4. The system starts with a feeder main into the sprinkler valve. The *riser* is the vertical piping to which the sprinkler valve, one-way check valve, fire department connection, alarm valve, main drain, and other components are attached. The *feed main* is the pipe connecting the riser to the cross mains. The cross mains directly service a number of branch lines on which the sprinklers are installed. Cross mains extend past the last branch lines and are capped to facilitate flushing. System piping decreases in size from the riser outward. The entire system is supported by hangers and clamps. All pipes in dry systems are pitched to help drain the system back toward the main drain.

Sprinklers

Sprinklers, often called sprinkler heads or automatic sprinklers, discharge water after the release of a

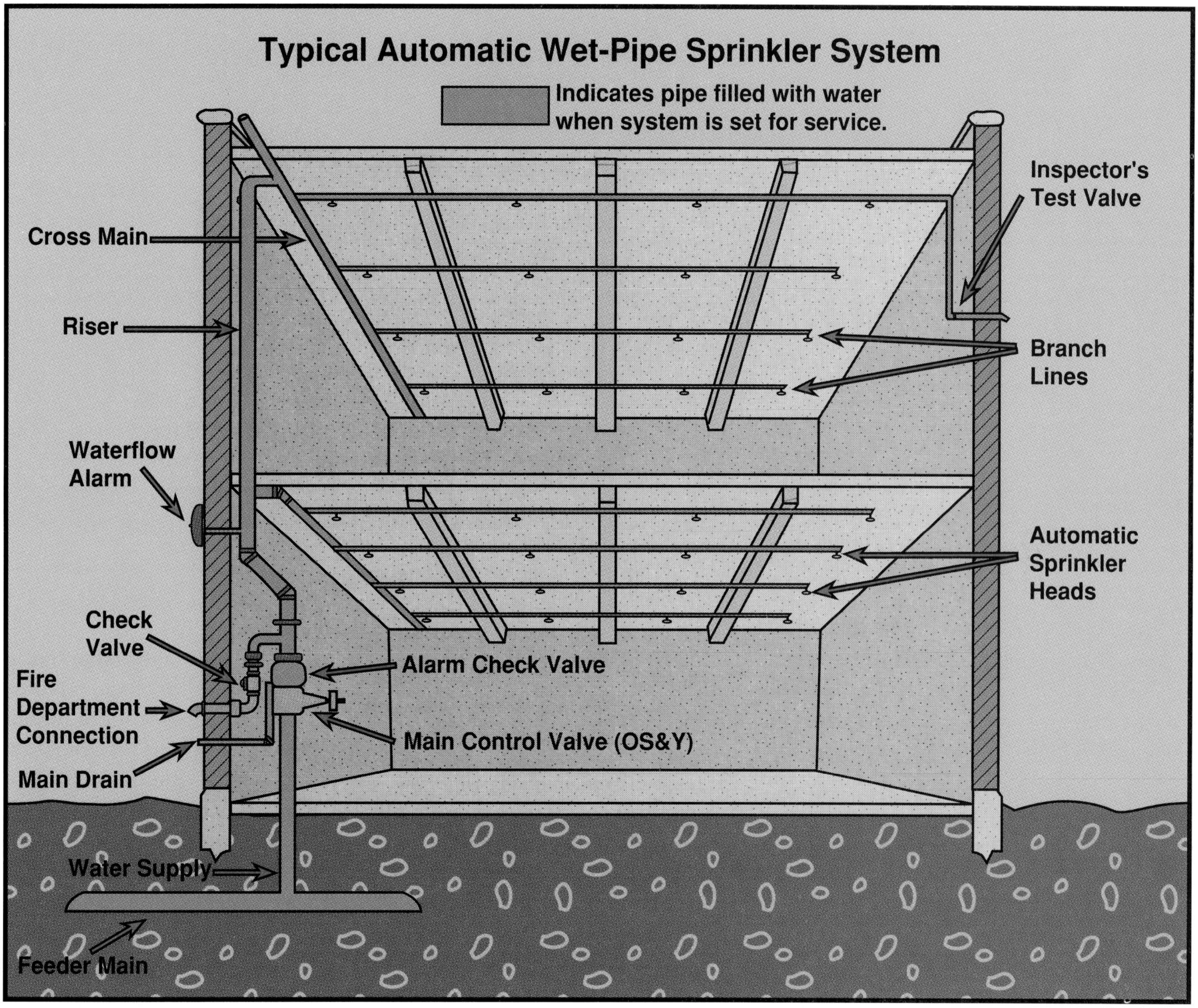

Figure 13.4 A typical wet-pipe system arrangement.

cap or plug that is activated by some heat-responsive element (Figure 13.5). This sprinkler may be thought of as a fixed spray nozzle that is operated individually by a thermal detector. There are numerous types and designs of sprinklers.

The sprinkler used for a given application should be based on the maximum temperature expected at the level of the sprinkler under normal conditions and anticipated rate of heat release produced by a fire in the particular area. The temperature rating is indicated by color coding the frame arms of the head, except for coated sprinklers and decorative heads (Table 13.1 on the next page). Coated sprinklers have colored frame arms, coating material, or a colored dot on the top of the deflector. Decorative sprinklers, such as plated or ceiling sprinklers, are not required to be color coded; however, some manufacturers use a dot on the top of the deflector (Figure 13.6).

Figure 13.5 Sprinkler heads discharge water after the release of a plug that is activated by some heat-responsive element.

Figure 13.6 Some manufacturers use a dot on the top of the deflector of decorative sprinklers.

TABLE 13.1
TEMPERATURE RATINGS, CLASSIFICATIONS, AND COLOR CODINGS

Max. Ceiling Temp. °F	Max. Ceiling Temp. °C	Temperature Rating °F	Temperature Rating °C	Temperature Classification	Color Code	Glass Bulb Colors
100	38	135 to 170	57 to 77	Ordinary	Uncolored or Black	Orange or Red
150	66	175 to 225	79 to 107	Intermediate	White	Yellow or Green
225	107	250 to 300	121 to 149	High	Blue	Blue
300	149	325 to 375	163 to 191	Extra High	Red	Purple
375	191	400 to 475	204 to 246	Very Extra High	Green	Black
475	246	500 to 575	260 to 302	Ultra High	Orange	Black
625	329	650	343	Ultra High	Orange	Black

Sprinklers are kept in a closed position by various means. Four of the most commonly used release mechanisms are fusible links, glass bulbs, chemical pellets, and quick-response mechanisms that fuse or open in response to heat.

FUSIBLE LINK

The design of a sprinkler that uses a fusible link involves a frame that is screwed into the sprinkler piping. Two levers press against the frame and a cap over the orifice in the frame holding the water back. The fusible link holds the levers together until the link is melted during a fire, after which the water pushes the levers and cap out of the way and strikes the deflector on the end of the frame (Figure 13.7). The deflector converts the standard ½-inch (12.5 mm) stream into water spray for more efficient extinguishment.

Figure 13.7 Two examples of a fusible-link sprinkler.

GLASS BULB

Some sprinklers use a small bulb filled with liquid and an air bubble to hold the orifice shut. Heat expands the liquid until the bubble is absorbed into the liquid. This increases the internal pressure until the bulb shatters at the proper temperature (Figure 13.8). The breaking temperature is regulated by the amount of liquid and the size of the bubble in the bulb. The liquid is color coded to designate the designed breaking temperature (refer back to Table 13.1). When the bulb shatters, the valve cap is released. The quantity of liquid in the bulb determines when it will shatter.

Figure 13.8 A glass-bulb sprinkler uses a small bulb filled with liquid and an air bubble to hold the orifice shut.

CHEMICAL PELLET

A pellet of solder, under compression, within a small cylinder melts at a predetermined temperature, allowing a plunger to move down and release the valve cap parts (Figure 13.9).

QUICK-RESPONSE

The quick-response mechanism was developed for life safety purposes. The specially designed fusible link offers increased surface area to collect the heat generated by a fire faster than in standard fusible link sprinklers (Figure 13.10). This results in a faster opening of the sprinkler and quicker extinguishment of the fire.

Figure 13.9 A chemical-pellet sprinkler.

Figure 13.10 A quick-response sprinkler.

Sprinkler Design

There are three basic designs for sprinklers: pendant, upright, and sidewall.

PENDANT

The pendant sprinkler extends down from the underside of the piping and is the most common type in use. This sprinkler sprays a stream of water downward into a deflector that breaks the stream into a hemispherical pattern (Figure 13.11).

Figure 13.11 The pendant sprinkler is the most common type of sprinkler in use.

UPRIGHT

The upright sprinkler sits on top of the piping and sprays water into a solid deflector that breaks it up into a hemispherical pattern that is redirected toward the floor. The upright standard sprinkler cannot be inverted for use in the hanging or pendant position, because in the inverted position the spray of water would be directed toward the ceiling (Figure 13.12).

SIDEWALL

The sidewall sprinkler extends from the side of a pipe and is used in small rooms where the branch line runs along a wall. It has a special deflector that creates a fan-shaped pattern of water (Figure 13.13).

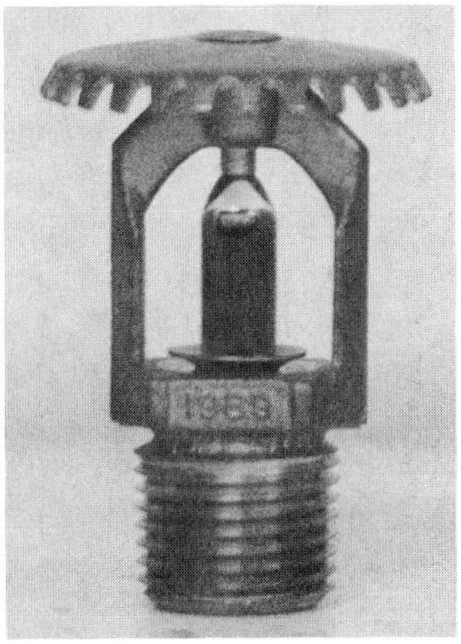

Figure 13.12 An upright sprinkler.

Figure 13.13 A sidewall sprinkler.

SPECIAL SPRINKLERS

Special sprinklers with corrosive-resistant coatings are available and should be installed in areas where chemical, moisture, or other corrosive vapors exist. Without this special coating, the operational parts of the sprinkler will corrode and may become inoperative in a very short time.

Sprinkler Storage

A storage cabinet for housing extra sprinklers and a sprinkler wrench should be installed in the area protected by the sprinkler system. Normally, these cabinets hold a minimum of six sprinklers for small systems (Figure 13.14). To prevent damage, use a sprinkler wrench when

Figure 13.14 A sprinkler storage cabinet.

changing sprinklers (Figure 13.15). Typically, this function should only be performed by representatives of the building's occupants.

Figure 13.15 Use a sprinkler wrench to change sprinklers.

Control Valves

Every sprinkler system is equipped with a main water control valve. Control valves are used to cut off the water supply to the system so that sprinklers can be replaced, maintenance can be performed, or operations interrupted. These valves are located between the source of water supply and the sprinkler system (Figure 13.16). The control valve is usually located immediately under the sprinkler alarm valve, the dry-pipe or deluge valve, or outside the building near the sprinkler system that it controls. The main control valve should always be returned to the open position after maintenance is complete. The valves should be secured in the open position or at least supervised to make sure that they are not inadvertently closed.

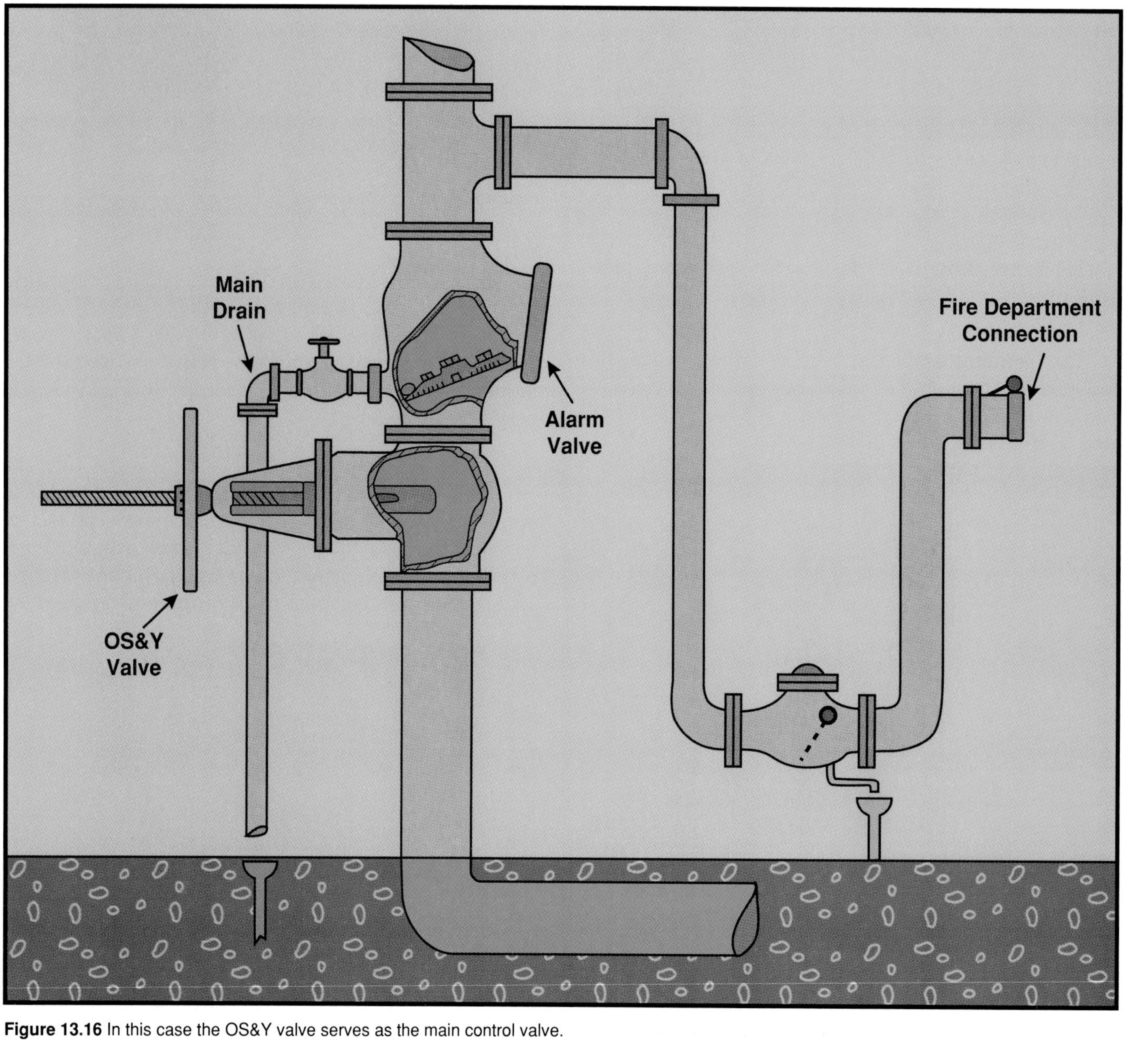

Figure 13.16 In this case the OS&Y valve serves as the main control valve.

Main water control valves are the *indicating* type and are manually operated. An indicating control valve is one that shows at a glance whether it is open or closed. There are four common types of indicator control valves used in sprinkler systems: OS&Y, post indicator, wall post indicator, and post indicator valve assembly.

OS&Y VALVE

The OS&Y valve is an outside screw and yoke valve. This valve has a yoke on the outside with a threaded stem that controls the opening and closing of the gate. The threaded portion of the stem is out of the yoke when the valve is open and inside the yoke when the valve is closed (Figures 13.17 a and b).

Figure 13.17a An open OS&Y valve.

Figure 13.17b A closed OS&Y valve.

POST INDICATOR VALVE (PIV)

The PIV is a hollow metal post that is attached to the valve housing. The valve stem is inside this post; on the stem is a movable target with the words *Open* and *Shut* at the opening. A PIV with the operating handle in the stored and locked position is shown in Figure 13.18.

WALL POST INDICATOR VALVE (WPIV)

A WPIV is similar to a PIV except that it extends through the wall with the target and valve operating nut on the outside of the building.

Figure 13.18 A PIV with the operating handle in the stored and locked position.

POST INDICATOR VALVE ASSEMBLY (PIVA)

A PIVA is similar to a PIV except the valve used is of the butterfly type (Figure 13.19), while the PIV and the WPIV use a gate valve.

Figure 13.19 A typical butterfly valve assembly.

Operating Valves

In addition to the main water control valves, sprinkler systems will employ various operating valves such as globe valves, stop or cock valves, check valves, and automatic drain valves. The alarm test valve is located on a pipe that connects the supply side of the alarm check valve to the retard chamber (Figure 13.20). This valve is
4-93 provided to simulate actuation of the system by allowing water to flow into the retard chamber and operate the waterflow alarm devices. An inspector's test valve is located in a remote part of the sprinkler system (Figure 13.21). The inspector's test valve is equipped with the same size orifice as one sprinkler and is used to simulate the activation of one sprinkler. The water from the inspector's test valve should drain to the outside of the building.

Figure 13.20 An alarm test valve.

Figure 13.21 The inspector's test valve is equipped with the same size orifice as one sprinkler and is used to simulate the activation of one sprinkler.

Waterflow Alarms

Actuation of fire alarms by sprinkler systems is accomplished by the operation of the alarm check valve, dry-pipe valve, or deluge valve. Sprinkler waterflow alarms are normally operated either hydraulically or electrically to warn that water is moving within the system. The hydraulic alarm is a local alarm used to alert the personnel in a sprinklered building or a passerby that water is flowing in the system (Figure 13.22). This type of alarm uses the water movement in the system to branch off to a water motor, which drives a local alarm gong. The electric waterflow alarm is also employed to alert building occupants, and in addition, it can be arranged to notify the fire department. With this type of alarm, the water movement presses against a diaphragm which in turn causes a switch to operate the alarm.

Figure 13.22 A waterflow alarm (also known as a water gong) warns that water is moving within the system.

WATER SUPPLY

Every sprinkler system should have a water supply of adequate volume, pressure, and reliability. In some instances, a second, independent water supply is required. A minimum water supply must be able to deliver the required volume of water to the highest sprinkler in a building at a residual pressure of 15 psi (100 kPa). The minimum flow is established by the hazard to be protected and is dependent upon the occupancy and fire loading conditions. A connection to a public water system that has adequate volume, pressure, and reliability is a good source of water for automatic sprinklers. This type of connection is often the only water supply. A gravity tank of the proper size also makes a reliable primary water supply. In order to give the minimum required pressure, the bottom of the tank should be at least 35 feet (11 m) above the highest sprinkler in the building (Figure 13.23).

Pressure tanks are another source of water supply, and they are used in connection with a secondary supply. Pressure tanks are normally located on the top floor or on the roof of the building. This type of tank is filled two-thirds full with water, and it carries an air pressure of at

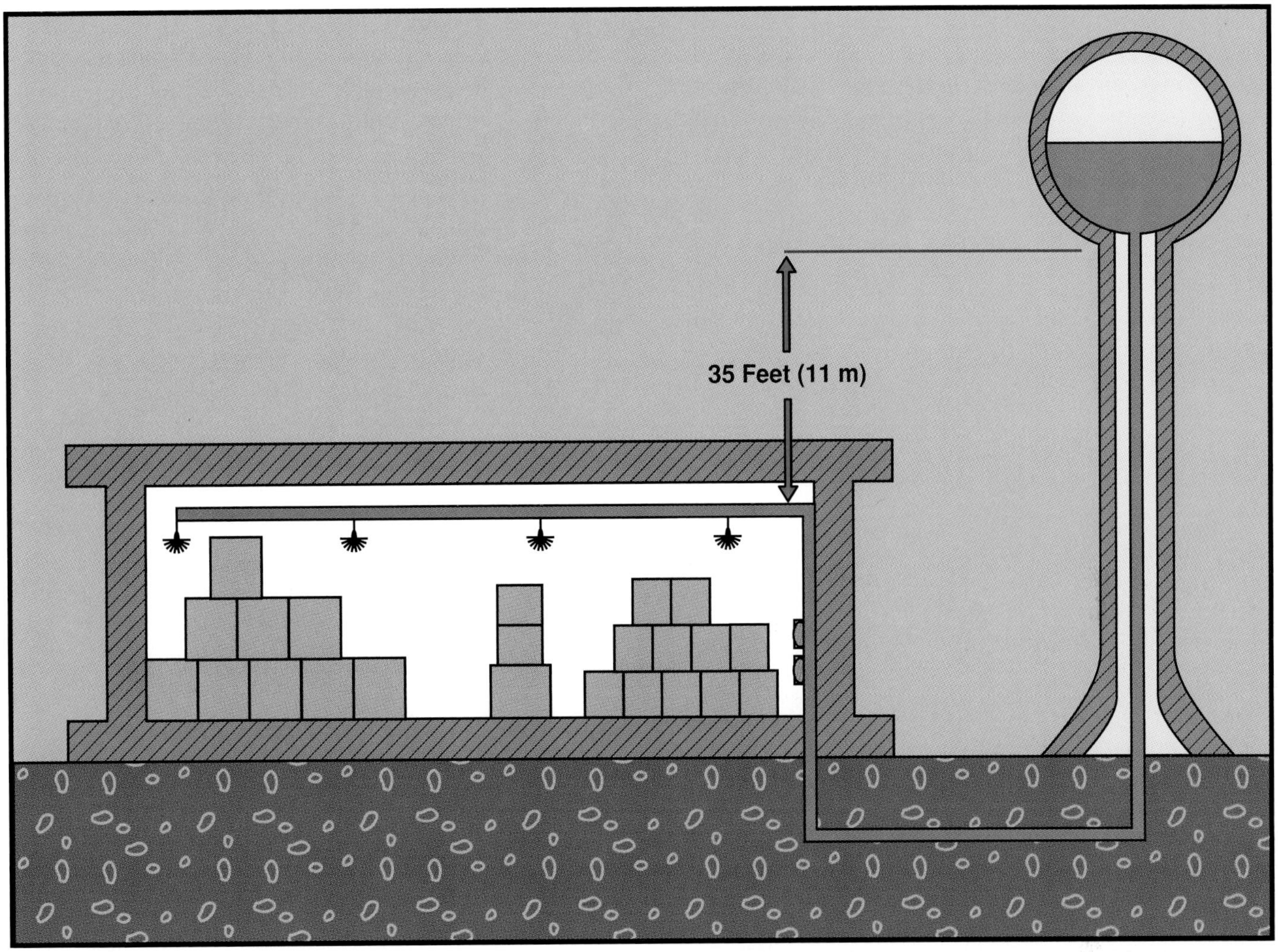

Figure 13.23 The water supply must be at least 35 feet (11 m) above the highest sprinkler.

least 75 psi (525 kPa). An adequate fire pump that takes suction from a static source, such as a large reservoir or storage tank, is used as a secondary source of water supply.

Fire Department Connections

As stated earlier, the water supply for sprinkler systems is designed to supply only a fraction of the sprinklers actually installed on the system. If a large fire should occur or if a pipe breaks, the sprinkler system will need an outside source of water and pressure to do its job effectively. This additional water and pressure can be provided by a pumper that is connected to the fire department sprinkler connection (FDC) (Figure 13.24). Fire department sprinkler connections consist of a siamese with at least two 2½-inch (65 mm) female connections or one large-diameter connection that is connected to a clappered inlet. As water flows into the system, it first passes through a check valve. This valve prevents water flow from the sprinkler system back into the fire department connection; however, it does allow

Figure 13.24 A fire department sprinkler connection.

water from the fire department connection to flow into the sprinkler system (Figure 13.25). The proper direction of water flow through a check valve may be denoted by arrows on the valve or by observing the appearance of the valve casting. A ball drip valve may also be installed at the check valve and FDC. This will keep the valve and connection dry and operating properly during freezing conditions.

Sprinkler connections should be supplied with water from pumpers that have a capacity of at least 1,000 gpm (4 000 L/min) or greater. A minimum of two 2½-inch (65 mm) hoses should be attached to the FDC. Whenever possible, fire department pumpers, other than those supporting the system, should operate from mains other than the primary water supply main for the system (Figure 13.26).

Fire department support of automatic sprinkler systems should include a standard plan of operation. The driver/operator should develop 150 psi (1 000 kPa) at the pumper and maintain this pressure if possible. Circumstances, such as those which exist in high-rise buildings or with deluge systems, may warrant a different

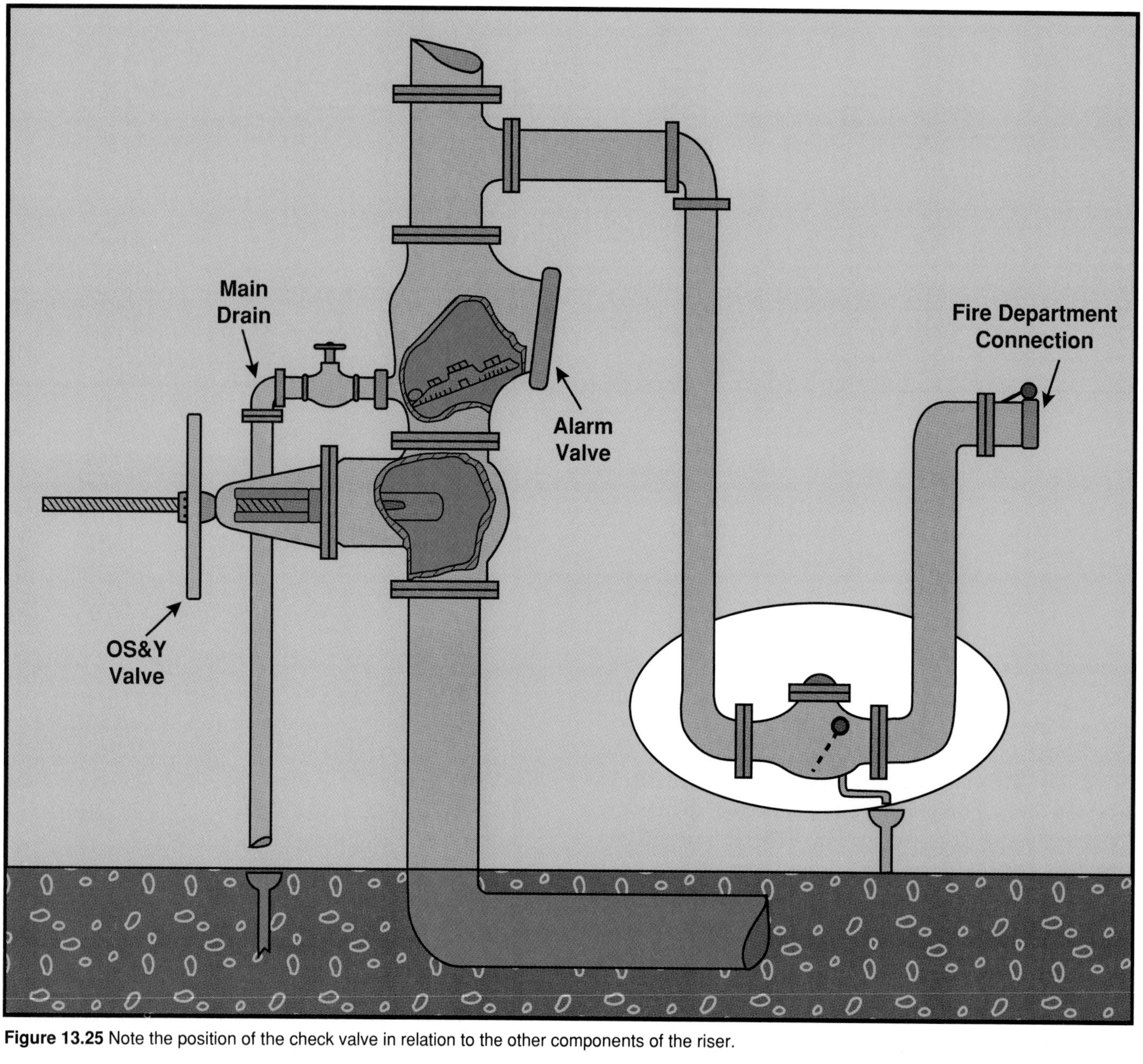

Figure 13.25 Note the position of the check valve in relation to the other components of the riser.

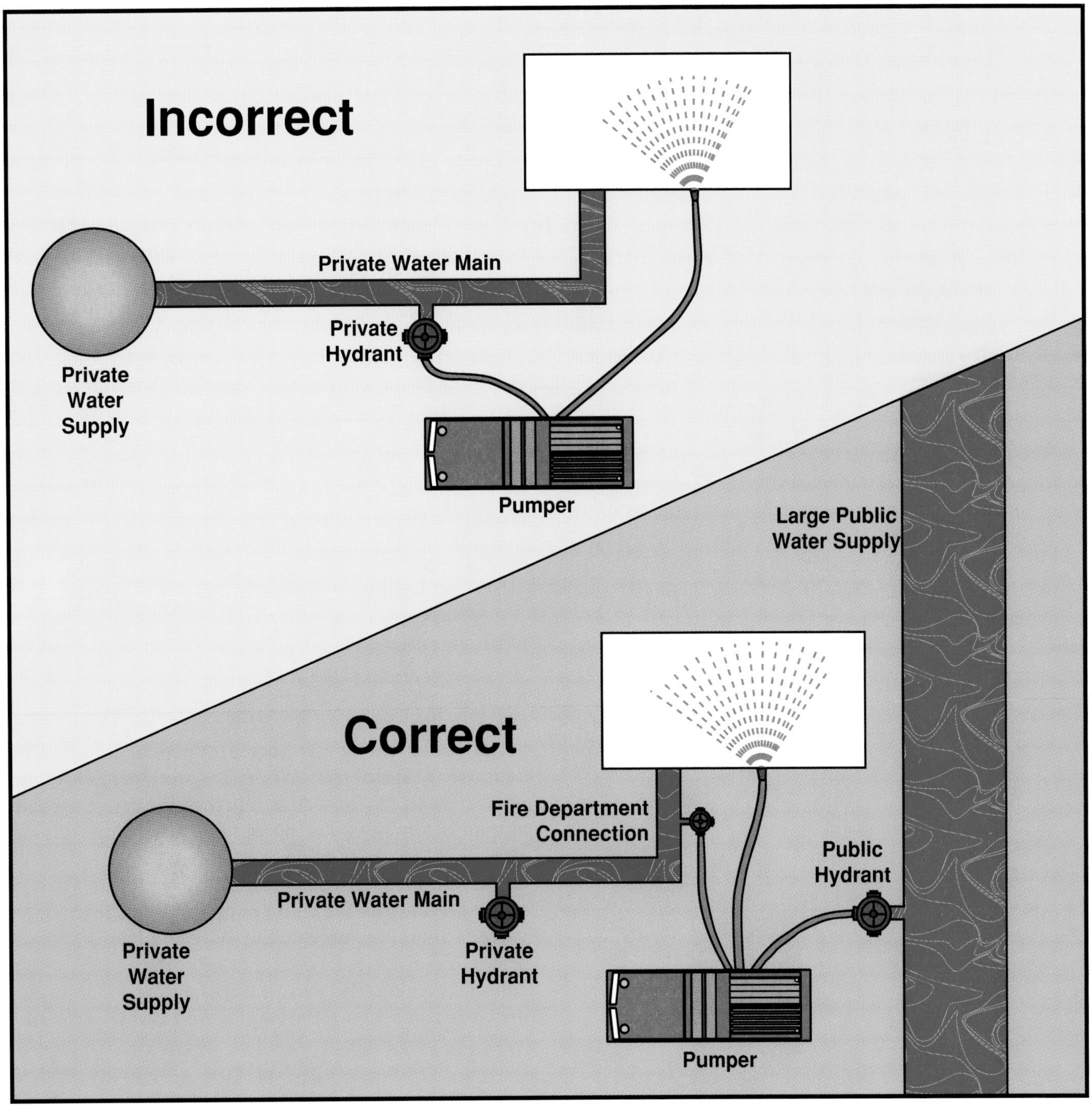

Figure 13.26 The correct and incorrect methods for boosting the water supply to a sprinkler system.

pump discharge pressure. Such a plan cannot be established until fire department personnel become familiar with sprinklered properties under their jurisdiction. A standard plan of operation should cover the buildings in the department's jurisdiction, including its type of occupancy, type of system, and the extent of the system. Therefore, an inspection survey is a prerequisite for the plan of operation (Figure 13.27 on the next page). A thorough knowledge of the water system is important, including volume and pressure.

TYPES OF SPRINKLER SYSTEMS

The following section highlights the five major types of sprinkler systems in use:

- Wet-pipe
- Dry-pipe
- Pre-action
- Deluge
- Residential

Firefighters should have a basic understanding of the operation of each type.

Figure 13.27 Fire department personnel conducting an inspection of a building's sprinkler system for their standard plan of operation.

Wet-Pipe System

The wet-pipe system is used in locations that will not be subjected to freezing temperatures. A wet-pipe system is the simplest type of automatic fire sprinkler system and generally requires little maintenance. This system contains water under pressure at all times and is connected to the water supply so that a fused sprinkler will immediately discharge a water spray in that area and actuate an alarm (Figure 13.28). This type of system is usually equipped with an alarm check valve that is installed in the main riser adjacent to where the feed main enters the building. To shut down the system, turn off the main water control valve and open the main drain. A pressure gauge on the riser should indicate system pressure. If it reads no pressure, chances are the system has been shut off at the main control valve.

Figure 13.28 A wet-pipe system.

RETARDING DEVICE

A wet-pipe sprinkler system may be equipped with a retarding device, commonly called a retard chamber, as part of the alarm check valve. The retard chamber catches excess water that may be sent through the alarm valve during momentary water pressure surges. This reduces the chance of a false alarm activation. The retarding chamber is installed between the alarm check valve and alarm signaling equipment. The retard chamber must be filled before the water can continue to the alarm equipment. Water drains through the small opening in the bottom of the chamber. A typical wet-pipe alarm check valve and retarding chamber in both the standby and fire positions are illustrated in Figure 13.29.

Dry-Pipe Sprinkler System

A dry-pipe sprinkler system should be used in locations where the piping may be subjected to freezing conditions. In a dry-pipe sprinkler system, air under pressure replaces water in the sprinkler piping above the dry-pipe valve. (A dry-pipe valve is a device that keeps water out of the sprinkler piping until a fire actuates a sprinkler.) When a sprinkler fuses, the pressurized air escapes first, and then the dry-pipe valve automatically opens to permit water into the piping system. A dry-pipe valve is designed so that a small amount of air pressure will hold back a much greater water pressure on the water supply side of the dry-pipe valve. This is accomplished by having a larger surface area on the air side of the clapper valve than on the water side of the valve. The valve will be equipped with an air pressure gauge above the clapper and a water pressure gauge below the clapper. Under normal circumstances, the air pressure gauge will read a pressure that is substantially lower than the water pressure gauge. If the gauges read the same, the system has been tripped and water has been allowed to enter the pipes. Dry systems are equipped with either electric or hydraulic alarm-signaling equipment. Figure 13.30(on the following page) illustrates the dry-pipe valve in both the standby and fire positions.

The required air pressure for dry systems usually ranges between 15 and 50 psi (100 kPa and 350 kPa). Air pressure that is needed to service a dry system may be derived from two different sources. These sources are either from a plant air service or from an air compressor and tank used exclusively for the sprinkler system.

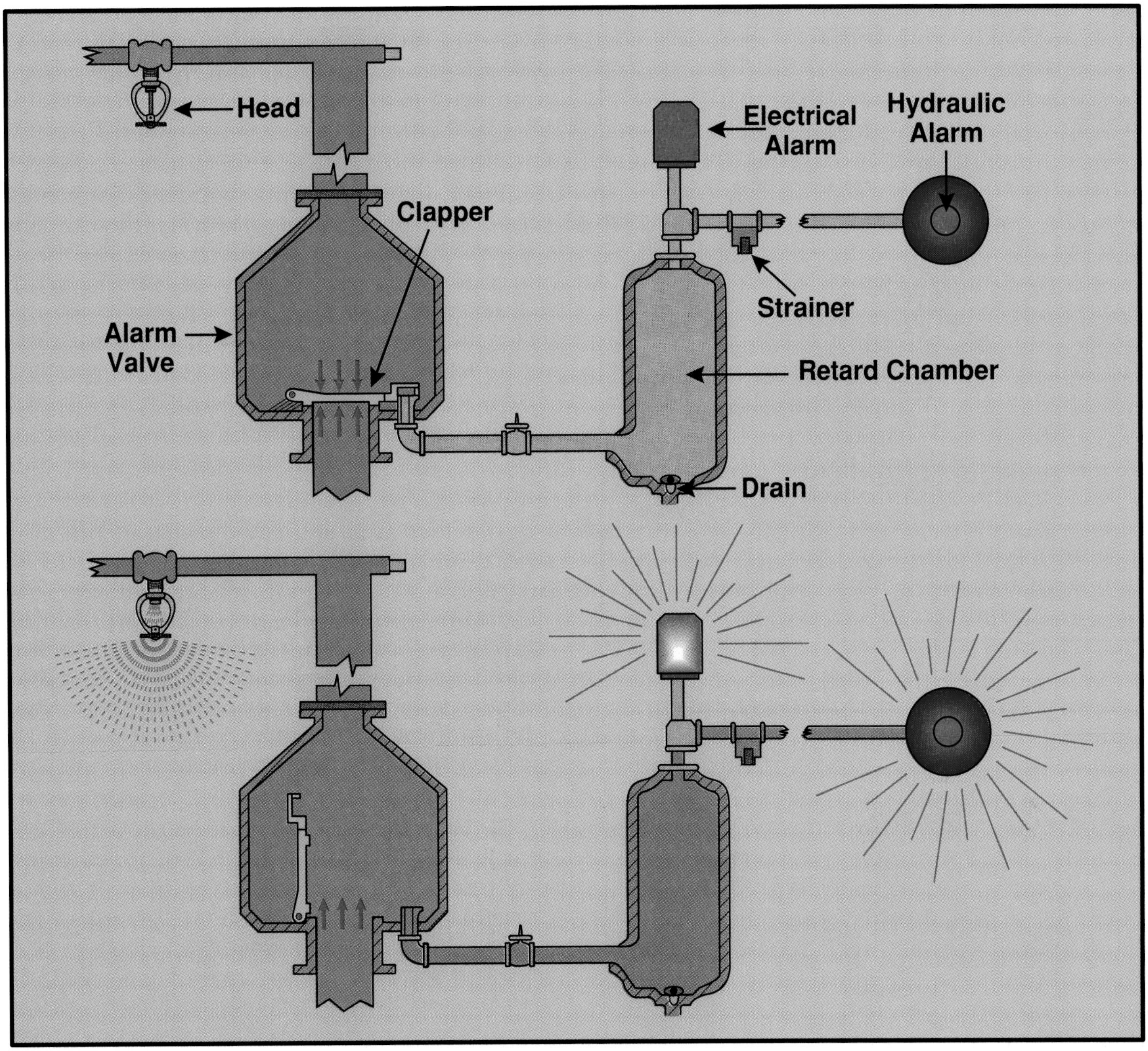

Figure 13.29 The retard chamber reduces the chance of false alarm system activation.

ACCELERATOR AND EXHAUSTERS

In a large, dry system several minutes could be lost while the air is being expelled from the system. Standards require a quick-opening device to be installed in systems that have a water capacity of over 500 gallons (2 000 L) (Figure 13.31 on the next page). Accelerators and exhausters are two types of quick-opening devices. The accelerator unbalances the differential in the dry-pipe valve causing it to trip more quickly, whereas the exhauster quickly expels the air from the system.

When a sprinkler is fused and air pressure in the accelerator-type system drops a few psi (usually 1 or 2 pounds [7 to 14 kPa]), a diaphragm in the accelerator becomes unbalanced. This unbalanced condition causes a valve to open, which permits the air pressure in the system to enter the intermediate chamber of the dry-pipe valve. As soon as air is equalized on both sides of the air clapper (normally 10 to 15 seconds), the valve is automatically tripped by water pressure. In the exhauster type, the fusing of a sprinkler causes a diaphragm to open a large valve. This action permits air pressure to quickly escape to the outside and the dry-pipe valve to trip.

Both of these devices are complicated mechanisms, and they require proper care and maintenance. They should be tested in the spring and fall by a competent individual. Although it will take longer, the dry-pipe valve will operate, even if the quick-opening devices do not operate.

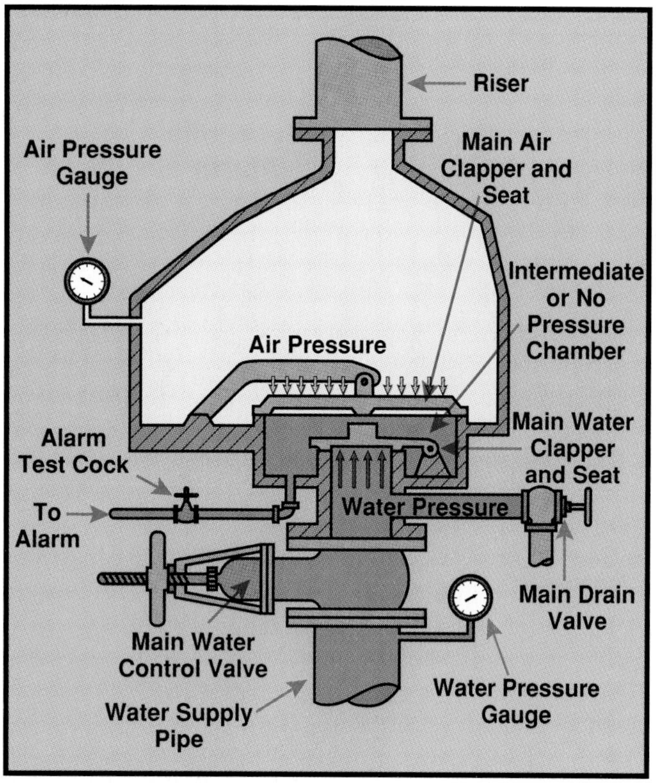

Figure 13.30 The parts of a dry-pipe valve.

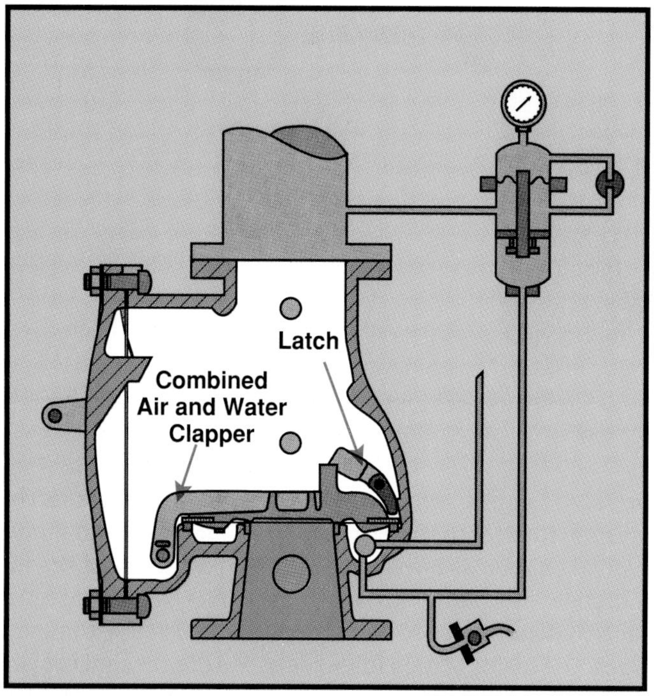

Figure 13.31 The quick-opening device is attached to the dry-pipe control valve.

Pre-Action System

A pre-action system is a dry system which employs a deluge-type valve, fire detection devices, and closed sprinklers. This type of system is used when it is especially important that water damage be prevented, even if pipes should be broken. The system will not discharge water into the sprinkler piping except in response to smoke or heat detection system actuation. A system that contains over 20 sprinklers must be supervised so that if the detection system fails, the system will still operate automatically.

Fire detection and operation of the system introduces water into the distribution piping before the opening of a sprinkler. In this system, fire detection devices operate a release located in the system actuation unit. This release opens the deluge valve and permits water to enter the distribution system so that water is ready when the sprinklers fuse. When water enters the system, an alarm sounds to give a warning before the opening of the sprinkler. Inspecting and testing the system will be essentially the same as that for a deluge system.

Deluge Sprinkler System

This system is ordinarily equipped with open sprinklers and a deluge valve. The purpose of a deluge system is to wet down the area where a fire originates by discharging water from all open sprinklers in the system. This system is normally used to protect extra hazardous occupancies. Many modern aircraft hangars are equipped with an automatic deluge system, which may be combined with a wet- or dry-pipe sprinkler system (Figure 13.32). A system using partly open and partly closed sprinklers is a variation of the deluge system.

Activation of the deluge system may be controlled by fire and heat detecting devices or smoke detecting devices plus a manual device. Because the deluge system is designed to operate automatically and the sprinklers do not have heat-responsive elements, it is necessary to provide a separate detection system. This detection system is connected to a tripping device that is responsible for activating the system. As there are several different modes of detection, there are also many different methods of operating the deluge valve. Deluge valves may be operated electrically, pneumatically, or hydraulically.

Residential Sprinkler Systems

Residential sprinkler systems are installed in one- and two-family dwellings for the protection of both lives and property (Figure 13.33). This type of sprinkler system is designed to prevent total fire involvement in the room of origin and to give the occupants of the dwelling a chance to escape. These systems may be wet or dry systems and are covered by NFPA 13D, *Standard for the Installation of Sprinkler Systems in One- and Two-Family Dwellings and Mobile Homes.*

Residential sprinkler systems employ "fast-response" sprinklers. This type of sprinkler is available in both

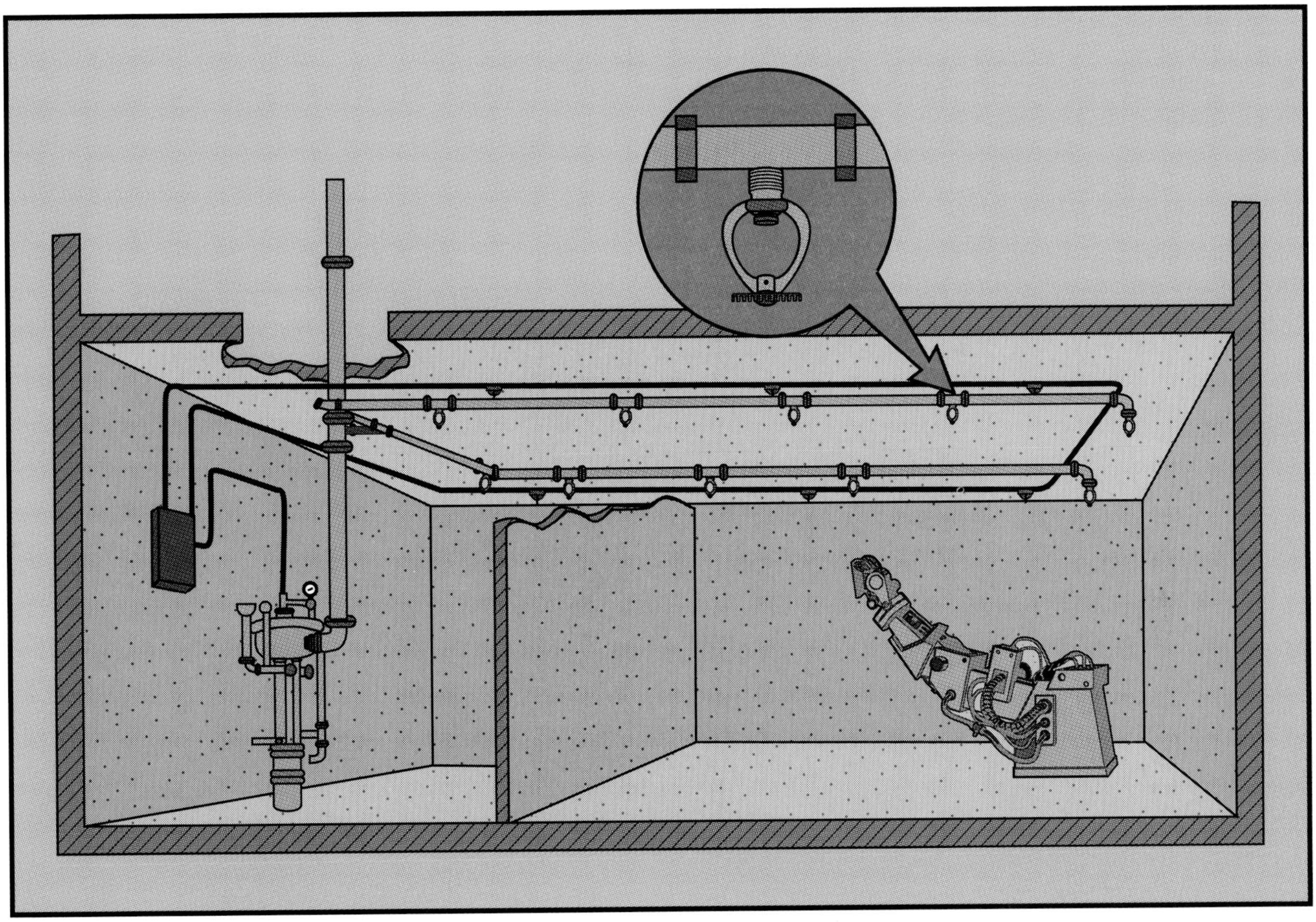

Figure 13.32 A typical deluge system arrangement.

Figure 13.33 Residential sprinkler systems are designed to prevent total fire involvement in one- and two-family dwellings.

conventional and decorative models. There are several types of piping systems that can be used on this type of system. Steel pipe is acceptable as used in other types of sprinkler systems. Plastic pipe is also very common in these systems. The advantage of plastic pipe is reduced cost, easier installation, and easier maintenance. The two types of plastic pipe that have been listed for use in residential sprinkler systems are chlorinated polyvinylchloride (CPVC) and polybutylene pipe. Residential sprinkler systems must also have a pressure gauge (to check air pressure on dry systems and water pressure on wet systems), a flow detector, and a means for draining and testing the system (Figure 13.34). These systems can either be connected directly to the city water supply or be tied in to the dwelling's domestic water system. A control valve is required to shut off the water to the sprinkler system and to the domestic water system if they are connected. If the sprinkler system is supplied separately from the domestic water system, the sprinkler control valve must be supervised in the open position.

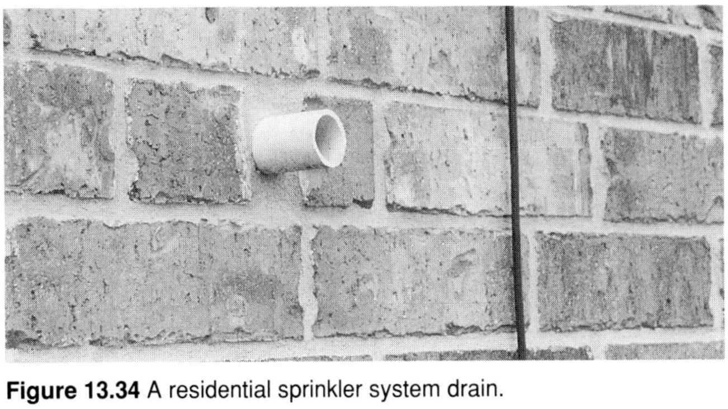

Figure 13.34 A residential sprinkler system drain.

Residential sprinkler systems operate in the same manner as other wet or dry systems. In the wet system, the sprinkler is fused by heat from the fire and water flow is initiated. In dry systems, the sprinkler is fused, air is exhausted, and water flow begins. The system alarm is actuated by the flow of water. The system is shut down by closing the sprinkler control valve (Figure 13.35). Some residential systems may be equipped with a fire department connection (usually a 1½-inch [38 mm] connection), while others will have no FDC.

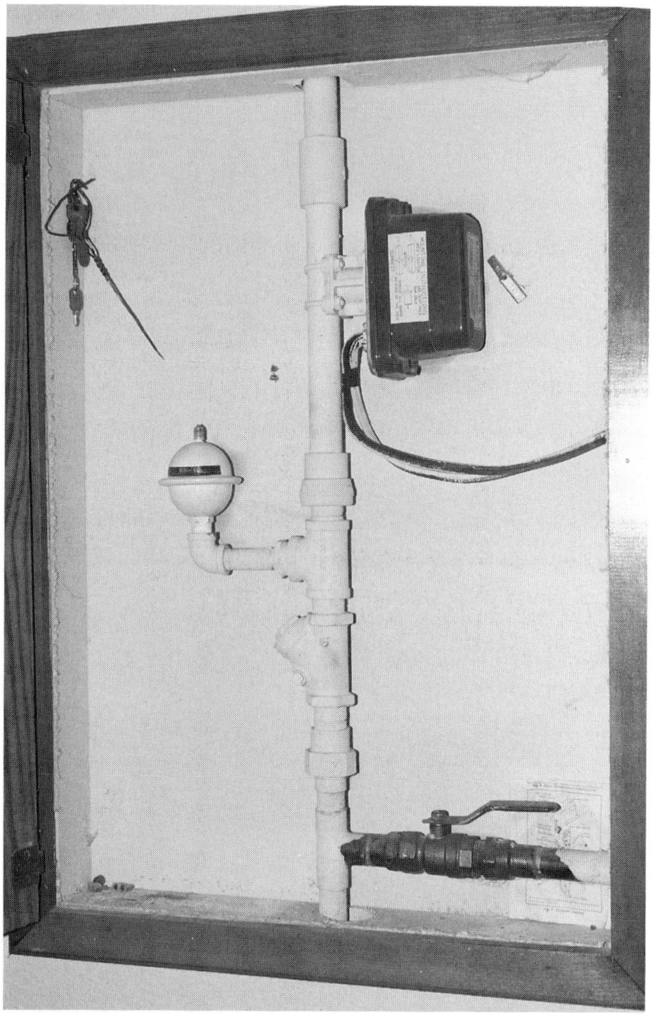

Figure 13.35 Residential sprinkler systems are shut down by closing the sprinkler control valve.

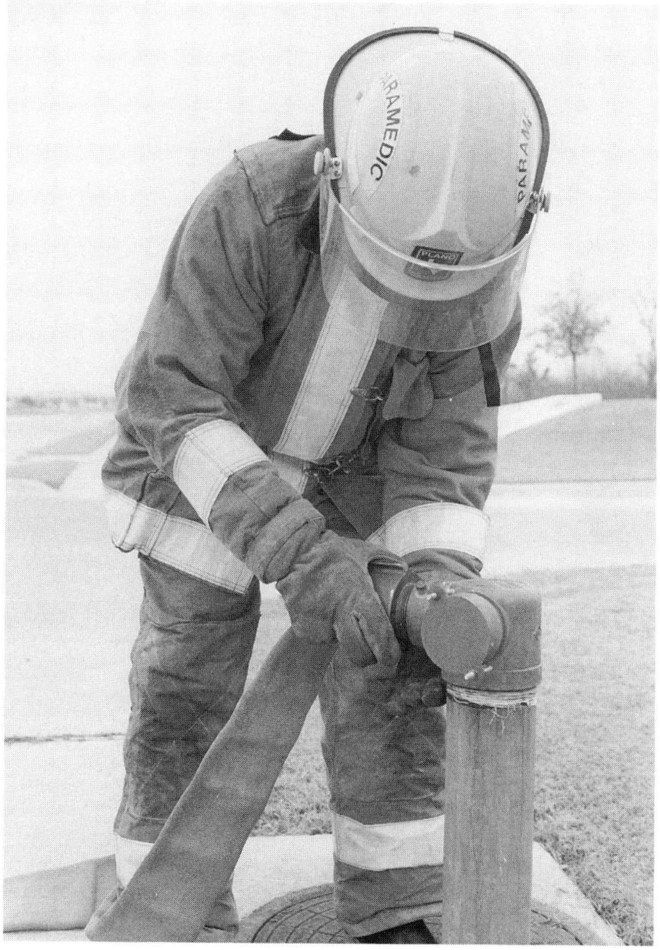

Figure 13.36 A firefighter is connecting the pumper to the fire department connection.

FACTORS TO CONSIDER DURING FIRES AT PROTECTED PROPERTIES

Several important considerations must be made when fighting fires in occupancies that have operating sprinkler systems:

- In addition to normal fire fighting operations, an early arriving pumper should connect to the FDC in accordance with the pre-incident plan (Figure 13.36). A maximum effort should be made to supply adequate water through a sprinkler system, and the water supply should be conserved for this purpose by limiting the use of direct hoselines. The discharge from sprinklers can be improved by increasing the pressure on the system. Pressure at the FDC should be in the range of 100 psi (700 kPa).
- Check control valves to see that they are open. Observe the discharge of sprinklers in the area of the fire and maintain pressure at the pumper to adequately serve the needs of the sprinkler system.
- Sprinkler control valves should not be closed until it is determined that the fire has been extinguished. However, they may be closed when fire officers are convinced that further operations will simply waste water, produce heavy water damage, or hamper the progress of final extinguishment by fire fighting personnel. Premature closure of the control valve could lead to a dramatic increase in the intensity of the fire. When a sprinkler control valve is closed, a firefighter with a portable radio should be stationed at the valve

in case it needs to be reopened should the fire rekindle. Pumpers should not be disconnected until after extinguishment has been determined by a thorough inspection.

- Sprinkler equipment should be restored to service before leaving the premises, if the weather and the extent of the damage permit. Fused sprinklers should be replaced with new sprinklers of proper temperature rating, and the control valve should then be opened. All sprinkler system maintenance should be performed by representatives of the occupant. For liability purposes, it is not recommended that fire department personnel service system components.

Stopping The Water Flow From A Sprinkler

On occasion, firefighters may be required to stop the flow of water from a single sprinkler that has been activated. This may even be necessary after the main water control valve has been closed, because the residual water in the system will continue to drain through the open sprinkler until the system is drained below that level. To stop the flow of water, small wooden wedges may be inserted between the discharge orifice and the deflector and tapped together by hand until the flow is stopped (Figure 13.37). Commercially made stoppers are also available that can be inserted to plug the orifice (Figure 13.38). Closing a sprinkler while water is flowing takes some practice, and all firefighters should be proficient in this skill.

Figure 13.37 To stop the water, insert small wooden wedges between the discharge orifice and the deflector, and tap together by hand until the flow stops.

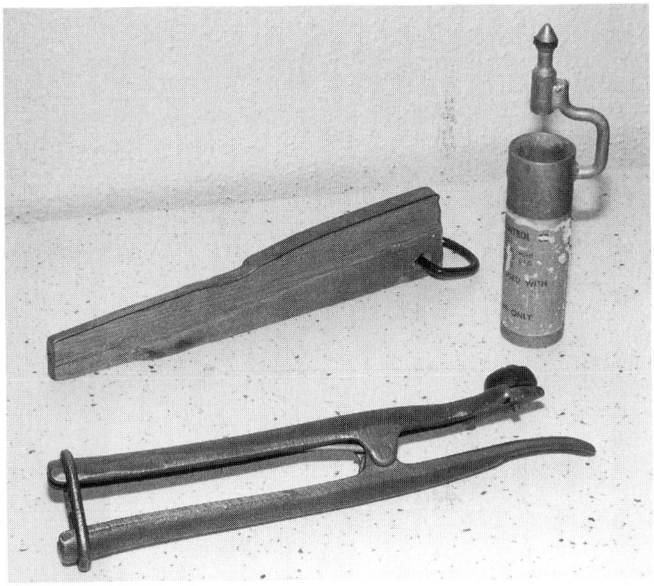

Figure 13.38 Commercially made stoppers are available.

14

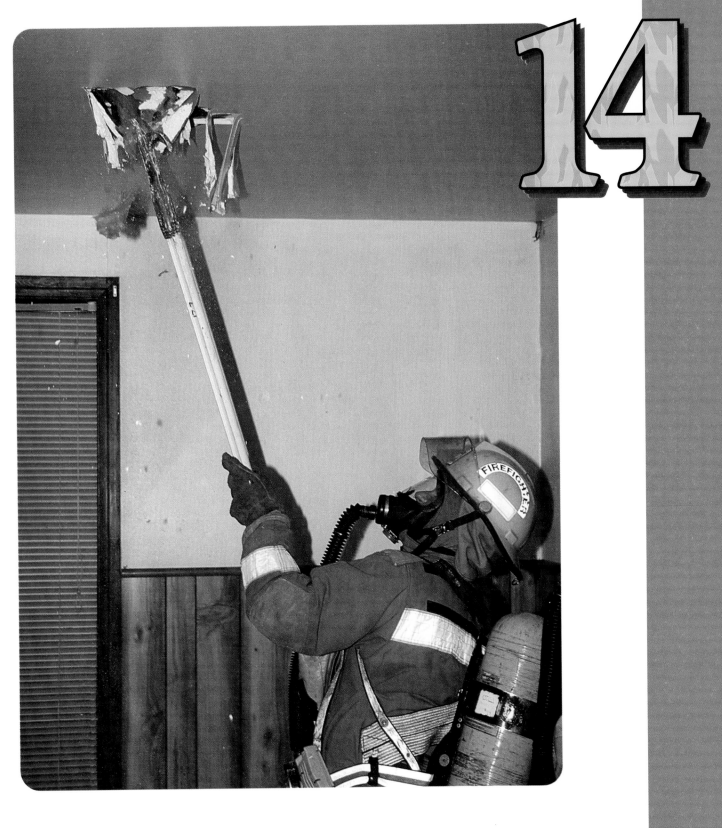

Salvage and Overhaul

This chapter provides information that addresses performance objectives described in NFPA 1001, *Standard for Fire Fighter Professional Qualifications* (1992), particularly those referenced in the following sections:

3-15 Salvage

3-15.1

3-15.2

3-15.3

3-15.4

3-15.5

3-15.6

3-15.7

3-15.8

3-16 Overhaul

3-16.1

3-16.2

3-16.3

3-16.4

3-16.5

4-16 Overhaul

4-16.1

4-16.2

4-16.3

4-16.4

Chapter 14
Salvage And Overhaul

Performing proper salvage and overhaul is one of the most effective means of building goodwill within the community. A fire department often receives words of appreciation and praise by the news media for good salvage and overhaul operations. This praise gives the firefighters a feeling of accomplishment — particularly when the appreciation comes from people who have had their belongings saved by the firefighters. It is common for fire officials to notice better morale and efficiency among firefighters who have significantly contributed to reducing fire loss by successfully practicing salvage and overhaul.

Salvage operations consist of those methods and operating procedures allied to fire fighting that aid in reducing fire, water, and smoke damage during and after fires. Some of these damages cannot be avoided because of the need to do forcible entry, apply water, vent the building, and search for fires throughout a structure. However, improved techniques in fire extinguishment plus prompt and effective use of good salvage procedures will minimize the total losses.

Overhaul operations consist of the search for and extinguishment of hidden or remaining fires; placing the building, its contents, and the fire area in a safe condition and protecting them from the elements; determining the area of origin and cause of the fire; and recognizing and preserving any evidence of arson. Salvage and overhaul may sometimes be performed simultaneously, but overhaul generally follows salvage operations. When effective salvage procedures precede a thorough and systematic overhaul, the result will have a significant effect upon reducing the extent of the loss and facilitate prompt restoration of the property to full use.

PLANNING FOR SALVAGE OPERATIONS

Efficient salvage operations require planning and training by fire officers and firefighters (Figure 14.1). Those people responsible for developing SOPs pertaining to salvage should review the department's salvage equipment inventory. NFPA apparatus standards can be used to obtain lists of recommended equipment to be carried on the apparatus. The training officer should see that all firefighters are adequately trained in salvage operations and in the use of salvage equipment. Chief officers should be instructed to give salvage operations a high priority. Salvage operations should begin before or at the same time the fire attack is initiated. First alarm assignments should be large enough to provide sufficient manpower to perform salvage operations, although fire officers should not hesitate to call additional help if required.

Figure 14.1 Firefighters must train regularly in salvage and overhaul techniques.

In some cases it may be prudent to delay the fire attack until salvage operations can be started. For instance, if an attic fire in an office building is allowed to burn another three or four minutes while the contents of the room(s) immediately below are covered, the fire is consuming some of the cheapest and most easily-replaced parts of the structure. The value of the contents can easily exceed the replacement cost of the roof/attic structure (Figure 14.2 on the next page). In this case, delaying pulling the ceilings while the contents are being covered is a sound tactic.

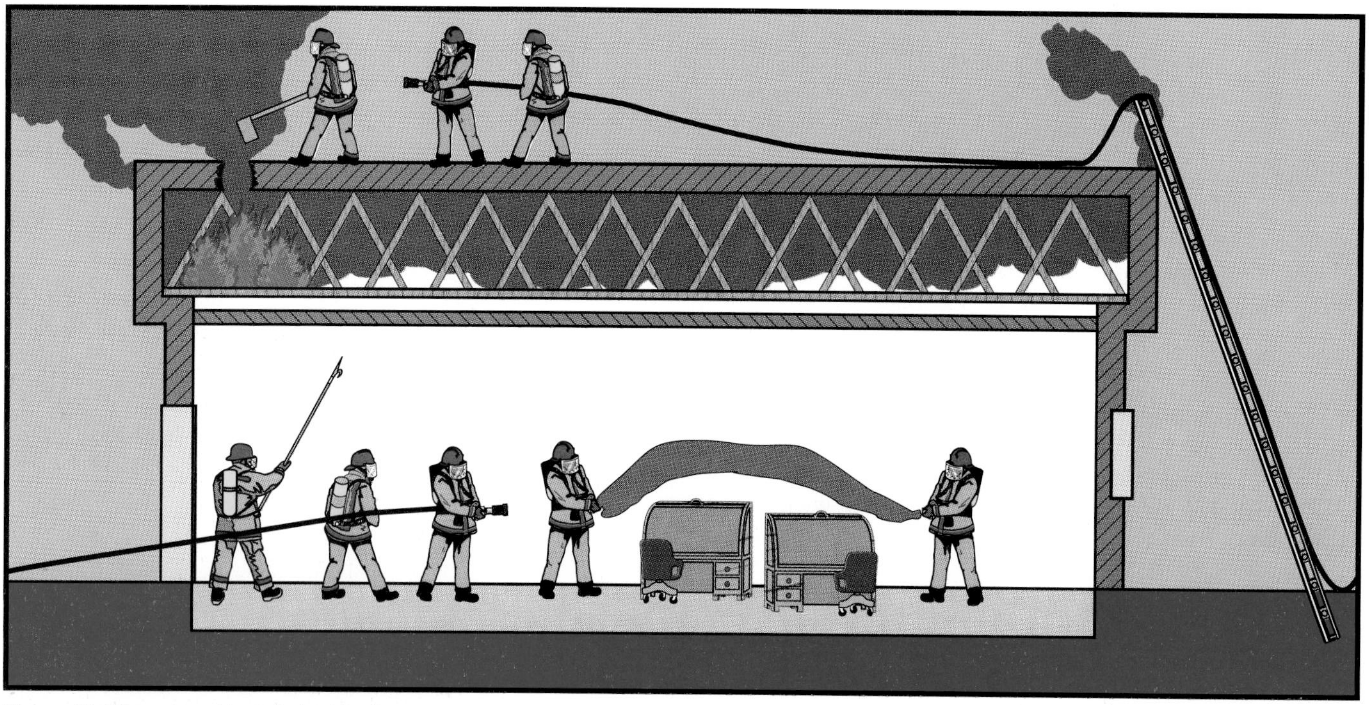

Figure 14.2 Covering objects below the fire area reduces the chance of damage caused by falling debris and water.

Arranging Contents To Be Covered

When possible, building contents should be arranged into close piles that can be covered with a minimum of salvage covers. This allows more contents to be protected than if the contents were covered in their original position.

When arranging household furnishings, group the furniture in the center of the room. If a reasonable degree of care is taken, one average-sized cover will usually protect the contents of one residential room (Figure 14.3). If the floor covering is a removable rug, slip the rug out from under the furniture as each piece is moved, and roll the rug for convenience. A dresser, chest, or high object may be placed at the end of the bed. If there is a roll of rug, place it on top to serve as a ridge pole. Other furniture can be grouped close by, and pictures, curtains, lamps, and clothing can be placed upon the bed. It may sometimes be necessary to place the cover into position before some articles are placed on the bed. In this event, bed and furniture will be protected until the other items are placed under the cover.

Figure 14.3 Gather furniture into the center of the room so that it can be covered.

Commercial occupancies provide challenges for firefighters who are trying to perform salvage functions (Figure 14.4). The actual arranging of contents to be covered may be limited when large stocks and display features are involved. Display shelves are frequently built to the ceiling and directly against the wall. This construction feature makes it difficult to cover shelving. When water flows down a wall, it will naturally come into contact with each shelf and wet the contents.

Figure 14.4 Some commercial occupancies, such as this grocery store, have large quantities of stock to be protected.

One common obstacle to efficient salvage operations is the lack of skids or pallets under stock susceptible to water damage. Some examples of contents that have perishable characteristics are flour, material in cardboard boxes, feed, paper, and other dry goods. Stock in basements should be placed on skids at least 5 inches (125 mm) above the floor, but stock on upper floors is reasonably safe with skids of lesser heights (Figure 14.5). If the contents are stacked too close to the ceiling, this also presents a salvage problem. There should be enough space between the stock and ceiling to allow firefighters to easily apply salvage covers. When salvage covers are limited, it is good practice to use available covers for water chutes and catchalls even though the water must be routed to the floor and cleaned up afterward. Firefighters must be extremely cautious of high-piled stock, such as boxed materials or rolled paper, that has become wet at the bottom. The wetness may reduce the strength of the material and cause the piles to collapse (Figure 14.6). Some rolls of paper can weigh a ton or more, and if these rolls of paper were to fall on firefighters, it could seriously injure or kill them.

SALVAGE COVERS AND EQUIPMENT

Traditionally, salvage covers are made of canvas materials that are waterproof. Some covers with a rubber coating have been used, but they are not common.

Figure 14.5 Placing boxes on skids provides them protection from small amounts of water on the floor.

Figure 14.6 Rolled paper storage can be hazardous to firefighters. If the bottom rolls become wet, the entire stack is likely to collapse. The rolls in this picture each weigh in excess of 2,500 pounds (1 130 kg).

Canvas salvage covers have reinforced corners and edge hems into which grommets are placed for hanging or draping (Figure 14.7). Canvas covers are found in various sizes, but the sizes ranging from 12 x 14 feet (3.7 m by 4.3 m) to 14 x 18 feet (4.3 m by 5.5 m) are most common.

In recent years the improvements in plastic technology have led to synthetic materials that are excellent for salvage covers. These synthetic covers are lightweight, easy to handle, economical, and practical for indoor and outdoor use (Figure 14.8). They are chemically inert,

Figure 14.7 Salvage cover grommet.

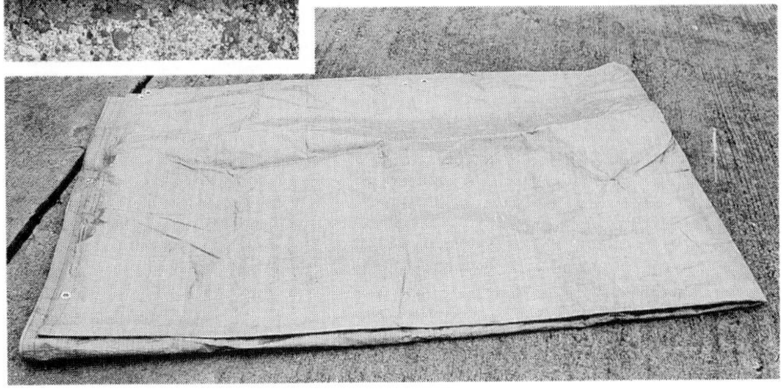

Figure 14.8 Synthetic salvage covers require less care than natural fiber covers.

available in colors, and are not generally affected by alkalines, oils, acids, caustics, or solvents. Synthetic covers are relatively unaffected by normal temperatures, remain flexible below zero degrees, and will not mold, mildew, or absorb moisture. Plastic covers have a tendency to slip from high-piled merchandise and are not well adapted to bagging by rolling the edges.

Salvage Cover Maintenance

Synthetic salvage covers do not require much maintenance. These covers need to be rinsed off when dirty and may be folded wet, but it is usually better to let them dry first so that they will not get a mildew smell (Figure 14.9). If torn, these covers may be patched. However, plastic covers are an economical item, and it may be preferable to replace them. They can be bought at most local hardware, automotive, or discount stores.

Proper cleaning, drying, and repairing of canvas salvage covers increases their span of service. Ordinarily, the only cleaning that is required for canvas salvage covers is showering with a hose stream and scrubbing with a broom (Figure 14.10). Covers that are extremely dirty and stained may be scrubbed with a detergent solution and then thoroughly rinsed. Permitting canvas salvage covers to dry while in a dirty condition is not a good practice. After carbon and ash stains have dried, a chemical reaction takes place that rots the covers. When dried, foreign materials are difficult to remove, even with a detergent. Canvas salvage covers should be completely dry before they are folded and placed in service. This practice is essential to prevent mildew and rot. There is no particular objection to outdoor drying of salvage covers except that wind tends to blow and whip the covers.

Figure 14.9 Synthetic salvage covers can be rinsed clean.

Figure 14.10 Extremely dirty salvage covers can be scrubbed with a detergent solution.

After salvage covers are dried, they should be examined for damage. Holes can be located by lining up three or four firefighters side-by-side along one end, and having them pick up the end and start passing it back over their heads while walking toward the other end, looking up at the underside of the cover (Figure 14.11). Light will show through even the smallest holes. Holes should be marked with chalk (Figure 14.12). The holes can be repaired by placing duct or mastic tape over them or by patching with iron-on or sew-on patches.

Figure 14.11 Firefighters are examining a salvage cover for damage.

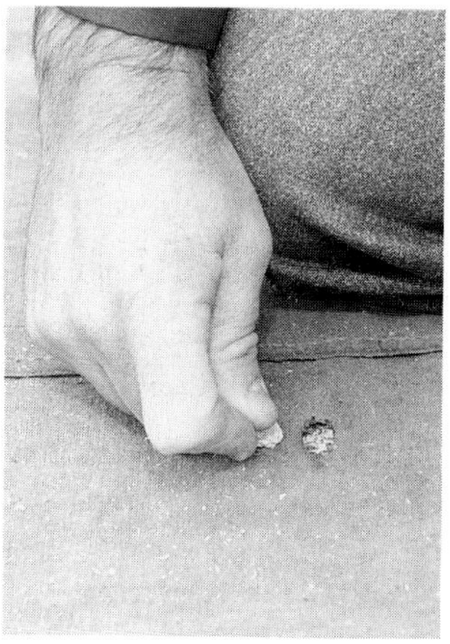

Figure 14.12 Any holes in the canvas should be marked with chalk.

Salvage Equipment

For conducting salvage operations at a fire, it is suggested that salvage equipment be located in one area on the apparatus (Figure 14.13). This common storage eliminates the need to search the apparatus for equipment. Smaller tools and equipment should be kept in a specially designated salvage tool box or other container to make them easier to carry.

The following is a list of typical salvage equipment to be carried on the apparatus. The use of this equipment, however, is not limited to salvage operations.

- Electrician's pliers
- Sidecutters
- Chisels
- Aviator tin snips
- Tin roof cutter (can opener)

Figure 14.13 Salvage equipment located in one compartment of the apparatus.

- Adjustable end wrenches
- Pipe wrenches
- Hammer(s)
- Sledgehammer
- Hacksaw
- Crosscut handsaw
- Heavy-duty stapler and staples
- Linoleum knife
- Wrecking bar
- Padlock and hasp
- Screwdriver(s)
- Applicable power tools
- Hydraulic jack
- An assortment of nails
- An assortment of screws
- Roofing or tar paper and roofing nails
- Plastic sheeting
- Wood lath
- Mops
- Squeegees
- Water scoops
- Scoop shovels
- Brooms
- Mop wringers with bucket
- Automatic sprinkler kit
- Water vacuum
- Submersible pump
- Sponges
- Chamois
- Assortment of rags
- 100-foot (30 m) length of electrical cable with locking-type connectors, 14-3 gauge or heavier
- Pigtail ground adapters, 2 wire to 3 wire, 14-3 gauge or heavier with 12-inch (300 mm) minimum length
- An approved ground fault interruption device
- Salvage covers
- Floor runners

Individual fire department policies dictate which apparatus the salvage equipment is carried on and who performs the salvage operations on the fire scene. It is common for truck (ladder) company personnel to perform salvage operations; however, engine company personnel may do salvage work in some jurisdictions.

Automatic Sprinkler Kit

The tools found in a sprinkler kit are needed when fighting fires in buildings protected by automatic sprinkler systems. These tools will be used to stop the flow of water from an open sprinkler. A flow of water from an open sprinkler can do considerable damage to merchandise on lower floors after the fire has been controlled. The following tools are suggested to form a sprinkler kit:

- Sprinkler wrench
- Sprinkler tong or stopper
- Sprinkler wedge

Carryalls (Debris Bags)

Carryalls, sometimes referred to as debris bags, are used to carry debris, catch falling debris, and provide a water basin for immersing small burning objects (Figure 14.14 on the next page). They can be made from an old or damaged salvage cover and are usually 6 or 8 feet (2 m or 2.5 m) square. Handholds are provided by lashing a sash cord through eyelets. If desired, the sash can be pulled tight, permitting the carryall to be puckered to form a bag.

Figure 14.14 A carryall can be used to haul debris from the building.

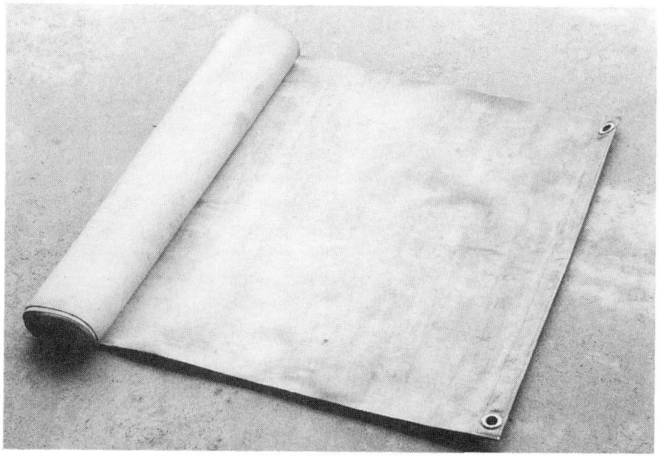

Figure 14.15 Floor runners are used to protect carpets and other floor coverings from foot traffic.

Figure 14.16 Portable pumps may be used to remove water from a structure.

Floor Runners

Costly floor coverings are sometimes damaged by mud and grime tracked in by firefighters. These floor coverings may be protected by using floor runners. Floor runners are usually 20 or 30 inches (500 mm or 750 mm) wide. They can be unrolled from an entrance to almost any part of a building. Commercially prepared vinyl-laminated nylon floor runners are lightweight, flexible, tough, heat and water resistant, and easy to maintain (Figure 14.15).

Portable Pumps

Portable water pumps are used to remove water from basements, elevator shafts, and sumps (Figure 14.16). Fire department pumpers should not be used for this purpose because they are intricate and expensive machines and are not intended to pump the dirty, gritty water found in such places. Trash-type pumps are best suited for salvage operations.

Various devices, known as jets or siphons, may be used for the removal of excess water. These devices can be moved to any point where a line of hose can be placed and an outlet for water can be provided.

Water Vacuum

One of the easiest and fastest ways to remove water, chemicals, acid solutions, and all other nonflammable liquids is the use of a water vacuum device. Dirt and small debris may also be removed from carpet, tile, and other types of floor coverings with this equipment. The water vacuum appliance consists of a tank (worn on the back or placed on wheels) and a nozzle. It has a suction powerful enough to extract liquids from deep pile carpeting. Backpack-type tanks normally have a capacity of 4 to 5 gallons (16 L to 20 L) and can be emptied by simply pulling a lanyard that empties the water through the nozzle or through a separate drain hose (Figure 14.17). Floor models on rollers may have capacities up to 20 gallons (80 L) and typically have to be dumped to be emptied (Figure 14.18).

Figure 14.17 Some water vacuums are designed to be worn on the back.

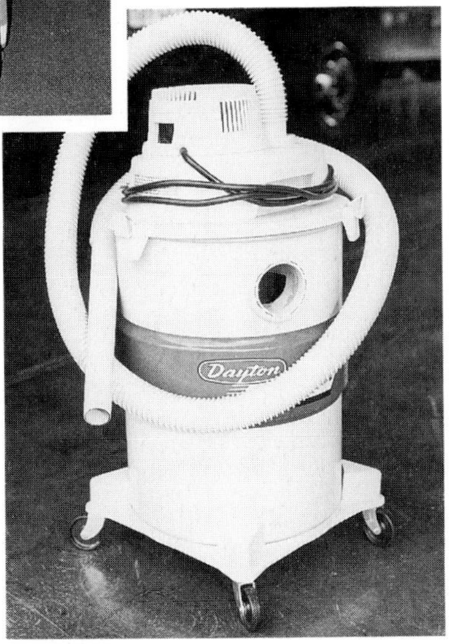

Figure 14.18 A floor model water vacuum.

Care And Maintenance Of Salvage Equipment

For the commonly used types of salvage equipment, use the following care and maintenance guidelines:

- Water Vacuum
  - —Inspect power cords for broken insulation
  - —Flush collection tank
  - —Clean nozzle
- Mops
  - —Clean with soap and water
  - —Dry thoroughly
- Tools
  - —Dry
  - —Lightly oil if needed
- Brooms
  - —Clean
  - —Sand handles if burred
- Buckets and tubs
  - —Clean
  - —Check for holes

METHODS OF FOLDING AND SPREADING SALVAGE COVERS

One of the key factors in successful salvage operations is the proper handling and deployment of salvage covers. The following sections highlight basic principles of salvage cover storage and deployment by one or two firefighters.

Rolling The Salvage Cover For A One-Firefighter Spread

The principal advantage of the one-firefighter salvage cover roll is that one person can quickly roll a cover across the top of an object and unfold it. To form this roll, two firefighters must make initial folds to reduce the width of the cover. Use the following procedure for making this roll. Note that Steps 1 through 6 are performed simultaneously by both firefighters on opposite sides of the cover. Steps 7 through 9 may be performed by both firefighters who are stationed at the same end of the roll.

Step 1: Grasping the cover with your outside hand midway between the center and the edge to be folded, place your other hand on the cover as a pivot midway between your outside hand and the center (Figure 14.19).

Step 2: Bring the fold over to the center of the cover (Figure 14.20). This will create an inside fold (center) and an outside fold.

Figure 14.19 Step 1: Grasping the cover with your outside hand midway between the center and the edge to be folded, place your other hand on the cover as a pivot midway between your outside hand and the center.

Figure 14.20 Step 2: Bring the fold over to the center of the cover.

Step 3: Grasping the cover corner with your outside hand, place your other hand as a pivot on the cover over the outside fold (Figure 14.21).

Step 4: Bring this outside edge over to the center, and place it on top of and in line with the previously placed first fold (Figure 14.22).

Step 5: Fold the other half of the cover in the same manner by using Steps 1 through 4. If the folds are not straight, they should be straightened. The partially completed fold is shown in Figure 14.23.

Step 6: Fold over about 12 inches (300 mm) at each end of the cover to make clean, even ends for the completed roll (Figure 14.24).

Figure 14.21 Step 3: Grasping the cover corner with your outside hand, place your other hand as a pivot on the cover over the outside fold.

Figure 14.22 Step 4: Bring this outside edge over to the center, and place it on top of and in line with the previously placed first fold.

Figure 14.23 Step 5: To fold the other half of the cover, use Steps 1 through 4.

Step 7: Start the roll by rolling and compressing one end into a tight compact roll, rolling toward the opposite end (Figure 14.25).

Step 8: As the roll progresses, tuck in any wrinkles that form ahead of the roll.

Step 9: The completed roll may be secured by inner-tube bands or Velcro® straps or tied with cords (Figure 14.26).

Figure 14.24 Step 6: Fold over about 12 inches (300 mm) at each end of the cover to make clean, even ends for the completed roll.

Figure 14.25 Step 7: Start the roll by rolling and compressing one end into a tight compact roll, rolling toward the opposite end.

Figure 14.26 Step 9: The completed roll may be secured by inner-tube bands or Velcro® straps or tied with cords.

One-Firefighter Spread With A Rolled Salvage Cover

A salvage cover rolled for a one-firefighter spread may be carried on the shoulder or under the arm. Use the following steps when one firefighter spreads a rolled salvage cover:

Step 1: Start at one end of the object to be covered. While holding the roll in your hands, unroll a sufficient amount to cover the end (Figure 14.27).

Step 2: Lay the roll on the object, and continue to unroll toward the opposite end (Figure 14.28).

Step 3: Let the rest of the roll either fall into place at the other end, or arrange it in position as shown in Figure 14.29.

Step 4: Stand at one end and grasp the open edges where convenient, one edge in each hand (Figure 14.30).

Figure 14.27 Step 1: Start at one end of the object to be covered. While holding the roll in your hands, unroll a sufficient amount to cover the end.

Figure 14.28 Step 2: Lay the roll on the object, and continue to unroll toward the opposite end.

Figure 14.29 Step 3: Let the rest of the roll either fall into place at the other end, or arrange it in position.

Figure 14.30 Step 4: Stand at one end and grasp the open edges where convenient, one edge in each hand.

Step 5: Open the sides of the cover over the object by snapping both hands up and out (Figure 14.31 on the next page).

Step 6: Open the other end of the cover over the object in the same manner, and tuck in all loose edges at the bottom (Figure 14.32 on the next page).

Note that in addition to covering stacks of objects, the rolled cover may be used as a floor runner. To do this, just unroll the cover and only spread it out as wide as necessary.

Salvage Cover Fold For A One-Firefighter Spread

Some departments prefer to carry salvage covers folded, as opposed to rolled. The following procedure highlights a fold for one-firefighter deployment. Again, two firefighters are needed to make this fold, and they will be performing the same functions simultaneously.

Figure 14.31 Step 5: Open the sides of the cover over the object by snapping both hands up and out.

Figure 14.32 Step 6: Open the other end of the cover over the object in the same manner, and tuck in all loose edges at the bottom.

Steps 1-5: The first five steps are the same as those described for the one-firefighter roll.

Step 6: With the cover folded to reduce width, grasp the same end of the cover, and bring it to a point just short of the center (Figure 14.33).

Step 7: Use one hand as a pivot, and bring the folded end over and on top of the previous fold (Figure 14.34).

Step 8: Continue, and once again, bring the folded ends over and on top of the previous fold at the middle (Figure 14.35).

Step 9: Fold the other end of the cover toward the center, leaving about 4 inches (100 mm) between the two folds (Figure 14.36).

Step 10: The space between the folds now serves as a hinge as you place one fold on top of the other for the completed fold (Figure 14.37).

Figure 14.33 Step 6: With the cover folded to reduce width, grasp the same end of the cover, and bring it to a point just short of the center.

Figure 14.34 Step 7: Use one hand as a pivot, and bring the folded end over and on top of the previous fold.

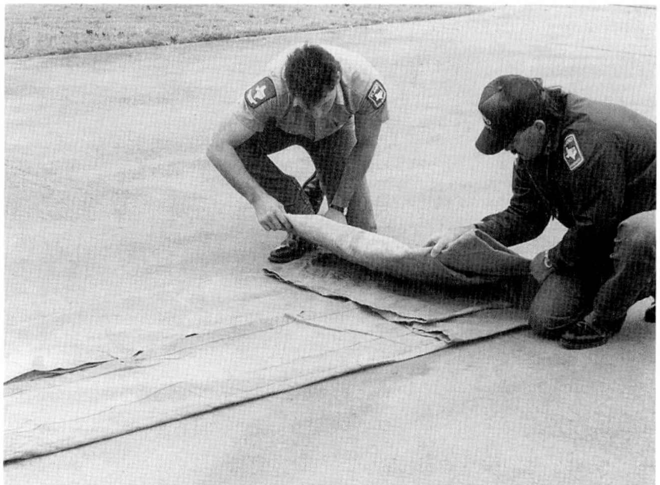

Figure 14.35 Step 8: Continue, and once again, bring the folded ends over and on top of the previous fold at the middle.

Figure 14.36 Step 9: Fold the other end of the cover toward the center, leaving about 4 inches (100 mm) between the two folds.

Figure 14.37 Step 10: The space between the folds now serves as a hinge as you place one fold on top of the other for the completed fold.

One-Firefighter Spread With A Folded Salvage Cover

A salvage cover folded for a one-firefighter spread may be carried in any manner. It is suggested, however, that this fold be carried on the shoulder for convenience. The following steps are to be used when one firefighter spreads a folded salvage cover:

Step 1: Lay the folded cover on top of and near the center of the object to be covered, and separate it at the first fold (Figure 14.38).

Step 2: Select either end, and continue to unfold the salvage cover by separating the next fold (Figure 14.39).

Step 3: Continue to unfold this same end toward the end of the object to be covered.

Step 4: Grasp the end of the cover near the center with both hands to prevent the corners from falling outward.

Step 5: Bring the end of the cover into position over the end of the object being covered (Figure 14.40).

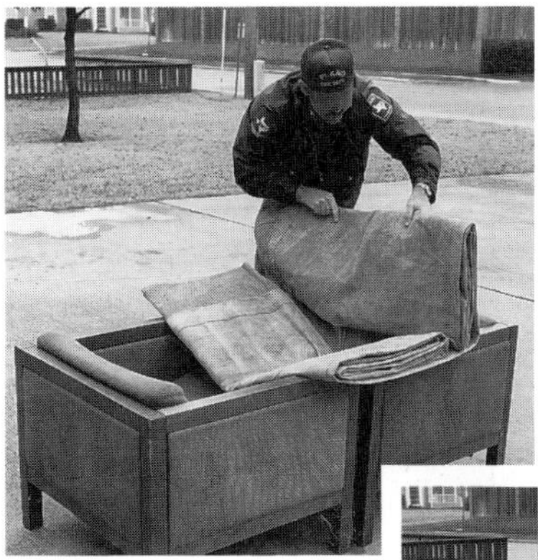

Figure 14.38 Step 1: Lay the folded cover on top of and near the center of the object to be covered, and separate it at the first fold.

Figure 14.39 Step 2: Select either end, and continue to unfold the salvage cover by separating the next fold.

Figure 14.40 Step 5: Bring the end of the cover into position over the end of the object being covered.

Step 6: Unfold the other end of the cover in the same manner over the object.

Step 7: Stand at one end and grasp the open edges where convenient, one edge in each hand.

Step 8: Open the sides of the cover over the object by snapping both hands up and out (Figure 14.41).

Figure 14.41 Step 8: Open the sides of the cover over the object by snapping both hands up and out.

Step 9: Open the other end of the cover over the object in the same manner, and tuck in all loose edges at the bottom (Figure 14.42).

Figure 14.42 Step 9: Open the other end of the cover over the object in the same manner, and tuck in all loose edges at the bottom.

Salvage Cover Fold For A Two-Firefighter Spread

Larger salvage covers cannot be easily handled by a single firefighter. Therefore, they must be folded for two-firefighter deployment. Both firefighters will be performing the same functions simultaneously. The following procedure can be used to make the two-firefighter fold:

Step 1: With the cover stretched lengthwise, grasp opposite ends of the cover at the center grommet and then pull the cover tightly between you. Raise this center fold high above the floor, and shake out the wrinkles to form the first half-fold (Figure 14.43).

Figure 14.43 Step 1: With the cover stretched lengthwise, grasp opposite ends of the cover at the center grommet and then pull the cover tightly between you. Raise this center fold high above the floor, and shake out the wrinkles to form the first half-fold.

Step 2: Spread the half-fold upon the floor, and smooth it flat to remove the wrinkles (Figure 14.44).

Step 3: Standing at each end of the half-fold and facing the cover, grasp the open-edge corners with the hand nearest to these corners. While in this position, place your corresponding foot at the center of the half-fold, thus making a pivot for the next fold (Figure 14.45).

Step 4: Stretch that part of the cover being folded tightly between the two of you. Make the quarter-fold by folding the open edges over the folded edge (Figure 14.46).

Figure 14.44 Step 2: Spread the half-fold upon the floor, and smooth it flat to remove the wrinkles.

Figure 14.45 Step 3: Standing at each end of the half-fold and facing the cover, grasp the open-edge corners with the hand nearest to these corners. While in this position, place your corresponding foot at the center of the half-fold, thus making a pivot for the next fold.

Figure 14.46 Step 4: Stretch that part of the cover being folded tightly between the two of you. Make the quarter-fold by folding the open edges over the folded edge.

Step 5: One firefighter should then stand on one end of the quarter-fold, while the other grasps the opposite end and shakes out all the wrinkles (Figure 14.47).

Step 6: The one holding the end of the cover carries this end to the opposite end, maintaining alignment of outside edges (Figure 14.48). Place the carried end on the opposite end, aligning all edges.

Step 7: Positioned at opposite ends, shake out all wrinkles and align all edges (Figure 14.49).

Step 8: Grasp the open ends, and use your inside foot as a pivot for the next fold (Figure 14.50).

Step 9: Bring these open ends over, and place them just short of the folded center fold (Figure 14.51).

Figure 14.49 Step 7: Positioned at opposite ends, shake out all wrinkles and align all edges.

Figure 14.47 Step 5: One firefighter should then stand on one end of the quarter-fold, while the other grasps the opposite end and shakes out all the wrinkles.

Figure 14.50 Step 8: Grasp the open ends, and use your inside foot as a pivot for the next fold.

Figure 14.48 Step 6: The one holding the end of the cover carries this end to the opposite end, maintaining alignment of outside edges.

Figure 14.51 Step 9: Bring these open ends over, and place them just short of the folded center fold.

Step 10: Continue this folding process by bringing the open ends over and just short of the folded end. During this fold your free hand may be used as a pivot to hold the cover straight (Figure 14.52).

Step 11: Complete the operation by making one more fold in the same manner. Bring the open ends over and to the folded end using your free hand as a pivot during the fold (Figure 14.53).

Figure 14.52 Step 10: Continue this folding process by bringing the open ends over and just short of the folded end.

Figure 14.53 Step 11: Complete the operation by making one more fold in the same manner. Bring the open ends over and to the folded end using your free hand as a pivot during the fold.

Carrying The Salvage Cover Folded For A Two-Firefighter Spread

Probably the most convenient way to carry this fold is on the shoulder with the open edges next to the neck. It makes little difference which end of the folded cover is placed in front of the carrier, because two open-end folds will be exposed. The cover should be carried so that the carrier can grab the lower pair of corners, and the second firefighter can grab the uppermost pair (Figure 14.54).

Figure 14.54 Proper method for carrying a cover for a two-firefighter spread.

Two-Firefighter Spread From A Folded Salvage Cover

The balloon throw is the most common method for two firefighters to deploy a salvage cover. The balloon throw gives better results when sufficient air is pocketed under the cover. This pocketed air gives the cover a parachute effect that floats it in place over the article to be covered. The following steps are for making the balloon throw:

Step 1: Stretch the cover along one side of the object to be covered, and separate the last half-fold by grasping each side of the cover near the ends (Figure 14.55).

Step 2: Make several accordion folds in your inside hand, and place your outside hand about mid-way down the end hem (Figure 14.56).

Step 3: Pull the cover tightly between you, and prepare to swing the folded part down, up, and out in one sweeping movement so as to pocket as much air as possible (Figure 14.57).

Step 4: When the cover is as high as you can reach, the accordion folds may either be pitched or carried across the object, an action which causes the cover to float over the object (Figure 14.58).

Step 5: As the cover is floated over the object, guide it into position and straighten for better water runoff (Figure 14.59).

Figure 14.55 Step 1: Stretch the cover along one side of the object to be covered, and separate the last half-fold by grasping each side of the cover near the ends.

Figure 14.56 Step 2: Make several accordion folds in your inside hand, and place your outside hand about midway down the end hem.

Figure 14.57 Step 3: Pull the cover tightly between you, and prepare to swing the folded part down, up, and out in one sweeping movement so as to pocket as much air as possible.

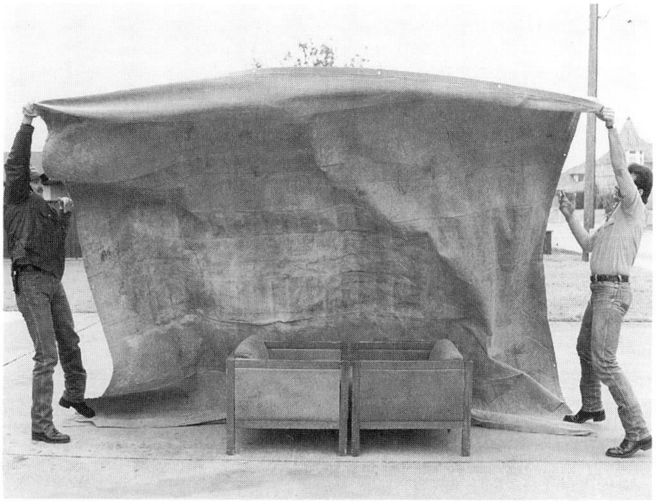

Figure 14.58 Step 4: When the cover is as high as you can reach, the accordion folds may either be pitched or carried across the object.

Figure 14.59 Step 5: As the cover is floated over the object, guide it into position and straighten for better water runoff.

IMPROVISING WITH SALVAGE COVERS

In addition to simply covering building contents, salvage covers may also be used to catch and route water from fire fighting operations. The following section details some of these special situations.

Removing Water With Chutes

Using a chute is one of the most practical methods of removing water that comes through the ceiling from upper floors. Water chutes may be constructed on the floor below fire fighting operations to drain runoff through windows or doors. Some fire departments carry prepared chutes, approximately 10 feet (3 m) long, as regular equipment, but it may be more practical to make chutes using one or more covers. Effective water chutes can be made with two pike poles and a salvage cover. This is accomplished by arranging the cover over the two pike poles that are extended between a window and a ladder or other support. The ladder or support should be elevated higher than the windowsill that will serve as the discharge point (Figure 14.60). Water chutes may also be constructed on stairways (Figure 14.61 on the next page). The sides of the cover should be rolled to prevent water from seeping off the edges of the cover.

Constructing A Catchall

A catchall is constructed from a salvage cover that has been placed on the floor to hold small amounts of water. The catchall may also be used as a temporary means to control large amounts of water until chutes can be constructed to route the water to the outside. Properly constructed catchalls will hold several hundred gallons

Figure 14.60 Ladders and tarps can be used to create water chutes that will channel excess water from the building.

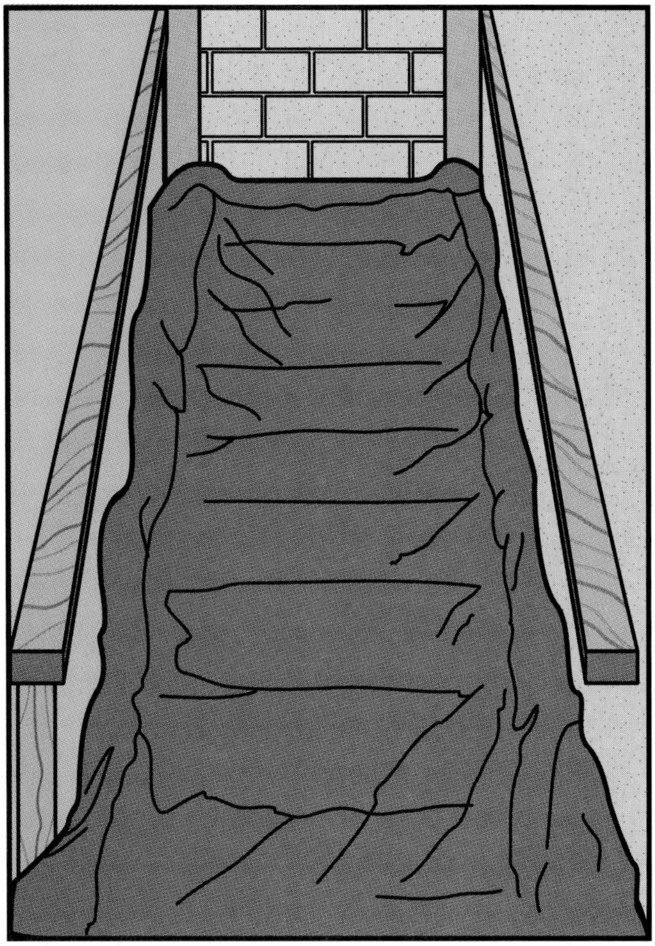

Figure 14.61 Tarps can be used to turn stairways into water chutes.

(liters) of water and often save considerable time during salvage operations. The cover should be placed into position as soon as possible, even before the sides of the cover are rolled. Two people are usually required to prepare a catchall to make more uniform rolls on all sides. The steps required to make a catchall are as follows:

Step 1: With the cover spread on the floor, roll the sides inward approximately 3 feet (1 m), and lay the ends of the side rolls over at a 90-degree angle to form the corners of the basin (Figure 14.62).

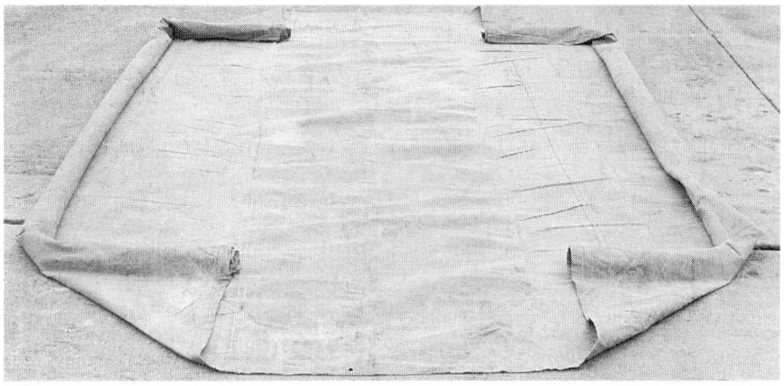

Figure 14.62 Step 1: With the cover spread on the floor, roll the sides inward approximately 3 feet (1 m), and lay the ends of the side rolls over at a 90-degree angle to form the corners of the basin.

Step 2: Roll one end into a tight roll on top of the side roll, and form a projected flap (Figure 14.63 a). Using your outside hand to lift the edge roll, tuck the end roll to lock the corners (Figure 14.63 b).

Step 3: Roll the other end in a like manner, and lock the corners. A completed catchall is shown in Figure 14.64.

Figure 14.63a Step 2: Roll one end into a tight roll on top of the side roll, and form a projected flap.

Figure 14.63b Using your outside hand to lift the edge roll, tuck the end roll to lock the corners.

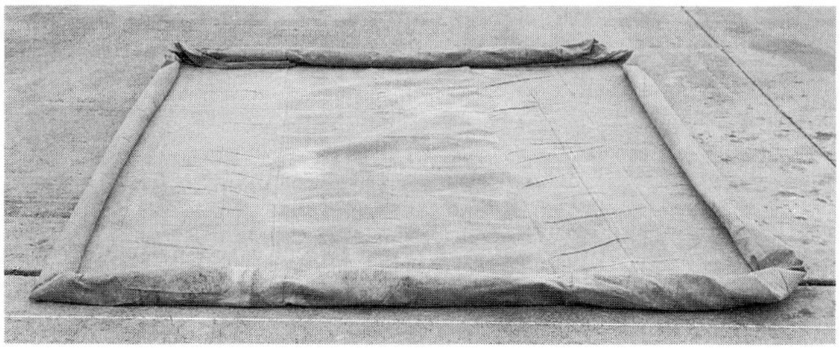

Figure 14.64 A completed catchall.

COVERING OPENINGS

One of the final parts of salvage operations is the covering of openings to prevent further damage to the property by weather. Cover any doors or windows that have been broken or removed with plywood, heavy plastic, or some similar materials to keep rain out (Figure 14.65). Cover openings in roofs with plywood, roofing paper, or tar paper. If roofing or tar paper is used, use appropriate roofing nails. Tack the edges down with laths between the nails and the paper.

Figure 14.65 Window openings may be covered with plastic to temporarily keep out the elements.

OVERHAUL

Overhaul is the practice of searching a fire scene to detect hidden fires or sparks which may rekindle and to note the possible point of origin and cause of fire (Figure 14.66). Afterwards, the building is to be left in as safe and habitable condition as possible. Salvage operations performed during fire fighting will directly affect any overhaul work that may be needed later. Many of the tools and equipment used for overhaul are the same as those used for other firefighter operations. Some of the tools and equipment used specifically for overhaul, along with their uses, may include:

- Pike poles, plaster hooks — Open ceilings to check on fire extension
- Axes — Open walls and floors
- Carryall, buckets, tubs — Carry debris or provide a basin for immersing smoldering material
- Shovels, hooks, forks — Move baled or loose materials

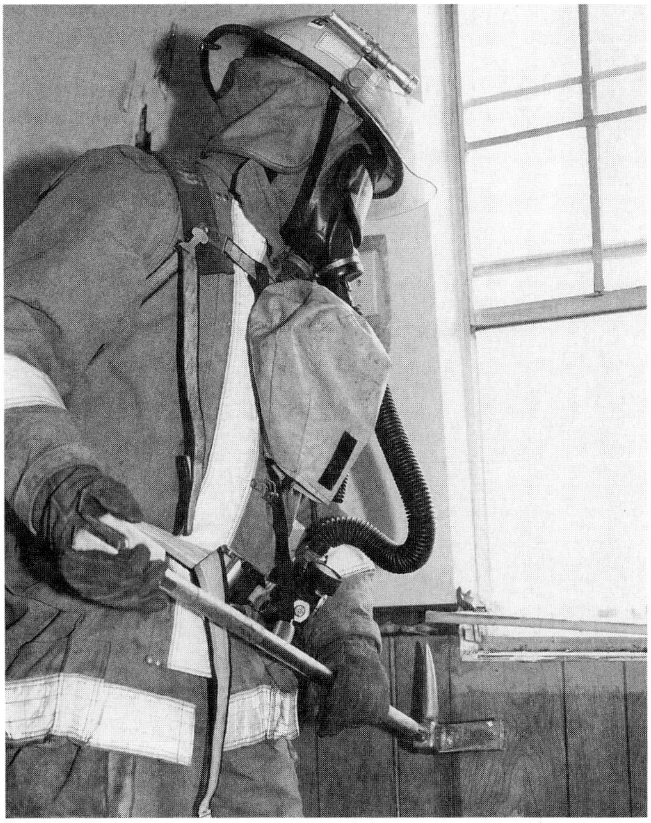

Figure 14.66 Proper overhaul reduces the chance of a rekindle fire.

Searching For Hidden Fires

Before starting a search for hidden fires, it is important to determine the condition of the building in the area to be searched. The intensity of the fire and the amount of water used for its control are two important factors that affect the condition of the building. The intensity of the fire determines the extent to which the structural members have been weakened, and the amount of water used determines the additional weight placed on floors and walls due to the absorbent qualities of the building contents. Consideration should be given to these factors for the protection of personnel during overhaul.

The firefighters should also be aware of other dangerous building conditions, such as the following:

- Weakened floors due to floor joists being burned away
- Concrete that has spalled due to the heat
- Weakened steel roof members (tensile strength is affected at about 500°F [260°C])
- Walls offset due to elongation of steel roof supports
- Weakened roof trusses due to burn-through of key members

- Mortar in wall joints opened due to excessive heat
- Wall ties holding veneer walls melted from heat

The firefighter can often detect hidden fires by sight, touch, sound, or electronic sensors (Figure 14.67). The following are some of the indicators for each:

- Sight
  - — Discoloration of materials
  - — Peeling paint
  - — Smoke emissions from cracks
  - — Cracked plaster
  - — Dried wallpaper
- Touch
  - — Feel walls and floors with the back of the hand (Figure 14.68)
- Sound
  - — Popping or cracking of burning
  - — Hissing of steam
- Electronic Sensors
  - — Detect heat variances in objects that may be in hidden locations through thermal imaging

Typically, overhaul begins in the area of actual fire involvement. The process of looking for extension should begin as soon as possible after the fire has been knocked down and the person responsible for determining the cause and origin of the fire says it is all right. The overhaul process can then be systematically moved away from the areas of heaviest involvement until firefighters are certain that no hidden fires remain. If it is found that the fire did extend to other areas, the firefighters must determine through what medium it traveled. When floor beams have burned at their ends where they enter a party wall, fire personnel overhaul the ends by flushing the voids in the wall with water. The far side of the wall should also be checked to see whether the fire or water has come through. Insulation materials should be thoroughly checked, because they can harbor hidden fires for a prolonged period. Usually, it is necessary to remove the material in order to properly check it or extinguish fire in it.

Figure 14.67 A firefighter checking for hidden fire with an infrared heat detector.

Figure 14.68 A firefighter using the back of his hand to feel for heat.

When the fire has burned around windows or doors, there is a possibility that there is fire remaining within the frames or casings. These areas should be opened to ensure complete extinguishment. These areas can be opened by simply pulling off the molding to expose the inner parts of the frame or casing (Figure 14.69). When

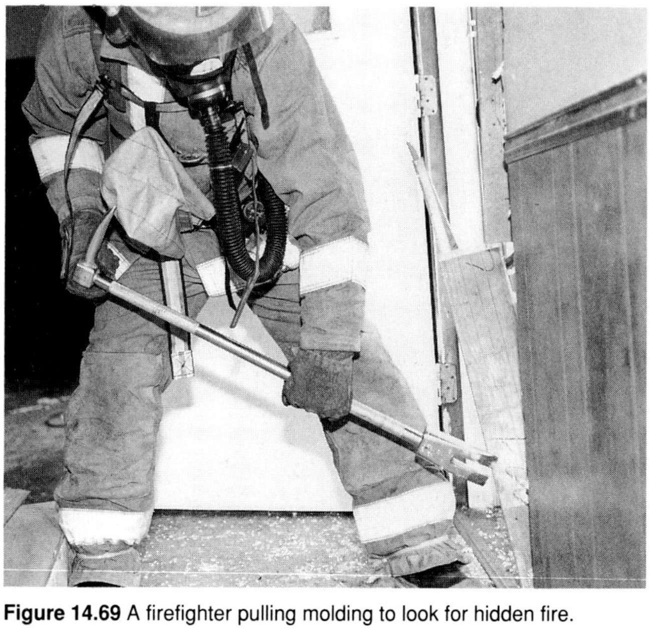

Figure 14.69 A firefighter pulling molding to look for hidden fire.

fire has burned around a combustible roof or cornice, it is advisable to open the cornice and inspect for hidden fires.

When concealed spaces below floors, above ceilings, or within walls and partitions must be opened during the search for hidden fires, the furnishings of the room should be moved to locations where they will not be damaged. Only enough wall, ceiling, or floor covering should be removed to ensure complete extinguishment. Weight-bearing members should not be disturbed.

The method of opening ceilings from below involves the use of either a pike pole or plaster hook. To open a plaster ceiling, the firefighter must first break the plaster and then pull off the lath. A pike pole is often used for this operation (Figure 14.70). Metal or composition ceilings may be pulled from the joist in a like manner. When pulling, the firefighter should not stand under the space to be opened. The firefighter should always be positioned between the area being pulled and the doorway to keep the exit route from being blocked with falling debris (Figure 14.71). The pull should be down and away to

Figure 14.70 A pike pole.

Figure 14.71 Pull ceilings to check for fire extension.

prevent the ceiling from dropping on the head of the firefighter. No firefighter should attempt to pull a ceiling without wearing full protective clothing, including eye and respiratory protection.

The plaster hook, made of tooled steel with a spear head and two sharp-edge blades, is useful in removing metal ceilings, plaster, or other obstructions to gain access to the fire (Figure 14.72). The blades fold downward when the spear is forced through an object and automatically open after penetration. This feature causes ceilings and wall materials to be pulled away when force is applied to the handle.

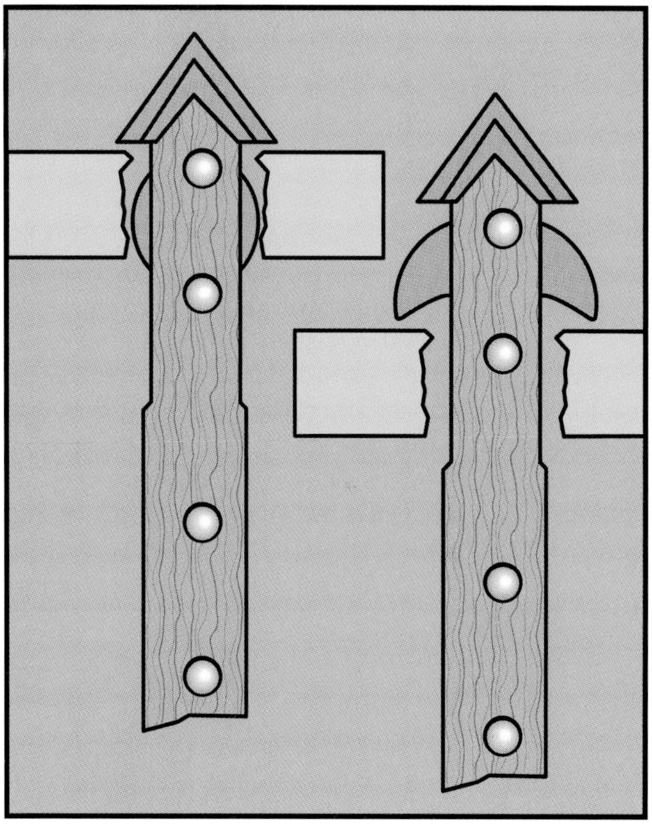

Figure 14.72 The method of operation for a plaster hook.

Extinguishing Hidden Fires

It is essential for firefighters to wear proper protective clothing including positive-pressure self-contained breathing apparatus while performing overhaul and extinguishing hidden fires (Figure 14.73 on the next page). Eye protection should be worn if it is safe to remove breathing apparatus.

Charged hoselines should always be available for the extinguishment of hidden fires; however, the same caliber of lines as were used to bring the fire under control are not always necessary. Fire department pumpers can often be disconnected from the hydrant; however, local policies may dictate that at least one supply

Figure 14.73 Full protective equipment, including SCBA, must be worn during overhaul operations.

line be left in place as a precaution. Typically, booster lines or 1½-inch (38 mm) attack lines are used for overhaul. During minor overhaul operations, air-presssurized water extinguishers or garden hoses may be used to extinguish small hidden fires. However, at least one attack line should still be available in the event it is needed. Regardless of the type of hose being used, the nozzle should be placed so that if it is accidentally opened, it will not cause additional water damage (Figure 14.74).

Quite frequently, small burning objects are uncovered during overhaul. Because of their size and condition, it is better to submerge the entire object in a container of water than to try drenching it with a stream of water. Bathtubs, sinks, lavatories, and wash tubs are all useful for this purpose. Larger furnishings, such as mattresses, stuffed furniture, and bed linens, should be removed to the outside where they can be easily and thoroughly extinguished. Firefighters should remember that all scorched or partially burned articles may prove helpful to an investigator in preparing an inventory or determining the cause of the fire.

The use of wetting agents is of considerable value when extinguishing hidden fires. The penetrating qualities of wetting agents usually permit complete extinguishment in cotton, upholstery, baled goods, and a large number of other materials. Special care should be taken to eliminate indiscriminate use and direction of hose streams.

Protecting And Preserving Evidence

Firefighters must keep two things clearly in mind regarding the protection and preservation of material evidence:

- Guard the evidence where it is found, untouched and undisturbed, to preserve the chain of custody (Figure 14.75).
- Properly identify, remove, and safeguard evidence that cannot be left at the scene of the fire (Figure 14.76).

Figure 14.74 To reduce the risk of excess water damage, leave the nozzle hanging out of a window until it is needed.

Figure 14.75 Do not touch or disturb the evidence found during overhaul operations. *Courtesy of Joseph J. Marino.*

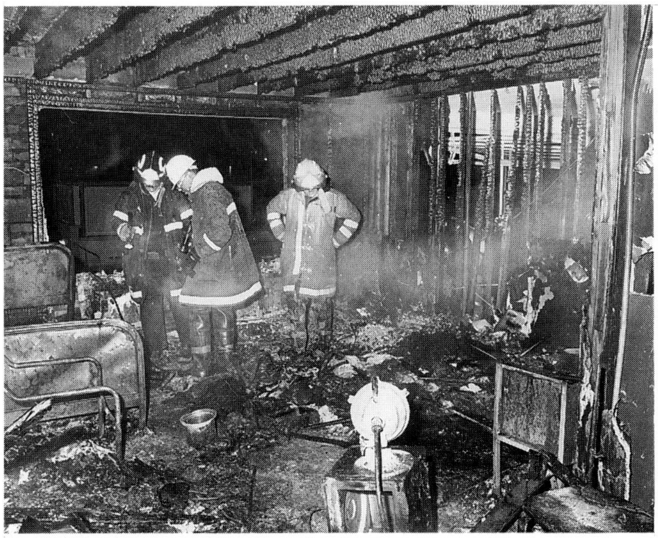

Figure 14.76 Properly identify, remove, and safeguard evidence that cannot be left at the scene of the fire. *Courtesy of Joseph J. Marino.*

No changes of any kind should be permitted in the evidence other than those absolutely necessary in the extinguishment of the fire. Photographs are excellent supporting evidence, if they are taken immediately. Firefighters should avoid trampling over possible evidence and obliterating it. The same precaution applied to the excessive use of water may help you avoid similar unsatisfactory results here. Human footprints must be protected to permit measurements of the prints, comparison of the prints, length of stride, position of feet, and any peculiarities in the gait (walk or run) of a suspect. Boxes placed over prints will prevent dust from blowing over otherwise clear prints and keep them in good condition for either photographs or plaster casts at a later time (Figure 14.77). Completely or partially burned papers found in a furnace, stove, or fireplace should be protected by immediately closing dampers and other openings.

All evidence collected by firefighters should be properly marked, tagged, identified, and preserved in clean containers. Careful notation should be made of the date, time, and place the evidence is found. Additional identifying marks on cans, bottles, and other articles, such as the initials of the person who collected the evidence, may also be noted. This procedure may establish unquestionable identity of the evidence. There must be a record of witnesses and of each person who has had or will have responsibility for care and preservation of the evidence.

When facilities for good protection are not available, partially burned paper and ash may be protected between layers of plastic or between pieces of window glass for the investigator and later transportation to a laboratory. Letters, documents, or bills demanding payment of money should be preserved to assist in establishing a person's financial condition that might indicate a motive for arson.

Wood suspected of containing paraffin or oil should be placed in a clear container and sealed until a chemical analysis can be made (Figure 14.78). All bottles should be labeled with stickers and identified. Objects, such as candlewicks and burned matches, should be placed in a bottle containing cotton to prevent breaking the evidence by jarring and handling. Samples of materials, such as cotton, wood, rayon, felt, and other fabrics, can be stored in clean, large-mouthed bottles. These bottles should be tightly sealed and properly marked. Volatile liquids, oil samples, oil-soaked rags, waste, or similar materials, should be put in tin cans and sealed. This procedure preserves odors as well as the materials themselves and allows them to be presented in court in their original condition. Any evidence of this sort should

Figure 14.77 Footprints may be covered with a box until they can be photographed or plaster casts can be made.

Figure 14.78 Samples can be kept in glass jars.

be sealed with wax so that when necessary, a laboratory technician may testify in court that the wax seal was unbroken until the proper time.

The firefighter who detects signs of possible arson and finds the evidence should be able to identify it later. When such material has been tagged, labeled, and properly marked, it is ready to be turned over to the proper authorities. Evidence should be kept under lock and key and as few people as possible should be permitted to handle it. A record of each person who has handled this evidence should always be kept.

After evidence has been properly preserved, debris may be cleaned up. Charred materials should be removed to prevent the possibility of rekindle and to help reduce the loss from smoke damage. Any unburned materials should be separated from the debris and cleaned. Debris may be shoveled into large containers, such as buckets or tubs, to reduce the number of trips back and forth to the fire area. It causes poor public relations to dump debris onto streets and sidewalks or to damage costly shrubbery. Rather, dump the debris in a backyard or alley that is not as visible (Figure 14.79). For more information, refer to IFSTA's **Fire Cause Determination** manual.

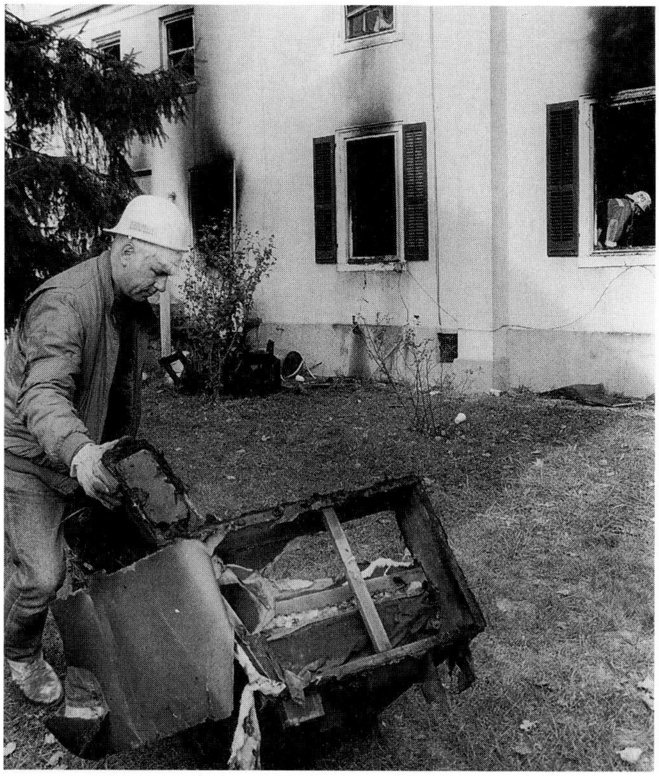

Figure 14.79 Debris may be piled outside after overhaul. Investigators may wish to examine it from this point. *Courtesy of Bob Esposito.*

15

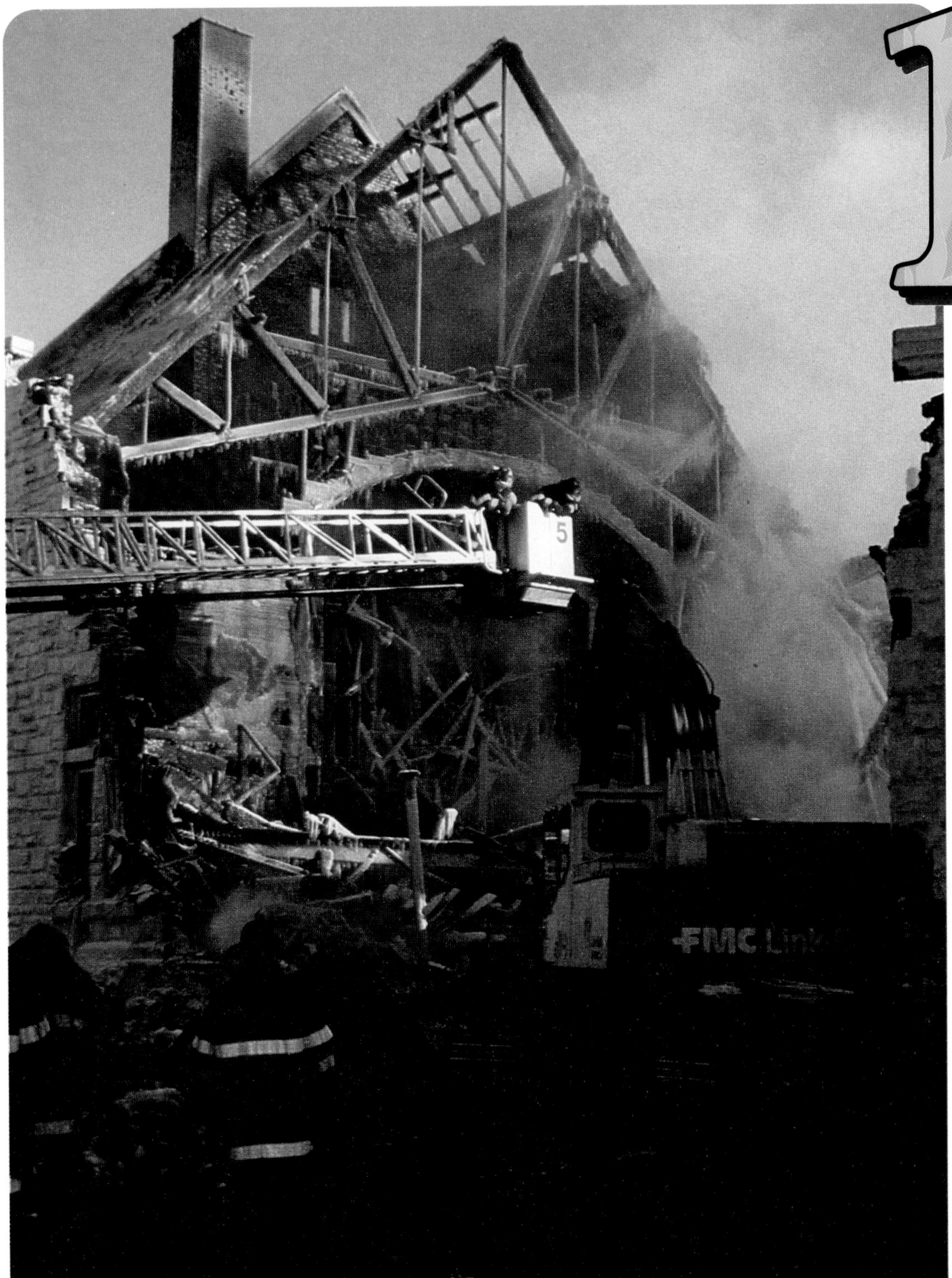

Building Construction

This chapter provides information that addresses performance objectives described in NFPA 1001, *Standard for Fire Fighter Professional Qualifications* (1992), particularly those referenced in the following sections:

4-23 Building Construction

4-23.1

4-23.2

4-23.3

4-23.4

4-23.5

4-23.6

4-23.7

Chapter 15
Building Construction

From a safety standpoint, all firefighters should have a basic knowledge of the principles of building construction. Knowledge of the various types of building construction and how fires react in each type gives the firefighter or fire officer an edge in planning for a safe and effective fire attack. History has proven that failure to recognize the potential dangers presented by a particular type of construction and the effects a fire has on it will lead to deadly results.

New technologies and designs are being used for building construction every day. Because of this expansive wave of change, as well as varied existing building types, it is impossible to highlight every conceivable situation in a chapter of this size. The purpose of this chapter is to introduce the firefighter to some of the most basic and common types of building construction and their characteristics to fire protection. The firefighter will also be introduced to common building construction terms and components. This chapter also covers some of the indicators that signify danger during fire fighting operations.

TYPES OF BUILDING CONSTRUCTION AND THE BASIC HAZARDS

Each building code classifies building construction in different terms. In general, construction classifications are based upon materials used in construction and upon hourly fire-resistance ratings of structural components. Most building codes have the same five construction classifications, but use different terms to name the classifications. For the purposes of this manual, the terms will be based on NFPA 220, *Standard on Types of Building Construction*. NFPA 220 uses Roman numerals to signify the five major classifications (Type I through V). Each classification is further broken down into subtypes using a three-digit Arabic number code or several letters (e.g. Type I-443):

- The first digit refers to the fire resistance rating (in hours) of the exterior bearing walls.
- The second digit refers to the fire resistance rating (in hours) of structural frames or columns and girders that support loads of more than one floor.
- The third digit indicates the fire resistance rating (in hours) of the floor construction.

A listing of construction types and the degree of fire resistance of each is shown in Table 15.1 on the next page. Construction classifications are further explained in the following sections.

Type I Construction

Type I construction is also known in some codes as fire-resistive construction. Type I construction has structural members, including walls, columns, beams, floors, and roofs, that are made of noncombustible or limited combustible materials (Figure 15.1). These buildings were originally intended to confine any fire and its resulting products of combustion to a given location. Because of the limited combustibility of the materials of construction, the primary fire hazards are the contents of the structure. The ability of Type I construction to confine the fire to a certain area can also be compromised by openings made in partitions and by improperly designed and dampered central heating and air-conditioning systems.

Figure 15.1 Type I construction. *Courtesy of Southwest Loss Control, Inc.*

TABLE 15.1
FIRE RESISTANCE REQUIREMENTS FOR TYPE I THROUGH V CONSTRUCTION

	Type 1		*Type II*			*Type III*		*Type IV*	*Type V*	
	443	*332*	*222*	*111*	*000*	*211*	*200*	*2HH*	*111*	*000*
EXTERIOR BEARING WALLS —										
Supporting more than one floor, columns or other bearing walls	4	3	2	1	0[1]	2	2	2	1	0[1]
Supporting one floor only	4	3	2	1	0[1]	2	2	2	1	0[1]
Supporting a roof only	4	3	1	1	0[1]	2	2	2	1	0[1]
INTERIOR BEARING WALLS —										
Supporting more than one floor, columns or other bearing walls	4	3	2	1	0	1	0	2	1	0
Supporting one floor only	3	2	2	1	0	1	0	1	1	0
Supporting a roof only	3	2	1	1	0	1	0	1	1	0
COLUMNS —										
Supporting more than one floor, bearing walls or other columns	4	3	2	1	0	1	0	H[2]	1	0
Supporting one floor only	3	2	2	1	0	1	0	H[2]	1	0
Supporting a roof only	3	2	1	1	0	1	0	H[2]	1	0
BEAMS, GIRDERS, TRUSSES & ARCHES —										
Supporting more than one floor, bearing walls or other columns	4	3	2	1	0	1	0	H[2]	1	0
Supporting one floor only	3	2	2	1	0	1	0	H[2]	1	0
Supporting a roof only	3	2	1	1	0	1	0	H[2]	1	0
FLOOR CONSTRUCTION	3	2	2	1	0	1	0	H[2]	1	0
ROOF CONSTRUCTION	2	1½	1	1	0	1	0	H[2]	1	0
EXTERIOR NONBEARING WALLS	0[1]	0[1]	0[1]	0[1]	0[1]	0[1]	0[1]	0[1]	0[1]	0[1]

Those members listed that are permitted to be of approved combustible material.

[1] Requirements for fire resistance of exterior walls, the provision of spandrel wall sections, and the limitation of protection of wall openings are not related to construction type. They need to be specified in other standards and codes, where appropriate, and may be required in addition to the requirements of this Standard for the construction type.

[2] "H" indicates heavy timber members; see text for requirements.

Type II Construction

4-93 Type II construction, also known as noncombustible or limited combustible construction, is similar to Type I except that the degree of fire resistance is lower. In some cases, materials with no fire-resistance ratings, such as untreated wood, may be used (Figure 15.2). Again, one of the primary fire protection concerns is the contents of the building. The heat buildup from a fire in the building can cause structural supports to fail. Another potential problem is the type of roof on the

Figure 15.2 Type II construction. *Courtesy of Southwest Loss Control, Inc.*

building. Type II construction buildings often have flat, built-up roofs. These roofs contain combustible felt, insulation, and roofing tar (Figure 15.3). Fire extension to the roof can eventually cause the entire roof to become involved and fail.

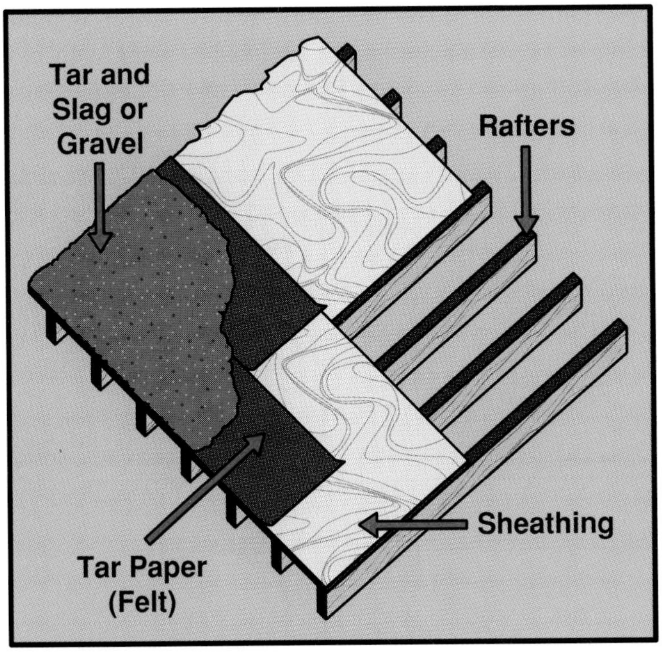

Figure 15.3 The components of a built-up roof.

Type III Construction

Type III construction, commonly referred to as ordinary construction, features exterior walls and structural members that are noncombustible or limited combustible materials. Interior structural members, including walls, columns, beams, floors, and roofs, are completely or partially constructed of wood (Figure 15.4). The wood used in these members is of smaller dimensions than required for Type IV, heavy timber construction. The primary fire concern specific to Type III construction is the problem of fire and smoke spread through concealed

Figure 15.4 Type III construction. *Courtesy of Southwest Loss Control, Inc.*

spaces. These spaces are between the walls, floors, and ceiling. Heat from a fire may be conducted to these concealed spaces through finish materials, such as dry wall, sheet rock®, or plaster, or the heat can enter the concealed spaces through holes in the finish materials. From there, the heat, smoke, and gases may be communicated to other parts of the structure. If enough heat is present, the fire may actually burn within the concealed spaces and feed on the combustible construction materials in the space. These hazards can be reduced considerably by placing fire stops inside these spaces to limit the spread of the combustion by-products.

Type IV Construction

Type IV construction is heavy timber construction. Heavy timber construction features exterior and interior walls and their associated structural members that are of noncombustible or limited combustible materials. Other interior structural members, including beams, columns, arches, floors, and roofs, are made of solid or laminated wood with no concealed spaces (Figure 15.5 on the next page). This wood must be of large enough dimensions to be considered heavy timber. These dimensions will vary depending on the particular code being used.

Heavy timber construction was used extensively in old factories, mills, and warehouses. It is rarely used today in new construction, other than occasionally in churches. The primary fire hazard associated with heavy timber construction is the massive amount of combustible contents (fire loading) presented by the structural timbers in addition to the contents of the building. Though the heavy timbers remain stable for a long period under fire conditions, they give off tremendous amounts of heat and pose serious exposure protection problems for firefighters.

Figure 15.5 Type IV construction. *Courtesy of Southwest Loss Control, Inc.*

Type V Construction

Type V construction, also called wood-frame construction, has exterior walls, bearing walls, floors, roofs, and supports made completely or partially of wood or other approved materials of smaller dimensions than those used for Type IV construction. Type V construction buildings present almost unlimited potential for fire extension within the building of origin and to nearby structures, and particularly if the nearby structures are also Type V construction (Figure 15.6). Firefighters must be alert for fire coming from doors or windows and extending to the exterior of the structure.

Figure 15.6 Type V construction. *Courtesy of Southwest Loss Control, Inc.*

THE EFFECTS OF FIRE ON COMMON BUILDING MATERIALS

All materials react differently when exposed to extreme heat or fire. Knowledge of how these materials react gives fire suppression personnel an idea of what to expect during fire fighting operations at a particular occupancy. This section reviews the common materials found in building construction and how they react to fire involvement.

Wood

Wood is used in various structural support systems. It may be used in load-bearing walls (those which support structural weight) or nonload-bearing walls (those that do not support structural weight). Most exterior walls are load bearing walls. An example of a load-bearing wall is a party wall. A party wall is a load-bearing wall that supports two adjacent structures (Figure 15.7). Some interior walls may also be load bearing, although this is often difficult to tell by just looking at them. This information should be obtained during pre-incident planning trips. An example of a nonload-bearing wall is a partition wall that simply divides two areas within a structure (Figure 15.8).

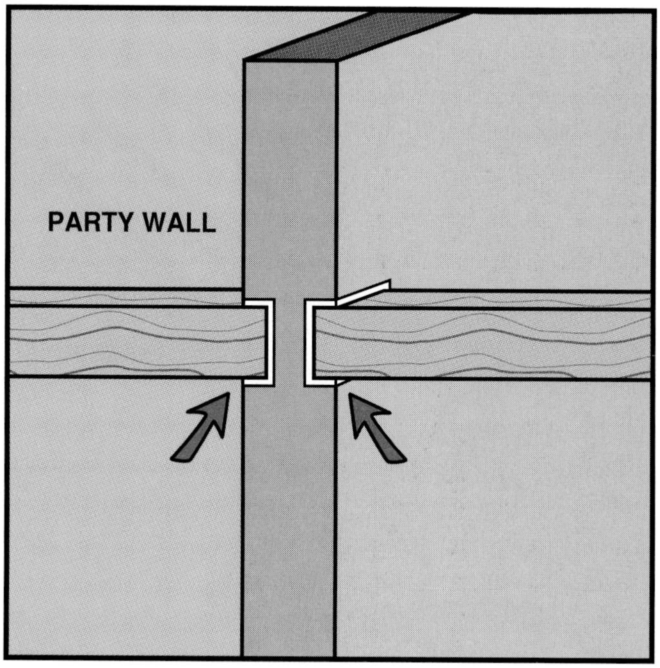

Figure 15.7 A diagram of a party wall.

The reaction of wood to fire conditions depends mainly on two factors: the size of the wood and the moisture content of the wood. The smaller the wood size, the more likely it is to lose structural integrity. Large pieces of wood, such as those used in heavy timber construction (Type IV), retain much of their original

Figure 15.8 Walls that separate offices or rooms are commonly nonload-bearing walls.

structural integrity, even after extensive fire exposure. Wood chars at a depth rate of about 1/40-inch (0.6 mm) per minute. Glued laminated structural members can be expected to perform in a similar manner (Figure 15.9). Smaller pieces of wood should be protected by dry wall or gypsum to increase their resistance to heat or fire.

Figure 15.9 Glue-lam beams are used in many types of occupancies.

The moisture content of the wood affects the rate at which it burns. Wood with a high moisture content (sometimes referred to as "green" wood) will not burn as fast as wood that has been cured or dried. In some cases, fire retardants may be added to wood to reduce the speed at which it will ignite or burn. Fire retardants are not always totally effective in reducing fire spread.

The application of water during fire fighting operations will not have a substantial effect on the structural integrity of the wood. In fact, the quicker water is applied to the wood, the less charring that occurs; this minimizes the reduction in the wood's strength. Firefighters should check wood studs and structural members to ascertain their structural integrity.

Masonry

Masonry includes bricks, blocks, stones, and unreinforced and reinforced concrete products (Figure 15.10). Masonry is used in a variety of wall types and is commonly used for fire walls. Fire walls separate two

Figure 15.10 The four main types of masonry construction are clockwise from the left: brick, block, stone, and poured concrete.

connected structures and prevent the spread of fire from one structure to the next. Fire walls can also be used to divide large structures into smaller portions and contain a fire to that particular portion of the structure. Fire walls are also used for cantilever walls. Cantilever walls are freestanding, and they are commonly found on large churches and shopping centers (Figure 15.11 on the next page). Block walls may be load-bearing walls; however, most brick and stone walls are used for veneer walls. Veneer walls are decorative and are usually attached to the outside of some type of load-bearing frame structure.

Figure 15.11 A cantilever wall.

Masonry is minimally affected by fire and exposure to high temperatures. Bricks rarely show any signs of loss of integrity or serious deterioration. Stones may spall or lose small portions of their surface when heated. Blocks may crack, but they usually retain most of their strength and basic structural stability. The mortar between the bricks, blocks, and stone may be subject to more deterioration and should be checked for signs of weakening (Figure 15.12).

Rapid cooling of masonry products, such as occurs when water is used to extinguish fire, may cause bricks, blocks, or stone to spall and crack. This is a common problem when water is used to extinguish chimney flue fires; the water causes the flue liner or fire bricks to crack. The masonry products should be inspected for signs of this damage after extinguishment has been completed.

Figure 15.12 Mortar is used between bricks, blocks, or stone.

Cast Iron

Cast iron is rarely used in modern construction. It will typically be found only in older buildings. It was commonly used as an exterior surface covering (veneer wall). These large sections were fastened to the masonry on the front of the building. The cast iron stands up well to fire and intense heat situations. The primary concern from a fire fighting standpoint is that the bolts or other connections that hold the cast iron to the building can fail, causing these large heavy sections of metal to come crashing down.

Steel

Steel is the primary material used for structural support in modern building construction (Figure 15.13). Steel structural members elongate when heated. A 50-foot (15 m) beam may elongate by as much as 4 inches (100 mm) when heated from room temperature to about 1,000°F (540°C). If the steel is restrained from movement at the ends, it will buckle and fail somewhere in the middle. For all purposes, the failure of steel structural members can be anticipated at temperatures near to or above 1,000°F (540°C). The temperature at which a specific steel member will fail depends on many variables, including the size of the member, the load it is under, the composition of the steel, and geometry of the member. For example, a lightweight, open-web truss will fail much quicker than a large I-beam.

Figure 15.13 A typical steel superstructure.

From a fire fighting perspective, firefighters must be cautious of the type of steel members that are used in a particular structure. Firefighters also need to determine how long the steel members have been exposed to heat; this gives an indication of when the members might fail. Another possibility for firefighters to consider is that elongating steel can actually push out load-bearing walls and cause a collapse (Figure 15.14). In this context, water does not have adverse effects on steel; if enough water is applied, it cools steel and reduces the chance of a collapse.

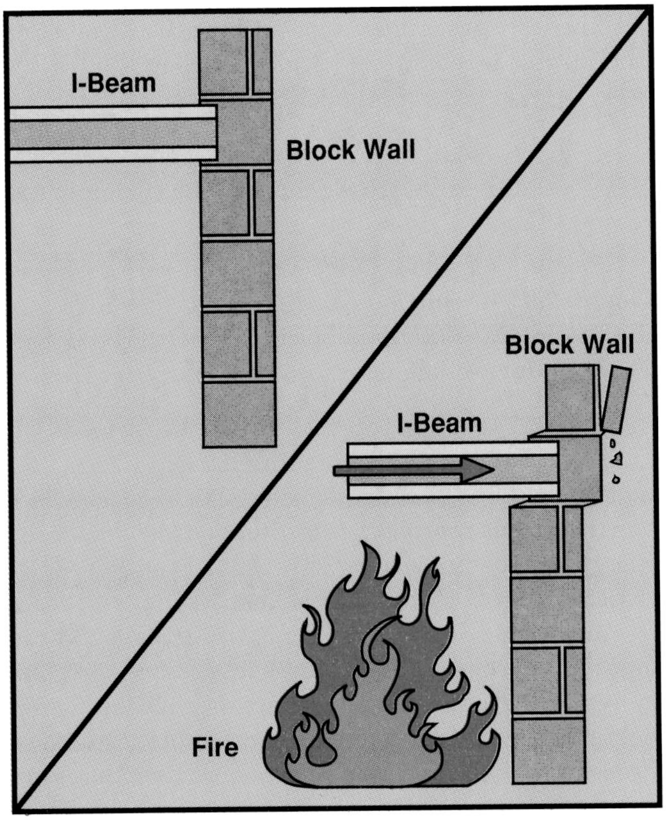

Figure 15.14 As beams expand they can push the wall outward, forcing a collapse.

Reinforced Concrete

Reinforced concrete is concrete that is internally fortified with steel reinforcement bars or mesh (Figure 15.15). This gives the material the compressive strength of concrete, along with the tensile strength of steel. Concrete does not perform particularly well under fire conditions; it loses strength and spalls. Heating may cause a failure of the bond between the concrete and the steel reinforcement. Firefighters should be cognizant of cracks and spalls in concrete surfaces. This is an indication that damage has occurred and that strength may be reduced.

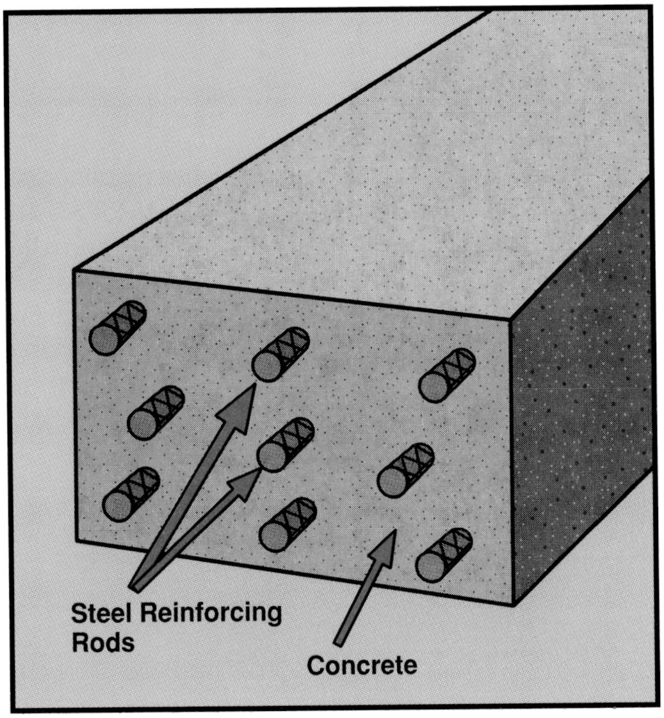

Figure 15.15 Reinforcing rods in concrete make the concrete stronger.

Gypsum

Gypsum is an inorganic product from which plaster and plaster boards are constructed (Figure 15.16). Gypsum has a high intrinsic water content, and the evaporation of this water takes a great deal of heat. The water content gives gypsum excellent heat-resistant, fire-retardant properties. Gypsum is commonly used to provide insulation to steel and wood structural members that are less adapted to high heat situations. Gypsum will gradually break down under fire conditions. In areas where the gypsum has failed, the structural members behind it will be subjected to higher temperatures and could fail as a result.

Figure 15.16 Gypsum board is used to cover interior walls.

Glass

Glass is not used for structural support; it is used in sheet form for doors and windows (Figure 15.17 on the next page). Fiberglass is also used for insulation purposes (Figure 15.18 on the next page). Though glass itself will not significantly add to the fire load, the resins used to bind fiberglass are combustible and can be tough to extinguish. Wire reinforced glass may provide some thermal protection as a separation, but for the most part glass is not an effective barrier to fire extension. Heated glass may crack when it is struck by a cold fire stream.

Figure 15.17 Some structures have large quantities of glass.

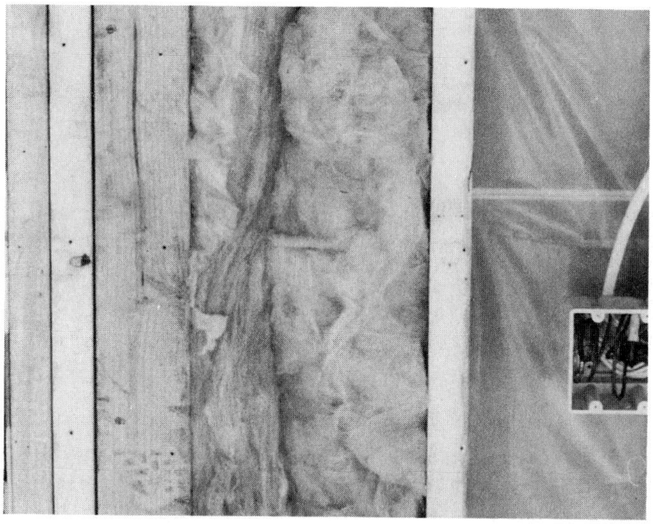
Figure 15.18 Fiberglass insulation.

FIREFIGHTER HAZARDS RELATED TO BUILDING CONSTRUCTION

The primary objective of understanding building construction and materials principles is being able to apply that information to the fireground. Firefighters should use their knowledge of these principles to monitor building conditions for signs of structural instability. Any problems that are noted should be reported to incident command personnel as quickly as possible. Even though a specific safety officer may be designated at the scene, it is the obligation of all personnel to constantly monitor for unsafe conditions. The following sections highlight some of the more critical issues related to building construction that affect firefighter safety.

Dangerous Building Conditions

Firefighters must be aware of the dangerous conditions created by the fire, as well as dangerous conditions that may be created by the firefighters trying to extinguish the fire. A potentially serious situation can be compounded if firefighters fail to recognize the seriousness of the situation and then take action that makes the situation worse.

There are two primary types of dangerous conditions that may be posed by a particular building:

- Conditions that contribute to the spread and intensity of the fire
- Conditions that make the building susceptible to collapse

These two conditions are obviously related. Conditions that contribute to the spread and intensity of the fire increase the likelihood of structural collapse. The following sections describe some of these conditions.

HEAVY CONTENT FIRE LOADING

Heavy content fire loading is the storing of combustible materials in high piles that are placed close together. (This provides a bountiful fuel load for a fire.) Heavy content fire loading is perhaps one of the most critical hazards in commercial and storage facilities (Figure 15.19). It is a critical hazard because the fire quickly overrides the capabilities of the fire sprinkler system (if present) and provides access problems for fire fighting personnel during manual fire fighting operations. Proper inspection and code enforcement procedures are the most effective defense against these hazards.

Figure 15.19 Commercial occupancies, such as this warehouse market, may have heavy content fire loading.

COMBUSTIBLE FURNISHINGS AND FINISHES

Combustible furnishings and finishes contribute to fire spread and smoke production (Figure 15.20). These two elements have been identified as major factors in the loss of many lives in fires such as the Coconut Grove fire in Boston (1942) and the Beverly Hills Supper Club fire in Southgate, KY (1977). Proper inspection and code enforcement procedures are the most effective defense against these hazards.

Figure 15.20 Theaters have large open areas and a considerable amount of upholstered furniture.

WOOD SHAKE SHINGLES

History has proven time and time again that wood shake shingles, even when treated with fire retardant, are not effective in stopping the spread of fire. This is a particular problem in wildland/urban interface situations, and the wood shake shingle has contributed regularly to conflagrations (Figure 15.21). Firefighters must use aggressive exposure protection tactics when faced with wood shake shingles.

Figure 15.21 Wood shake shingles are major contributors to fire spread.

WOODEN FLOORS AND CEILINGS

Combustible structural components, such as wood framing, floors, and ceilings, will also contribute to the fire loading in the building. Prolonged exposure to fire may weaken them and increase the chances of collapse.

LARGE, OPEN SPACES

The large, open spaces in buildings contribute to the spread of fire throughout the facility. Such spaces may be found in warehouses, churches, large atriums, common attics or cocklofts, and theaters (Figure 15.22). In these facilities, proper vertical ventilation is essential for slowing the spread of fire.

Figure 15.22 Buildings with large open spaces burn fast, and in many cases, they collapse quickly.

BUILDING COLLAPSE

Many firefighters have been seriously injured or killed by a structural collapse at fire fighting operations. The collapse results from damage caused by the fire or fire fighting operations to the structural system of the building. Some buildings, based on their construction and age, are more inclined to collapse than others. Knowledge of, and the ability to recognize, the type of construction is important to fire fighting operations. For example, buildings featuring lightweight or truss construction will succumb to the effects of fire much quicker than will a heavy timber building. The age of the building is also a determining factor. Older buildings that have been exposed to much weather and use are often more likely to collapse than are newer buildings, although this is not always true (Figure 15.23). Wooden structural components in older structures may also have dehydrated to the point that their ignition temperature has decreased and their flame spread characteristics have increased. This information should be obtained when conducting inspections and pre-incident planning.

Figure 15.23 Old buildings can become very dangerous during fires.

The longer a fire burns in a building, the more likely it is that the building will collapse. Fire weakens the structural support system until it becomes incapable of holding the weight of the building and collapses. The time it takes for this to happen varies with the fire severity, type of construction, the presence or absence of heavy machinery on upper floors or on the roof, and general condition of the building. The following are indicators of potential building collapse that all firefighters should be aware of and be on the lookout for at every fire:

- Cracks or separations in walls, floors, ceilings, and roof structures (Figure 15.24)
- Evidence of existing structural instability such as the presence of tie rods and stars that hold the walls together (Figure 15.25)
- Loose bricks, blocks, or stones falling from the buildings
- Deteriorated mortar between the masonry
- Walls that appear to be leaning in one direction or another

Figure 15.24 Note that these reinforcement stars were pulled so tight that the walls cracked around them.

Figure 15.25 Reinforcement stars on a building are an indication that the building was already in bad shape before the fire started.

- Structural members that appear to be distorted or pulling away from walls
- Fires beneath floors that support heavy machinery or other extreme weight loads
- Prolonged fire exposure to the structural members
- Unusual creaks and cracking noises

Fire fighting operations may also increase the risk of building collapse. Improper vertical ventilation techniques that result in the cutting of structural supports could weaken the structure. The amount of water used to extinguish the fire adds extra weight to the structure and can weaken it. This is particularly true of "surround and drown" master stream operations. Water only a few inches deep over a large area can add many tons of weight to an already weakened structure.

If fire personnel believe that the collapse of a building is imminent, or even possible, immediate safety precautions should be taken. First, all personnel who are operating within the building should be immediately be pulled out. A collapse zone should be set up around the perimeter of the building (Figure 15.26). The collapse zone should be equal to the height of the building. No personnel or apparatus should be allowed to operate in the collapse zone. If it is necessary to operate fire streams from within the collapse zone, unmanned master streams should be used. Once the streams have been placed, the personnel should immediately retreat to an area outside the collapse zone.

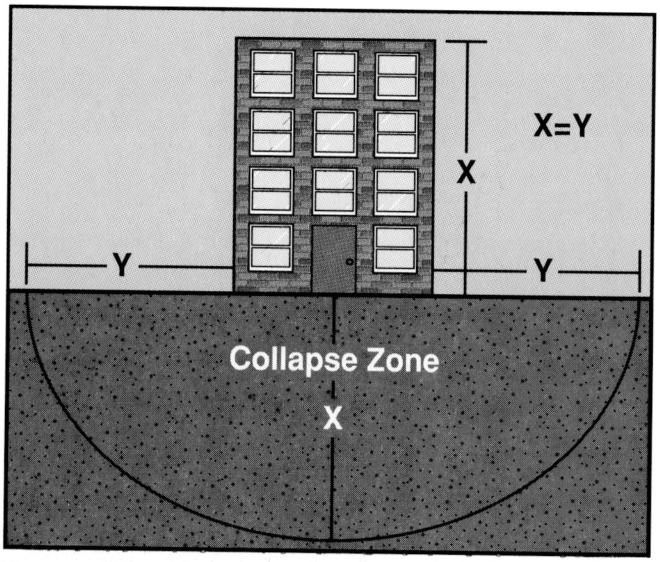

Figure 15.26 A collapse zone should be equal to the height of the building.

Lightweight And Truss Construction Hazards

One of the most serious building construction hazards facing firefighters today is the increased use of lightweight and trussed support systems (Figure 15.27).

Figure 15.27 Lightweight truss construction can be extremely hazardous to firefighters during fire fighting operations.

Lightweight construction is most common in regions that are not subject to snowfall; therefore, the roof need only be strong enough to support itself. Lightweight construction is most commonly found in houses, apartments, and small commercial buildings. The two most common types found are lightweight metal and lightweight wood trusses. Lightweight steel trusses are made from a long steel bar that is bent at a 90-degree angle with flat or angular pieces welded to the top and bottom (Figure 15.28). Lightweight wood trusses are constructed of 2 x 3 or 2 x 4-inch (50 mm by 75 mm or 50 mm by 100 mm) boards that are connected together by gusset plates (Figure 15.29). Gusset plates are small metal plates (usually 18 to 22 gauge metal) with prongs that penetrate about ⅜-inch (9 mm) into the wood.

Figure 15.28 Unprotected lightweight steel trusses fail quickly when exposed to high heat.

Figure 15.29 Firefighters should note wood truss roofs that are being constructed in their response area.

Other types of trusses, such as bowstrings, are found in virtually every community. They are used in buildings, such as car dealerships, bowling alleys, factories, and supermarkets, that have large open spaces. Bowstrings are often easily denoted by their rounded appearance, though many will appear otherwise (Figure 15.30).

All trusses are designed to work as an integral unit. Some members are in tension, and others are in compression. The strength is directly proportional to the number of members in the truss. One thing common to all types of trusses is that if one member fails, the entire truss is likely to fail. Once an entire truss fails, usually the truss next to it will fail, and the domino principle soon takes over until a total collapse has occurred.

Figure 15.30 The telltale rounded shape of a bowstring truss roof may be hidden by square-end parapets.

Experience has shown that lightweight metal and wood trusses will fail after 5 to 10 minutes of exposure to fire. For steel trusses, the 1,000°F (540°C) discussed earlier in this chapter is the critical temperature. Research indicates that the gusset plates in wood trusses will fail when the char depth on the wood is only ¼-inch (6 mm) deep. Although the trusses may be protected to give longer protection, most are not protected.

Some lightweight construction assemblies use I-beams that are fabricated from wooden 2 x 3 or 2 x 4-inch (50 mm by 75 mm or 50 mm by 100 mm) boards (Figure 15.31). These I-beams are just as hazardous as the wood trusses and should be treated with similar precautions.

It is important that firefighters know what buildings in their district have truss roofs or floors. Truss-containing buildings that have been exposed to fire conditions for 5 to 10 minutes (usually they have been exposed that long before the fire department arrives) should not be entered and crews should not go onto the roof.

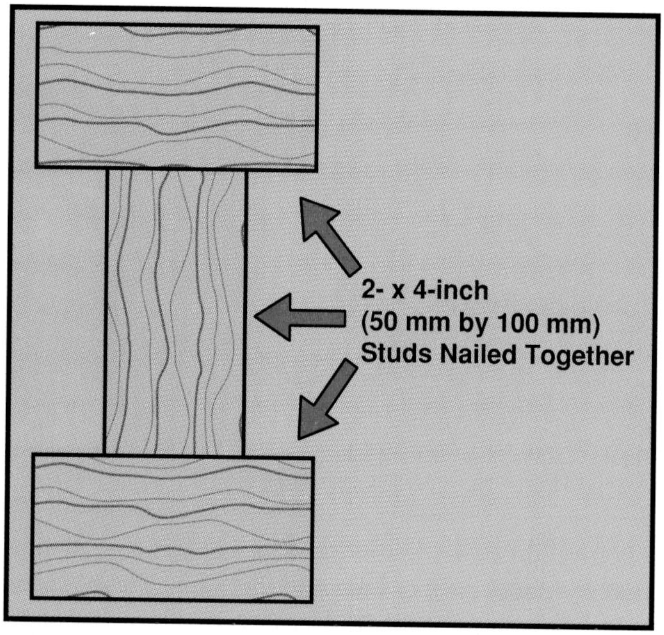

Figure 15.31 Lightweight wooden I-beams are made by nailing 2 x 4-inch (50 mm by 100 mm) boards together.

16

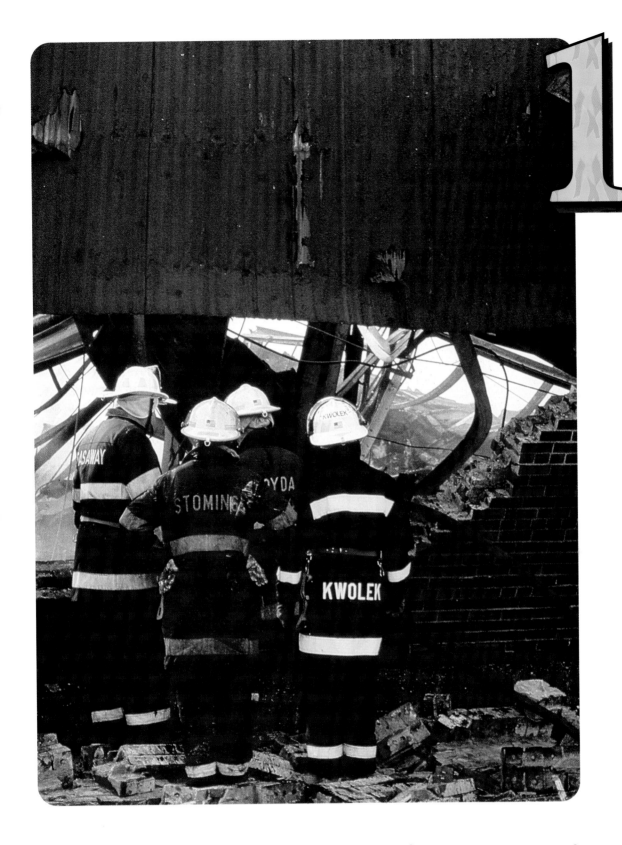

Firefighters' Responsibility in Fire Cause Determination

TABLE 16.1
COLOR OF SMOKE PRODUCED BY VARIOUS COMBUSTIBLES

Combustible	*Smoke Color*
Hay/Vegetable compounds	White
Phosphorous	White
Benzine	White to gray
Nitrocellulose	Yellow to brownish yellow
Sulfur	Yellow to brownish yellow
Sulfuric acid, nitric acid, hydrochloric acid	Yellow to brownish yellow
Gunpowder	Yellow to brownish yellow
Chlorine gas	Greenish yellow
Wood	Gray to brown
Paper	Gray to brown
Cloth	Gray to brown
Iodine	Violet
Cooking oil	Brown
Naphtha	Brown to black
Lacquer thinner	Brownish black
Turpentine	Black to brown
Acetone	Black
Kerosene	Black
Gasoline	Black
Lubricating oil	Black
Rubber	Black
Tar	Black
Coal	Black
Foamed plastics	Black

Figure 16.8 Check doors for signs of forced entry to the structure. Note the pry marks on this door jamb. *Courtesy of Dennis Cranor, Kansas City (KS) Fire Department.*

- *Discarded containers.* Containers found inside or outside the structure could have held flammable liquids (Figure 16.9).
- *Discarded burglary tools.* These tools could be evidence that the fire was set to cover up another crime.
- *Familiar faces.* Look for familiar faces in the crowd. Arsonists often stay at or return to the scene to admire their work. Firefighters should note people that they see at more than one scene. They might be fire buffs, or they might be habitual firesetters.

Figure 16.9 Sometimes flammable liquid containers are found near the structure. *Courtesy of Bob Esposito.*

During Fire Fighting

Firefighters should continue to observe any conditions that may lead to the determination of the fire cause.

- *Unusual odors.* Gasoline, kerosene, paint thinner, and other common accelerants often can be smelled and identified by the firefighters. Perfumes, deodorants, and ammonia are sometimes used by arsonists in attempts to disguise the smell of accelerants.

  CAUTION: Self-contained breathing apparatus (SCBA) should always be worn during fire fighting operations. Thus, odors may not be detectable until after extinguishment when a hydrocarbon detector can be used.

- *Abnormal behavior of fire when water is applied.* Flashbacks, reignition, several rekindles in the same area, and an increase in the fire's intensity when ordinary combustibles seem to be the only fuel involved are indications of flammable liquid.

- *Items not normally found at that particular location.* Such items could include containers, timing devices, residues, and electrical equipment.
- *Obstacles hindering fire fighting.* An example is furniture placed in doorways and hallways (Figure 16.10).
- *Structural alterations.* Alterations to the structure include plaster being removed to expose wood; holes made in ceilings, walls, and floors to spread fire or to injure firefighters who may fall through them; and fire doors being secured open.
- *Incendiary devices.* Any contrivance designed and used to start a fire. Most incendiary devices leave evidence of their existence, especially the metal parts of electrical or mechanical devices (Figure 16.11). More than one device may be used, and sometimes a faulty device can be found.

Figure 16.10 Furniture may be placed in front of doors to block the firefighter's entry to the structure.

Figure 16.11 Not all incendiary devices work as planned; some remain intact. *Courtesy of Dennis Cranor, Kansas City (KS) Fire Department.*

- *Trailer.* A trailer is a combustible material (for example, rolled rags, blankets, newspapers, or flammable liquid) used to spread fire from one point to another (Figure 16.12). It will usually leave char or burn patterns and may be used with incendiary devices.
- *Burn patterns.* Areas of uneven burning, local heavy charring, or charring in unusual places should be carefully noted. This charring indicates that an accelerant may have been spread at these locations.
- *Heat intensity.* High heat intensity could indicate flammable liquids.
- *Rate of spread.* Unusually fast spreading fire could indicate flammable liquids were used.
- *Availability of documents.* Sudden availability of insurance papers, inventory lists, deeds, or other legal documents could indicate plans were made for a fire in advance.
- *Fire detector systems.* If fire protection systems and devices are inoperable, they should be checked for evidence of tampering or intentional damage (Figure 16.13).

Figure 16.12 Note the trail of flammable liquid from the container to the structure. *Courtesy of Dennis Cranor, Kansas City (KS) Fire Department.*

Figure 16.13 A smoke detector may be disabled to delay the detection of the fire.

- *Intrusion alarms.* Intrusion alarms should also be checked to see if they have been tampered with or damaged intentionally.
- *Site of fire.* Fire in unusual places can be a source of suspicious activity.
- *Personal possessions.* Look for absence or shortage of clothing, records, furnishings, appliances,

and similar costly items; absence of personal possessions such as diplomas, financial papers, and toys; items of sentimental value such as photo albums, wedding pictures, and heirlooms; absence of pets that would ordinarily be in the structure. (**NOTE:** Do not read too much into the lack of material possessions; some people just do not have as much as others.)

- *Equipment or stock.* Look for absence of stock or used stock substituted for new, fixtures, display cases, equipment, raw materials, or records.
- *Business records.* Are important business records out of their normal place and left where they would be endangered by fire? File cabinet drawers left open are often a sign of intent to damage the records.
- *Household items.* Do major household items appear to be replaced with junk? Do other items appear to be replaced with things of inferior quality?

After The Fire

All facts concerning the fire should be reported to the officer in charge as soon as possible. If the fire is of suspicious origin, each firefighter should write a chronological account of important circumstances personally observed (not hearsay or conjecture). Such an account will aid investigators and be invaluable if the firefighter must testify in court. Cases often come to trial long after the incident, and a written account will stand up better in court than testimony from memory.

After the fire, the incident commander, or someone charged by the incident commander, interviews occupants, owners, and witnesses for information needed to complete the fire report. If the cause is not already clear, this person may also conduct the preliminary cause investigation.

Salvage and overhaul are probably the pivotal operations in determining fire cause. Some departments take great pride in their salvage and overhaul work; some boasting that they leave a building neater, cleaner, and more orderly than it was before the fire began. Admirable as this is, it may destroy the evidence that shows how the fire started. Debris should not be moved more than is necessary and especially in the area of origin, because the investigation will be hampered. Neither should debris be thrown outside in a pile — evidence is buried this way. Such thorough salvage and overhaul work must be left until after the fire investigator has completed the work needed to gather all the appropriate evidence. Once the fire investigator has gathered all the required evidence and information, the thorough overhaul work can be done. However, even at this point, firefighters should continue to be watchful for any further evidence.

CONDUCT AND STATEMENTS AT THE SCENE

Although the firefighters and officer should obtain all information possible pertaining to the fire, there should be no attempt to cross-examine a potential arson suspect — that is the job of the trained investigator, not of the firefighter or officer. If they are inclined to do so, allow the owners or occupants of the property to talk freely. Some valuable information is often gathered this way. From the moment one suspects a particular person of arson, a trained investigator must be called to conduct an interview (Figure 16.14).

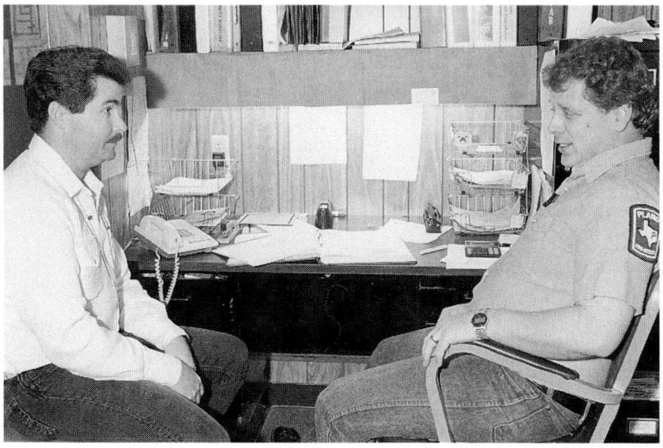

Figure 16.14 Carefully interview all potential suspects.

Do not make statements of accusation, personal opinion, or probable cause to anyone until the investigator arrives. Then, make these statements only to the investigator. Any statement regarding the fire cause should be made only after the investigator and ranking fire officer or incident commander have agreed to its accuracy and validity and have given permission for it to be made public.

Statements made in jest should never be made at the scene. These statements can easily be overheard by the property owner, a news reporter, or other bystanders, all of whom could consider such statements fact. Most reporters are eager, and a microphone could be anywhere. Careless, unauthorized, or premature remarks that are published or broadcast can be very embarrassing to the fire department. Many times these remarks will impede the efforts of the investigator to prove malicious intent as the fire cause. "The fire is under investigation" is a sufficient reply to any question concerning cause.

SECURING THE FIRE SCENE

The most efficient and complete efforts to determine the cause of a malicious and incendiary fire are completely wasted unless the building and premises are properly secured and guarded until the investigator has finished evaluating the evidence exactly as it appears at the scene. Firefighters should take care not to contaminate the scene while operating power tools, hoselines, or other equipment.

If an investigator is not immediately available, the premises should be guarded and kept under the control of the fire department until all evidence has been collected (Figure 16.15). All evidence should be marked, tagged, and photographed at this time, because in many instances a search warrant or written consent to search will be needed for further visits to the premises. This duty might be given to the police, depending on local policies and manpower availability, but whenever possible it should be carried out by fire department personnel.

CAUTION HAZARDOUS MATERIAL

Figure 16.15 Fire line tape can be used to cordon off the scene. *Courtesy of Tulsa (OK) Fire Department.*

The fire department has the authority to bar access to any building during fire fighting and as long afterward as is deemed reasonably necessary (Figure 16.16). Fire department personnel should be aware, however, of any local laws pertaining to the right of access by owners or occupants. Fire department authority ends as soon as the last firefighter leaves the scene. Further visits to the scene will require a search warrant if the owner does not give permission to enter the building.

No one, including the property owner or occupant should be allowed to enter the premises for any reason unless accompanied by a fire officer or responsible firefighter. A recorded log of any such entry should be kept, showing the person's name, times of entry and departure, and a description of any items taken from the scene.

The premises can be secured and protected in several ways with only a few personnel. In fenced areas, gates may be locked and possibly watched by one person.

POSTED
POSTED
KEEP OUT
CRIMINAL INVESTIGATION IN PROGRESS
VIOLATORS WILL BE PROSECUTED
BY ORDER OF
Pennsburg Fire
Marshal's Office
Date ________

Figure 16.16 Signs should be posted to keep people out of the fire building.

Areas may be cordoned off and marked by signs. Goods and materials may be piled around the entrance to a small business or plant to discourage entry. At large manufacturing plants, a full-time guard force is often employed, so they could handle the situation. In some extreme instances, all doors, windows, or other entrances could be completely closed with plywood or similar material.

LEGAL CONSIDERATIONS

As previously stated, firefighters may remain on location as long as necessary, but once they leave they may be required to get a search warrant to reenter the scene. This is based on the case of Michigan vs. Tyler (436 U.S. 499, 56 L.Ed. 2d 486 [1978]). The U.S. Supreme Court held in that case that "once in a building [to extinguish a fire], firefighters may seize [without a warrant] evidence of arson that is in plain view ... [and] officials need no warrant to remain in a building for a reasonable time to investigate the cause of a blaze after it has been extinguished."

The Court agreed, with modification, with the Michigan State Supreme Court's statement that "[if] there has been a fire, the blaze extinguished and the firefighters have left the premises, a warrant is required to re-enter and search the premises, unless there is consent ...."

The import of these decisions seems to be that if there is evidence of possible arson, the fire department should leave at least one person on the premises until the investigator arrives. To leave the premises, return later without a search warrant, and make a search might be enough to make prosecution impossible or for an appellate court to overturn a conviction.

Each department should learn the legal opinions that affect its jurisdiction in this regard. These opinions or interpretations can be obtained from such persons as the district attorney or state attorney general. The fire department should write a standard operating procedure around these opinions.

17

Fire Alarms and Communications

This chapter provides information that addresses performance objectives described in NFPA 1001, *Standard for Fire Fighter Professional Qualifications* (1992), particularly those referenced in the following sections:

3-4 Fire Alarm and Communications

3-4.1

3-4.2

3-4.3

3-4.4

3-5.5

4-22 Fire Prevention, Public Fire Education and Fire Cause Determination

4-22.8

Chapter 17
Fire Alarms And Communications

The expedient handling of fire alarms or calls for help is a significant factor in the successful outcome of any incident. History has proven time and time again that failure to quickly communicate the need for help will result in large and tragic losses. Firefighters must be knowledgeable in all aspects of communications within their particular department.

Fire department communications include all of the methods by which the public can notify the communications center of an emergency. They also include all of the methods by which the center can notify the proper fire fighting forces and relay information between all personnel involved at the scene. Firefighters must know how to handle routine communications such as telephone calls for business purposes.

THE COMMUNICATIONS CENTER

The communications center can be described as the brain of the fire department's central nervous system. The communications center is the point through which nearly all information flows, is processed, and then is acted upon. The communications center houses the equipment and personnel to receive alarms and dispatch resources. Depending on the size and communication requirements of the department, the communications center may be located in a fire station or in a separate building (Figure 17.1). In some jurisdictions, the fire communications center will be a part of a larger, joint communications center for all emergency services (Figure 17.2).

The building or portion of the building used as the communications center should not be located near high-risk fire hazards. Combustible materials should not be permitted in its construction except for a floor covering laid directly upon a noncombustible base. All lights should not be on a single, branch-line fuse. Adequate emergency power is essential, including a reliable secondary source independent of other sources that can apply its power to the line within 10 seconds.

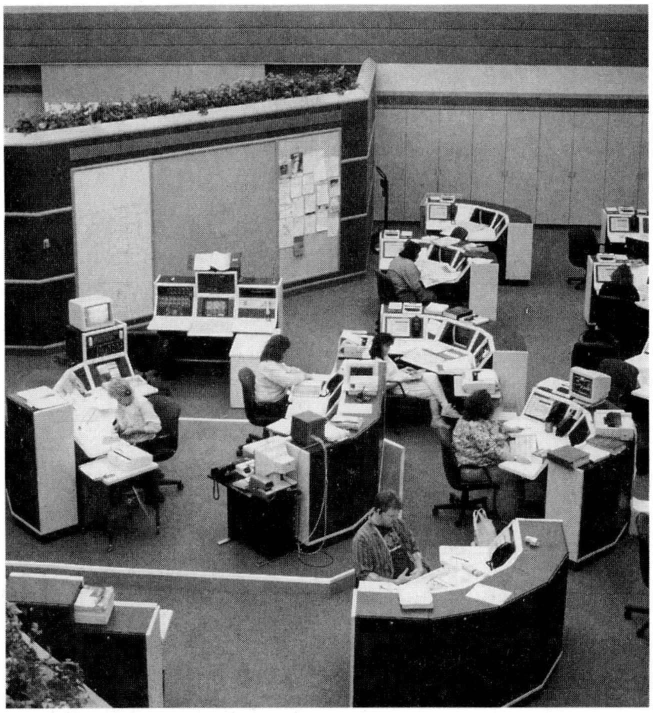

Figure 17.1 Large communication centers may require a staff of many personnel.

Figure 17.2 The fire dispatchers may be part of a much larger joint dispatch operation.

For municipalities receiving more than 2,500 alarms per year, NFPA standards require at least two fully trained operators on duty at all times (Figure 17.3). Actually, more may be needed. In smaller areas where only one operator is required, a major incident may overload the abilities of the single operator. This same possibility also exists in large municipalities with two or more operators.

Figure 17.3 Only one dispatcher is required if the volume of calls is below 2,500 per year.

The communications center may be equipped with a variety of communications equipment, depending on local capabilities. Some of the more common equipment includes:

- Two-way radio equipment for communicating with line personnel
- Tone-alert equipment for dispatching resources (Figure 17.4)
- Telephones for handling routine and emergency phone calls (Figure 17.5)

Figure 17.4 Typical tone-alert equipment found at a dispatch console.

Figure 17.5 The phone must be convenient for the dispatcher to use.

- Direct-line phone equipment for communications with hospitals, utilities, other response agencies, etc.
- Computer equipment for dispatch information and communications to line personnel (Figure 17.6)
- Tape-recording equipment to record phone calls and radio traffic (Figure 17.7)
- Alarm receiving equipment for municipal alarm box systems and private fire alarm systems

All communications equipment should be attached to the auxiliary power supply to ensure continued operation during public service power interruptions.

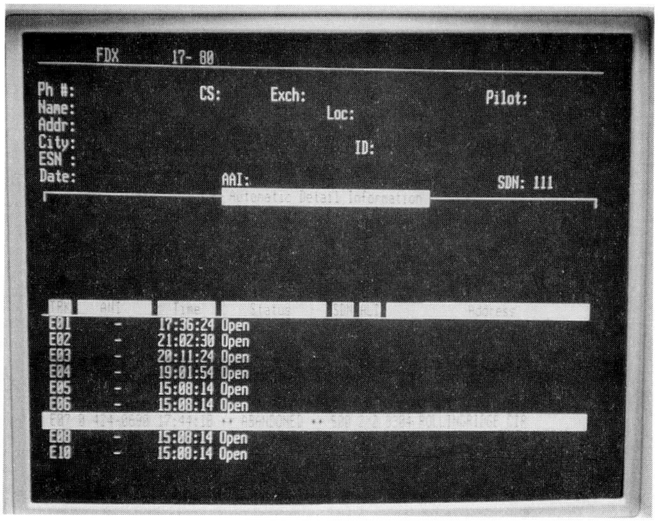

Figure 17.6 Computers make the dispatch process more efficient.

Figure 17.7 All radio transmissions and phone calls must be recorded.

REPORTING A FIRE

Adequate information must be taken from a caller reporting an emergency. This information provides a correct response by emergency personnel in the shortest amount of time possible.

A department's fire safety education program should include information on how to report a fire correctly. The public must be trained to report a fire in the following manner:

Reporting An Emergency By Telephone

- Dial the appropriate number.
  - — Fire department 7-digit number
  - — 9-1-1
  - — "0" perator
- Give address, with cross street or landmark if possible.
- State your name and location.
- Give the telephone number you are calling from.
- State the nature of the emergency.
- Stay on the line if requested to do so by the dispatcher.

Reporting An Emergency From A Fire Alarm Pull Box

- Send signal as directed on the box (Figure 17.8).
- Stay at the box until fire personnel arrive so that you can provide the exact location of the emergency to them.

Figure 17.8 A municipal fire alarm box.

Reporting An Emergency From A Local Alarm Box

- Send signal as directed on the box (Figure 17.9).
- Notify the fire department by telephone using the guidelines given earlier.

Figure 17.9 A local fire alarm pull station that is commonly found in many occupancies.

METHODS OF RECEIVING ALARMS FROM THE PUBLIC

Fire alarms may be received in two different ways: public alerting systems (telephone, radio, people walking into a fire station, etc.) and private alarm systems. The following sections highlight each of these methods.

Public Alerting Systems

Public alerting systems are those systems that may be used by anyone to report an emergency. These systems include telephones, two-way radios, radio alarm boxes, wired telegraph circuit boxes, telephone fire alarm boxes, and walk-in reports.

TELEPHONE

The public telephone system performs a very valuable function in reporting fires. It is, by far, the most commonly used means of communication for reporting emergencies. NFPA 1221, *Standard for the Installation, Maintenance and Use of Public Fire Service Communication Systems*, says that the alarm communication center should have at least one telephone line, and in larger municipalities, additional lines depending upon the traffic handled.

Depending on the phone system capabilities, the fire department emergency number may be a 7-digit number or 9-1-1. Emergency telephone number stickers that may be placed directly on the telephone help reduce time delays when calling the fire department (Figure 17.10). When equipped with either of these systems, it is important that the alarm operator get the correct location, caller's name, the phone number where the caller is, the nature of the incident, and any other pertinent information. The dispatcher must also be prepared to provide the caller with helpful directions or advice.

Figure 17.10 Every phone should be equipped with an emergency sticker.

Some municipalities are equipped with Enhanced 9-1-1 (E-9-1-1) systems. E-9-1-1 systems combine telephone and computer equipment to provide the dispatcher with instant information on the location from which the call is being made. As soon as the dispatcher picks up the phone, the computer screen shows from where the call is being made. This system allows help to be sent even if the person on the other end of the line is incapable of giving proper information.

Many business calls will also come in on the public telephones. For this reason, it is important to know the correct procedure for processing business calls. The following list describes some of these procedures:

- Answer calls promptly.
- Be pleasant and identify the department or company and yourself. For example, "Good morning, Pennsburg Fire Company, Chief Seip speaking."
- Be prepared to take messages.
- Take accurate messages by including date, time, name of caller, caller's number, message, and your name.
- Never leave the line open or someone on hold for an extended period of time.
- Post the message or deliver the message promptly to the person it is intended for.
- Terminate calls courteously. Always hang up last.

RADIO

On occasion, the report of an emergency may be received via radio. This type of report will most likely come from fire department personnel who are already on the streets for some reason and happen upon an emergency. The firefighter or dispatcher in the station monitoring the radio should take the same kind of information that would be taken from a telephone caller. Once all the appropriate information is taken, additional resources should be dispatched if required.

Some fire departments also monitor citizen's band (CB) radio frequencies for reports of emergencies. The universal frequency for reporting emergencies, and the one most commonly monitored by emergency providers, is CB channel 9. Reports taken via the CB radio should be handled the same as those taken by telephone; however, in place of the call-back number, the caller's radio handle or designation should be taken.

WALK-INS

Occasionally, a civilian will walk into a fire station and report an emergency that has just occurred in the vicinity of the station (Figure 17.11). Firefighters in the station should get the location and type of incident from the person, as well as the person's name and address if possible. Once the information is obtained, local policy will dictate the next step. Some departments require the person taking the complaint to first notify the dispatch center by phone before taking further action. Other jurisdictions will allow the personnel in the station to immediately start their response and radio the dispatcher with information on the incident to which they are responding.

Figure 17.11 Firefighters must know what information to get from a civilian who walks into the station to report an emergency.

WIRED TELEGRAPH CIRCUIT BOX

Wired telegraph circuit boxes were commonly used, particularly in metropolitan and heavy industrial areas, to provide a means for people on the street to summon the fire department (Figure 17.12). The system is limited in that the only information transmitted is the location, and malicious false alarms are a problem. For this reason, they have been eliminated in many localities. This alarm system is operated by pressing a lever in the alarm box that starts a wound-spring mechanism. The rotating mechanism transmits a code by opening and closing the circuit. Each box transmits a different code to specify its location. Wiring may be overhead or underground.

TELEPHONE FIRE ALARM BOX

A telephone is installed in the fire alarm box for direct voice contact, allowing for exchange of more information on the type of response needed. If the num-

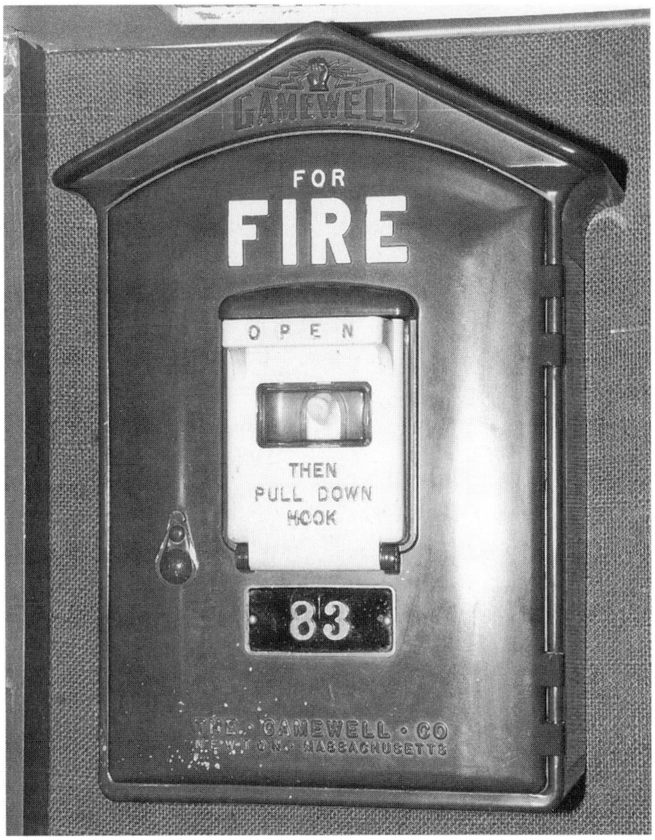

Figure 17.12 Older fire alarm systems use telegraph boxes.

ber of required boxes is unusually large, causing an increased panel size, some departments have used the combination telegraph and telephone-type circuits. This would give the best of both systems. The basic pull-down hook is used to send the coded signal, and a telephone is included for additional use (Figure 17.13 on the next page).

RADIO FIRE ALARM BOX

A radio alarm box contains an independent radio transmitter with a battery power supply (Figure 17.14 on the next page). Solar recharging is available for some systems. Others feature a wound-spring alternator to provide power when the operating handle is pulled.

There are different types of radio boxes. Activating the alarm in radio boxes alerts the fire department dispatcher by an audible signal, visual light indicator, and a printed record indicating the location. Some models have, besides a red alarm-light indicator, a different colored light indicator that shows a test or tamper signal. By using a time clock within the box, it can test itself every 24 hours. If the box pole is struck or tampered with, the tamper light comes on and gives the box location. Some boxes are numbered and this number also appears on the display panel, informing the dispatcher of the box involved and its location.

Some printing systems, when activated by the incoming radio signal, print the day of the year, time of

Figure 17.13 Some localities have emergency telephones.

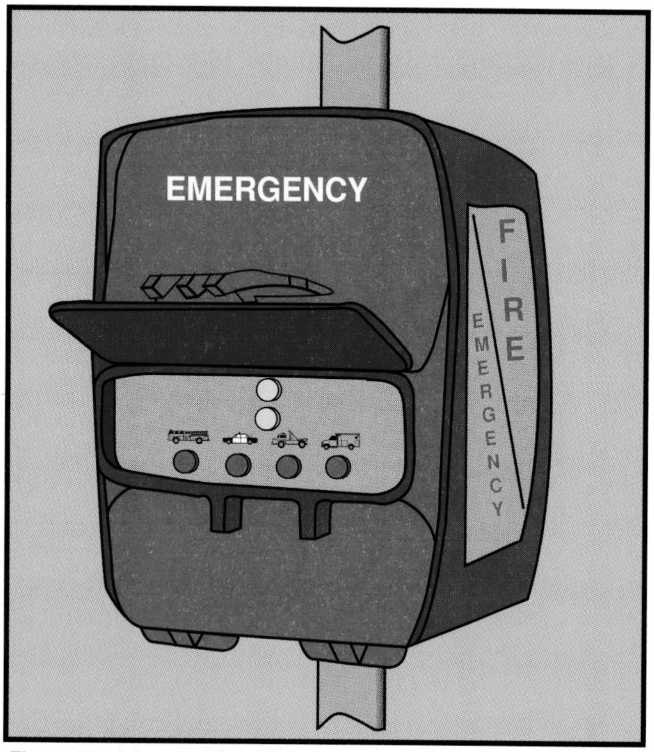

Figure 17.14 A radio alarm box.

day in 24-hour time, message sent by the box, box number, and a coded signal that indicates the strength of the battery within the box. Some radio alarm boxes are designed so that a person can select fire, police, or ambulance service. Some radio alarm boxes have two-way communications capabilities. Dispatchers answering these radio alarm box reports should take the same information as they would take by phone.

Private Fire Alarm Signaling Systems

Besides public means of notifying the fire department of an emergency, private protective signaling systems are used to detect and transmit alarms to a fire department communications center. These systems can be used to:

- Notify occupants to evacuate the premises.
- Summon the fire department or other organized assistance.
- Supervise extinguishing systems to ensure their operability when needed.
- Supervise industrial processes to warn of abnormalities that may contribute to a fire hazard.
- Supervise personnel to ensure performance of assigned duties.
- Actuate fire control equipment.

Before examining the types of systems, it is important to have a basic understanding of the various alarm initiating devices found on private fire alarm signaling systems.

ALARM INITIATING DEVICES

Alarm initiating devices are the nerve endings of the system. It is through these devices that the alarm signal is originated. The basic types of initiating devices are as follows:

- Manually activated
- Thermal sensitive
- Visible products-of-combustion detectors
- Invisible products-of-combustion detectors
- Flame detectors
- Waterflow detectors

Manually Activated Devices

Manually activated devices are commonly called pull stations (Figure 17.15). These devices require that people first recognize that a fire emergency exists and then act by using a pull station or, in some cases, a telephone fire alarm device.

Figure 17.15 A typical manual pull station.

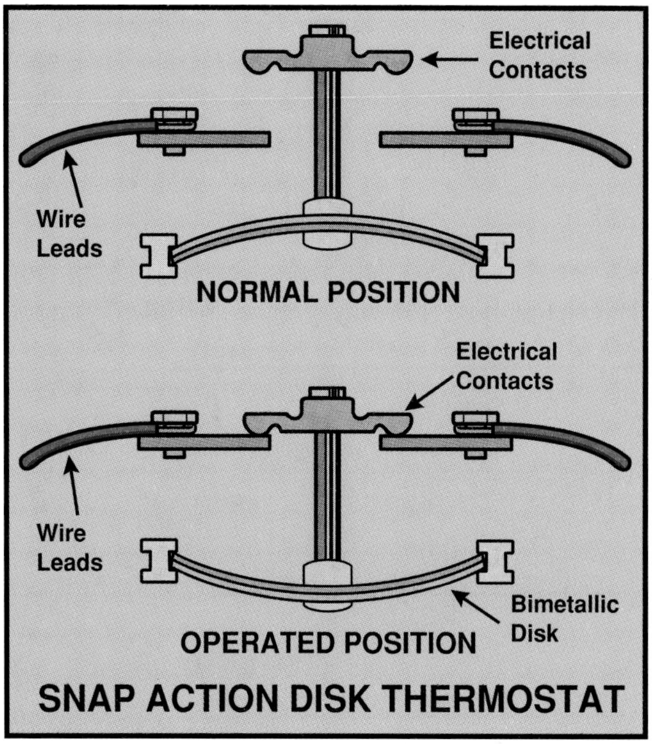

Figure 17.16 A bimetallic heat sensor.

Thermal Sensitive Devices

Several different types of heat detection devices, such as fixed-temperature devices and rate-of-rise detectors, employed in private fire alarm signaling systems are discussed in this section. Fixed-temperature fire alarm initiating devices activate at a predetermined temperature. Three basic types of fixed-temperature alarm initiating devices are used:

- A device using a bimetallic strip or disk
- A device using a soft metal alloy or thermoplastic resins that melt
- A device using expansion of heated solvents

Bimetallic strips consist of two metals or metal alloys that are bonded together; these metals have different expansion ratios when heated (Figure 17.16). When the strip is heated, the strip with the higher expansion ratio arches toward the strip made of the metal with the lower expansion ratio. This action causes an electrical contact to be opened or closed, thus initiating an alarm.

Fixed-temperature devices that melt employ soft metal alloys or thermoplastic resins to hold together a two-piece link or latching mechanism. A frequently used metal alloy is solder. When the solder melts, the link separates or the latch is released, causing an alarm signal to sound.

Another type of fixed-temperature device actuates through the expansion of heated solvents. In this type of device, small glass bulbs contain solvents; the bulbs are manufactured to break at predetermined pressures. When the solvent is heated, it vaporizes. The resulting vapor pressure breaks the glass bulb, which has been holding two electrical contacts apart. The contacts close, causing an alarm signal to be sent.

Rate-of-rise detectors respond to quick changes in temperature rather than activating at a fixed temperature. Most rate-of-rise detectors have a small chamber filled with air. A small vent from the chamber allows for slow changes in temperatures. If rapid heating occurs, however, the air inside the chamber expands. The small vent cannot relieve the pressure, which forces a diaphragm out. The movement of the diaphragm either opens or closes a set of electrical contacts that initiate an alarm signal (Figure 17.17). A second type

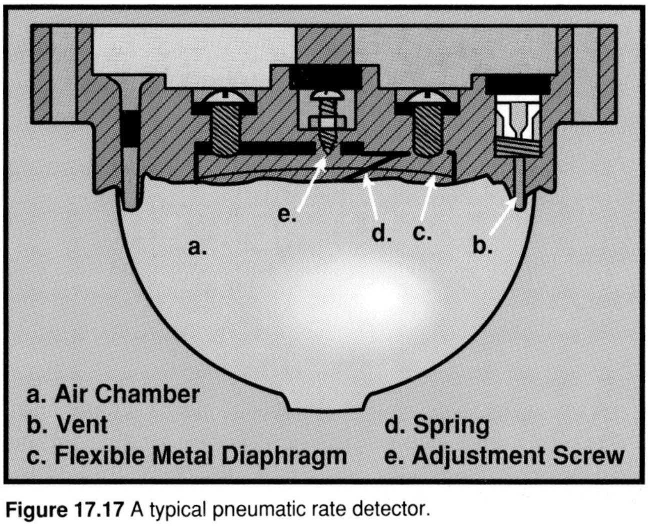

Figure 17.17 A typical pneumatic rate detector.

uses bimetallic strips (Figure 17.18), and a third type uses thermoelectric sensors to detect rapid changes in temperature. Rate-of-rise detectors tend to react more quickly than fixed-temperature detectors, but they are not quite as reliable.

Combination rate-of-rise/fixed-temperature detectors feature the quicker reaction rate of the rate-of-rise detector coupled with the higher dependability of the fixed-temperature unit (Figure 17.19).

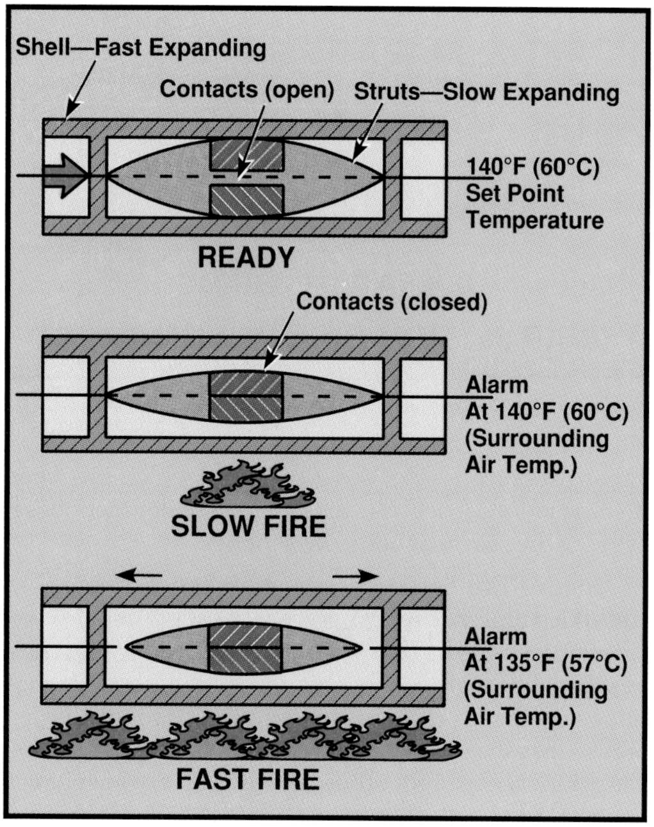

Figure 17.18 A bimetallic rate-of-rise detector.

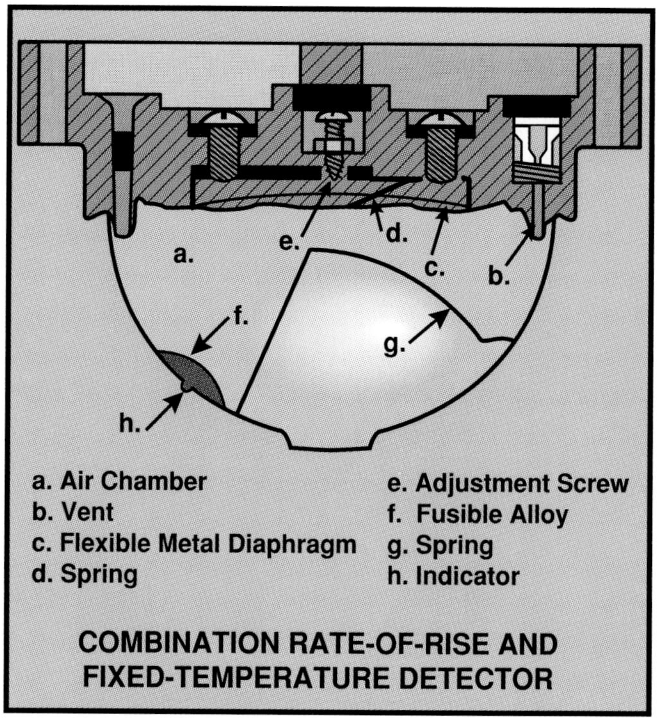

Figure 17.19 This detector is a combination fixed-temperature/rate-of-rise detector.

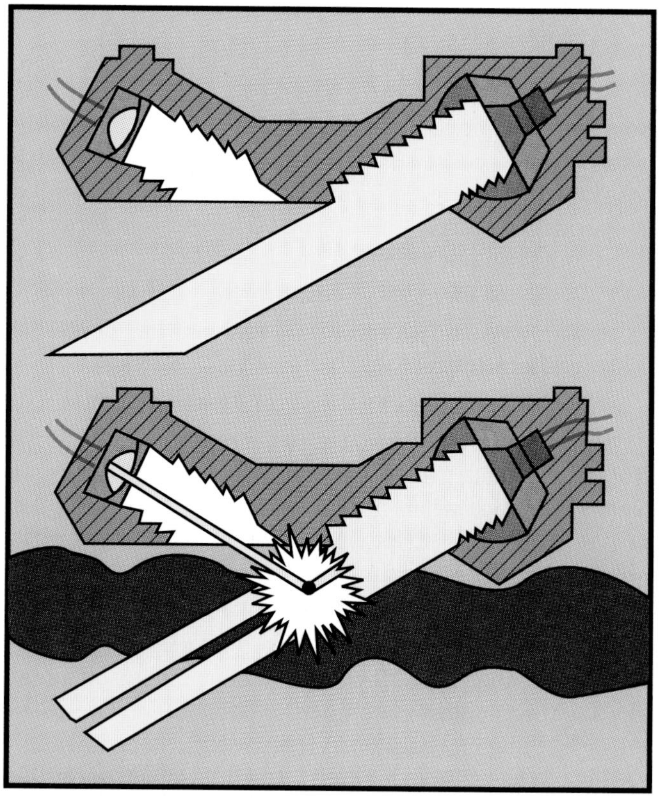

Figure 17.20 This detector is activated when a foreign object interrupts the beam of light.

Visible Products-Of-Combustion Detectors

Visible products-of-combustion smoke detectors use a photoelectric cell coupled with a specific light source. The photoelectric cell functions in two ways to detect smoke: beam application and refractory application.

The beam application uses a beam of light focused across the area being monitored onto a photoelectric cell. The cell constantly converts the beam into current, which keeps a switch open. When smoke blocks the path of the light beam, the current is no longer produced, the switch closes, and an alarm signal is initiated (Figure 17.20).

The refractory photocell uses a light beam that passes through a small chamber at a point away from the light source. Normally, the light does not strike the photocell, and no current is produced. When a current does not flow, a switch in the current remains open. When smoke enters the chamber, it causes the light beam to be refracted (scattered) in random directions. A portion of the scattered light strikes the photocell, causing current to flow. This current closes the switch and activates the alarm signal (Figure 17.21).

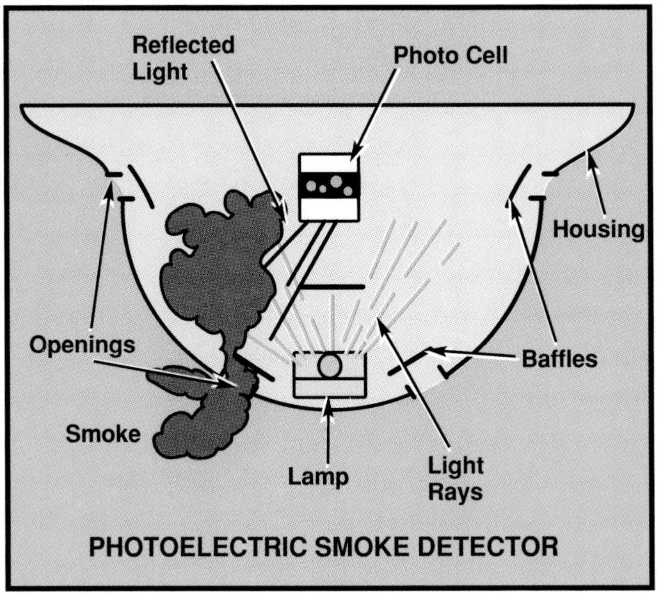

Figure 17.21 The operating principle of a photoelectric smoke detector.

Invisible Products-Of-Combustion Detectors

During a fire, molecules ionize as they undergo combustion. The ionized molecules have an electron imbalance and tend to steal electrons from other molecules. The operation of the invisible products-of-combustion detector uses this phenomenon, and for this reason they are commonly called ionization detectors.

The detector has a sensing chamber that samples the air in a room. A small amount of radioactive material (usually americium) adjacent to the opening of the chamber ionizes the air particles as they enter. Inside the chamber are two electrical plates: one positively charged and one negatively charged. The ionized particles free electrons, which travel to the positive plate. Thus, a minute current normally flows between the two plates. When ionized products of combustion enter the chamber, they pick up the electrons freed by the radioactive ionization. The current between the plates ceases and an alarm signal is initiated (Figure 17.22).

Flame Detectors

Flame detectors are sometimes also called light detectors. There are two basic types: those that detect light in the ultraviolet wave spectrum (UV detectors) and those that detect light in the infrared wave spectrum (IR detectors).

Ultraviolet detectors can give false alarms when they are in contact with sunlight and arc welding.

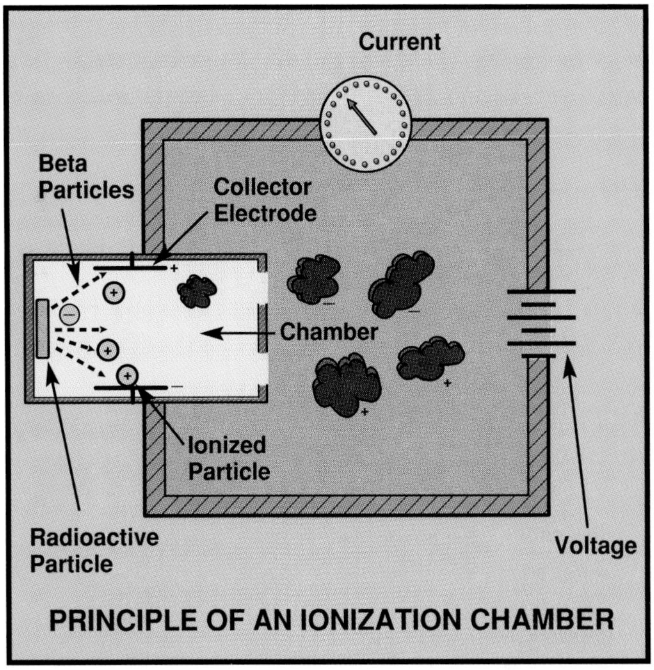

Figure 17.22 The operating principle of an ionization smoke detector.

Therefore, their use is limited to areas where these and other sources of ultraviolet light can be eliminated. Infrared detectors are effective in monitoring large areas. To prevent activation from infrared light sources other than fires, infrared detectors require the flickering action of a flame before they activate to send an alarm.

Waterflow Detectors

A waterflow detector recognizes movement of water within the sprinkler or standpipe system. Once movement is noted, the detector gives a local alarm and/or may transmit the alarm. The intent of a waterflow alarm is to transmit an alarm whenever a sprinkler opens or a standpipe valve is opened. This device is usually attached to the main sprinkler riser. Waterflow detectors are susceptible to false alarms caused by pressure fluctuations within the water supply system.

TYPES OF PRIVATE FIRE ALARM SIGNALING SYSTEMS

Private fire alarm signaling systems are a primary defense against large-scale fires in all types of structures. Their primary benefit is the early detection of fire conditions and the corresponding transmittal of an alarm to fire personnel. Early detection and reporting of fires is a key factor in preventing a large loss situation. The following section highlights some of the more common types of private fire alarm signaling systems.

Central Station Protective Signaling Systems

A central station system is a commercial fire alerting system or group of systems that signal the alarm to a central station. This agency has trained and experienced personnel continuously on duty to receive signals, retransmit fire alarms to the fire department, and take whatever action the supervisory signals indicate is necessary (Figure 17.23).

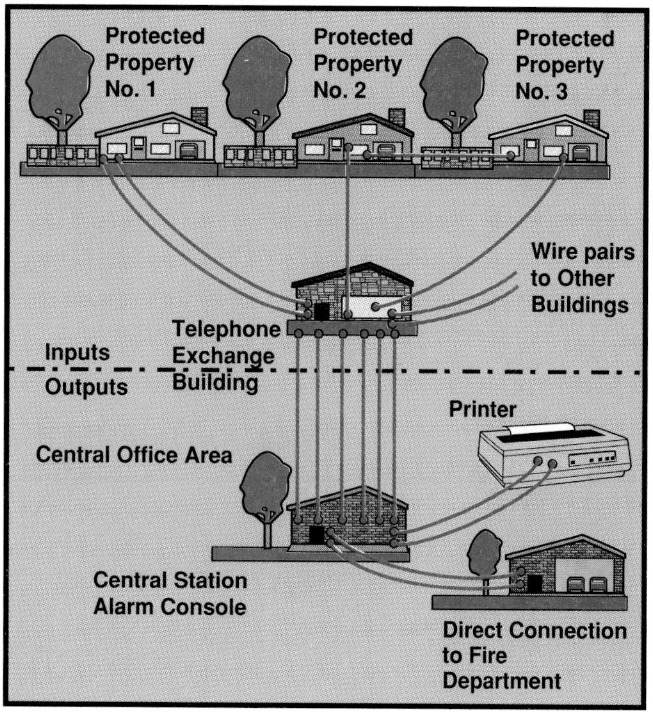

Figure 17.23 A typical central station alarm system.

Local Protective Signaling Systems

An alarm or supervisory signal operates on the protected premises and is primarily for the notification of occupants and supervisory personnel. If fire department response is required, the department must be called by a building occupant.

Auxiliary Protective Signaling Systems

An auxiliary alarm system is, by itself, insufficient for notifying the fire department, but in combination with a public system, it does summon a response. The alarm devices are usually owned and maintained by the property owner, but the connecting facilities between the protected property and the fire department are part of the public alerting system.

Visual indicators are provided at a location acceptable to the fire department to further direct firefighters to the precise location of an alarm (Figure 17.24). This provision is especially necessary in a large complex of warehouses, where, at times, no personnel are on duty to direct the firefighters. Indicators connected to all the alarm signal devices must show the following:

- Particular building of the complex
- Which floor of the building
- Which section of that floor

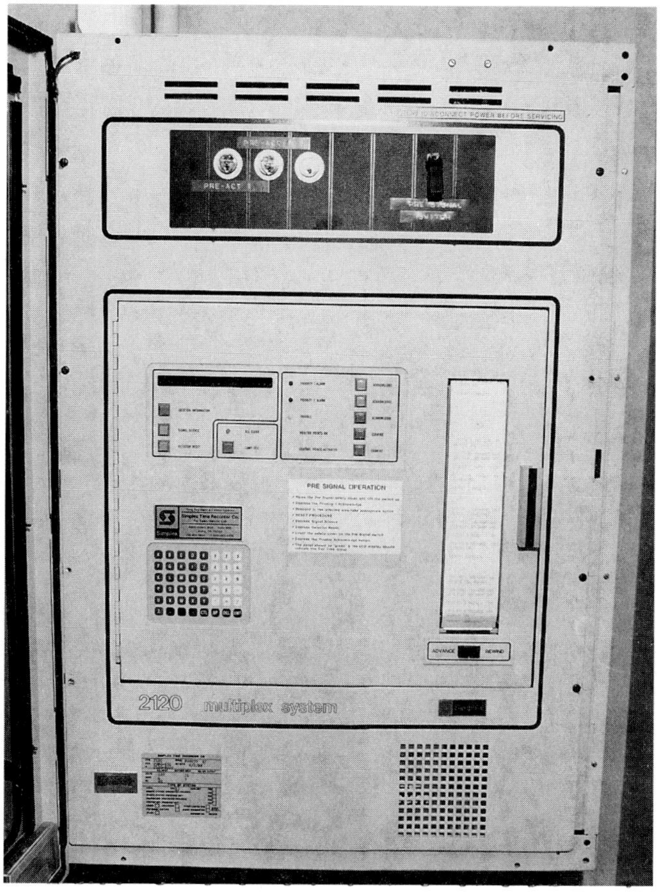

Figure 17.24 A typical annunciator panel.

Remote Station Protective Signaling Systems

Remote station protective signaling systems are installed to protect private premises (Figure 17.25). Generally, these systems are installed and maintained by a commercial central station company. Instead of the alarm going to the commercial central station, it goes by a direct circuit to a panel at a public fire alarm office or fire station where operators are available to act on the alarm.

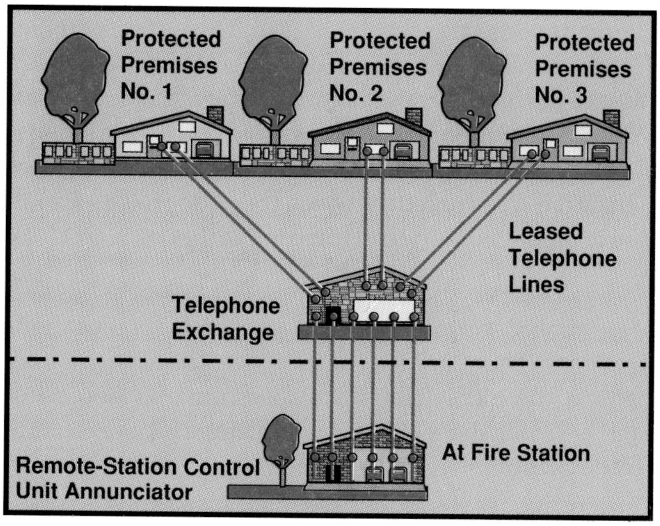

Figure 17.25 A remote station system.

Proprietary Protective Signaling Systems

These are systems protecting large plants and installations that maintain their own proprietary central supervising station at the protected property (Figure 17.26). From this supervising station, alarms are retransmitted manually or automatically to the fire department alarm office or fire station. Personnel in the supervising station can call the department, after investigation, to give information about gate number, type of fire, or building number.

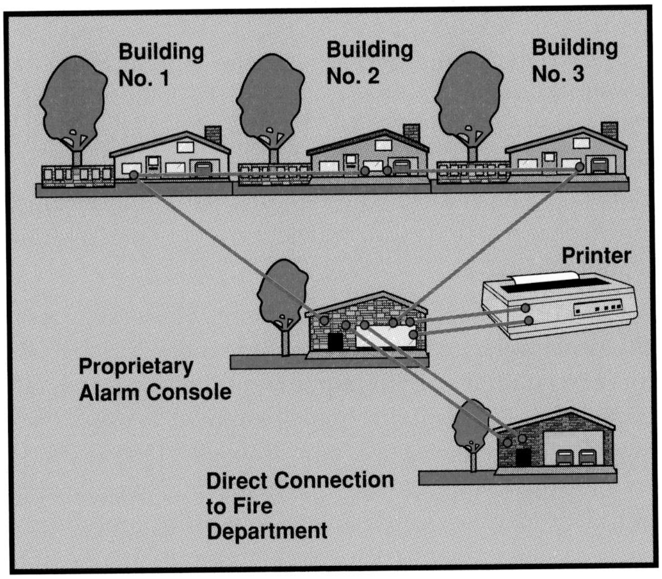

Figure 17.26 A proprietary system.

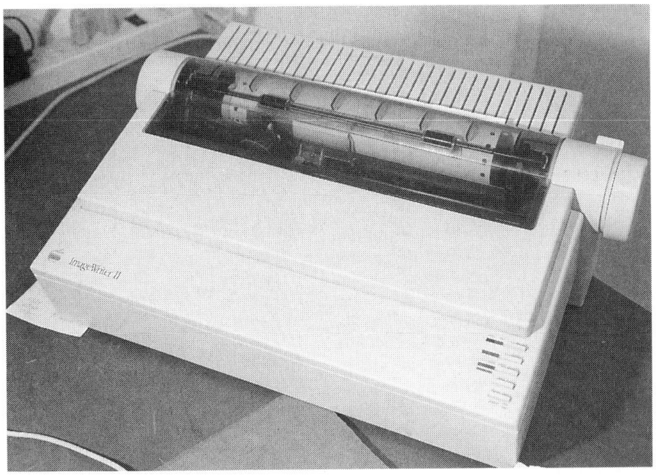

Figure 17.27 Computerized fire departments may have printers in the fire station that give hard copy of the address of the emergency, as well as other pertinent information.

Figure 17.28 A sign board displays which unit is being alerted to respond. *Courtesy of Phoenix (AZ) Fire Department.*

METHODS OF ALERTING FIRE DEPARTMENT PERSONNEL

There are a variety of methods for alerting firefighters of an emergency. The particular way that fire stations and personnel are alerted depends on whether or not the station is manned.

Alerting Manned Stations

Technological advances have brought about new, modern alerting systems to accompany the more traditional types. These types of new alerting systems include the following:

- Computerized line printer or terminal screen (Figure 17.27)
- Vocal alarm
- Teletype
- House bell or gong
- House light (Figure 17.28)
- Telephone from dispatcher
- Telegraph register
- Radio with tone alert

When planning or redesigning station alerting systems, the alerting methods employed should be effective but not startling to personnel. Extremely loud audible devices or bright lights that come on in the middle of the night can be somewhat shocking to the firefighter's system and can raise stress and anxiety levels. Dull lights and reasonable audible devices are much better.

Alerting Unmanned Stations

In order to facilitate as quick a response as possible from unmanned stations, some method of simultaneously notifying all personnel must be used. These systems include:

- Pagers (Figure 17.29)

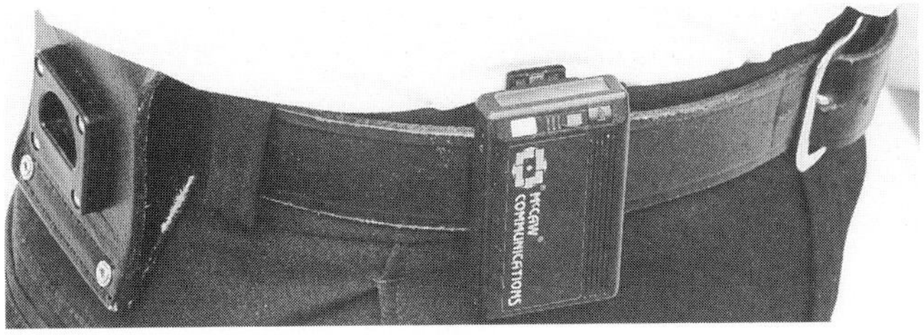

Figure 17.29 Many volunteer firefighters rely on pagers to alert them to respond to an emergency.

- Home electronic monitors (Figure 17.30)
- Telephones
- Sirens (Figure 17.31)
- Whistles or air horns

Pagers and home electronic monitors are activated by tone signals that are sent over radio waves. The advantage of pagers is that firefighters can carry them wherever they go. Home monitors and telephones require the firefighter to be at home to be aware of the need to respond. Sirens, whistles, and air horns are employed most commonly in small communities. These devices make a loud noise that all members of the community can hear. This noise is both an advantage and a disadvantage. Civilians will be aware that emergency traffic may be on the streets; however, many may also be inclined to follow the apparatus and congest the emergency scene.

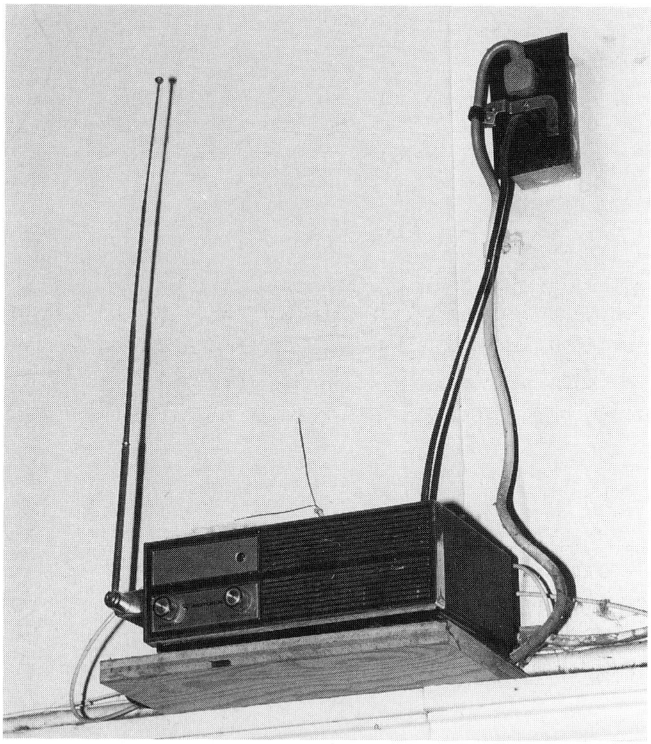

Figure 17.30 Home tone-alert monitors alert volunteers to respond.

Figure 17.31 Fire sirens are still used in many municipalities.

RADIO PROCEDURES

All radio communication in the United States is under authorization from the Federal Communications Commission (FCC). Fire departments that operate radio equipment must hold radio licenses from the FCC. Depending on the radio system in a particular locality, one license may cover several departments that operate a joint system.

There should be a certain amount of instruction on the local level in the correct use of the radio. Firefighters must be familiar with the equipment they have and how it operates. Firefighters should be taught to use correct voice procedures when talking on the radio. The firefighter should speak slightly louder than normal, but in a slow and clear manner. Departmental operating procedures, such as department codes, test procedures, and time limits on radios, need to be established. Ten-codes were popular in the early days of radio equipment because of poor transmission and reception. Advances in radio technology have reduced the need for ten-codes and many departments have converted to "clear text" or simple English. Local department rules should specify who is authorized to transmit on the radio. It is a federal offense to send personal or nonfire service messages over the radio.

Other important considerations that should be remembered include the following:

- Avoid unnecessary transmissions. Be brief, accurate, and to the point.
- Do not transmit until determining if the air is clear.
- Any unit working at a fire or rescue scene has priority over any other transmission.
- Do not use profane or obscene language on the air.
- Hold the microphone 1 to 2 inches (25 mm to 50 mm) from your mouth at a 45-degree angle (Figure 17.32).
- Speak calmly, clearly, and distinctly in a natural conversational rhythm at medium speed.

Figure 17.32 Hold the mike 1 to 2 inches (25 mm to 50 mm) from the mouth at a 45-degree angle.

- Avoid laying the microphone on the seat of the vehicle, because the switch may be pressed and cause interference.
- Do not touch the antenna when transmitting. Radio frequency burns might result.

At times, it may be necessary to broadcast emergency traffic (urgent message) over the radio. Emergency traffic is necessary in situations such as when additional resources are needed or detailed instructions must be relayed through the dispatcher. When the need occurs to transmit emergency traffic, the person transmitting the message should make the urgency clear to the dispatcher. For example:

"Dispatch from Chief 65, emergency traffic."

At that point, the dispatcher should give an attention tone if used in that system, advise all other units to stand by, and then advise the caller to proceed with the emergency traffic.

Evacuation signals are used when command personnel decide that all firefighters should be pulled from within a burning building or other hazardous area because conditions have deteriorated beyond the point of reasonable safety. All firefighters should be familiar with their department's method of sounding an evacuation signal. There are several ways this communication may be done. The two most common ways are to broadcast a radio message ordering them to evacuate and to sound the audible warning devices on the apparatus at the fire scene for an extended period of time. The radio broadcast of an evacuation signal should be handled in a manner similar to that described for emergency traffic. The message should be broadcast several times to make sure that everyone hears it. The use of audible warning devices on apparatus, such as sirens and air horns, will work in small structures, but may not be heard by everyone when working in a large building.

Arrival And Progress Reports

Whether or not codes or verbal descriptions are used, first-arriving companies should use the radio to provide a description of the conditions found at the scene. All firefighters should know how to provide an accurate report of the conditions they see as they arrive on the scene. This process is often referred to as making a size-up. Each department should establish its own format for size-up reports. A good size-up report establishes a time of arrival and allows other responding units to anticipate what actions might be taken upon their arrival. A typical size-up report might sound like the following:

"Engine 65 to Dispatch, we are on location at Long Alley and 5th Street with light smoke showing from a two-story, wood-frame storage barn. We are stretching a handline at this time, and Engine 65 is establishing Long Alley Command."

Or like this:

"Engine 65 to Dispatch, we are on location at 68 Main Street with heavy fire showing from the north side, second floor of a two-story, ordinary construction boarding house. Engine 38 is laying a 5-inch line at the rear. Have Engine 71 lay a line to the front. Trucks 65 and 15 should start rescue and ventilation, Engine 65 is Main Street Command on side one of the fire building. Dispatch the second alarm and have them stage at Front and Main."

When giving a report of conditions upon arrival, the following information should be included:

- Address, particularly if other than the one reported
- Building and occupancy description
- Nature and extent of fire
- Attack mode selected
- Rescue and exposure problems
- Instructions to other responding units
- Location of incident command position

Once fire fighting operations have begun, it is important that the communications center be kept advised of the actions taken at the emergency scene. Such progress reports should indicate the following:

- Name of the command officer
- Change in command location
- Progress (or lack of it) in situation control

LOCATION Valley Shoe Store	**BOX** 65-103	**OCCUPANCY TYPE** Commercial/Mercantile
ADDRESS 334 Main Steet	**HYDRANTS** 3rd and Main Heimbach Bros. Hardware Store Reigner's Parking Lot	**SPECIAL HAZARDS** Truss Roof Structure

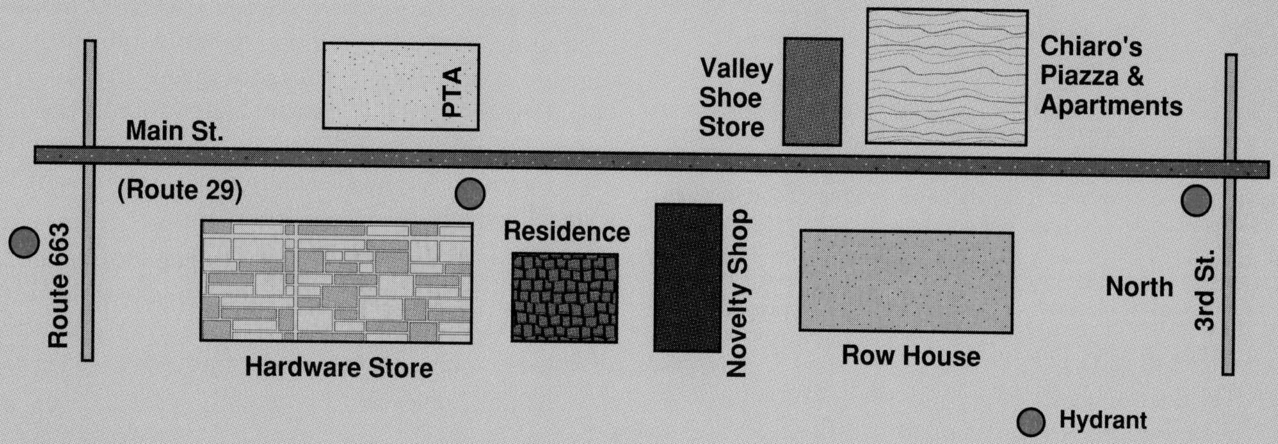

ALARM	ENGINE	TRUCK	TANKER	SQUAD	COVER STATION 65	MEDIC
1	65-22 65-21	65-11	—	65-61	—	369 (1 unit)
2	38-21 38-22	BE-15	—	38-81	—	
3	71-21 42-21	BU-18	42-31	71-81 42-6	BU-57	369 (2 units)
4	BE 58 BU 57	—	BE 58 BU 57	BU 57	BE 14	—
5	—	—	—	—	—	—

SPECIAL CALL	UNIT
Cascade	38,BU58,64
Canteen	41
Light/Power	BE15,69-81

Legend
BE - Berks Co.
BU - Bucks Co.

Figure 17.33 A typical phantom box assignment.

- Direction of fire spread.
- Exposures by direction, height, occupancy, and distance
- Any problems or needs
- Anticipated actions — holding, doubtful

Calls For Additional Response

At some fires, it may be necessary for additional units to be called. Normally, only the incident commander at the fire may order multiple alarms or additional response. Depending on who arrives first, the incident commander may be a company officer, chief officer, or a line firefighter.

All firefighters need to know the local procedure for requesting additional alarms (Figure 17.33). They must also be familiar with the alarm signals, such as for a multiple or special alarm, and what to do when they are received. Personnel should know the number and types of units that will respond to these alarms.

When multiple alarms are given for a single fire, maintaining communications with each unit becomes more difficult as the radio traffic increases. To reduce the load on the communications center, a mobile, radio-equipped, command vehicle can be used at large fires (Figure 17.34).

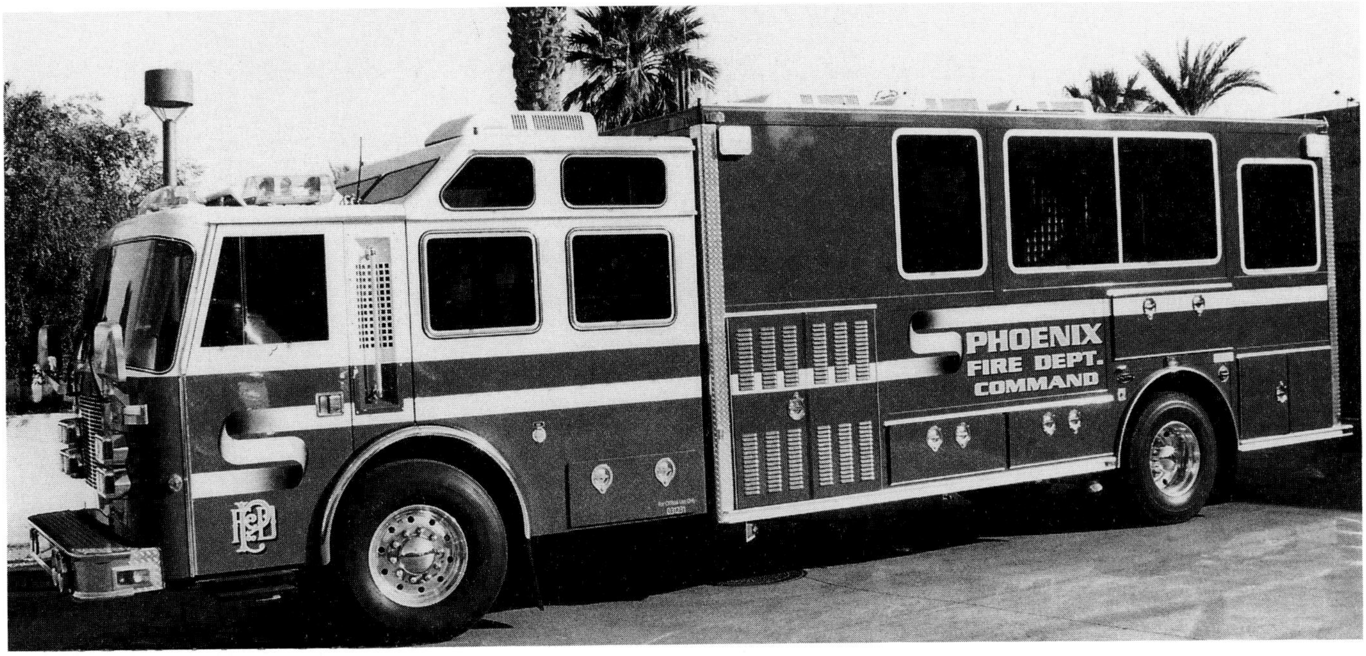

Figure 17.34 This mobile command post and communications vehicle can be used at large incidents.

18

Fire Prevention and Public Fire Education

- To gain valuable on-site information for the pre-incident plan
- To enforce fire and building codes

The fire department administration should specify how the firefighters are to conduct the building inspection. Although the firefighters may not have much input into its development, they must be familiar with the department's inspection policy. This will give the firefighters a better understanding of the whole inspection process, and they will be better prepared to answer questions that may be asked during the inspection.

Upon completion of the on-site visit, the firefighters are responsible for writing and forwarding reports to the fire department administration (Figure 18.3). Chief officers use the reports to develop information resource systems (files or records that may be in print and/or on computer) and strategic plans for the pre-incident plan.

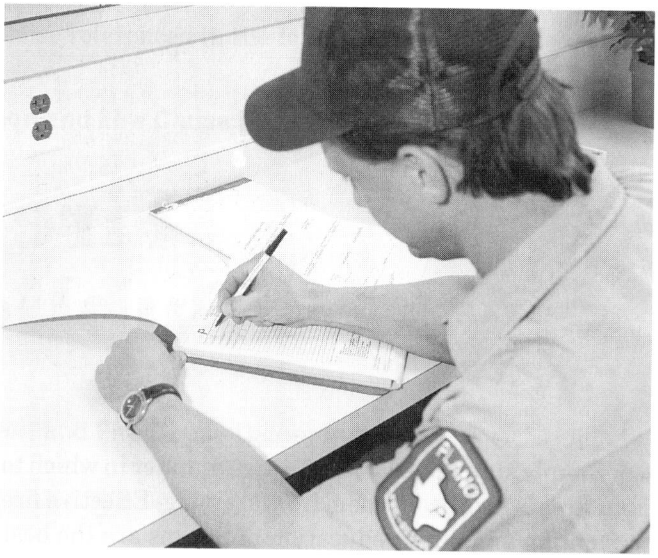

Figure 18.3 A firefighter must prepare accurate reports for record keeping.

Pre-incident planning has several characteristics that make the process a positive, effective tool in reducing fire and life loss. It is the result of a collective effort from all levels of the fire department, beginning with the firefighters performing the on-site inspection. The information gathered allows the coordination of resources to meet the special demands placed upon the fire department by the buildings within its jurisdiction. The success of the pre-incident planning process depends on the ability of the firefighters to perform adequate building inspections and to forward complete, accurately written reports.

PERSONAL REQUIREMENTS FOR FIREFIGHTERS WHO PERFORM INSPECTIONS

To the public, the uniform and badge worn by firefighters indicate that they are qualified to discuss fire protection and can give reliable advice on how fire hazards can be corrected. However, firefighters must realize their limitations in this area. Firefighters meeting the fire prevention and public fire education objectives found in NFPA 1001 will have a basic understanding of these principles; however, they are by no means fully qualified inspectors or public fire educators. Firefighters who perform inspections are expected to recognize basic hazards and report them through the appropriate channels within their organization. They may offer corrective advice, especially in dwelling inspections; however, they should not tackle situations that they are not trained to handle. Any person who is expected to take a more authoritative role in inspections should be trained to meet the objectives found in NFPA 1031 *Standard for Professional Qualifications for Fire Inspector*. Public fire educators should meet the requirements of NFPA 1035 *Standard For Professional Qualifications for Public Fire Educator*.

Firefighters will most commonly perform inspections as part of a company. In many cases, the company officer will have more inspection training and experience than the firefighters. A prerequisite for anyone making inspections is confidence in one's ability. The firefighter must have the ability to meet the public, to make a favorable impression, and to judge conditions (Figure 18.4). The ability to judge conditions can be developed only by training and using available reference material. A firefighter's confidence in his or her ability to transpose visual information into written reports and sketches can be accomplished through practice. It is extremely important for the firefighter to have confidence in the officer in charge of the fire prevention section and to be able to assist when unfamiliar hazards

Figure 18.4 The firefighter must make a positive impression on the occupant.

and conditions are encountered. If firefighters have confidence in themselves, the public will have confidence in them.

When performing fire prevention activities, the firefighter should project a neat appearance. The firefighter's uniform should be clean and in good condition, and the firefighter should be well groomed. A neat appearance will demand more respect from the civilians being dealt with and will bolster the fire department's public image.

INSPECTION EQUIPMENT

The equipment needed by a firefighter to adequately perform an inspection may be divided into equipment needed at the fire station and equipment used at the place of inspection. A recommended list of equipment that may be needed includes the following:

At The Inspection Site

- Neat coveralls for crawling into attics and concealed spaces
- Clipboard, inspection forms, sketching, or engineering paper, and standard plan symbols
- Pen or pencil
- 50-foot tape
- Flashlight—most concealed spaces are not lighted
- Camera equipped with flash attachment
- Pitot tube and gauges, when water flow tests are required (Figure 18.5)
- Reference books

Figure 18.5 A pitot tube is used for waterflow tests.

At the Fire Station

- Inspection forms
- Inspection reports
- Inspection manuals
- Adequate records
- Inspection file
- Drawing materials
- Drawing board

FIRE COMPANY INSPECTION PROCEDURES

Fire department administrators must consider the number of demands competing for the firefighter's time. Firefighters cannot choose the time to do fire fighting, but they have the prerogative of selecting the hour of the day, the day of the week, and even the season of the year to perform inspection services. Because of this choice, the fire department administration should set up a schedule for inspection activities.

Even though the fire department has days scheduled for inspections, this schedule may not always coincide with the availability of the building occupants. Thus, the company officer should call the occupant ahead of time to arrange for the inspection (Figure 18.6). The company officer should inform the occupant of the purpose of a fire prevention inspection and find out what day and time would be suitable for the occupant. This procedure enables fire inspections to be scheduled at a time that will not inflict a hardship on either the occupant or the company. Inspections are usually made during normal business hours, but night inspections are sometimes necessary. Occupancies, such as theaters, night clubs, and halls, may require a part of the inspection to be done in the daytime and completed at night while the building is occupied (Figure 18.7).

Figure 18.6 Call ahead so that the inspection will not be a surprise to the occupant.

Figure 18.7 Theaters may have to be inspected at night to observe actual working conditions.

Approaching The Occupant

Certain observations should be made before the firefighter enters the property. The firefighter should look for the location of fire hydrants, fire alarm boxes, and exposures (Figure 18.8). The condition of the streets and the general housekeeping of the area surrounding the occupancy are extremely important from the standpoint of fire apparatus response. Before entering the property, the firefighter should record in the preliminary notes the type of buildings, occupancy, and general appearance of the neighborhood. The firefighter should also check the following:

- Address numbers for visibility
- All sides of the building for accessibility
- Forcible entry problems posed by barred windows or high-security doors
- Overhead obstructions that would restrict aerial ladder operation (Figure 18.9)

Figure 18.8 Look for potential exposure problems around the building in question.

Figure 18.9 Look for large overhangs or trees that might pose an obstacle to aerial apparatus.

An earnest effort by the firefighters to create a favorable impression upon the owner will help to establish a courteous and cooperative relationship. The firefighters should enter the premises at the main entrance and obtain permission to make the inspection. An inspection should never be attempted without the proper permission. If necessary, firefighters may have to wait to see the proper authority, because this person may be busy with other important matters. The company officer should introduce the crew and state their business. If the owner has been informed of the purpose of the inspection in advance, this introduction will be much easier. A representative of the occupant should accompany the firefighters during the entire inspection. Such a guide will help obtain ready access to all areas of the building and provide answers to questions.

Making The Inspection

To make a thorough inspection, firefighters should take sufficient time to make notes and sketches of all important features. Additional time to discuss fire protection problems with the owner will pay dividends. A complete set of notes and a well-prepared sketch of the building provide dependable information from which a complete report can be written (Figure 18.10).

From an inspection point of view, it does not matter whether the inspection is started on the roof and worked downward or started in the basement and worked upward. From a practical standpoint, however, many firefighters find it helpful and less confusing to start on the roof. In either procedure, the route should be planned so that the firefighters can systematically inspect each floor. In large or complicated buildings, it may be necessary to make more than one visit to complete the inspection.

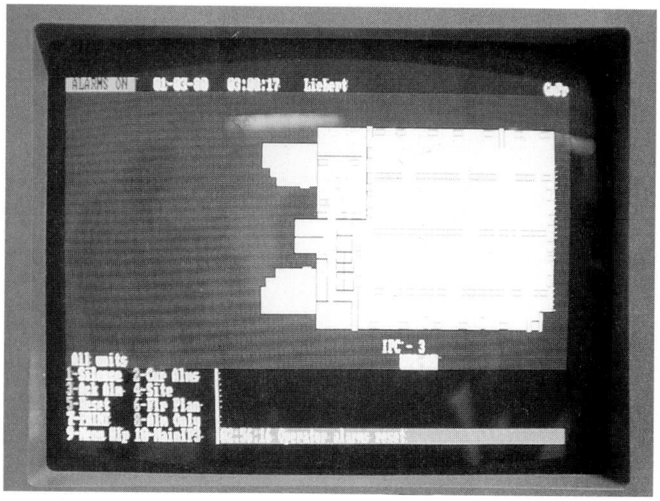

Figure 18.10 Some occupancies may have their building layout on computer. A hard copy of this information would save firefighters from having to sketch the layout.

After permission to inspect the property has been obtained, the firefighter should return to the outside of the building and first conduct an inspection of the exterior (Figure 18.11). This procedure makes the inspection of the interior easier and provides the necessary information for drawing the exterior walls on the floor plan sketch. When the survey of the exterior is completed, the inspection team should go directly to the roof or basement and proceed with a systematic inspection. Each floor should be inspected in succession and all areas on each floor should be inspected. The firefighter should ask that all locked rooms or closets be opened and tactfully explain why it is necessary to see these areas. For example, if the guide says, "There is nothing in this locked room," the firefighter may say, "Yes, we understand, but it is not particularly the material in the room that we must see. The size, shape, and construction of the room are important features." If admission to an area or room is refused because of scientific research or a secret process, the firefighter should suggest that a cover or a screen be set up to permit the inspection. Secret areas from which the inspection crew is barred should be reported to the fire chief or fire marshal for appropriate action.

Figure 18.11 Check fire department connections for obstructions or damage.

If the property includes several buildings, each should be inspected separately. It is a good idea to start on the roof of the highest building from which the firefighter can get a general view. A sketch of each floor should be completed before proceeding to the next floor. If a floor plan that was made on a previous inspection is used, the inspection can proceed more rapidly. Make sure, however, to record any changes that have been made and change the floor plan sketch accordingly.

The Final Interview

Reporting to the person with authority, after one or two days of intensive inspection, can do much to maintain a cooperative attitude of the owner toward matters that need attention (Figure 18.12). To leave the premises without consulting the person with authority might give the impression that the inspection did not take long and that it was an unimportant venture. During this interview, a firefighter or the company officer should first comment favorably on the good conditions that were found, and let it be known that these good points will certainly be contained in the inspection report. Unfavorable conditions may be discussed in general, but the firefighter should avoid technicalities and direct conclusions at that time. Explain that such conditions will be studied more fully and that recommended solutions will be submitted in a written report. Too much talk concerning unfavorable conditions will give the impression of petty faultfinding and may lead to an argumentative discussion. A final interview also gives the firefighter an opportunity to express thanks for the courtesies extended to the fire department and opens the way to explain how firefighters will study these reports from the standpoint of fire fighting procedures.

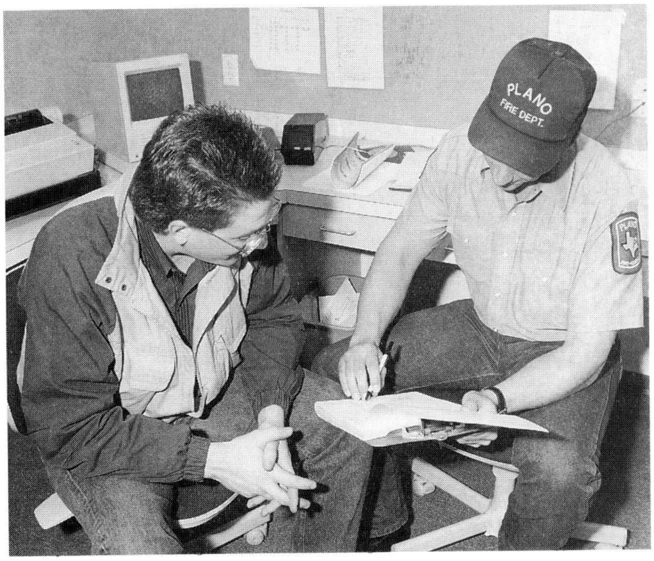

Figure 18.12 Following the inspection, review the results with the occupant.

Follow-Up

A follow-up system is one of the best ways to impress the importance of inspections upon occupants or owners. The follow-up should not necessarily be a surprise visit. The occupant should be told during the final interview of the inspection that a representative will return. This follow-up should be made within a few weeks after the occupant or the owner has received the written report. It should be conducted much in the same way as the original inspection. The follow-up representative should first go to the person to whom the report was written to discuss the inspection. Entering the property without this contact may give the impression that the firefighter is snooping. Being fair and honest in all dealings will prove more beneficial in the end.

Map And Sketch Making

Maps that convey information relative to construction, fire protection, occupancy, special hazards, and other details of buildings in legible form are an asset to fire prevention and fire suppression personnel. Larger occupancies or complexes may already have maps that were prepared by map companies. These maps use standard map symbols that are used almost universally (Figure 18.13).

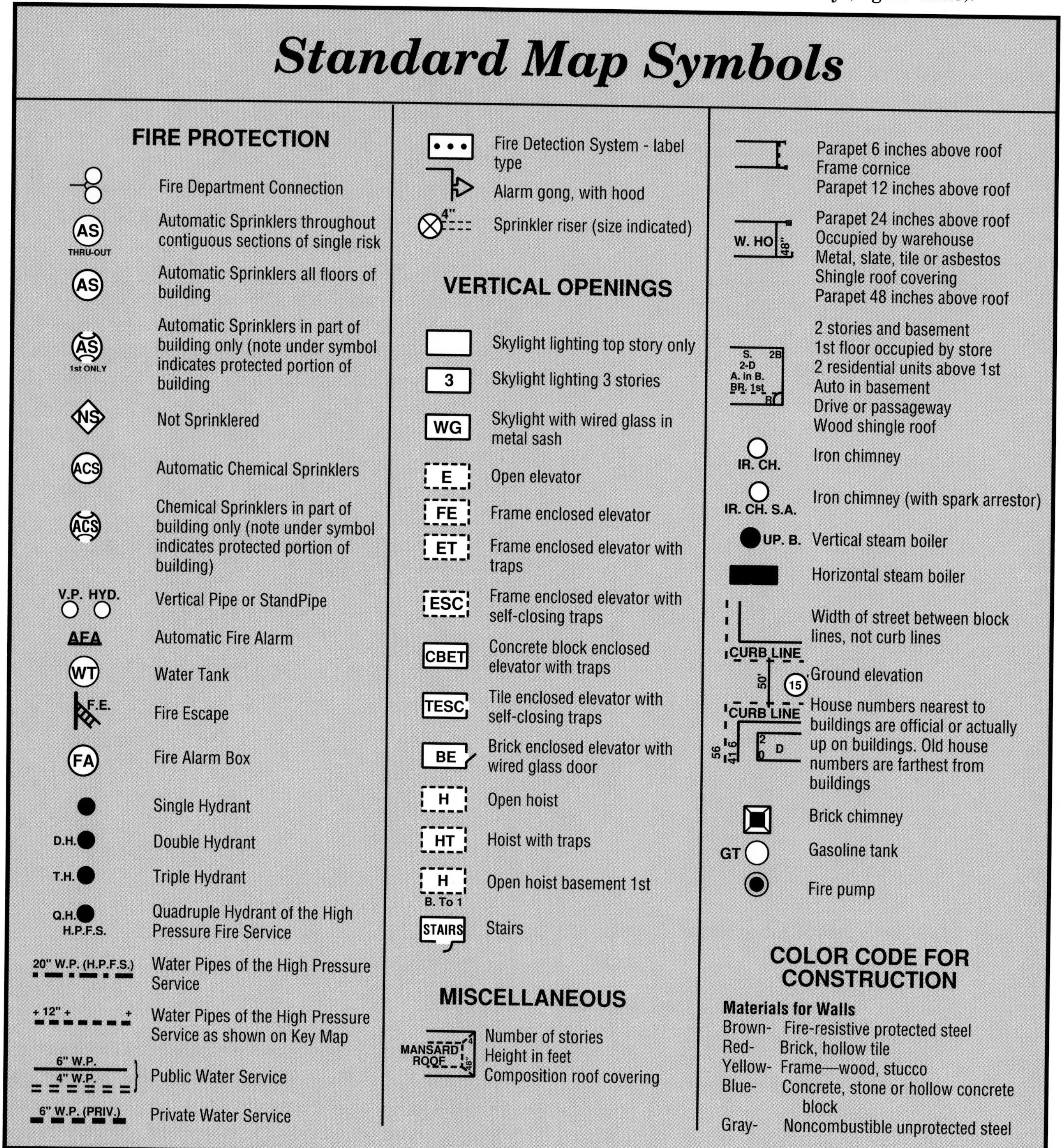
Standard Map Symbols
FIRE PROTECTION
Fire Department Connection
AS THRU-OUT
Automatic Sprinklers throughout contiguous sections of single risk
AS
Automatic Sprinklers all floors of building
AS 1st ONLY
Automatic Sprinklers in part of building only (note under symbol indicates protected portion of building
NS
Not Sprinklered
ACS
Automatic Chemical Sprinklers
ACS
Chemical Sprinklers in part of building only (note under symbol indicates protected portion of building)
V.P. HYD.
Vertical Pipe or StandPipe
AFA
Automatic Fire Alarm
WT
Water Tank
F.E.
Fire Escape
FA
Fire Alarm Box
Single Hydrant
D.H.
Double Hydrant
T.H.
Triple Hydrant
Q.H. H.P.F.S.
Quadruple Hydrant of the High Pressure Fire Service
20" W.P. (H.P.F.S.)
Water Pipes of the High Pressure Service
+ 12" + +
Water Pipes of the High Pressure Service as shown on Key Map
6" W.P.
4" W.P.
Public Water Service
6" W.P. (PRIV.)
Private Water Service
Fire Detection System - label type
Alarm gong, with hood
4"
Sprinkler riser (size indicated)
VERTICAL OPENINGS
Skylight lighting top story only
3
Skylight lighting 3 stories
WG
Skylight with wired glass in metal sash
E
Open elevator
FE
Frame enclosed elevator
ET
Frame enclosed elevator with traps
ESC
Frame enclosed elevator with self-closing traps
CBET
Concrete block enclosed elevator with traps
TESC
Tile enclosed elevator with self-closing traps
BE
Brick enclosed elevator with wired glass door
H
Open hoist
HT
Hoist with traps
H B. To 1
Open hoist basement 1st
STAIRS
Stairs
MISCELLANEOUS
MANSARD ROOF 48"
Number of stories
Height in feet
Composition roof covering
Parapet 6 inches above roof
Frame cornice
Parapet 12 inches above roof
W. HO 48"
Parapet 24 inches above roof
Occupied by warehouse
Metal, slate, tile or asbestos
Shingle roof covering
Parapet 48 inches above roof
S. 2B 2-D A. in B. BR. 1st B
2 stories and basement
1st floor occupied by store
2 residential units above 1st
Auto in basement
Drive or passageway
Wood shingle roof
IR. CH.
Iron chimney
IR. CH. S.A.
Iron chimney (with spark arrestor)
UP. B.
Vertical steam boiler
Horizontal steam boiler
CURB LINE
Width of street between block lines, not curb lines
50'
15
Ground elevation
CURB LINE
56 41 6 2 0 D
House numbers nearest to buildings are official or actually up on buildings. Old house numbers are farthest from buildings
Brick chimney
GT
Gasoline tank
Fire pump
COLOR CODE FOR CONSTRUCTION
Materials for Walls
Brown- Fire-resistive protected steel
Red- Brick, hollow tile
Yellow- Frame—wood, stucco
Blue- Concrete, stone or hollow concrete block
Gray- Noncombustible unprotected steel

Figure 18.13 Standard map symbols that firefighters may choose to use.

For buildings where existing maps are not accurate or available, fire department personnel should include some sort of sketch with their inspection report to show the general arrangement of the property with respect to streets, other buildings, and important features that will help determine fire fighting procedures. This sketch is commonly called a plot plan of the area. Property sketches that have been made during an inspection do not need to be drawn again during a reinspection except to make changes or additions. A firefighter's sketch of an area frequently constitutes the most informative part of an inspection report and should be made with neatness and accuracy.

A sketch that is made during an inspection may be done freehand with the aid of a clipboard and a 6-inch (160 mm) rule. Data should be recorded by using standard plan symbols as much as possible. Engineering or graph paper can make the process somewhat simpler.

The floor plan consists of an outline of each floor of the building. By using standard symbols, the firefighter can show the type of construction, thickness of walls, partitions, openings, roof types, parapets, and other important features. In addition to these features, protection devices, water mains, valves, and other miscellaneous items can be included (Figure 18.14).

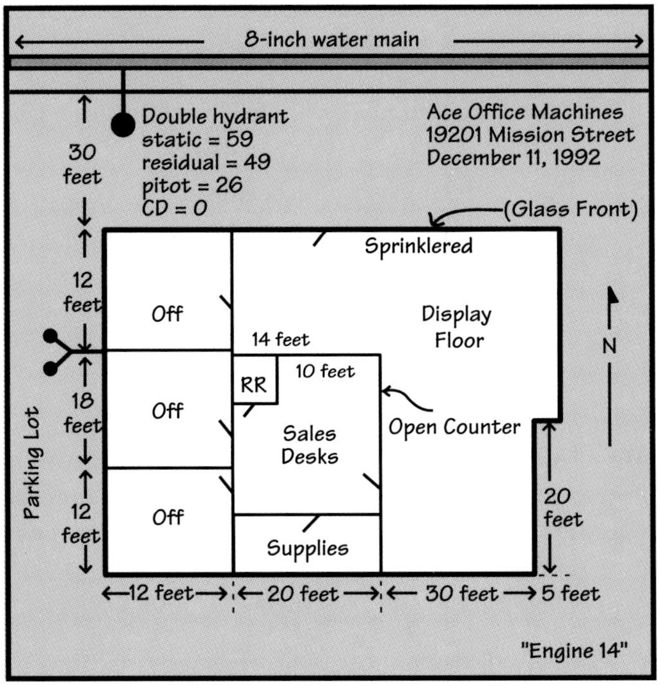

Figure 18.14 A basic floor plan sketch.

A sectional elevation sketch of a structure, consisting of a cross section or cutaway view of a particular portion of a building along a selected imaginary line, may be needed to show elevation changes, mezzanines, balconies, or other structural features (Figure 18.15). The easiest sectional view to portray is to establish the imaginary line along an exterior wall. This location theoretically removes the exterior wall and exposes such features as roof construction, floor construction, parapets, basements, attics, and other items that are difficult to show on a floor plan. Establishing the imaginary line along an exterior wall may not always show the section of the building that is desired. In this case, it may be better to divide the building near the center or along a line where a separate wing is attached to the main structure.

From the firefighter's sketch and notes, a permanent drawing can be made to be filed for future reference and classroom study. The permanent drawing should be drawn to scale.

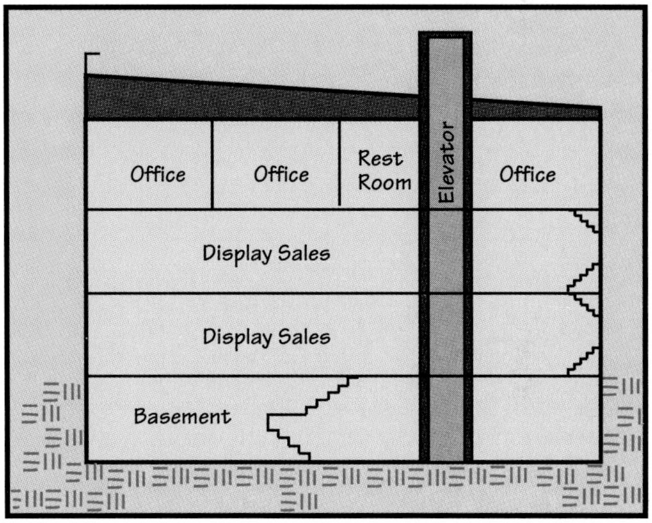

Figure 18.15 Sectional views show a cross section of the occupancy from top to bottom.

Photographs

Photographs show worthwhile detail for inspection reports, especially if they can be taken from more than one angle. One view that is especially good from a fire fighting standpoint is from an elevated position (Figure 18.16). An adjoining building or elevated tower can be

Figure 18.16 When possible, get an elevated view of the building.

used for this purpose. Interior and close-up photographs are very effective aids in making a complete report. A good camera, equipped with flash attachment, should be considered standard equipment for fire department inspectors. Photographs should not be taken if they make the occupant uncomfortable.

Written Reports

Written reports not only serve as records of the inspection, but also can be used as the basis for legal action. Without written evidence of an inspection, no proof exists that the firefighter or inspector gave the owner notice of hazardous conditions or corrective measures to be taken.

Firefighters must record every fire inspection, but a formal report is not always necessary. On the majority of inspections, the firefighter can easily record data and violations on the inspection form or checklist. Such forms often provide a means by which the firefighter can make recommendations to the owner (Figure 18.17). Situations that require formal reports are those that involve life-threatening hazards, major renovations to comply with codes, or an extensive list of minor violations.

The formal inspection report provides information to the owner/occupant. In the contents of the report, the firefighter should inform, analyze, and recommend. It is sometimes necessary for another person to complete an inspection or to conduct follow-up inspections; therefore, it is very important that the report be complete.

In an inspection report, firefighters are generally concerned with presenting the facts and evidence to prove a point, draw a conclusion, or justify a recommendation. This information must be presented in a businesslike manner. The report should not be opinionated, biased, emotional, or unfair.

The formal report should be written in letter form, including the following information:

- Name of business
- Type of occupancy
- Date of inspection
- Name of people conducting the inspection
- Name of business owner/occupant
- Name of property owner
- Edition of applicable code, as reference for future inspections
- List of violations and their locations stated in specific terms (Code Section numbers should be referenced)

PLANO FIRE DEPARTMENT
Fire Inspection Report

Location ____________ Date of Inspection ____________
Name of Business/Type of Occupancy ____________ No. Floors ______
Construction ____________

Inspection Type:
Regular .... A ____
Complaint .... B ____
Cert. of Occupancy .... C ____
System .... D ____
Home .... E ____

Violations requiring corrective action are marked below with a cross (X).
Inspection will be made within________days.

1. Aisles
   - Blocked .... A ___
   - Inadequate .... B ___
   - Not lighted .... C ___
2. Alarm System
   - None .... A ___
   - Inadequate .... B ___
   - Defective .... C ___
   - Other .... D ___
3. Building Repairs
   - Dangerous Conditions ..... A ___
   - Other .... B ___
4. Burning
   - No Permit .... A ___
   - Out of Hours .... B ___
   - Illegal Container .... C ___
   - Too Close to Building .... D ___
5. Chemicals
   - Improper Storage .... A ___
   - No Permit .... B ___
6. Combustibles
   - Excessive Storage .... A ___
   - Improper Storage .... B ___
7. Doors
   - Blocked .... A ___
   - Not Self-Closing .... B ___
   - Unapproved .... C ___
   - None as Required .... D ___
8. Ducts
   - No Fire Damper .... A ___
   - Others .... B ___
9. Electricity
   - Overloaded .... A ___
   - Defective .... B ___
   - Extension Cords .... C ___
   - Other .... D ___
10. Elevators
    - Emergency Tools/Key ..... A ___
    - Improper Enclosure .... B ___
    - Pit Area .... C ___
    - Other .... D ___
11. Exit Lights/Signs
    - Out .... A ___
    - Needed .... B ___
    - Wrong Type/Location .... C ___
12. Extinguishers
    - Recharge .... A ___
    - Hang .... B ___
    - Repair or Test .... C ___
    - Inadequate .... D ___
    - Wrong Type .... E ___
13. Exits
    - Insufficient. .... A ___
    - Not Indicated/Lights .... B ___
    - Blocked or Locked .... C ___
    - Wrong Type .... D ___
    - Wrong Hardware .... E ___
14. Explosives/Ammunition
    - Improper Storage .... A ___
    - No Permit .... B ___
    - Other .... C ___
15. Fire Door
    - Blocked/Inoperative .... A ___
    - No Closing Devices .... B ___
    - Inadequate .... C ___
    - None .... D ___
16. Fire Escape
    - Defective/Repairs .... A ___
    - None as Required .... B ___
    - Inadequate or Wrong .... C ___
17. Flameproofing
    - Curtains/Drapes .... A ___
    - Decorations .... B ___
18. Flammable Liquids
    - Improper Storage .... A ___
    - No Safety Cans .... B ___
    - Too Close to Heat .... C ___
    - Improper Dispensing .... D ___
    - No Permit .... E ___
19. Housekeeping
    - Area/Yards/Buildings .... A ___
    - Improper Disposal .... B ___
    - Excessive Storage of Waste .... C ___
20. Firewalls/Enclosures
    - None .... A ___
    - Incomplete .... B ___
    - Door Not to Code .... C ___
    - Openings Unprotected ..... D ___
21. Open Flame Devices
    - Open Top/Sides .... A ___
    - Use Not to Code .... B ___
22. Paint Spraying
    - Vented to Outside. .... A ___
    - Booth Not to Code/None ... B ___
    - No Sprinkler Heads .... C ___
    - Other .... D ___
23. Fixed Ext. Systems
    - Inadequate Coverage ...... A ___
    - Needed Service .... B ___
    - Other .... C ___
24. Heating
    - Repair .... A ___
    - Combustion Air .... B ___
    - Remove Combustibles ..... C ___
    - Room Enclosure .... D ___
    - Other .... E ___
25. Sprinklers
    - Defective Heads .... A ___
    - Incomplete Coverage ...... B ___
    - Storage Too Close to Head . C ___
    - Other .... D ___
26. Stairs
    - No Handrail .... A ___
    - Blocked .... B ___
27. Standpipes-Hose Cabinets
    - Threads/Hose .... A ___
    - Repair/Other .... B ___
28. Welding
    - No Permit .... A ___
    - No Fire Curtain .... B ___
    - Too Close to Combustibles .. C ___
29. Address
    - Not Posted .... A ___
30. Other Violations
    - See REMARKS below ..... A ___

A copy of this notice will be on file in the office of the Bureau of Fire Prevention for further action.

REMARKS

____________ Owner/Agent/Manager/Representative ____________ Inspector

Emergency Contact #1 ____________ Emergency Contact #2 ____________

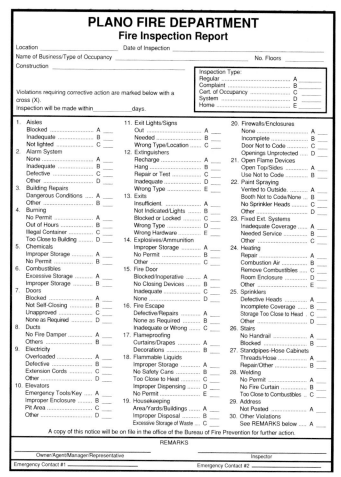

Figure 18.17 A sample inspection form.

- Specific recommendations for correcting each violation
- Date of the follow-up inspection

In the final portion of the report, firefighters should answer any questions they can and refer the owner/occupant to the fire marshal's office for further assistance. Reports can be sent by registered mail to ensure that they are received.

Proper filing techniques are also an important aspect of record keeping. A report is an accurate account used to describe a specific state or condition. The report becomes record when it is stored and is capable of being retrieved upon request. All records should be kept in orderly files so that they can be found when needed.

FIRE HAZARDS

A *fire hazard* may be defined as a condition that will encourage a fire to start or will increase the extent or severity of the fire (Figure 18.18). In order to prevent a situation from becoming hazardous, the fuel supply, heat source, and oxygen supply (fire triangle) must be considered. If any one of these can be eliminated, a fire

Figure 18.18 Spray booths are one example of hazards that firefighters should take special note of. *Courtesy of Des Plaines (IL) Fire Department.*

cannot occur. The oxygen supply hazard is normally present in air, and very little control can be maintained over the oxygen supply except in special cases.

Fuel supply and heat source hazards are more readily controlled than the oxygen supply hazard. If heat sources are kept separated from fuel supplies, the condition will be safe indefinitely. Some fuel supplies may be easily ignited, and they may have some characteristics that make them dangerous from a fire safety standpoint. Any heat source may be dangerous. Some common fuel supply and heat source hazards are included in the following lists.

Fuel Supply Hazards

- Ordinary combustibles such as wood, cloth, or paper
- Flammable and combustible liquids such as gasoline, oils, lacquers, or alcohol (Figure 18.19)
- Flammable and combustible gases such as natural gas, liquefied petroleum gas (LPG), and compressed natural gas (CNG)
- Chemicals such as nitrates, oxides, or chlorates
- Dusts such as grain, wood, metal, or coal
- Metals such as magnesium, sodium, or potassium
- Plastics, resins, and cellulose

Heat Source Hazards

- Chemical heat energy — Heat of combustion, spontaneous heating, heat of decomposition, heat of solution. This occurs as a result of material being improperly used or stored. The materials may come in contact with each other and react, or they may decompose and generate heat.

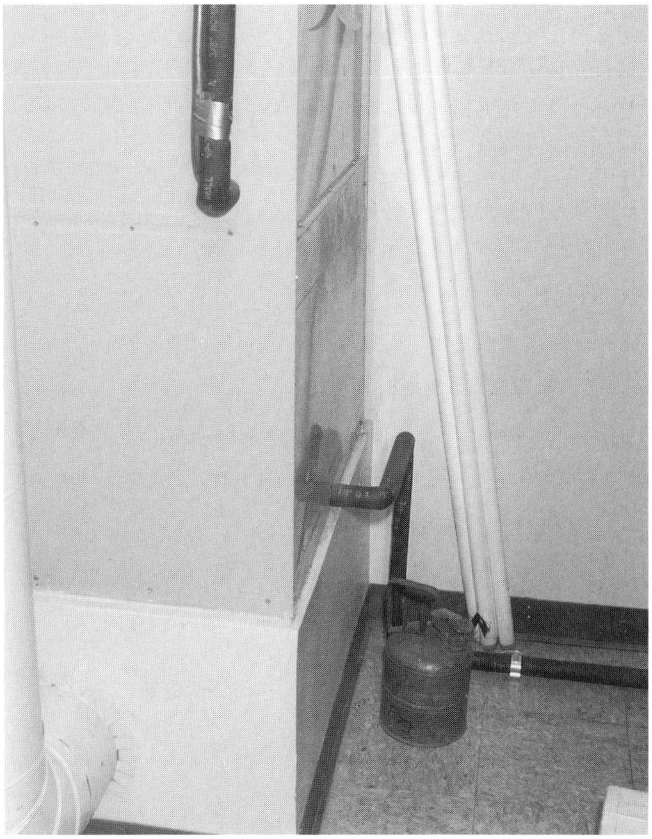

Figure 18.19 Look for obvious hazards such as this flammable liquid container stored next to the heater.

- Electrical heat energy — Resistance heating, dielectric heating, heat from arcing, heat from static electricity. Poorly maintained electrical appliances, exposed wiring, and lightning are sources of electrical heat energies.
- Mechanical heat energy — Friction heat and heat of compression. Moving parts on machines, such as belts and bearings, are a source of mechanical heating (Figure 18.20).
- Nuclear heat energy — Heat created by fission is not commonly encountered by most firefighters.

Figure 18.20 Many types of machines, even automobiles, have belts and bearings that can overheat and start a fire if not operating properly.

Common Hazards

The term "common" could be misleading to some individuals. It refers to the probable frequency of the hazard being found and not to the severity of the hazard. A common fire hazard is a condition that is prevalent in almost all occupancies and will encourage a fire to start. Common fire hazards are listed in contrast to special hazards, which are usually characteristic of a given industry. Some common hazards are poor housekeeping; heating, lighting, and power equipment; floor cleaning compounds; packing materials; fumigation substances; and other flammable and combustible liquids.

Firefighters should be alert for the dangers posed by these common hazards. Poor housekeeping can make maneuvering through an area difficult. Poor housekeeping also increases the fire load in an area and increases the chance that a flammable or combustible material may come in contact with an ignition source. It also hides fire hazards in the clutter. Improperly functioning heating, lighting, or other electrical equipment can provide an ignition source for nearby combustibles. Floor cleaning compounds, fumigation substances, and other flammable and combustible liquids that are improperly used and stored can provide a volatile fuel source should an ignition source be present.

Firefighters must also make sure that exits are adequate in number and in reasonable condition (Figure 18.21). Firefighters should check to make sure that all storage is removed from exit aisles, corridors, or passageways; check the operation of all exit doors (especially those equipped with panic hardware) and make sure that any chains, deadbolts, or other extra locking devices are removed *immediately*, or the operation will be shut down until they are removed.

Personal hazards are probably the most serious of all common hazards. The term *personal hazard* covers all individual traits, habits, and personalities of the people who work, live, or visit the property or building in question. Personal hazards may be considered intangible, but they are always present. A combined program of public fire education and good fire prevention practices can minimize personal hazards.

Special Hazards

A special fire hazard may be defined as one that arises from the processes or operations that are characteristic of the individual occupancy. Commercial, manufacturing, and public assembly occupancies will each have their own particular hazards. Some of these hazards are:

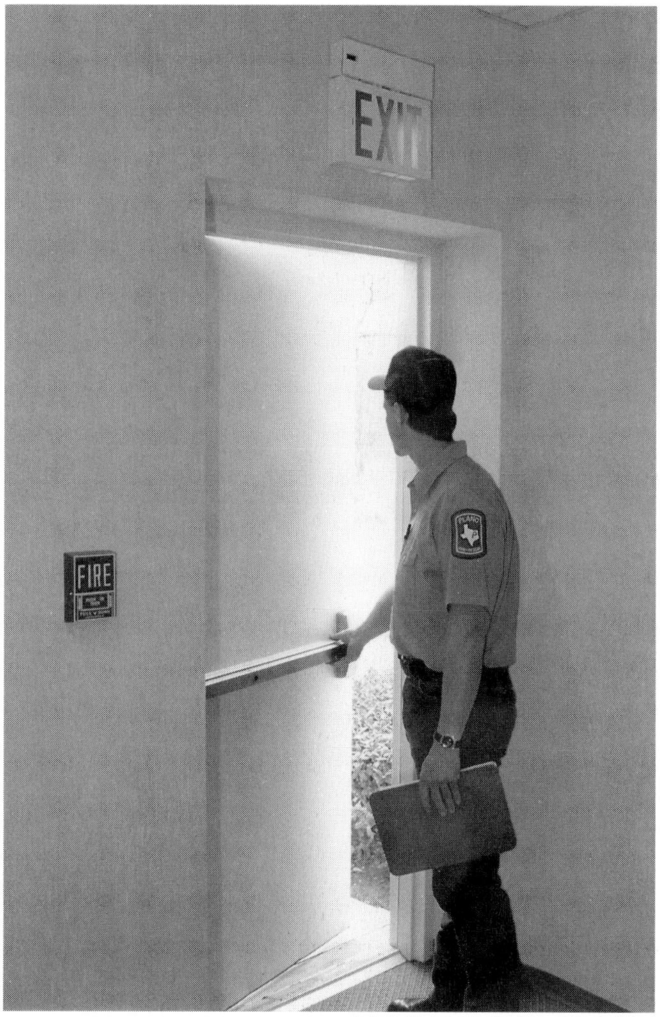

Figure 18.21 Make sure that all exits work properly.

Commercial Occupancies

- Large amounts of contents
- Mixed variety of contents
- Difficulties in entering during closed periods
- Common attics and cocklofts in many multiple occupancies (Figure 18.22)

Manufacturing

- Flammable liquids in dip tanks, ovens, and driers; plus those used in mixing, coating, spraying, and degreasing processes
- High-piled storage of combustible materials (Figure 18.23)
- Vehicles, such as fork trucks and other trucks, inside the building
- Large, open areas
- Large-scale use of flammable and combustible gases

Figure 18.22 Strip shopping centers often have common cocklofts that promote the spread of fire throughout the building.

Figure 18.23 High-piled storage produces extreme fire loads.

Public Assembly

- Large numbers of people present
- Insufficient, blocked, or locked exits
- Highly combustible interior finishes
- Storage of materials in paths of egress

Target Hazard Properties

A target hazard may be defined as a facility or process that could produce or stimulate a fire that could cause a large loss of life or property. Some examples of target hazards are lumberyards, bulk oil storage, area shopping centers, hospitals, theaters, nursing homes, rows of frame tenements, and schools (Figure 18.24). These occupancies should receive special attention in inspection programs.

Figure 18.24 Pay particular attention to target hazards such as this hospital.

INSPECTING FIRE PROTECTION SYSTEMS

Firefighters should pay particular attention to fire protection systems in the occupancies they inspect. Depending on departmental policies, firefighters making inspections may or may not actually test the system. However, regardless of whether or not the system is tested, firefighters should inspect each of the following systems.

Fixed Extinguishing Systems

The firefighter should check for:

- Open water supply valves
- Appropriate pressures on the gauges of automatic sprinkler systems (Figure 18.25)

Figure 18.25 Check the gauges to make sure that appropriate pressure is available.

- Damage or tampering of the components of the system
- Sprinklers that have been painted or otherwise damaged
- Special agent extinguishing systems that are charged and ready for service

Portable Extinguishers

The firefighter should make sure that:

- The portable extinguisher is a suitable type for its location
- Extinguishers are charged to the appropriate level indicated on the gauge
- The occupant replaces any extinguishers that show signs of damage or tampering
- The travel distances between extinguishers meet applicable code requirements
- Access to the extinguisher is not obstructed (Figure 18.26)

Figure 18.26 Objects should not be piled in the access area to fire extinguishers.

Standpipe Systems

The firefighter should inspect:

- Standpipe systems to ensure that they are ready for service
- Outlets for workable hose threads
- Outlets for foreign objects that may have been inserted into them
- Standpipe cabinet to see if hose and nozzles are there and to check their condition (Figure 18.27)
- Pressure reducing valves (PRV) to make sure that they are set at a workable level

Fire Detection And Alarm Systems

The firefighter should check:

- Fire detection and alarm systems to make sure that they are operable
- Annunciator panels to make sure that the power supply is on and that the system is not in a trouble mode (Figure 18.28)
- Entire system to make sure that no components, such as detector heads, are missing

DWELLING INSPECTIONS

Dwelling inspections, also called home fire safety surveys, may be part of a house-to-house fire prevention program, or they can be done when requested on an individual basis. They should be performed not only to provide an inspection, but to perform an educational and

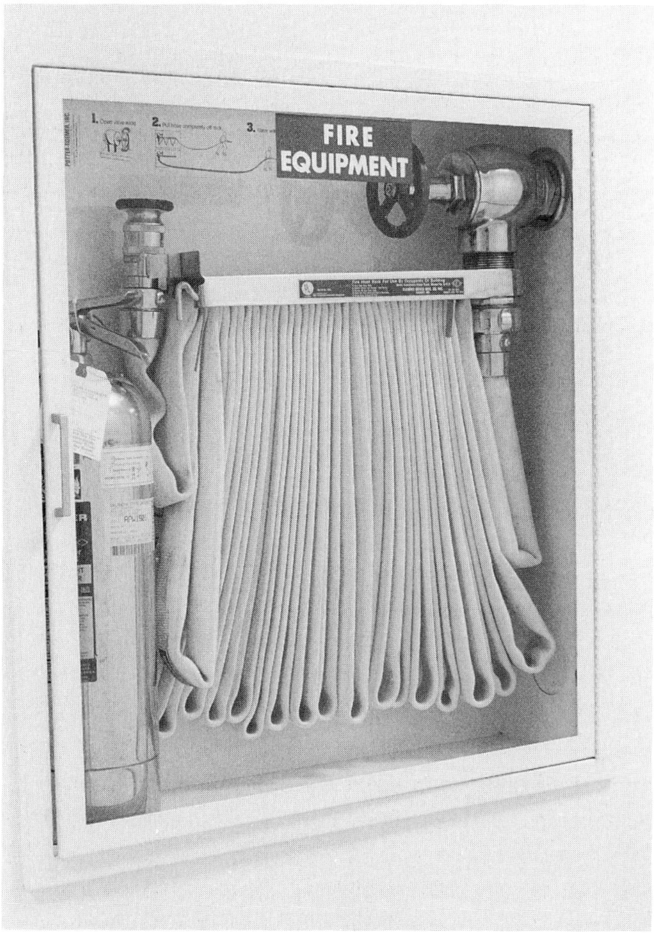

Figure 18.27 Hose cabinets or standpipe reel lines should be in good order and free of debris.

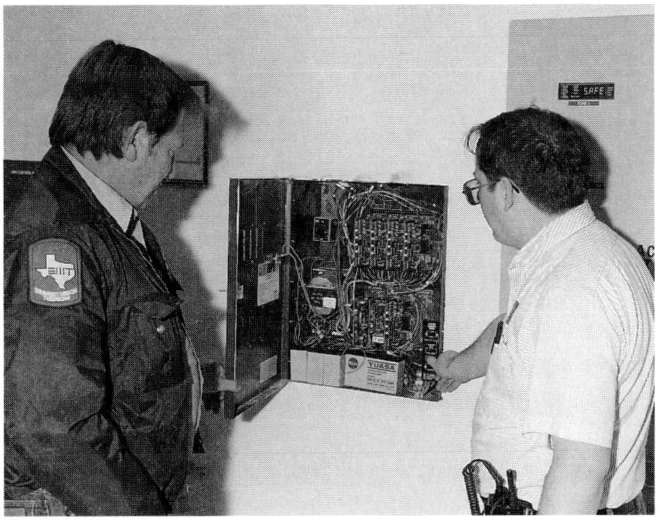

Figure 18.28 Make sure that annunciator panels are functioning properly.

advisory service as well. When dwelling inspections are done as part of an organized program, a great deal of advanced planning and publicity is necessary to gain full acceptance by the community. It must be understood that the program is a fire prevention activity and not a code enforcement activity. In other words, the firefighter is to look for hazards, not code violations. The objectives of home fire inspection programs are as follows:

- To obtain proper life safety conditions
- To keep fires from starting
- To help the owner or occupant understand and improve existing conditions

In addition to reducing loss of life and property damage, the fire department will realize other important benefits. Dwelling inspections give the firefighters a chance to impress people with the programs and activities of the fire department. The citizens who support the fire department will feel that they are getting "more for their money." Inspections give them a complete service, not just emergency service, and they will become more familiar with the duties and responsibilities of firefighters. Personal visits by firefighters to households for dwelling inspections generally result in an improvement in community support of the department.

Dwelling inspections are also a good way to distribute fire prevention literature, home fire drill information, child or invalid markers, telephone stickers, and other fire safety information (Figure 18.29). The firefighter can explain each item in the literature and possibly tie in a "local angle" of fire experience. Many fire departments also print special cards or slips to compliment the homeowner when the dwelling is found to be in a fire safe condition. Other cards are used to notify

Figure 18.29 Firefighters can provide homeowners with printed fire prevention materials during the inspection.

absent households that the firefighters visited the property for inspection.

Firefighters also gain valuable information when performing dwelling inspections. Firefighters who participate in dwelling inspections become acquainted with streets, hydrants and water supply locations, area development, and home construction, which is a part of pre-incident planning. Notes on these items and other useful information should be made and discussed during training sessions. Using the fire apparatus will also improve driver proficiency. While these fringe benefits are helpful, the primary reason for making inspections is to reduce fire and life hazards.

Firefighter Responsibilities

Every firefighter must fully understand that a dwelling inspection campaign is a fire department effort to reduce the number of fire deaths and home fires. It is the firefighter's responsibility to represent his or her fire department and meet the citizens with dignity and pride. The public has every right to expect firefighters to be fully qualified to advise on matters pertaining to fire prevention.

The approach to each home should be made on sidewalks or a path (never cut across the lawn), and shoes should be cleaned before entering the house (Figure 18.30 on the next page). If no one is at home and the firefighter is to leave some material, the mail box should

Figure 18.30 Make sure to clean your shoes before entering the house.

not be used; leave it between the doors or partially beneath the doormat. The firefighter should introduce himself or herself to the householder and state the purpose of the dwelling inspection service. The firefighter should then ask for permission to make the inspection. The firefighter should be courteous at all times, even when the home owner declines the inspection. When hazards are found, corrections should not be ordered. The hazard and its potential danger should be explained, and the proper method for correction suggested. The firefighter should not argue with the householder and should be complimentary when favorable conditions are found. The inspection should only include the basement, attic, utility room, storage rooms, kitchen, and garage. However, other rooms of the house may be inspected at the request of the householder or if necessary to access fuse boxes or breaker panels located in bedroom closets. It is important that the homeowner be sold on the idea of carrying out suggestions. For a successful dwelling inspection, the firefighter should use the following guidelines:

- Maintain a courteous attitude on all inspections.
- Thank the owner or occupant for the invitation into their home.
- Remember that the primary interest is preventing a fire that could take the lives of the occupants and destroy the home.
- Make constructive comments regarding the elimination of hazardous conditions.
- Keep the inspection confidential.
- Do not gossip.
- Never make notes from the inspection available to an insurance carrier, repair service organization, sales promotion groups, or any publicity group that would identify a given home.

What To Look For

There are several items firefighters should look for when making inspections of homes. It may be helpful to fill out an inspection form for each dwelling. The form can serve as a guide for firefighters making the inspection, or it can be used to make summaries of the inspections. A copy of this form can be given to the occupant.

Firefighters should be alert for signs of the most common causes of fires. These include the following:

- Heating appliances (Figure 18.31)
- Cooking procedures
- Smoking materials
- Electrical distribution
- Electrical appliances

It is critical that firefighters know these common causes in order to conduct inspections and to educate the public. An inspection provides the property owner with a valuable consulting service and is a means by which firefighters can more effectively carry out their responsibility of protecting lives and property. The following sections highlight some of the other items that should be checked.

ATTICS

When permission is obtained to go into the attic, firefighters should check for faulty electrical equipment, wiring, and other hazards. Attics should not be used for storage, and occupants should be encouraged to clean out all combustible materials.

ITEMS TO BE CHECKED FROM OUTSIDE

Condition of roof. Roofing that is old and warped is easily ignited by sparks and flying brands.

Condition of chimneys. Many types of chimneys are found in residential structures (Figure 18.32). Chimneys supported on wooden posts or brackets are likely to crack from settling and allow sparks or hot flue gases to set fire to woodwork. Loose bricks, open joints, and cracks that can be seen indicate that similar defects may exist in other parts of the chimney where they might start a fire. In such cases, a thorough investigation should be made. Improperly installed prefabricated chim-

Figure 18.31 Check the home heating unit for obvious problems.

Figure 18.32 Masonry chimneys are found on many homes with fireplaces.

Figure 18.33 Many newer chimneys are of the metal, prefabricated type.

neys can be hazardous if they are not sufficiently insulated or are in contact with combustible materials such as wood studs in walls (Figure 18.33). Chimneys should be equipped with spark arrestors to prevent sparks from exiting the chimney and starting a fire somewhere else (Figure 18.34).

Figure 18.34 A chimney should have a spark arrestor to prevent flying brand fires.

Condition of yard. Dry grass, leaves, paper, boards, branches of trees, and other combustible waste materials in yards and under porches and houses are readily ignited and are a fire hazard to buildings. Tall grass and vegetation should be kept at least 30 feet (10 m) from the house (Figure 18.35). Trees should not be allowed to hang over the roof line.

Waste burners. The location and arrangements for outdoor burning of waste should be noted for conformity to local regulations. Burn barrels should have metal screens over them to prevent embers from spreading the fire outside the barrel.

Condition of garages and sheds. Cleanliness and good maintenance are important precautions against fire that apply to sheds and garages as well as other buildings.

Flammable liquids. Gasoline and other similar flammable liquids should not be kept in dwellings.

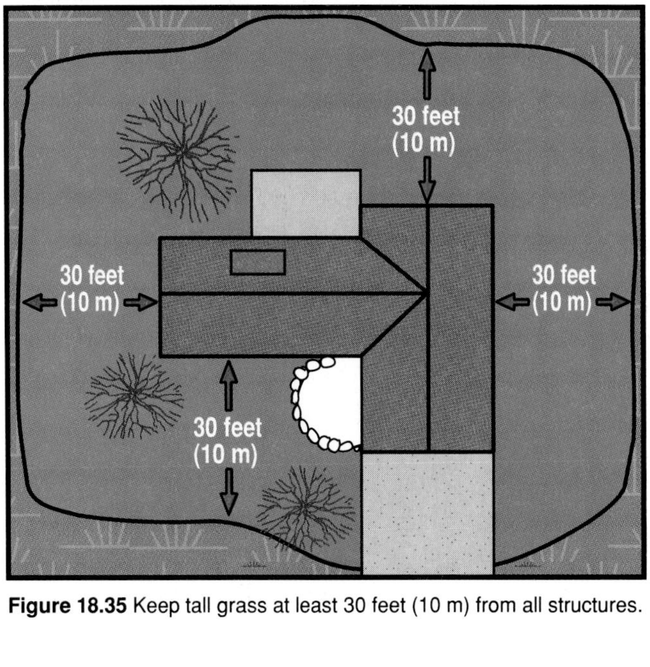

Figure 18.35 Keep tall grass at least 30 feet (10 m) from all structures.

Flammable liquids should be stored in a safety-type can in an outside storage area (Figure 18.36). Flammable liquids should never be used for home dry cleaning or for other purposes that would expose the dwelling to their explosive vapors.

Figure 18.36 Flammable and combustible liquids should be stored in tool sheds.

ITEMS TO BE CHECKED IN BASEMENT

Accumulations of waste and discarded material. Waste paper, discarded furniture, and partially filled paint cans constitute a fire hazard. A suggestion from a firefighter may provide the necessary incentive to dispose of such accumulations. Rags soiled with oil or paint are especially hazardous because of the danger of spontaneous ignition and because of the fuel that they add to a fire, regardless of its origin.

Furnace and stove vent pipes. Firefighters should be familiar with local regulations governing the installation of vent pipes from heating and cooking appliances. Charring of wood and blistering of paint indicate exposure to excessive temperatures.

Gas appliances. Corroded piping and rubber tubing may result in gas leaks. Automatic gas devices, without provisions for automatically cutting off the supply of gas when the pilot flame is extinguished, may produce an explosion. Devices that do not automatically come with an automatic gas control device should be retrofitted with one. A separate shutoff should be provided on the supply line to every appliance and the shutoff should be accessible without having to move the appliance.

Oil burning installations. Oil burners, supply tanks, and piping need to be properly installed to avoid danger of fire. Installations should be checked against local regulations and with NFPA 31, *Standard for the Installation of Oil Burning Equipment.*

Work rooms. Removal of shavings from work benches and the orderly storage of flammable liquids are features to be commented on by firefighters to the occupant.

HOME FIRE SAFETY

In addition to performing an inspection of the premises, firefighters should also provide home fire safety information to the occupants. Home fire safety should include escape plans in case of fire. These plans should be carefully reviewed and *practiced*, especially with children, at *regular intervals.* Some basic home fire safety rules include:

- Have two (or more) escape exits from every room.
- Windows should be easily opened by anyone and doors should remain closed (Figure 18.37).
- Always stay low if awakened by smoke; do not raise up (Figure 18.38).
- Have a whistle by every bed to alert other family members if awakened by the smell of smoke.
- Roll out of bed and crawl to the door. Feel the door; if it is warm, use the window for escape (Figure 18.39).

Figure 18.37 Children should be instructed on window operation.

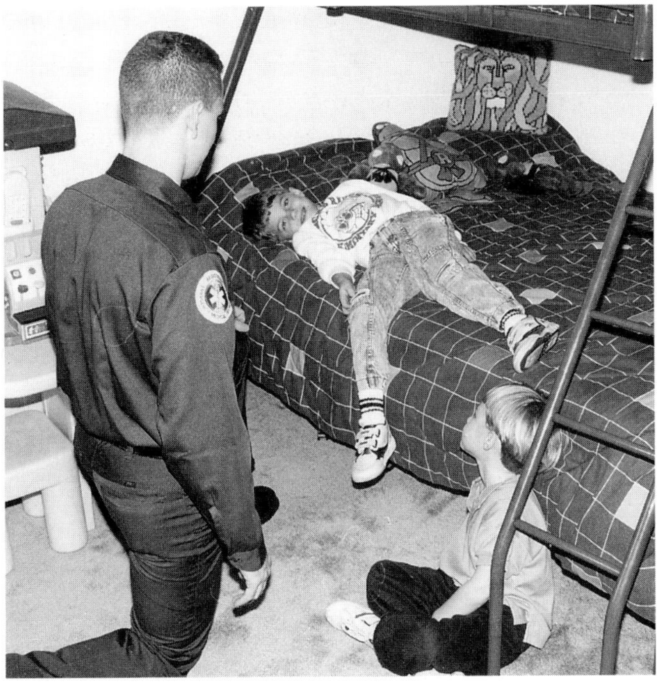

Figure 18.38 Emphasize that children should roll out of bed in a fire situation. They should not sit or stand up.

Figure 18.39 Children should be instructed to feel the door before opening it.

- Never return to the house once outside.
- A fire escape ladder should be kept by the window in a second-story house, and all family members should practice descending it.
- A meeting place outside the home should be agreed upon so that all members can be accounted for after escaping.
- Ask a neighbor to call the fire department.

STOP, DROP, AND ROLL

Firefighters should inform people of the actions to take if their clothing catches on fire. Impress upon people that they should not run — this action will only fan and worsen the flames. If your clothes catch on fire, you should STOP immediately where you are, DROP to the ground, and ROLL around until the flames are smothered (Figure 18.40). If someone else's clothes catch on fire, assist them in dropping to the ground and smothering the flames. A blanket or heavy coat can be used to help smother the flames. Once the fire is out, remove the burned clothes if possible, and cool the area with cold water if available. Summon emergency medical assistance immediately.

Figure 18.40 Practice the stop, drop, and roll technique for extinguishing clothing fires.

SMOKE DETECTORS

An important part of conducting home inspections is the firefighter's knowledge of residential smoke detectors. The firefighter must have the ability to educate the homeowner of the advantages and proper use of smoke detectors (Figure 18.41 on the next page). Smoke detectors greatly reduce the response time to a fire situation, especially during sleeping hours. This factor is a key to increasing the resident's chance of safely exiting the home in a fire situation.

Fire fighting is the most hazardous occupation in the United States, and search and rescue is one of the most dangerous services the firefighter performs. While an early warning is often credited with saving the lives of the home's occupants, do not forget that the smoke detector's warning may have also protected a firefighter from having to enter the structure. When smoke detectors are not in the home, the firefighter is then forced to go in the home and rescue the occupants.

Figure 18.41 All homes need smoke detectors.

Types Of Smoke Detectors

Smoke detectors operate on one of two basic principles: ionization or photoelectric. For maximum protection, the user should understand the advantages and disadvantages of both types.

IONIZATION

The ionization detector uses a small amount of radioactive material to make the air within a sensing chamber conduct electricity. When very small smoke particles enter the sensing chamber, they interfere with the conduction of electricity, reducing the current and triggering the alarm. The particles to which the detector responds are often smaller than can be seen with the human eye. Because the greatest number of these invisible particles are produced by flaming fires, ionization detectors respond slightly faster to open flaming fires than do photoelectric detectors. The radiation source in ionization detectors is not a hazard to the home's occupants. The U.S. Nuclear Regulatory Commission (NRC) performs a radiation safety analysis to determine that detectors meet safety requirements.

PHOTOELECTRIC

The photoelectric detector uses a small light source — either an incandescent bulb or a light-emitting diode (LED) — that shines its light into a dark sensing chamber. The sensing chamber also contains an electrical, light-sensitive component known as a photocell. The light source and photocell are arranged so that light from the source does not normally strike the photocell. When smoke particles enter the sensing chamber of the photoelectric detector, the light is reflected off the surface of the smoke particle, allowing it to strike the photocell and increase the voltage from the photocell. (This reflection of light is the same means by which we see smoke in the air; that is, light from the room strikes the smoke and reflects it into our eyes.) When the voltage reaches a predetermined level, the detector activates. Smoke particles that scatter visible light are larger in diameter than those that an ionization detector senses. Because smoldering fires produce these larger smoke particles in their greatest numbers, photoelectric detectors respond slightly faster to smoldering fires than ionization detectors.

Even though the average particle size changes considerably with temperature, all fires produce a broad range of particle sizes. Therefore, both types of detectors will detect most fires. While there will be some variation in the detector response time, the differences are fairly small when compared to the amount of escape time the detector provides. Numerous field tests have shown that either type of detector, when correctly installed, will provide adequate warning for escape.

Power Sources

Batteries or household current can power residential smoke detectors. Battery-operated detectors offer the advantage of easy installation — a screwdriver and a few minutes are all that are needed (Figure 18.42). Battery models are also independent of house power circuits and will operate during power failures. It is critical that only the specific battery recommended by the detector manufacturer be used for replacement. In many instances, a unit with the wrong battery will not respond to smoke even though its test button will function. Additionally, some detectors require special batteries that may be difficult to find in all but large metropolitan areas or are available only by mail order. While some detectors on the market take special batteries, many detectors require common batteries. If battery-powered detectors are preferred, residents of relatively isolated areas should purchase detectors using batteries that are readily available. The batteries should be changed at least twice a year or more if necessary. A good way to remember when to change the batteries is in the spring and fall when the clocks are set forward and back.

Field data indicate that many purchasers of battery-operated detectors have not immediately replaced worn-out batteries. Consequently, many codes requiring detectors in newly constructed homes specify 110-volt,

Figure 18.42 The most common type of residential smoke detector is battery operated.

hard-wired units, because people who did not voluntarily purchase detectors are assumed to be less knowledgeable about battery replacement. Therefore, the detector powered by household current is usually the most reliable mechanism. However, in some rural areas and areas with high thunderstorm occurrence, power failures may be more frequent, and battery-operated units may be more appropriate.

Several manufacturers produce electrically powered units with stand-by batteries for use during power failures. However, most stand-by battery systems do not last any longer than the primary batteries in battery-operated detectors. While possibly providing peace of mind, these units can be inoperable during power failures if the owners have not replaced a worn-out, stand-by battery.

Smoke Detector Location

A smoke detector in every room will provide the fastest detection times. However, the modest increase in escape time may not be enough to justify the additional expense in some family budgets.

Extensive field tests show that installing a smoke detector on every level of the living unit provides good all-around protection for the least investment (Figure 18.43). When providing this "every level" detection, the user should consider locations such as bedrooms, hallways, stairways, and normal exit routes.

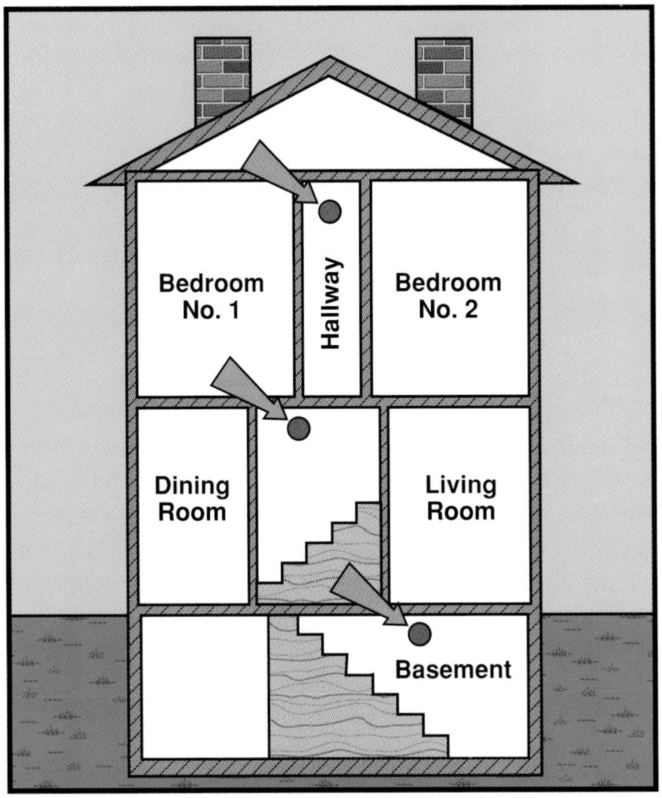

Figure 18.43 Smoke detectors should be located on each level of the structure.

At the minimum, users should install a smoke detector in the hallway outside each sleeping area and between the sleeping area and other rooms in the house (Figure 18.44). The detectors should be close enough to the bedrooms so that the alarm can be heard when the bedroom door is closed. It is most desirable to mount the detectors on the ceiling. However, if the ceiling mount is not possible, position them as high on the walls as possible.

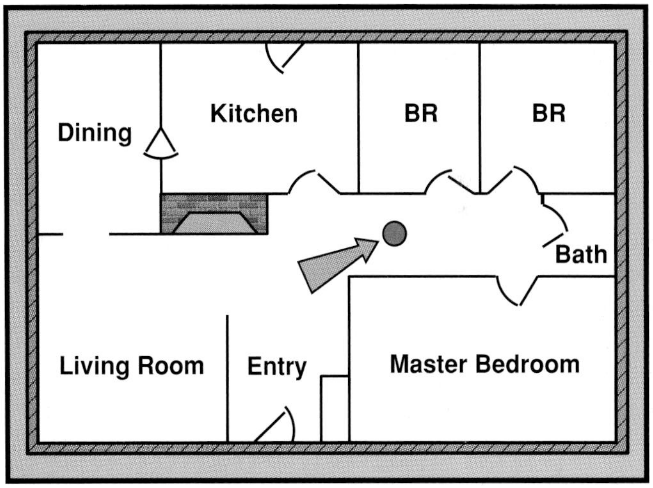

Figure 18.44 Locate smoke detectors outside the sleeping area of the home.

Testing

The smoke detector should be tested weekly according to the manufacturer's instructions. It is important that owners test their detectors at least once a month with smoke or an aerosol product specifically designed for this purpose, because some detectors have test buttons that only check the device's horn circuit. The owner of the detector may not realize this. Only when the detector incorporates a test button that simulates smoke or checks the detector's sensitivity can the "smoke test" be eliminated (Figure 18.45 on the next page). Smoke detectors with test buttons that simulate smoke or check sensitivity are recommended in homes where "smoke testing" is not likely to be conducted such as a handicapped or elderly person's home.

Testing with smoke is simple, but safe ways of producing the smoke should be used. Cigarette smoke is most often used for testing. Smoke from burning incense or a small piece of smoldering cotton rope or string in an ashtray can be used, especially when testing photoelectric detectors. Ionization detectors can also be tested by blowing over the top of the flame on a wood or paper match, directing the invisible or visible smoke particles toward the openings of the smoke detector. Canned smoke is available and has an advantage of not adhering to and obscuring the sensing device of the detector.

Figure 18.45 Test the smoke detector to make sure that it functions properly.

FIRE EXIT DRILLS FOR SCHOOLS

Fire exit drills are a matter of great importance when inspecting school buildings. Some states have laws that establish where the responsibility lies for holding school fire drills, their frequency, and other details. Firefighters should be aware of the specific requirements of such regulations. Where no law, code, or ordinance exists that requires exit drills, the fire department should encourage voluntary adoption of a regular drill program which includes drills at least twice per school year: once shortly after classes begin in the fall and once again after the holidays to reinforce the procedures and to involve newly enrolled students.

The purpose of fire exit drills is to ensure the efficient and safe use of the available exit facilities. Proper drills ensure orderly exit under controlled supervision and prevent panic, which has been responsible for many fatalities in numerous major fire disasters. Order and control are the primary purposes of the drill. Speed in emptying buildings, while desirable, is not in itself an objective and should be made secondary to keeping proper order and discipline.

Fire exit drills should be held with sufficient frequency to familiarize all occupants with the drill as a matter of established routine. Drills should be held at different times and under varying conditions to simulate the unusual conditions experienced during a fire. Some of these periods are during the changing of classes, when the school is at assembly, during the recess or gymnastic periods, or during other special events. If a drill is called during the time classes are changing, students should immediately proceed to the nearest available exit in an orderly manner. Instruction cards should be conspicuously posted that describe the procedure of the drills.

Responsibility for the planning and conducting of drills should be assigned only to those who are competent and qualified to exercise leadership. Emphasis should be placed upon orderly evacuation under proper discipline rather than upon speed. Drills should include procedures that ensure all persons in the building, or all persons subject to the drill, participate. If a fire exit drill is considered merely as a routine exercise from which some persons may be excused, there is grave danger that in an actual fire, the drill will fail in its intended purpose.

Each class or group should proceed to a predetermined point outside the building and remain there while a check is made to see that all are accounted for and that the building is safe to reenter. Such points should be sufficiently far from the building to avoid danger from fire in the building, interference with fire department operations, or confusion between classes or groups. No one should be permitted to reenter until after the drill is completed.

All fire exit drill alarms should be sounded on the fire alarm system and not on the signal system used to dismiss classes. Whenever any of the school authorities determine that an actual fire exists, they should immediately notify the local fire department. It should be a duty of principals and teachers to inspect all exit facilities daily. Particular attention should be given to keeping all doors unlocked, the paths of egress unblocked, and all stairs and fire escapes free from all obstructions. Any condition likely to interfere with safe exit should be immediately corrected or reported to the appropriate authorities.

Firefighters should check on the frequency of exit drills and the time required to vacate the building. A stopwatch should be considered as standard equipment for the evacuation practice of school buildings. The building fire alarm system should be examined to see if the alarm can be heard in all portions of the building and that it can be activated from each floor. Arrangements for promptly notifying the fire department should also be investigated.

FIRE STATION TOURS

Firefighters are frequently required to give tours of the fire station to civilians (Figure 18.46). These may be either spur-of-the-moment visits from people who walk in off the street or organized citizens' groups. Fire prevention week tours for groups of children are common. Firefighters should use these tours as an opportunity to impress the public. It is important to provide a good image for the fire department. During the time that the civilians are in the station, all firefighters should be dressed appropriately and performing appropriate duties. Citizens should not be witness to firefighters who are sleeping or engaged in unproductive activities (Figure 18.47). Do not hesitate to show the people all of the equipment in the station. You may wish to allow them to don some equipment to see how it feels. Answer all questions courteously and to the best of your ability. Fire safety and prevention information can also be passed on to visitors at this time.

Never allow visitors, especially children, to roam around the fire station unescorted. Visiting groups should be met by an assigned firefighter or officer who carefully explains what steps to take in case of an alarm (Figure 18.48). Special care should be taken to protect curious children from sliding poles. All groups should be kept together and, if necessary, arranged into smaller groups with a firefighter assigned to each group.

Equipment and apparatus should be demonstrated with considerable caution to ensure that no one gets into a dangerous position. Place a firefighter at each corner of an apparatus to prevent any youngster from slipping into, onto, or under the apparatus. Putting visitors on elevating platforms and ladders should be prohibited. Exercise caution when blowing the siren in the presence of children. There should be a cover plate at the face of the siren to keep curious children from putting their fingers into it.

Station mascots (dogs, cats, etc.) can be potential safety and liability hazards (Figure 18.49). The wisdom of keeping them around fire stations should be carefully weighed. Excited animals have been known to strike out and bite visitors and firefighters. If animals are kept in the fire station, they should be cared for by a veterinarian and should receive all the necessary inoculations to assure good health.

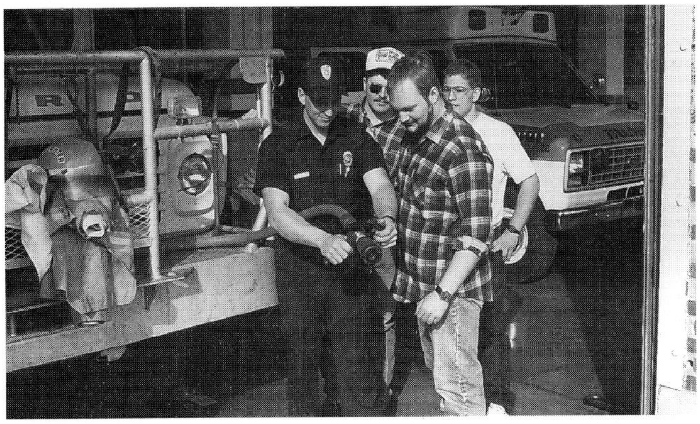

Figure 18.46 Firefighters commonly give station tours to civilians.

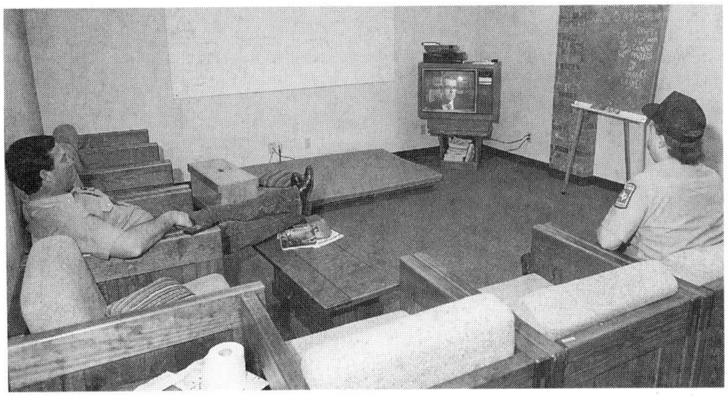

Figure 18.47 It is poor public relations to allow civilians to see lounging firefighters.

Figure 18.48 At the beginning of the tour, instruct the civilians where they should stand if an alarm comes in during the tour.

Figure 18.49 The decision to keep a mascot (such as a dog) in the station should be given careful consideration, because a dog can pose a potential hazard to firefighters and visitors.

19

Firefighter Safety

This chapter provides information that addresses performance objectives described in NFPA 1001, *Standard for Fire Fighter Professional Qualifications* (1992), particularly those referenced in the following sections:

3-3 Safety

3-3.1

3-3.2

3-3.3

3-3.4

3-3.5

3-3.6

3-3.7

3-3.8

3-3.9

3-3.10

3-3.11

4-3 Safety

4-3.1

4-3.2

Chapter 19
Firefighter Safety

Fire fighting is one of the world's most dangerous jobs, and accidents in this profession can result in costly losses — the greatest loss being the death of a firefighter (Figure 19.1). Other losses may include lost manpower (due to injuries), damaged equipment (which is expensive to repair or replace), and legal expenses. In order to prevent these losses, it is necessary to prevent the accidents that cause them. Reducing accidents will save lives and money.

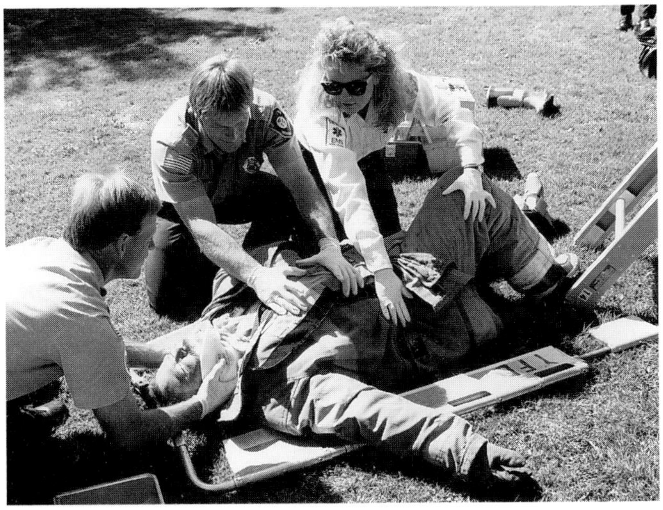

Figure 19.1 A firefighter receiving medical attention.

There are two basic factors that motivate accident control efforts within the fire fighting profession: the humane factor and the economic factor (Figure 19.2). The humane factor, while interrelated with economics, stems from the natural desire to conserve human resources to prevent needless suffering from physical pain or emotional stress. The economic factor includes legal expenses and expenses caused by the loss of manpower, apparatus, equipment, tools, property, or systems.

Yet despite these two strong motivational factors, firefighters have traditionally accepted injuries and related losses as part of their vocation. Knowing their job to be one of the most hazardous, many firefighters are resigned to occupational accidents, injuries, and fatalities; and this is compounded by the stereotypical image of the firefighter as heroic and fearless in the face of danger. The image we should be fostering is one of the firefighter who is too smart and too professional to take unnecessary risks. Most firefighter injuries are a direct result of preventable accidents.

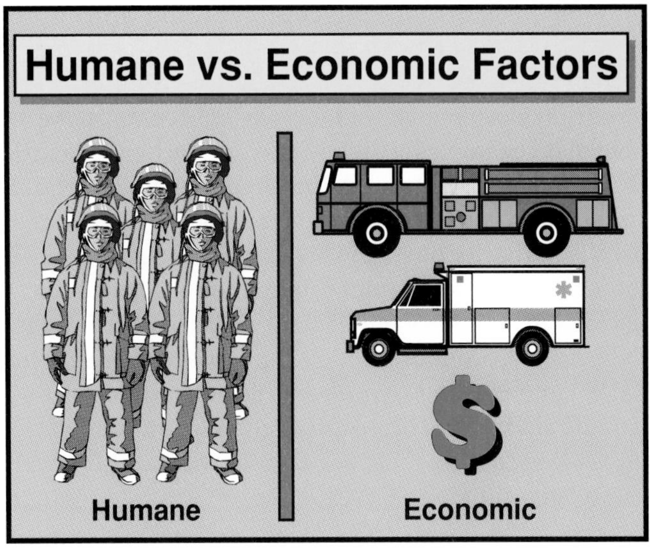

Figure 19.2 Safety is motivated by humane and economic factors.

An *accident* is defined as an unplanned, uncontrolled event resulting from unsafe acts and/or unsafe occupational conditions, either of which can result in injury. An *injury* can be defined as a hurt, damage, or loss sustained as a result of an accident.

The relationship between an accident and an injury is readily apparent. The injury can be the unfortunate result of the accident, although not every accident results in an injury. It is important to realize, however, that injuries will not be prevented unless the accidents that cause them are stopped.

Only by analyzing and understanding the causes of accidents is it possible to prevent them from happening. Safety reduces the number and severity of accidents.

The function of a safety program is to develop an objective procedure to carefully analyze, evaluate, and correct those factors that cause accidents and subsequent injuries.

SAFETY STANDARDS FOR THE FIRE SERVICE

Obviously, each fire department will have to tailor a safety program to meet its special needs. This process may be made easier by using existing safety standards as guidelines for program development. Before discussing these standards, it is important to understand the difference between laws and standards. *Laws* are rules of conduct that are adopted and enforced by an authority having jurisdiction. Anyone operating within the jurisdiction is bound to obey these laws or is subject to penalty for not doing so (Figure 19.3). *Standards* are criterion documents that are developed to serve as models or examples of desired performance or behaviors. No one is required to meet the requirements set forth in standards, unless those standards are legally adopted by the authority having jurisdiction, in which case they become law. For example, if your fire department is located in a state or municipality that has not adopted NFPA 1500, *Standard on Fire Department Occupational Safety and Health Program*, you do not have to meet the standard. However, the courts may hold you accountable for failing to recognize and comply with a well-known consensus standard.

Figure 19.3 Citizens who do not obey the laws are subject to penalty.

NFPA 1500, *Standard on Fire Department Occupational Safety and Health Program*

NFPA 1500, *Standard on Fire Department Occupational Safety and Health Program*, was originally approved in 1987. The standard contains the minimum requirements and procedures for a safety and health program. The standard may be applied to a fire department or similar organization, public or private. It calls upon the fire department to recognize safety and health as official objectives of the department and to provide as safe and healthy a work environment as possible. The basic concept of NFPA 1500 is to apply the same degree of safety throughout the fire service regardless of individual status or type of organization. Because it is a minimum standard, none of the objectives are intended to restrict a department or jurisdiction from exceeding the requirements specified by the standard.

GOALS AND OBJECTIVES

When the safety policy is adopted by the fire department, the goals and objectives must be determined. A *goal* is a broad, general, nonmeasurable statement of desired achievement. The main goals of any good safety program should be to:

- Prevent human suffering, deaths, injuries, illnesses, and exposures to hazardous atmospheres and contagious diseases.
- Prevent damage/loss of equipment.
- Reduce the incidence and severity of accidents and hazardous exposures.

Objectives are different from goals. An *objective* is a specific, measurable, achievable statement of intended accomplishment. Objectives always include a certain time frame. It is important to understand the difference in goals and objectives in order to successfully plan and implement a safety program. Once the goals of the program have been established, the objectives must be determined. NFPA 1500 should serve as a guide in establishing objectives.

In most cases, the fire chief, in conjunction with fire department management staff, will set the objectives of the safety program. However, in some cases, the authority having jurisdiction over the fire department will set the objectives.

EMPLOYEE INTEREST

The success of a safety program will begin at the top of the fire department administrative chain (Figure 19.4). The administration's attitude toward safety is invariably reflected in the attitude of the supervising officers, which in turn affects firefighters. An effective safety program becomes a matter of developing, promoting, and practicing an ongoing attitude of involvement throughout the organization. *Nothing can damage a safety program more than an officer who openly rejects or ridicules the program.* The participation of every supervising officer, particularly company officers who are in daily contact with the firefighters, should be enlisted to ensure the effectiveness of the overall safety program.

Figure 19.4 Safety begins at the top of the fire department command structure.

Safety requires effort on the part of everyone. If one person does not participate or abide by the rules of the program, chances are that others will follow this bad example. Because of their leadership role, officers must provide a good example and follow all safety rules. It is not enough to teach safety practices; they must be practiced and enforced. Breaking bad habits will not be easy for some, and once the new procedures are established, they must be maintained by *everyone*. If not, people will revert back to the old procedures.

The remainder of this chapter addresses various issues related to firefighter health and safety. Many of these topics are addressed in NFPA 1001, *Standard for Fire Fighter Professional Qualifications.*

PHYSICAL FITNESS AND HEALTH CONSIDERATIONS

Fire fighting is one of the most physically demanding and dangerous of all professions. Firefighters must be in good physical condition so that they are capable of handling these extreme demands. Firefighters need strength to perform such tasks as rescue victims, place ladders, handle hoselines, and force entry with heavy tools (Figure 19.5). Aerobic endurance is required to move rapidly down hallways, climb ladders, or fight fires on steep hillsides. Flexibility is needed to reach for equipment, tilt a ladder, and move a victim onto a ladder (Figure 19.6). Firefighters must perform these activities with little or no warm-up exercises and with 40 or more pounds (18 or more kg) of protective equipment. In addition, these difficult activities must often be performed in adversely hot, cold, or humid atmospheres filled with smoke and deadly gases.

Based on these job demands, it is obvious why such a large percentage of firefighter deaths are caused by heart attacks. Such activity places extreme stress on the body, especially the heart. Firefighters must be physically and mentally prepared for the job, or this stress will take its toll. Physical fitness and health programs can help reduce accidents, injuries, heart attacks, illnesses, and mental stress caused by the rigors of the job. Healthy, physically fit firefighters are able to perform their duties better, longer, and more safely than unfit firefighters (Figure 19.7 on the next page).

Figure 19.5 Raising a ladder requires strength and flexibility.

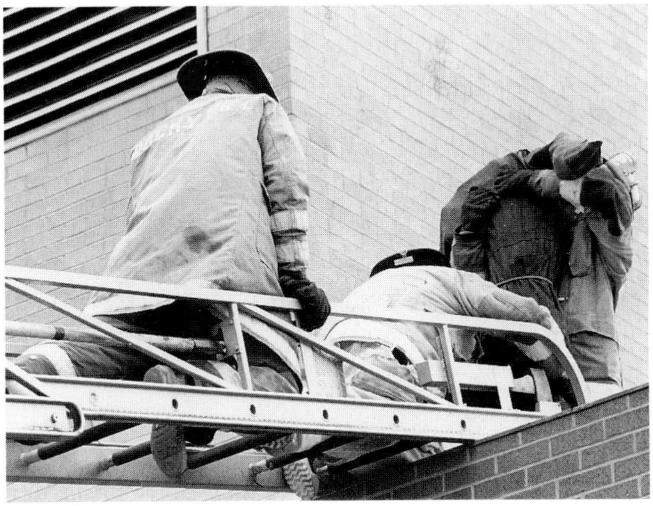

Figure 19.6 Moving a victim onto a ladder is an extremely physically demanding task.

Figure 19.7 Firefighter fitness is a key in long-term health maintenance.

A reason for determining the health of firefighters upon hiring involves possible future occupational discharges. For example, a firefighter may be found to have serious lung damage. Was the damage caused by the job? Was it unrelated to the firefighter's occupation? Did the person have the condition before entering the department? Only a thorough medical examination at the beginning of employment and a continuing diagnostic program can ascertain job-related injuries or illnesses.

The fire service is starting to recognize the importance of occupational safety. This recognition is reflected in the creation of NFPA 1500. Good health and physical fitness will never eradicate all deaths and injuries. However, many problems, especially heart attacks, can be reduced by good health and fitness. Body strength and flexibility can reduce falls and sprains, preventing damage to bone and tissue. Good lung capacity can reduce relative carbon dioxide levels and sustain the oxygen input into the bloodstream.

Firefighter physical fitness should be an ongoing maintenance program. It is the responsibility of the department to ensure that measures are taken to limit the number of stress-related accidents and illnesses. Physical fitness and health programs are a good way of fulfilling this responsibility.

Physical Fitness Programs

Many departments have recognized the value of a continuing fitness program. Since the implementation of these programs, departments have experienced lower injury rates, reduced severity of injuries, and fewer service-connected disability retirements. Firefighters participating in such fitness programs have enjoyed better general health and attained a higher work capacity (Figure 19.8).

Figure 19.8 All members should participate in physical fitness activities.

A physical fitness program will help reduce stress-related injuries and fatalities by strengthening the cardiovascular system. A healthy body can take more stress. Indeed, exercise is often a good way to relieve stress. Exercise will improve the health of the muscles, heart, and lungs, thereby reducing the chances of heart attack and other stress-related injuries and illnesses.

The following are characteristics of a successful physical fitness program:

- Total support of top fire department management
- Financial commitment
- Total participation in the program from chief to rookie based on job requirements
- Union and employee approval and input
- Program designed to improve and maintain cardiovascular endurance, strength, and flexibility
- Individual records to show improvements
- Initial physical exam for all involved
- Relative improvement emphasized as top priority

CARDIOVASCULAR TRAINING

The ability of the lungs, heart, and circulatory system to deliver oxygen can be greatly increased by physical training. To train the cardiovascular system, it must be placed under a fairly constant load. Studies have shown that to improve cardiovascular efficiency, training should be carried out two to three times a week at a level sufficient to raise the pulse rate within a specific range (60 to 85 percent of maximum heart rate), depending on current fitness levels.

Taking the pulse is not difficult and can be done at the radial artery in the wrist (Figure 19.9). The pulse should be taken within 5 seconds after stopping exercise and recorded for 10 seconds and multiplied by 6 (or recorded 15 seconds and multiplied by 4) to give the rate for 1 minute. Table 19.1 lists the 10-second pulse rates for various age groups.

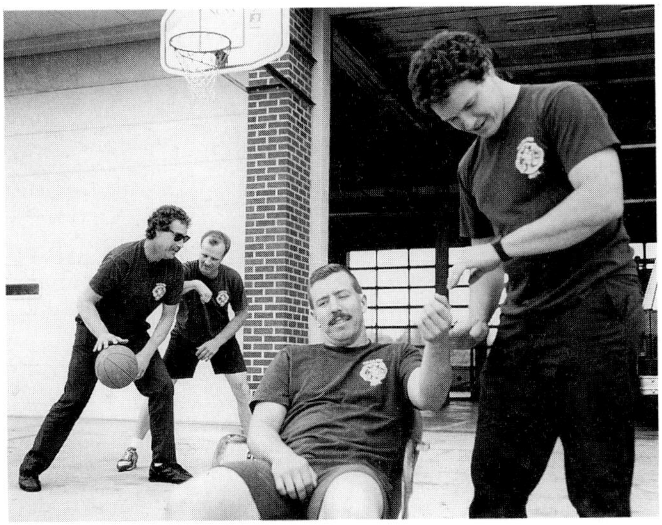

Figure 19.9 The pulse should be checked following physical exercise.

As training progresses, the increased efficiency of the heart can be measured by a drop in pulse rate for a given load. Likewise, the resting heart rate will decrease. As the work load is increased and the individual becomes conditioned, the resting heart rate will decrease. The work load must then be increased to achieve the same training effect. Once a basic level of cardiovascular fitness has been achieved, it can be maintained indefinitely with 20 minutes or more of elevated heart rate at least three times per week. As this occurs, the amount of blood being pumped by each stroke of the heart is increased and more efficiently moves the needed oxygen to the muscles of the body.

MUSCLE TRAINING

A second and equally important aspect of a physical fitness program is the strengthening of muscles and joints of the back, abdomen, legs, and arms. As the body's structural strength is improved, it is made more resistive to injury during any demanding activity (Figure 19.10).

It is important to note that whatever types of physical activities are employed in the physical fitness program, all exercising should be preceded by a 5- to 10-minute warm-up (Figure 19.11 on the next page). This warm-up should include stretching and moderate activity. The workout should also be followed by a 5- to 10-minute cool-down period.

BACK INJURY PREVENTION

One area that should be included in any fitness program is back injury prevention. The costs due to back

TABLE 19.1
TARGET PULSE RATES DURING EXERCISE
AGES 20-60

Age	*Target Pulse Rate** 60%	85%	Max
20 - 22	20	28	33
23 - 25	20	28	33
26 - 28	19	27	32
29 - 31	19	27	32
32 - 34	19	26	31
35 - 37	18	26	31
38 - 40	18	26	30
41 - 43	18	25	30
44 - 46	18	25	29
47 - 49	17	24	29
50 - 52	17	24	28
53 - 55	17	24	28
56 - 58	16	23	27
59 - 61	16	23	27

*Taken for 10 seconds

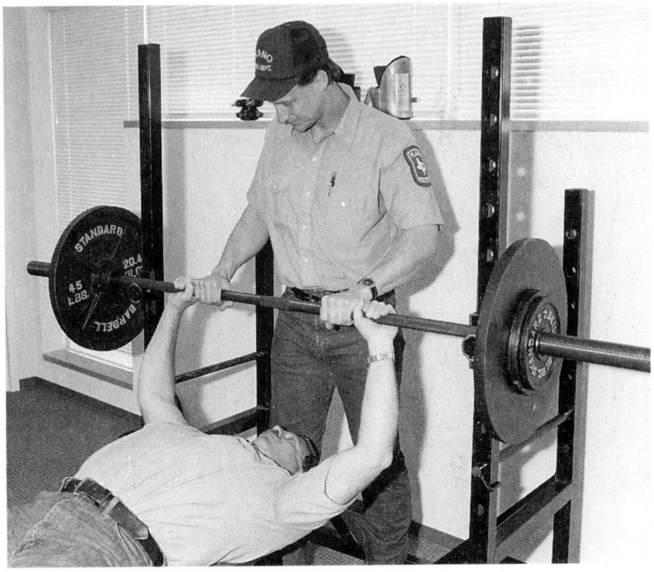

Figure 19.10 Strength training is important for firefighters.

Figure 19.11 All exercising should be preceded by stretching.

injuries are staggering, not to mention the pain and anguish. The average amount of lost time comes to 26 days per injury. Lower back pain accounts for approximately 20 percent of all medical expenses and disability pensions paid by the fire service. Back injuries (not to include traumatic injuries) are the result of the cumulative effects of wear and tear on the spine. This inevitable wear and tear is accelerated by inattention to numerous day-to-day activities such as lifting, carrying, and posture.

The amount of pressure exerted on the fulcrum or spine is directly related to the distance the mass is from the fulcrum, as can be seen by examining a cantilevered beam. The force required to lift a 10-pound weight at the end of a weight arm 10 times longer than the power arm is approximately 100 pounds, or 10 times the force required if the mass were located at a distance from the fulcrum equal to the power arm (Figure 19.12). What this example indicates in practical terms is that the pressure on the spine is greatly increased as the object being lifted is held at a distance out from the discs of the lower back.

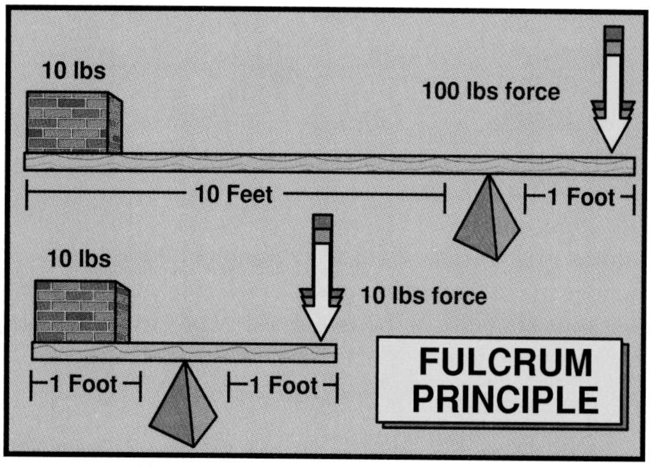

Figure 19.12 Understanding the fulcrum principle can make lifting easier.

The solution for proper lifting is so simple when compared to the magnitude of the problem. It can be summed up in three words: BEND YOUR KNEES. By bending your knees when lifting an object, the load can be more closely approached and will come closer to your spine (fulcrum). With your knees bent, your spine will assume a more comfortable and natural position (Figure 19.13). The load will then be lifted by your thighs, the most powerful muscle group of the body. At the beginning, it may take quite a bit of conscious effort to remember to bend your knees, but with practice, it will become natural.

Figure 19.13 Using proper lifting techniques helps to avoid back injuries.

Obviously, not all operations will allow firefighters to bend their knees properly because of the size, shape, and position of the load. Loads that cannot be lifted using the natural technique of bending the knees cannot be lifted safely and will put excessive stress on the spine as well as accelerate wear and tear. When situations allow, large loads should be handled by sufficient numbers of people to allow safe lifting (Figure 19.14).

Figure 19.14 Use two firefighters to carry heavy objects.

Nutrition

A quick glance around some fire stations — or in the mirror — should be all that is necessary to convince most firefighters that obesity is a frequent problem. In those rare cases where it is not, poor nutrition often is. A better understanding of nutrition should be one of the goals of any physical fitness program. The evidence is clear that a change in the average diet will significantly contribute to a healthier, safer firefighter.

Several factors have led to people being overweight in today's society:

- A significant drop in the average amount of personal energy output since the beginning of this century has created a surplus of energy (calories). A surplus of 3,500 calories equals a 1-pound (1,600 calories equal 1 kg) gain in weight.
- Increasing amounts of refined sugars in the diet have replaced the naturally occurring complex carbohydrates (starches). Complex carbohydrates include fruits, vegetables, and grains (including breads and pasta). Natural complex carbohydrates are more nourishing and were once a staple in our diet. Unlike complex carbohydrates, the sugars cannot be stored for energy.
- The consumption of fats in our diets has increased by about 30 percent since 1910. Fats can be stored as a source of energy; however, the human body was not designed to efficiently carry around excessive fat buildup.

In 1977, the Senate Select Committee on Nutritional and Human Needs made the following recommendations to improve our diet:

1. To avoid being overweight, consume only as much energy (calories) as needed; if overweight, decrease caloric intake and increase exercise. The rule of thumb used in exercise circles is: consume no more than 15 calories per day for every pound of body weight you wish to maintain.
2. Increase the consumption of complex carbohydrates and "naturally" occurring sugars from about 28 percent (the present level) to about 48 percent of caloric intake. Complex carbohydrates supply the digestive tract with the high levels of fiber necessary for proper bowel function.
3. Reduce the consumption of refined and processed sugar by about 45 percent to account for about 10 percent of caloric intake.
4. Reduce overall fat consumption from about 40 percent to about 30 percent of total calories. Product labels list fat in grams, so it is necessary to multiply the listed fat by 9 (9 calories per gram of fat) to convert the fat content to calories. You can then compute the percentage of total calories made up of fat.
5. Reduce saturated fat consumption to account for about 10 percent of total caloric intake and balance that with polyunsaturated and monosaturated fats, which should account for 10 percent of caloric intake each. Saturated fats include coconut oil, butter, lard, meat and chicken fat, and cheese.
6. Reduce cholesterol consumption based on the recommendation of a physician or other medical professional knowledgeable on the subject. Cholesterol is a controversial subject that has raged for ages. Most recommendations for reducing cholesterol levels start with a reduction of egg intake to two to four eggs a week.
7. Limit the intake of sodium by reducing the intake of salt to about 5 grams per day. Heavy users of salt are more likely to develop high blood pressure. (The actual nutritional requirement for salt is only about 5 percent of the 5 grams recommended.)

The goals cited by the committee are not to be taken as absolutes but rather as recommendations for improving nutrition and health. These recommendations are subject to change as our understanding of the body's nutritional requirements improve. However, it is essential that nutritional considerations are included in any physical fitness program.

The Effects Of Cigarette Smoking On Firefighters

It is a well-established fact that cigarette smoking is harmful to the lungs and to the respiratory tract, which is responsible for filtering impurities out of the air we breathe. After prolonged exposure to toxic substances contained in tobacco smoke, such as nitrogen oxides, hydrocarbons, and carbon monoxide, the protective mechanisms of the lungs fail and cannot be restored. The firefighter who smokes is more prone to lung injury than the average citizen because of double exposure by cigarette smoke and products of combustion found on the fire scene (Figure 19.15).

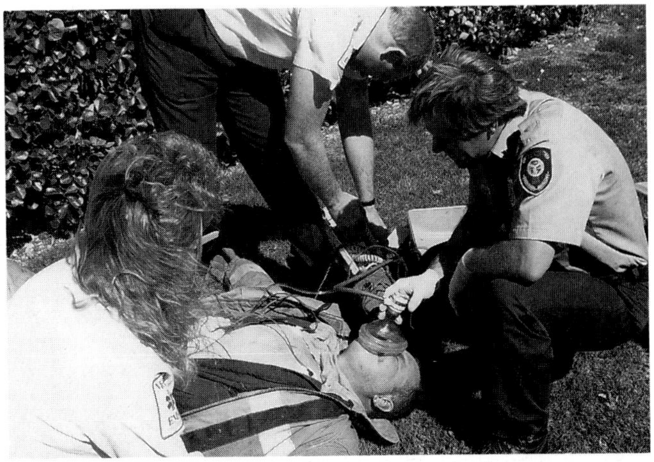

Figure 19.15 A downed firefighter receiving oxygen therapy.

Excessive exposure to carbon monoxide associated with cigarette smoking increases the blood's carbon monoxide level and decreases the blood's oxygen level. These two factors can have very serious complications for the smoking firefighter. Smoking causes a marked decrease in endurance and in cardiovascular function.

The firefighter who smokes is more prone to advanced arteriosclerosis (thickening of the walls of the arteries) that might impede the flow of blood. Smoking can also lead to heart attack or stroke. A smoker is two to three times more likely to have a heart attack and four to seven times more likely to die from that heart attack than the nonsmoker. The smoker also has a significantly greater chance of developing lung cancer.

The effects of smoking can directly affect a firefighter's ability to physically perform at a safe and effective level. Because of this, many departments have successfully implemented smoking policies (Figure 19.16). These policies range from no smoking while on duty, to more comprehensive policies that require firefighters to sign a statement saying that they will no longer smoke — on or off the job. Failure to obey the department smoking policy can result in disciplinary actions that range from reprimand to dismissal. To date, the courts have upheld dismissal based on the department smoking policy. Fire departments that adopt no-smoking policies should provide smoking members with the counseling or treatment needed to quit their smoking habit.

Figure 19.16 Many fire departments have instituted no-smoking policies in their fire stations.

FIREGROUND SAFETY

The greatest number of firefighter injuries and fatalities occur at the emergency scene (Figure 19.17). It is here that all training, planning, inspecting, and learning become vital. Being able to assimilate all information concerning the scene is of paramount importance to safe action. The behavior of a fire is not totally unpredictable if adequate information about the structure, fuels, and any special hazards involved is readily available. Limited information and knowledge, underestimating the fire potential, and miscalculating the structural integrity of the building can cause injuries and fatalities (Figure 19.18). Disorder, confusion, and chaos at the emergency scene can also lead to accidents. To avoid injuries, the emergency scene should be well structured with clearly defined roles and responsibilities. The officer in command of the incident must recognize potential hazards and conduct operations to minimize potential accidents and injuries.

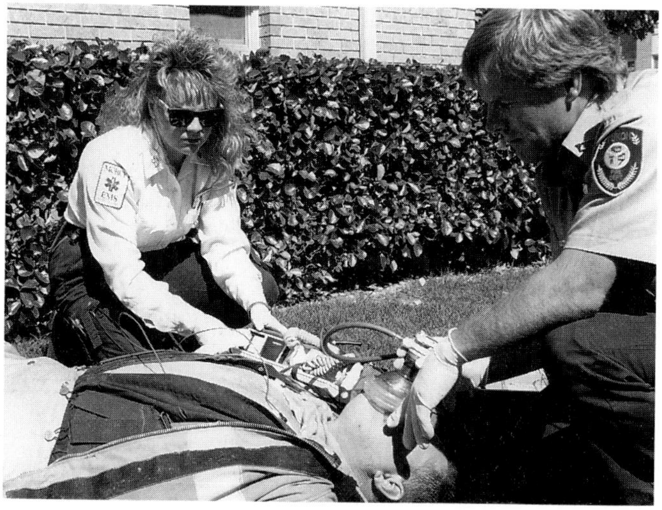

Figure 19.17 An injured firefighter at the emergency scene.

Figure 19.18 Fire personnel must understand how a structure reacts to the stresses of fire. *Courtesy of Harvey Eisner.*

Even though there are obvious variations in fires, most fires have more similarities than differences. These similarities are the basis for standard operating procedures (SOPs). Standard operating procedures are departmental policies that require and assign specific procedures to be carried out at any given incident. These procedures provide a standard set of actions that are the core of every fire fighting incident plan. The incident commander knows the SOPs and can base a plan of action upon them. Examples of SOPs are as follows:

1. The first unit on the scene assumes command.
2. The first-arriving engine attacks the fire.
3. The second-arriving engine will lay a supply line(s) to the first engine.
4. The first-arriving ladder truck will perform necessary forcible entry, search, rescue, and ventilation.

These SOPs will be performed with crews wearing complete protective clothing and SCBA.

When standard operating procedures are followed, chaos on the fire scene is reduced. All resources can be used in a coordinated effort to rescue victims, control the fire, and conserve property. Operational procedures that are standardized, clearly written, and mandated to each department member establish accountability and increase command and control effectiveness. When the firefighters of individual units are trained properly in SOPs, confusion is lessened. Firefighters will understand their duties and require a minimum of direction. SOPs will also help prevent duplication of effort and uncoordinated operations because all positions will be assigned and covered.

Safety should be a top priority when designing SOPs. Requiring SCBA for all crews is an example of a safety consideration. SOPs should be applied to all situations, including medical responses. SOPs should be designed to limit personnel exposure to contagious diseases. For example, SOPs may require personnel to use pocket masks when performing mouth-to-mouth resuscitation. SOPs may also require personnel to wear rubber gloves and safety glasses to prevent contact with patients' body fluids during medical emergencies (Figure 19.19). The assumption and transfer of command, communications procedures, and tactical procedures are other areas that must be covered by the SOPs.

The Officer's Role

Officers are the keys to safe fireground operations. All fire officers should meet the requirements specified in NFPA 1021, *Standard for Fire Officer Professional Qualifications*. Officers should be trained to evaluate the safety of structures and to deploy equipment and personnel. The first officer on the scene should be able to evaluate the situation and begin making decisions without waiting for another officer. If officers lack the ability to accurately evaluate a fire situation, they can make decisions that adversely affect the safety of firefighters. The situation at a fire can change frequently, and officers should be able to reevaluate and communicate their findings to the incident commander.

Figure 19.19 An SOP may dictate that firefighters responding to an EMS call wear protective rubber gloves.

Officers must assist incident commanders by keeping them informed of approaching dangers and hazardous conditions. All departmental officers and chiefs should recognize their technical limitations and make contingency plans to promptly provide the sources of technical advice necessary.

Incident Command And Accountability

To avoid chaos and confusion at the emergency scene, a unifying system must be employed to establish roles and responsibilities. Confusion can cause personnel to work against each other instead of against the fire. If a firefighter is confused about orders, he or she could hesitate in performing a specific duty at a vital time. Conflicting orders could cause personnel to open up the roof before interior crews are ready. A lack of command could result in crews pushing the fire into unburned portions of the structure, or worse yet, into other crews. Coordination of all personnel at the scene is absolutely necessary for operations to be conducted safely.

Many systems exist throughout the nation for the command and control of resources at emergency incidents. The National Fire Academy (NFA) has adopted the Incident Command System (ICS) as its base for teaching the concepts of incident command.

ICS is a system that is well documented and has been successfully used in managing available resources at a variety of emergency operations (Figure 19.20). All procedures will not perfectly fit all departments nor will the system necessarily need to be fully implemented for all situations the fire service will encounter, particularly routine responses.

Figure 19.20 Large fires require sound incident command procedures. *Courtesy of Bob Norman, Elkton (MD) V.F.D.*

The system consists of procedures for controlling personnel, facilities, equipment, and communications. It is designed to expand as needed from the time an incident occurs until the requirements for management and operations no longer exist. The ICS should be staffed and operated by qualified personnel from any emergency services agency and may involve personnel from a variety of agencies.

The system can be used for any type or size of emergency, ranging from a minor incident involving a single unit to a major emergency involving several agencies. The Incident Command System allows agencies to communicate using common terminology and operating procedures and allows for the timely combining of resources during an emergency.

COMPONENTS OF THE ICS

The Incident Command System has a number of components. The following interactive components provide the basis for an effective ICS operation:

- Common terminology
- Modular organization
- Integrated communications
- Unified command structure
- Consolidated action plans
- Manageable span-of-control
- Predesignated incident facilities
- Comprehensible resource management

ORGANIZATION AND OPERATIONS

All members of the department should be trained in the Incident Command System, and standard operating procedures should be designed in accordance with the system. This procedure enables personnel to know what is expected of them and what role and responsibilities they have at any given incident. The integration of SOPs with the five major functional areas of the ICS will eliminate confusion at the emergency scene and will help operations run smoothly and safely. The five major functional areas of the Incident Command System are command, operations, planning, logistics, and finance (Figure 19.21).

Command

An operation cannot be implemented unless someone is in charge. The person in charge of the operation is the incident commander. Command is responsible for all incident activities, including the development and implementation of strategic decisions. The responsibility of incident activities also includes dealing with the results of the strategic decisions. Command is the authority over the ordering and releasing of resources. There may be a limited number of staff positions within the command area such as the safety officer, liaison, and the public information officer.

Operations

This area is responsible for the management of all operations directly applicable to the primary mission. Its function is to direct the organization's tactical operations to meet the strategic goals developed by command (Figure 19.22). The operations area is divided into branches that may be further divided into groups or divisions. Groups are teams that are assigned to perform a particular task, such as ventilation or rescue.

Planning

This area is responsible for the collection, evaluation, dissemination, and use of information concerning the development of the incident. Planning is also responsible for maintaining the status of resources. Command will use the information compiled by planning to develop strategic goals and make alternate plans. Specific entities that may be placed under planning include the following:

- Resource unit
- Situation status unit
- Demobilization unit
- Technical specialist

Logistics

This section is responsible for providing the facilities, services, and materials necessary to support the incident. The two branches within logistics are the support branch and the service branch. The support branch includes

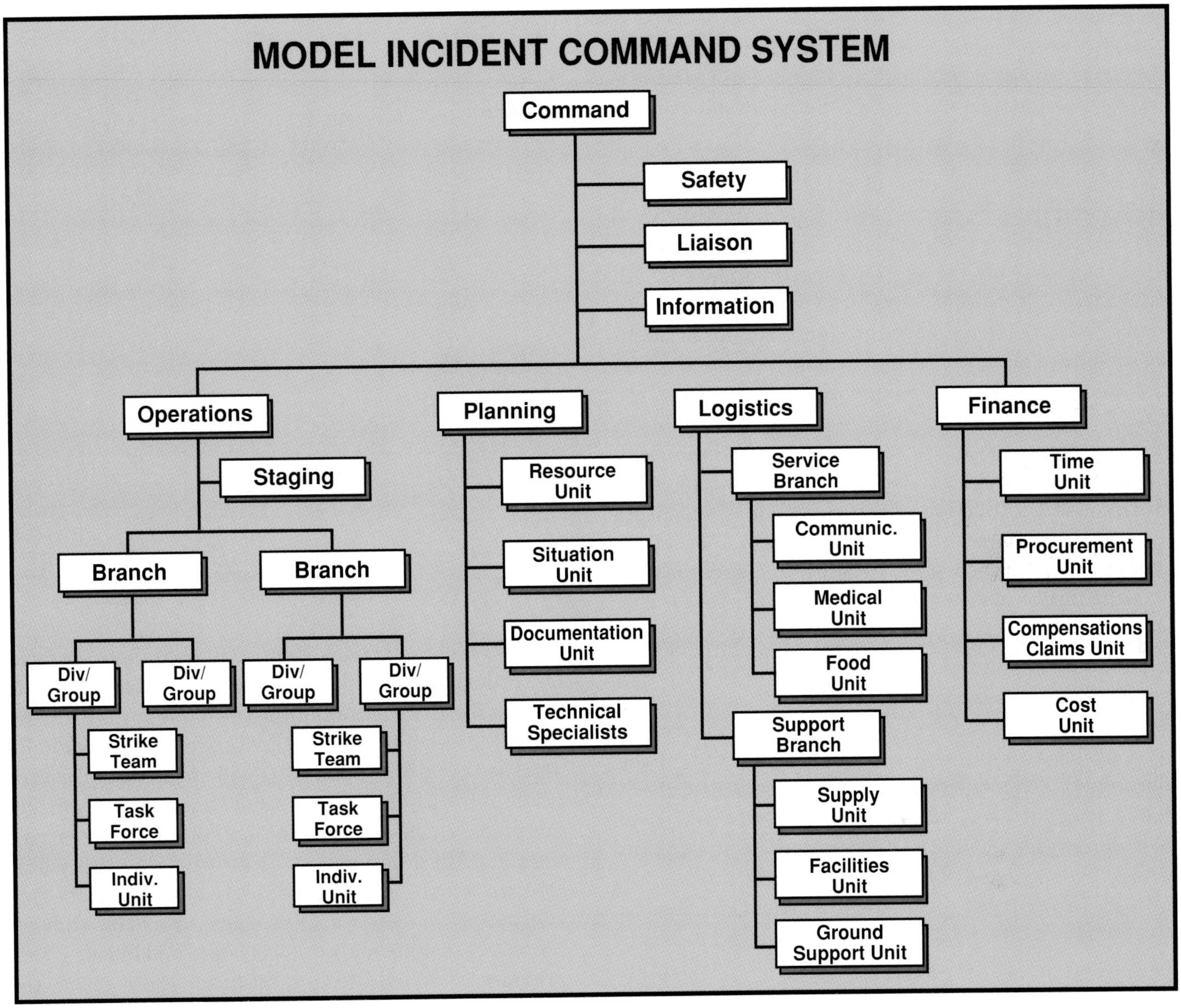

Figure 19.21 A model incident command chart.

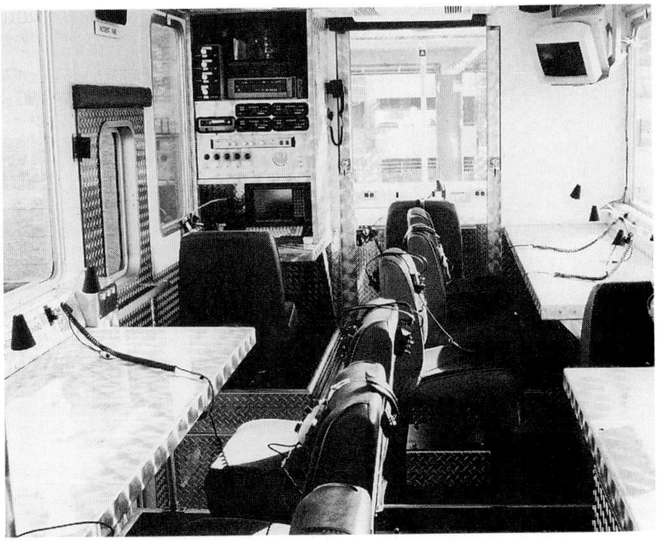

Figure 19.22 This command post is used to direct tactical operations.

medical, communications, and food services (Figure 19.23). The service branch includes supplies, facilities, and ground support (vehicle services).

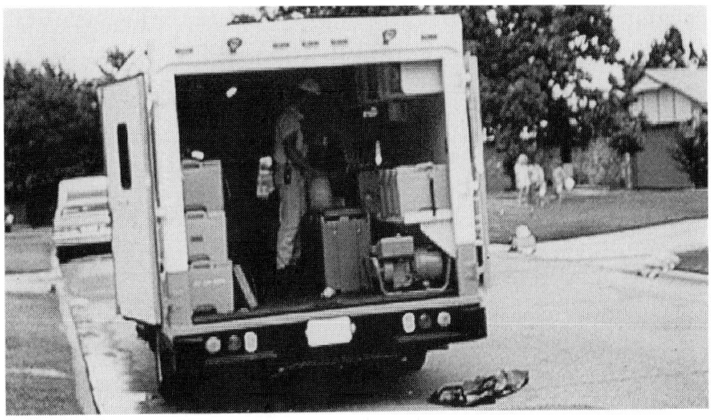

Figure 19.23 Canteen units provide food and drinks for emergency personnel.

Finance

This area has the responsibility for all costs and financial aspects of the incident. Finance will generally be of concern only at large-scale, long-term incidents.

PERSONNEL ACCOUNTABILITY SYSTEM

Each department must develop its own system of accountability. The system should be standardized so that it is used at every incident. The system can be as simple or as complex as the department needs. All personnel should be familiar with the system. Accountability is vital in the event of a serious accident or structural collapse. If the incident commander does not know who is on the fireground and where the firefighters are located, it is impossible to determine who and how many may have been trapped inside or injured. Flashover and backdraft may trap or injure firefighters. SCBAs can malfunction or run out of air. Firefighters can get lost in mazes of rooms and corridors. Too many firefighters have died because they were not discovered missing until it was too late.

Although the proper use of an incident command system will keep track of personnel on the emergency scene, a separate personnel accountability system may also be used. The incident commander should have an on-duty list to know exactly who is operating on the fireground. The incident commander should also have a list or some means of identifying the location and job of each person at all times. Officers must be in constant contact with personnel by head count, communication, or sight. One copy of the on-duty list should be kept by the company officer, and one copy should be in the apparatus. By taking a head count at every incident, it is possible to account for all personnel on the fireground.

Tag System

A simple tag system can aid in accounting for personnel within the fireground perimeter. Personnel can be equipped with a personal identification tag (Figure 19.24). Upon entering the fireground perimeter, firefighters leave their tags at a given location or with a designated person (command post, apparatus compartment, company officer, control officer, or sector officer) (Figure 19.25). Tags can be attached to a control board or personnel ID chart for quick reference. Upon leaving the fireground perimeter, the firefighters collect their tags. By using this system, it is possible for officers to know exactly who is operating on the fireground.

SCBA Tag System

An SCBA tag system provides closer accountability for personnel inside the structure. All personnel entering a hazardous atmosphere must be required to wear full protective clothing with SCBA. These firefighters must be trained and certified for SCBA use. Each SCBA is provided with a tag containing the name of the user and the air pressure (Figure 19.26). Having individually assigned breathing apparatus will ensure that the user is familiar with the apparatus and has a properly fitting facepiece. Upon entering the building, personnel give their tags to a designated supervisor. The supervisor records time of entry and expected time of exit. This supervisor also does a brief check to ensure that all protective equipment is properly used and in place. This provides complete accountability for those inside the structure and ensures that they are in proper gear.

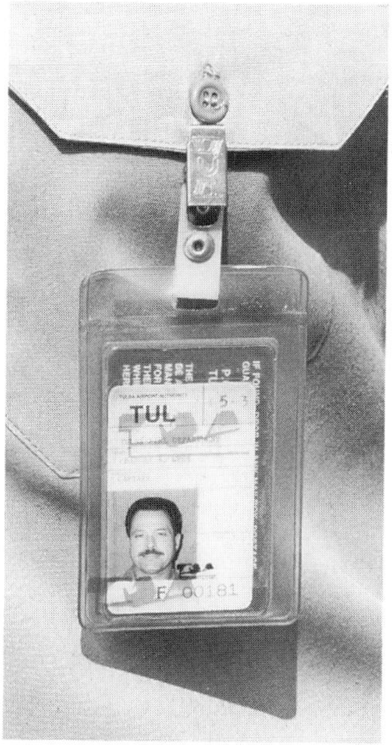

Figure 19.24 Identification badges may be worn on both the station uniform and the turnout clothing.

Figure 19.25 A firefighter identification tag on apparatus hook.

Figure 19.26 Name tags can be attached to the SCBA.

Firefighters leaving the danger area take back their tags so that the control officer knows who is safely outside and who is still inside the structure or danger area. Relief crews are sent in before the estimated time of the sounding of the low-pressure alarms.

Trapped Or Disoriented Firefighters

Even with the best incident command or accountability system in place, unusual circumstances can lead to a firefighter, or a group of firefighters, becoming trapped or disoriented within a burning structure. Unexpected structural collapse, doors closing behind crews, or firefighters straying from a hose or lifeline are all ways that this scenario may evolve.

Firefighters that become disoriented should try to remain calm. Becoming overly excited reduces the firefighter's ability to think and react quickly. Excitement or disorientation also causes firefighters to expend their air supply faster than normal. If possible, the firefighters should try to retrace their steps to their present location. If retracing is not possible, the firefighters should try to seek an exit from the building or at least from the area that is on fire. The firefighters should shout for help every once in a while so that other personnel who may be in the area will hear them. They can also activate their PASS device to produce a loud, distinctive noise. While moving around, the firefighters should be alert for a hoseline or safety line. If one is found, they should follow it (Figure 19.27). The male coupling will be the one with lugs on the shank (Figure 19.28). The male coupling signifies the exit direction. If the firefighters are not having any success finding their way out, they should find a place of relative safety and activate their PASS device to send out a help signal.

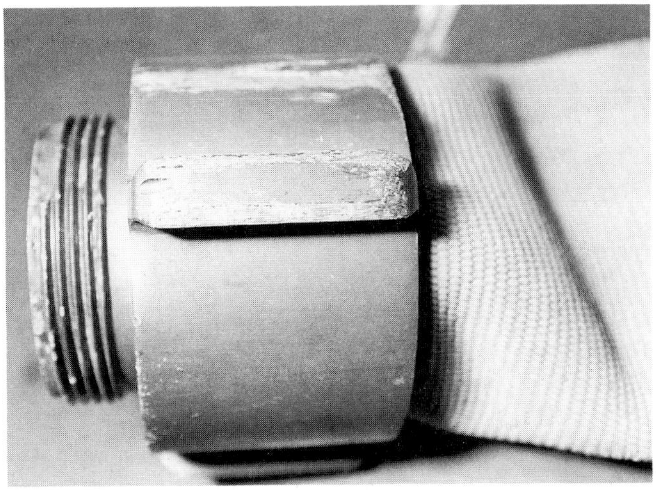

Figure 19.28 The male shanks have lugs on them. This tells the firefighter which direction to follow for exiting the building.

Firefighters who become trapped by a structural collapse or suffer some sort of injury that prevents them from moving about do not have all the options that a disoriented firefighter has. Firefighters may call out so that other personnel in the area may hear them; this will help personnel locate the area the firefighters are in. The firefighters should immediately activate their PASS devices. The firefighters should try to maintain their composure to maximize the air supply.

If trapped or disoriented firefighters have a radio, they should try to make radio contact as quickly as possible with other personnel on the emergency scene. The firefighters should try to describe their location as accurately as possible to narrow down the search area for rescuers.

Firefighters searching for a lost or disoriented firefighter should first try to quickly obtain an idea of the

Figure 19.27 Follow the hoseline to safety in emergency situations.

last location of the firefighter. When performing the search, the rescuers should stop every so often and become perfectly quiet. This may allow the rescuers to hear calls for help or the downed firefighter's PASS device tone.

Controlling Building Utility Services

In order to maximize safety on the fireground and to prevent excess damage to the fire building, it is often necessary for firefighters to shut off one or more utility services to the building. All firefighters must be knowledgeable in the proper methods for shutting off each type of utility.

The safest way for firefighters to disconnect electrical service to a building is to shut off the main breakers at the electrical service box or panel (Figure 19.29). Fire personnel should not physically cut outside wires or pull electrical meters. Cutting the drop wires into a building can be hazardous if proper equipment is not used. Pulling the electrical meters can result in a blinding or explosive electrical arc if not done properly. In some instances, pulling the meter may not even stop the flow of electricity to the building.

Natural gas service to a building may be halted by turning the petcock on the gas meter to the off position,

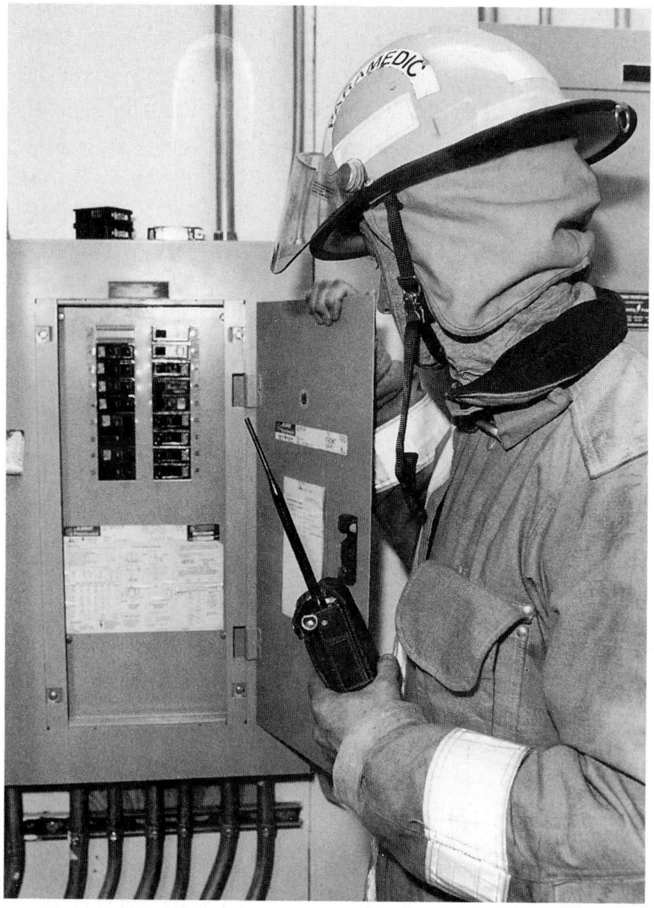

Figure 19.29 A typical electrical panel.

which is when the petcock is perpendicular to the pipe. The petcock may be turned with a spanner wrench, adjustable wrench, or a gas meter wrench (Figure 19.30). Turn all visible valves to the off position to ensure that gas service is stopped. Never extinguish a gas-fed fire until the gas can be shut off. Extinguishing the fire before controlling the flow of gas may result in a reflash fire of disastrous proportions.

Figure 19.30 A spanner wrench may be used to close most natural gas petcocks.

Some buildings may be served by liquefied petroleum gas (LPG) or compressed natural gas (CNG). These gases are stored in small cylinders or tanks outside the building (Figure 19.31). Both types of containers are equipped with valves that can be closed in the event the supply of gas needs to be halted. It may be necessary to play water on these tanks if they are exposed to fire from the structure.

Water service to a building may be stopped by one of several methods. The easiest method is to close the valve on the feeder main to the building. This valve may be located either inside or outside the building. Inside valves are usually in the form of a screw-type valve or quarter-turn valve. Outside valves may be on a post similar to a hydrant, or they may be below ground. Those below ground will require a long instrument, called a key tool, to reach down in the hole and close the valve (Figure 19.32).

Figure 19.31 Many buildings are supplied by LPG or CNG tanks.

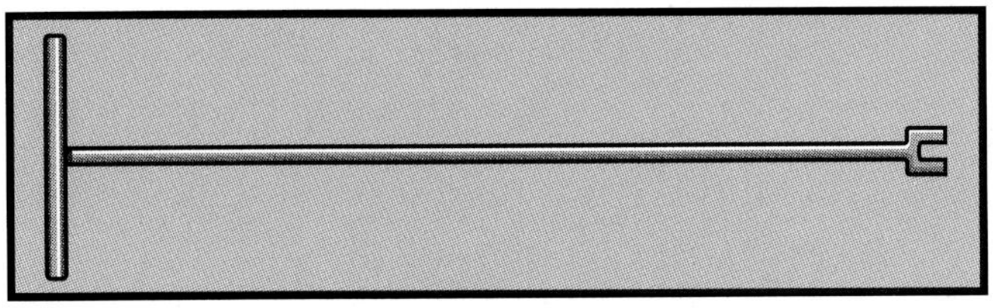

Figure 19.32 A key tool for shutting off the water supply.

Regardless of which service the fire department discontinues to the building, fire department personnel should never turn the service back on. That job should be done by the utility authority or representatives of the owner or occupant. This reduces the fire department's liability should a problem arise from the restoration of services.

Handling Electrical Emergencies

Considering the variety of electrically powered conveniences used in homes and businesses and the vast electrical distribution system, firefighters inevitably will be exposed to the hazards of electricity during a fire, vehicle accident, or rescue operation (Figure 19.33). Serious or fatal injuries can occur from electrical shocks or burns.

There needs to be departmental policy governing the handling of electrical emergencies. Planning should include getting advice and support from power utility personnel who can be instrumental in establishing procedures for electrical emergencies. Arrange for the utility personnel's prompt assistance during an emergency. Planning will not only help the incident commander evaluate the scene, but can prevent the firefighters from endangering themselves and others. A planned procedure in which all fire department personnel are well trained will lessen the likelihood of dangerous mistakes during electrical emergencies.

Figure 19.33 Overhead wires can fall on vehicles and energize the entire area.

A typical departmental policy for handling electrical exposure can include the following points. Depending on local preferences, however, the policy may be more expansive.

- Notify power company personnel as soon as there is an electrical hazard.
- Let power company personnel handle energized electrical equipment if possible. The exception to this policy is when a life is in immediate danger, a rescue must be performed, and the rescuer has the proper knowledge and equipment. This equipment must be certified to be dielectric by an approved testing agency, and the firefighter must be thoroughly trained in its use.
- Do not use solid or straight hose streams when a possible electrical hazard exists.
- Let pole top or cross arm fires burn until utility personnel shut down the power, unless part of the pole or cross arm is in danger of falling. In this case a fire extinguisher rated for Class C fires may be used to extinguish the fire. A firefighter will most likely have to use a fire department aerial device to reach the fire.
- Exercise extreme caution when using all ladders around electrical hazards.
- When possible, avoid parking apparatus under overhead wires (Figure 19.34).

Figure 19.34 Avoid parking apparatus under overhead wires.

- OSHA 29 CFR 1926.550 states that aerial devices or ground ladders shall be kept a minimum of 10 feet from lines rated 50 kV (50,000 volts) or lower (Figure 19.35). For lines over 50 kV, minimum clearance shall be 10 feet plus 0.4 inches for each 1 kV over 50 kV, or use twice the length of line insulator but never less than 10 feet.
- Let only power company personnel cut electrical wire.
- Treat all wires as "live" high-voltage wires.
- Establish a danger zone of at least one span in either direction from downed power wires (Figure 19.36).
- Wear full protective clothing when electrical hazards exist.

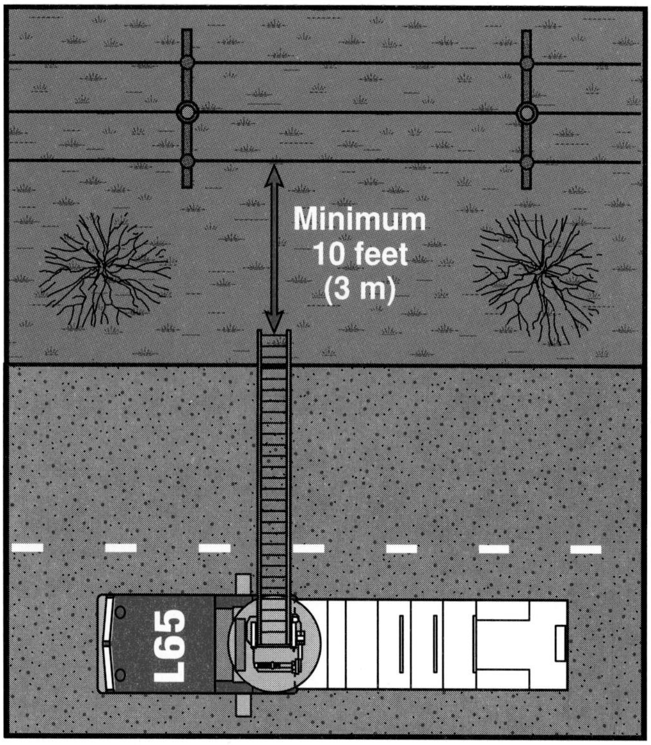

Figure 19.35 Keep ladders at least 10 feet from all overhead wires.

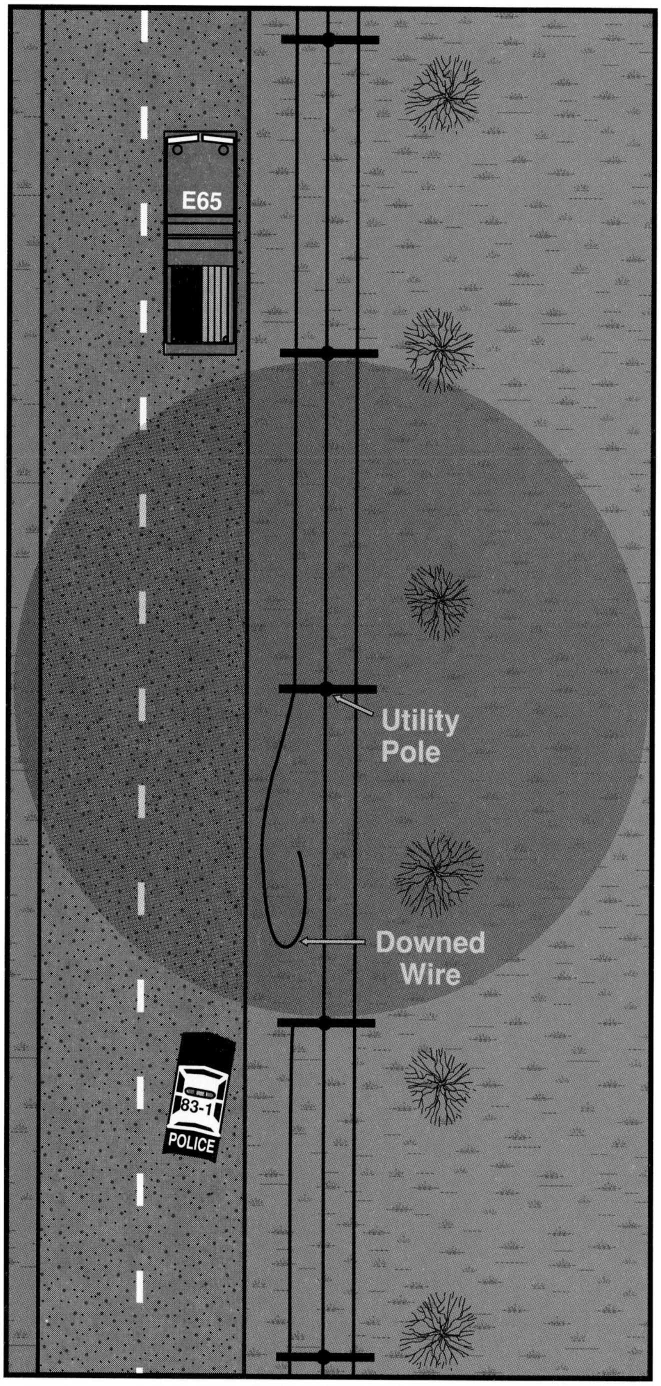

Figure 19.36 Set up at least a one-span safety zone around downed wires.

TOOL AND EQUIPMENT SAFETY

Tools and equipment are vital to a firefighter's job. However, accidents can happen if the firefighter is not properly trained in the use and care of tools and equipment. Poorly maintained tools and equipment can be very dangerous and can result in costly accidents to firefighters in the station and at the emergency scene. NFPA 1500 stresses the importance of safety in every aspect of tool and equipment design, construction, purchase, usage, maintenance, inspection, and repair.

When working in a station shop or on the emergency scene, firefighters must use personal protective equipment (PPE). Using PPE is fundamental for safe work practices. Although PPE does not take the place of good tool engineering, design, and use practices, it does provide personal protection against hazards that exist.

Eye And Face Protection

Eye and face protection is necessary when (Figures 19.37 a through c):

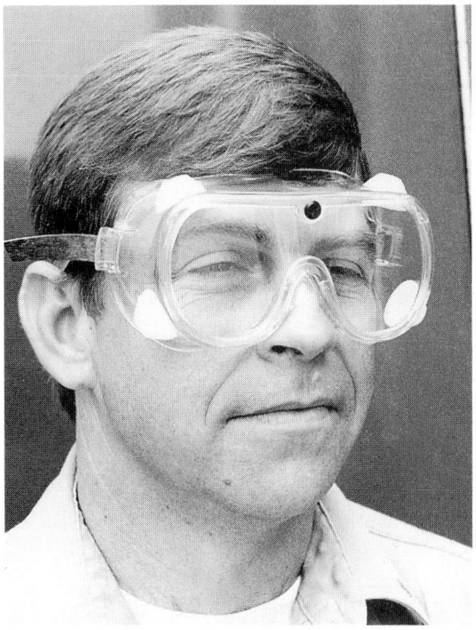

Figure 19.37a Any time there is a danger to the eyes, safety glasses or goggles must be worn.

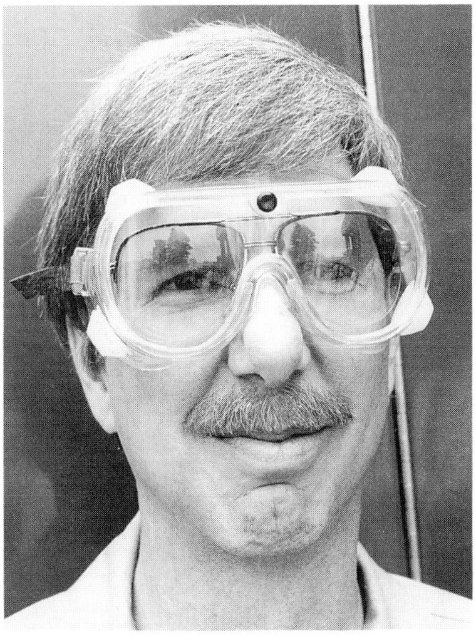

Figure 19.37b Goggles that fit over prescription glasses are available.

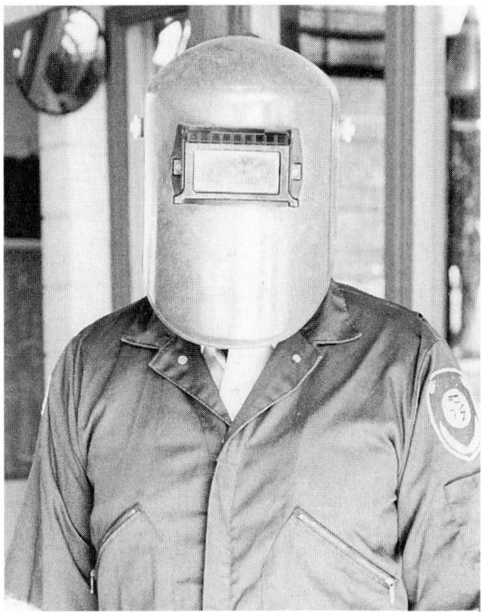

Figure 19.37c Shields that reduce ultraviolet radiation should be worn during welding operations.

- Working with any tool that has the potential to create airborne fragments
- Working with chemicals that may be dangerous if splashed into the eyes (including blood and other bodily fluids)
- Working with welding equipment that may cause damage via bright arcs

A Bureau of Labor statistics study found that about 60 percent of workers who suffered eye injuries were not wearing eye protection. Design, construction, testing, and use of eye and face protection must be in accordance with ANSI Z87.1

Ear Protection

Exposure to high noise levels can cause permanent hearing loss. Because there is no cure for noise-induced hearing loss, it is imperative that preventive measures be taken. Appropriate hearing protection, in the form of personal earplugs or earmuffs, should be provided in areas where the noise level exceeds the permissible levels established by OSHA (Figure 19.38). Table 19.2 lists the permissible noise levels and their gdurations.

Shop Tools

The most widely used tools in the station shop are hand tools and small power tools. Observe the following guidelines when using hand and power tools:

- Wear appropriate personal protective equipment.
- Remove jewelry, including rings and watches.
- Select the appropriate tool for the job.

Figure 19.38 Earmuffs reduce noise to tolerable levels.

TABLE 19.2
PERMISSIBLE NOISE EXPOSURES

Duration per day, hours	*Sound Level dBA slow response*
8	90
6	92
4	95
3	97
2	100
1½	102
1	105
½	110
¼	115

From OSHA 1910.95

- Know the manufacturer's instructions and follow them.
- Inspect tools before use to determine their condition. If a tool has deteriorated or is broken, replace it.
- Provide adequate storage space for tools, and always return them promptly to storage after use. Whenever necessary, inspect and clean the tools before storing.
- Consult with and secure the approval of the manufacturer before modifying the tool.
- Use spark-resistant tools when working in flammable atmospheres such as around a vehicle's fuel system.

HAND TOOLS

Inspect all tools before each use to ensure that they are in good condition. This inspection may prevent an accident caused by tool failure. Homemade tool-handle extensions or "cheaters" are sometimes incorrectly used to provide extra leverage for wrenches, pry bars, and similar tools (Figure 19.39). The use of a cheater can overload the tool beyond its designed capabilities. This overloading can cause the tool to break suddenly, not only while the cheater is attached but also later when the weakened tool is being used normally.

Figure 19.39 Misusing tools, such as adding a cheater bar to this pry bar, may result in tool failure and/or injury to rescuers or victims.

POWER TOOLS

Grinders, drills, saws, and welding equipment are commonly found in fire stations (Figure 19.40). Improperly used, these tools can do a lot of damage — fast! Whether the tool is driven by air or electricity, it will have a specific, safe method of operation that must be understood and followed. Only those firefighters who have read and who understand the tool manufacturer's instructions should be allowed to use power tools. It is important for instructions to be accessible to the firefighters.

Figure 19.40 Power tools, such as a grinder, are found in most fire stations.

Repairs should always be made by someone trained and authorized to properly repair the damaged tool. Depending on the department, this person may be someone within the fire department or an outside equipment dealer and repair agent. Keeping accurate records of repairs can help spot misuse before the tool causes an accident.

Choose electrically powered tools carefully. Look for the label of a nationally recognized testing laboratory such as Underwriters Laboratories. This label means the tool has met safety requirements set by the testing laboratory. However, it does not mean that the tool will be safe regardless of how it is abused or misused.

Make sure that the outlets used to power electrical tools have a good ground. Contact qualified personnel to test electrical systems if you are unsure of the ground. Good wiring can break down over time; therefore, it may be wise to check outlets periodically. Install a ground fault circuit interrupter (GFCI) in the circuit. If a tool's insulation fails, the GFCI will shut off the electricity very quickly to prevent or reduce the severity of shock to the user. GFCI units can be wired into circuits like a regular circuit breaker or outlet, or they can be plugged into a properly grounded outlet. Whichever type you use, follow the manufacturer's instructions closely.

Any electrical tool not marked "double insulated" should have a three-prong plug (Figure 19.41). For firefighter safety, the third prong must connect to a good

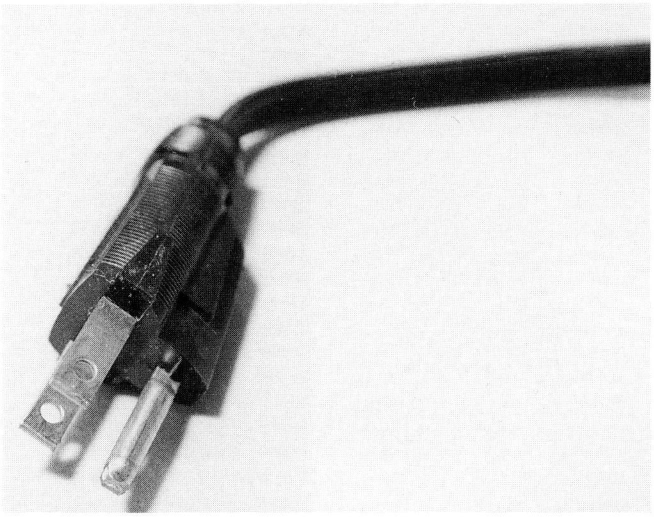

Figure 19.41 A three-prong plug should be used when a tool is not double insulated.

ground while the tool is in use. Bypassing the ground plug in any way opens the door to injuries or deaths from unpredictable electrical shorts.

POWER SAWS

The most common types of power saws used by firefighters are rotary saws and chain saws. When improperly operated or maintained, a power saw can be the most dangerous type of tool that a firefighter will use.

Rotary saws may be found in the station or on the emergency scene. Rotary saws found in the station are usually bench- or table-mounted (Figure 19.42). Rotary saws used on the emergency scene are generally of the rescue or forcible entry design (Figure 19.43).

Following a few simple safety rules when using power saws will prevent most typical accidents. These rules are as follows:

- Match the saw to the task and the material to be cut. Never push a saw beyond its design limitations.
- Wear proper protective equipment, including gloves and eye protection. Avoid wearing loose, dangling clothing that may become entangled in the saw.
- Have hoselines in place when forcing entry into an area where fire is suspected or when performing vertical ventilation. Hoselines are also essential when cutting materials that generate sparks.
- Avoid the use of all saws when working in a flammable atmosphere or near flammable liquids. Sparks generated by the saw or the saw's hot muffler are ignition sources for the vapors.

Figure 19.42 Rotary saws used in the station are usually mounted on a table or a bench.

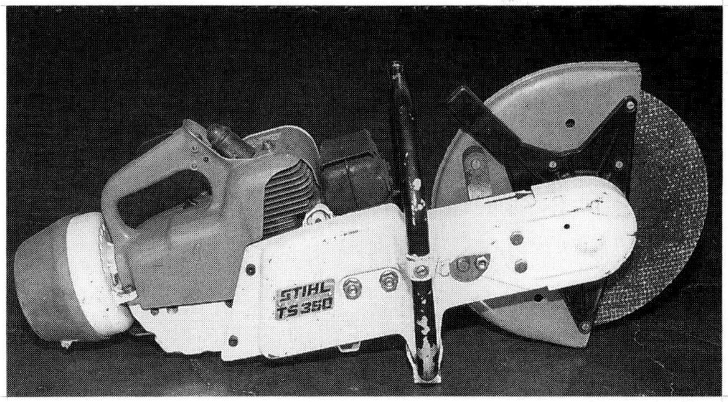

Figure 19.43 A rotary rescue saw.

- Keep unprotected and unessential people out of the work area.
- Follow manufacturer's guidelines for proper saw operation.
- Use caution to avoid igniting gasoline vapors when refueling a hot, gasoline-powered saw. It is best to allow the saw to cool before refueling.
- Keep blades and chains well sharpened. A dull saw is far more likely to cause an accident than a sharp one.

EMERGENCY POWER AND LIGHTING EQUIPMENT

It would certainly make operations easier if all emergency situations occurred during daylight hours. Unfortunately, this does not always happen. Many incidents occur during the evening, which creates the need to artificially light the scene. Lighting the scene provides a safer, more efficient atmosphere in which to work. Firefighters must be knowledgeable of when and how to properly and safely operate the emergency power and lighting equipment.

Power Plants

There are two basic types of power plants: inverters and generators.

INVERTERS

Inverters (alternators) are used on emergency vehicles when large amounts of power are not necessary (Figure 19.44). The inverter is a step-up transformer that converts the vehicle's 12- or 24-volt DC current into 110- or 220-volt AC current. Advantages of inverters are fuel efficiency and low or nonexistent noise during operation. Disadvantages include small capacities and limited mobility from the vehicle.

Figure 19.44 Some inverters are mounted in the compartment of the apparatus. *Courtesy of Stillwater (OK) Fire Department.*

GENERATORS

Generators can be either portable or fixed to the apparatus. They are the most common power source used for emergency services. Portable generators are powered by small gasoline or diesel engines and generally have 110- and/or 220-volt capacities (Figure 19.45). Most portable generators are designed to be carried by either one or two people. They are extremely useful when electrical power is needed in an area that is not accessible to the vehicle-mounted system.

Figure 19.45 Portable generators provide limited power supply.

Vehicle-mounted generators usually have a larger capacity than portable units (Figure 19.46). In addition to providing power for portable equipment, vehicle-mounted generators are responsible for providing power for the floodlighting system on the vehicle. Vehicle-mounted generators can be powered by gasoline, diesel, or propane engines or hydraulic or power take-off systems. Fixed floodlights are usually wired directly to the unit through a switch, and outlets are also provided for other equipment. These power plants generally have 110- and 220-volt capabilities with capacities up to 50 kW and occasionally greater. However, mounted generators are noisy, and it is difficult to talk and hear around them.

Figure 19.46 Vehicle-mounted generators may have capacities of 50 kW or more.

Lighting Equipment

Lighting equipment can be divided into two categories: fixed and portable. Portable lights are used when fixed lights are not within reaching distance or when additional lighting is necessary. Portable lights generally range from 300 to 1,000 watts (Figure 19.47). They may be supplied by a cord from the power plant, or they may have a self-contained power unit. The lights usually have handles for safe carrying and large bases

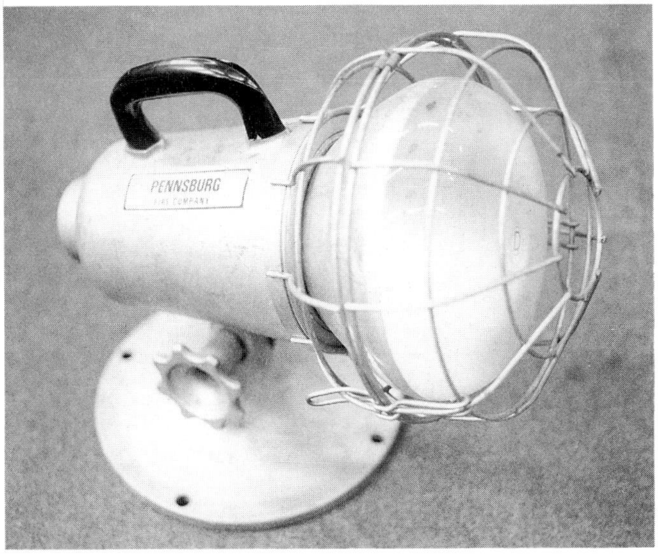

Figure 19.47 Portable floodlights should have protective enclosures over the bulbs.

for stable setting and placement. Some portable lights are connected to telescoping stands, which eliminate the need for personnel to hold the portable lights or to find something to set them on (Figure 19.48). Care must be exercised when using portable floodlights; this equipment operates at a very high temperature, and even brief contact with a person can result in serious burns.

Figure 19.48 Portable floodlights on telescoping stands provide an elevated light source that does not require anyone to hold it.

Fixed lights are mounted to the vehicle, and their main function is to provide overall lighting of the emergency scene. Fixed lights are usually mounted so that they can be raised, lowered, or turned to provide the best possible lighting. Often, these lights are mounted on telescoping poles that allow this movement (Figure 19.49). More elaborate setups include hydraulically operated booms with a bank of lights (Figures 19.50 a and b). A bank of

Figure 19.49 Many apparatus have equipment with telescoping floodlights.

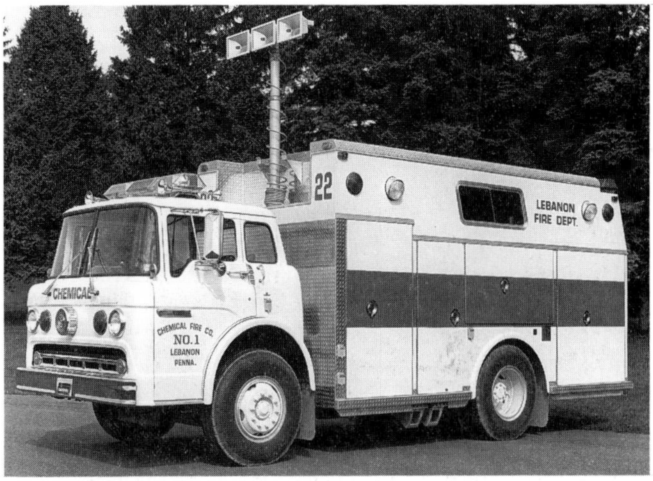

Figure 19.50a Emergency vehicles may be equipped with light towers capable of raising a bank of lights 40 feet (13 m) or more. *Courtesy of Will-Burt Company, Telescoping Mast Division.*

Figure 19.50b Another style of emergency vehicle equipped with light towers. *Courtesy of Saulsbury Fire Apparatus Co.*

lights generally has a capacity of 500 to 1,500 watts per light. The amount of lighting should be carefully matched with the amount of power available from the power plant. Overtaxing the power plant will give poor lighting, may damage the power generating unit or the lights, and will restrict the operation of other electrical tools.

Auxiliary Electrical Equipment

A variety of other equipment may be used in conjunction with power plants and lighting equipment. Electrical cables or extension cords are necessary to conduct electric power to portable equipment. The most common sized cable is the 12-gauge, 3-wire type. The cord may be stored in coils, on portable cord reels, or on fixed automatic rewind reels (Figures 19.51 a through c). Twist-lock receptacles provide secure, safe connections (Figure 19.52). Electrical cable must be adequately insulated, waterproof, and have no exposed wires.

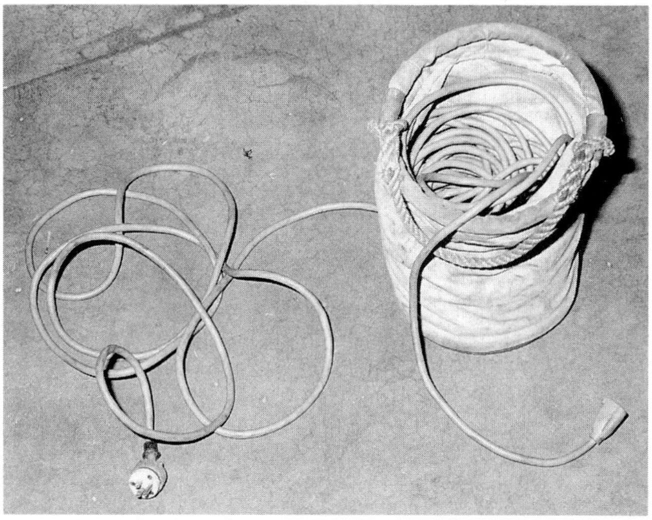

Figure 19.51a A coiled electric cord in a canvas bag. *Courtesy of Stillwater (OK) Fire Department.*

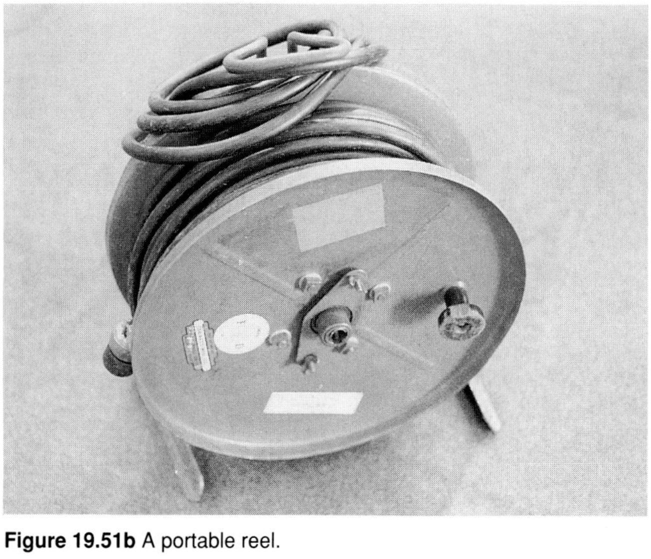

Figure 19.51b A portable reel.

Figure 19.51c A power reel. *Courtesy of East Greenville (PA) Fire Co. No. 1.*

Figure 19.52 Twist-lock connections are the safest and most effective for fire and rescue operations.

Junction boxes may be used when multiple connections are needed (Figure 19.53). The junction is supplied by one inlet from the power plant and is fitted with several outlets.

In situations where mutual aid departments frequently work together and have different sizes or types of receptacles (for example, one has two prongs, the other has three), adapters should be carried so that equipment can be interchanged. Adapters should also be carried to allow rescuers to plug their equipment into standard electrical outlets.

Figure 19.53 Junction boxes allow one electrical supply line to be branched into several different lines.

Maintenance And Servicing Of Power Plants And Lighting Equipment

The servicing and maintenance of portable power plants and lighting equipment are essential for their reliable operation. The following list contains general guidelines for the servicing and maintenance of power plants and lighting equipment. (These guidelines are not designed to replace the owner's manual that comes with the equipment.)

- Run power plants at least once a week, for at least 20 minutes, while powering an electrical device.
- Check fluid levels weekly. Check gas and oil levels after every use. Drain the power plant of all fluids if it is not to be in service for an extended period.
- Wear gloves when changing quartz bulbs. Normal hand oil can cause a bulb to explode when it is energized.
- Inspect electrical cords at weekly intervals to ensure that the insulation is not damaged.
- Inspect the spark plug, spark plug wire, and carburetor at weekly intervals. A spare spark plug should also be readily accessible.
- Test electrical devices for operating status while the power plant is running.
- Change extra gasoline approximately every three weeks to ensure freshness.

SAFETY ON THE APPARATUS

The most common, dangerous situation that the firefighter experiences is riding the apparatus to and from emergency calls. All firefighters are exposed to the same dangers, regardless of the size of their community.

Passengers and driver/operators of emergency vehicles shall not dress while the apparatus is in motion. Preferably, all firefighters should ride within a fully enclosed portion of the cab (Figure 19.54). Firefighters that are not riding in enclosed seats should wear helmets and eye protection. If sirens and noise levels exceed 90 decibels, firefighters should also wear hearing protection (Figure 19.55).

Figure 19.54 All new apparatus must have fully enclosed cabs.

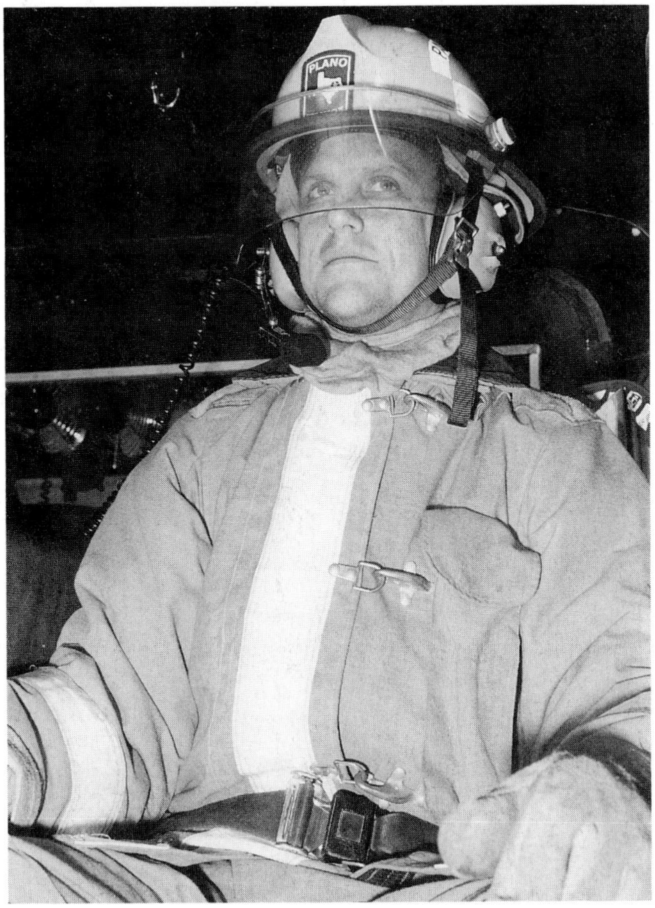

Figure 19.55 Firefighters riding in open jump seats should wear appropriate ear and eye protection, as well as their seat belt.

All firefighters must be seated with their seat belts fastened when the vehicle is in motion, regardless of whether the firefighters are riding in an enclosed or unenclosed portion of the cab (Figure 19.56). Seat belt usage prevents or lessens two major causes of injury or death in vehicular accidents. Seat belt usage prevents passengers from being thrown from the vehicle and also keeps them in their seats so that they are not thrown against part of the vehicle's interior.

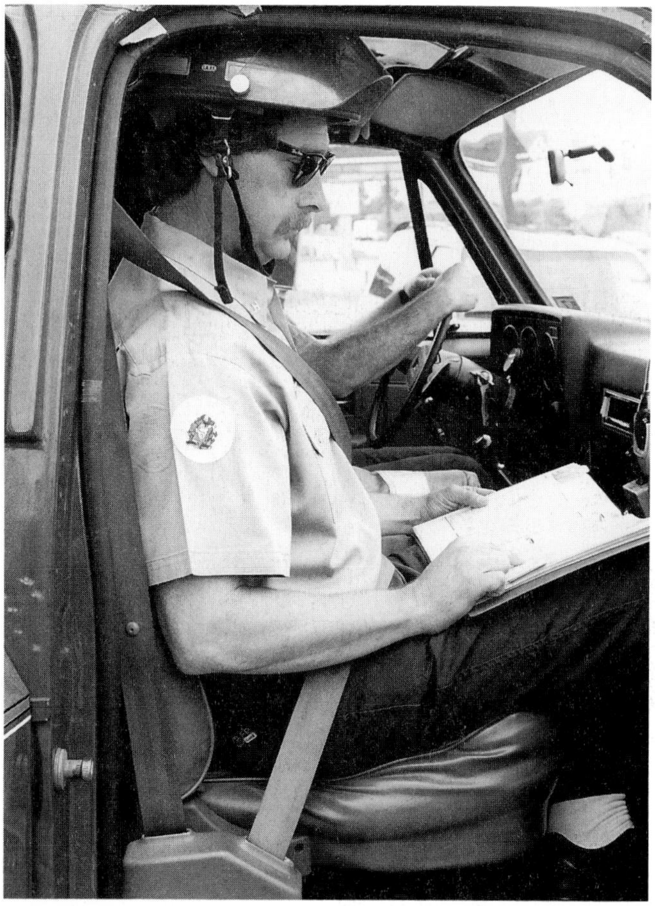

Figure 19.56 All firefighters must wear their seat belts.

Firefighters who oppose wearing seat belts, because of the time spent in fastening them, lack a sound argument. The difference in the speed of response is very small, and that extra second may save a firefighter's life. All apparatus should be purchased with seat belts. Older apparatus can be retrofitted with seat belts. In many cases, better design and fastening hardware can reduce the fastening time. Cross-chest belts should be permanently fastened to the lap belt so that only one metal-to-metal fastener is required (Figure 19.57). Fastening these belts can be done with one swing of the arm if the belt has been properly adjusted.

When new apparatus is purchased, it should meet the specifications of the appropriate NFPA apparatus design specification standard (1901, 1902, 1903, or 1904), including enclosed seats and seat belts large enough to accommodate a firefighter in full protective clothing. Firefighters should NOT be in a standing position anywhere on the apparatus.

Figure 19.57 Three-point, cross-chest seat belts are the most desirable type.

WARNING

Do not ride on the tailboard. Many firefighters have been killed falling from tailboards. This practice must be discontinued.

If it is absolutely necessary to ride in an unenclosed jump seat, safety bars are available that may prevent a firefighter from falling out (Figure 19.58). These bars are not a substitute for safety procedures that require firefighters to ride in safe, enclosed positions wearing their seat belts; however, they do have value as an additional barrier between the firefighter and the road. When the bars are installed, they should be secured with bolts strong enough to withstand the full weight of a firefighter falling upon them. Safety bars that are held in an upright position by straps or ropes provide no extra

Figure 19.58 Safety bars provide very limited protection to occupants of the jump seat area.

security for the firefighter riding in the jump seat. Safety gates are more effective than safety bars; they reduce the opening size to prevent a person from slipping out of the jumpseat area (Figure 19.59). Many apparatus are equipped with half-doors, which accomplish the same goal as safety gates.

Figure 19.59 Safety gates are more effective than bars.

The officer in charge should signal the driver/operator when all firefighters are fully clothed and in secure positions. Driver/operators should never move the apparatus until they are certain all personnel are in a seated and secured position. If firefighters sit facing to the rear with a seat belt on, they are better protected from head-on accidents involving the apparatus.

Firefighters should always use handrails when mounting or dismounting the apparatus (Figure 19.60). Using handrails reduces the chance of firefighters accidentally slipping and falling from the apparatus. However, firefighters should not use the handrail when dismounting an apparatus that has an aerial device extended close to electrical wires. If the aerial device contacts the charged lines and the firefighter is in contact with the apparatus and the ground at the same time, there is a chance the firefighter will be electrocuted. Always jump clear of an apparatus that might be electrically energized (Figure 19.61).

SAFETY IN THE FIRE STATION

Most firefighters' duties and activities center around the station, and a significant portion of their on-duty

Figure 19.60 Use handrails when entering or exiting the apparatus.

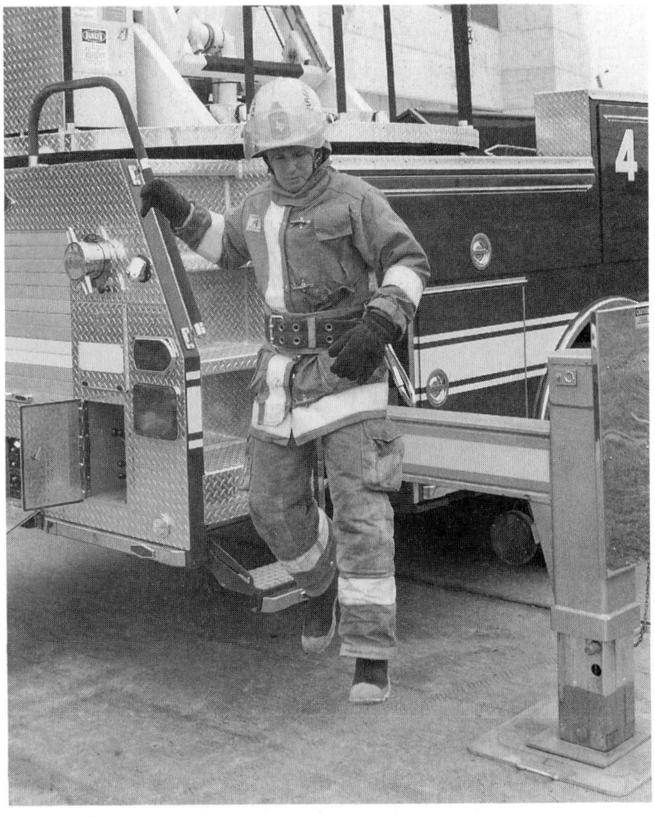

Figure 19.61 Jump clear of aerial apparatus that may be in contact with energized electrical wires.

time is spent there. A department's safety officer should examine station conditions and work procedures as closely as fireground procedures. Hazards in the fire station not only endanger firefighters but can also endanger visitors who enter the station. Visitors are the responsibility of the fire department while they are in the building. Therefore, safe conditions must exist to limit the possibility of accidents and injuries.

There are certain safety hazards common to any fire station. There are also certain types of accidents that are not limited to any specific location within a station. Improper lifting techniques and slip-and-fall accidents are two of the most common accidents that result in injury.

Although back strains are the most common injuries related to improper lifting techniques, bruises, sprains, and fractures can also result from the improper lifting and carrying of materials. Improper lifting techniques not only result in personal injury, but they may also result in damage to equipment if it is dropped or improperly handled in the process of lifting. Back injuries have been statistically proven to be the most expensive single type of accident in terms of workers compensation, and they occur with surprising frequency.

Every firefighter should be instructed in the correct method of lifting (Figure 19.62). Only with assistance should a firefighter attempt to lift or carry an object that is too bulky or heavy for one person to safely handle. Lifting and carrying heavy or bulky objects without help can result in unnecessary strains and injuries. If the object is big, bulky, or heavy, get help to lift it.

Figure 19.62 When working around the station, use plenty of help to lift heavy loads.

Another type of common accident is the slip, trip, and/or fall. Numerous factors contribute to these accidents. A slip, trip, or fall generally results from poor footing. This can be caused by slippery surfaces, objects or substances on surfaces, inattention to footing on stairs, uneven surfaces, and similar hazards. These accidents can easily result in minor and serious injuries as well as damaged equipment. To prevent such accidents, it is important to stress good housekeeping. For example, floors must be kept clean and free from slipping hazards such as loose items and spills (Figure 19.63). Aisles must be unobstructed and stairs should be well lighted. In addition to walking surfaces (floors, stairs, and aisles), items, such as handrails, slide poles, and slides, must also be maintained in a safe condition.

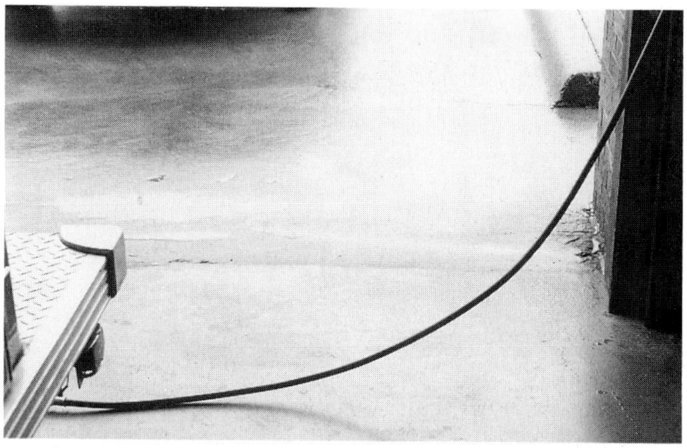

Figure 19.63 Eliminate tripping hazards around the station.

SAFETY IN TRAINING

NFPA 1500 requires that all personnel who may engage in structural fire fighting participate in training at least monthly. Ideally, this monthly training will serve to reinforce safe practices until they become automatic. Other types of training are required on an "as needed" basis. For example, training is required when new procedures or equipment are introduced. There should be at least two training sessions of this type per year.

All personnel participating in training at a drill site should be fully clothed in protective gear. Raising ladders, laying hose, or performing any other activity that simulates actual fire scene conditions requires the use of protective gear.

If trainees have colds, severe headaches, or other symptoms indicating physical discomfort or illness, they should not continue training until a medical examination can determine their fitness. Some trainees might feel uncomfortable telling an instructor they are physically unable to continue training. Older trainees and trainees who apparently are not in good physical condition should be watched closely for signs of fatigue, chest pains, or unusually labored breathing during heavy exercise. Physical discomfort or illness can lead to accidents; however, accidents can be prevented by determining the physical condition of all participating personnel before training.

Horseplay during training must be forbidden, because it can lead to accidents and injuries (Figure 19.64). If the trainees want to play pranks, the instructor will need to find out why. Boredom will often cause trainees to fidget and release their energy in horseplay. Boredom can develop easily in a training program that does not allow everyone to see the demonstration or to participate in the activities because of the size of the group.

Figure 19.64 Horseplay must not be tolerated.

Maintaining And Servicing Equipment

Equipment used for fire training evolutions must be in excellent condition. Items used frequently for training will often wear out sooner than those used routinely in the fire station. Examples of frequently used training items are ropes, straps, buckles, and other harness parts that must be tied, fastened, and unfastened repeatedly. Tools with wooden handles can quickly become worn and splintered when used repeatedly by trainees. All tools and equipment should be inspected before each drill to ensure their reliability (Figure 19.65). Records must also be maintained on all equipment used for training. Training equipment, like all fire fighting equipment, should be tested according to manufacturer's instructions and applicable standards.

Figure 19.65 Inspect all equipment before its use in training exercises.

Appendix A

The following objectives are reprinted from NFPA 1001, *Standard for Fire Fighter Professional Qualifications* (1992). These objectives do not represent the standard in its entirety. For more details, check the standard itself. The beginning of each chapter lists the objective numbers that apply to the chapter.

CHAPTER 3 FIRE FIGHTER I

3-1 General. The Level I Fire Fighter shall meet the job performance requirements defined in sections 3-2 through 3-23 of this Standard.

3-2 Fire Department Organization.

3-2.1 Describe the organization of the fire department.

3-2.2 Explain the Fire Fighter I's role as a member of the organization.

3-2.3 Explain the mission of the fire service and of the local fire department.

3-2.4 Explain the function of a standard operating procedure.

3-2.5 Explain fire department rules and regulations that apply to the position of fire fighter.

3-2.6* Explain the basic components of incident management and the fire fighters' role within the local incident management system.

> **A-3-2.6** An incident management system is an organized plan of roles, responsibilities, and standard operating procedures used to direct and control an emergency incident. An incident management system identifies the roles and responsibilities of all members involved in emergency operations. (See NFPA 1561, *Standard on Fire Department Incident Management System.*)

3-2.7 Explain the role of other agencies that may respond to emergencies.

3-2.8 Describe the components of a member assistance program.

3-3* Safety.

> **A-3-3** NFPA Fire Fighter death and injury statistics confirm fire fighter accidents and injuries are excessive. It is important to increase the awareness of the hazards associated with fire fighting. The locations listed in this section are those used by NFPA to identify where injuries to fire fighters occur.

3-3.1 Describe the responsibilities of a fire fighter within the occupational health and safety policy adopted of the Authority Having Jurisdiction (AHJ) and NFPA 1500, *Standard on Fire Department Occupational Safety and Health Program.*

3-3.2 Describe the elements of a personnel accountability system and demonstrate the application of the system at an incident.

3-3.3 Identify dangerous building conditions created by fire.

3-3.4 Demonstrate procedures for action when trapped or disoriented in a fire situation or in a hostile environment.

3-3.5 Explain hazards related to electrical emergencies.

3-3.6 Demonstrate safety procedures when using fire service lighting equipment, given the following:

(a) Power supply (portable or mounted)

(b) Lights

(c) Cords

(d) Connectors

(e) Ground-fault interrupter (GFI)

3-3.7 Demonstrate the use of seat belts, noise barriers and other safety equipment provided for protection while riding on apparatus.

3-3.8 Demonstrate safety procedures when mounting, dismounting and operating around fire apparatus.

3-3.9 Shut off the utility services to a building.

3-3.10 Identify a minimum of three common types of accidents and/or injuries, and their causes, which occur in the following locations:

(a) Fire ground

(b) Responding and returning

(c) Training

(d) Non-fire emergencies

(e) Other on-duty locations

3-3.11 Identify safety procedures for ensuring a safe station/facility environment.

> **A-3-3.11** Due to the large number of injuries which occur in the fire station, personnel need to be aware of methods for eliminating accidents and injuries. The following is a partial list of safety concerns relating to the provision of a safe work facility.
>
> (a) Back injuries are a frequent cause of lost work time. Whenever possible, personnel should redesign tasks to eliminate lifting and lowering, and use lifting devices. Demonstration of proper lifting procedures should be included as routine safety training.
>
> (b) Personnel should be aware of unsafe conditions which could lead to falls, such as unguarded pole holes; to tripping, such as tools and equipment not put away after use and in the main paths of travel; and to slipping, such as wet floors.
>
> (c) The importance of eliminating "horseplay" should be stressed, as it can cause needless injuries.
>
> (d) Personnel should be encouraged to request assistance in performing tasks or duties which can be more safely performed with additional personnel.
>
> (e) Good physical fitness should be encouraged, as it will enable personnel to perform job tasks and functions with less chance of injury.
>
> (f) All personnel should be aware of proper methods of disinfecting themselves and equipment after use.

3-3.12 Identify potential long-term consequences of exposure to products of combustion.

3-4 Fire Alarm and Communications.

3-4.1 Explain the procedures for a citizen to report a fire or other emergency.

3-4.2 Explain the procedures for receiving an alarm from dispatch or a report of an emergency from the public, and demonstrate appropriate action.

3-4.3 Define the purpose and function of all alarm-receiving instruments and personnel-alerting equipment provided to the department and its members.

3-4.4 Identify procedures required for receipt and processing of business and personal calls.

3-4.5 Define and demonstrate prescribed fire department radio procedures including:

(a) Routine traffic

(b) Emergency traffic

(c) Emergency evacuation signals

3-5 Fire Behavior.

3-5.1 Define fire.

3-5.2 Define the fire triangle and tetrahedron.

3-5.3 Recognize the following conditions, explain their associated hazards and appropriate actions:

(a) Incipient fire

(b) Rollover

(c) Hot smoldering fire

(d) Flashover

(e) Steady state burning

(f) Backdraft

3-5.4 Identify three products of combustion commonly found in structural fires which create a life hazard.

3-5.5 Define the three methods of heat transfer.

3-5.6 Define the three physical stages of matter in which fuels are commonly found.

3-5.7 Define the relationship of the concentrations of oxygen to combustibility and life safety.

3-5.8* Describe the process of thermal layering that occurs in structural fires and how to avoid disturbing the normal layering of heat.

> **A-3-5.8** Thermal layering is the process by which heat will accumulate during fires in a confined space. The hottest air and gases will accumulate at the highest (typically ceiling) level. The levels of heat will decrease down towards the floor level, where heat should be at the minimum. It is important for fire fighters to understand this concept because it reinforces their need to stay low during interior fire fighting operations. It is also important for fire fighters to realize that they must avoid disturbing this normal layering of heat by improper ventilation or hose stream techniques. Upsetting the normal layering process could result in driving the harmful heat levels down to the floor level, thus seriously endangering the fire fighters or victims in the area.

3-6 Portable Extinguishers.

3-6.1 Identify the classification of types of fire as they relate to the use of portable extinguishers.

3-6.2 Define the portable extinguisher rating system.

3-6.3 Given a group of differing extinguishers, identify the appropriate extinguishers and application procedures for the various classes of fire.

3-6.4 Extinguish class A and B fires using the appropriate portable fire extinguishers.

3-7 Personal Protective Equipment.

3-7.1* Identify the function of the following articles of protective equipment:

(a) Helmet (with eye shield)

(b) Hood

(c) Boots

(d) Gloves

(e) Protective coat

(f) Protective trousers

(g) Self-contained breathing apparatus (SCBA)

(h) Personal alert safety system (PASS)

(i) Eye protection

> **A-3-7.1** Fire fighters should understand the importance of the type of clothing worn under personal protective equipment. See NFPA 1975, *Standard on Station/Work Uniforms for Fire Fighters*, for additional information.

3-7.2 Identify and demonstrate the care, use, inspection, maintenance and limitations of the protective clothing and equipment assigned or available for use.

3-7.3 Demonstrate the donning and doffing of the protective equipment specified in section 3-7.1.

3-7.4 Identify the hazardous environments requiring the use of respiratory protection.

3-7.5 Identify the physical requirements of an SCBA wearer.

3-7.6 Describe the uses and limitations of SCBA.

3-7.7 Identify each component and safety feature of the SCBA.

3-7.8 Describe the function of each component of the SCBA.

3-7.9 Demonstrate donning and doffing the SCBA while wearing protective clothing.

3-7.10 Demonstrate that the SCBA is in a safe condition for immediate use.

3-7.11* Demonstrate the use of SCBA in conditions of obscured visibility.

> **A-3-7.11** When training exercises are intended to simulate emergency conditions, smoke generating devices that do not create a hazard are required. Several accidents have occurred when smoke bombs or other smoke generating devices that produce a toxic atmosphere have been used for training exercises. All exercises should be conducted in accordance with the requirements of NFPA 1404, *Standard for a Fire Department Self-Contained Breathing Apparatus Program*.

3-7.12 Demonstrate the use of SCBA in conditions of restricted passage.

3-7.13* Demonstrate the following emergency procedures to be used in the event of SCBA failure:

(a) Use of the emergency by-pass or purge valve

(b) Conservation of air

(c) Breathing from the breathing tube or regulator in the event of a face piece failure.

> **A-3-7.13** Follow the manufacturers recommendations for specific procedures.

3-7.14 Demonstrate procedures for maximizing the air capacity of an SCBA under work conditions.

3-7.15 Demonstrate replacement of an expended cylinder on an SCBA assembly with a full cylinder.

3-7.16 Demonstrate and document routine maintenance for SCBA, including inspection, cleaning, sanitizing, and cylinder recharging.

3-7.17* Demonstrate rescue procedures for the following, without compromising the rescuer's respiratory protection:

(a) A fire fighter with functioning respiratory protection

(b) A fire fighter without functioning respiratory protection

(c) A civilian without respiratory protection.

> **A-3-7.17** SCBA emergency procedures should be an integral part of any SCBA program, with written policies for the removal of victims, both civilian and fire service, from hazardous atmospheres without compromising the rescuer's SCBA for any reason.

3.8 Forcible Entry.

3-8.1 Identify materials and construction features of doors, windows, and walls and the dangers associated with forcing entry through each.

3-8.2 Force entry through at least three (3) different types of doors, windows and walls.

3-8.3 Identify materials and construction features of door and window locking devices.

3-8.4 Identify the method and demonstrate procedures of through the lock entry for doors and windows.

3-8.5 Identify methods and procedures for cleaning, maintaining, and inspecting hand tools used for forcible entry.

3-8.6 Identify and safely carry at least one (1) of the following:

(a) Cutting tool

(b) Prying tool

(c) Pulling tool

(d) Striking tool

3-9 Ventilation.

3-9.1 Define the principles of ventilation, and identify the advantages and effects of proper ventilation.

3-9.2 Identify the safety considerations and precautions to be taken in performing ventilation.

3-9.3 Describe the advantages and disadvantages of the following types of ventilation:

(a) Vertical

(b) Horizontal

(c) Trench/strip

(d) Mechanical

(e) Mechanical pressurization

(f) Hydraulic

3-9.4 Identify the signs, causes and effects of backdraft explosions.

3-9.5 Identify methods of preventing a backdraft explosion.

3-9.6 Identify the types of tools used during ventilation.

3-9.7 Recognize the characteristics of, and list necessary precautions when ventilating at least the following roof types:

(a) Flat

(b) Shed

(c) Pitched

(d) Arched

3-9.8 Demonstrate determining the integrity of a roof system by sounding.

3-9.9 Describe how the following factors are used to determine the integrity of a roof system:

(a) Construction

(b) Visual observation

(c) Elapsed time of fire

3-9.10 Define procedures for the types of ventilation referred to in section 3-9.3.

3-9.11 Demonstrate opening various types of windows from inside and outside, with and without the use of tools.

3-9.12 Demonstrate breaking window or door glass, and removing obstructions.

3-9.13 Using both hand and power tools, demonstrate the ventilation of both pitched and flat roofs.

3.10 Ropes.

3-10.1* Explain the uses of and tie a bowline knot, a clove hitch, figure of eight on the bight, a becket or sheet bend, overhand safety knot, and half hitch, given the proper size and amount of rope.

> **A-3-10.1** The recommendations of the rope manufacturer for the proper knots to be used with their product should be followed

3-10.2 Tie an approved knot, and hoist any selected forcible entry tool, pike pole/hook, ground ladder, hose line, extinguisher or appliance to a height of at least 12 ft (3.7 m), given the proper rope.

3-10.3 Demonstrate the procedures of inspecting, maintaining, and storing rope.

3-10.4 Use a rope to tie ladders, hose, and other objects to secure them.

3-10.5 Identify the reasons for placing a rope out of service.

3-10.6 Distinguish between life safety and utility ropes.

3.11 Ladders.

3-11.1 Identify and describe the use of the following types of ladders:

(a) Folding/attic

(b) Roof

(c) Extension

(d) Straight/wall

(e) Aerial devices

3-11.2* Carry, position, raise, and lower the following ground ladders:

(a) 14 foot single or wall ladder

(b) 24 foot extension ladder

(c) 35 foot extension ladder

(d) Folding/attic

> **A-3-11.2** Comparable lengths of ladders of each type as specified by the authority having jurisdiction may be substituted for the stated lengths.

3-11.3 Demonstrate the procedures of working from

ground or aerial ladders with tools and appliances, with and without a safety harness.

3-11.4 Climb the full length of each type of ground and aerial ladder available to the authority having jurisdiction and demonstrate:

(a) Carrying fire fighting tools or equipment while ascending and descending

(b) Bring an injured person down the ladders.

3-11.5 Demonstrate the deployment of a roof ladder on a pitched roof.

3-12 Fire Hose, Appliances and Streams.

3-12.1 Describe the application of each size and type of hose on a pumper as required to be carried by NFPA 1901, *Standard for Pumper Fire Apparatus*, in Section 7-1.2 and 7-2.

3-12.2 Demonstrate the use of nozzles, adaptors, and hose appliances and tools on a pumper as required to be carried by NFPA 1901, *Standard for Pumper Fire Apparatus*, in Section 7-2.

3-12.3 Advance uncharged and charged attack lines of two different sizes, 1½ inch (38 mm) or larger, from a pumper, given the necessary equipment and operating as a member of a team, for the following evolutions:

(a) Into a structure

(b) Up a ladder to a second floor landing

(c) Up an inside stairway to an upper floor

(d) Up an outside stairway to an upper floor

(e) Down an inside stairway to a lower floor

(f) Down an outside stairway to a lower floor

(g) To an upper floor by hoisting.

3-12.4 Given fire hose used for fire attack (minimum of 1½ inches, 38 mm) and water supply (minimum of 2½ inches, 65 mm) demonstrate:

(a) Three types of hose loads and finishes

(b) Three types of hose rolls

(c) Coupling and uncoupling two lengths

(d) Two hose carries

(e) Extending hose lines

(f) Replacing burst sections of hose

3-12.5 Demonstrate operation of a charged attack line 1½ inches (35 mm) or larger from a ground ladder.

3-12.6 Demonstrate carrying a 100 foot (30 m) attack line of 1½ inches (38 mm) or larger, into a building, connecting it to a standpipe and advancing the line from the standpipe.

3-12.7 Demonstrate a hand lay of 300 feet (90 m) of supply line 2½ inches (65 mm) or larger from a pumper to a water source.

3-12.8 Define a fire stream.

3-12.9 Define water hammer and at least one method for its prevention.

3-12.10 Demonstrate how to open and close a nozzle and how to adjust its stream pattern and flow setting, when applicable.

3-12.11 Given a selection of nozzles and tips, shall identify the type, design, operation, required nozzle pressure, and flow of each.

3-12.12* Define the following methods of water application:

(a) Direct

(b) Indirect

(c) Combination.

A-3-12.12 When applying water to interior fires the fire fighter must be aware of the effects each type of stream may have on the hot gas layer within the space. The term "thermal balance" has been used to describe the stratification of temperatures within the space. The improper application of the fire stream may disrupt this stratification and cause the hot gases and products of combustion to a lower level in the space (thermal imbalance).

3-12.13 Identify precautions to be followed while advancing hose lines to a fire.

3-12.14 Identify three (3) observable results that are obtained when the proper application of a fire stream is accomplished.

3-13 Foam Fire Streams.

3-13.1 Assemble and operate a foam fire stream arrangement given the appropriate equipment.

3-13.2 Demonstrate the methods for applying a foam stream.

3-14* Fire Control.

A-3-14 Live fire evolutions should be conducted in accordance with the requirements of NFPA 1403, *Standard on Live Fire Evolutions in Structures*. It is further recommended that prior to involvement in interior live fire evolutions, the fire fighter demonstrate the use of SCBA in smoke and elevated temperature conditions.

3-14.1* Control and/or extinguish the following live fires using appropriate protective equipment, fire fighting tools, extinguishing agents and working as a member of a team:

(a) Pile/stacks of class A combustible materials (exterior)

(b) Open pans of combustible liquids (exterior)

(c) Vehicle fires

(d) Storage containers (exterior dumpster/trash bin)

(e) Class A combustible materials within a structure (interior attack)

A-3-14.1 In areas where environmental or other concerns restrict the use of fuels required in this section. Properly installed and monitored gas fueled fire simulators or other modifications to the Requirement may be substituted.

3-14.2 Explain the procedures for extinguishing ground cover fires.

3-15 Salvage.

3-15.1 Identify the purpose of salvage, and its value to the public and the fire department.

3-15.2 Demonstrate two (2) folds and rolls for salvage covers.

3-15.3 Demonstrate two (2) methods of deploying salvage covers to cover property.

3-15.4 Demonstrate the construction and use of a water chute.

3-15.5 Demonstrate the construction and use of a water catch-all.

3-15.6 Demonstrate the covering or closing of building openings, including doors, windows, floors and roofs.

3-15.7 Demonstrate the removal of debris, and removal and routing of water from a structure.

3-15.8 Demonstrate the procedures of inspection, cleaning, and maintaining salvage equipment.

3-16 Overhaul.

3-16.1 Identify the purpose of overhaul.

3-16.2 Recognize at least four (4) indicators of hidden fires.

3-16.3 Expose hidden fires by opening ceilings, walls, floors, and pulling apart burned materials.

3-16.4 Separate, remove and relocate charred material to a safe location while protecting the area of origin for determination of cause.

3-16.5 Define duties of fire fighters left at the fire scene for fire and security surveillance.

3-17 Emergency Medical Care.

3-17.1 Define the principles of infection control and Universal Blood and Body Fluid Precautions as prescribed for public safety workers by the Centers for Disease Control in *Guidelines for Prevention of Transmission of Human Immunodeficiency Virus and Hepatitis B Virus to Health-Care and Public-Safety Workers*.

3-17.2 Demonstrate the use, decontamination, disinfection, and disposal of personal protective equipment used for protection from infection.

3-17.3* Perform the following procedures as defined in the Journal of the American Medical Association, *Standards and Guidelines for Cardiopulmonary Resuscitation (CPR) and Emergency Cardiac Care (ECC)*:

(a) Single rescuer CPR

1. Adult
2. Child
3. Infant

(b) Two rescuer CPR on an adult

(c) Management of an obstructed airway

1. Conscious and unconscious adult
2. Conscious and unconscious child
3. Conscious and unconscious infant

A-3-17.3 This requirement may be met with courses such as provided by the American Red Cross in its Basic Life Support for the Professional Rescuer course and comparable courses offered by the American Heart Association. Certification by either organization at the appropriate skill level would constitute meeting this requirement.

3-17.4 Demonstrate the use of a resuscitation mask in the performance of single and two rescuer CPR.

3-17.5 Demonstrate a primary survey for life-threatening injuries.

3-17.6 Identify three (3) types of external bleeding, and the characteristics of each type.

3-17.7 Demonstrate three (3) procedures for controlling external bleeding.

3-17-8 Identify characteristics and emergency medical care of thermal burns according to degree and severity.

3-17.9 Identify the emergency medical care for chemical burns, including chemical burns of the eyes.

3-17.10 Identify symptoms and demonstrate emergency medical care of traumatic shock.

3-17.11 Identify the symptoms and demonstrate emergency medical care for ingested poisons and drug overdoses.

3-17.12 Identify the method of contacting the poison control center serving the region the Authority Having Jurisdiction is located in.

3-18 Rescue.

3-18.1 Define and demonstrate primary and secondary search procedures under fire conditions:

(a) With a rope or hose line.

(b) Without a rope or hose line.

3-18.2 Don a life safety harness that meets the requirements of NFPA 1983, *Standard on Fire Service Life Safety Rope, Harness, and Hardware*.

3-18.3 Inspect a life safety harness and identify the conditions that would require its removal from service.

3-18.4 Demonstrate the removal of injured persons from an immediate hazard by the use of carriers, drags and stretchers.

3-19 Water Supplies.

3-19.1 Connect a supply hose to a hydrant, and fully open and close the hydrant.

3-19.2 Demonstrate hydrant to pumper hose connections for forward and reverse hose lays.

3-19.3 Assemble and connect the equipment necessary for drafting from a static water supply source.

3-19.4 Describe the deployment of a portable water tank.

3-19.5 Describe the assembling of equipment necessary for the transfer of water between portable water tanks.

3-19.6 Describe loading and off-loading of tanks on mobile water supply apparatus.

3-20 Sprinklers.

3-20.1 Define the value of automatic sprinklers in providing safety to the occupants of a structure.

3-20.2 Identify a fire department sprinkler connection and water motor alarm.

3-20.3 Connect hose line(s) to a fire department connection of a sprinkler or standpipe system.

3-20.4 Explain how the automatic sprinkler head activates and releases water.

3-20.5 Stop the flow of water from a sprinkler head using a wedge or stopper.

3-20.6 Identify the main control valve on an automatic sprinkler system.

3-20.7 Operate a main control valve on an automatic sprinkler system from "open" to "closed" and then back to "open."

3-21 Response to Hazardous Materials Incidents.

3-21.1 Meet the requirements defined in NFPA 472, *Standard for Professional Competence of Responders to Hazardous Materials Incidents*, Section 2-2, First Responder Awareness Level.

3-22 Fire Prevention, Public Fire Education and Fire Cause Determination.

3-22.1 Identify five (5) common causes of fires and their prevention.

3-22.2 Define the importance of inspection and public fire education programs to fire department public relations and the community.

3-22.3 Demonstrate inspection procedures for private dwellings.

3-22.4* Present a prepared program to an identified audience, given a lesson plan, time allotment and instructional materials, for the following topics:

(a) Stop, Drop and Roll

(b) Crawl Low in Smoke

(c) Escape Planning

(d) Alerting Others

(e) Calling the fire department

(f) Fire Station Tour

(g) Residential Smoke Detector Placement and Maintenance

> **A-3-22.4** Home escape plans as described in "Operation EDITH" (Exit Drills in the Home) or in NFPA's *Learn Not to Burn Curriculum* should be used to meet this requirement.

3-22.5 Document the presentation of a program covered in 3-22.4, given a reporting form, that includes:

(a) Program Title

CHAPTER 4 FIRE FIGHTER II

(b) Number of Participants

(c) Evaluations

3-23 Building Construction.

No Job Performance Requirements in Fire Fighter I

4-1 General. The Level II Fire Fighter shall meet the job performance requirements defined in sections 4-2 through 4-23 of this Standard.

4-2 Fire Department Organization.

4-2.1 Explain the Fire Fighter II's role as a member of the organization.

4-2.2 Explain the responsibilities for the fire fighter in assuming and transferring command within an incident management system.

4-3 Safety.

4-3.1 Identify applicable local, state/provincial and federal laws and regulations related to occupational health and safety.

4-3.2 Demonstrate the service and maintenance of portable power plants and lighting equipment.

4-3.3 Safely operate a total of twelve (12) types of hand and power tools used for forcible entry, rescue and ventilation.

4-4 Fire Alarm and Communications.

4-4.1 Define the policy and demonstrate the procedures concerning the ordering and transmitting of multiple alarms of fire and calls for special assistance from the emergency scene.

4-4.2 Identify supervisory alarm equipment provided in the fire station and the prescribed action to be taken upon receipt of designated signals.

4-4.3 Identify fire location indicators provided to direct fire fighters to specific locations in protected public or private properties.

4-5 Fire Behavior.

4-5.1 Define the following units of heat measurement:

(a) British Thermal Unit (BTU)

(b) Fahrenheit (°F)

(c) Celsius (°C)

(d) Calorie (C).

4-5.2 Define the hazard of finely divided fuels as they relate to the combustion process.

4-5.3 Define flash point, fire point, and ignition temperature.

4-5.4 Identify two (2) chemical, mechanical, and electrical energy heat sources.

4-6 Portable Extinguishers.

No Job Performance Requirements in Fire Fighter II

4-7 Personal Protective Equipment.

No Job Performance Requirements in Fire Fighter II

4-8 Forcible Entry.

No Job Performance Requirements in Fire Fighter II

4-9 Ventilation.

4-9.1* Identify the manual and automatic venting devices found within structures.

A-4-9.1 The following classes are examples:

(a) Curtain boards

(b) Roof monitors

(c) Continuous gravity vents

(d) Unit type vents

(e) Automatic heat and smoke vents

(f) Sawtooth roof skylights

(g) Exterior wall openings

4-9.2 Describe the operations and considerations necessary to control the spread of smoke and fire through duct systems including:

(a) Determine location and routing of ducts

(b) Shutting down systems to prevent spread of heat and smoke

(c) Examine duct system after thorough ventilation

(d) Check false ceilings or framing enclosing duct systems

(e) Check duct system outlets

(f) Determine if duct system has openings, smoke dampers, or smoke detectors.

4-9.3 Identify considerations that must be made when determining the location and size of a ventilation opening including:

(a) Availability of natural openings

(b) Location of the fire

(c) Direction in which the fire will be drawn

(d) Type of building construction

(e) Wind direction

(f) Progress of the fire

(g) Condition of the building

(h) Obstructions

(i) Relative efficiency of large vs. small openings

4-9.4 Identify the location of the opening, the method to be used, and the precautions to be taken when ventilating a basement.

4-9.5* Identify fire ground situations where forced ventilation procedures may be required.

> **A-4-9.5** Forced ventilation in this requirement includes both positive and negative pressure methods.

4-10 Ropes.

4-10.1 Given a fire fighting or rescue task requiring the use of rope, select the appropriate size, strength, type and length of rope to accomplish the task.

4-10.2 Select an appropriate knot, given a fire fighting or rescue task requiring the use of rope.

4-11 Ladders.

4-11.1 Identify the materials used in ladder construction.

4-11.2 Identify the load capacities established by NFPA 1931, *Standard on Design of and Design Verification Tests for Fire Department Ground Ladders* and NFPA 1904, *Standard for Aerial Ladder and Elevating Platform Fire Apparatus* for ground and aerial ladders.

4-11.3 Demonstrate the procedures of cleaning ladders.

4-11.4 Demonstrate inspection and maintenance procedures for different types of ground and aerial ladders.

4-11.5 Describe the annual service test for ground ladders.

4-12 Fire Hose, Appliances, and Streams.

4-12.1 Given three (3) fire situations, select the proper nozzle and hose for fire attack.

4-12.2 Select adapters and appliances to be used in three (3) specific fire ground situations.

4-12.3 Demonstrate the procedures for cleaning and maintaining fire hose, couplings, and nozzles; and inspecting for damage.

4-12.4 Demonstrate an annual service test for fire hose.

4-12.5 Describe and demonstrate the operation of fog and solid stream nozzles.

4-12.6 Identify the rate of water flow necessary to control fire in a room of specified volume.

4-12.7 Describe the advantages and disadvantages of solid and fog streams.

4-13 Foam Fire Streams.

4-13.1 Define the four (4) methods by which foam prevents or controls a hazard.

4-13.2 Define the principle by which foam is generated.

4-13.3 Define common causes for the poor generation of foam and identify the procedures for correcting each.

4-13.4 Define the difference between hydrocarbon and polar solvent fuels and identify the type of foam concentrate required for each fuel.

4-13.5 Define the advantages, characteristics, and precautions for use of the following types of foam:

(a) Protein

(b) Fluoroprotein

(c) Film forming fluoroprotein (FFFP)

(d) Aqueous film forming foam (AFFF)

(e) Hazardous materials vapor mitigating foam

(f) Medium-and high-expansion foam

(g) Class A foams

4-13.6 Define the precautions that must be taken when using high expansion foam to attack structural fires.

4-14 Fire Control.

4-14.1* Control and/or extinguish the following live fires using appropriate protective equipment, fire fighting tools, extinguishing agents and working as a member of a team:

(a) An exterior combustible liquids fire of at least 100 square feet (9 m^2), using a foam fire stream.

(b) A fire in an elevated location within a structure (e.g. upper level floor, attic).

(c) A hidden fire within a structure (e.g. within walls, crawl spaces).

(d) A fire involving energized electrical components.

(e) A fire involving flammable gas cylinder (exterior).

(f) A fire in a below-grade area or other location requiring initial attack from above.

> **A-4-14.1** In areas where environmental or other concerns restrict the use of fuels required in this section, properly installed and monitored gas fueled fire simulators or other modifications to the Requirement may be substituted.

4-15 Salvage.

No Job Performance Requirements in Fire Fighter II

4-16 Overhaul.

4-16.1 Identify the procedures and safety precautions to follow during overhaul.

4-16.2 List five (5) indicators of structural instability.

4-16.3 Identify and preserve evidence of fire cause and origin.

4-16.4 Identify the procedures for restoration of the premises after a fire.

4-17 Emergency Medical Care.

No Job Performance Requirements for Fire Fighter II

4-18 Rescue.

4-18.1* Describe the procedures and safety procedures as they apply to the following rescue activities:

(a) Structural collapses

(b) Trench collapses

(c) Caves and tunnels

(d) Water and ice emergencies

(e) Elevators and escalators

(f) Emergencies involving energized electrical lines

(g) Industrial accidents

(h) Other hazards particular to the local jurisdiction

A-4-18.1 Each of the activities listed in this requirement constitutes a major component of the rescue field. The Committees intent is that this requirement develop an awareness of the procedures and safety requirements involved.

4-18.2 Demonstrate the use of the following rescue tools:

(a) Cribbing and shoring material

(b) Block and tackle

(c) Hydraulic devices

(d) Pneumatic devices

(e) Ratchet device

4-18.3 Demonstrate the following evolutions, which may be required to extricate an entrapped victim of a motor vehicle accident, by displacing:

(a) Vehicle roof

(b) Vehicle door

(c) Vehicle windshield

(d) Steering wheel

(e) Steering column and dashboard

4-18.4 Raise and lower a person a minimum of 20 vertical feet (6 m) with a rope rescue system.

4-19 Water Supplies.

4-19.1 Identify the water distribution system, and other water sources in the local community.

4-19.2 Identify the following parts of a water distribution system:

(a) Distributors

(b) Primary feeders

(c) Secondary feeders

4-19.3 Explain the operation of a:

(a) Dry-barrel hydrant

(b) Wet-barrel hydrant

4-19.4 Define the following terms as they relate to water supply:

(a) Static pressure

(b) Normal operating pressure

(c) Residual pressure

(d) Flow pressure

4-19.5 Identify the following types of water main valves:

(a) Indicating

(b) Nonindicating

4-19.6 Describe how the following conditions reduce hydrant effectiveness:

(a) Obstructions to use of hydrant

(b) Direction of hydrant outlets to suitability of use

(c) Mechanical damage

(d) Rust and corrosion

(e) Failure to open the hydrant fully

(f) Susceptibility to freezing

4-19.7 Identify the apparatus, equipment and appliances required to provide water at rural locations by relay pumping or a mobile water supply apparatus shuttle.

4-19.8 Identify and explain the four (4) fundamental components of a modern water system.

4-19.9 Given a Pitot tube and gauge, read and record flow pressures from three different sized orifices.

4-19.10 Identify the pipe sizes used in water distribution systems for residential, business, and industrial districts.

4-19.11 Identify two (2) causes of increased resistance or friction loss in water mains.

4-20 Sprinklers.

4-20.1 Identify the sources of water supply for sprinkler systems, including:

(a) Public water systems

(b) Gravity tank

(c) Pressure tanks

(d) Pumps

(e) Fire department connections

4-20.2 Identify how the direction of water flow through

a fire department connection check valve can be determined, including:

(a) Arrows

(b) Pivot casting

4-20.3 Identify the location and appearance of the control and operating valves of a sprinkler system including:

(a) Outside screw and yoke (OS&Y)

(b) Post indicator

(c) Wall post indicator

4-20.4 Identify the main drain valve on an automatic sprinkler system.

4-20.5 Open and close a main drain valve on an automatic sprinkler system.

4-20.6 Identify and define the dangers of premature closure of sprinkler main control valve, and of using hydrants to supply hose streams when the same water system is supplying the automatic sprinkler system.

4-20.7 Identify the difference between an automatic sprinkler system that affords complete coverage and a partial sprinkler system.

4-20.8 Describe the following types of sprinkler systems:

(a) Wet

(b) Dry

(c) Deluge

(d) Residential

4-20.9 Read and record the indicated pressures on all gauges provided on a standard wet automatic sprinkler system and identify each gauge.

4-20.10 Read and record the indicated pressures on all gauges provided on a standard dry pipe automatic sprinkler system and identify each gauge.

4-20.11 Define the reliability of automatic sprinkler systems, and give eight (8) reasons for unsatisfactory performance.

4-21 Response to Hazardous Materials Incidents.

4-21.1 Meet the requirements defined in NFPA 472, *Standard for Professional Competence of Responders to Hazardous Materials Incidents*, Section 2-3, First Responder Operational Level.

4/93

4-22 Fire Prevention, Public Fire Education and Fire Cause Determination.

4-22.1 Prepare a prefire plan that includes diagrams or sketches of a building to record the location of items of concern.

4-22.2* Complete a basic fire incident report and describe the importance of this information.

A-4-22.2 A basic fire report should contain information as recommended by NFPA 902M, *Fire Reporting Field Incident Manual.*

4-22.3 Conduct a building fire safety survey and prepare a written report summarizing the results.

4-22.4 Identify school exit drill procedures.

4-22.5 Identify life safety programs for the home.

4/93

4-22.6 Identify common fire hazards and make recommendations for their correction.

4-22.7 Identify responsibilities of the fire fighter in determining the point of origin, cause, and protection of evidence in fires.

4-22.8 Inspect fire protection standpipe systems for readiness, including visual inspection of hose (where provided), nozzles, hose outlet threads, and fire department connections.

4-22.9 Identify smoke, flame and heat-detection alarm systems.

4-22.10 Identify the fire hazards commonly found in manufacturing, commercial, residential, and public assembly occupancies.

4-22.11 Identify standard types of chimneys and flues, and recognize deficiencies likely to cause fires.

4-23 Building Construction.

4-23.1 Describe the basic structural characteristics of the following types of building construction:

(a) Wood frame

(b) Ordinary

(c) Heavy timber

(d) Non-combustible

(e) Fire resistant

4-23.2 Identify the general fire behavior expected with each type of building construction, including the spread of fire and the safety of the building, occupants, and fire fighters.

4-23.3 Describe at least three (3) hazards associated with truss and lightweight construction.

4-23.4 Identify dangerous building conditions created by fire and fire suppression activities.

4-23.5 Identify five (5) indicators of building collapse.

4-23.6 Describe the effects of fire and fire suppression activities on the following building materials:

(a) Wood

(b) Masonry (brick, block, stone)

(c) Cast iron

(d) Steel

(e) Reinforced concrete

(f) Gypsum wall board

(g) Glass

(h) Plaster on lath

4-23.7 Define the following terms as they relate to building construction:

(a) Veneer wall (exterior)

(b) Party wall

(c) Fire wall

(d) Partition wall

(e) Cantilever or unsupported wall

(f) Load bearing

Index

I

IFSTA MANUALS AND FPP PRODUCTS

For a current catalog describing these and other products call or write your local IFSTA distributor or Fire Protection Publications, IFSTA Headquarters, Oklahoma State University, Stillwater, OK 74078-0118. Phone: 1-800-654-4055

FIRE DEPARTMENT AERIAL APPARATUS
includes information on the driver/operator's qualifications; vehicle operation; types of aerial apparatus; positioning, stabilizing, and operating aerial devices; tactics for aerial devices; and maintaining, testing, and purchasing aerial apparatus. Detailed appendices describe specific manufacturers' aerial devices. 1st Edition (1991), 416 pages, addresses NFPA 1002.

STUDY GUIDE FOR AERIAL APPARATUS
The companion study guide in question and answer format. 1st Edition (1991), 152 pages.

AIRCRAFT RESCUE AND FIRE FIGHTING
comprehensively covers commercial, military, and general aviation. All the information you need is in one place. Subjects covered include: personal protective equipment, apparatus and equipment, extinguishing agents, engines and systems, fire fighting procedures, hazardous materials, and fire prevention. Over 240 photographs and two-color illustrations. It also contains a glossary and review questions with answers. 3rd Edition (1992), 272 pages, addresses NFPA 1003.

BUILDING CONSTRUCTION RELATED TO THE FIRE SERVICE
helps firefighters become aware of the many construction designs and features of buildings found in a typical first alarm district and how these designs serve or hinder the suppression effort. Subjects include construction principles, assemblies and their resistance to fire, building services, door and window assemblies, and special types of structures. 1st Edition (1986), 166 pages, addresses NFPA 1001 and NFPA 1031, levels I & II.

CHIEF OFFICER
lists, explains, and illustrates the skills necessary to plan and maintain an efficient and cost-effective fire department. The combination of an ever-increasing fire problem, spiraling personnel and equipment costs, and the development of new technologies and methods for decision making requires far more than expertise in fire suppression. Today's chief officer must possess the ability to plan and administrate as well as have political expertise. 1st Edition (1985), 211 pages, addresses NFPA 1021, level VI.

SELF-INSTRUCTION FOR CHIEF OFFICER
The companion study guide in question and answer format. 1st Edition, 142 pages.

FIRE DEPARTMENT COMPANY OFFICER
focuses on the basic principles of fire department organization, working relationships, and personnel management. For the firefighter aspiring to become a company officer, or a company officer wishing to improve management skills, this manual helps develop and improve the necessary traits to effectively manage the fire company. 2nd Edition (1990), 278 pages, addresses NFPA 1021, levels I, II, & III.

COMPANY OFFICER STUDY GUIDE
The companion study guide in question and answer format. Includes problem applications and case studies. 1st Edition (1991), 256 pages.

ESSENTIALS OF FIRE FIGHTING
is the "bible" on basic firefighter skills and is used throughout the world. The easy-to-read format is enhanced by 1,500 photographs and illustrations. Step-by-step instructions are provided for many fire fighting tasks. Topics covered include: personal protective equipment, building construction, firefighter safety, fire behavior, portable extinguishers, SCBA, ropes and knots, rescue, forcible entry, ventilation, communications, water supplies, fire streams, hose, fire cause determination, public fire education and prevention, fire suppression techniques, ladders, salvage and overhaul, and automatic sprinkler systems. 3rd Edition (1992), addresses NFPA 1001.

STUDY GUIDE FOR 3rd EDITION OF ESSENTIALS OF FIRE FIGHTING
The companion learning tool for the new 3rd edition of the manual. It contains questions and answers to help you learn the important information in the book. 1st Edition (1992).

PRINCIPLES OF EXTRICATION
leads you step-by-step through the procedures for disentangling victims from cars, buses, trains, farm equipment, and industrial situations. Fully illustrated with color diagrams and more than 500 photographs. It includes rescue company organization, protective clothing, and evaluating resources. Review questions with answers at the end of each chapter. 1st Edition (1990), 400 pages.

FIRE CAUSE DETERMINATION
gives you the information necessary to make on-scene fire cause determinations. You will know when to call for a trained investigator, and you will be able to help the investigator. It includes a profile of firesetters, finding origin and cause, documenting evidence, interviewing witnesses, and courtroom demeanor. 1st Edition (1982), 159 pages, addresses NFPA 1021, Fire Officer I, and NFPA 1031, levels I & II.

FIRE SERVICE FIRST RESPONDER
provides the information needed to evaluate and treat patients with serious injuries or illnesses. It familiarizes the reader with a wide variety of medical equipment and supplies. **First Responder** applies to safety, security, fire brigade, and law enforcement personnel, as well as fire service personnel, who are required to administer emergency medical care. 1st Edition (1987), 340 pages, addresses NFPA 1001, levels I & II, and DOT First Responder.

FORCIBLE ENTRY
reflects the growing concern for the reduction of property damage as well as firefighter safety. This comprehensive manual contains technical information about forcible entry tactics, tools, and methods, as well as door, window, and wall construction. Tactics discuss the degree of danger to the structure and leaving the building secure after entry. Includes a section on locks and through-the-lock entry. Review questions and answers at the end of each chapter. 7th Edition (1987), 270 pages, helpful for NFPA 1001.

GROUND COVER FIRE FIGHTING PRACTICES

explains the dramatic difference between structural fire fighting and wildland fire fighting. Ground cover fires include fires in weeds, grass, field crops, and brush. It discusses the apparatus, equipment, and extinguishing agents used to combat wildland fires. Outdoor fire behavior and how fuels, weather, and topography affect fire spread are explained. The text also covers personnel safety, management, and suppression methods. It contains a glossary, sample fire operation plan, fire control organization system, fire origin and cause determination, and water expansion pump systems. 2nd Edition (1982), 152 pages.

FIRE SERVICE GROUND LADDER PRACTICES

is a "how to" manual for learning how to handle, raise, and climb ground ladders; it also details maintenance and service testing. Basic information is presented with a variety of methods that allow the readers to select the best method for their locale. The chapter on Special Uses includes: ladders as a stretcher, a slide, a float drag, a water chute, and more. The manual contains a glossary, review questions and answers, and a sample testing and repair form. 8th Edition (1984), 388 pages, addresses NFPA 1001.

HAZARDOUS MATERIALS FOR FIRST RESPONDERS

provides basic information on hazardous materials with sections on site management and decontamination. It includes a description of various types of materials, their characteristics, and containers. The manual covers the effects of weather, topography, and environment on the behavior of hazardous materials and control efforts. Pre-incident planning and post-incident analysis are covered. 1st Edition (1988), 357 pages, addresses NFPA 472, 29 CFR 1910.120 and NFPA 1001.

STUDY GUIDE FOR HAZARDOUS MATERIALS FOR FIRST RESPONDERS

The companion study guide in question and answer format also includes case studies that simulate incidents. 1st Edition (1989), 208 pages.

HAZARDOUS MATERIALS: MANAGING THE INCIDENT

takes you beyond the basic information found in **Hazardous Materials for First Responders**. Directed to the leader/commander, this manual sets forth basic practices clearly and comprehensively. Charts, tables, and checklists guide you through the organization and planning stages to decontamination. This text, along with the accompanying workbook and instructor's guide, provides a comprehensive learning package. 1st Edition (1988), 206 pages, helpful for NFPA 1021.

STUDENT WORKBOOK FOR HAZARDOUS MATERIALS: MANAGING THE INCIDENT

provides questions and answers to enhance the student's comprehension and retention. 1st Edition (1988), 176 pages.

INSTRUCTOR'S GUIDE FOR HAZARDOUS MATERIALS: MANAGING THE INCIDENT

provides lessons based on each chapter, adult learning tips, and appendices of references and suggested audio visuals. 1st Edition (1988),142 pages.

HAZ MAT RESPONSE TEAM LEAK AND SPILL GUIDE

contains articles by Michael Hildebrand reprinted from *Speaking of Fire*'s popular Hazardous Materials Nuts and Bolts series. Two additional articles from *Speaking of Fire* and the hazardous material incident SOP from the Chicago Fire Department are also included. 1st Edition (1984), 57 pages.

EMERGENCY OPERATIONS IN HIGH-RACK STORAGE

is a concise summary of emergency operations in the high-rack storage area of a warehouse. It explains how to develop a pre-emergency plan, what equipment will be necessary to implement the plan, type and amount of training personnel will need to handle an emergency, and interfacing with various agencies. Includes consideration questions, points not to be overlooked, and trial scenarios. 1st Edition (1981), 97 pages.

HOSE PRACTICES

reflects the latest information on modern fire hose and couplings. It is the most comprehensive single source about hose and its use. The manual details basic methods of handling hose, including large diameter hose. It is fully illustrated with photographs showing loads, evolutions, and techniques. This complete and practical book explains the national standards for hose and couplings. 7th Edition (1988), 245 pages, addresses NFPA 1001.

FIRE PROTECTION HYDRAULICS AND WATER SUPPLY ANALYSIS

covers the quantity and pressure of water needed to provide adequate fire protection, the ability of existing water supply systems to provide fire protection, the adequacy of a water supply for a sprinkler system, and alternatives for deficient water supply systems. 1st Edition (1990), 340 pages.

INCIDENT COMMAND SYSTEM (ICS)

was developed by a multiagency task force. Using this system, fire, police, and other government groups can operate together effectively under a single command. The system is modular and can be used to meet the requirements of both day-to-day and large-incident operations. It is the approved basic command system taught at the National Fire Academy. 1st Edition (1983), 220 pages, helpful for NFPA 1021.

INDUSTRIAL FIRE PROTECTION

is designed for the person charged with the responsibility of developing, implementing, and coordinating fire protection. A "must read" for fire service personnel who will coordinate with industry/business for pre-incident planning. The text includes guidelines for establishing a company policy, organization and planning for the emergency, establishing a fire prevention plan, incipient fire fighting tactics, an overview of interior structural fire fighting, and fixed fire fighting systems. 1st Edition (1982), 207 pages, written for 29 CFR. 1910, Subpart L, and helpful for NFPA 1021 and NFPA 1031.

FIRE INSPECTION AND CODE ENFORCEMENT

provides a comprehensive, state-of-the-art reference and training manual for both uniformed and civilian inspectors. It is a comprehensive guide to the principles and techniques of inspection. Text includes information on how fire travels, electrical hazards, and fire resistance requirements. It covers storage, handling, and use of hazardous materials; fire protection systems; and building construction for fire and life safety. 5th Edition (1987), 316 pages, addresses NFPA 1001 and NFPA 1031, levels I & II.

STUDY GUIDE FOR FIRE INSPECTION AND CODE ENFORCEMENT

The companion study guide in question and answer format with case studies. 1st Edition (1989), 272 pages.

FIRE SERVICE INSTRUCTOR

explains the characteristics of a good instructor, shows you how to determine training requirements, and teach to the level of your class. It discusses the types, principles, and procedures of teaching and learning, and covers the use of effective training aids and devices. The purpose and principles of testing as well as test construction are covered. Included are chapters on safety, legal considerations, and computers. 5th Edition (1990), 325 pages, addresses NFPA 1041, levels I & II.

LEADERSHIP IN THE FIRE SERVICE

was created from the series of lectures given by Robert F. Hamm to assist in leadership development. It provides the foundation for getting along with others, explains how to gain the confidence of your personnel, and covers what is expected of an officer. Included is information on supervision, evaluations, delegating, and teaching. Some of the topics include: the successful leader today, a look into the past may reveal the future, and self-analysis for officers. 1st Edition (1967), 132 pages.

FIRE SERVICE ORIENTATION AND INDOCTRINATION

relates the traditions, history, and organization of the fire service. It includes operation of the fire department, responsibilities and duties of firefighters, and the function of fire department companies. This exciting and informative text is for anyone dealing with the fire service who needs a basic understanding and overview. The perfect book for new or prospective members, buffs, your congressman or council members, fire service sales personnel, and industrial brigades. 2nd Edition (1984), 187 pages, addresses NFPA 1001.

PRIVATE FIRE PROTECTION AND DETECTION

introduces the firefighter, inspection personnel, brigade/safety member, insurance inspector/investigator, or fire protection student to fixed systems, extinguishers, and detection. It covers the various types of equipment, their installation, maintenance, and testing. Systems discussed are: wet-pipe, dry-pipe, pre-action, deluge, residential, carbon dioxide, Halogen, dry- and wet-chemical, foam, standpipe, and fire extinguishers. 1st Edition (1979), 170 pages, addresses NFPA 1001 and NFPA 1031, levels I & II.

PUBLIC FIRE EDUCATION

provides valuable information for ending public apathy and ignorance about fire. This manual gives you the knowledge to plan and implement fire prevention campaigns. It shows you how to tailor the individual programs to your audience as well as the time of year or specific problems. It includes working with the media, resource exchange, and smoke detectors. 1st Edition (1979), 169 pages, helpful for NFPA 1021 and 1031.

FIRE DEPARTMENT PUMPING APPARATUS

is the Driver/Operator's encyclopedia on operating fire pumps and pumping apparatus. It covers pumpers, tankers (tenders), brush apparatus, and aerials with pumps. This comprehensive volume explains safe driving techniques, getting maximum efficiency from the pump, and basic water supply. It includes specification writing, apparatus testing, and extensive appendices of pump manufacturers. 7th Edition (1989), 374 pages, addresses NFPA 1002.

STUDY GUIDE FOR PUMPING APPARATUS

The companion study guide in question and answer format. 1st Edition (1990), 100 pages.

FIRE SERVICE RESCUE PRACTICES

is a comprehensive training text for firefighters and fire brigade members that expands proficiency in moving and removing victims from hazardous situations. This extensively illustrated manual includes rescuer safety, effects of rescue work on victims, rescue from hazardous atmospheres, trenching, and outdoor searches. 5th Edition (1981), 262 pages, addresses NFPA 1001.

RESIDENTIAL SPRINKLERS A PRIMER

outlines United States residential fire experience, system components, engineering requirements, and issues concerning automatic and fixed residential sprinkler systems. Written by Gary Courtney and Scott Kerwood and reprinted from *Speaking of Fire*. An excellent reference source for any fire service library and an excellent supplement to **Private Fire Protection.** 1st Edition (1986), 16 pages.

FIRE DEPARTMENT OCCUPATIONAL SAFETY

addresses the basic responsibilities and qualifications for a safety officer and the minimum requirements and procedures for a safety and health program. Included in this manual is an overview of establishing and implementing a safety program, physical fitness and health considerations, safety in training, fire station safety, tool and equipment safety and maintenance, personal protective equipment, en route hazards and response, emergency scene safety, and special hazards. 2nd Edition (1991), 396 pages, addresses NFPA 1500, 1501.

SALVAGE AND OVERHAUL

covers planning salvage operations, equipment selection and care, as well as describing methods and techniques for using salvage equipment to minimize fire damage caused by water, smoke, heat, and debris. The overhaul section includes methods for finding hidden fire, protection of fire cause evidence, safety during overhaul operations, and restoration of property and fire protection systems after a fire. 7th Edition (1985), 225 pages, addresses NFPA 1001.

SELF-CONTAINED BREATHING APPARATUS

contains all the basics of SCBA use, care, testing, and operation. Special attention is given to safety and training. The chapter on Emergency Conditions Breathing has been completely revised to incorporate safer emergency methods that can be used with newer models of SCBA. Also included are appendices describing regulatory agencies and donning and doffing procedures for nine types of SCBA. The manual has been thoroughly updated to cover NFPA, OSHA, ANSI, and NIOSH regulations and standards as they pertain to SCBA. 2nd Edition (1991), 360 pages, addresses NFPA 1001.

STUDY GUIDE FOR SELF-CONTAINED BREATHING APPARATUS

The companion study guide in question and answer format. 1st Edition (1991).

FIRE STREAM PRACTICES

brings you an all new approach to calculating friction loss. This carefully written text covers the physics of fire and water; the characteristics, requirements, and principles of good streams; and fire fighting foams. **Streams** includes formulas for the application of fire fighting hydraulics, as well as actions and reactions created by applying streams under a variety of circumstances. The friction loss equations and answers are included, and review questions are located at the end of each chapter. 7th Edition (1989), 464 pages, addresses NFPA 1001 and NFPA 1002.

GASOLINE TANK TRUCK EMERGENCIES

provides emergency response personnel with background information, general procedures, and response guidelines to be followed when responding to and operating at incidents involving MC-306/DOT 406 cargo tank trucks. Specific topics include: incident management procedures, site safety considerations, methods of product transfer, and vehicle uprighting considerations. 1st Edition (1992), 60 pages, addresses NFPA 472.

FIRE VENTILATION PRACTICES

presents the principles, practices, objectives, and advantages of ventilation. It includes the factors and phases of combustion, flammable liquid characteristics, products of combustion, backdrafts, transmission of heat, and building construction considerations. The manual reflects the new techniques in building construction and their effects on ventilation procedures. Methods and procedures are thoroughly explained with numerous photographs and drawings. The text also includes: vertical (top), horizontal (cross), and forced ventilation; and a glossary. 6th Edition (1980), 131 pages, addresses NFPA 1001.

FIRE SERVICE PRACTICES FOR VOLUNTEER AND SMALL COMMUNITY FIRE DEPARTMENTS

presents those training practices that are most consistent with the activities of smaller fire departments. Consideration is given to the limitations of small community fire department resources. Techniques for performing basic skills are explained, accompanied by detailed illustrations and photographs. 6th Edition (1984), 311 pages.

WATER SUPPLIES FOR FIRE PROTECTION

acquaints you with the principles, requirements, and standards used to provide water for fire fighting. Rural water supplies as well as fixed systems are discussed. Abundant photographs, illustrations, tables, and diagrams make this the most complete text available. It includes requirements for size and carrying capacity of mains, hydrant specifications, maintenance procedures conducted by the fire department, and relevant maps and record keeping procedures. Review questions at the end of each chapter. 4th Edition (1988), 268 pages, addresses NFPA 1001, NFPA 1002, and NFPA 1031, levels I & II.

TEACHING PACKAGES

LEADERSHIP

This teaching package is designed to assist the instructor in teaching leadership and motivational skills. Cause and effect, behavior and consequences, listening and communications are themes throughout the course that stress the reality of the job and the people one deals with daily. Before each lesson is a title page that gives an outline of the subject matter to be covered, the approximate time required to teach the material, the specific learning objectives, and the references for the instructor's preparation. Sources for suggested films and videotapes are included.

CURRICULUM PACKAGE FOR IFSTA COMPANY OFFICER

A competency-based Teaching Package with lesson plans and activities to teach the student the information and skills needed to qualify for the position of Company Officer. Corresponds to **Fire Department Company Officer**, 2nd Edition.

The Package includes the Company Officer Instructor's Guide (the how, what, and when to teach); the Student Guide (a workbook for group instruction or self-study); and 143 full-color overhead transparencies.

ESSENTIALS CURRICULUM PACKAGE

A competency-based teaching package with lesson plans and activities to teach the student the information and skills needed to qualify for the position of Fire Fighter I or II. Corresponds to **Essentials of Fire Fighting**, 3rd Edition.

The Package, with Instructor's Guide, Student Guide, and more than 400 transparencies, is scheduled for publication in 1993.

TRANSLATIONS

LO ESENCIAL EN EL COMBATE DE INCENDIOS

is a direct translation of **Essentials of Fire Fighting**, 2nd edition. Please contact your distributor or FPP for shipping charges to addresses outside U.S. and Canada. 444 pages.

PRACTICAS Y TEORIA PARA BOMBEROS

is a direct translation of **Fire Service Practices for Volunteer and Small Community Fire Departments**, 6th edition. Please contact your distributor or FPP for shipping charges to addresses outside U.S. and Canada. 347 pages.

OTHER ITEMS

TRAINING AIDS

Fire Protection Publications carries a complete line of videos, overhead transparencies, and slides. Call for a current catalog.

NEWSLETTER

The nationally acclaimed and award winning newsletter, *Speaking of Fire*, is published quarterly and available to you free. Call today for your free subscription.

COMMENT SHEET

DATE ____________________ NAME ____

ADDRESS ____

ORGANIZATION REPRESENTED ____

CHAPTER TITLE ____ NUMBER ______________

SECTION/PARAGRAPH/FIGURE ______________________________ PAGE ______________

1. Proposal (include proposed wording, or identification of wording to be deleted), OR PROPOSED FIGURE:

2. Statement of Problem and Substantiation for Proposal:

RETURN TO: IFSTA Editor
Fire Protection Publications
Oklahoma State University
Stillwater, OK 74078

SIGNATURE ______________________________

Use this sheet to make any suggestions, recommendations, or comments. We need your input to make the manuals as up to date as possible. Your help is appreciated. Use additional pages if necessary.